Freshwater Microbiology

Freshwater Microbiology

Biodiversity and Dynamic Interactions of Microorganisms in the Aquatic Environment

David C. Sigee
University of Manchester, UK

JOHN WILEY & SONS, LTD

Telephone   (+44) 1243 779777

Email (for orders and customer service enquiries): cs-books@wiley.co.uk
Visit our Home Page on www.wileyeurope.com or www.wiley.com

Reprinted July 2005

Other Wiley Editorial Offices

John Wiley & Sons Inc., 111 River Street, Hoboken, NJ 07030, USA

Jossey-Bass, 989 Market Street, San Francisco, CA 94103-1741, USA

Wiley-VCH Verlag GmbH, Boschstr. 12, D-69469 Weinheim, Germany

John Wiley & Sons Australia Ltd, 33 Park Road, Milton, Queensland 4064, Australia

John Wiley & Sons (Asia) Pte Ltd, 2 Clementi Loop #02-01, Jin Xing Distripark, Singapore 129809

John Wiley & Sons Canada Ltd, 22 Worcester Road, Etobicoke, Ontario, Canada M9W 1L1

Wiley also publishes its books in a variety of electronic formats. Some content that
appears in print may not be available in electronic books.

Library of Congress Cataloging-in-Publication Data

Sigee, D. C.
   Freshwater microbiology: biodiversity and dynamic interactions of microorganisms in the
freshwater environment / David C. Sigee.
      p. : cm.
   Includes bibliographical references and index.
   ISBN 0-471-48528-4 (cloth : alk. paper) – ISBN 0-471-48529-2 (pbk. : alk. paper) 1. Freshwater
microbiology.
   [DNLM: 1. Fresh Water – microbiology. QW 80 S574f 2004] I. Title.
QR105.5.S545 2004
579'.176–dc22   2004021738

British Library Cataloguing in Publication Data

A catalogue record for this book is available from the British Library

ISBN 10: 0-471-48529-2 (P/B)
ISBN 13: 978-0-471-48529-2 (P /B)
ISBN 10: 0-471-48528-4 (H/B)
ISBN 13: 978-0-471-48528-5 (H/B)

Typeset in 10/12 pt Times by Thomson Press, New Delhi, India.
Printed and bound in Great Britain by Antony Rowe Ltd, Chippenham, Wiltshire.
This book is printed on acid-free paper responsibly manufactured from sustainable forestry
in which at least two trees are planted for each one used for paper production.

Contents

3 Algae: the major microbial biomass in freshwater systems — 105

5   Inorganic nutrients: uptake and cycling in freshwater systems

9 Grazing activities in the freshwater environment: the role of protozoa and invertebrates   **401**

Preface

Although freshwater microorganisms are not as readily apparent as macroscopic fauna (invertebrates, fish) and flora (higher plants, large algae), they are universally present within aquatic habitats and their ecological impact is of fundamental importance. This book examines the diversity and dynamic activities of freshwater microbes including micro-algae, bacteria, viruses, actinomycetes, fungi, and protozoa.

On an environmental scale, the activities of these organisms range from the microlevel (e.g., localized adsorption of nutrients, surface secretion of exoenzymes) through community dynamics (interactions within planktonic and benthic populations) to large-scale environmental effects. These include major changes in inorganic nutrient concentrations, formation of anaerobic hypolimnia, and stabilization of mudflats. Thus, although freshwater microorganisms are strongly influenced by their physical and chemical environment, they can in turn exert their own effects on their surroundings.

In a temporal dimension, the biology of these biota embraces activities which range from femtoseconds (e.g., light receptor parameters) to seconds (light responsive gene activity), diurnal oscillations, seasonal changes (phytoplankton succession), and transitions over decades and centuries (e.g., long-term response to acidification and nutrient changes).

The study of freshwater microorganisms involves all major disciplines within biology, and I have attempted to bring together aspects of taxonomy, molecular biology, biochemistry, structural biology, and classical ecology within this volume. A complete study of freshwater microorganisms must also clearly relate to other freshwater biota, and I have emphasized links in relation to general food webs – plus specific interactions such as zooplankton grazing, algal/higher plant competition, and the role of fish and macrophytes in the biomanipulation of algal populations.

In addition to being an area of intrinsic biological interest, freshwater microbiology has increasing practical relevance in relation to human activities, population increase, and our use of environmental resources. In this respect microorganisms play a key role in the deterioration of freshwater environments that results from eutrophication, and management of aquatic systems requires a good understanding of the principles of freshwater microbiology. The biology of these biota is also important in relation to other applied aspects such as the breakdown of organic pollutants, the spread of human pathogens, and the environmental impact of genetically-engineered microorganisms.

Inevitably, a book of this type relies heavily on previously published work, and I would like to thank holders of copyright (see separate listing) for granting permission to publish original diagrams and figures. I am also grateful to colleagues (particularly Professor A.P.J. Trinci and Dr R.D. Butler) for commenting on sections of the manuscript, and for the helpful, detailed comments of anonymous reviewers. Finally, my thanks are due to the numerous research colleagues who I have worked with over the years, including Jo Abraham, Martin Andrews, Ed Bellinger, Karen Booth, Ron Butler, Sue Clay, Andrew Dean, Ebtesam El-Bestawy, Robert Glenn, Harry Epton, Larry Kearns, Vlad Krivtsov, Eugenia Levado, Christina Tien, and Keith White.

David Sigee
Manchester, 2004.

Copyright acknowledgements

I am very grateful to the following individuals for allowing me to use previously unpublished material – Drs Nina Bondarenko, Andrew Dean, Harry Epton, Robert Glenn, Eugenia Levado, Ms Elishka Rigney, Mr Richard Sigee, Mrs Rosemary Sigee, Dr Christina Tien, Professor James Van Etten and Dr Keith White.

I also thank the following copyright holders for giving me permission to use previously published material – The American Society for Microbiology, Biopress Ltd., Blackwell Publishing Ltd., Cambridge University Press, Elsevier Science, European Journal of Phycology, Inter-research Science, John Wiley & Sons, Parthenon Publishing, Kluwer Academic Publishers, The Linnean Society of London, NRC Research Press, Scanning Microscopy Inc. and Springer-Verlag.

1

Microbial diversity and freshwater ecosystems

1.1 General introduction

This book explores the diversity, interactions and activities of microbes (microorganisms) within freshwater environments. These form an important part of the biosphere, which also includes oceans, terrestrial environments and the earth's atmosphere.

1.1.1 The aquatic existence

It is now generally accepted that life originated between 3.5 and 4 billion years ago in the aquatic environment, initially as self-replicating molecules (Alberts *et al.*, 1962). The subsequent evolution of prokaryotes, followed by eukaryotes, led to the existence of microorganisms which are highly adapted to aquatic systems. The biological importance of the physical properties of water is discussed in Section 2.1.2.

Life in the aquatic environment (freshwater and marine) has numerous potential advantages over terrestrial existence. These include physical support (buoyancy), accessibility of three-dimensional space, passive movement by water currents, dispersal of motile gametes in a liquid medium, minimal loss of water (freshwater systems), lower extremes of temperature and solar radiation, and ready availability of soluble organic and inorganic nutrients.

Potential disadvantages of aquatic environments include osmotic differences between the organism and the surrounding aquatic medium (leading to endosmosis or exosmosis) and a high degree of physical disturbance in many aquatic systems. In undisturbed aquatic systems such as lakes, photosynthetic organisms have to maintain their position at the top of the water column for light availability. In many water bodies (e.g., lake water column), physical and chemical parameters show a continuum – with few distinct microhabitats. In these situations, species compete in relation to different growth and reproductive strategies rather than specific adaptations to localized environmental conditions.

1.1.2 The global water supply – limnology and oceanography

Water covers seven tenths of the Earth's surface and occupies an estimated total volume of $1.38 \times 10^9 \, km^3$ (Table 1.1). Most of this water occurs between continents, where it is present as oceans (96.1 per cent of global water) plus a major part of the atmospheric water. The remaining 3.9 per cent of water (Table 1.1 shaded boxes), present within continental boundaries (including polar ice-caps), occurs mainly as polar ice and ground water. The latter is present as freely exchangeable (i.e., not

Freshwater Microbiology: Biodiversity and Dynamic Interactions of Microorganisms in the Aquatic Environment   David C. Sigee
© 2005 John Wiley & Sons, Ltd   ISBNs: 0-471-48529-2 (pbk) 0-471-48528-4 (hbk)

Table 1.1 Global distribution of water (adapted from Horne and Goldman, 1994)

Site	Volume (km^3)	% of water within continents
Oceans	1 322 000 000	
Polar ice caps and glaciers	29 200 000	54.57
Exchangeable ground water	24 000 000	44.85
Freshwater lakes	125 000	0.23
Saline lakes and inland seas	104 000	0.19
Soil and subsoil water	65 000	0.12
Atmospheric vapour	14 000	0.026
Rivers and streams	1 200	0.022
Annual inputs		
Surface runoff to ocean	37 000	
Ground water to sea	1 600	
Precipitation		
Rainfall on ocean	412 000 000	
Rainfall on land and lakes	108 000 000	

chemically-bound) water in subterranean regions such as aquifers at varying depths within the Earth's crust. Non-polar surface freshwaters, including soil water, lakes, rivers and streams occupy approximately 0.0013 per cent of the global water, or 0.37 per cent of water occurring within continental boundaries. The volume of saline lakewater approximately equals that of freshwater lakes. The largest uncertainty is the estimation of ground water volume. Annual inputs by precipitation are estimated at 5.2×10^8 km^3, with a resulting flow from continental (freshwater) systems to oceans of about 38 600 km^3.

The distinction between oceans and continental water bodies leads to the two main disciplines of aquatic biology – oceanography and limnology.

- Oceanography is the study of aquatic systems between continents. It mainly involves saltwater, with major impact on global parameters such as temperature change, the carbon cycle and water circulation.

- Limnology is the study of aquatic systems contained within continental boundaries, including freshwater and saltwater sites.

The study of freshwater biology is thus part of limnology. Although freshwater systems do not have the global impact of oceans, they are of major importance to biology. They are important ecological features within continental boundaries, have distinctive groups of organisms, and show close links with terrestrial ecosystems.

The two main sites (over 99 per cent by volume) of continental water – the polar ice-caps and exchangeable ground water – are extreme environments which have received relatively little microbiological attention until recent years. Although most limnological studies have been carried out on lakes, rivers, and wetlands the importance of other water bodies – particularly the vast frozen environments of the polar regions (see below) – should not be overlooked. Microbiological aspects of snow and ice environments are discussed in Sections 2.17 and 3.12, and the metabolic activities of bacteria in ground water in Section 2.14.2.

Although there are many differences between limnological (inland) and oceanic (intercontinental) systems, there are also some close similarities. The biology of planktonic organisms in lakes, for example, shows many similarities to that of oceans – and much of our understanding of freshwater biota (e.g., the biology of aquatic viruses (Chapter 7) comes from studies on marine systems.

1.1.3 Freshwater systems: some terms and definitions

Freshwater microorganisms

Microorganisms may be defined as those organisms that are not readily visible to the naked eye, requiring a microscope for detailed observation. These biota have a size range (maximum linear dimension) up to $200 \, \mu m$, and vary from viruses, through bacteria and archea, to micro-algae, fungi and protozoa. Higher plants, macro-algae, invertebrates and vertebrates do not fall in this category and are not considered in detail, except where they relate to microbial activities. These include photosynthetic competition between higher plants and micro-algae (Section 4.8) and the role of zooplankton as grazers of algae and bacteria (Section 9.8).

Freshwater environments: water in the liquid and frozen state

Freshwater environments are considered to include all those sites where freshwater occurs as the main external medium, either in the liquid or frozen state. Although frozen aquatic environments have long been thought of as microbiological deserts, recent studies have shown this not to be the case. The Antarctic sub-continent, for example, is now known to be rich in microorganisms (Vincent, 1988), with protozoa, fungi, bacteria, and microalgae often locally abundant and interacting to form highly-structured communities. New microorganisms, including freeze-tolerant phototrophs and heterotrophs, have been discovered and include a number of endemic organisms. New biotic environments have also been discovered within this apparently hostile environment – which includes extensive snow-fields, tidal lakes, ice-shelf pools, rock crystal pools, hypersaline soils, fellfield microhabitats, and glacial melt-water streams. Many of these polar environments are saline, and the aquatic microbiology

of these regions is considered here mainly in terms of freshwater snow-fields in relation to extreme aquatic environments (Section 2.17) and the cryophilic adaptations of micro-algae (Section 3.12).

This book deals with aquatic systems where water is present in the liquid state for at least part of the year. In most situations (temperate lakes, rivers, and wetlands) water is frozen for only a limited time, but in polar regions the reverse is true. Some regions of the ice-caps are permanently frozen, but other areas have occasional or periodic melting. In many snow-fields, the short-term presence of water in the liquid state during the annual melt results in a burst of metabolic activity and is important for the limited growth and dispersal of snow microorganisms and for the completion of microbial life cycles (Section 3.12).

Freshwater and saline environments

Within inland waters, aquatic sites show a gradation from water with a low ionic content (freshwater) to environments with a high ionic content (saline) – typically dominated by sodium and chlorine ions. Saline waters include estuaries (Sections 2.12 and 2.13), saline lakes (Section 2.15.3) and extensive regions of the polar ice-caps (Vincent, 1988). The high ionic concentrations of these sites can also be recorded in terms of high electrical conductivity (specific conductance) and high osmotic potential.

The physiological demands of saline and freshwater conditions are so different that aquatic organisms are normally adapted to one set of conditions but not the other, so they occur in either saline or freshwater conditions. The importance of salinity in determining the species composition of the aquatic microbial community was demonstrated in a recent survey of Australian saline lakes (Gell and Gasse, 1990), where distinct assemblages of diatoms were present in low salinity (oligosaline) and high salinity (hypersaline) waters. Some diatom species, however, were present over the whole range of saline conditions, indicating the ability of some microorganisms to be completely independent of salt concentration and ionic ratios. Long-term adaptability to different saline conditions is also indicated by the ability of some organisms to migrate from saltwater to freshwater sites, and establish themselves in their new conditions. This appears to be the case for various littoral red algae of freshwater lakes (Section 3.1.3), which were originally derived from marine environments (Lin and Blum, 1977).

Differences between freshwater/saltwater environments and their microbial communities, are particularly significant in global terms (Chapter 2), where the dominance of saline conditions is clear in terms of area coverage, total biomass, and overall contribution to carbon cycling.

Lentic and lotic freshwater systems

Freshwater environments show wide variations in terms of their physical and chemical characteristics, and the influence these parameters have on the microbial communities they contain. These aspects are considered in Chapter 2, but one important distinction needs to be made at this stage – between lentic and lotic systems. Freshwater environments can be grouped into standing waters (lentic systems – including ponds, lakes, marshes and other enclosed water bodies) and flowing waters (lotic systems – rivers, estuaries and canals). The distinction between lentic and lotic systems is not absolute, and almost all water bodies have some element of through-flow. Key differences between lentic and lotic systems in terms of carbon availability and food webs are considered in Section 1.8.

1.1.4   *The biology of freshwater microorganisms*

In this book the biology of freshwater microorganisms is considered from five major aspects:

- Microbial diversity and interactions within ecosystems (Chapter 1); these interactions include temporal changes in succession and feeding (trophic) interactions.

- Variations between different environmental systems, including lakes, rivers, and wetlands (Chapter 2). Each system has its own mixture of microbial communities, and its own set of physical and biological characteristics.

- Characteristics and activities of the five major groups of microbial organisms – algae (Chapter 3), bacteria (Chapter 6), viruses (Chapter 7), fungi (Chapter 8), and protozoa (Chapter 9).

- The requirement of freshwater microorganisms for two major environmental resources – light (Chapter 4) and inorganic nutrients (Chapter 5). These are considered immediately after the section on algae, since these organisms are the major consumers of both commodities.

- The microbial response to eutrophication. Environmental problems associated with nutrient increase are of increasing importance and reflect both a microbial response to environmental change and a microbial effect on environmental conditions.

A.   BIOLOGICAL DIVERSITY IN THE FRESHWATER ENVIRONMENT

1.2   Biodiversity of microorganisms

1.2.1   *Domains of life*

With the exception of viruses (which constitute a distinct group of non-freeliving organisms) the most fundamental element of taxonomic diversity within the freshwater environment lies in the separation of biota into three major domains – the Bacteria,

Archaea, and Eukarya. Organisms within these domains can be distinguished in terms of a number of key fine-structural, biochemical, and physiological characteristics (Table 1.2).

Cell organization is a key feature, with the absence of a nuclear membrane defining the Bacteria (Figure 3.2) and Archaea as prokaryotes. These prokaryote domains also lack complex systems of membrane-enclosed organelles, have

Table 1.2   The three domains of life* in freshwater environments (adapted from Purves *et al.*, 1997 – shaded areas: distinctive prokaryote/eukaryote features). The four major kingdoms within the Domain Eukarya simply group organisms with broadly similar features, and do not imply any phyllogenetic interrelationships. The Protista (protozoa, algae, sline moulds) and Fungi are particularly heterogeneous assemblages, containing organisms that have arisen via diverse evolutionary routes (polyphyletic)

Characteristic	Domains		
	Bacteria	Archaea	Eukarya
Kingdoms	Eubacteria	Archaebacteria	Fungi Protista Plantae Animalia
Level of cellular organisation	prokaryote	prokaryote	eukaryote
Membrane-enclosed nucleus	Absent	Absent	Present
Membrane-enclosed organelles	Absent	Absent	Present
Ribosomes	70s	70s	80s
Peptidoglycan cell wall	Present	Absent	Absent
Membrane lipids	Ester-linked Unbranched	Ether-linked Branched	Ester-linked Unbranched
Initiator tRNA	Formyl-methionine	Methionine	Methionine
Operons	Yes	Yes	No
Plasmids	Yes	Yes	Rare
RNA polymerases	One	Several	Three
Sensitive to chloramphenicol & streptomycin	Yes	No	No
Ribosomes sensitive to diphtheria toxin	No	Yes	Yes
Some are methanogens	No	Yes	No
Some fix nitrogen	Yes	Yes	No
Some conduct chlorophyll – based photosynthesis	Yes	No	Yes

*Excluding viruses, which are not free-living organisms.

70s ribosomes and have genetic systems which include plasmids and function by operons. Prokaryote features also include a unicellular or colonial (but not multicellular) organization, and a small cell size ($<5 \mu m$ diameter). The domain Bacteria is widely represented in all freshwater environments and contains a single kingdom, the Eubacteria, which includes bacteria, actinomycetes and blue-green algae (see Figure 3.2). In contrast to this, members of the domain Archaea (single kingdom Archaebacteria) tend to be restricted within freshwater environments to extreme situations (Section 2.15). Further subdivisions within this diverse group are shown in Table 2.13 and Figure 2.20.

The domain Eukarya takes its name from the fundamental eukaryote organization of its cells and may be conveniently divided (Purves *et al.*, 1997) into four kingdoms – the Protista (unicellular and colonial organisms), Plantae (multicellular photosynthetic eukaryotes), Fungi (see Chapter 8) and Animalia (multicellular heterotrophic eukaryotes). The Protista include two important groups of freshwater microorganisms – the micro-algae and protozoa.

Taxonomic characteristics of the five major groups of freshwater microorganisms – algae, bacteria, viruses, fungi (with actinomycetes), and protozoa – are detailed in the respective chapters on these organisms.

Table 1.3   Size range of major freshwater planktonic organisms (the shaded area falls within the category of dissolved organic carbon (<0.2 μm))

Size Group	Maximum linear dimension (μm)	Viruses, fungi, and bacteria	Algae	Protozoa	Zooplankton
Femtoplankton	<0.2	Viruses and small bacteria			
Picoplankton	0.2–2	Some large, linear viruses (*Inoviridae*) Most of the bacterial population	Unicellular blue-green algae		
Nanoplankton	2–20	Larger bacteria Fungal zoospores and hyphae (diameter)	Unicellular eukaryotic algae	Nanoflagellates	
Microplankton	20–200		Unicellular and colonial algae	Most planktonic protozoa	Rotifers Copepod larvae
Macroplankton	>200		Large colonial blue-green algae		Crustacea

1.2.2   Size range

Size is an important parameter for all freshwater microorganisms, affecting their location within the freshwater environment, their biological activities, and their removal by predators. The importance of size and shape in planktonic algae is considered in some detail in Section 3.4, and includes aspects such as predation by other organisms, sinking rates, and potential growth rates.

In the case of free-floating (planktonic) organisms, the maximum linear dimension ranges from <0.2 μm–>200 μm, with separation of the biota into five major size categories (Table 1.3), from femtoplankton to macroplankton.

Femtoplankton (<0.2 μm)

The distinction between particulate (insoluble) and non-particulate (soluble) material in freshwater systems is usually defined in terms of retention by a 0.2 μm pore-size filter membrane. On this basis, the smallest size group, the Femtoplankton (<0.2 μm) falls within the non-particulate category and the constituent viruses and small bacteria are strictly

part of the dissolved organic material (DOM) or dissolved organic carbon (DOC) of the freshwater environment.

Picoplankton (0.2–2 μm)

This group is almost entirely composed of prokaryotes (bacteria and blue-green algae) with potentially rapid growth rates and the ability to carry out rapid colonization of freshwater environments. These organisms have negligible sinking rates, and are subject to significant predation by small rotifers, protozoa, and filter-feeding crustaceans. Some large linear viruses (family *Inoviridae*) also fit into this size category (Figure 7.14) and have been linked to the infection of bacterial populations (Section 7.8).

Nanoplankton (2–20 μm)

Typically eukaryote flagellated unicellular organisms, this group includes fungal zoospores, algae, and protozoa. These organisms are the principal food of micro- and macrozooplankton, and have low sinking rates and high potential growth rates.

Within this assemblage, unicellular algae, along with picocyanobacteria, are particularly important in the short-term development of algal blooms which may occur during brief growing seasons or at various points in a more prolonged seasonal sequence.

Microplankton (20–200 μm)

Larger microplankton are retained by traditional ~70 μm mesh size phytoplankton nets, and are highly prone to sinking in the absence of buoyancy aids. These organisms are consumed by larger crustacea, and are also the principal food of pelagic and benthic omnivorous fish. Growth rates are moderate to low.

Macroplankton (>200 μm)

These have similar biological features to the larger microplankton, and are characterized by the colonial blue-green algae and by the multicellular zooplankton (rotifers and crustacea). The biology of meso-trophic and eutrophic lakes in temperate climates are typically dominated by these size categories over the summer growth period, with separate population peaks of colonial algae (diatoms, blue-greens) and zooplankton (crustacea) at different times of year. Although macroplanktonic organisms are characteristically slow-growing, they typically make the greatest contribution to biomass under conditions of adequate nutrient supply. Differences between picoplankton and macroplankton in terms of growth rate, short-term colonization and long-term domination of freshwater environments reflect fundamental differences in evolutionary selection strategy and the way these biota are adapted to different environmental conditions. The distinction between small size (r-strategist) and large size (K-strategist) organisms is considered more fully below.

1.2.3   Autotrophs and heterotrophs

Freshwater microorganisms may be divided according to their feeding activity (trophic status) into two major groups:

- Autotrophs – synthesize their complex carbon compounds from external CO_2. Most also obtain their supplies of nutrient (e.g., nitrogen and phosphorus) from simple inorganic compounds. These phototrophic microorganisms include microalgae and photosynthetic bacteria, and are the main creators of biomass (primary producers) in many freshwater ecosystems. This is not always the case, however, since photosynthetic microorganisms are outcompeted by larger algae and macrophytes in some aquatic systems, particularly wetland communities (Section 4.8).

- Heterotrophs – use complex organic compounds as a source of carbon. By far the majority of freshwater microorganisms (most bacteria, protozoa, fungi) are heterotrophic. Even within the algae, various groups have evolved the capacity for heterotrophic nutrition (Section 3.11) and many organisms currently included in the protozoon assemblage have probably evolved from photosynthetic ancestors.

Heterotrophic nutrition involves a wide diversity of activities (Table 1.4) with microorganisms obtaining their organic material in three main ways – saprotrophy, predation, and in association with living organisms. Saprotrophic organisms obtain their nutrients from non-living material. This may be assimilated in three main ways: direct uptake as soluble compounds (chiefly bacteria), indirect uptake by secretion of external enzymes (exoenzymes) followed by absorption of the hydrolytic products (bacteria and fungi), and ingestion of particulate matter by phagocytosis (protozoa). Predation is carried out by protozoa, and involves capture, ingestion, and internal digestion of other living organisms such as bacteria and algae. Protozoa can capture their prey either by active motility or, as sedentary organisms, by the use of filter feeding processes. The third major aspect of heterotrophy involves associations with living organisms and includes parasitism and symbiosis. Parasitism almost invariably involves a strict dependence of the parasite on the host organism as part of the parasitic life cycle, though in some cases the benefits to the parasite are not entirely clear. The common

Table 1.4 Heterotrophic nutrition in freshwater microorganisms

Mode of nutrition	Group of organisms	Characteristics and ecological importance	Book reference
Saprophytic Uptake of nonliving biomass			
(a) Osmotrophy: direct uptake of soluble organic compounds	Planktonic heterotrophic bacteria	Assimilation of dissolved organic carbon (DOC) secreted by phytoplankton basis of microbial loop in planktonic systems	Section 4.7.2 Section 6.10 Figures 6.13 and 6.16
	Osmotrophic algae and protozoa	Uptake of soluble organic compounds in high nutrient environments	Algae: Section 3.11 Protozoa: Section 9.6.3 and 9.6.4
(b) External digestion of insoluble biomass	Bacteria, actinomycetes, and fungi	Secretion of exo-enzymes, uptake of digestion products; typical of benthic environments of lakes and streams where organic detritus accumulates	
	Bacteria	Divide and penetrate biomass as invasive populations	Section 6.6.3
	Actinomycetes and fungi	Penetrate biomass via extension of mycelial system	Actinomycetes: Section 8.2.3 Fungi: Section 8.5
(c) Detritus feeders; ingestion of dead particulate material	Protozoa	Engulph dead particulate biomass by phagocytosis; typical of benthic environments	Section 9.6
Predation (carnivory); ingestion and killing of live organisms	Protozoa	Catch and ingest living microorganisms by phagocytosis; important in the control of bacterioplankton populations by heterotrophic nanoflagellates (HNFs) and in benthic environments	HNFs: Sections 6.10.2; and 9.5 Benthic: Section 9.6
	Heterotrophic algae	Catch and ingest bacteria and other algae; strategic response to limitations in photosynthesis	Section 3.11
Association with living organisms			
(a) Parasitism (unilateral nutritive benefit)	Viruses	Major parasites of all freshwater biota; important in the control of phytoplankton and bacterioplankton populations	Chapter 7: Phycoviruses Cyanophages Bacteriophages
	Fungi	Important parasites of a wide range of organisms, including: (a) Phytoplankton; chytrid fungi have major importance in limiting planktonic algal populations (b) Invertebrates	Sections 8.6 and 8.7 Section 8.6.2
	Bacteria	Carried out by specialized bacteria such as *Bdellovibrio*	Section 6.12.2

Table 1.4   (*Continued*)

Mode of nutrition	Group of organisms	Characteristics and ecological importance	Book reference
	Protozoa	Opalinid protozoa commonly occur as parasites in the gut of freshwater amphibia[a]	
(b) Symbiosis (mutual nutritive benefit)	Bacteria	Association with blue-green algal heterocyst	Section 6.13.3 Figures 6.23 and 6.24
(c) Commensalism (no direct nutritive benefit)	Bacteria, protozoa, algae	Many microorganisms have a casual epiphytic association with the surfaces of other biota; the phycosphere of colonial algae may have a diverse population of such organisms	Section 6.13.1 Figure 6.23

The above table contains examples of the major heterotrophic interactions and associations occurring in the freshwater environment and is not intended to be a complete listing.
[a] Sleigh, 1973.

presence of Opalinid protozoa in the gut of adult amphibia, for example, is thought to involve nourishment from mucous secretions and gut fluids, but there is no penetration of host tissues or any apparent adverse effects on the host (Sleigh, 1973). As with other freshwater parasites, this organism has a well-defined life cycle that fits in with the aquatic biology of the host. Commensal microorganisms such as epiphytic bacteria, protozoa, and algae derive no direct nutrient from their associated host, though there may be indirect nutritive benefits. These are seen in the phycosphere community that occurs in association with mucilaginous colonial algae (Figure 6.23), where a close coupling occurs between the local growth of bacterial populations and the trophic activities of bacterivorous protozoa.

Within the nutritive diversity shown in Table 1.4, particular heterotrophic activities stand out as having special ecological significance. These include the uptake of DOC by planktonic bacteria, breakdown of detritus by benthic organisms (bacteria, fungi, and protozoa) and limitation of phytoplankton populations by parasitic activity (viruses and fungi). The importance of heterotrophic nutrition in the freshwater environment runs as a theme throughout the book, with specific examples discussed in appropriate sections (listed in Table 1.4).

Autotrophic and heterotrophic activities within communities

The overall balance between autotrophic and heterotrophic activities within freshwater communities is an important aspect of ecosystem function (Pradeep Ram *et al.*, 2003). This determines the net exchange of carbon with the surrounding atmosphere (which influences global warming) and can be expressed as the *net ecosystem production* (*NEP*), where

$$NEP = P - R \qquad (1.1)$$

and *P* = primary productivity (carbon uptake during photosynthesis) *R* = community respiration (carbon loss by respiratory breakdown).

NEP will have a positive value in ecosystems that are net autotrophic (*P* > *R*), and a negative value where the community is net heterotrophic (*P* < *R*).

The balance between *P* and *R* is determined by the relative metabolic contributions of autotrophic and heterotrophic organisms, which in turn relates to environmental parameters such as light availability (promoting phototrophy) and availability of externally-derived (allochthonous) carbon-promoting heterotrophy.

Benthic communities vary from net heterotrophy (e.g., profundal sediments, dominated by anaerobic heterotrophic bacteria) to net autotrophy (e.g., shallow lake sediments, dominated by benthic algae – Figure 2.4). Pelagic communities typically have greater light exposure than benthic systems, and (in lentic water bodies) often have a greater autotrophic contribution. The balance of available nutrients is important, and low nutrient (oligotrophic) lakes are typically net heterotrophic while high nutrient (eutrophic) lakes show net autotrophy (see Tables 2.6 and 10.3). This difference arises because oligotrophic conditions support only limited carbon uptake by phototrophy (low inorganic nutrient availability) but significant heterotrophy can still occur due to input of allochthonous carbon. In eutrophic lakes, both modes of nutrition are increased, but the increased availability of soluble inorganic nutrients promotes autotrophy to a greater extent.

The pelagic communities of lotic systems may also show great variation in net ecosystem production. This is seen particularly well in some estuaries (Section 2.12), where the value for *NEP* shows marked seasonal fluctuation with river inflow (Pradeep Ram *et al.*, 2003).

1.2.4 Planktonic and benthic microorganisms

Freshwater organisms can be divided into two main groups, according to where they spend the major part of their growth phase – pelagic organisms (present in the main body of water) and benthic organisms (associated with the sediments). Pelagic biota can be further subdivided into nekton (strongly swimming organisms such as fish) and plankton (free-floating). The latter tend to drift within the water body, though they may have limited motility and achieve vertical movement. Pelagic microorganisms are essentially plankton, so the distinction within this assemblage is between planktonic and benthic states.

Although freshwater organisms can be categorized as planktonic or benthic, most species have both planktonic and benthic phases within their life cycle. Biofilm bacteria, for example, have a dynamic equilibrium between attached and planktonic states, and freshwater algae also show clear planktonic/benthic interactions in lake and stream environments (Section 3.6.1). Planktonic and benthic phases of the same species typically show considerable differences in metabolic activity. This is particularly the case for freshwater bacteria, where quorum-sensing mechanisms lead to the induction of stationary-phase physiology in biofilm communities (Figure 1.7).

Planktonic microorganisms

These biota frequently have specialized mechanisms to migrate or maintain their position vertically within the water column, and are particularly well represented by micro-algae and bacteria. Planktonic organisms can be further divided into holoplanktonic forms (where the organism is present in the water column for a major part of the annual cycle) and meroplanktonic organisms (where the planktonic phase is restricted over time) biota. The distinction between holoplanktonic and meroplanktonic species is particularly well shown by the algae, where the meroplanktonic state can be viewed as an adaptation for short-term competition (Section 3.10.1).

Benthic microorganisms

Present at the surface and within sediments, these are dominated by biota such as fungi, protozoa, and bacteria that are able to break down sedimented organic debris. The diversity of benthic life forms is particularly well represented by the protozoa (Section 9.6), which include both attached and freely-motile organisms, with a variety of feeding mechanisms.

A number of key terms have been used to define particular groups of benthic organisms, including one major group – the periphyton. This term is used to describe all the 'plant-like' microorganisms (microflora) present on substrata, including microscopic algae, bacteria and fungi. This term excludes 'animal-like' organisms, such as micro-invertebrates and protozoa, but includes filamentous algae (e.g., *Cladophora*, *Spirogyra*, *Chara*, and *Vaucheria*).

Differences occur within the periphyton in terms of the nature of the substrate, which may be living (e.g., plant surfaces) or non-living (organic or inorganic, with different particle sizes).

- *Epiphytic organisms* are associated with the surfaces of higher plants and macroalgae. The substratum in this case is often metabolically active, and the epiphytic association may be important in terms of competition for light, metabolic exchange and nutrient availability.

- *Epilithic, epipsammic and epipelic microorganisms* grow on non-living substrates which differ in particle size. Epilithic biota occur on hard, relatively inert substrata such as gravel, pebbles and large rocks. Decrease in particle size leads to epipsammic organisms (present on sand) and epipelic organisms (present on fine sediments such as mud). Relatively few of the larger microorganisms, such as algae, live in sand, since particles are too unstable and may crush them. Quite large algae are present on fine sediments, however. including large motile diatoms, motile filamentous algae and large motile flagellates such as *Euglena*.

A further group of plant-like benthic organisms includes the metaphyton, which are loosely associated (but not attached) to the substratum. These are characterized by clouds of filamentous green algae such as *Spirogyra*, *Mougeotia* or *Zygnema*, which become loosely aggregated and accumulate in regions of substratum that are free from currents and waves. Metaphyton usually arises from other substrata, such as higher plant surfaces (epiphyton).

1.2.5 Metabolically active and inactive states

Although freshwater microorganisms are often thought of as dynamic, metabolically active biota, this is not always the case. All species have inactive or inert periods for at least part of their life cycle; these may occur as temporary dormant phases (sur-

viving adverse conditions) or terminal phases of senescence, leading to death.

Dormant phases

In temperate climates, particularly, many microorganisms overwinter on the sediments in a dormant state. The formation of resistant spores typically occurs as environmental conditions deteriorate (overcrowding, reduced nutrient and light availability, accumulation of toxic metabolites, reduced temperature), and may be preceded by sexual reproduction. Figure 1.1 shows the active and dormant states of the protozoon *Actinophrys*, and dormant spores (akinetes) of the blue-green alga *Anabaena* can be seen in Figure 5.14. In some cases, dormancy appears to relate to a specific environmental change, such as oxygen concentration. In the water column of lakes, obligate anaerobes such as photosynthetic purple bacteria are metabolically active in the anaerobic hypolimnion during stratification, but overwinter in flocs of organic material once the lake becomes oxygenated at autumn overturn (Eichler and Pfennig, 1990).

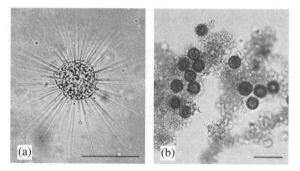

Figure 1.1 Metabolically active and inactive phases in *Actinophrys*, a heliozoon amoeba: (a) active pelagic or trophic phase – arms (axopodia) radiate out from a central body and capture prey (mainly flagellates and other protozoa) which make passive contact; (b) metabolically-inactive benthic phase – low power view of group of dense, resistant cysts within sedimentary debris, the pelagic phase transforms to a cyst when the food supply becomes exhausted and sinks to the bottom of the water body (scales = 50 µm)

In bacteria and viruses, metabolic inactivity is also nutrient-related. Comparisons of total and viable bacterial counts suggest that only a small fraction of aquatic populations are metabolically-active. Under certain environmental conditions, bacteria remain inert until they encounter appropriate growth conditions – particularly in relation to nutrient supply. Viruses are also metabolically inactive within the water medium, and are only activated when they encounter a healthy host cell (Section 7.2). In the inert state, they occur as free particles (virions) within the water medium, and are vulnerable to inactivation and loss processes (Figure 7.1).

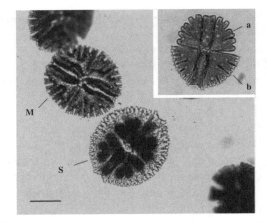

Figure 1.2 Cycles of death and regeneration in an algal population (scale = 100 μm)

Senescence

Senescence and cell death are just as much a part of population development as growth processes, and play an important role in the dynamics of microbial communities. In some situations, death is induced by specific environmental factors such as nutrient limitation (Section 5.2.4), solar irradiation (photo-inhibition, Section 4.9) and parasitism (Chapters 7 and 8). In other situations, senescence may occur due to internal changes, particularly after a long sequence of divisions and probably due to programmed cell death.

The occurrence of growth and senescence within a single-species population is shown in Figure 1.2, which is taken from a stationary phase in a laboratory monoculture of the green alga *Micrasterias*. Mature cells (M) of this placoderm desmid have two halves (semicells) of equal size, each with a single chloroplast. Cells that are undergoing senescence (S) show condensation of the chloroplasts to the centre of the cell, leading to degeneration and the formation of colourless dead cells that just have the remains of the cell wall. Actively growing cells that have recently divided (inset) have one normal sized semicell (b) derived from the mother cell, plus a new semicell (a) that will attain full size on completion of the growth cycle. Although cell differences in growth cycle and the onset of senescence are particularly clear in populations of this organism, they are also characteristic of other populations where they are not so easily observed. Within phytoplankton populations at the top of the water column, for example, senescence and cell death occurring in a wide range of species results in a continuous rain of mixed organisms to the lower part of the water column and the sediments. These can be collected in sediment traps suspended in the water column and subsequently analysed.

1.2.6 Evolutionary strategies: r-selected and K-selected organisms

In the population of cells shown in Figure 1.2, regeneration (formation of new organisms) and death continuously occur. The growth of populations that include continuous cycles of regeneration and death is best described by 'models of continuous growth' (Begon *et al.*, 1996), and leads to a consideration of the importance of competition in population increase and the designation of two opposing evolutionary strategies – adopted by r-selected and K-selected organisms.

Exponential and sigmoidal models of continuous growth

The speed with which a single-species population (N) increases with time will be denoted by dN/dt.

The increase in size of the whole population is the sum of the growth contribution of all individuals, so the rate of increase per individual (r) will be:

$$r = \frac{dN}{dt}\left[\frac{1}{N}\right] \qquad (1.2)$$

or

$$rN = \frac{dN}{dt}$$

The parameter r is also referred to as the 'per capita rate of increase' and describes the fundamental ability of individual organisms to grow and reproduce. The above equation describes population growth in conditions of unlimited resources and generates an exponential increase in population with time (dashed line, Figure 1.3). In practice, intra-specific competition for resources (particularly nutrients and space) must also be taken into account, and the population growth curve then becomes sigmoidal (solid line, Figure 1.3). In this situation, the population (N) rises to a maximum value (K), which is the maximum value (the carrying capacity) that the environment can support.

The growth equation which describes the sigmoidal curve is referred to as the 'logistic equation' and has the form:

$$\frac{dN}{dt} = rN\left[\frac{K-N}{K}\right] \qquad (1.3)$$

This sigmoidal curve is well-known to microbiologists, where the growth of microorganisms in batch culture follows a sequence of lag, logarithmic and stationary phases. In the lower part of the curve (lag phase), when population is sparse, the population is beginning to colonize the new environment and relatively little competition occurs between cells. As the population approaches the carrying capacity (K), resources (nutrients, space) become limited and high levels of competition occur between cells.

r-selected and K-selected organisms

The conditions seen in the single-species growth curves of organisms cultured under laboratory conditions (Figure 1.3) can also be applied to freshwater environments, where mixed populations occur. Some environments have low population densities, and are dominated by organisms that are adapted for high rates of growth and rapid colonization, while others have high population densities and

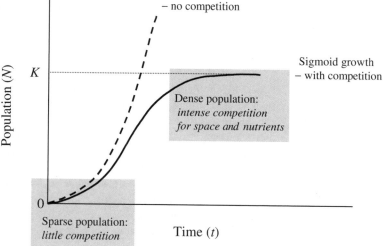

Figure 1.3 Effect of intra-specific competition on the population growth of a single species (based on a figure from Begon *et al.*, 1996)

Table 1.5 Biological characteristics of environments that are dominated by r-selected and K-selected organisms. Many of these characteristics (shaded boxes) directly relate to differences in population level

Environmental parameter	Domination by r-selected organisms	Domination by K-selected organisms
Example	• Lake Clearwater phase • Early biofilm	• Lake autumn bloom • Maturating biofilm
Population density	**Low**	**High**
Biological diversity	Low to moderate	Typically high
Competition	Low	High
Accumulation of metabolites in water medium	Low	High – including DOC and blue-green toxins
Resource availability	High availability of space and/or nutrients	Limited
Liability of a species population to parasitic (virus, fungal) attack	Typically low, but may be high in high-stress situations	High (high host population density)
Population stability	Low – populations liable to sudden change	High
Stress level	May be low (biofilm colonization) or high (lake Clearwater phase)	May be low (normal autumn algal bloom) or high (intense blue-green dominance)

are dominated by organisms that are adapted to survive in these highly competitive conditions.

This fundamental distinction between two major types of adaptation was first described by MacArthur and Wilson (1967) in relation to differences in selection pressure at different phases in the colonization of oceanic islands. These authors separated organisms into r-strategists, which were the initial colonizers of the island environment, and K-strategists – adapted to the more crowded, later conditions. In uncrowded environments, organisms able to grow and reproduce rapidly, with high productivity, will be most suited to dominate the environment. They are subsequently replaced by organisms which are more suited to a crowded community that is approaching its population limit and are referred to as K-strategists. The terms r and K are derived from the logistic equation for population growth, which specifies that the *per capita* rate of increase (r) is maximized under sparse conditions. With increasing crowding, a decline in the *per capita* rate of increase occurs until the population density equilibrates to its upper level or carrying capacity (K).

The adaptation of r- and K-strategists to environments of differing population density has a range of biological implications (Table 1.5), including differences in biological diversity, competition, accumulation of metabolites, resource availability, and liability to parasitic attack. The incidence of ecological stress (see Section 1.6) can be high or low in either situation, depending on the particular circumstances. The adaptation of r-strategists to uncrowded conditions also makes them suitable to unstable environments, where growth is limited to short periods of time and high population levels cannot become established. Under such conditions the ability of these organisms to grow rapidly and exploit growth opportunities as they become available gives them a competitive advantage. In line with this, r-strategists are typically small, with a short life cycle and high growth rate. In contrast, K-strategists are adapted to stable environmental conditions, where dense populations develop and rapid growth is not an advantage (Andrews, 1992). All groups of freshwater microorganisms contain r- and K-selected species, and all aquatic environments have situations (during colonization, and at different times in the mature community) when *r*- and *K*-strategists are respectively adapted to the prevailing situation. Planktonic bacteria are particularly good examples of *r*-strategists (Section 6.1.3), existing in a metabolically inert form for much of their life cycle, but able to grow and multiply within a short space of time when nutrients become available. Phytoplankton contain well-adapted examples of both types of organism, with changes in dominance of *r*- and *K*-selected algae during the seasonal progression. One particular group of algae, the dinoflagellates, are perhaps the best example of all microbial *K*-strategists (see Section 3.10.2).

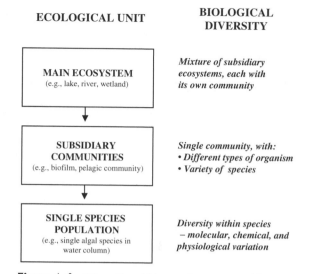

Figure 1.4   Diversity within species, communities, and groups of communities at different levels of organisation in freshwater systems

1.3   Biodiversity in ecosystems, communities, and species populations

Individual microorganisms are part of larger groups or communities, and any consideration of biodiversity must take into account the wider community and the environment in which it occurs.

A community is a naturally occurring group of organisms living in a particular location, such as a lake or stream. The occurrence and interactions of organisms living within a discrete environment constitutes a functional unit, referred to by Tansley (1935) as an ecological system or 'ecosystem'. This was considered by Tansley to consist of two major components – the biome or ecosystem community (the entire group of organisms) and the habitat (physical environment). Each particular ecosystem can be regarded to some extent as self-contained and as a basic unit of ecology. Ecosystems are themselves part of a larger geographic or global unit, the biosphere, which is the sum of all ecosystems within a particular zone.

The nature and evaluation of community diversity partly depends on environmental scale and can be considered as a hierachy of three levels – main ecosystems, subsidiary communities, and species populations (Figure 1.4).

1.3.1   Main ecosystems

Major aquatic environments such as lakes, rivers and wetlands form a discrete ecological unit, with their own characteristic community of organisms; they can be referred to as main ecosystems. Each of these main ecosystems contains a diverse array of distinctive groups of organisms (subsidiary communities), each in their own particular environment, forming subsidiary ecosystems.

As an example of this, a typical temperate lake (main ecosystem) can be divided into two major regions – the peripheral shoreline (littoral) zone and the central zone (Figure 2.4), each with its own subsidiary communities and ecosystems. The central zone can be separated into pelagic and benthic groups of organisms, with further resolution within the pelagic zone of neuston, phycosphere and photosynthetic bacterial communities (Table 1.6). These occur in discrete small-scale environments, and are entirely composed of microorganisms.

The littoral zone of the lake is dominated by attached organisms, which in eutrophic lakes can be divided into three major groups – macrophyte, upper

Table 1.6 Subsidiary communities within the central and littoral zones of a stratified temperate lake (main ecosystem)

Main part of lake	Subsidiary community	Physical characteristics	Book reference
Central zone	Pelagic community	Varied throughout water column – epilimnion and hypolimnion	Section 3.7.2
	Neuston	Lake surface microlayer	Section 2.16 Figure 2.21
	Phycosphere community	Colony mucilage – attachment and matrix for biota	Section 6.13 Figure 6.23
	Population of photosynthetic bacteria	Top of hypolimnion – anaerobic plus limited light	Section 6.7.3
	Benthic community	Bottom of water column and sediments; anaerobic, nutrient-rich, no light	Section 6.6.3
Littoral zone	Macrophyte community	Exposed to high light	Section 3.6.2
	Upper periphyton (eulittoral) community	High sunlight and temperature, with wave action in some lakes	Figure 3.11
	Lower periphyton (sublittoral community)	Thermally stable region, with low light and turbulence	Figure 3.11
	Epiphytic community on macrophyte leaves	Attached to plant surface; high light, oxygenated, water motion	Figure 3.10
	Algal and bacterial biofilms on exposed rock surfaces	Thin surface layer on hard substrate	Sections 3.7.2 and 6.11

Microbial communities are indicated by shading.

periphyton and lower periphyton communities (see Figure 3.11). Within these, various distinctive micro-communities occur including epiphytic communities on macrophyte leaves, and algal and bacterial biofilms on exposed rock surfaces.

Although these various groups of organisms have a degree of autonomy, they all interrelate within the main ecosystem. The pelagic and benthic communities, for example, may appear very distinct during the summer growth phase of eutrophic lakes but the benthic community depends on continuous biomass input from the pelagic zone by sedimentation, and populations of planktonic organisms arise by recruitment from the benthic zone.

1.3.2 Diversity within subsidiary communities

Different communities have their own distinctive pattern of organisms with their own level of biodiversity. Within these communities, this diversity can be measured in various ways including variation in size of organisms, presence of mucilaginous and non-mucilaginous types, attached and free-floating biota and proportions of autotrophic and heterotrophic species. In practice, diversity is often determined simply in terms of species content. This is illustrated in Section 3.13. where the variety of algae within phytoplankton is considered in terms of indices of species diversity. Significant changes occur in these indices during seasonal development of phytoplankton (Figure 3.28), indicating fundamental changes in community diversity throughout the growth season.

1.3.3 Biodiversity within single-species populations

Phenotypic and genetic diversity within the ecosystem populations of individual species represents an important but relatively unexplored area of biological diversity in freshwater systems. One of the problems in studying variability at this level is that classical biochemical (test-tube) techniques are difficult to apply to single species within mixed populations.

The use of analytical microscopical techniques, however, has gone some way to overcoming this problem since these have sufficient spatial resolution to determine the chemical and molecular composition of single microorganisms within mixed populations. Some of these techniques are discussed in relation to algal populations (Section 3.2.2) and include the use of light microscope infrared spectroscopy to study the vibrational states of different molecular groups (Figure 3.3; Sigee *et al.*, 2002) and the use of scanning electron microscope X-ray micro-analaysis to determine variations in elemental composition (Figure 5.2; Sigee *et al.*, 1993).

Both of these approaches have demonstrated considerable variability within the single-species populations that make up the mixed phytoplankton assemblage, indicating that intra-specific variation in planktonic algae is an important feature of the pelagic environment.

Molecular techniques also have considerable potential for looking at biodiversity within single species populations, and have been used to study various aspects of genetical variation within natural populations. These include sub-species (strain) variation in blue-green algae (Section 3.3.1) and variations in plasmid-borne resistance in aquatic bacteria (Section 6.5.1). Differences between planktonic and attached (biofilm) bacteria in terms of the expression of stationary phase genes (Section 1.4.1) also constitute an important an important intra-specific variable within benthic systems.

B. ECOSYSTEMS

Ecosystems vary in size and composition, and contain a wide range of organisms which interact with each other and with the environment. Individual ecosystems have a number of important properties (McNaughton, 1993; Berendse, 1993):

- a distinct pattern of interactions between organisms,
- defined routes of biomass formation and transfer,
- maintenance of the internal environment,
- interactions with the external environment.

The range of organisms present in aquatic communities define and characterize the system concerned, and are involved in the generation and transfer of biomass. They have distinct roles and interactions within the ecosystem, occupying particular trophic levels and forming an interconnected system of feeding relationships (the food web). The balance of individual species within the food web is determined primarily by resource (light, nutrient availability) and competition. This in turn affects

variety in the range and proportions of the different organisms (biodiversity), with important implications for the overall functioning of the system. Community structure is closely related to ecosystem stability and physical stress levels in the environment (see Section 1.6). Interactions between ecosystems and their surrounding environment occur in various ways. One example of this is the net exchange of carbon between the aquatic ecosystem and the adjacent atmosphere, which can be quantified in relation to net ecosystem production (NEP) (see Section 1.2.3.)

This section considers the dynamic properties of freshwater communities within two ecosystems where microorganisms have a key role – the microbial biofilm and the pelagic ecosystem of lakes. The microbial biofilm is a small-scale ecosystem composed entirely of microorganisms, while the pelagic ecosystem is a large-scale functional unit containing a wide range of biota. Summary diagrams of ecosystem functions for these two systems are respectively given in Figures 1.6 and 1.8. Contrasts between the diagrams to some extent reflect diversity in research approach rather than fundamental ecosystem differences. In the case of biofilms, for example, a lot of information has been obtained on genetic interactions, but relatively little is known about food webs. The reverse is true for pelagic ecosystems.

1.4 The biofilm community: a small-scale freshwater ecosystem

Microbial biofilms provide a useful model system for considering fundamental aspects of community interactions and ecosystem function (Allison *et al.*, 2000). Their small scale makes them amenable to laboratory as well as environmental experimentation, and the close proximity of organisms within the biofilm leads to high levels of biological inter-

action. Within the context of this book, the exclusively microbial composition of biofilms also makes them particularly appropriate for consideration.

Microbial biofilms occur as discrete communities within a gelatinous matrix, and are present as a surface layer on rocks and stones in lakes and rivers. The biological composition of biofilms varies with environmental conditions, including factors such as ambient light intensity, water flow rate and prior colonization history. In some cases they are entirely bacterial (Section 6.11), while in others they initiate with the settlement of diatoms and develop into larger scale periphyton communities (Figure 9.17). The biofilm shown in Figure 1.5 is a mature mixed biofilm, such as might occur on a river substratum under conditions of limited light, and consists of a balanced community composed mainly of bacteria, with algae, protozoa, and fungi also present. The

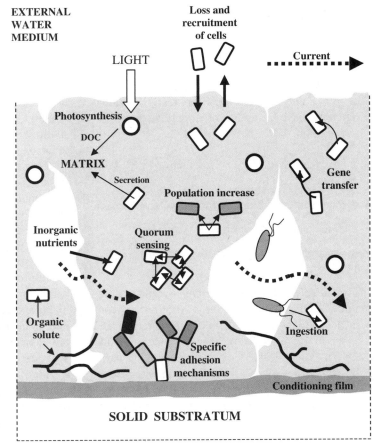

Figure 1.5 Biological interactions within a mixed biofilm: A small-scale microbial ecosystem. Found as a surface layer on rocks and stones in rivers and the littoral zone of lakes, the thickness of the biofilm ranges from a few micrometers (μm) to several millimetres (mm). Different microorganisms within the biofilm (not drawn to scale) are indicated by the following symbols: Bacteria - ▭ Algae ○ Protozoa ⬭ Fungi ∿ In a mature biofilm, the gelatinous matrix ( ) is typically highly structured, occurring as columns with interstitial spaces through which water can percolate (•••►). Water flow also occurs across the surface of the biofilm. Solid arrows simply relate organisms to particular activities (see also Figure 6.17)

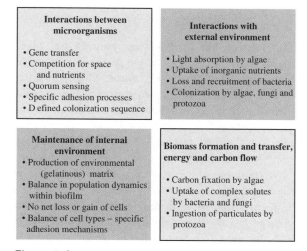

Interactions between microorganisms
• Gene transfer • Competition for space and nutrients • Quorum sensing • Specific adhesion processes • Defined colonization sequence

Interactions with external environment
• Light absorption by algae • Uptake of inorganic nutrients • Loss and recruitment of bacteria • Colonization by algae, fungi and protozoa

Maintenance of internal environment
• Production of environmental (gelatinous) matrix • Balance in population dynamics within biofilm • No net loss or gain of cells • Balance of cell types – specific adhesion mechanisms

Biomass formation and transfer, energy and carbon flow
• Carbon fixation by algae • Uptake of complex solutes by bacteria and fungi • Ingestion of particulates by protozoa

Figure 1.6 Major ecosystem functions in a mixed microbial biofilm community

organisms are largely present within a gelatinous matrix, which defines the extent of the ecosystem. The matrix typically has a columnar appearance, with channels or pores between the columns through which water percolates.

The microbial biofilm shares with other small-scale microbial (e.g., neuston and phycosphere) ecosystems and with larger ecosystems the four key aspects of ecosystem function listed above – interactions between organisms, biomass transfer, homoestasis and interactions with the external environment (Figure 1.6).

1.4.1   Interactions between microorganisms

Microbial interactions illustrated in Figure 1.6 occur both within the main population of bacteria and in relation to other microorganisms. Bacterial interactions determine the structure and diversity of the biofilm and include gene transfer, quorum sensing and specific adhesion processes. Trophic (feeding) interactions occur between different groups of biota and can be considered in terms of food webs.

Gene transfer

Microbial biofilms are particularly important in relation to gene transfer between bacterial cells,

since they are a part of the aquatic environment where the transfer process is optimized due to the close proximity of the organisms concerned. The importance of genetic diversity and spread of novel genes is discussed in Section 6.3, and the mechanisms of transfer by transformation, transduction or conjugation, in Section 6.4. Gene transfer between bacterial cells has been studied under both laboratory and environmental conditions.

Quorum sensing

The physiological activities of bacteria vary considerably in relation to population density (number of cells per unit volume of medium). The pattern of gene activity, in particular, differs markedly within a single bacterial species between the sparse planktonic populations that occur in the general water medium and the dense community of cells present in the microbial biofilm. The mechanism underlying this difference in gene expression is referred to as 'quorum sensing' and depends (in Gram-negative bacteria) on the release of the signal molecule acyl homoserine lactone (AHL) into the water medium (Davies, 2000).

At low population density, release of AHL results in the formation of low signal concentrations in the water medium, and no quorum response. At high population density, the concentration of AHL in the water reaches a critical level, activating DNA transcription factors (by binding to them) and triggering the activation of AHL-responsive genes (Figure 1.7). These AHL-responsive genes have been shown to be present in a wide range of biofilm bacteria (including *Pseudomonas aeruginosa*) and to operate in environmental (stream) biofilms (McLean *et al.*, 1997). They encode a variety of cell functions, including population density regulation, pseudomonad virulence factors and stationary-phase characteristics. The induction of stationary-phase physiology is an important feature of biofilms, distinguishing these organisms from their planktonic counterparts. In *Pseudomonas aeruginosa*, the stationary phase state is caused by induction of a stationary phase sigma (transcription) factor known as RpoS (Suh *et al.*, 1999), which activates a wide

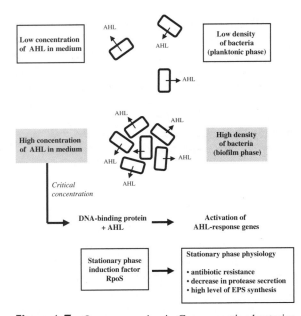

Figure 1.7   Quorum sensing in Gram-negative bacteria: induction of stationary phase physiology at high population density in biofilms (AHL – acyl homoserine lactone signal molecule)

spectrum of stationary phase characteristics. These include greater antibiotic resistance, decrease in protease secretion, reduced motility and higher levels of extracellular polysaccharide (EPS) synthesis. These changes promote the formation of the biofilm (EPS) matrix, and the retention of cells within the surface film (reduced motility), and are thus crucial in forming and maintaining the biofilm ecosystem.

Specific adhesion processes

Specific adhesion mechanisms are important in the secondary colonization and maturation of bacterial biofilms (Section 6.11). During biofilm development, different species of bacteria enter the community at different times, and there is a need for specific recognition systems to maintain the necessary sequence and hierarchy of association. This is achieved by specific co-aggregation, which involves receptor-mediated recognition and specific binding between different species of bacteria (Figure 6.18).

Trophic interactions

Ingestion of biofilm bacteria, diatoms and blue-green algae is carried out by protozoa that move over exposed submerged surfaces. These protozoa include hypotrich ciliates, hypostome ciliates (with ventral mouths) plus bodonid and euglenoid flagellates (Section 9.3). Ingestion of biofilm microorganisms is the first stage in the sequence of biomass transfers that forms the food web of the biofilm ecosystem, but also connects with the food web of the wider lake or river ecosystem.

In addition to protozoa, surface biofilms are also grazed by larger invertebrates such as snails and larvae of Ephemeroptera and Trichoptera (Section 9.13). These larger organisms are extraneous to the biofilm community.

1.4.2   Biomass formation and transfer

Relatively little is known about biomass formation and transfer in aquatic biofilms. The presence of autotrophic algae such as diatoms and blue-green algae generates fixed carbon by photosynthesis, and some release of dissolved organic carbon (DOC) would be expected. Biomass transfer occurs via ingestion of cellular and other particulate material by protozoa and invertebrates (see above), and also by assimilation of secreted organic material such as DOC and matrix by bacteria and fungi. Extraneous organic material may also enter biofilms and become part of the biomass transfer.

1.4.3   Maintenenace of the internal environment

The microbial biofilm illustrated in Figure 1.5 is a mature film that has arisen from a sequence of colonization processes (Section 6.11). This has resulted in a stable and balanced internal environment which has two major components:

- The gelatinous matrix, secreted by the bacteria, and forming a complex architecture that includes internal spaces (pores and channels) with a water circulation connecting to the outside medium.

- Populations of different microorganisms that are in a state of balanced equilibrium.

Once the biofilm has become established, the internal environment will be maintained and controlled by the resident microorganisms. This involves the following processes.

1. *Continued production of gelatinous matrix.* Continued secretion of the extracellular polysaccharide (EPS) matrix by bacterial cells tends to balance the loss caused by detachment of pieces of biofilm at the water surface and entrainment in the current.

2. *Controlled population growth.* Growth rates of bacteria within the biofilm are controlled by the quorum sensing system that acts as a negative feedback process. High population densities trigger the induction of stationary phase characteristics, including reduced rates of cell division. If the population of bacteria becomes depleted (e.g., due to ingestion) the quorum control will cease to operate, and division will increase.

3. *Balance of mixed populations.* The balance of different organisms will be determined by differential growth rates, stable food webs and specific adhesion mechanisms. The latter is particularly important during biofilm development and maturation, when the changing pattern of species composition that accompanies the colonization process is determined by differential recognition and adhesion characteristics. These features may also control the entry of bacteria into the mature biofilm and their retention within the community.

1.4.4 Interactions with the external environment

Although microbial biofilms operate as a functional (and to some extent self-contained) unit, they are not separate from either their physicochemical or biological surrounding environment. Important physicochemical characteristics include light (required for photosynthesis), inorganic nutrients, soluble organic nutrients and dissolved oxygen. The entry and exit of soluble components into and out of the biofilm is promoted by internal water circulation, and leads to concentration gradients within the gelatinous matrix. These concentration gradients may be important determinants of local biofilm physiology and microstructure. The potential importance of nutrients is indicated by laboratory studies on monospecies biofilms (Molin *et al.*, 2000), which have suggested that biofilm structure is greatly influenced by the concentration and quality of nutrients.

The external biotic environment also has direct influences on biofilms. These are exerted through invertebrate grazing activities, and also entry and loss of resident microbes at the water interface. The entry of particulate (by sedimentation) or soluble (water flow) organic materials into the biofilm is a further biotic effect that provides important substrates for heterotrophic nutrition.

1.5 The pelagic ecosystem: a large-scale unit within the lake environment

Pelagic ecosystems occupy the main water body of the lake and contain the largest subsidiary community within the main lake ecosystem, encompassing all the free floating (planktonic) and strongly swimming (nektonic) organisms. Pelagic ecosystems of lakes occupy the largest volume of all freshwater environments (excluding snow and ice systems), and show close similarities to the marine pelagic ecosystems of seas and oceans.

Characteristics of lake pelagic ecosystems in relation to the four key parameters of ecosystem function are summarized in Figure 1.8.

1.5.1 Interactions between organisms

A range of defined interactions occur between microorganisms in pelagic lake ecosystems, including competition for resources, antagonism, trophic

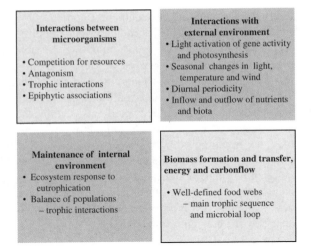

Interactions between microorganisms

- Competition for resources
- Antagonism
- Trophic interactions
- Epiphytic associations

Interactions with external environment

- Light activation of gene activity and photosynthesis
- Seasonal changes in light, temperature and wind
- Diurnal periodicity
- Inflow and outflow of nutrients and biota

Maintenance of internal environment

- Ecosystem response to eutrophication
- Balance of populations
  – trophic interactions

Biomass formation and transfer, energy and carbonflow

- Well-defined food webs
  – main trophic sequence and microbial loop

Figure 1.8   Major ecosystem functions in a lake pelagic community

interactions and epiphytic associations. Some of the major microbial interactions are listed in Table 1.7.

Competition for resources such as light and inorganic nutrients (silicate, phosphate, nitrate) are an important contributory factor in determining the relative ability of different species populations to grow and dominate the pelagic environment. In cases where competition for more than one nutrient is involved, a change in algal dominance may occur with shift in nutrient balance (Figure 5.26).

Antagonistic interactions may also occur as part of the competition for resources, allowing the population of the successful antagonist to access the resources and also generating nutrients from the target organism. Freshwater bacteria in particular, which are known to produce a range of anti-algal metabolites, may be important as antagonists in the termination of algal blooms, and are also potentially useful as biological control agents. Dominance of blue-green algae in bloom conditions may also be regarded as antagonistic activity, since the low CO_2/high pH micro-environment created by these organisms inhibits the growth of eukaryote algae in the top part of the water column.

Trophic interactions in the pelagic zone are often very specific, and may result in close coupling of particular microbial populations. This is the case for heterotrophic bacteria and phytoplankton, which are linked by algal DOC production. A trophic link also occurs between bacteria and predatory heterotrophic nanoflagellate (HNF) protozoa, with a buildup of HNF populations in some lakes at times of high bacterial count. Parasitic viral and fungal infections of lake bacteria and phytoplankton provide examples of highly specific microbial interactions and are ecologically important in limiting host population growth and productivity.

Epiphytic associations are common in the pelagic environment, with many unicellular organisms (bacteria, protozoa and algae) occurring on (or within) the surface of larger biota such as colonial algae and zooplankton. With the exception of blue-green algal heterocyst bacteria, little is known about the possible exchange of nutrients between epiphyte and host. Whether this occurs or not, epiphytic sites such as colonial blue-green algae are trophically significant in providing microcosms of microbial activity, each with its own localized food web. Epiphytic sites also provide a point of attachment in a part of the lake that is otherwise devoid of substratum, and allow planktonic organisms the opportunity to exist in equilibrium with a sedentary (non-planktonic) phase.

Seasonal changes in a temperate lake

The importance of microbial interactions in the pelagic ecosystem of temperate lakes is seen in the seasonal progression of biomass and populations that occurs during the growth season (Figure 3.14). This sequence is considered in the section on algae, since it is driven primarily by the response of phytoplankton populations to environmental alteration, and all other lake biota show correlated changes. The role of other lake biota within the seasonal transition is important and includes:

- control of phytoplankton spring and autumn blooms by fungal and viral epidemics;

- control of phytoplankton populations by protozoon and zooplankton grazing, with the occurrence of a clear-water phase;

Table 1.7 Interactions between microorganisms in lake pelagic ecosystems

Type of interaction	Interaction between organisms	Major parameter involved or result of interaction
1. **Competition for resources**	Competition between phytoplankton species	Light (Section 4.3) Inorganic nutrients (Section 5.2)
	Nutrient competition between algae and bacteria	Inorganic nutrients (Section 6.12)
2. **Antagonism**	Repression of eukaryote algae by blue-greens under bloom conditions	CO_2/pH equilibrium in blooms (Sections 4.7 and 10.6)
	Destruction of blue-green algae by antagonistic bacteria	Anti-microbial compounds (Section 6.12)
3. **Trophic interactions**	Association between bacteria and blue-green algae	Heterocyst symbiotic association (Section 6.13)
	Coupling of productivity between phytoplankton and bacterioplankton	DOC production and uptake (Section 6.10)
	Ingestion of bacteria by HNF protozoa	Control of bacterial populations (Section 9.5)
	Infection of phytoplankton by Cyanophage and phycovirus Viruses	Parasitic control of algal populations (Sections 7.6 and 7.7)
	Infection of phytoplankton populations by chytrid fungi	Parasitic control of algal populations (Sections 8.6 and 8.7)
	Infection of bacterial populations by bacteriophages	Parasitic control of bacterial populations (Section 7.8)
4. **Epiphytic associations**	Associations between protozoa, bacteria, and algae with colonial blue greens (phycosphere ecosystem)	• Spatial association – evidence of direct nutrient interaction with host (Section 6.13.1) • Local microcosm; food web with diverse microbial population (Figure 6.23)

- peaks in total bacterial count that correlate (slightly out of phase) with phytoplankton blooms, in relation to DOC production and accumulation.

1.5.2 Trophic connections and biomass transfer

Biomass is created by the photosynthetic activity of autotrophic organisms and is subsequently consumed by heterotrophs, passing from one group to another in a defined sequence of trophic progression. The interconnections and transfer processes that are involved in this sequence can be considered at three main levels of complexity:

- the food chain – a linear sequence describing the progression in terms of major groups of organisms;

- the ecological pyramid – a linear sequence, in which organisms are grouped in terms of ecological role, and dynamics of biomass transfer are quantified – diagrammatic representations have a pyramidal shape, hence the name;

- the food web – showing detailed interactions between particular groups and species, often including details of biomass or energy transfer within the network.

The pelagic food chain

In pelagic environments, interactions between organisms can be considered in relation to two major routes of biomass transfer – the major trophic sequence and the microbial loop.

The major trophic sequence   The major trophic or grazing sequence is the principal system of mass transfer in the aquatic environment (Figure 1.9(a)). It involves production of biomass by photosynthetic algae (primary producers) followed by a succession of grazing and ingestion. Ingestion of algae is carried out by herbivores (e.g., zooplankton), followed by a series of carnivores (e.g., zooplankivorous then piscivorous fish). Further details of the grazing sequence are given in later sections in reference to specific lake systems (Section 1.7).

Until quite recently, trophic interactions within freshwater systems were considered almost entirely in relation to this classic sequence, involving organisms which could be caught in plankton nets or were clearly visible as planktonic or benthic biota. This has now changed, with the realization that:

- in many pelagic systems, primary production is also carried out by algae which pass through phytoplankton nets and form major populations of actively photosynthetic picoplankton;

- in addition to algae, primary production is also carried out by photosynthetic bacteria and by some protozoa (containing symbionts);

- many planktonic and benthic algae also have heterotrophic capabilities, able to assimilate complex carbon compounds in addition to their autotrophic activities;

- primary productivity is also channelled into a second major trophic sequence which involves bacterial and protozoon populations and is referred to as the microbial loop.

The microbial loop   The existence of a second major trophic sequence, the microbial loop, was realized when it became clear that:

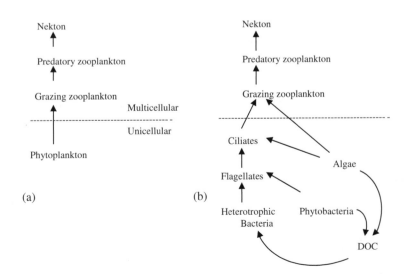

Figure 1.9 Incorporation of microbial loop into the classical pelagic food chain: (a) classical pelagic food chain; (b) incorporation of various microbial components (based on a figure from Dobson and Frid, 1998)

- a substantial part of primary productivity does not enter the grazing sequence, but is released as dissolved organic carbon (DOC) into the environment;

- large populations of viruses, bacteria and protozoa are present in many aquatic systems and play a major role in the release and utilization of algal DOC.

In the microbial loop (Figure 1.9(b)), DOC released by microalgae is routed back to the grazing sequence via a succession of microbial organisms. DOC is metabolized by bacteria, which are then ingested in sequence by protozoa and smaller zooplankton, leading back into the main sequence. The microbial loop is particularly well seen in planktonic systems, where stable populations of protozoa, algae and bacteria may occur under balanced (steady-state) conditions. At other times, populations are not in equilibrium, and many studies have confirmed a temporal sequence of microorganisms in such circumstances, with bacteria succeeding phytoplankton blooms and protozoa succeeding the bacteria.

The ecological pyramid

At any point in time, the biota contained within the pelagic zone of a lake may be considered to be in a state of equilibrium – with a balance between major groups in terms of numbers, biomass and energy content. These broad interrelationships are typically illustrated as ecological pyramids, as shown in Figure 1.10 for the pelagic community of a temperate lake during winter and summer. The main conclusions that emerge from this concept are as follows.

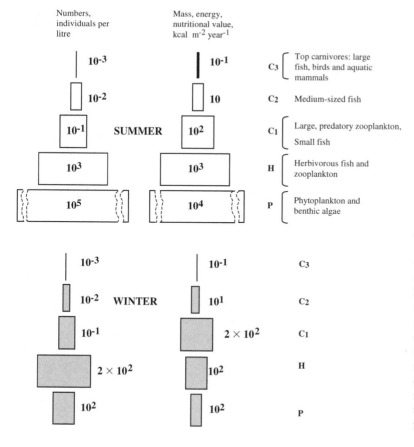

Figure 1.10 Ecological pyramid for the pelagic ecosystem of a typical temperate lake. Population levels for different trophic levels within the lake water column are shown as total counts on the left side, and biomass (also equivalent to energy and nutritional value) on the right. The patterns of population counts and biomass have a clear pyramidal shape during the summer growth phase (top diagram) but not during winter (bottom diagram) – when many of the organisms occur on the sediments in a non-metabolic resistant state. Symbols for different trophic levels are: P (primary producers), H (herbivores), C1 to C3 (carnivores) (based on a figure from Horne and Goldman, 1994)

- Summer populations show a clear pyramid of biomass, with phytoplankton occupying the base of the pyramid (10^8 individuals l^{-1}) leading to herbivores, primary carnivores and secondary carnivores (10^{-3} individuals l^{-1}, or one individual every $1000\,l$ of lake water).

- Winter conditions do not permit active growth of the primary producer, and the overall pyramidal structure is lost. At this time of year, the population of primary producer is down to about 10^2 organisms l^{-1} and herbivorous fish and zooplankton are the main population.

The classical concept of the ecological pyramid is based on growth of the primary producer, with a succession of dependent groups of primary and secondary consumers. Although it provides a useful overview of population levels at different stages in the trophic sequence, there are a number of limitations.

- Organisms are considered either as producers or consumers. Recent studies have shown that this is not the case with a number of algal groups – where organisms can exist either as autotrophs or as heterotrophs (Section 3.11).

- The classical ecological pyramid deals only with the main trophic sequence and ignores the microbial loop. A substantial amount of primary productivity is diverted into the production of dissolved organic carbon (DOC), and is then routed into the microbial community via the loop system, supporting a large population of heterotrophic bacteria and protozoa (Figure 1.9(b)).

The food web

Trophic interactions between individual species or particular groups of biota can be considered in more detail as a two-dimensional interconnected flow diagram (food web).

Generalized food web   A generalized food web for the major groups of biota within the pelagic lake system is shown in Figure 1.11. This diagram includes the five major groups of microorganisms – viruses, fungi, bacteria, protozoa and algae – and again emphasizes the distinction between the main trophic sequence (ingestion) and the microbial loop. Dissolved organic carbon (DOC) plays a key role in linking the productivity of phytoplankton and zooplankton to bacteria, and also linking the pelagic ecosystem to the external environment via exogenous DOC.

The main trophic sequence involves input of energy (light), CO_2 and inorganic nutrients into

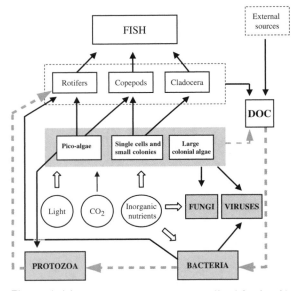

Figure 1.11   Simplified scheme (generalized food web) for trophic interactions in the pelagic freshwater environment. Microorganisms are indicated by shaded boxes. Solid arrows indicate direction of carbon flow, including main trophic sequence (⟶) and microbial loop (- - ▶). Carbon flow from protozoa is part of the main sequence (ingestion of pico-algae) and also the microbial loop (ingestion of bacteria). The sequence of organisms (individual or in groups) from top carnivore (fish) to primary producers (algae) is referred to as the trophic cascade. The zooplankton assemblage is represented by three major groups (rotifers, copepods, and Cladocera), and phytoplankton by pico-algae, small colonial and single-called micro-algae, and large colonial forms. DOC is dissolved organic carbon

phytoplankton, which can be separated into a gradation of subgroups. These are then differentially consumed by protozoa and different zooplankton groups, which are then consumed by zooplanktivorous fish – and ultimately piscivorous fish and other vertebrates as top predators. Trophic interactions between groups are indicated as carbon flow. This relatively simple diagram illustrates some of the complexities of the food web. These include multiple connections – picoalgae, for example, are consumed by protozoa, rotifers and copepods – and the fact that protozoa have a key role in both the main trophic sequence and the microbial loop. Interconnections between the different groups give the appearance of a cascade of organisms, and this generalized food web is sometimes referred to as the 'trophic cascade' (Carpenter and Kitchell, 1993). One other important aspect that emerges from this diagram is the question of control of phytoplankton and zooplankton productivity. Is this determined primarily by the supply of light and nutrients (bottom-up control) or by fish predation (top-down control)? The question of bottom-up and top-down control is considered later in relation to particular microbial groups, and also in relation to the control of blue-green algal blooms resulting from eutrophication.

Patterns of ingestion and biomass transfer

Patterns of ingestion (i.e., which organisms are eating which) within aquatic food webs can be determined by gut content analysis, though this may fail where soft-bodied organisms are ingested or where advanced digestion has occurred (Stübing *et al.*, 2003). In most situations, food web interactions may also be inferred from known predator–prey relationships, particularly in relation to ingestion of defined size ranges by zooplankton (Section 9.8). Other approaches may also be useful, such as the use of X-ray microanalysis (Section 5.1.2) to detect silicon in groups of zooplankton that are feeding on diatoms, and the use of trophic biomarker compounds and stable isotope analyses.

Trophic biomarker compounds. The trophic biomarker concept is based on observations that particular dietary components pass from one organism to another, and are incorporated without any chemical change. This has been used particularly in relation to lipid composition, where fatty acid profiles of primary producers, herbivores and carnivores have revealed useful information on aquatic food web relationships (Stübing *et al.*, 2003).

Stable isotope analysis. Uptake of stable carbon and nitrogen isotopes by primary consumers, and selective retention during biomass transfer, provides useful information on patterns of ingestion in aquatic food webs.

Analysis of carbon and nitrogen stable isotopes in Lake Baikal (Russia), for example, has shown that the pelagic food web has an isotopically ordered structure (Yoshii *et al.*, 1999), where the concentrations of $\delta^{13}C$ and $\delta^{15}N$ within individual species both showed a well-defined relationship to trophic level (Figure 1.12). $\delta^{15}N$ levels in particular had a clear trend of stepwise enrichment with progression from primary producer to zooplankton, pelagic fish and freshwater seal. Further details on the food web in this lake are given in Case Study 1.2.

The technique can also be used to demonstrate changes in food web dynamics caused, for example, by eutrophication. The resulting transition in shallow Greenland lakes, from benthic to pelagic primary production, is matched by a change in the feeding habits of primary consumers (Vadeboncoeur *et al.*, 2003). The use of carbon stable isotope analysis in this situation to detect changes in resource use is possible because phytoplankton discriminate against $\delta^{13}C$ more than benthic algae (periphyton), and consumers conserve these differences. During the eutrophication process, the $\delta^{13}C$ content of benthic grazers (amphipods, isopods and snails feeding on benthic algae) changes to that of benthic filter feeders (mussels and chironomids consuming phytoplankton). This transition in the $\delta^{13}C$ signature of grazers results from a decrease in periphyton availability and an increase in phytoplankton biomass, which settles out on the sediments.

Biomass and energy flow

The flow of biological material within the food web involves a sequence of ingestion, breakdown and synthesis of

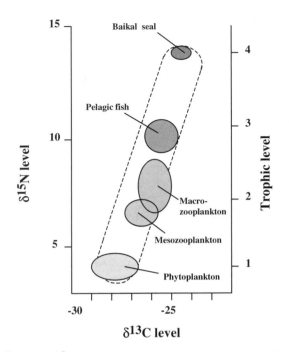

Figure 1.12 Stable isotope transitions in the pelagic food web of Lake Baikal (Russia). The levels of two stable isotopes ((δ15N and (δ13C) are shown within individual species at four trophic levels:

(1)   Phytoplankton: *Aulacoseira*

(2)   Zooplankton: *Epischura* (meso-) and *Macrohectopus* (macro-)

(3)   Pelagic fish: *Coregonus*

(4)   Freshwater seal: *Phoca*

Isotope levels are expressed as arbitrary units in reference to non-biological standards, and are shown as the area around the mean ($\pm$ one standard deviation) (based on a figure from Yoshii *et al.*, 1999)

new biomass. During the transition from one organism to another there are progressive changes in the chemical composition of biomass (e.g., alterations in isotope content, see Figure 1.12) and also loss of potential energy (biomass) due to inefficiencies in the conversion process and respiratory breakdown of some of the ingested material.

Incorporation of data on mass or energy content of different groups provides quantitative information on the flow of materials within the food web

and the efficiency of energy conversion within the ecosystem. Information is provided later on these aspects in relation to two case studies (Section 1.7).

1.5.3   Maintenance of the internal environment

Maintenance of the internal environment (homeostasis) is an important property of all ecosystems and is particularly well seen in the pelagic environment. In practical terms, homeostasis involves the ability of the system to overcome short-term physical, chemical and biological perturbations which may be imposed from outside (e.g., severe climatic change, pollutant inflow), or arise internally (e.g., random increase in a microbial population). Homeostasis operates to restore the ecological balance that occurs between biota and to maintain the biodiversity of the ecosystem.

In addition to its importance for natural ecological processes, homeostasis has a key role in the community response to human-mediated nutrient-enrichment (eutrophication) of the freshwater environment. The adverse environmental effects that result from eutrophication arise due to a breakdown in ecosystem homeostasis, and the recovery of freshwater systems requires the restoration of a balanced community and the return of homeostatic mechanisms. The role of homeostasis in eutrophication is discussed in Chapter 10, and the relationship between homeostasis and ecosystem stability in Section 1.6.

1.5.4   Interactions with the external environment

Interactions with the external environment have a key effect on both parts (Tansley, 1935) of the pelagic ecosystem, influencing both the community of organisms and the habitat that they occupy. These interactions have particular importance for the lake pelagic ecosystem.

- A large surface area of the pelagic system is directly exposed to the atmosphere, and is thus rapidly affected by external physico-chemical change.

- The pelagic ecosystem is the major area in the lake for penetration of light and the generation of biomass through photosynthesis (primary productivity). The pelagic system acts as a source of biomass for other systems such as the deep benthic environment, and for littoral microsystems such as bacterial (non-mixed) biofilms.

- External changes lead to seasonal stratification (and subsequent de-stratification) of the pelagic environment, with major impact on the biology of lake organisms.

During daytime hours, penetration of light into the pelagic environment is continuously changing, with correlated changes in the response of phytoplankton cells. These changes include activation of light-responsive genes (Section 3.3.2) and alterations in the rate of photosynthesis. Longer-term changes in light level are also important, particularly in relation to seasonal changes and diurnal periodicity.

Seasonal changes. The impact of external effects is seen particularly clearly in seasonal changes in lake biota (Figure 3.14), where alterations in surface light, temperature and wind trigger the onset of the spring diatom bloom, cause stratification in the water column, and promote a succession of changes in phytoplankton and other lake organisms.

Diurnal periodicity. In addition to seasonal effects, external changes in light intensity also promote a diurnal periodicity in phytoplankton activities (Section 4.10). These involve entrainment of phytoplankton dynamic activities (vertical migration), physiological processes (photosynthetic activity, nitrogen fixation) and gene transcription (expression of clock genes). Other biota such as zooplankton are also directly affected by daily changes in light intensity, and diurnal periodicity promotes the vertical migration of these organisms and their interactions with phytoplankton. These include the timing and position (within the water column) of zooplankton grazing of phytoplankton and the recycling of phosphorus from zooplankton to phytoplankton.

The dynamics of lake hydrology may also exert an important external effect on the pelagic ecosystem, with inflow and outflow of aquatic biota, particulate matter and soluble nutrients. Towards the end of the growth season, for example, inflow may be particularly important in the supply of soluble inorganic nutrients to a nutrient-depleted system, while the outflow of biota such as planktonic algae may influence bloom formation. Flushing the pelagic system is an important management option with many freshwater systems in the control of late summer blue-green algal blooms. The loss and recruitment of biota may also occur via the surface of the lake. This does not appear to have been documented for lake microorganisms, but insects such as mosquitoes and chironomid flies enter the system as eggs, have an aquatic larval stage, then exit the ecosystem on metamorphosis to the imago stage.

1.6   Homeostasis and ecosystem stability

All biological systems are subject to environmental changes that may impair function. These are referred to as 'stress factors' and can operate at the level of ecosystems (this section), individual organisms (e.g., physiological influence of high light, Section 4.9) and molecular systems. Prokaryotes respond to physicochemical stresses such as salinity, high temperature and acute nutrient deprivation by the induction of special transcription factors (sigma proteins) that lead to the expression of a range of stress genes (Sections 3.3.2 and 6.6.4).

Ecosystems are frequently subject to external or internal environmental effects that tend to bring about changes in the biological community. These effects can be defined as 'environmental stress' when they impair the structure or function of food webs and other dynamic aspects of the biological system. In this section, the response of aquatic ecosystems (including microbial communities) to stress is considered in relation to general ecosystem theory, observed stress responses, assessment of ecosystem stability, changes in community structure and biological response signatures.

1.6.1 Stress factors

Examples of external environmental stresses which affect aquatic environments include major environmental change (floods, alteration of land usage), climatic change (droughts, temperature, wind disturbance), local physical parameters (rate of water flow) and chemical effects (toxic pollutants, nutrient enrichment). Biological perturbations, such as the introduction of new species, may also be important stress factors. The ecology of many lakes in the UK, for example, is affected by the introduction and spread of non-native plants such as the New Zealand pygmyweed. Originally brought into the country for use as a garden pond plant, this readily colonizes other freshwater habitats – spreading across water bodies, eliminating native plants and reducing habitat diversity.

Stress factors affecting microbial communities

Various examples are given in this volume of the effects of external stress on aquatic microbial communities. The rate of water flow is an important parameter in lotic systems, for example, affecting the development of biofilm and epilithon communities. The flow rate of effluent into activated sludge (sewage) systems also provides a good example of external stress. At high flow rates the protozoon community is limited to rapidly reproducing species, with all the features typical of r-selected organisms (Chapter 9). In lakes and other standing waters, inflow of inorganic nutrients also constitutes a major stress factor, leading to eutrophication and the breakdown of ecosystem homeostasis (Chapter 10).

Stress factors developing internally within aquatic environments may also be important. In some cases these internal stresses are secondary effects of external factors, such as the effects of blue-green algal blooms resulting from eutrophication. Another example of an internal stress that may result from external factors is the induction of anoxic conditions by the bacterial bloom that develops as a response to organic pollution. This has an important effect on the composition and activities of the emergent protozoon community (Chapter 9).

In other cases, internal stress factors are not directly linked to external perturbations. In lakes, they may occur as part of the seasonal cycle – and include the substantial increase in zooplankton grazing activity that occurs during the clear-water phase, which has a major influence on the phytoplankton population. Development of late summer intense blue-green algal blooms in natural eutrophic waters may also impose stress on the ecosystem, affecting the growth of other algae and in some cases limiting the development of zooplankton.

1.6.2 General theoretical predictions: the community response

Theoretical considerations suggest that a number of general responses may be expected in ecosystem communities in relation to external stress (Odum, 1985; Cairns, 2002). The principles that underlie these responses apply equally to combinations of communities as well as single communities (e.g., biofilms, plankton populations) within complex ecosystems. These responses are summarized in Figure 1.13, and include changes in energetics, nutrient cycling, community structure and general system characteristics.

- *Energetics.* Energy changes would be expected as an early response to external stress, with an increase in respiration (R) as organisms cope with the disorder caused by disturbance. Chemical energy within the system is directed from biomass production (productivity P) to respiration, the P/R ratio becomes unbalanced and the R/B (respiration/biomass) ratio increases. The drain of productive energy within the system means that auxiliary energy (from outside) becomes increasingly important for continued survival. Disturbance also results in less efficient use of primary production, with an increase in unused resources.

- *Nutrient cycling.* Increased nutrient turnover and decreased cycling frequently appear in stressed ecosystems. The decreased nutrient cycling results from changes in community structure

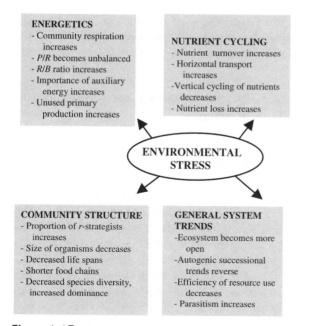

ENERGETICS
- Community respiration
  increases
- *P/R* becomes unbalanced
- *R/B* ratio increases
- Importance of auxiliary
  energy increases
- Unused primary
  production increases

NUTRIENT CYCLING
- Nutrient turnover increases
- Horizontal transport
  increases
- Vertical cycling of nutrients
  decreases
- Nutrient loss increases

**ENVIRONMENTAL
STRESS**

COMMUNITY STRUCTURE
- Proportion of *r*-strategists
  increases
- Size of organisms decreases
- Decreased life spans
- Shorter food chains
- Decreased species diversity,
  increased dominance

**GENERAL SYSTEM
TRENDS**
- Ecosystem becomes more
  open
- Autogenic successional
  trends reverse
- Efficiency of resource use
  decreases
- Parasitism increases

Figure 1.13 Stress responses expected in aquatic systems, based on general ecosystem theory: *P* – energy for biomass production, *R* – respiration, *B* – biomass (based on a figure from Odum, 1985)

(Figure 1.14), with increased horizontal transport but reduced vertical cycling. Together, these changes in turnover and cycling result in nutrient accumulation which, as with unused primary production, may be lost from the system.

- *Community structure.* Under stressed conditions, we would expect an increase in species that are able to grow rapidly and exploit temporary advantages. Such opportunistic (*r*-selected) organisms typically have a small size and decreased life spans. The competitive strategy of *r*-selected organisms and their adaptation to uncrowded and unstable environments has been considered in Section 1.2.6. Disturbance to the food web also results in shorter food chains and decreased species diversity as particular species attain temporary dominance.

- *General system-level trends.* Under stress conditions, ecosystems tend to become more open and existing (autogenic) successional changes become

reversed. The efficiency of resource use decreases due to unused primary production and increased nutrient loss. As individual organisms become physiologically stressed, their susceptibility to parasitic infection increases.

1.6.3 Observed stress responses: from molecules to communities

Observed responses to stress range from molecular events (e.g., light inactivation of proteins) to alterations in physiology, balance of populations (selective mortality), ecological dynamics (reduced recruitment) and community composition (reduced diversity). In terms of the aquatic community, many of the observed stress responses have been predicted on the basis of general ecosystem theory (previous section), which applies equally to aquatic and terrestrial systems.

Aquatic ecosystems show many of the stress responses noted in Figure 1.13. Cattaneo *et al.* (1988), for example, demonstrated a general decrease in size in a range of lake biota (diatoms, thecamoebians and cladocerans) under chronic and progressive conditions of metal pollution. Havens and Carlson (1998), investigating plankton community structure in a range of lakes of different pH level (7.3 to 4.2), showed that reductions in pH correlated with reductions in food web complexity and species diversity. Schindler (1987), in a study of lake community responses to anthropogenic stress, demonstrated changes in the proportion of *r*-strategists, organism life spans, species diversity and the relative openness of the ecosystem.

1.6.4 Assessment of ecosystem stability

The ability of ecosystems to resist stress and return to, or maintain, their biological integrity is an example of the general principle of homeostasis or maintenance of the internal environment. In ecosystems, the internal environment comprises the biological, chemical and physical characteristics of the system – with particular emphasis on the species composition and dynamic interactions of the constituent organisms. The extent to which an ecosystem is altered by any of the above stress factors (Section 1.6.1) is a measure of its stability – the greater the stability, the more able is the system to resist external change.

Although concepts of ecosystem stability and the mechanisms that control it have been developed mainly in relation to terrestrial environments, many of the principles may also be applied to aquatic systems. The stability of ecosystems has been defined and discussed in relation to three interrelated aspects.

- *Species composition.* Stable systems show constancy in overall species numbers or relative proportions of individual species under stress. This has been defined as 'no-oscillation stability' and implies predictability and continuity of species composition within the ecosystem (Dunbar, 1973).

- *Biodiversity and complexity of the food web.* It is generally accepted that ecosystem stability is partly a function of trophic structure and the overall diversity of food webs (see Elton, 1958). This diversity of trophic interconnections means that oscillations within the ecosystem are reduced and that there is increased resilience to outside influences. MacArthur (1955) has proposed a direct quantitative relationship between ecosystem stability and the number of links in the food web, and suggested that a large number of interactive feeding links allows a wide variety of adjustments to stress within the system and also provides alternative channels for energy flow.

- *The ability of the system to maintain (or return to) its original state under external stress.* This property has been referred to as 'stability resistance' (Regier and Cowell, 1972). This property of environmental homeostasis is particularly relevant when considering the impact of blue-green algal blooms on the lake ecosystem.

Some freshwater bodies have developed very stable ecosystems with high levels of biodiversity. This is particularly the case for ancient lakes such as Lake Baikal (Russia), where a diverse assemblage of species has evolved in a relatively stable environment over a long period of time (Mazepova, 1998).

1.6.5 Ecosystem stability and community structure

Although some freshwater systems have evolved relatively stable and complex communities, the concept that ecosystem stability relates primarily to biotic interactions and food web diversity applies particularly to established terrestrial systems such as tropical rain forests. These typically conform to the community structure model developed by Hairston *et al.* (1960), where physical disturbance is of minimal importance. In this model (Figure 1.14(a)), competition occurs mainly between primary producers (for space, nutrients, light) and between predators (competition for limited food resource). Populations of herbivores are unable to limit the growth of primary producers, and are kept below the theoretical carrying capacity of their available food supply by predation.

Freshwater ecosystems frequently differ from terrestrial ones in having greater physical disturbance and greater physical heterogeneity. These systems relate more to the community structure model developed by Menge and Sutherland (1987), where varying levels of physical stress (disturbance) cause major alterations in competition and predation (Figure 1.14(b)). Highest stress levels (continual disturbance) do not allow any species to establish, but as stress is reduced the community becomes more complex until, at low stress levels, it is equivalent to that envisaged by Hairston *et al.* (1960). Variations in stress level and community structure can be seen in both lotic and lentic aquatic systems.

Streams and rivers

Turbulent water disturbance in fast-moving rivers provides a good example of a high-stress environment, corresponding to stages (3) and (4) of the Menge–Sutherland model (Figure 1.14(b)). In this situation, unattached organisms such as

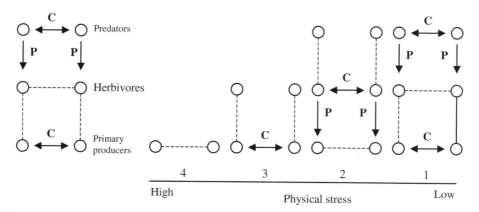

Figure 1.14 Alternative models of community structure showing interactions between the main groups of aquatic organisms under varying conditions of physical stress: (a) in the Hairston–Smith–Solobodkin model, predator and primary producer populations are controlled by competition, but herbivores are controlled by predation; (b) in the Menge–Sutherland model, community structure varies with stress level. With increasing stress (level 1 to level 4), community structure becomes simpler and moves from the Hairston model, to situations where there is a progressive decrease in the effect of predators (level 2), then herbivores (level 3). The model is related to specific ecological examples in Section 1.6 (figure adapted and redrawn from Dobson and Frid, 1998, based on original papers by Hairston *et al.*, 1960, and Menge and Sutherland, 1987)

invertebrates and free-moving microorganisms are not able to establish themselves and are swept away by the current. Only attached microorganisms such as periphyton communities and biofilms are able to remain in such conditions, with varying degrees of competition between organisms colonizing exposed surfaces. Even these attached communities may be limited by environmental stress. In highly disturbed ecosystems such as head-water streams, algal succession may show little progression beyond diatom colonization of exposed surfaces (Section 3.7.2, Figure 3.15).

Lakes

The microbiology of lakes is influenced by stress factors in both natural and derived situations.

Under natural conditions, variations in community structure and stress levels are seen during the seasonal cycle (Section 3.13.2). High stress levels occur during the clear-water phase, where competition between herbivores (zooplankton) is high and phytoplankton populations are low and show rapid change (Figure 1.14.(b), stage (2)). Only rapidly growing algae (*r*-selected organisms) are able to survive the adverse conditions at this time. In contrast to this, the summer bloom period is typically a phase of minimal stress levels, where zooplankton grazing pressure is reduced and high levels of competition occur between established populations of different phytoplankton species (Figure 1.14(b), stage (1)).

An example of induced stress occurs where lakes are subject to high nutrient pollution (eutrophication) as a result of human activities. This has major effects on the microbial community and is discussed in Chapter 10. The growth of blue-green algae results in the suppression of both herbivores (zooplankton) and predators (fish), with strong competition within the phytoplankton community leading to blue-green dominance (Figure 1.14(b), stage (3)).

1.6.6 Biological response signatures

Different types and degrees of stress result in different community responses. These responses can be used to evaluate the source, type and impact of stress on a particular aquatic system, and have been referred to as 'biological response signatures' (Yoder and Rankin, 1995; Simon, 2002). Biological response signatures may be defined as discernable patterns in the response of aquatic communities, allowing the investigator to discriminate between different types of stress.

In their paper describing the selective effects of discrete environmental disturbances on rivers and streams in Ohio (USA), Yoder and Rankin (1995) were able to segregate various impacts into nine categories of disturbance. These comprised complex toxic release, conventional municipal discharge,

sewer overflows, channelization, diffuse agricultural pollution, flow alteration, impoundment, combined sewer overflow with toxic discharge and livestock access.

The use of biological signatures normally requires a broad assessment of the aquatic community rather than looking at individual 'indicator species.' This assessment involves the determination of community indices using multiple indicator groups, detailed taxonomic resolution and standardized sampling procedures. Most programmes of biological assessment use macro-scale indicators such as fish, invertebrates and higher plant communities (Simon, 2002). There is considerable scope, however, for using microbial communities – including benthic diatom assemblages to assess nutrient impacts on flowing waters and phytoplankton communities to evaluate eutrophication of standing waters.

C. FOOD WEBS IN LENTIC AND LOTIC SYSTEMS

The ecosystems of standing water (lentic) and running water (lentic) systems show a number of important differences. These reflect differences in the biota (e.g., relative proportions of pelagic and benthic organisms), the environment (e.g., absence and presence of water motion), and in the input of external (allochthonous) material into the aquatic system. Lentic food webs are considered in Section 1.7 by reference to pelagic communities, and comparisons are made in Section 1.8 to lotic systems.

1.7 Pelagic food webs

The generalized pelagic food web shown in Figure 1.11 was based on major groups of orga-

nisms, and becomes highly complex when individual species – each showing multiple connections to other species – are considered. For this reason investigations involving food webs tend to focus on particular aspects of the lake ecosystem, such as particular groups of biota (e.g., excluding the major microbial component), particular parts of the lake (e.g., the pelagic ecosystem), or particular time sequences (e.g., during a particular algal bloom). These different approaches will be illustrated by two case studies from lake ecosystems – the transient microbial food web associated with a spring algal bloom and a generalized annual food web present within the water column of an oligotrophic lake (Lake Baikal, Russia).

Case study 1.1   Microbial food web associated with an algal bloom

Classical studies on species interactions in aquatic systems tend to overlook the role of smaller microorganisms, such as bacteria and fungi. It is now clear that these are of major importance in such ecosystems, not only in controlling populations of other biota (e.g., role of chytrid fungi in limiting phytoplankton populations) but also themselves being a significant part of the biomass transformations.

Weisse et al. (1990) investigated microbial interactions during the intensive phytoplankton Spring bloom of 1988 in Lake Constance (Germany), looking at changes in biomass, productivity and grazing rates of various microorganisms (phytoplankton, free-living bacteria, ciliate protozoa) and zooplankton. Biomass within particular groups of biota was determined as the amount of carbon contained within a m^2 water column extending to a depth of 20m from the lake surface. The carbon budget is summarized in Figure 1.15, and can be divided into three aspects – growth and development of *phytoplankton (bloom formation), the classic pelagic food web, and the microbial loop.*

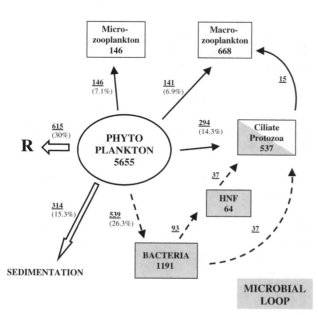

CLASSIC PELAGIC FOOD WEB

Figure 1.15 Pelagic food web during the spring phytoplankton bloom in Lake Constance (Germany), figure based on data from Weisse *et al.* (1990). Major routes of carbon flow are indicated by arrows. The Pelagic food web has two main components – the classic food web (solid arrows: ingestion of phytoplankton by zooplankton and protozoa) and the microbial loop (broken arrows: involving bacteria, heterotrophic nanoflagellates (HNF) and protozoa). The mean biomass (mg C m^{-2}) of each group of organisms is given within their compartment, and the rate of biomass transformation (carbon flux in mg C m^{-2} d^{-1}) against the arrow. All data are derived as an average from the top 20m of the water column, over the 29 day study period. In addition to the classic food web and microbial loop, phytoplankton biomass is also lost from the pelagic system via respiration (*R*) and sedimentation

The phytoplankton bloom

Increases in light intensity and temperature, combined with lake stratification, lead to the development of a spring phytoplankton bloom. This is composed mainly of unicellular algae, with small (diameter 0.9 μm) coccoid blue-green algae (pico-cyanobacteria) being the most abundant autotrophic organisms present. Diatoms (e.g., *Stephanodiscus hantzschii*), green algae (e.g., *Chlorella sp.*), and small cryptophytes were also abundant in the mixed phytoplankton.

The phytoplankton population reached the peak of the bloom (chlorophyll-a concentration >40 μg l^{-1}) in mid-April, followed by a rapid decline. Phytoplankton growth during over the bloom period correlated with the removal of soluble phosphorus, nitrogen, and silicon from the lake water, depleting the nutrients to such low levels that continued growth became limited and the bloom was terminated. The marked increase in N/P ratio during bloom development suggests that phosphorus limitation is particularly important.

Algal biomass built up during photosynthesis was lost or transformed in various ways. Determination of carbon fluxes (Figure 1.15) indicated major losses due to respiration (30 per cent), sedimentation (15 per cent), ingestion by zooplankton and protozoa (28 per cent) and uptake into bacteria (microbial loop, 26 per cent).

The classical pelagic food web

Ingestion of phytoplankton is carried out by three main groups of organisms – ciliate protozoa (particularly the oligotrichs *Strombidium* and *Strobilidium*), microzooplankton (copepod nauplius larvae, rotifers), and macrozooplankton (chiefly copepods). The increase in ciliate protozoon towards the end of the algal bloom is partly due to ingestion of algae and partly due to ingestion of heterotrophic nanoflagellates as part of the microbial loop.

The microbial loop

The role of heterotrophic bacteria and protozoa in the consumption and recycling of algal biomass has been noted previously, and forms a particularly important route of carbon transfer during the Spring phytoplankton bloom of Lake Constance. This is indicated by the following points.

- *Carbon flux.* A large fraction (over 26 per cent) of the particulate phytoplankton carbon flux is channelled through the microbial loop, in addition to the assimilation of algal exudates that is also routed mainly through the bacterial population. Taken together, these heterotrophic activities of planktonic bacteria add up to more than 50 per cent of net primary production.

- *The bacterial biomass.* This is the highest biomass of all the heterotrophic organisms and is about 20 per cent of the algal biomass, averaged over the sampling period.

- *Dynamics of bacterial populations.* Bacterial populations show a substantial increase during the latter part of the algal bloom. This increase in bacterial count is consistent with mass transfer from one population to another, and parallels increased levels of algal exudate followed by increased particulate material derived from breakdown of the algal bloom.

Within the microbial loop, bacteria are ingested mainly by small colourless non-photosynthetic flagellated organisms (heterotrophic nanoflagellates or HNF), rather than by ciliate protozoa. Although there is a substantial carbon flux into HNF populations, these show little increase (Figure 1.15) due to intensive grazing by ciliate protozoa.

Case study 1.2   General food web in the water column of Lake Baikal (Russia)

Although the biota of Lake Baikal as a whole is diverse, the community present in the open water has evolved within a homogeneous environment and comprises relatively few species and trophic groups (Mazepova, 1998). Biotic interactions within the water column (pelagic zone) of ancient lakes such as Lake Baikal are thus comparatively simple, and provide a useful model for considering the general properties of an aquatic food web.

Major pattern of interactions

Long-term studies carried out by the Russian Academy of Sciences (Limnological Institute) have established the major pattern of species interactions and energy flow within the pelagic zone. In this system, primary production is carried out by about 10 major species of phytoplankton, with separate blooms of diatoms and picophytoplankton at different times of year (Figure 2.9). The main secondary production derived from this phytoplankton biomass is carried out by bacteria, protozoa, zooplankton, fish, and a mammal. The simplified food web shown in Figure 1.16 emphasizes some of the major interconnections within this system and consists of phytoplankton, microbial loop (bacteria), and four heterotrophic levels. These comprise a major zooplankton herbovore (H) and three levels of carnivore – C_1 (zooplankton), C_2 (fish) and C_3 (mammal) – as illustrated in Figure 1.17.

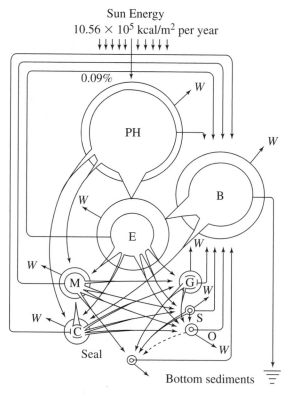

Figure 1.16 Annual energy flow (kcal/m² per year) in the main trophic chains of the pelagic ecosystem of Lake Baikal (Russia) (figure reproduced from Mazepova, 1998, with permission from Elsevier). The amounts of energy are represented by spheres of corresponding volume: outer sphere – assimilation (A), inner sphere – net production P). W = metabolic expenditure.

		A	P	W	
PH	Phytoplankton	972	875	97	P
B	Bacteria	602	315	263	
E	*Epischura*	326	81	245	H
M	*Macrohectopus*	18	5	13	H, C1
C	*Cyclops*	13	3	9	H, C1
G	Golomjankas	6	3	4	C1, C2
S/O	Sculpin, Omul	2	1	2	C1, C2
Seal	Seal	1	0	1	C3

Trophic designations are given (far right) in relation to the classic food web.

- *Major herbivore.* The main herbivore is represented by one endemic species of calanoid copepod, *Epischura baicalensis.* As in other ancient lakes the zooplankton is dominated by copepods rather than cladocerans, which in Baikal are not a constant component of the plankton

- *Carnivore – C_1.* The second heterotrophic level in ancient lakes is represented by a small and large predator. In the case of Baikal, these are a copepod (*Cyclops kolensis*) and an endemic amphipod (*Macrohectopus branickii*), respectively.

- *Carnivore – C_2.* The third heterotrophic level is represented by zooplanktivorous fishes, including two species of Compephoridae plus *Cottocomephorus* and *Coregonus*.

- *Carnivore – C_3.* Top predator in Lake Baikal is the endemic freshwater seal (*Phoca baicalensis*).

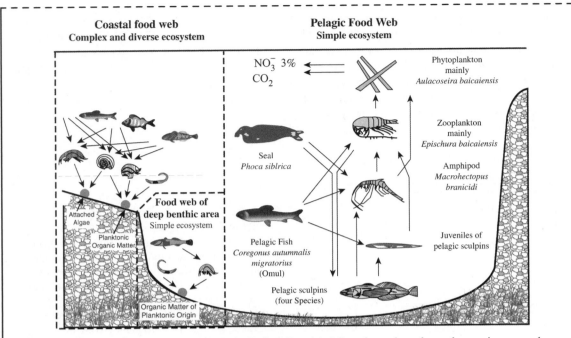

Figure 1.17 Diversity of food webs in Lake Baikal (Russia). Microalgae play a key role as primary producers in the form of living phytoplankton (pelagic food web), phytoplankton detritus (deep benthic area, coastal zone) and attached algae (coastal zone). Arrows indicate source of food or inorganic substrate. Figure adapted and redrawn from Yoshii *et al.*, 1999.

The seasonal and long-term dynamics of this food web have been studied in terms of the biomass of the above organisms. The flow of energy (kcal) under 1 m^2 of lake surface per year for the pelagic zone as a whole has been estimated on the basis of the amount of organic carbon in the organisms. Values for some of the principal energy transformations are shown in Figure 1.16, and indicate the following.

- At each level in the food web, the initial conversion of energy by assimilation leads to the generation of biomass (net production) and the expenditure of energy for metabolism. The proportion of assimilated energy that is used for metabolism is higher for fish and zooplankton (60–75 per cent) compared to bacteria (40 per cent) and phytoplankton (10 per cent).

- The primary conversion of light to potential energy by phytoplankton is highly inefficient. The primary assimilation of 972 kcal/m^2 per year from incident light energy of 10.56×10^5 kcal represents a conversion factor of 0.09 per cent.

- In terms of net productivity, energy levels show a marked decrease throughout the system. During the initial part of the trophic succession, these levels fall by a factor of 10 per cent, with net productivity of phytoplankton (875 kcal) translating to 80 kcal (herbivores) and 8 kcal (primary carnivores).

- Within the microbial loop, bacteria use approximately 60 per cent of the potential energy assimilated during primary production (phytoplankton). This compares to a value of about 40 per cent for the herbivores, and emphasizes the importance of the bacterial metabolism to the pelagic food web.

Food web complexities and interactions

Even with relatively simple food webs such as the Baikal pelagic system there are complexities in terms of species interactions. These complexities operate both within a single food web and between separate food webs.

Complexities within the food web   Within the food web, various factors make calculations of biomass and energy flow very difficult. These include the following.

- Omnivorous habits of individual species: in Lake Baikal, for example, the copepod *Epischura* ingests bacteria as well as phytoplankton, the primary carnivore *Macrohectopus* ingests phytoplankton and Cyclops in addition to *Epischura* and zooplanktivorous fish ingest all types of zooplankton.

- Temporal and spatial variations in biota can also complicate energy flow analyses. Populations of individual species show both seasonal and annual variation. Variation in phytoplankton is particularly marked, with the diatom *Aulacoseira baicalensis* forming major blooms in some years but not others. The abundance of *Cyclops* is also highly variable, being barely detectable in some years but reaching 80–90 per cent of the zooplankton biomass in others.

  Spatial distribution of *Cyclops* within the lake is also very patchy, leading to localized food webs where it is the dominant member of the zooplankton and others where it is much less important.

Interactions between food webs   Within a particular ecosystem, different food webs interact in time and space.

- Within the water column of Lake Baikal, the simplified food web shown in Figure 1.16 is derived from a succession of separate pelagic food webs that occur throughout the annual cycle. This sequence occurs due to changes in phytoplankton composition. During the early part of the season the lake is dominated by diatoms (Figure 2.9) with *Epischura* as the main herbivore (Figure 1.17). In late summer, picophytoplankton (particularly blue-green algae) dominate the lake, leading to a food web in which flagellate protozoa become the predominant herbivore.

- The pelagic environment is not an isolated system within the lake in terms of biotic interactions, and the food web that characterizes this part of the lake also has connections with the food webs in the coastal (littoral) and deep benthic zones of the lake (Figure 1.17).

  Intermixing of pelagic organisms between the pelagic and littoral zones occurs at various times of the year. At the end of summer, a few species of Palaearctic derivation (Section 2.4.2) temporarily penetrate the open water from the peripheral zone and shallow bays, but are not able to complete their life cycles.

  A more permanent interaction occurs between the pelagic and benthic food webs, since death and sedimentation of pelagic organisms results in the continuous addition of biological material to the bottom of the lake – where it becomes the primary source of organic carbon for the benthic food web.

Changes in the nutrient status of lakes can also influence interactions between pelagic and benthic food webs. Recent studies by Vadebonceur *et al.* (2003) on shallow oligotrophic lakes of Greenland have shown a shift in productivity from benthic to pelagic food webs with increased nutrient status, accompanied by a breakdown in the linkage between the two habitats in terms of food web interconnections.

1.8   Communities and food webs of running waters

In standing water (lentic) systems, the ecology is dominated by phytoplankton and the entry of carbon into the food chain mainly occurs internally (autochthonous origin) via photosynthesis. In contrast to this, running waters are dominated by attached (benthic) organisms and the entry of carbon is principally from external sources (allochthonous origin) as complex organic compounds. As a result of these differences, lentic systems often have a net autotrophic metabolism, while lotic systems typically show net heterotrophy (Section 1.2.3). At the microbial level, biofilm communities (Figure 1.5) are particularly important in rivers and streams, and are dominated by either bacteria or algae (periphyton) – depending on stage of development and location.

The lotic ecosystem will be considered in relation to the mainly allochthonous origin of the organic carbon, the relative importance of pelagic and benthic communities and microbial food webs.

1.8.1   Allochthonous carbon: dissolved and particulate matter in river systems

Although there is some production and release of autochthonous carbon by riverine phytoplankton (Section 1.8.2), benthic algae, and higher plants, most of the carbon in streams comes from the catchment area or flood plain. The dependence of flowing waters on allochthonous sources of carbon involves the entry of both dissolved organic carbon (DOC) and particulate carbon (POC) into the aquatic system, as reflected in the changes in concentration of these components during the seasonal river cycle.

Dissolved organic carbon (DOC)

The amount – and characteristics – of DOC that enters streams is controlled by both biotic (vegetation type, human activity) and physicochemical (geochemistry, hydrology) factors of the catchment

area. Hydrological characteristics such as the flow path and residence time of water in soil horizons (differing in organic matter content and sorption capacities) are of particular importance. The molecular composition of DOC in streams is diverse, including both highly stable (refractory) and rapidly changing, unstable (labile) components. Only part of this can be broken down by bacteria, with relatively little degradation (<1 per cent) of the more refractory components such as humic acids and high molecular weight DOC but considerably more breakdown (>50 per cent) of the more labile low molecular weight DOC compounds, particularly those of human origin (Meyer, 1994). Studies on the elemental composition of DOC in relation to bio-availability (Meyer, 1994) have suggested that the atomic ratios of H/C and O/C in DOC can be used to predict the ability of bacteria to degrade these compounds. The H/C ratio in particular is directly proportional to readily degradable aliphatic compounds and inversely proportional to less degradable aromatic material. Because of the variation in DOC bioavailability, the amount of DOC entering the system is not a direct measure of DOC supporting the microbial food web.

Particulate organic carbon

Particulate organic matter (diameter $>0.2\,\mu m$) ranges from finely dispersed material (including bacteria) to large particulate matter such as leaf litter and other plant debris. Much of this material is directly deposited into the flowing water from surrounding vegetation.

Large particulate matter such as leaf litter is an important carbon source for the microbial food web, serving as a substrate for fungal (Section 8.5.2) and bacterial growth and as a source of DOC. On entering a stream or river, leaf litter releases an initial pulse of rapidly leachable, water soluble material, followed by a slow release of DOC due to microbial degradation (Meyer, 1994). The breakdown and release of organic material from leaf litter is accelerated by the feeding activities of invertebrates.

Changes in the concentrations of dissolved and particulate carbon during the seasonal cycle

Concentrations of dissolved and particulate carbon show wide fluctuations in many streams and rivers during the annual cycle, with recorded DOC values generally in the range 1–10 mg l^{-1}, but reaching much higher levels in some rivers (Burney, 1994). In most cases, this annual variation reflects the allochthonous derivation of these compounds and the seasonal fluctuations of entry into the river system.

The large rivers of Southern Asia, for example, have large changes in DOC concentration, with close correlations between seasonal flow patterns and the timing of DOC maxima and minima (Table 1.8). The dissolved organic carbon levels in the Indus and Ganges–Brahmaputra Rivers reach a maximum near the end of rising water levels, due to overflow and entrainment from highly productive flood plains. The pulse of DOC then rapidly declines as water levels recede due to mixing, metabolic removal, and dilution. In the upper Mississippi River (USA), the very high autumn allochthonous DOC levels are derived not from soil, but from leaching of leaf litter.

Although many rivers show elevated allochthonous DOC concentrations at times of flood and terrestrial runoff (summarized in Burney, 1994), wide differences in seasonal patterns can occur. The Shetucket River (USA), for example, is unusual in showing minimal DOC levels during high inflow winter months, but maximum levels during the low-flow summer period (Klotz and Matson, 1978). The summer maximum was attributed to the generation of autochthonous carbon by secretion and senescence of benthic algae. In winter, the allochthonous DOC input was overridden and diluted by high discharge due to ice melt and heavy rain.

Seasonal patterns in some rivers are at least in part driven by *in situ* planktonic primary production. In the Gambia River (West Africa), allochthonous DOC concentrations reach a maximum at the time of maximum discharge into the system, but accumulation of autochthonous DOC occurs during the low-flow period. This was linked to phytoplankton production and was associated with elevated river water pH levels (Lesack *et al.*, 1984).

Table 1.8 Varying importance of allochthonous and autochthonous dissolved organic carbon (DOC) in different river systems (taken from Burney, 1994)

River	Maximum concentration (mg l^{-1})	Minimum concentration (mg l^{-1})	DOC origin
Allochthonous input			
Ganges	9.3	1.3	Rising water from flood plains
(Bangladesh)	(July)	(June)	
Brahmaputra	6.5	1.3–2.6	
(Bangladesh)	(July)	(rest of year)	
Mississippi	21	6	Leaching of leaf litter in autumn
(USA)	(November)	(March)	
Autochthonous input			
Shetucket	6.2–10	2–4	Production by benthic algae
(USA)	(May and September)	(January–April)	during low-flow period
Gambia	3.7	1.3	Mainly allochthonous but some
(West Africa)	(September)	(December)	low-water production by phytoplankton

Although DOC in flowing waters is principally derived from external (allochthonous) origin, internal (autochthonous) sourcing may be important in rivers such as the Shetucket (USA) and Gambia (West Africa). In each case the peak in DOC concentration corresponds to the timing and origin of carbon input.

1.8.2 Pelagic and benthic communities

Lotic systems differ considerably in the extent to which pelagic and benthic communities are able to develop. This depends particularly on size and flow, with a major distinction between large, slow-flowing rivers and small, turbulent streams.

Large rivers: the development of a phytoplankton community

In most lotic systems, phytoplankton are simply displaced by the current, and are not able to form standing populations. Because of this continuous displacement, net production within a defined section can only occur when local growth rates exceed downstream losses. Phytoplankton growth in lotic systems tends to be limited by ambient light intensity (overhanging foliage), turbidity, and circulation within the water column (no stratification), while downstream loss is largely a function of current velocity. Phytoplankton production in riverine systems, and the development of a pelagic community, is thus largely regulated by light availability in combination with hydrological processes.

Recent studies by Sellers and Bukaveckas (2003) have demonstrated that phytoplankton production may be significant in large rivers. Local biomass accumulation occurs particularly in shallow reaches during peaks of low discharge and turbidity, when phytoplankton experiences prolonged exposure to favourable light conditions. Observations on a large navigational pool in the Ohio River (USA), for example, showed that at times of high discharge, phytoplankton productivity within the pool was <10 per cent of phytoplankton input from upstream and tributary sources. At times of low discharge, phytoplankton production in the pool exceeded external algal sources (Sellers and Bukaveckas, 2003).

Small rivers and streams: development of a benthic community

Most lotic systems are dominated by benthic communities, with little development of pelagic food webs. Solid surfaces are rapidly colonized by microbial organisms, leading to the development of bacterial and algal biofilms. These permanent microbial communities are an important aspect of the lotic environment, with algal biofilms making a significant contribution to primary productivity and both algal and bacterial biofilms forming part of the ecosystem food web. Bacterial biofilms typically produce copious amounts of extracellular matrix, creating a distinctive micro-environment that limits the extent of the community (Figure 1.5).

Benthic bacteria   Bacteria present on the surface, and in subsurface regions, of stream-bed sediments are involved in a number of key ecosystem processes – including the breakdown (mineralization) of organic matter (Section 6.6), assimilation of inorganic nutrients, and acting as a food source for consumer organisms (Section 9.13). While quantitative aspects of the supply of organic matter clearly influence the abundance and productivity of sediment bacteria, qualitative aspects are also important (Findlay *et al.*, 2003). These include delivery, particle size distribution, and chemical composition – all of which affect the spatial and temporal composition of bacterial communities. Chemical composition is particularly important, and recent studies by Findlay *et al.* (2003) have emphasized differences between labile (easily assimilable) and recalcitrant (poorly assimilable) carbon sources in promoting bacterial community responses such as oxygen consumption, productivity, extracellular enzyme activity, and community composition (Section 6.2).

The growth of heterotrophic bacterial populations in benthic environments relates to both productivity (availability of organic carbon, inorganic nutrients, terminal acceptors) and loss processes such as grazing (Section 9.13) and viral infection (Section 7.8.2).

Ecological pyramids and food webs   The comparatively low level of phytoplankton and zooplankton in lotic systems means that ecological pyramids from rivers and streams look very different from those of standing waters. Food webs of lotic systems (Figure 1.19) are also very different. The grazing (phytoplankton, zooplankton, fish) food web that dominates lakes is of much reduced importance, and the low level of internal (autochthonous) DOC production by phytoplankton and other photosynthetic organisms means that the pelagic microbial loop – recycling carbon released from phytoplankton back into the macrobiota – has little application. The metabolic coupling and correlation in populations between planktonic algae and bacteria seen in lakes (Section 6.10) is not a feature of lotic systems.

The microbial loop of lotic systems

Although the pelagic microbial loop has little relevance in most lotic systems, the benthic microbial community is important in cycling externally-derived DOC into multicellular organisms within the stream environment. This benthic microbial loop (Figure 1.18) is initiated by the arrival of externally-derived carbon into the river system as dissolved organic carbon (DOC) and particulate organic matter. The latter is broken down by invertebrates (conversion of coarse to fine particulate matter) then digested by fungal and bacterial exoenzymes to form DOC, with subsequent uptake by these two groups of organisms. Protozoa are involved in the direct uptake of particulate material and the ingestion of bacteria. The context of the Microbial loop within the overall lotic food web is shown in Figure 1.19.

Comparison between pelagic and benthic microbial loops (Table 1.9) emphasizes a number of key differences in addition to the internal or external origins of the carbon supply. Non-microbial biota have a greater impact on the benthic microbial loop, carrying out the initial shredding or breakdown of the coarse particulate organic (CPOM) matter and generating fine particulate material and DOC which are ingested by microorganisms. Connections with the overall food web also differ (Figure 1.19), with relatively good linkage between benthic and planktonic biota in lotic systems, but fewer biomass transfers between microorganisms and top carnivores within the river ecosystem.

1.8.3   The microbial food web

The microbial loop of flowing water is completed by the assimilation of bacterial carbon, which then passes into multicellular biota (Figure 1.18). Bacteria form the main microbial biomass and are present mainly within biofilms and associated with organic debris such as leaf litter.

They are ingested by three main groups of benthic organisms – protozoa, insect larvae, and meiofauna. These differ in the efficiency with which they ingest bacteria, as determined by bacterial carbon uptake per unit weight of the organism (Table 1.10).

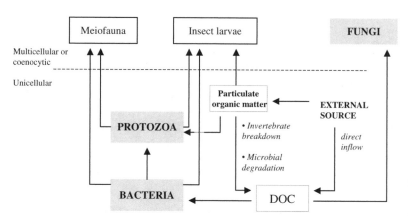

Figure 1.18 Benthic microbial loop in flowing waters, showing the major activities of benthic microorganisms (shaded boxes) in the retrieval of soluble carbon streams and rivers. The context of the microbial loop within the overall lotic food web is shown in Figure 1.19

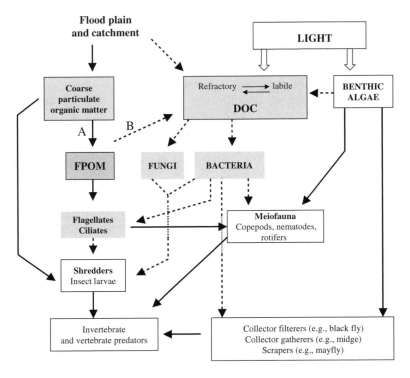

Figure 1.19 Generalized food web for fast-flowing waters: some of the main linkages within the lotic food web are shown, emphasizing the importance of the generation of biomass from external carbon via the microbial loop (- - - - ►); Coarse particulate matter is initially cut up by invertebrates (A) to fine particulate organic matter (FPOM), then degraded by enzyme action (B) to dissolved organic carbon (DOC). In fast-flowing streams, light acts primarily on benthic algae and DOC, with little generation of autochthonous carbon via phytoplankton (not shown)

Protozoa (flagellates and ciliates)

This group of unicellular organisms are important grazers of biofilms. They have significant impact on bacterial populations, with assimilation rates of the order of 10^{-1} to 10^{-2} μg bacterial carbon mg^{-1} protozoon biomass d^{-1}.

Insect larvae

Aquatic insect larvae ingest bacteria that are present mainly on organic debris such as leaf litter. They are both predators and competitors of microbial organisms, directly consuming bacteria and fungi and also digesting the organic detritus that is an impor-

Table 1.9   Key differences between the microbial loops of flowing and standing waters

	Standing waters (lentic systems)	Flowing waters (lotic systems)
Main source of DOC	Autochthonous – phytoplankton	Allochthonous – catchment area
Main location of bacteria in water column	Plankton	Benthos
Ingestion of bacteria	Ingested as single dispersed organisms	Often ingested with debris or in biofilms
Linkage between benthic and planktonic groups	Poor	Good
Number of transfers between bacteria and top consumers	Many	Few

Table 1.10 Assimilation of bacterial carbon by consumers in streams measured using radiolabelled bacteria. Shaded areas – values for carbon assimilation >1 (taken from Meyer, 1994)

Taxonomic group	Species or family	Assimilation of bacterial carbon*
Insect larvae	Stoneflies (Peltoperlidae)	1.2×10^{-4}
	Craneflies (*Tipula*)	5×10^{-4}
	Mayflies (*Stenonema*)	2
	Black flies (*Simulium*)	16–267
Isopod crustacea	*Lirceus*	0.7–6×10^{-3}
Copepod crustacea	Harpacticoid (*Atthyella*)	1–36
Protozoa	Ciliates	6×10^{-2}
	Flagellates	2.8×10^{-1}

*Measured as μg Bacterial C per mg animal per day.

tant basis for the bacterial food chain. Some insect larvae (e.g., stoneflies and craneflies) ingest bacteria simply as part of the litter that they collect, resulting in relatively low bacterial assimilation rates (about 10^{-4} μg mg^{-1} d^{-1}). Other larvae have specialized bacterial collection strategies, including filtration (black fly) and surface scraping (mayfly), resulting in much higher bacterial assimilation rates (Table 1.10).

Meiofauna

These are animals inhabiting the bottom of a lake or river that are just visible to the namost important ked eye and include copepods, nematodes and rotifers. These are generally regarded as the most important bacterial predators in lotic systems, and include both filter feeders and biofilm grazers. Their bacterial carbon assimilation rates are typically 1–4 orders of magnitude greater than that of other bacterial consumers (Meyer, 1994).

The microbial loop in running waters is part of a more complex food web that also involves photosynthetic carbon production by algae and higher plants, saprophytic and parasitic activites of fungi and viruses and important roles for the large invertebrate and vertebrate predators (Figure 1.19).

Primary production by benthic algae (periphyton) varies in importance in different systems (Table 1.8), depending partly on the depth of the water column

and light penetration to the substratum. The development of periphyton communities is discussed in Section 3.7.2, and the lack of photoinhibition in Section 4.9.5. Light is also important in the water column in relation to degradation of non-assimilable (refractory) DOC. UV-irradiation, in particular, has been shown to have a major effect in converting humic acids to more labile forms of DOC (DeHaan, 1993).

The ecological importance of bacteria, protozoa, and fungi in lotic food webs has been mentioned in relation to the microbial loop (Figure 1.18). Further information is subsequently given in relation to the breakdown of organic matter by benthic bacteria (Section 6.6.3), saprophytic activities of benthic fungi (Section 8.5), and the role of protozoa in the ingestion of both living and non-living particulate matter (Section 9.6).

The transfer of biomass from microbes to top carnivores does not have such a defined route as pelagic foodwebs, and the domination of herbivorous activities in the water column by crustaceans (zooplankton) does not occur in the lotic food web. Herbivory in running waters occurs mainly at the sediment surface and involves the ingestion of either unattached or attached microorganisms. The consumption of microorganisms in lotic communities is discussed further in Section 9.13 and involves ingestion of:

- free-moving biota such as bacteria and protozoa, present as localized populations around organic debris, by meiofauna and protozoa (Figure 9.7);

- microorganisms, particularly fungi and bacteria, that are present within organic debris such as leaf litter – this is carried out by invertebrate shredders, gougers, and collector gatherers, prominent amongst which are insect larvae (Table 1.10);

- biofilm and other microorganisms that are attached to solid substratum – this is carried out by shredding, scraping and rasping (Figure 9.17).

The fragmentation and ingestion of leaf litter by invertebrates requires partial breakdown of this material by fungal activity (Section 8.5.2). The substrate colonization, invasion, and macerating activity of these organisms thus promotes their own ingestion during the final stages of leaf processing (Figure 8.4).

2

Freshwater environments: the influence of physico-chemical conditions on microbial communities

This chapter considers the range and diversity of freshwater environments, emphasizing the influence of physical and chemical conditions on the composition and activities of the microbial communities that they contain. More detailed aspects of the importance of light (Chapter 4) and inorganic nutrients (Chapter 5) are considered later, and the environmental implications of eutrophication (nutrient enrichment) are discussed in Chapter 10.

A. INTRODUCTION

2.1 The aquatic medium: water, dissolved and particulate components

At the micro level, the aquatic medium surrounding freshwater biota is a heterogeneous mixture of three main constituents – particulate material, soluble components, and water matrix. Soluble constituents are discussed in Sections 5.1 (inorganic components) and 4.7.2 (dissolved organic carbon – DOC).

2.1.1 Particulate Matter

Particulate matter is normally defined as comprising all solid material with a diameter (longest axis) greater than 0.2 μm, and can be further divided into coarse particulate matter (diameter >1 mm) and fine particulate matter (0.2 μm–1 mm). This particulate material includes both inorganic and organic components, with further separation of the latter into living (e.g., planktonic organisms) and non-living (e.g., leaf debris, dead planktonic organisms) groups (Table 2.1). The arbitrary designation of dissolved organic matter (DOM) as including all material with a size <0.2 μm leads to the anomalous inclusion of both particulate material (e.g., virus particles, fine organic debris) and genuinely soluble material (e.g., small MW organic molecules, inorganic anions and cations) in this category.

With these size definitions, microorganisms are major components of all three categories of organic material – CPOM, FPOM, and DOM (Table 2.1). The size ranges of planktonic organisms are considered in further detail in Sections 1.2.2 and 3.4.2. The dissolved organic material (DOM) of aquatic systems can be measured as total mass or more

Freshwater Microbiology: Biodiversity and Dynamic Interactions of Microorganisms in the Aquatic Environment   David C. Sigee
© 2005 John Wiley & Sons, Ltd   ISBNs: 0-471-48529-2 (pbk) 0-471-48528-4 (hbk)

Table 2.1 Size classification of organic material in aquatic systems

Size category	Size range	Non-living material	Living material
Coarse particulate organic material (CPOM)	>1 mm	Breakdown products of higher plants (e.g., leaf litter) and plankton debris	Wide range of planktonic organisms
Fine particulate organic material (FPOM)	0.2 μm–1 mm		
Dissolved organic material (DOM)[*]	<0.2 μm	Fine debris Algal exudates Humic acids	Viruses

[*]Dissolved organic material (DOM) is sometimes measured as dissolved organic carbon (DOC) and is the same biomass.

typically in terms of carbon content, when it is expressed as dissolved organic carbon (DOC). On the above size definitions, DOC also includes some particulate as well as truly soluble material.

2.1.2 Aquatic Matrix

In the freshwater environment, water is important both as a major internal constituent of biota and also as the environmental matrix. Because of this dual role, the physical/chemical properties of water are central to the physiology and ecology of freshwater organisms. The properties of water in the liquid state are considered in this section (Table 2.2) and water in the frozen state in Section 2.17 (Table 2.15 – snow ecosystems).

Physical properties of water

Water is unusual in comparison with other molecules of a similar structure (H_2S, NH_3, HF) in that it exists as a liquid rather than a vapour at the earth's surface, and is one of only two inorganic liquids (together with mercury) that can exist in this state under ambient temperature and pressure. Many of the unique properties of water result from its unusual molecular structure, with the separation of charges between electronegative oxygen and electropositive hydrogen atoms. This charge asymmetry leads to weak hydrogen bonding between molecules, resulting in molecular complexes – $(H_2O)_n$ – with a semi-crystalline structure. It is these chemical prop-

erties that determine the liquid state of water at room temperature, along with the other physical properties summarized in Table 2.2.

Water has a wide temperature range and high values for viscosity, specific heat capacity, latent heat of evaporation, dielectric constant, and surface tension. Thermal conductivity, light absorption, and scattering coefficients are low. These properties have a major effects on the physical characteristics of the aquatic environment, including temperature changes at the water surface, ice formation, stratification, mixing and circulation of water, sedimentation of particles within the water column, and the light regime of lake and rivers (Table 2.2). Some physical properties, such as the coefficient of thermal expansion and viscosity, vary considerably with temperature – so the environmental effects are also strongly temperature-dependent. Coefficients of light absorption and scattering vary with spectral characteristics and are wavelength-dependent (Kirk, 1994).

Influence of the physical properties of water on the biology of aquatic microorganisms

The physical properties of water exert fundamental and wide-ranging influence on the biology of freshwater microorganisms (Figure 2.1) through their effects on the surrounding aquatic environment (Table 2.2). Some of these effects are general to all water bodies, while others are more specific to lotic or lentic systems, respectively.

Table 2.2 Physical properties of water (data mainly from Ellis, 1984, and Dean, 1999)

Property	Value (SI units)	Environmental implications
Temperature range (θ)	**Wide range** 273–373 K (0–100°C)	Water is in liquid form for most of the year in temperate to tropical lakes
Coefficient of cubic thermal expansion (γ) (at 293 K)	2.1×10^{-4} K^{-1}	Surface heating leads to decrease in density and retention of heated water at the lake surface
Density (ρ) (at 298 K)	1.0 g cm^{-3}	• Close similarity in density of water and aquatic biota limits the rate of sedimentation – Stokes' equation. • Density/temperature variations create thermal stratification in standing waters
Absolute viscosity (η) (at 298 K)	**High** (8.91×10^{-4} N s m^{-2})	High viscosity limits: • mixing of water in lakes and streams • sedimentation of particulates – Stokes' equation • velocity of flow in lotic systems (Reynolds number)
Thermal conductivity (λ)	**Low** (0.6 W m^{-1} K^{-1})	Poor thermal conductivity and high specific heat result in: • a relatively slow response to changes in air temperature
Specific heat capacity (C_p)	**High** ($4.2 \times$ J g^{-1} K^{-1})	• localized surface heating and stratification in standing waters
Latent heat of fusion (Δh_m) (at T_m and 1 atm)	3.3×10^5 J kg^{-1}	Important in determining the rate of ice formation and release of heat at the water body surface
Latent heat of evaporation (Δh_b) (at T_b)	**High** (2.2×10^6 Jkg^{-1})	Important in determining the rate of evaporation and loss of heat at the water body surface
Dielectric constant (ε) (at 293 K)	**High** (80.10 – no units)	Determines the ability of ions of opposite charge to stay in solution
Surface tension (σ)	**High** (7.28 N m^{-1})	High surface tension is important in the accumulation of fine particulate material at the water surface
Light absorption coefficient (a)[*] (over λ range of 280–550 nm)	**Low** 0.0239–0.064 (fraction absorbed) m^{-1}	Low light absorption and scattering lead to high levels of light penetration within the water column, reaching sediments that occur within the photic zone
Light scattering coefficient (b)[*] (over λ range of 400–500 nm)	**Low** 0.058–0.0019 (fraction scattered) m^{-1}	

[*]Information on light absorption and scattering coefficients, over specific wavelength (λ) ranges, from Kirk (1994).

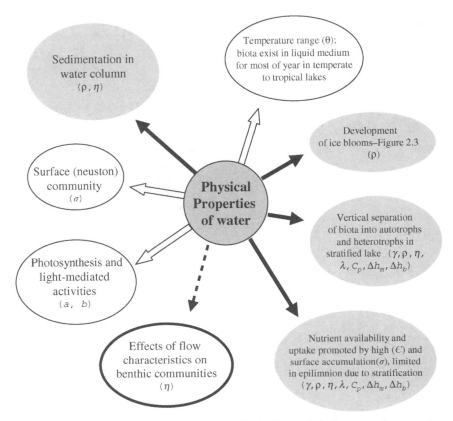

Figure 2.1 Influence of physical properties of water on the biology of freshwater microorganisms: the physical properties of water influence the biology of microorganisms separately in lakes (➡), rivers (--➤) and in both types of environment (⇨); the quantitative physical parameters determining each of these biological aspects are indicated by symbols (see Table 2.2).

General effects Temperature characteristics of water are generally important, by modulating the influence of external changes in atmospheric temperature. Low thermal conductivity and high heat capacity modulate diurnal and seasonal changes in temperature. This has wide-ranging implications, including relatively high rates of decomposition (compared with terrestrial systems) by fungi (Section 8.5) and other saprophytes. Other characteristics of general importance include high surface tension (development of surface neuston communities) and low light absorption/scattering. Deep penetration of the water column and irradiation of sediments promotes photosynthesis and other light-mediated activities in a wide range of environments.

Lotic environments The bulk flow of water in lotic systems can be quantified in relation to the dimensionless Reynolds number (Re), where:

$$Re = \frac{\bar{U}L}{\eta} \qquad (2.1)$$

and $\bar{U}$ is the velocity of the fluid (m s^{-1}), L is characteristic length scale (m), and η is kinematic viscosity (Kirk, 1994). The Reynolds number is characteristic for a particular bulk-flow situation, and can be used to distinguish different types of flow which range from laminar to turbulent. At low Re values (<500), viscous forces predominate and flow is laminar. Increases in Reynolds number (500–10^3) lead to transitional flow, with values over 10^3 signifying turbulent conditions. The Reynolds number thus quantifies the ratio of inertial to viscous forces within a liquid, and emphasizes the importance of viscosity in determining the flow characteristics of a stream or river. The relatively high viscosity of water reduces the value of *Re* and favours laminar flow conditions.

The velocity and type of water flow is of considerable significance to lotic benthic communities in relation to dispersal, nutrient acquisition, competition, predator-prey interactions, and the creation of microenvironments (Section 2.10).

Lentic environments The physical properties of water are important to lake microorganisms in a number of specific ways including lake stratification (Section 2.5.1), development of ice-blooms, and the establishment of surface biofilm (neuston) microbial communities.

Formation of ice-blooms. Variations in density with temperature are particularly important in maintaining thermal stratification within standing waters, with temperature normally decreasing with depth in the water column. The density of water, however, reaches a maximum at 3.95°C, resulting in a reverse stratification below ice (temperature increasing with depth) as surface water approaches freezing point. The anomalous temperature – density curve of water (Figure 2.2) is particularly important in ice-covered water bodies such as Lake Baikal (Russia), where localized heating below the ice layer in late

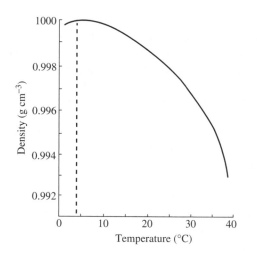

Figure 2.2 Changes in the density of pure water with temperature: water has a maximum density at 3.95°C, and shows an almost exponential decrease at higher temperatures

spring results in the generation of weak convection currents. These are sufficient to maintain microbial communities (particularly diatoms) in suspension below the ice ((Figure 2.3) at a time of year when light intensity is increasing, resulting in algal blooms. If the density of water simply decreased with a rise in temperature from 0°C, no downdraft would occur and circulation below the ice would not take place.

Surface biofilm communities. The high surface tension of water is important in the establishment of a

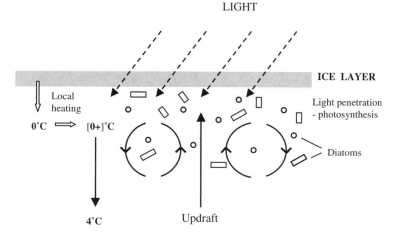

Figure 2.3 Formation of ice-blooms due to the anomalous temperature-density relationship of water. Under appropriate ice conditions, light penetration leads to turbulent suspension of diatoms (*Aulacoseira baicalensis*) and supports photosynthesis. Local heating raises the temperature of water below the ice (from 0 to 4°C), increasing density and promoting a downdraft. This causes sub-surface water circulation, generating updrafts which carry phytoplankton up into the light zone, where they remain suspended in the turbulent conditions, leading to a diatom bloom below the ice.

surface biofilm, with an associated community of surface microorganisms. The lowering of surface tension by organic molecules leads to their spontaneous adsorption at the air – water interface, creating a high-nutrient microlayer (Section 2.16). This acts as a site for microbial attachment and nutrient supply, leading to the development of dense surface populations of bacteria, algae, and protozoa. These surface microorganisms are part of a larger community of surface biota (neuston), which include various invertebrates. Many of these, such as the water beetle (*Notonecta*) and mosquito larvae, exploit the high surface tension of water to attach to the surface.

2.2 Freshwater environments

Freshwater environments show wide variation in terms of size, permanence, and complexity. They also vary in terms of the state of the water medium (frozen or liquid) and in a wide range of other physico-chemical conditions imposed on the living organisms that inhabit them.

The distinction between standing waters (lentic systems) and flowing waters (lotic systems) has been made previously. In this chapter, freshwater systems are considered under four major headings – lakes, wetlands (lentic systems), rivers, and estuaries (lotic systems). Although the majority of these environments allow a wide range of microorganisms to survive and grow under 'normal' conditions, other environments are more restricted in their support of freshwater biota. Such 'extreme' environments severely limit the metabolic activities of aquatic organisms, and are considered in the final section.

Table 2.3 summarizes the main microbial communities that dominate these major freshwater systems, and the main physical and biological constraints that affect them. These are considered further in the appropriate sections, but in each case involve complex interactions between hydrology (flow, turbulence), external factors (light, temperature), and biological constraints (competition for nutrients, colonization, grazing).

Table 2.3 Physical and biological factors affecting the microbial communities in different freshwater environments

Environment	Major microbial community	Major physical characteristics	Major biological constraints
Lake	Plankton	Stratification Wind-generated turbulence	Nutrient competition Grazing Parasitism
Wetlands (a) Flood plain	Plankton Epiphytic biota	Periodic desiccation	Competition between algae and macrophytes
(b) Permanent	Benthos	Periodic flooding	
River	Benthos	Linear flow Flow-generated turbulence	Colonization competition Biofilm grazing
Estuary (a) Mudflats	Epipelic biofilms	Desiccation High Light Exposure to salt water	Competition and grazing at mud surface
(b) Outflow	Plankton	Mixing with saltwater Turbidity	Grazing in water column

B. LAKES

Lakes vary in size from small ponds to the great lakes of North America, Africa and Russia. The world's 10 largest lakes (Table 2.4) comprise massive aquatic environments, ranging in size from 28 000 to 350 000 km^2 and varying from freshwater to marine bodies.

The biology of lakes is dominated by free-floating or planktonic organisms – including bacteria, algae and invertebrates which inhabit the main part of the lake and form the major biomass. The growth and development of these and other lake organisms are determined by local chemical (e.g., nutrient concentrations) and physical (e.g., light regime) conditions, which in turn are determined by more general geomorphological and geographic features, including the following.

- Lake morphology and hydrology: this includes the overall structure of the lake (morphology), water flow into and out of the lake (hydrology), and interactions of lakes with their surrounding environment.

- The climatic environment: weather conditions influence various physical parameters of lakes including light regime, temperature, and water movement.

The long-term evolution of lake biota is strongly influenced by another geographic feature of lakes – their isolation and independent development. These broad aspects are considered in relation to the biology of lake biota, with particular emphasis on the microorganisms.

2.3   Lake morphology and hydrology

2.3.1   Lake morphology

Lake morphology refers to the overall shape of the lake basin and is important in terms of water flow, nutrient accumulation, light penetration (lake depth), mixing of the water column, and separation of the

Table 2.4 The worlds 10 largest lakes – ancient lakes are in bold type (adapted from Horne and Goldman, 1994, Martens, 1997 and Kalff, 2002)

Lake	Area (km^2)	Volume (km^3)	Salinity	Age of lake[*] (My)	Endemism[†] (%)	Mean water retention time (years)
Caspian (former USSR/Iran)	**374 000**	**78 200**	**Mostly marine**	**2–3**	**27**	**210**
Superior (USA, Canada)	82 100	12 200	Fresh			191
Victoria (East Africa)	**68 500**	**2700**	**Fresh**	**0.75**	**–**	**123**
Aral (Russia)	64 100	1000	Mostly marine			
Huron (USA/Canada)	59 500	3500	Fresh			22
Michigan (USA)	57 800	4900	Fresh			99
Tanganyika (Africa)	**32 900**	**18 900**	**Mostly fresh**	**9–12**	**56**	**6000**
Baikal (Russia)	**31 500**	**23 000**	**Fresh**	**25–30**	**54**	**327**
Great Bear (Canada)	31 300	3400	Fresh			131
Malawi/Nyasa (Africa)	**30 800**	**8400**	**Fresh**	**4.5–8.6**	**–**	**1225**

[*]Age of lake, based on paleohydrology, sediment fossils, and estimates from molecular data (molecular clock).
[†]Approximate figures for the percentage of endemic fauna and flora; no comprehensive summaries are available for Lakes Victoria and Malawi.

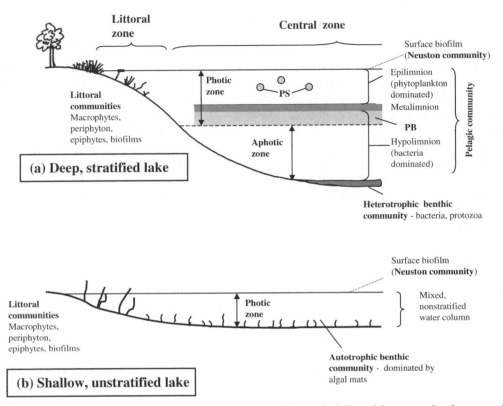

Figure 2.4   Lake morphology and major communities – deep lake and shallow lakes may develop very different communities (figures not drawn to scale)

(a) *Deep, stratified lake.* Mesotrophic and eutrophic deep (>20 m) lakes typically show clear distinction between littoral and central communities. In summer, the stratified water column is clearly differentiated into autotrophic (epilimnion) and heterotrophic (hypolimnion) populations, and the lack of light penetration to the sediments results in the development of a heterotrophic (bacterial-dominated) benthic community. The stratified water column supports summer growths of colonial blue-green algae (PS – phycosphere communities) in the epilimnion, and a distinct layer of photosynthetic bacteria (PB) under reducing conditions at the base of the photic zone

(b) *Shallow, unstratified lake.* The distinction between littoral and central zones may be less pronounced, with some macrophyte development in central regions. In lakes with depths <6 m, the water column may be too shallow for stratification to develop, with turbulent mixing occurring throughput the growth period. This tends to maintain diatoms in suspension, suppresses the development of colonial blue-green algae and prevents the development of communities that require reducing conditions, e.g., anaerobic photosynthetic and heterotrophic bacteria. Light penetration to the sediments allows the growth of extensive algal mats, forming an autotrophic benthic community

lake into two main regions – a littoral zone (edge of the lake) and a central (pelagic) region (Figure 2.4).

Littoral and central regions

Most lakes have clear separation of littoral and central regions, each with their own distinctive communities. The development of higher plant (macrophyte) communities as part of the littoral flora is typical of mesotrophic and eutrophic (Figure 2.6(b)) water bodies, but not oligotrophic ones such as Wastewater, UK (Figure 2.6(a)) and Lake Tahoe, USA (Figure 4.25). In ancient lakes such as Lake Baikal (Russia), the presence of littoral and central communities also provides an important separation between non-endemic and endemic groups of organisms (Section 2.4).

Littoral zone The edge of the lake is a shallow region characterized by distinct physical (high light penetration, wave action), chemical (influx of allochthonous DOC, see below), and biological characteristics. The shallow water in the littoral zone allows light penetration to the sediments, which provide substratum for rooting macrophytes (mainly higher plants), which are able to survive strong wave action and tend to dominate meso-trophic and eutrophic littoral communities.

In many lakes the littoral zone also has a rich microbial community, with filamentous and unicel-lular algae present on sediments, where they form part of the periphyton community (Sections 1.2.2, 3.6). Some algae also occur as free floating masses (metaphyton) and as epiphytes on the leaves and stems of higher plants. Bacteria, fungi, and protozoa are also well represented in this part of the lake, occurring (with algae) as mixed biofilm commu-nities. The microbial community of the littoral zone differs from that of the pelagic zone in terms of taxonomic composition, domination of benthic organisms and heterotrophic consumption of mainly allochthonous DOC. The littoral region of the lake is also a major site for recruitment of benthic algae to form planktonic populations (Brunberg and Blomqvist, 2003; Section 3.10.1).

Allochthonous DOC. Inflow of externally-derived dissolved organic carbon (allochthonous DOC) from the surrounding terrestrial environment has an important influence on the littoral community. The level of DOC that enters the edge of the lake is regulated primarily by rainfall and wetland flushing (Vinebrooke and Leavitt, 1998), and is dependent on climatic conditions. In general, entry of DOC is enhanced by wet (increased flushing) and warm climates (promoting development of terrestrial vege-tation and soil fertility). Alpine lakes near to the tree line are particularly sensitive to climatically-induced changes in DOC influx (Leavitt *et al.*, 1997), since they normally receive significantly less DOC (mean value 0.9 mg Cl^{-1}) compared with subalpine lakes (4.7 mg Cl^{-1}). The potential effects of allochthonous DOC on the littoral microbial community can be separated into three main aspects (Vinebrooke and Leavitt, 1998). These are summarized in Figure 2.5, and include the following.

- *Substrate.* Allochthonous DOC provides an organic substrate for heterotrophic microorgan-isms, and contains associated anions (PO_4, NO_3) and cations which may be released in a bio-available form. The extent to which the DOC can be broken down varies considerably, with some materials (e.g., humic acids) requiring photo-induced conversion from a refractory to a labile state for microbial assimilation. Solar

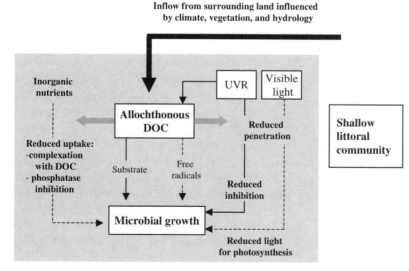

Figure 2.5 Potential effects of al-lochthonous DOC on the shallow lit-toral community occurring at the edge of lakes. Direct and indirect effects of DOC influx can either promote (——▶) or inhibit (--▶) growth of mi-crobial communities, which include both heterotrophic and photoauto-trophic organisms.
Allochthonous DOC: externally-derived dissolved organic carbon.
UVR: ultraviolet radiation

radiation (particularly UVR) can also convert soluble organic materials to toxic compounds such as hydrogen peroxide, carbon monoxide, and free radicals.

- *Light penetration.* High levels of DOC influence both the spectral quality and the intensity of light that penetrates the littoral water column, and may reduce the depth of the photic zone to such a degree that the substratum is no longer illuminated. Reduced penetration of photosynthetically-active radiation (PAR) potentially inhibits microbial growth, while reduced penetration of ultraviolet radiation (UVR) would be expected to promote it.

- *Inorganic nutrient uptake.* Influx of DOC can cause large-scale removal of free anions and cations, and thus their availability to littoral microorganisms. Humic acids in particular are rich in electronegative ligands (mainly oxygen groups) that bind cations, and may cause severe depletion of limiting elements such as Fe^{3+}. Phosphates and nitrates may also be complexed, limiting their microbial uptake. Deactivation of phosphatase exo-enzymes, required for the hydrolysis of soluble organic phosphates (Section 5.8.3), may also result from the DOC influx.

Studies by Vinebrooke and Leavitt (1998) have shown that allochthonous DOC is an important resource for littoral food webs in oligotrophic alpine lakes. The precise effects of DOC influx will vary with individual littoral communities, but appear to be highly selective in relation to particular groups of biota. In the above study, for example, addition of DOC to littoral enclosures significantly increased the biomass of benthic microorganisms growing on hard substrata (epilithon), but had no effects on the abundance of those living on fine sediments (epipelon), benthic heterotrophic bacteria, protozoa, and phytoplankton. Natural UV irradiance significantly enhanced the positive effect of DOC on epilithon, and directly increased the abundance of epipelon.

Pelagic zone The main part of the lake (central or pelagic zone) has limited light penetration and is dominated by free-floating organisms (plankton). This part of the lake consists of a vertical water column with an air – water interface at the top and lake sediments at the base.

The air – water interface exhibits some features of an extreme environment and is considered in detail in Section 2.16. This part of the lake has its own community of organisms (neuston), which include an extensive microflora (forming a mixed biofilm) plus higher animals (insect larvae, surface beetles such as *Notonecta*) and plants. Some of these may be blown about by the wind (referred to as 'pleuston'), including floating higher plants such as *Pistia stratiotes* (water cabbage) – a major pest of tropical lakes.

The major community within the lake in terms of area cover and total biomass is the pelagic community (Section 1.5), which occurs throughout the water column. Within this part of the lake, actively swimming animals such as fish and some mammals are able to locate themselves at any depth by their own activity and are referred to as nekton. In contrast to this, planktonic organisms are less actively motile and have much less ability to adjust their position within the water column.

The water column of lakes is closely linked to the sediment in terms of nutrient cycling and interchange of aquatic populations. This part of the lake is inhabited by benthic organisms (collectively referred to as benthos), with bacteria being particularly important in sediments that have no light (aphotic zone) or oxygen (anoxic).

Shallow lakes

Lake depth is an important feature of lake morphology, influencing light penetration to the sediments and mixing of the water column, both of which can have a major affect on the development of lake communities (Figure 2.4(b)).

Hollingworth Lake, UK (Figure 10.16), provides a good example of a temperate, shallow water body with depths not exceeding 10 m. The water column of this lake is too shallow for the establishment of a stable temperature gradient and, with the added influence of frequent wind activity (exposed upland position), is in a non-stratified frequently-mixed (polymictic) state. This influences the phytoplankton population, and although the lake is nutrient-rich (mesotrophic/eutrophic status), continuous turbulence maintains high populations of diatoms throughout the year. The expected summer development of colonial blue-greens in a lake of this nutrient status (Table 10.4) does not materialize, though these algae may form problematic blooms of *Oscillatoria* later in the year (Section 10.10.1). Penetration of light to the sediments allows an algal population to become established, and the benthic community is primarily autotrophic. Continuous mixing in the water column of shallow lakes prevents the development of anaerobic conditions and the occurrence of populations of anaerobic photosynthetic and heterotrophic bacteria that are so prominent in deep eutrophic lakes.

The development of an extensive autotrophic benthic community is a prominent feature of shallow low-nutrient water bodies. In shallow, oligotrophic Greenland lakes, for example, benthic algae (periphyton) are responsible for 80–90 per cent of primary production (Vadeboncoeur *et al.*, 2003). With increasing eutrophication, benthic algae become out-competed by phytoplankton, and the balance of productivity changes. A decrease in the depth of the water column of shallow lakes may also lead to a wetland situation, in which benthic algae also have to compete with macrophytes.

2.3.2 Lake hydrology and the surrounding terrestrial environment

The hydrology of a lake involves all aspects of water flow, including the lake itself and the inflow and outflow of water. Although interest in lakes tends to focus on the main water body, they are clearly part of a wider catchment area and have important interactions with this – as illustrated by the potential effects of terrestrial DOC on the lake littoral community (Figure 2.5).

Much of the water contained in lakes is derived as inflow from the surrounding land mass, and the chemistry of the lake is largely determined by the chemistry of the inflow water. Entry of inorganic nutrients (particularly phosphates and nitrates), dissolved organic material (e.g., humic acids), dissolved cations, and suspended solids determines the trophic status of the lake and has a major influence on the development and growth of micro-organisms.

Hydrology and trophic status

The importance of the catchment area to the biology of the lake is particularly well illustrated by comparison of upland and lowland lakes. Lakes in mountainous areas typically have a relatively infertile catchment area, resulting in oligotrophic (low nutrient) status. Lowland lakes, in contrast, collect their water from nutrient-rich (often cultivated) land and are typically eutrophic. This distinction is seen particularly well within the English Lake District, where mountain lakes such as Wastewater are much less nutrient-rich and productive than lowland water bodies such as Esthwaite (Figure 2.6). The effect of location and hydrology on nutrient availability in lakes determines a wide range of biological and related characteristics – including phytoplankton biomass (primary productivity), growth of bacteria and zooplankton (secondary productivity), light penetration and oxygenation within the water column. The general features of eutrophic and oligotrophic lakes are summarized in Table 2.5, and discussed more fully in Chapter 10.

The presence and type of vegetation in the catchment area may also be important. This is seen in the case of low nutrient (oligotrophic) lakes that are surrounded by dense woodland (Hoyos and Comin, 1999). Lake Sanabria (Spain), for example, has low inorganic nutrient concentrations throughout the year, but has inflow of dissolved organic fractions from the surrounding forested catchment area. The colour of the lake (20–40 mg Pt l^{-1}) is indicative of the relatively high amount of externally-derived (allochthonous) organic material dissolved in the waters of the lake. This input of organic material has an important impact on the biology of what would otherwise be a very nutrient-limited oligotrophic lake. Blooms of blue-green are able to develop, and the deeper penetration of red light

Figure 2.6 Oligotrophic and eutrophic lakes, showing two contrasting lakes from the same geographic area (Lake District, UK), differing in nutrient status and microbial productivity; (a) Wastewater – an oligotrophic mountain lake, reaching a depth of 91 m; (b) Esthwaite – a lowland eutrophic lake with well-developed plankton populations, fish stocks and littoral reed beds

compared with other wavelengths is relevant to pigment composition and photosynthetic activity. Wood fires (whether natural or deliberate) are frequent in the catchment area, leading to areas of deforestation – with increased runoff of rain and increased nutrient flux into the lake. The occurrence of summer fires around Lake Sanabria is particularly important in years of low rainfall, when the accumulation of flushed organic material is enhanced due to a high water retention time and extensive development of blue-green algal blooms can develop (Chapter 10).

Inflow, outflow, and retention times

Most natural lakes have an inflow and an outflow, with gradual and continuous replacement of the lake water. The average time taken to refill a lake basin with water if it were emptied is referred to as the mean hydraulic water retention time (WRT in years), and provides a measure of the circulation of water within the aquatic system. WRT (also known as the flushing time or simply the water residence time) has the symbol τ_w and can be calculated from the basin volume (V in m^3) and the average annual water outflow (Q in m^3 yr^{-1}), where:

$$\tau_w = \frac{V}{Q} \qquad (2.2)$$

This value clearly relates to the balance between basin morphology and the hydrology of the catchment area. The retention time of most lakes and reservoirs is about 1–10 years. Some of the world's largest lakes have values considerably in excess of this (Table 2.4), with extreme values of 6000 and 1225 years for lakes Tanganyika (Africa) and Malawi (Nyasa). Size is not an overriding factor, however, since one of the longest residence times (about 700 years) has been recorded for Lake Tahoe (USA) – a lake of only moderate size (156 km^3).

Equation (2.1) describes flushing characteristics in terms of the whole lake, and provides a reasonable estimate for the time taken by an inflow water molecule to pass though the system in unstratified, well-mixed water bodies. In stratified lakes, however, inflow molecules may simply enter and leave via the epilimnion, with a much shorter retention time. Lake Constance (Europe), for example, receives most of its water from melting glaciers during stratification. The epilimnion WRT for this lake is just 0.13 years (1.5 months) compared with an average whole-lake WRT of about 4 years.

Hydraulic retention time has implications for lake microorganisms in two main ways.

- *Short-term accumulation of biomass*. Lakes and reservoirs with a very short retention time tend to have a rapid flow-through of pelagic biomass,

Table 2.5 General features of eutrophic and oligotrophic lakes

Parameter	Oligotrophic lakes	Eutrophic lakes
Morphology and hydrology	Often deep highland lake with infertile undisturbed catchment area	Often shallow lowland lake with cultivated, disturbed or naturally fertile catchment area
Nutrients		
Availability	Low (limiting) levels of at least one major nutrient (N, P, or Si)	High winter levels. High supply rates
Presence in water column	Minor loss from epilimnion. Reduced compounds become oxidized throughout	Major depletion in epilimnion Reduced compounds in hypolimnion
General productivity	Low primary and secondary productivity	High primary (algae) and secondary (all other biota) productivity
Phytoplankton species	Species adapted to low nutrient lakes: desmids and chrysophytes	Species adapted to high nutrient lakes: centric diatoms and colonial blue-greens
Light penetration	High, due to transparent water. Often reaching below thermocline. Secchi depth 8–40 m	Low, due to cloudy water. Often not reaching thermocline. Secchi depth 0.1–2 m
Oxygenation	Saturation in epilimnion, with little variation through water column	Great variation in water column Supersaturated (100–250%) in epilimnion, low levels (0–100%) in hypolimnion
Macrophyte vegetation	Poorly developed or absent	Water lilies and sedges well developed in littoral zone

A more detailed view of biological changes with trophic status is given in Table 10.3.

avoiding accumulation of phytoplankton and the development of standing algal blooms. This may be important in the control of blue-green algal blooms in other water bodies, where dense accumulations of these organisms can be removed by simply flushing out the system (i.e., effectively reducing the residence time).

• *Long-term accumulation of nutrients.* Lakes with a long retention time act as an almost permanent sink for nutrients, leading to long-term accumulation and eutrophication. In lakes with a shorter retention time (e.g., Lake Windermere, UK of 0.75 years) the accumulation is much less. In Lake Marion (Canada), which has a retention time of a only few days (Horne and Goldman,

1994), nutrients are flushed out of the system almost as soon as they have entered.

In addition to lake morphology and the hydrology of the catchment area, hydraulic retention time may also be influenced by rainfall. In Lake Sanabria (Spain), for example, annual water retention time ranges from <0.5 years in times of high rainfall to >1 year during extended periods of drought (Hoyos and Comin, 1999). This has considerable influence on eutrophication and the development of algal blooms in this lake.

Some lakes (e.g., Dead Sea, Israel) have inflow but no outflow. These lose water entirely by evaporation, leading to salt accumulation and the development of a high-saline environment. Other lakes, such as volcanic crater lakes (also known as 'cone lakes'), lack both inflow and outflow, and are controlled in size by basin morphology and the balance between precipitation and evaporation. A considerable number of such lakes occur in Africa along the Uganda–Zaire border and are often very deep, with small drainage basins.

Changes to lake hydrology: climatic and anthropogenic effects

Lakes are fragile environments, susceptible to both climatic and human (anthropogenic) influences. These may affect the net inflow of water into lakes, with further affects on lake size, nutrient accumulation and general biology.

The Aral Sea (Russia) has been dramatically altered by human intervention, with diversion of inflow water for agricultural irrigation, leading to a 50 per cent reduction in the area of the lake over the last 10 years. Other lakes have been much influenced by climatic change. Lake Chad (West Africa) is a shallow lake that has experienced alternate cycles of drought and high rainfall, leading to a cycle of changes in the depth and area of the lake.

2.4   Lakes as isolated environments

To some degree, inland lakes may be regarded as independent and enclosed aquatic systems, and show a number of important features in relation to this:

2.4.1   Isolated development

Many lakes show independent and separate evolution of biota. The extent to which this can occur depends on the isolation and age of the lake. Most lakes today are about 10 000 years old, and will disappear within the next 100 000 years. However, about 10 lakes are much older, between 1 million and 20 million years old. These ancient lakes include the Caspian Sea (former USSR/Iran) plus Lakes Victoria (East Africa), Tanganyika (Africa),

Baikal (Russia) and Malawi/Nyasa (Africa) (see Table 2.4).

Ancient lakes are of particular biological interest since, in contrast to more recent (mostly post-glacial) lakes, they have exceptionally high species diversity and levels of endemicity in their fauna and flora (Table 2.4). The high diversity is partly a direct result of the relative constancy of physical conditions over very long periods of time. During this period, highly stable, complex and buffered assemblages of organisms have developed, in which physical and biological stress have become less important in determining the evolutionary success or failure of individual species (Sanders, 1968). Complex processes of immigration, speciation, and extinction have also contributed to this diversity.

The endemic fauna and flora of ancient lakes represent survival from ancient populations which have become extinct elsewhere. These different aspects are clearly exhibited by Lake Baikal, which is situated in Eastern Siberia and is estimated at being 20–25 million years old.

2.4.2   Lake Baikal: an ancient lake with a diverse and unique fauna and flora

Lake Baikal (Russia) is the world's oldest lake, and an area of major biological interest (Kozhov, 1963). It is also the deepest (>1600 m) and the largest lake by area (31 500 km^2) and by volume (23 000 km^3), holding approximately one fifth of the global unfrozen freshwater. This lake is oligotrophic and completely oxygenated, allowing a variety of animal and plant species to inhabit the lake right down to the bottom of the water column (Kozhova and Matveev, 1998). The lake has two major inflow rivers (the Selenga and Angara Rivers), and is sub-divided into three main basins, of which the middle and south basins have the greatest depth (Figure 2.7).

Two distinct and essentially separate groups of biota occur in the lake. One group of organisms (palaearctic biota) is typical of the fauna and flora currently present throughout much of Siberia, while the other group (Baikalian biota) is endemic and entirely restricted to this lake. Endemic species in

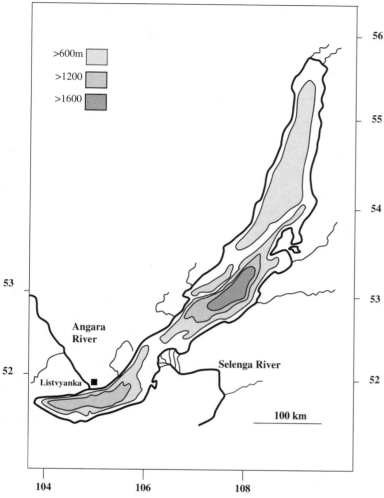

>600m

>1200

>1600

Angara
River

Selenga River

Listvyanka

100 km

Figure 2.7 Lake Baikal (Russia): owing to its continental climate and high latitude (51–55°N), the lake is frozen from January to May, despite the large water mass. About 80 per cent of the surface area occurs as a deep pelagic zone (>250 m depth) with only a short period (July–September) when surface temperatures become warm (10–15°C). Primary productivity is highest during this short warm period (based on a figure from Nagata *et al.*, 1994)

this lake are thought to have survived from populations which were widespread across the arctic region shortly after the lake was formed (early Tertiary period), and include the filamentous diatom *Aulacoseira baicalensis* (Figure 2.8). This dominates the phytoplankton during the early part of the seasonal cycle, to be replaced by the blue-green unicellular alga *Synechocystis* during the later warm period (Figure 2.9).

According to recent listings (Kozhova and Izmest'eva, 1998) there are over 2500 recognized species of freshwater biota in Lake Baikal, of which at least 50 per cent are endemic. Although studies of endemic biota have concentrated particularly on

macrofauna such as flatworms, annelids, arthropods, molluscs, and chordates, microfauna and flora are also well represented (Table 2.6). Approximately 16 per cent of the algae found in the lake are endemic – including dinoflagellates (five species), diatoms (six) and green algae (34). A substantially higher proportion (35 per cent) of the microfauna are unique, with rotifers, copedod, and cladoceran crustacea all having species that are endemic to this lake. These numbers of endemic species must be regarded as conservative, since information on many of the taxonomic groups is incomplete and no listings are currently available for groups such as free-living ciliates, fungi, and bacteria.

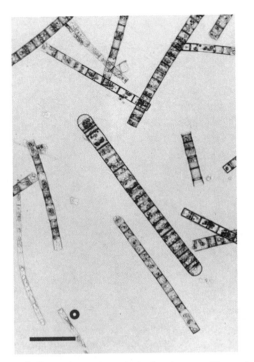

Figure 2.8 *Aulacoseira baicalensis,* a filamentous diatom endemic to Lake Baikal. In this sample the filaments are of two sizes – the thicker ones being derived from resistant propagules (auxospores). Photograph by Dr Nina Bondarenko, Limnological Institute, Irkutsk, Russia (Scale = 100 μm)

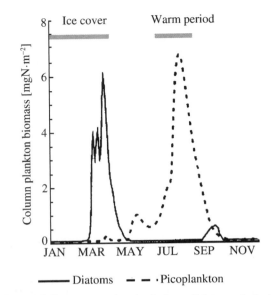

Figure 2.9 Changes in physical conditions and phytoplankton dominance in Lake Baikal (Russia). The major diatom bloom (*Aulacoseira baicalensis*) occurs under ice from January to May, and is followed by a bloom of blue-green algae – of which *Synechocystis limnetica* is the most abundant. Growth of these picocyanobacteria accounts for about 60 per cent of the total annual primary production of the lake and occurs in July/September during the period of maximum water temperature (10–15°C). Adapted from a simulation of the phytoplankton annual cycle (based on remote sensing) by Semovski *et al.,* 2000

The palaearctic and Baikalian fauna occur in different parts of the lake, with non-endemic species being confined to peripheral regions (down to a depth of 20 m) – while endemic species occur in the central deeper regions. The latter are present within the water column at depths of 15–1600 m, and have a vertical region of overlap (ecotone) with palaearctic species at 15–20 m (Sanders, 1968). Some of the major endemic planktonic and nektonic species in Lake Baikal are illustrated in Figure 1.17. The pelagic community occurring in the main body of the lake constitutes a relatively simple ecosystem (Case Study 1.2) and is dominated by these endemic species, with relatively little impact from the more recent palaearctic arrivals.

2.5 Climatic influences on lakes

Climatic conditions affect the physical characteristics of lakes in relation to the duration, intensity, and spectral quality of sunlight (light regime), ambient temperature (affecting ice cover, water temperature), and wind action (causing turbulent mixing). Rainfall is important in terms of inflow, lake volume, and hydraulic residence time (Equation (2.1)), and the combined effects of solar heating and wind determine water loss through evaporation. The effects of wind action in causing vertical mixing of lakes is important for redistribution of nutrients and maintenance of heavy phytoplankton cells (particularly diatoms) in suspension. The

Table 2.6 Total and endemic species of algae and micro-fauna in Lake Baikal (based on Kozhova and Izmest'eva, 1998)

Classification	Total species	Number of endemic species	Unknown
Algae			
Cyanophyta	80	2	6
Dinophyta	48	5	13
Bacillariophyta	49	6	1
Chlorophyta	109	34	9
Chrysophyta	2	1	0
Charophyta	5	0	0
Xanthophyta	5	0	0
Total	**(298)**	**48**	**29**
%		**16%**	**10%**
Microfauna			
Rotifera	198	32	0
Copepoda calanoida	5	1	0
Copepoda cyclopoida	43	24	0
Copepoda harpacticoda	72	63	0
Cladocera	48	7	0
Total	**366**	**127**	**0**
% Total		**35%**	

majority of temperate lakes are shallow to moderately deep, and show mixing throughout the water column (holomictic). Other lakes (particularly deep mountain lakes) are never completely mixed throughout the water column and are termed meromictic. Lake Tanganyika (East Africa), with a maximum depth of 556 m is a good example of a meromictic lake. Lack of complete mixing within the water column in oligotripohic lakes can result in long-term nutrient accumulation in the hypolimnion, with infrequent release causing major eutrophication problems (Case Study 10.6).

Climatic effects on lakes are important in terms of seasonal changes within a particular lake system and also in determining geographic differences between standing waters. These geographic differences are considered in relation to temperate, tropical, and polar lakes.

2.5.1 Temperate lakes – seasonal variations and lake stratification

The majority of studies on freshwater lakes have been carried out on water bodies in temperate regions. Under these climatic conditions there is a clear seasonal succession, with alternation of ice cover (winter) and ice-free conditions (summer) in many cases. Such lakes typically show a limited thermal stratification (layering) in winter, pronounced stratification from late spring to autumn, with two intervening periods of mixing – the spring and autumn overturns. Lakes showing this double annual mixing are termed dimictic. Lakes which lack ice cover show mixing throughout winter (monomictic) and lakes which have permanent ice cover have no complete mixing (amictic).

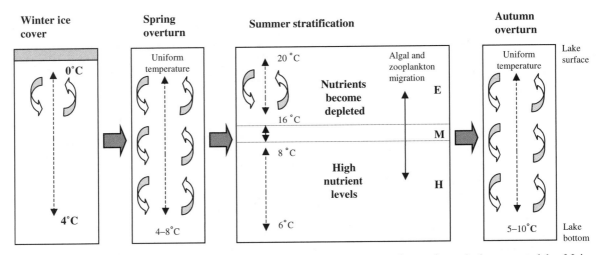

Figure 2.10 Seasonal changes in turbulence and stratification in the water column of a typical temperate lake. Major phases are shown as separate blocks, temperature gradients as broken arrows and turbulence as curved arrows. E – epilimnion, M – metalimnion, H – hypolimnion. The lake shown in this diagram is holomictic (mixes throughout the water column) and dimictic (two periods of complete mixing, the spring and autumn overturns)

The four seasonal phases of dimictic temperate lakes can be summarized (Figure 2.10) as follows.

Winter stratification

Ice cover in winter prevents wind-generated turbulence and complete mixing of the water column. The anomalous density – freezing curve of water leads to inverse stratification, with the least dense body of water (0–4°C) at the top of the water column, and temperature increasing with depth to a maximum of 4°C. As spring approaches, solar heating of water below the ice results in a limited circulation, which may be important in maintaining diatoms in suspension and supporting a sub-ice algal bloom (see Figure 2.3).

Spring overturn

Melting of the ice exposes the lake water to wind action, creating turbulence and mixing within the water column.

Summer stratification

As spring progresses, increases in solar energy cause localized heating at the top of the lake which results in a surface layer of warm water (the epilimnion). Mixing occurs within the surface layer due to moderate wind activity. The epilimnion remains at the top of the lake due to its low density, and there is minimal heat transfer to the lower part of the water column (hypolimnion) due to poor thermal conductivity. The volume of surface water that is heated, and the resulting depth of the epilimnion, is limited by the high thermal capacity of water. The resulting temperature profile through the water column (Figure 2.11) is relatively uniform within the epilimnion, decreases rapidly with depth in the transitional layer (metalimnion or thermocline), then remains low in the hypolimnion.

Autumn overturn

Decreases in light intensity and air temperature lead to a reduction in surface heating of the lake, which,

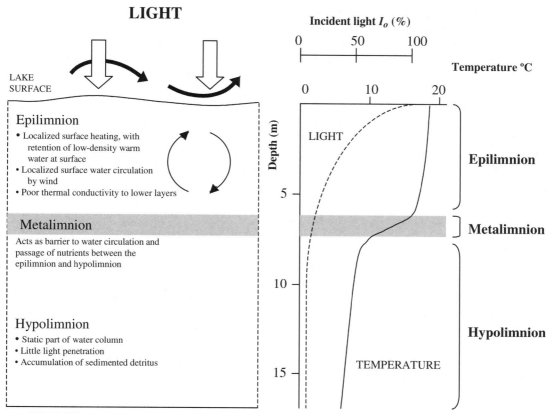

Figure 2.11 Development of stratification in the water column of a temperate lake. In late spring, increased solar energy (⇩) and mild wind activity (➡) creates a surface layer of warm water, the epilimnion. This becomes separated from a cooler lower layer (hypolimnion) by a transition zone – the metalimnion or thermocline. The stratified water column shows vertical differentiation in relation to light intensity, temperature and water circulation. The effects of stratification on oxygenation and the distribution nitrates, nitrites and ammonia are shown in Figure 5.9

coupled with increased wind activity, causes a breakdown in stratification and a mixing of lake contents. Wind activity also accelerates this process by contributing to the reduction in surface temperatures as a result of increased evaporation (heat loss as latent heat of vaporization). Freezing of the lake surface subsequently returns the lake to the period of winter stratification.

2.5.2 Biological significance of stratification

Although stratification has been described in relation to temperate lakes, it is not restricted to this climatic category and occurs wherever surface heating, wind action, and lake depth are appropriate. Stratification has a major impact on the biology of lake organisms for the following reasons.

- Primary production of biomass and uptake of soluble inorganic nutrients mainly occur in the epilimnion. In many lakes, phytoplankton growth in this part of the water column leads to severe nutrient depletion, with very limited replenishment from the hypolimnion. Lack of water mixing and nutrient exchange across the metalimnion promotes and emphasizes the separation between epilimnion (high biomass, low inorganic nutrients) and hypolimnion (low biomass, high nutrients). The effect of stratification on the distribution of inorganic nutrients within the water column is discussed in Case Study 5.1 and Section 5.3, and the penetration of light in Section 4.1.3.

- Seasonal changes in surface nutrients are an important aspect of the temporal sequence in phytoplankton populations (see Section 3.7.2) and related populations of other lake biota. Towards the end of the stratified period, algal migration within the water column becomes increasingly important as a means to access the high nutrient levels occurring in the lower part of the lake.

- Death of biota in the epilimnion leads to sedimentation of organic debris, which passes into the hypolimnion. In this part of the water column, where there is no mass removal of nutrients by phytoplankton, the continuous influx of debris further increases nutrient levels and supports large populations of heterotrophic bacteria. These in turn create a high oxygen demand, further influencing the physical and chemical conditions with the lake. Nutrient levels in the hypolimnion are also enhanced by the release of nutrients, particularly phosphates, from sediments. All of these changes depend on the overall nutrient status of the lake, as summarized (for nitrogen compounds) in Figure 5.9.

Summer stratification of the lake thus results in a marked compartmentalization of the pelagic environment, with clear separation of physical (light, temperature), chemical (nutrient levels, oxygen concentrations), and biological (autotrophs in epilimnion, heterotrophs in hypolimnion) characteristics.

2.5.3 Tropical lakes

Lakes outside the temperate climatic zone show differences from the above sequence as the winter period is reduced (decreasing latitude, tropical lakes) or increased (increasing latitude, polar and sub-polar lakes)

Tropical lakes (Ryding and Rast, 1989) lack the freeze–thaw cycle of temperate water bodies, but do show seasonality in terms of rainfall. Some of the major features of tropical lakes are summarized in Table 2.7. Because of the more favourable

Table 2.7 Comparison of tropical and temperate lakes

	Temperate zone	Tropical lakes and reservoirs
Rainfall	Throughout year	Very seasonal, mainly in summer
Temperature	Well-defined annual cycle	Limited annual cycle - range 10°C - mean 25°C
Growing season	Main growth restricted to summer period	Throughout whole year
Lake stratification	Typically well defined, marked physico-chemical variation within water column	- Less pronounced than temperate lakes - Oxygen depletion occurs irrespective of trophic status
Nutrient cycling	Determined largely by physical mixing	Determined largely by biological re-cycling
Productivity	Moderate	High
For mesotrophic lake:		
Mean primary productivity ($gC\ m^{-2}d^{-1}$)	1.0	2–3
Algal blooms	Typically spring and autumn	Any time of year
Dominant algal types	Diatoms	Blue-green algae

(information from Ryding and Rast, 1989, and Gardner *et al.*, 1998).

year-round growing season, tropical lakes usually have a higher productivity compared to those of temperate climates. Phytoplankton blooms can occur at any time of year, irrespective of the annual cycle. Tropical lakes often develop very low N/P ratios, favouring dominance of blue-green algae.

Some features of tropical lakes resemble those of temperate lakes in summer, but contrasts are greater during other parts of the year (Gardner *et al.*, 1998). Fundamental differences occur between the two systems in terms of physical and biological dynamics, particularly in reference to nutrient cycling – which may be dominated either by physical (temperate lakes) or biological (tropical lakes) factors. Nutrient cycling in temperate lakes is determined largely by physical mixing or circulation of the water column at the beginning and end of the growth season, while in tropical lakes the biological process of recycling is more important.

Circulation of nutrients in large temperate lakes occurs due to temperature-driven spring and autumn mixing of the water column, and results in the transfer of oxygen to hypolimnetic regions and nutrients to the euphotic zone. In contrast, tropical lakes typically do not show strong seasonal water column mixing patterns, and tend to develop more permanent high nutrient anoxic hypolimnia. If nutrient substrates are available, water column nutrients tend to recycle more rapidly in tropical lakes due to higher temperatures enhancing biological activity and enzyme reactions.

2.5.4 Polar and sub-polar lakes

With increasing latitude, the duration of ice cover during the annual cycle increases and the summer period of ice-free water decreases.

The presence of ice tends to limit algal growth, particularly when covered by snow, due to decreased light penetration (Figure 2.12). If the ice cover is thin, however, and snow is not present, light penetration may be sufficient to support algal growth and may even promote bloom formation (Kelley, 1997). This is the case with Lake Baikal (Russia), which is frozen from January to May, and where annual blooms of the diatom *Aulacoseira*

Figure 2.12 Frozen mountain lake, Rogaland, Norway: the thick layer of ice and snow severely limits light penetration and growth of phytoplankton for much of the year. Photograph by Richard Sigee

baicalensis are regularly observed below the ice – accounting for about 40 per cent of the annual primary productivity. Under-ice populations of this diatom have been monitored by remote sensing (Semovski *et al.*, 2000) and show an initial increase in February–March, when water temperatures are 0.1–0.2°C. Populations reach a maximum in April–May when water temperature reaches 1–3°C and ice cover is beginning to break up (Figure 2.9). The occurrence of diatom blooms under ice shows strong interannual variability, and is remarkable both in terms of the limited light conditions and the ability of the algae to stay in suspension. As with other diatoms, *Aulacoseira baicalensis* is substantially heavier than water and would sink out of the water column (sedimentation rate 2.8 m d^{-1}) unless maintained in suspension by water movement. Wind-generated water turbulence is not possible below the ice, but convection currents caused by solar heating may occur and appear to be the major factor in maintaining these diatom populations in suspension. The mechanism of these convection currents is unique to freshwater systems, and depends on the fact that pure water has a maximum density at 4°C. The water column under ice shows an inverse temperature gradient from 0°C (immediately below the ice) to a maximum of 4°C lower down (Figure 2.3). Solar heating of the water below the ice causes an increase in density, resulting in gravitational instability and the generation of convection currents which carry the diatoms into the surface water.

With further increase in latitude, the period of ice cover becomes even more extensive. Polar lakes, such as Lake Fryxell in southern Victoria Land (Antarctic), are frozen all year and are permanently stratified – without any mixing (amictic). These

Antarctic lakes are among the most hostile freshwater environments in the world. Light penetration is non-existent in winter (permanent darkness) and is very limited in summer (1–4 m of ice and snow cover), yet there is still a substantial population of phototrophic planktonic micro-algae (up to 10^7 cells l^{-1}) and associated biota (McNight *et al.*, 2000).

C. WETLANDS

Wetlands can be defined as shallow aquatic systems in which the water column is typically no greater than 1–2 m in depth. The microbiology of wetlands shows major similarities to that of shallow lakes (with which they form an environmental continuum) and includes substantial and competing communities of planktonic, epiphytic and benthic microorganisms (Table 2.3). Many wetlands differ from shallow lakes in showing greater extremes of seasonal change (with periodic desiccation and flooding) and are particularly liable to more permanent alterations and loss of habitat due to human interference. Problems of maintenance and sustainability are key aspects of wetland management (Kvet *et al.*, 2002).

2.6 General characteristics

Although solar radiation typically penetrates to the sediment, supporting the growth of epipelic algae, light levels may be less than expected due to the following.

- Resuspension of sediments – numerous studies of lakeshore wetlands and shallow lakes have demonstrated sediment resuspension by wind and increased turbidity of the water column.

- Presence of macrophytes – growth of higher plants (macrophytes) leads to interception of light by leaves and stems and may seriously limit light availability to the different groups of algae.

Productivity of different algal groups in wetlands is considered in Section 4.4, and competition

between macrophytes and algae in Section 4.8. Wetlands are amongst the most endangered environments on the planet, liable to loss by drainage or flooding (dam creation), and also frequently exposed to eutrophication. The effects of eutrophication (and remedial action) on the Broads Wetland area (UK) is discussed in Case Study 10.4.

2.6.1 Wetland diversity and global scale

Wetlands occupy a continuum between terrestrial and fully aquatic environmental systems. They are typically present as regions of waterlogged landscape, composed of shallow lakes, water meadows, marshes, and shallow meandering rivers. Many wetlands, such as the seasonal wetlands of North West Australia (Figure 4.20), show considerable fluctuation in the height of the water table, with alternation between wet and dry periods. Such variations may arise due to period flooding, seasonal alternations in rainfall, and activities of man, such as regular drainage activities.

Most wetlands have a water depth >2 m, and occupy as much as six per cent of the global land area. Regional amounts of wetland may be much higher than this. In Canada, for example, wetlands occupy about 14 per cent of the total landmass – due to the extensive area of peatlands in the northern part of the country. Wetlands occur in every continent except Antarctica.

2.6.2 Unifying features of wetlands

Although wetlands may be defined in various ways, they all have a number of features in common

(Pokorny *et al.*, 2002a):

- presence of standing water over the soil surface for at least part of the annual cycle, causing waterlogging;

- characteristic features of wetland soil and surface water – including lack of oxygen and the presence of both organic and inorganic reducing substances;

- presence of organisms which are tolerant of, or are unable to survive without, periodic flooding;

- substantial development of higher plants (macrophytes), which are frequently dominant.

Wetlands typically have a complex spatial structure. This applies both at the geographic scale (Figure 2.13) and at field level. Many wetland habitats appear as a mosaic of pools and macrophyte vegetation, with considerable heterogeneity in both biological and physical composition. The importance of macrophyte vegetation in this habitat is emphasized by the characterization of wetlands according to the different plant types that predominate – including marshes (emergent macrophytes), swamps (trees), and bogs (*Sphagnum* moss). Although wetlands are typically seen as entirely natural habitats, they often incorporate man-made features such as fishponds, shallow water bodies derived from peat excavations (e.g., Broads Wetland Area, UK), and constructed reed-beds for wastewater treatment.

2.6.3 The role of wetlands in energy and material flow

Wetlands exert a strong regional importance in terms of energy and material flow and are ecologically important in relation to stability and biodiversity (Pokorny *et al.*, 2002a). The stabilizing influence of wetlands is mediated in several ways as follows.

- Wetlands act as an energy buffer, with most of the energy of incoming solar radiation being dissipated in the form of latent heat of vaporization.

- They act as sinks for atmospheric CO_2 and retard its atmospheric increase. This arises because organic carbon accumulation in such systems prevails over carbon mineralization, with the low rates of decomposition leading to the formation of peat or the accumulation of organic detritus.

- As zones of nutrient accumulation, wetlands act as a buffer between land and water – protecting surface waters from increases in nutrient input.

- In areas of heavy metal discharge, the accumulation of metals in wetlands buffers any major increase in adjacent open waters.

- As regions of high water-retaining capacity, wetlands buffer water discharge from catchment areas and make an important contribution to flood control.

2.7 Wetland habitats and communities

Studies on the microbiology of wetlands have concentrated largely on algae, which form the major microbial biomass at these sites. The algae may be grouped into four main assemblages, some of which are illustrated in Figure 4.21:

- epipelic algae – living on and within soft sediments;

- epiphytic algae – composed of prostrate, erect, and heterotrichous algae growing on the external surfaces of submerged and emergent vascular/non-vascular plants;

- metaphyton – consisting of cohesive floating and sub-surface mats, usually composed of filamentous green algae – the mats originate as epiphyton, which detach due to turbulence and may float due to the presence of trapped gases;

- phytoplankton – these are free-floating algae present in the water column, and occur either as permanent residents in this part of the water body or are derived secondarily from epipelic and epiphytic algae.

Table 2.8 General characteristics of four major states of wetland environments (Adapted from Goldsborough and Robinson, 1996)

Property	Dry-state	Open-state	Sheltered-state	Lake-state
Dominant algae	**Epipelon**	**Epiphyton**	**Metaphyton**	**Phytoplankton**
Algal primary production	**Low**	**Medium**	**High**	**Variable**
Water level	Low	Medium	Medium	High
Water column disturbance	Rare	Frequent	Rare	Frequent
Light penetration through water column	**High**	**Variable**	**High**	**Low**
Nutrients	–	Medium	High	High
Aquatic macrophytes	Few	Abundant	Medium	Few
Herbivory	Low	High	Low	Low
Secondary production	Low	High	Low	Variable

Shaded boxes show microbial (algal) properties of the different wetland states.

Epipelic and epiphytic algae form part of the general benthic algal (periphyton) assemblage (Section 1.2.4). Goldsborough and Robinson (1996) have suggested that wetlands can be divided into four quasi-stable states, which are separately dominated by these four algal assemblages. These different wetland states are also characterized by differences in algal primary production, presence of aquatic macrophytes, degree of herbivory and level of secondary production (Table 2.8).

1. *Dry-state wetlands.* Characterized by exposed moist sediments, and dominated by epipelic algae. This occurs where very low water levels occur as part of a seasonal cycle (e.g., drought period), drawdown of an existing wetland, or early flooding of a new wetland. Biomass of true aquatic macrophytes is low, with replacement by 'wet meadows' of grasses and sedges.

2. *Open-state wetlands.* These are characterized by a turbulent water column containing abundant macrophytes. Open-state wetlands may originate as dry wetlands fill with water, or because of biomanipulation of a phytoplankton-dominated wetland. Epiphyton biomass is high due to the presence of a large macrophyte surface area and high nutrient levels resulting from the water inflow. Maintenance of epiphyton dominance paradoxically depends on a high grazing pressure by benthivorous fish and other fauna. This prevents over-accumulation of epiphyton biomass to the point that it causes macrophyte decline. If epiphyton growth occurs unchecked, progressive shading of macrophytes leads to their replacement by phytoplankton and the development of a lake-state wetland.

3. *Sheltered-state wetland.* Under conditions of high nutrients, stable water column, and shelter by macrophytes or bordering vegetation, epiphyton may develop and detach to form dense mats of metaphyton. These carpet the sediment, and may also form a surface biomass. The further development of metaphyton mats, which are composed of green algae such as *Cladophora*, depends on high nutrient status, high irradiance, alkaline pH, high calcium, and low N/P ratios.

4. *Lake-state wetland.* This is characterized by high water and nutrient levels, with a turbid water column containing abundant phytoplankton. Epiphyton, metaphyton, and epipelon are not able to develop due to low light penetration and limited macrophyte substratum. Lake-state wetland may develop if rapid water input occurs, leading to loss of macrophytes and allowing the rapid development of phytoplankton.

2.8   Case studies on wetland areas

Two case studies are included on particular wetland sites. One of these relates to eutrophication and the development of blue-green algae in the Broads Wetland Area, UK (Case Study 10.4). The other relates to recent studies on the Třeboň Biosphere Reserve. Both of these examples emphasize the importance of human activity on both the origin and development of wetlands.

Case study 2.1   Třeboň basin biosphere reserve

The Třeboň Biosphere Reserve, in South Czechoslavakia, is one of about 60 biosphere reserves which are recognized by the Ramsar Convention as being a wetland area of major international importance. Recent studies (Kvet *et al.*, 2002) have emphasized the biological diversity and ecological importance of this site. The reserve (*Třeboň basin*) is a shallow and large depression which has been modified by human activities for more than eight centuries. The result is a diverse semi-natural countryside which includes three main types of wetland – fishponds (over 500), water meadows, and peat bogs, interspersed with forest, arable land, and habitations (Figure 2.13).

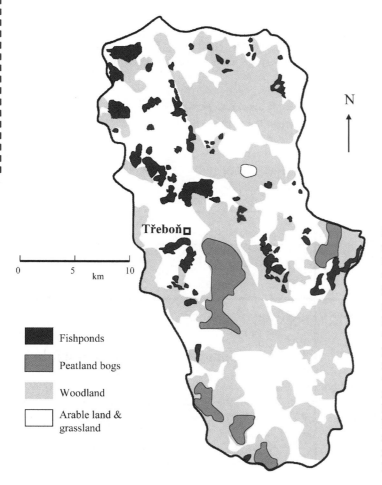

Figure 2.13 Key role of aquatic habitats in the Třeboň Biosphere Reserve (Czech Republic). As with many other wetland areas, the reserve is characterized by a mosaic of shallow open water (fishponds), water meadows, peatland bogs, and drier regions of grassland and woodland. In this wetland area the fishponds are largely man-made, and are managed as high-nutrient (eutrophic) environments. The microbiology of the fishponds is dominated by intense populations of microalgae, and competition between algae and macrophytes (figure redrawn from Kvet *et al.*, 2002 and reproduced with permission of Parthenon Publishing)

Studies on biological diversity within the wetland areas have concentrated mainly on macrophytes, fish, waterfowl, and insects – with relatively little information on microorganisms. The main exception to this is the micro-algal flora, which is diverse, and an important component of the various ecosystems.

Fishponds and peat bogs have their own distinctive algal flora (Table 2.9), with further subdivisions in the case of the peat bogs in relation to the microhabitats (drainage channels, acid pools, and submerged sphagnum moss communities) of this complex wetland environment. Although algae are an important component of the flora in both of these habitats, they contribute only a fraction of the primary production – with higher plants (macrophytes) and sphagnum mosses respectively taking a dominant role in the two situations. Of all aquatic habitats, wetlands exhibit the greatest level of competition between algae and higher macrophytes (Section 4.8).

Table 2.9   Microbial diversity in a wetland area; the table shows variation in algal communities (dominated by microalgae) in fishpond and peat bog wetlands of the Třeboň Biosphere Reserve

Wetland	Algal community
Fishponds[*]	
(a) Oligotrophic and mesotrophic ponds	Dominated by *Chrysophyceae* & *Dinophyceae*
(b) Eutrophic ponds	Dominated by green algae (*Chlorococcales* and *Volvocales*) diatoms and blue-green algae
Peat Bog[†]	
(a) Drainage channels	Green algae (*Microspora*)
(b) Acid pools	Green flagellates
	Green filamentous algae
(c) Submerged Sphagnum moss	*Cylindrocystis*
	Klebsormidium
	Microthamnion
	Euglena

Information taken from: [*]Pechar *et al.* (2002), [†]Jenik *et al.* (2002).

D.   STREAMS AND RIVERS

The lotic or running water environment comprises mainly streams and rivers, though situations of high water flow in estuaries and flooding of marshlands also come within this category. The aquatic environment of both streams and rivers is dominated by continuous unidirectional flow. Rivers differ from streams in being considered as larger, faster moving, and often warmer water bodies.

2.9   Comparison of lotic and lentic systems

As with the lake environment, the lotic environment consists of a water column lying over a bed or substratum. Differences between lentic and lotic systems in relation to food webs and derivation of assimilable carbon have been discussed in

Section 1.8. In terms of physical characteristics, fast-flowing streams differ from lakes in various ways as follows.

- The water body is transient (low retention time) at any particular site. It is also typically highly turbulent, with no thermal or chemical stratification. Both of these factors combine to limit the development of a planktonic community (phytoplankton and zooplankton), which may be significant in larger rivers but does not normally make a major contribution to the stream community.

- The substratum (bed) of streams is typically well aerated and exposed to light, and is the major site of algae and associated biota.

- Inflow from the catchment area is typically periodic, giving rise to marked fluctuations in water level and flow velocity (high-flow and low-flow periods). Concentrations of dissolved and particulate matter, which are mainly derived from the catchment area, typically follow this periodicity in inflow.

Because of these differences, the ecosystem of streams is dominated by the benthic organisms (Figure 1.19), with algal and bacterial biofilms (Figure 1.5) being particularly important at the microbial level. Benthic algae, including single-celled (e.g., diatoms) and colonial (e.g., filamentous green and blue-green algae) are the most successful photosynthetic organisms to exploit streams as habitat, and are ecologically important in primary production, nutrient transformation, sediment stabilization, and habitat provision for other benthic organisms (Section 3.6). Benthic algae can also proliferate in nutrient-enriched, stable-flowing streams, causing water management problems. Because of their hydrology, and the importance of terrestrial input, streams are particularly liable to variations in water quality. Benthic algae show a rapid response to such changes, and are potentially useful in monitoring pollution effects in water lotic situations (Biggs, 1996).

Some of the larger rivers occupy an intermediate position in the distinction between lotic and lentic environments, with limited movement of the water body, restriction of turbulence to the upper layers and sediments that are exposed to minimal water flow and are in permanent darkness. In this situation, the river may resemble lakes in showing thermal and chemical stratification and develop substantial populations of phytoplankton and zooplankton. Recent studies have shown that phytoplankton abundance in some rivers is comparable to moderately productive lentic systems with well-established pelagic food webs (Wehr and Thorp, 1997), and modelling studies have also simulated high lotic phytoplankton productivity under appropriate hydrological and irradiation conditions (Sellers and Bukaveckas, 2003).

2.10 River flow and the benthic community

In flowing water systems, the physical world is governed by water in motion (Paterson and Black, 1999). In models of river ecology, water flow is typically the dominant forcing function (or 'master variable') to which all other river processes and patterns can be traced – including changes in river morphology, distribution of organisms in time and space, and rates of energy transfer and material cycling. The primary importance of water flow in limiting the development of plankton (by displacement) and promoting dominance of benthic communities has already been mentioned. Microorganisms are present in the water column, however, and the flow characteristics of the main water body and the benthic region present distinct environments in the lotic system.

2.10.1 Flow characteristics of lotic systems

Main water flow

Flow of water within the water column has two main parameters of biological significance – velocity and type of motion (laminar or turbulent). The most useful quantitative index of water motion is the Reynolds number (Re) which relates water velocity to viscosity (Equation (2.1)), and can be used to indicate whether water flow will be laminar ($Re < 500$), transitional (500–10^3), or turbulent ($> 10^3$). The relatively high viscosity of water tends to reduce the Reynolds number and promote laminar flow (Section 2.1.2).

In most rivers and streams, water flow in the main channel is turbulent (Allan, 1995). Laminar flow requires shallow, slow-moving water – with conditions becoming transitional (leading to turbulence) when velocities exceed 10 cm s^{-1} and depth exceeds 0.1 m. This turbulence is biologically significant in maintaining non-motile particulate organisms (such as diatoms) in suspension, and also in carrying fine particulate allochthonous carbon matter along the course of the system (Figure 2.17).

Benthic flow: the boundary layer

Friction between moving fluid and stationary surfaces (or vice versa) results in a region of reduced flow, or boundary layer. This is important as a general environmental zone at the base of the water column (Figure 2.14) and is also relevant at the micro-level in relation to individual algal cells and nutrient availability (Section 5.2.3, Figure 5.7).

The vertical thickness of the benthic boundary layer is considered to extend from the sediment surface (where velocity is zero) to a point where velocity is 90 per cent of main water flow. The dynamics of this layer have been most easily studied in the laboratory, where factors such as water flow and roughness of the solid surface can be controlled (Allan, 1995). These studies show the zone of reduced water flow to be quite thin. Under laboratory conditions, main velocities of 20 cm s^{-1} or less result in boundary layers of about 5 mm. Under these conditions, algae and bacteria on exposed surfaces would experience water velocities less than 1 per cent of the main current. At flow speeds greater than 20 cm s^{-1}, turbulence increases dramatically and boundary layers markedly reduce in thickness – with estimated values of 0.5–1 mm when the main current is 20–50 cm s^{-1}. Whatever the exact value for water flow at benthic surfaces, it is clear that the boundary layer forms an important

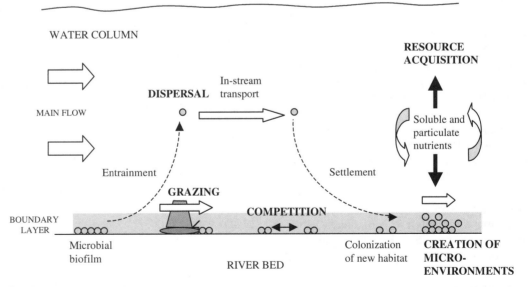

Figure 2.14 Effects of river flow on benthic microorganisms. The majority of benthic microorganisms occur in a zone of reduced flow, the boundary layer (shaded), where most of the microbial interactions occur. This layer extends to a thickness of about 5 mm, depending on main flow rate and surface roughness. Dispersal of microorganisms takes place within the main current ($\Rightarrow$), to which large grazing invertebrates are also exposed

micro-environment – with major significance for the biology of lotic microorganisms.

2.10.2 Influence of water flow on benthic microorganisms

Water flow can affect bottom-living organisms in a number of ways (Figure 2.14) – including dispersal, nutrient acquisition, competition, grazing (predator–prey interactions), and the creation of microenvironments. These factors affect the whole variety of benthic organisms, ranging from fish and large invertebrates to smaller invertebrates (meiofauna) and microorganisms present in bacterial and algal biofilms.

Dispersal, emigration, and immigration processes

Dispersal of benthic microorganisms by water flow involves three main phases – entry into the water column (entrainment), in-stream transport and delivery to the river bed (settlement). Local populations of many benthic (e.g., biofilm) organisms are affected by loss (emigration) and entry (immigration) processes (Figure 2.15), often involving an intervening dispersal over large distances. These organisms are detached and carried downstream via water currents, providing a significant supply of immigrants to newly-available substrates. Such drifting organisms are rarely adapted to planktonic life, but are simply in transit from one site to another. Some benthic organisms appear to exploit flow as an active opportunity for dispersal – including algae, which use flow-mediated dispersal to vacate unfavourable habitats. Bothwell *et al.*, (1989) demonstrated that benthic diatoms were able to selectively emigrate from experimentally-darkened sites by altering their buoyancy or form resistance and becoming entrained in the river current.

The immigration of organisms into benthic communities such as biofilms occurs by selective uptake from the planktonic phase. Biofilm formation (Figure 2.16) involves selective adsorption of macromolecules, followed by bacteria, a mixture of

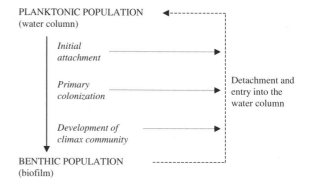

Figure 2.15 Role of emigration and immigration in the development of a biofilm community. River flow carries planktonic microorganisms such as bacteria and diatoms to exposed surfaces, where they enter the biofilm (immigration) and are involved in the colonization sequence – from initial attachment to climax community. Organisms detach from the biofilm (emigration) at all points along the colonization sequence, becoming entrained into the water column and available for further biofilm colonization downstream

microorganisms then invertebrates. Studies on the colonization of new substrates by benthic diatoms (McCormick and Stevenson, 1991; Stevenson, 1996) have demonstrated an inverse relationship between velocity of river flow and immigration rates. This is true for both early and late colonizers of algal biofilms (see Figure 3.15, Table 3.7), suggesting that flow rates affect the establishment of new organisms throughout the whole succession. The rate of immigration can be determined by dividing the 24 h substrate population (cells mm^{-2}) by the water column population (cells ml^{-1}). Reduction in the rate of immigration by high water flow may arise either due to a reduction in the rate of cell delivery, or to higher rates of post-contact removal. Over the longer term, cell growth and division in the biofilm are also important, and are also influenced by current.

Nutrient acquisition

The growth and reproduction of benthic microorganisms requires a constant supply of nutrients.

Figure 2.16 Selective uptake of macromolecules and microorganisms from the water column during biofilm development. Initial adsorption of macromolecules onto freshly-exposed surfaces (physicochemical phase) leads to the formation of a surface-conditioning film. This is then followed by selective adhesion and immigration (into the biofilm) of bacteria, other microorganisms and invertebrates (biological phase)

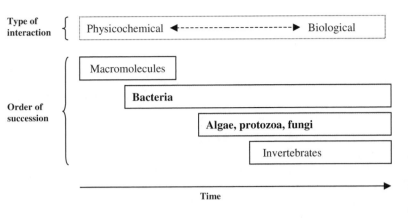

These are mainly derived from the water column and include dissolved inorganic material and particulate organic food. River flow can affect the acquisition of such resources by influencing their delivery to benthic communities from the water column and by affecting the conditions of uptake within the benthic microenvironment.

• *Nutrient delivery from the water column.* The concentration of dissolved nutrients in the water column is affected by the magnitude of turbulent mixing, while the distribution of suspended particles is determined by both turbulence and particle settling velocity. Conditions of high turbulence will ensure delivery of soluble nutrients to the benthic communities, but reduce the availability of particulate material which will remain in suspension and not settle out.

• *Nutrient absorption within the microenvironment.* The absorption of soluble nutrients by microbial biofilms is affected at the microlevel in two main ways – circulation within the community and restrictions in flow at the water/community interface.

Circulation of water within biofilms is important for nutrient access and also for the exchange of gases such as O_2 and CO_2. Both bacterial (Figures 1.5 and 6.17) and algal (Figure 3.15) biofilms have complex patterns of water circulation which determine concentration gradients and availability within the community.

Restrictions in water flow occur at all surfaces within the aquatic environment. Studies on nutrient uptake by benthic algae and other aquatic plants have demonstrated the importance of a surface water shell (Figure 5.7), referred to as the laminar sub-layer (Stevenson, 1990). This surrounds all organisms in the aquatic environment and is a thin layer of static water across which all soluble materials have to diffuse to enter or exit the organism. The thickness of the laminar sub-layer is inversely related to current velocity, leading to higher rates of nutrient absorption and growth by attached algae in fast-flowing conditions.

The relationship between algal biomass and current velocity follows a unimodal relationship (Biggs *et al.*, 1998). At the lowest velocities, uptake of nutrients by algae is limited by the thickness of the static sub-layer and by low levels of turbulent flux. Increased velocity leads to greater nutrient uptake (reduced sub-layer thickness), but loss of biomass occurs beyond a critical velocity due to increased losses by sloughing and entrainment. Maximum levels of algal biomass in benthic mats thus occur at intermediate current velocities as a result of this subsidy–stress relationship.

• *Uptake of particulate material.* Water flow has important effects on the ingestion of particulate material by filter feeders such as benthic protozoa. Although a low flow rate is beneficial in allowing sedimentation from the water column, it has a reverse effect at the microlevel since collection rates depend on a high flux of particulate material in contact with the feeding structures. An increase in flow rate increases particulate contact, but if the level becomes excessive the performance of the feeding structures is impaired by the high drag forces that occur under such fast-flowing conditions. As with soluble nutrient uptake, the increase in water flow has initial benefits (nutrient subsidy) followed by adverse effects (environmental stress). This subsidy-stress situation can be represented graphically as a unimodal relationship between water velocity and feeding activity.

Competition

Flow rate can affect both short-distance and long-distance competition between microorganisms. Short-distance effects operate within communities such as diatom biofilms, where variations in flow rate can affect the establishment and growth of particular species in different ways, thus having a major influence on competition. Competitive interactions between algae normally continue into the mature periphyton community. In conditions of very high water flow and turbulence, however, algal succession may not proceed beyond the diatom biofilm stage, leading to a limitation in the competitive interactions that develop.

The role of water flow in controlling the supply rate of limiting resources gives it the potential to generate long-distance competitive interactions. Upstream organisms can reduce resource availability to those located downstream (but not vice versa) in one of two ways – exploitation (where resources are directly consumed by upstream animals) and interference (where upstream individuals alter flow characteristics, reducing velocity and downstream nutrient uptake).

Predator–prey interactions

Water flow can affect predator–prey interactions by influencing encounter rates or by altering the ability of the prey to escape following an encounter. Various studies on invertebrates (reported in Hart and Finelli, 1999) have demonstrated reduced predator impact with increasing current velocities, and the same principles may apply to microbial predators such as protozoa.

Interactions between grazers and benthic algae may also be influenced by water flow (Hart and Finelli, 1999). Laboratory and field studies have shown that grazing of benthic algae by snails, caddis fly larvae, and crayfish was high at low water velocities, but decreased with increased flow rates. Reductions in grazing rate with velocity probably arose due to the high drag forces operating on the large predator organisms, and the resulting tendency for them to be dislodged. The growth of many microbial benthic communities is thus enhanced in regions of high water flow (flow-mediated refuges), where invertebrate grazing activity is permanently limited.

Creation of microenvironments

Flowing water interacts with benthic communities to create microhabitats which are important for the development of the lotic community (Mulholland, 1996). The extent of these microenvironments depends on the spatial development of the benthic biomass and the conditions of water flow in that part of the river.

The accumulation of biomass on the bottom of the stream can lead to major changes in the hydraulic characteristics of the benthic environment. Zones of low-velocity or stationary water develop within and around bacterial biofilms, algal biofilms, and periphyton communities, forming a variety of supplementary boundary zones in addition to the main benthic boundary layer noted previously. These boundary zones range from the thin laminar sublayer (or viscous sublayer) that occurs on surfaces directly exposed to the current, to large regions of quiescent water contained within the internal spaces of benthic communities.

The zones of quiescent water form important regions within which unattached organisms such as algae, protozoa and small invertebrates can exist (see, for example, Sebater and Munoz, 2000). Quiescent boundary zones prevent downstream displacement of biota by the strong advectic forces of flowing water, thus counteracting one of the fundamental problems of life in lotic ecosystems. Such microenvironments are a general feature of attached benthic communities, and are important in maintaining biodiversity and nutrient cycling. The local regeneration of inorganic nutrients by this process may be a major factor in maintaining nutrient supply to primary producers, and sustaining high rates of primary productivity in benthic systems (Mulholland, 1996).

2.11   River hydrology

The overall characteristics of particular streams and rivers are determined by water input from the catchment area and by the underlying geology. More detailed characteristics of the river environment vary along the course of the water body – with a succession of meanders or an alternation of gravel beds and deep pools leading to a variety of different microenvironments and associated communities.

Although a river may be described as a succession of distinct habitats and communities, it can also be considered as a single continuum – in which changes in biological and physical characteristics occur along a linear gradient. The river continuum concept (RCC) was developed by Vannote *et al.* (1980) to relate the activities of primary producers, heterotrophs and carbon availability along the entire course of the river. The RCC:

- characterizes lotic systems as an interconnected system of streams with a continuum of longitudinally linked physical, chemical, and biological parameters;

- describes biological changes in the invertebrate, microbial, and macroalgal biota in relation to external carbon availability and the opportunity for primary productivity;

- links the activity of organisms upstream to physico-chemical conditions and communities downstream.

In the upper reaches of the river (Figure 2.17), there is a high influx of external (allochthonous) carbon, and invertebrate and microbial activity is mainly concerned with physical breakdown and biochemical degradation. Primary production by benthic algae does not contribute much carbon to the system due to heavy shading by surrounding vegetation. System level respiration (R) exceeds primary production (P), where $P/R < 1$. Further down the stream, degradation of external organic matter becomes reduced and primary production by benthic algae increases due to higher light levels ($P/R > 1$). Many lotic systems do not progress beyond this point, but in large rivers (greater than sixth order) there is development of phytoplankton and zooplankton populations. Even though the river basin is open to full sunlight in this situation, high levels of fine particulate matter from upstream, combined with greater depth and resuspended sediments limit phytoplankton productivity. The water column is often turbid, with limited light penetration, and primary production again falls below system level respiration ($P/R < 1$).

The river continuum concept has stimulated increased research in lotic systems and led to various concepts such as nutrient recycling. This regular and periodic activity along the length of the stream is referred to as nutrient spiralling (Elwood *et al.*, 1981), and is discussed in Case Study 5.1.

Variation in current along the course of the river leads to differences in sedimentation and accumulation of organic material, including 'drift' and 'detritus'. The term 'drift' is used to describe living benthic invertebrates and algae that have released or lost their substrate attachment. These are swept downstream, where they may find another favourable site for attachment or food availability. 'Detritus' consists of dead organic particles with associated epiphytic bacteria, fungi, protozoa, and rotifers. Fish and invertebrate feeders on drift and detritus are distributed to make optimum use of this food supply.

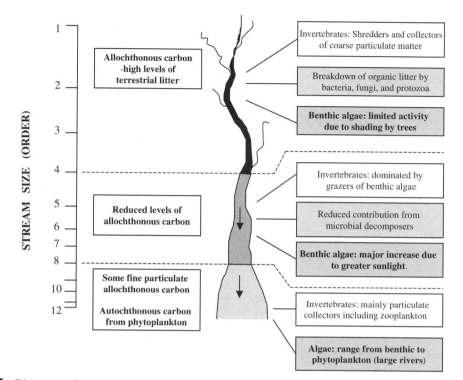

STREAM SIZE (ORDER)

1
2
3
4
5
6
7
8
10
12

Allochthonous carbon
-high levels of
terrestrial litter

Reduced levels of
allochthonous carbon

Some fine particulate
allochthonous carbon

Autochthonous carbon
from phytoplankton

Invertebrates: Shredders and collectors
of coarse particulate matter

Breakdown of organic litter by
bacteria, fungi, and protozoa

**Benthic algae: limited activity
due to shading by trees**

Invertebrates: dominated by
grazers of benthic algae

Reduced contribution from
microbial decomposers

**Benthic algae: major increase due
to greater sunlight.**

Invertebrates: mainly particulate
collectors including zooplankton

**Algae: range from benthic to
phytoplankton (large rivers)**

Figure 2.17 Diagrammatic representation of the river continuum concept. Different orders of stream size are designated in relation to size of channel and flow characteristics. The stream illustrated is 12th order, with a three phase transition (shaded regions). Continuous flow of water delivers high levels of external (allochthonous) carbon to the upper reaches (orders 1–4). This is gradually degraded, with increased proportion of carbon (autochthonous) from benthic algae (orders 4–8) then phytoplankton (orders 8–12). Microorganisms are initially dominated by heterotrophs, then benthic algae, with phytoplankton becoming an important component in the lower reaches of large rivers, river flow, (➡) (based on a figure from Vannote *et al.*, 1980)

E.   ESTUARIES

Estuaries and their associated wetlands differ from other aquatic environments considered in this book in being only partially freshwater. They are, by reason of their topographical location, semi-enclosed bodies of sea water which are diluted with freshwater from terrestrial watersheds. Inflow of fresh water from the river catchment area is periodic, varying mainly with rainfall, and results in estuarine communities which are freshwater or brackish for at least part of the year.

Water movements within the estuary depend on tidal currents, estuarine circulation, river discharge, and inflow from groundwater (Figure 2.18). The extent of these movements depends to a large extent on the morphometry of the estuary basin. Although daily tidal variation in the height of water is normally within 1–2 m, this may rise to over 10 m in funnel-shaped estuaries such as the Severn estuary in the UK and the Bay of Fundy in Newfoundland, Canada.

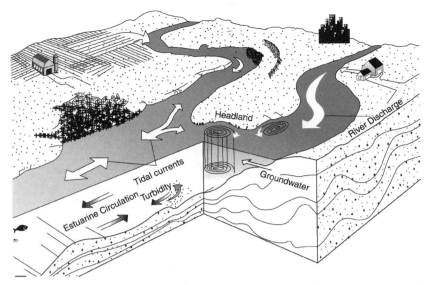

Figure 2.18 Schematic diagram of the upper reaches of an estuary, emphasizing freshwater input into the system. A number of physical processes contribute to the temporal and spatial variability of the estuarine system. In addition to tidal currents and river discharge, which interact to determine periodic bulk flow, water movement is also caused by vertical estuarine circulation (within the water column), tide-induced residual circulations at the headlands, groundwater inflow, and run-off from surrounding land at times of high rainfall. Anthropogenic input (agricultural, industrial, and domestic) contributes to the river discharge. Physical and climatic processes controlling water movement modulate the freshwater/ saltwater exchange with fringing marshes, mudflats and the adjoining coastal region. Biological transport within the pelagic zones can be passive (e.g., phytoplankton) or active (e.g., fish migration). Mudflats and marshes provide important sites for benthic microorganisms, including bacterial and diatom populations (based on a figure from Geyer *et al.*, 2000)

2.12 River inflow: water mixing, estuarine productivity, and eutrophication of coastal areas

River inflow is important for the mixing of fresh and saltwater systems, productivity within the estuarine system and eutrophication of surrounding coastal areas.

2.12.1 Mixing of fresh and saltwaters

Mixing of high-density salt water and freshwater results in a region of brackish water, the 'salt wedge', which tends to underlie the freshwater and to extend upstream due to its higher density. This salt wedge gives a layered structure to the estuarine water body, which shows some similarity to the lower thermal layer (hypolimnion) of freshwater lakes. The salt wedge is highly mobile, moving upstream with the tide, and varies in extent during the year in relation to the inflow of stream water.

The region of mixing of salt water and freshwater provides an important habitat for organisms that are specialized for living in brackish environments. Organisms more suited to the marine environment may be introduced into freshwater conditions. This includes, for example, the toxic dinoflagellate *Pfiesteria piscicida* which has recently been implicated in extensive fish kills in the Pocomoke and Chicamaconico Rivers of the Chesapeake Bay estuarine system, USA (Section 10.1.4, Silbergeld *et al.*, 2000). The region of saline and freshwater mixing is also a region of high biotic diversity, occurring as a result of the high productivity for both phytoplankton and benthos. This arises due to 'salting out' or flocculation of small particles of dissolved organic matter present in the freshwater. These organic particles tend to collect in the bottom layer of the saline wedge and may be returned upstream in the high tide.

2.12.2 High productivity of estuarine systems

Recent studies on the estuarine environment have given particular emphasis to the importance of river inflow to the nutrient status of estuaries, and the effect of the freshwater input on estuarine biota (Hobbie, 2000). This river inflow carries with it sediments, organic matter, and inorganic nutrients from terrestrial sources, some of which is retained in estuarine sediments (Burdige and Zheng, 1998). The high rates of autotrophic and heterotrophic growth that result from inflow of soluble and particulate material result in increased productivity in all of the communities which make up the estuary system (Simenstad *et al.*, 2000).

Within the main channel of the estuary, overall productivity is largely the result of microbial activities in the water column rather than the sediments, since the bulk of nutrient transfer occurs within the pelagic system. The periodic influx of organic compounds from the river tends to promote peaks of heterotrophic productivity, and seasonal oscillations between net autotrophic growth (phytoplankton activity) and heterotrophic growth (mainly bacterioplankton) are a feature of particular interest. Seasonal variations in such ecosystem function are particularly prominent in monsoon climates and have been studied, for example, in the Mandovi-Zuari system of southwestern India (Pradeep Ram *et al.*, 2003).

Within this estuarine system, the balance between autotrophic and heterotrophic activity was expressed as the net ecosystem production (NEP), which reflects the difference between primary production (P) and community respiration (R) (see Section 1.2.3). During the monsoon period, high inflow of river water into the estuary resulted in a negative value (-46 mmol C m^{-2} d^{-1}) for NEP, indicating a situation of net heterotrophy. This results from a combined decrease in phytoplankton productivity and an increase in bacterioplankton

productivity. Decreased productivity of the phytoplankton population can largely be attributed to removal (advection) by the fast-flowing current (water retention time reduced to 5–6 days) and to reduction in light levels caused by increased turbidity. Heterotrophic bacterial productivity shows an increase due to influx of allochthonous soluble and particulate carbon and an increase in the generation of soluble nutrients from the sediments. In non-monsoon periods (pre- and post-monsoon phases) the water column showed net autotrophy, with a positive NEP value (49 mmol C m^{-2} d^{-1}). The switch to phytoplankton dominance during these periods reflects a reverse in the parameters described previously for monsoon conditions.

Comparison of productivity in estuarine and lake ecosystems

The high productivity of estuarine systems can be considered in terms of energy subsidies (Odum, 1971), and differs from highly productive lakes in a number of key respects. In estuaries:

- nutrients and food are brought to many of the resident organisms by river and tidal flow, saving energy that would otherwise be required for searching or capture;

- mixing of seawater and freshwater causes flocculation of fine suspended matter into larger particles used by zooplankton and benthic filter feeders (see previously);

- heating of shallow water and exposed mud flats at low tide increases nutrient recycling via bacterial decay and accelerated growth of benthic organisms;

- particulate matter is resuspended by tidal action and has continued availability for uptake – this is in contrast to stratified lakes, where particulate material is lost from the region of major productivity (epilimnion) by sedimentation.

- the presence of adjacent sea tends to limit extremes in temperature.

2.12.3   Eutrophication of surrounding coastal areas

Freshwater inflow into major estuarine systems such as Chesapeake Bay, USA (Boynton and Kemp, 2000) and the Mississippi basin, USA (Rabalais *et al.*, 2000) results in eutrophication of the bay and surrounding coastal areas. In the case of Chesapeake Bay, on the east coast of the USA, long-term data sets and related modelling studies (Vorosmarty and Peterson, 2000) have demonstrated a clear linking between aquatic input of the major tributary rivers and biological parameters in the estuarine (bay) area. Linear relationships occurred between primary production, benthic–pelagic coupling and nutrient recycling in relation to freshwater inflow and nutrient loading rates.

The role of human activities in increasing the nutrient loading of estuaries and surrounding coastal regions has been the cause for much concern. Various studies have emphasized the related occurrence of declining fisheries, increased incidence of toxic algal blooms and loss of submerged aquatic vegetation communities in deeper coastal waters due to the development of hypoxia and (in some cases) anoxia.

Eutrophication of coastal regions around the Mandovi–Zuari estuary system of southwestern India (Pradeep Ram *et al.*, 2003) has resulted in negative NEP values, with excess organic matter from estuarine outflow promoting net heterotrophy throughout the year.

2.13   Habitats and communities

Estuaries have well-defined zonation, resembling lakes and rivers in possessing clear pelagic, benthic, and littoral zones. They also frequently have extensive mud flats and tidal marshes, which are exposed at low tide. Extensive intermixing of soluble nutrients and organic particles between these zones is mediated by tidal action, resulting in a highly productive environment (see previously) with a wide range of biota ranging from high bacterial populations to fish, shellfish, and birds.

Estuarine systems are characterized by two main groups of primary producers – phytoplankton (in the pelagic zone of the drainage channels) and microphytobenthos (on submerged sediments and exposed mud flats).

2.13.1   Pelagic zone

The pelagic community of estuaries contains the full range of biota seen in lakes. It is dominated by phytoplankton that is grazed by larger zooplankton (mainly copepods), smaller organisms such as juvenile copepods and heterotrophic protozoa such as flagellates and ciliates (Underwood and Kromkamp, 1999). Benthic organisms also feed on the algae and are especially important in estuarine systems, which are typically shallow and are characterized by intense benthic–pelagic coupling.

In many estuaries (Burney, 1994), the periodic development of an extensive phytoplankton community results in annual fluctuations of dissolved organic carbon (DOC) that relate more to variations in algal productivity (autochthonous DOC) than to inflow from catchment sources (allochthonous DOC). In these estuaries, concentrations of DOC follow the seasonal cycle of primary production, peaking in summer or associated with spring or autumn phytoplankton blooms. Seasonal changes in allochthonous carbon have been noted previously.

Salinity is an important factor in determining the composition of estuarine phytoplankton, with mainly freshwater species in the upper regions – dominated by diatoms. Increasing numbers of marine algae occur towards the sea, with dinoflagellates and prymnesiophytes (often *Phaeocystis*) being particularly abundant. Some algae, such as the diatom *Skeletonema costatum* have a wide salt tolerance (euryhaline) and are very common throughout the estuarine system.

2.13.2   Sediments and mudflats

Communities of algae (microphytobenthos) occur at the surface of submerged sediments within the water column and on exposed mudflats, forming a discrete algal biofilm (Underwood and Kromkamp, 1999; Smith and Underwood, 2000).

Ecological importance of benthic biofilms

Algal biofilms are important in various ways, including the exchange of materials between sediment and water column, sediment stabilization, and estuarine productivity.

Sediment–water exchange   In shallow estuarine conditions, where there is active photosynthesis at the bottom of the water column, algae affect sediment–water interchange in two main ways (Nedwell *et al.*, 1999). First, algae may effectively scavenge upwardly transported nitrogen and phosphorus, reducing the effluxes of ammonium and phosphate ions into the water body. Second, generation of oxygen extends the oxygenated layer to the bottom of the water column and may influence the processes of nitrification, denitrification and exchange of dissolved organic phosphorus.

Sediment stabilization   Almost all sediments in natural waters show evidence of biological activity, and the production of cell exudates by a wide range of biota is of great importance in sediment stabilization. Benthic diatoms appear to be particularly important in this process, as observed by Holland *et al.* (1974) who noted that stabilization was more pronounced by diatoms that produced copious amounts of polymer (mucilage). In addition to diatoms, bacteria and blue-green algae are also of major importance (Wetherbee *et al.*, 1998).

Estuarine productivity   Benthic algae make a major contribution to carbon fixation in the estuarine environment, contributing up to 50 per cent of total productivity in some situations (Underwood and Kromkamp, 1999).

Variation in algal biofilms

The species content and biomass of these communities varies considerably with differences in relation to depth (within the substratum), spatial distribution, and temporal variation.

Depth variation   Although algal benthic communities are mainly present within the top few millimeters of the surface, they do extend into deeper layers. These algae are present in many mudflat and intertidal soft-sediment habitats, and are dominated by epipelic (mud-dwelling) diatoms. Epipelic diatoms are motile and can migrate through the sediment in response to tidal and diurnal rhythms, appearing at the surface of the sediment during periods of emersion and migrating down when the sediment becomes immersed. Epipelic diatoms move through the production and extrusion of carbohydrate-rich, long-chained heteropolymers. Under high light conditions, a well-defined zonation occurs within the epipelic biofilm, with defined populations of *Euglena*, epipelic diatoms, and filamentous blue-greens positioning themselves in relation to ambient irradiation (Figure 2.19(b)).

Spatial variation   In addition to vertical heterogeneity, benthic algal communities also show considerable spatial heterogeneity over the surface of sediments – both in terms of total biomass and species composition (Underwood and Kromkamp, 1999). These spatial variations are complex and depend on a range of inter-related factors, including sediment particle size, nutrient content, water content, degree of salinity and substratum mobility. Exposed estuarine sediments are liable to extended periods of desiccation, interspersed with periods of both freshwater and seawater flooding – creating a very complex environment. This is seen in Figure 2.19, which illustrates changes across the shoreline up into the region of mudflats. Mudflats are composed of fine, cohesive, high-nutrient sediments and typically support large populations of epipelic diatoms. Transition to a higher particle sized sandy substratum in the intertidal zone results in a change from epipelic, motile diatoms species to small, virtually immobile diatoms which produce extracellular polymeric materials for attachment.

Total biomass and productivity in the epipelic biofilm reaches a maximum value at an intermediate position between the high mudflats and inter-tidal zone (Figure 2.19). This position is a compromise between the growth-limiting conditions of low nutrient, reduced light (frequent immersion) in the inter-tidal zone, and prolonged desiccation plus intense irradiation in the high mudflats.

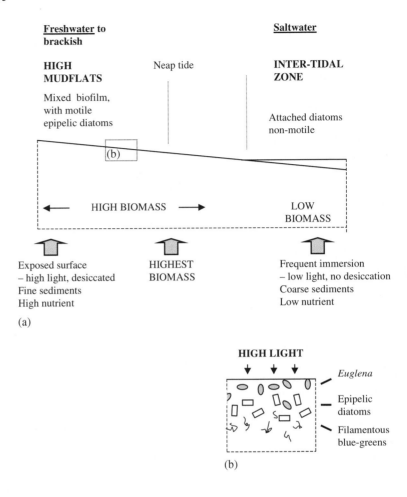

Figure 2.19 Occurrence of microphytobenthos in estuarine inter-tidal mudflats: (a) growth of diatoms may be limited in the high mudflats due to desiccation and high light levels, so highest biomass tends to occur lower down towards the inter-tidal zone; (b) detail from the surface biofilm (based on information from Underwood and Kromkamp, 1999)

Temporal variation   Benthic algal communities show considerable variation in biomass both within and between annual cycles, with reported chlorophyll-a concentrations ranging as widely as 0.1–460 $\mu g\ g^{-1}$ sediment. Although various studies of these communities have shown increases in epipelic diatom biomass during summer months, peaks of biomass do occur at other times of year – varying with external factors such as periods of desiccation and flooding. In general, it appears that epipelic diatom assemblages are less directly influenced by seasonal changes than phytoplankton communities (Underwood and Kromkamp, 1999).

F. ADVERSE AND EXTREME CONDITIONS IN FRESHWATER ENVIRONMENTS

Many freshwater environments are characterized by adverse physical and chemical conditions, where microbial growth and metabolic activity are limited within all or part of the habitat and for varying periods of time within the seasonal cycle. In some cases, these adverse conditions become extreme, and the 'abnormal' environment is incompatible with most forms of life.

This section considers the range of adverse and extreme environmental conditions that aquatic biota living in inland waters may experience, then deals in detail with two specific environments. The first of these, the air–water interface, is an example of potentially extreme conditions at the micro-level. The second example considers microbial adaptations to snow and ice and relates to adverse conditions which may occur over a wide environmental area.

2.14   Adverse conditions as part of the environmental continuum

Many adverse aspects of the freshwater habitats are not extreme, and may simply be viewed as part of the normal environmental continuum. These include variations in oxygen concentration, nutrient availability, and the inhibitory effects of solar radiation. A wide range of autotrophic and heterotrophic micro-organisms show adaptations that allow them to survive these often localized and transient conditions.

2.14.1   Variations in oxygen concentration

Microorganisms fall into two main categories in relation to oxygen requirements – aerobic organisms (able to grow in the presence of O_2) and anaerobes (able to grow in its absence). These groups are further divided into obligate and facultative organisms, depending on whether the situation is an absolute requirement or not. Further subdivisions within the categories also occur, including the environmentally important groups – the 'microaeriophiles' – which are damaged by the normal atmospheric level of oxygen (20 per cent) but do require it at lower concentrations (2–5 per cent) for growth. Microaerophiles tend to be restricted to limited zones between aerobic (with oxygen) and anaerobic (no free oxygen) regions and include bacteria (e.g., *Campylobacter*) and protozoa such as *Loxodes* (Figure 9.6).

Although life originated and evolved in anaerobic environments, most of the biosphere (including freshwater environments) today is aerobic. Many anaerobic niches exist, however, containing organisms living under conditions which may resemble

the anoxic environment of the earth until approximately two billion years ago.

Although the absence of oxygen might be regarded as an extreme situation within the current biosphere, anoxic niches are widespread and often occur in close proximity to oxygenated conditions. This close association is important since aerobic and anaerobic processes come together in the biogeochemical cycling of carbon, nitrogen, phosphorus, sulphur and other elements (Chapter 5). Bacterial anaerobic decomposers (Sections 6.6 and 6.8) are particularly important in these cycles, and the two worlds rely and depend on each other.

The presence of oxygen may also present problems to freshwater biota, and excessive build-up may constitute an extreme environment. Some biochemical processes, such as the fixation of nitrogen via the nitrogenase enzyme, appear to have arisen under anaerobic conditions, and are inhibited by the presence of even small amounts of oxygen. Various strategies have evolved to facilitate the continuation of this process under potentially hostile aerobic conditions (Section 5.6). Excessive levels of oxygen may occur in lakes during algal bloom formation, where evolution of oxygen at high rates of photosynthesis leads to super-saturation of surface waters (Section 10.7).

2.14.2   Nutrient availability

Limitations in the external concentration of inorganic nutrients such as nitrogen (Section 5.3), phosphorus (Section 5.7), silicon (Section 5.10), and trace elements (Section 5.11) occur widely in freshwater environments, restricting the growth of both autotrophic (algae) and heterotrophic (bacteria) micro-organisms.

Limitations in organic substrate for heterotrophic microbes are also widespread, and many bacteria are carbon-limited in freshwater systems. In the water column of most lakes and rivers, for example, the majority of bacteria are in a starvation-induced state of dormancy, as indicated by differences in total and viable counts (Section 6.9.2). Well-defined molecular and physiological mechanisms have evolved (Section 6.6.4) to allow bacteria to survive low nutrient availability until high nutrient conditions return. These trigger a burst of metabolic activity, growth and population increase – allowing the bacteria to exploit the growth resource for as long as it lasts.

Although most aquatic environments exhibit some degree of nutrient limitation, this becomes acute in certain situations, such as snow/ice systems (Section 2.17) and subterranean ground water (aquifers).

Aquifer microorganisms

Microbial communities in subterranean ground waters exist in various aquatic compartments, from shallow to deep (>200 m) systems, and show considerable physiological diversity. Ground water microorganisms are mainly present within biofilms at the surface of sediment particles, and include both lithotrophic and heterotrophic organisms (Frederickson *et al.*, 1989). Lithotrophic organisms include manganese (Mn IV)- and iron (FeIII)-reducing bacteria in low-oxygen conditions, with sulphate-reducing and methanogenic bacteria occurring in completely anoxic environments.

In the majority of aquifers, the predominant microorganisms are thought to be aerobic or facultatively-aerobic heterotrophs, mostly in the genus *Pseudomonas* (Kazumi and Capone, 1994). These bacteria may have to experience a range of adverse conditions (Table 2.11), including very low nutrient concentrations, and are adapted to grow and survive at extremes of organic carbon availability. Evidence for extreme nutrient limitation of bacteria in aquifers is provided by estimates of inter-related bacterial abundance, nutrient supply, and metabolic activities (Kazumi and Capone, 1994):

- Total bacterial counts in sediments from a range of depths within a pristine (non-contaminated) aquifer have been shown to be directly correlated to total organic carbon levels.

- Values for bacterial abundance and metabolic activity (glucose uptake, total thymidine incorporation per gram dry weight of sediment) in aquifers are generally 10 to 1000-fold lower than soil or sediments of surface aquatic systems.

- The efficiency of glucose uptake is particularly high in aquifer bacteria, with only 10 per cent of added label being recovered as CO_2.

2.14.3 Solar radiation

High levels of solar radiation must be regarded as a normal aspect of many freshwater environments, and aquatic organisms have evolved a wide range of adaptations to minimize or avoid the harmful effects of exposure to such conditions (Section 4.9). Although excessive radiation may be regarded as part of the normal situation, there are some environments where solar radiation (including ultraviolet UV-B radiation) reaches extreme and very damaging levels. These include the micro-environment at the air–water interface (Section 2.16), exposed snowfields (Section 2.17), and the entire water column of some alpine lakes (Section 2.5.4).

At the other end of the radiation continuum, some aquatic environments experience chronically limiting light levels. Many polar lakes are permanently frozen, and planktonic algae have adopted strategies of heterotrophic nutrition to supplement the limited photosynthetic carbon and related inorganic nutrient uptake (Section 3.11).

2.15 Extreme environmental conditions

Our understanding of biodiversity in the microbial world has recently grown with the realization that many microorganisms are able to grow in extreme environments (Seckbach, 2000). Organisms living in such environments are referred to as 'extremophiles' (Seckbach and Oren, 2000) and are able to tolerate a wide range of 'hostile' external conditions. Some of these are listed in Table 2.11, which also emphasizes the point that extreme environments typically have multiple adverse conditions. These include:

2.15.1 Temperature

Temperature is one of the most important environment factors that determines the growth and survival

Table 2.10 Temperature ranges for microbial growth – colour coding: ▨ psychrophiles ☐ mesophiles ▨ thermophiles ▨ hyperthermophiles (data from Prescott *et al.*, 2002)

Microorganism	Cardinal temperatures (°C)		
	Minimum	Optimum	Maximum
Nonphotosynthetic prokaryotes			
Bacillus psychrophilus	−10	23–24	28–30
Micrococcus cryophilus	−4	10	24
Pseudomonas fluorescens	4	25–30	40
Thermoplasma acidophilum	45	59	62
Thermus aquaticus	40	70–72	79
Sulfolobus acidocaldarius	60	80	85
Pyrodictium occultum	82	105	110
Photosynthetic prokaryotes			
Rhodospirillim rubrum	ND	30–35	ND
Synechococcus eximius	70	79	84
Eukaryotic algae			
Chlamydomonas nivalis	−36	0	4
Fragilaria sublinearis	−2	5–6	8–9
Chlorella pyrenoidosa	ND	25–26	29
Skeletonema costatum	6	16–26	>28
Cyanidium caldarium	30–34	45–50	56
Fungi			
Mucor pusillus	21–23	45–50	50–58
Protozoa			
Amoeba proteus	4–6	22	35
Paramecium caudatum	ND	25	28–30
Tetrahymena pyriformis	6–7	20–25	33

ND – not determined.

of microorganisms. Each organism has a characteristic range of growth with minimum, optimum, and maximum temperature values (Table 2.10). Microorganisms can exist in environments from subzero temperatures (minimum −5°C) to the boiling point of water and above (maximum 113°C), and may be classified according to their temperature requirement into four major groups as follows.

- *Psychrophiles* grow well at 0°C, with an optimum temperature of 15°C or lower. These organisms are readily isolated from Arctic and Antarctic habitats, and include algae (e.g., *Chlamydomonas nivalis*), bacteria (wide range of species), and Archaea (e.g., *Methanogenium*). Psychrophiles have enzymes, transport systems, and protein synthetic systems which function well at low temperatures. Cell membranes remain semi-

liquid when cold due to high levels of unsaturated fatty acids, but become disrupted at higher temperatures (>20°C), leaking cell constituents into the external medium.

- *Mesophiles* have growth optima around 20–45°C, with a growth range of about 15–45°C. Most freshwater microorganisms fall within this category.

- *Thermophiles* can grow at temperatures >55°C, with growth optima typically between 55–65°C. The great majority of thermophiles are prokaryotes, though some algae and fungi are also thermophilic (Table 2.10). The prokaryotes include Archaea, which are the most thermophiloic of all microorganisms. Thermophiles have heat-stable enzymes and protein synthetic

Table 2.11 Multiple adverse conditions in some extreme aquatic environments

Environment	Typical micro-organism	Temperature	pH	Oxygen	Water potential	Salinity	Osmotic Pressure	Hydro-static P	Solar radiation	Inorganic nutrients	Organic pollutans	Heavy metals
Sulphur springs	**Thiobacillus thiooxidans** **Cyanidium**	H	L	L								
Hydrothermal vents (deep lakes)	**Thermococcus**	H	L	L				H				
Deep water: non-hydrothermal	**Colwellia hadaliensis**	L						H	L			
Soda lakes	**Spirochaeta** **Spirulina**		H	L	L	H	H					
Subterranean water. Aquifers	**Pseudomonas spp.**	H	H	L	L	H	H	H	L	L		
Snow/ice Ecosystem	*Chlamydomonas nivalis* *Chloromonas*	L	L		L				H/L	L		
Polar lakes (permanently frozen)	*Amphidinium cryophilum*	L							L			
Neuston biofilm	**Bacteria**, unicell algae & protozoa			H					H		H	
Industrial discharges Acid mine drainage	*Picrophilus oshimae* *Klebsormidium flaccidum*		L									H
Fertilised fishponds. Sewage polluted lakes	*Chlorococcalian algae*	H	H/L	H						H	H	

Particular extreme aquatic environments are characterized by high (H) and low (L) levels of different environmental parameters. Listed organisms and groups include prokaryotes (Archaea, bacteria, blue-green algae – bold type) and eukaryote algae (plain type). In snow/ice systems, solar radiation can be very high at the snow surface but rapidly attenuates with depth. Extreme eutrophic systems such as fertilised fishponds typically have very high oxygen concentrations in the epilimnion during periods of high algal growth, becoming anoxic with depth. Organic pollutants in these systems can arise as secondary environmental effects and include decomposing organic material and algal toxins.

systems that function well at high temperatures. Cell membranes remain intact and functional at high temperatures due to high levels of saturated lipids, with high melting points.

- *Hyperthermophiles* can grow at 90°C or above, and may have maxima above 100°C. Hyperthermophiles have growth optima in the range 80–113°C, and include the bacteria *Pyrococcus abysii* and *Pyrodictium occultum* which growth in deep ocean vents.

Low-temperature organisms occur in habitats that are frozen for most of the year. The absolute requirement for liquid water means that these organisms are only able to grow in water pockets (surrounded by ice) or in melting ice that arises during the short summer period. Recently, microorganisms have been isolated from ice cores of up to 300 m depth from Vostock, the Russian ice station in Antarctica. These cores contain bacteria, eukaryote algae, blue-green algae, and fungi, in a state of permanently-frozen dormancy. Below the station lies a vast freshwater lake, which may also contain microbiota. A more detailed account of snow and ice microorganisms is given in Section 2.17, with information on adaptations of snow algae in Section 3.12.

High-temperature environments originate as a result of solar heating, geothermal activity, combustion processes, and human activity (industrial discharges, power plants), and often also carry extremes of pH, oxygen concentration and hydrostatic pressure.

2.15.2 pH

Microbes can exist in environments ranging from extremely acid (down to ∼pH 0) to very alkaline (as high as pH 12), and fall into three main groups – acidophiles (growth range pH 0–5.5), neutrophiles (pH 5.5–8.5), and alkalophiles (pH 8.5–12.0) (see Figure 2.20). Strongly acid environments also frequently correlate with heavy metal pollution (Section 2.14.6).

Acidophiles

Low pH conditions often originate as a result of biogeochemical activities of bacteria and are typical of hydrothermal vents, sulphur springs, and acid mine drainage. Acidophiles living in such environments comprise the following.

- Eukaryotes – including extreme acidophilic fungi (belonging to the genera *Acontium*, *Cephalosporium*, and *Trichosporon*), acidophilic green algae (e.g., *Klebsormidium*), and acidothermophilic red algae (belonging to the *Cyanidiaceae*).

- Prokaryotes – prominent in this group are organisms in the domain *Archaea*, many of which are also anaerobic and thermophilic.

Recent studies (Sabater *et al.*, 2003) on an extremely acid and heavy-metal-polluted stream (the Rio Tinto, Spain) have emphasized the role that benthic algae may have in the cycling of organic matter and energy flow in such extreme, heterotroph-dominated conditions. The benthic community of this stream was dominated by mats of the acidophilic green alga *Klebsormidium* – with fungal hyphae, iron bacterial sheaths, diatoms (*Pinnularia*), and mineral particles all part of the mixed biofilm.

Neutrophiles

Although most freshwater microorganisms are neutrophiles, some differences do occur between the major groups. Bacteria and protozoa tend to occur in environments close to neutrality, while fungi prefer slightly acid surroundings (pH 4–6). Algae show a wide variation, with different phytoplankton species being adapted to acid (often oligotrophic) or alkaline (often eutrophic) waters. Under bloom conditions in nutrient-rich waters, the lake water may become highly alkaline (pH 11), which only colonial blue-green algae are able to tolerate (Section 10.7).

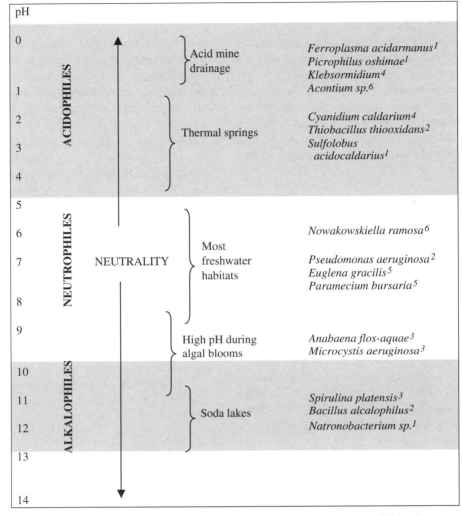

Figure 2.20 Range of pH conditions and microorganisms in aquatic environments of inland waters; the range of microorganisms includes archaeans[1], bacteria[2], blue-green – algae[3], eukaryotic algae[4], protozoa[5], and fungi[6]

Alkalophiles

Alkalophilic microorganisms grow optimally at pH values above 8.5, and live in soda lakes, hot springs, and subterrestrial environments; alkalinity can rise to values as high as pH 12. These organisms include the archaeans, bacteria, and blue-green algae (Figure 2.20).

Over the wide range of environmental pH conditions to which freshwater microorganisms are exposed (Figure 2.20), internal (intracellular) pH

is maintained close to neutrality. This pH homeostasis appears to be maintained in several ways (Prescott *et al.*, 2002). With many organisms, the plasma membrane may be relatively impermeable to H^+ ions (protons), and internal buffering mechanisms reduce internal pH change. Specific exchange mechanisms may also operate. Neutrophiles appear to export potassium for protons using an antiport transport system, and extreme alkalophiles such as *Bacillus alcalophilus* maintain a neutral pH by exchanging internal sodium ions for external protons. When internal pH does alter, other meta-bolic processes come into play. The bacterium *Escherichia coli* has been shown to produce an array of new proteins when intracellular pH falls to pH 5.5–6 as part of an acidic tolerance response. When pH falls below this, polypeptide denaturation is reduced by the synthesis of chaperones such as acid shock and heat shock proteins.

2.15.3 Conditions of low water availability: saline environments

The amount of free water available to microorgan-isms can be reduced by interaction with solute molecules (the osmotic effect), adsorption to solid surfaces (the matric effect), or because water is in a solid rather than a liquid state (the phase effect). Water availability is greatly reduced in saline envir-onments due to the osmotic effect, and in frozen habitats (Section 2.17) where it is present in the solid state.

Water activity

Water availability in saline environments can be expressed in terms of the water activity (a_{w}), which relates the vapour pressure of the saline solution (P_{soln}) to the vapour pressure of pure water (P_{water}), where:

$$a_{\mathrm{w}} = \frac{P_{\mathrm{soln}}}{P_{\mathrm{water}}} \qquad (2.3)$$

In natural waters, availability (a_{w}) ranges from values of about 1 (freshwater) to 0.75 (saturated salt lakes). Survival at low a_{w} requires the main-tenance of a high internal solute concentration to retain water. Some organisms are able to do this over a range of external molarities (osmotolerant), and in the case of saline environments, are referred to as halophiles.

Halophiles

Most halophiles grow in salt solutions ranging from 2–20 per cent of a saturated NaCl solution, with extreme halophiles occuring in hypersaline brines from 20 per cent NaCl up to complete saturation. Such organisms occur in inland salt lakes, soda lakes, and subterrestrial subsurface brines, and include archeans (*Halobacterium* sp.), bacteria, and algae (e.g., *Dunaliella viridis*). Among the organisms living in hypersaline environments are aerobic and anaerobic heterotrophs, sulphate redu-cers, methanogens, and phototrophs.

The archean *Halobacterium* provides a good example of an extreme halophile (Prescott *et al.*, 2002), and can be isolated from inland waters such as the Dead Sea (Jordan/Israel) and The Great Salt Lake, Utah (USA) – where salt concentrations approach saturation. This organism contains enor-mous quantities of intracellular potassium ions, reaching concentrations of 4–7 M. The high con-centration of K^+ ions not only promotes osmotic balance with the external medium, but is also important in stabilizing enzymes, ribosomes, and transport proteins and in maintaining their activity. Significant modifications occur to the structure of proteins and cell membranes, and the plasmalemma and cell wall of *Halobacterium* are further stabi-lized by high concentrations of Na^+ ions. Extreme halophilic bacteria such as *Halobacterium* have become highly adapted for survival in an environ-ment that would destroy most other microorgan-isms. In doing so, they have lost ecological flexibility, and are now restricted to a very limited aquatic habitat.

Saline lakes and wetlands

Saline lakes and wetlands have a high ion concen-tration, which in most cases is dominated by Na^+, Cl^- and can be measured in terms of straight concentration, molarity, or electrical conductivity (specific conductance). Sodium chloride lakes are particularly common in Australia, South America, and Antarctica. The ionic composition of saline lakes varies considerably, and these water bodies should be more properly characterized by reference to the dominant and subdominant ions, e.g., Lake Bogoria, Kenya ($NaCO_3Cl$), Lake Van, Turkey ($NaClCO_3$) and the Dead Sea, Israel/Jordan

(NaMgCl). Regions such as East Africa contain lakes that are dominated by $NaCO_3$. These 'soda lakes' arise because the fresh inflow waters are disproportionely rich in HCO_3, but relatively poor in Ca and Mg. In all of these saline lakes, microorganisms are surrounded by water which has an ionic balance different from cell contents, has a high osmotic pressure, and in many cases is highly alkaline.

Salinity range and microbial communities

The salt concentration of saline lakes ranges from conductance values of 1000–270 000 microSiemens (μS) cm^{-1}, compared with values of 20–650 μScm^{-1} for freshwater bodies. Over the above range for saline lakes, aquatic microorganisms tend to be adapted to particular windows of salinity value. This was demonstrated in a recent diatom survey of Australian saline lakes (Gell and Gasse, 1990), where distinct diatom communities were identified over a six-step sequence (oligosaline to hypersaline lakes) of increasing salinity (Table 2.12). The extreme limiting environment of high saline (eusaline to hypersaline) lakes was indicated by the restricted group of diatom species present, resulting in a low overall biodiversity level within the diatom community. Although most diatom spe-

cies were present in salinity-related assemblages, not all species were restricted to a particular salinity range. Four diatom species occurred over a wide range of saline conditions – demonstrating their independence from ionic constraints.

2.15.4 Conditions of low water availability: ice and snow environments

Although it might seem unlikely that low water availability would be a problem in aquatic environments (rich in H_2O), this is certainly the case for long-term frozen regions such as glaciers and snow fields, where liquid water may not occur or may be restricted to a brief melt once a year. A decrease in H_2O content (desiccation) may occur in such environments due to the twin effects of low water availability and loss of H_2O in the frozen state due to sublimation.

Environmental constraints in frozen environments are discussed in Section 2.17, and the adaptations of micro-algae to such conditions in Section 3.12. These organisms are adapted to long periods of dormancy for most of the year, with brief phases of growth and completion of their life cycle during the short summer period.

Table 2.12 Diatom communities in a range of Australian saline lakes (Gell and Gasse, 1990)

Category of saline lake	Conductivity range (μS cm^{-1})	Diatom community (representative species)
Hypersaline	70–195 000	*Achnanthes brevipes* *Berkeleya rutilans* *Amphora coffaeformis*
Metasaline	40–70 000	*Navicula elegans* *Nitzschia sp.*
Eusaline	30–40 000	
Polysaline	15–30 000	*Anomoeneis sphaerophora*
Mesosaline	5–15 000	*Tabularia parva* *Cymbella pusilla*
Oligosaline	1–5000	*Eunotia lunaris,* *Achnanthes minutissima* *Stauroneis pachycephala* *Navicula augusta*

The salt content of saline lakes depends on dilution or concentration of existing levels in relation to climatic conditions (balance between precipitation/ evaporation) and input of salts from the atmosphere, saline crusts and groundwater. Eusaline to hypersaline lakes (shaded area) represent extreme aquatic environments which have restricted species (adapted to high saline conditions) and low biodiversity.

2.15.5   Variations in hydrostatic pressure

Most freshwater microorganisms live in surface waters, where hydrostatic pressure ranges from about 1–10 atmospheres. The hydrostratic pressure within the depths of a water body may rise well above this range, however, increasing from the surface level of 1 atmosphere by a 1 atmosphere increment for every 10 m depth in water column. Many bacteria present in deep waters are barotolerant (unaffected by increased hydrostatic pressure) and some are barophiles – growing more readily at high pressures (100–1000 atmospheres). So far, barophilic organisms have been identified in several genera of bacteria (e.g., *Photobacterium*, *Shewanella*, *Colwellia*) and Archaea (e.g., *Pyrococcus*, *Methanococcus*).

Obligate barophiles have an absolute requirement for high pressure for optimal growth, and have been found in deep-sea samples at depths below 10 000 m. Such deep-dwelling microbes also grow at low temperatures ($-4°C$) and are termed barophilic psychrophiles. Although such marine environments are outside the scope of this book, barophiles may also occur in freshwater locations such as Lake Baikal (Russia), which reaches depths of over 1600 m. Barophilic psychrophiles have been detected in deep ice near the Vostock Station in Antarctica, and may also be present in the freshwater lake covered by thousands of meters of ice below this.

In contrast to deep sea and deep ice high-pressure environments which are cold, the subterranean environment is characterized by 20–30°C temperature increases for every km of depth. Archaea and bacteria have been isolated from water bodies deep below the Earth's surface, some of which have been shown to be hyperthermophilic (Seckbach and Oren, 2000). It seems likely that some of these organisms are also barophilic.

2.15.6   Organic and inorganic pollution

Contamination of freshwaters by inorganic pollutants (e.g., heavy metals) and organic pollutants (e.g., pesticides) has led to many extreme environmental situations. The adverse effects of contamination with organic pollutants are described in Section 9.6.3.

Heavy metal pollution

Pollution with heavy metals (particularly zinc, copper, and lead) often accompanies conditions of high acidity and typically arises as a result of industrial contamination or acid mine discharge. Lake D'Orta (Italy) provides a good example of industrial pollution, with a history of initial copper contamination ($\sim$1930–1960), followed by mixed pollution with Cu, Zn, Ni, and Cr plus acidification. The microbial response to these extreme conditions was investigated over the time period from sediment cores (Cattaneo *et al.*, 1998) in relation to two main groups – diatoms (algae) and thecamoebians (protozoa). In both cases, the stress response involved a decrease in individual mean size, which reflected a shift in the taxonomic composition of the microbial assemblages. Within the diatoms, communities dominated by *Cyclotella comensis* (biovolume 100–1000 μm^3) and *Cyclotella bodanina* (1000–10 000 μm^3) were replaced by assemblages dominated by *Achnanthes minutissima* ($<100 \mu m^3$). In the thecamoebians, small species (e.g., *Difflugia proteiformis*) came to dominate assemblages that had previously been largely populated by larger species such as *Difflugia globulus*. Changes in mean size within these microbial populations not only reflected a shift in species composition but also a significant change within the size range of individual taxa.

In the case of acid mine drainage, acidification is largely caused by the creation of sulphuric acid by chemical oxidation and the activity of sulphur-oxidizing bacteria. The activity of these organisms also generates strongly oxidizing conditions that finally result in heavy metal metal leaching from the drained strata, causing pollution. The case of extreme river acidification quoted earlier (Section 2.14.2, Sabater *et al.*, 2003) provides a good example of heavy metal acid mine drainage.

Archaean microbes and extreme environments

Microorganisms within the domain Archaea are quite distinct from other living organisms (Table 1.1) and are particularly characteristic of extreme aquatic habitats. These organisms are prominent within most of the extreme environments discussed previously, and include thermophiles, acidophiles, alkalophiles, halophiles, and barophiles (Seckbach, 2000; Prescott *et al.*, 2002). The most thermophilic of all microbes occur within the archaean domain, and include *Pyrobolus fumari*, which grows up to a temperature of 113°C and is able to survive within an autoclave (121°C) for up to 1 hour. The most acidophilic prokaryote known is *Picrophilus oshimae*, a thermophilic archaean whose pH optimum is 0.07. This organism is stable in hot (optimum 60°C) and extremely acidic (below pH 1) media.

Although Archaea form a distinctive taxon, there is great diversity within the assemblage. The major groups of archaeans are summarized in Table 2.13, which also emphasizes the wide-ranging adaptations shown by this group to extreme aquatic environments. Many of these organisms are adapted to multiple adverse factors.

Table 2.13 Characteristics of major groups within the Domain Archaea (adapted from Prescott *et al.*, 2002)

Group	Taxonomic characteristics	Environmental characteristics	Representative genera
Methanogenic archaea	Methane is the major end product. S^0 may be reduced to H_2S without energy formation. Cells possess coenzyme M, factors 420 and 430 and methanopterin	Strictly anaerobic	*Methanobacterium* *Methanococcus* *Methanomicrobium* *Methanosarcina*
Archaea sulphate reducers	Irregular gram-negative coccoid cells. H_2S formed from thiosulphate and sulphate. Autotrophic or heterotrophic. Possess factor 420 and methanopterin but not coenzyme M or factor 430	Extremely thermophilic. Strictly anaerobic	*Archaeoglobus*
Extremely halophilic archaea	Coccoid or irregularly-shaped cells. Gram-negative or gram-positive. Mainly aerobic chemoorganotrophs. Form red colonies.	Extreme halophiles. Require high sodium chloride concentrations (>1.5 M) for growth. Neutrophilic or alkalophilic. Mesophilic or slightly thermophilic	*Halobacterium* *Halococcus* *Natronobacterium*
Cell wall-less archaea	Pleomorphic cells lacking a cell wall. Chemoorganotrophic. Plasma membrane contains a mannose-rich glycoprotein and a lipoglycan	Thermoacidophilic. Facultative anaerobes	*Thermoplasma*
Extremely thermophilic S^0-metabolizers	Gram-negative rods, filaments or cocci. Autotrophic or heterotrophic. S^0 reduced to H_2S anaerobically, H_2S or S^0 oxidized to H_2SO_4 aerobically	Obligately thermophilic (optimum temperature between 70–110°C). Usually strict anaerobes. Acidophilic or neutrophilic	*Desulfurococcus* *Pyrodictium* *Pyrococcus* *Sulfolobus* *Thermococcus* *Thermoproteus*

2.16 A potentially extreme microenvironment: the air–water surface

A surface film occurs at the air – water interface of all aquatic systems. This interface provides an example of a microenvironment which is liable to experience extreme chemical and physical conditions, including localized concentrations of organic and inorganic toxins, fluctuating temperatures, and extreme exposure to solar radiation.

The air–water interface is important both in global terms (occupying over 70 per cent of the world's surface) and in relation to individual aquatic systems where it modulates the interchange of energy and materials between the atmosphere and the water body. In standing and running waters, dissolved and particulate matter accumulate within the surface layer from both aerial and water column sources, creating a distinct environment – the surface microlayer. This surface region has its own physical and chemical characteristics, and is associated with a distinct community of organisms – the neuston (Maki and Hermansson, 1994). This includes bacteria, micro-algae, and protozoa (Figure 2.21).

2.16.1 Chemical composition of the surface microlayer

Most organic chemicals reduce the surface tension of water, releasing free energy at the liquid surface. Because of this, these molecules have a tendency to adsorb to the air–water interface and remain within the surface microlayer – reaching high concentrations. This surface layer has a thickness of about 0.1–10 μm, depending on the water body, and has a chemical composition quite different from the rest of the water column.

The surface microlayer contains a wide range of both organic and inorganic chemicals (Table 2.14). The organic component occurs as both particulate organic material (POM) and dissolved organic material (DOM or DOC). Fatty acids, lipids, and hydrocarbons play an important role at the water surface. These molecules are referred to as 'dry surfactants' and are very surface-active. The hydrocarbon assemblage contains poly-aromatic and chlorinated hydrocarbons, including polychlorinated biphenyls (PCBs) and pesticides. PCBs and pesticides can become enriched in surface films from atmospheric deposition, reaching concentrations high enough to cause stress to microorganisms and the reproductive stages of fish and shellfish

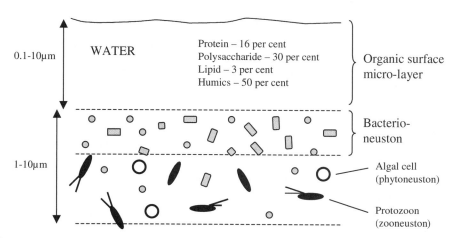

Figure 2.21 Microbial populations associated with the air-water surface biofilm. The bacterioneuston layer varies from single-cell (monolayer) to several-cell (multilayer) thickness. Individual bacterial cells may be free-floating (planktonic) or attached to macromolecules at the edge of the surface microlayer

Table 2.14 Chemical composition of surface microlayer (Neuston community). Information taken from Maki and Hermansson, 1994

Chemical	Physical properties	Derivation[*] and chemical properties
Organic Components		
Fatty acids, lipids, and hydrocarbons	'Dry surfactants' – very active surface molecules Amphiphilic	Mainly autochthonous Concentration of PCBs and pesticides from aerial (allochthonous) deposition
Proteins and polysaccharides	'Wet surfactants' anchor to surface by hydrophobic regions.	Autochthonous Detected by infrared spectroscopy
Phenolic and humic material	Liable to photo-oxidation at lake surface. Strong UV absorption.	Allochthonous. Potential inhibitors of microbial activity
Inorganic Components		
Inorganic nutrients – phosphate, ammonia, nitrate, and nitrite ions	Non-adsorptive accumulation: do not reduce surface tension	Autochthonous Potential nutrients for surface microbes
Metals	Particulate and dissolved Frequently complexed with organic molecules. 10^3–10^4 × concentration factor.	Autochthonous and allochthonous (pollution). Potential inhibitors of microbial activity

[*]Derivation is either autochthonous (from inside the aquatic system) or allochthonous (external derivation).

associated with surface films. Proteins and polysaccharides have been identified in surface films by infrared spectroscopy and other analytical techniques (Maki and Hermansson, 1994) and make up a major part of the organic complement. These 'wet surfactants' probably anchor to the water surface via their hydrophobic portions, unfolding as the interface is penetrated. The presence of phenolic and humic material in surface microfilms has been demonstrated by ultraviolet (UV) absorbance. In freshwater systems, these molecules are external (allochthonous) in derivation. These strongly UV-absorbing molecules are susceptible to photo-oxidation at the water surface, and may act as natural inhibitors of microbial activity.

In addition to dissolved organic material, various inorganic molecules also accumulate in the surface microlayer – including phosphate, ammonia, nitrate, and nitrite ions. These molecules do not lower the surface energy at the air–water interface, so do not adsorb in the same way as organic components. Their presence at the water surface is due to other factors such as excretion and lysis by microfilm organisms, transport in rising air bubbles, and asso-

ciation with organic matter by complexation. The inorganic composition of lake surface biofilms varies with nutrient status, with greater surface film enrichment of phosphate and ammonia nitrogen occurring in eutrophic lakes compared to oligotrophic sites. Accumulation of metals (Cr, Cu, Fe, Pb, Hg, and Zn), present both as particulate and dissolved material, has also been demonstrated in surface films. Although metals appear to have a relatively brief residence time (minutes to hours) in surface films, they are potentially important as toxins for microorganisms and other lake biota.

2.16.2 Physical processes and transformations in the surface biofilm

The surface film is subject to various exchange (input/output) and transformation processes (Figure 2.22). Input of dry deposition and rain from the atmosphere is counteracted by aerosol formation and evaporation. Aerosol droplets, generated by air turbulence and local heating (insolation), are normally in the range of 2–10 µm diameter, and can

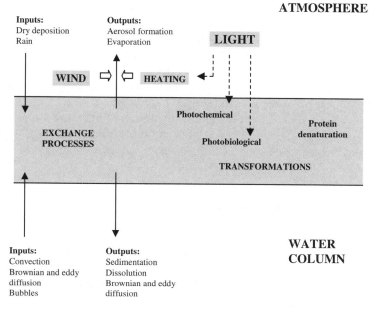

Figure 2.22 Physico-chemical exchange and transformation processes in the surface biofilm. Light damage occurs to isolated molecules (photochemical) and whole organisms (photobiological). The surface biofilm is indicated as the central shaded area

transport microorganisms (e.g., bacteria) and chemicals (e.g., algal toxins) away from the water surface. Input of material from the water column is compensated by loss of particulate and soluble matter, mainly by sedimentation and dissolution. Surface biofilms also act as a static barrier to gas exchange, and may influence the temperature at the air–water interface by retarding evaporation or by affecting water movement at the water surface.

Various biological transformations occur within the surface film, including photodegradation and protein denaturation.

Photodegradation

During daylight hours, the air–water interface is in a state of considerable photochemical and photobiological activity. Both visible and UV light have direct effects on chemical and biological (microorganism) components, and also have indirect effects by generating short-lived oxygen radicals in the presence of free oxygen. These radicals are highly reactive and lead to further degradation of organic matter.

Aromatic and sulphur-containing components of macromolecules are particularly susceptible to photodegradation by UV absorption. Damage to amino acids and proteins involves a range of processes, including cleavage of N–C bonds to form NH_4^+, generation of H_2S from cysteine and breakage of glycoside linkages (by hydrolysis) in ribose (but not deoxyribose) nucleotides (Wangersky, 1976). Damaging effects of light also have potentially serious effects on surface microorganisms (Section 2.16.3).

Protein degradation

Studies by Graham and Phillips (1979) on model systems have demonstrated irreversible adsorption and denaturation of proteins at water surfaces. The adsorption process follows a defined sequence of events:

- diffusion of proteins to the water surface, leading to random collision with the air–water interface;

- a reversible phase of adsorption, in which some proteins are lost (desorbed) while others remain or are recruited – molecular competition is a key aspect of this phase;

• permanent adsorption, with irreversible changes in protein configuration – this leads to the formation of a primary protein monolayer, with a thickness of 5–6 nm and an area loading of 2.6–3.3 mg protein cm^{-3};

• continued adsorption, leading to a maximum protein film thickness of >10 nm.

2.16.3 Microbial community at the air–water interface

Bacteria (bacterioneuston), algae (phytoneuston), and protozoa (zooneuston) all contribute to the surface film community (Figure 2.21).

Bacterioneuston

High bacterial counts at air–water interfaces have been recorded in numerous reports (see Maki and Hermansson, 1994), suggesting that this is a regular feature of surface biofilms. These organisms probably occur both as single cells and microcolonies, with the bacterial layer in different biofilms ranging from a monolayer to multilayer structure. Some of the bacteria are freely motile (planktonic) while others are attached to the molecular matrix within the surface microlayer. Studies involving model systems suggest that bacterial adhesion to the surface microlayer has similar characteristics to adhesion to solid surfaces, with bacterial surface hydrophobicity and charge strongly influencing surface biofilm adhesion (Hermansson et al., 1982).

The growth and metabolic activity of bacteria within the surface biofilm is influenced by opposing environmental factors. These aspects are promoted by high nutrient levels, but may be limited by unstable temperature conditions, localized accumulation of organic toxins and heavy metals, and intense levels of UV and visible solar radiation. In spite of this environmental stress, studies on nutrient uptake and respiration suggest that a major proportion of the bacterial population is metabolically active and physiologically-adapted to the local conditions (Maki and Hermansson, 1994). The bacterial population that develops at the water surface is probably initially selected for in terms of ability to adhere to the surface microlayer.

After this initial colonization, further growth and survival depend on ability to utilize the organic substrate and to tolerate adverse environmental conditions.

Phytoneuston

Bacteria are generally considered to be the primary colonizers of the water surface film, with algae and protozoa entering the system as secondary populations and extending the biofilm below the zone of attached bacteria (Figure 2.21).

The presence of algae in surface biofilms has been demonstrated by direct observation and by determination of chlorophyll-a. Surface populations of these organisms, however, appear to be highly variable, and the greatest enrichment with phytoneuston appears to occur in discrete patches with high concentrations of associated organic material. These localized accumulations of surface algae increase the thickness and complexity of the biofilm. Algae may enter the surface community in various ways, and may interchange with phytoplankton populations in the water column as part of a diurnal oscillation. As residents of surface films, algae share the same environmental constraints as bacteria – with solar destruction of photosynthetic complexes being additionally important. In spite of these potential limitations, analysis of primary productivity at the water surface suggests higher values than lower down in the main water column (Hardy and Apts, 1989).

Zooneuston

Amoebae and ciliate protozoa have been reported at the air–water interface, with a correlation between their population level and the density of bacterioneuston. These secondary colonizers form an important connection between prokaryotes and higher trophic levels, and link with higher organisms that associate with the water surface – including copepods and various larvae.

2.17   Microbial communities of snow and ice: life in the frozen state

This section considers habitats where fresh water occurs as the main environmental medium, but is in a frozen state for most of the year. The water is present either as snow or ice, and the habitats include snow fields, glaciers, lake ice, sea ice, and the relatively freshwater surfaces of ice shelves. The physicochemical features and biology of frozen environments are current areas of major research interest (Jones *et al.*, 2001), with particular emphasis on the microbial ecology of snow habitats (Hoham and Duval, 2001).

2.17.1   Snow and ice as an extreme environment

Many of these frozen freshwater environments present extremes in terms of temperature, acidity, radiation levels, minimal nutrients, and desiccation. In spite of such adverse conditions, snow and ice environments frequently support actively-growing populations of bacteria, fungi, algae, and some invertebrates. Adaptations of these organisms to the physicochemical constraints of the frozen environment have been well documented, particularly in relation to snow algae (Section 3.12).

The effects of the frozen environment on the biology of snow and ice organisms is matched by their own influence on the biogeochemistry of snow and ice. These organisms are closely involved in the primary production, nutrient cycling, respiration, decomposition, metal accumulation, and food webs which are characteristic of such habitats. The presence of microbes may also affect physical properties of the environment. Snow algae and other microbes in snow, for example, may reduce short-wave reflectance by as much as 50 per cent (Pomeroy and Brun, 2001).

2.17.2   Requirement for water in the liquid state

Desiccation may seem an unusual feature in situations where water, albeit in a frozen state, is present in abundance. In all of the above habitats, however, the growth, reproduction, and dispersal of microbial organisms depends on the presence of liquid (free) water for at least part of the year.

Some freshwater environments are totally and permanently frozen and do not support life. Thus, even though viable microbial cells and spores are present, the great Antarctic polar plateau and the Greenland ice sheet are virtually lifeless because of the lack of water in the liquid phase. Free water is periodically present in other frozen environments, however, including:

- the bottom surfaces of Antarctic glaciers, where basal melting generates an important habitat for microbes;

- sea ice of the Arctic ocean, where summer meltwater present on the ice surface often contains blooms of snow algae;

- snowfields, where seasonal melting of the ice supports abundant microbial life.

Snow ecosystems have been studied in considerable detail and are discussed below.

2.17.3   Snow ecosystems

Snow ecosystems typically consist of a layer of snow (snow cover) lying over ground, with microbial populations present within the snow and at the ground surface (Figure 2.23). Physical factors affecting the system are mediated via the overlying atmosphere (incoming solar and infrared radiation, rain and snowfall, wind action) and the underlying ground layer (melt run-off, ground heat flux).

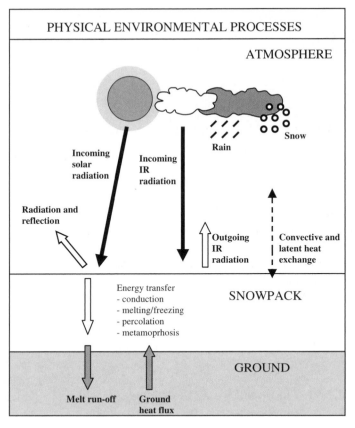

Figure 2.23 Conceptual diagram of mass and energy fluxes controlling the energetics of snow cover. *Interactions between snow and atmosphere*: energy exchanges include incoming and reflected solar and infrared (IR) radiation, convection (turbulent heat flow), latent heat exchange (evaporation, sublimation, refreezing) at the snow surface. *Physical changes within snowpack*: energy and mass changes are associated with heat conduction, melting/refreezing, water percolation, and snow metamorphosis. *Interactions between snow and ground*: energy enters the snowpack as heat flux and mass loss occurs as melt run-off (figure redrawn from Pomeroy and Brun (2001) and reproduced with permission from Cambridge University Press)

Various studies have been carried out on snow ecosystems, investigating the physical properties of snow (Pomeroy and Brun, 2001), snow chemistry and nutrient cycling (Tranter and Jones, 2001) and microbial ecology (Hoham and Duval, 2001). The global importance of snow as a freshwater environment arises because of the large winter land coverage at high latitude and high altitude, with more than half the northern hemisphere land mass being covered each year (Groisman and Davies, 2001).

Snow ecosystems can be considered in relation to the snow matrix (physical properties) and the constituent biota.

2.17.4 *The physical properties of snow*

Snow cover interacts strongly with the global climate system, being formed as a result of local weather characteristics and also conversely influencing climatic conditions. The physical properties of snow are summarized in Table 2.15 and are important both in terms of snow as a static environment medium and in relation to the dynamic changes (seasonal and non-seasonal) occurring as a response to weather fluctuations.

The physical properties of snow have an overriding influence on the potential for biological colonization and growth in a number of ways (Pomeroy and Brun, 2001).

Energy bank

Snow stores and releases energy that enters the ecosystem. Potential energy is stored as latent heat of fusion and vaporization and as crystal bonding forces. Where clear seasonal changes occur, the periodic uptake and loss of energy at different times of the year makes snow a highly variable habitat, particularly in relation to the balance between frozen and liquid states.

Table 2.15 Physical properties of snow (after Pomeroy and Brun, 2001)

Property	Value (SI units)	Environmental implications
Melting point (θ)	273 K (0°C)	The liquid water content of a snowpack declines rapidly below this temperature
Latent heat of vaporization	2.83 MJ kg^{-1} of snow	The high value means that snow is not readily lost by sublimation, and remains as a long-term habitat
Latent heat of fusion (Δh_m) (at T_m and 1 atm)	3.3×10^5 J kg^{-1}	The high energy level required to convert snow to water means that it is not readily lost by melting
Thermal conductivity (λ)	0.045 W m^{-1} K^{-1} (value for dry snow with a density of 100 kg m^{-3})	The very low thermal conductivity means that here is very low heat transfer into the snowpack to raise the internal temperature
Reflectance of short-wave (solar) radiation	80–90% for fresh, continuous snow cover	Low absorption of solar radiation minimizes solar heating and light energy available for photosynthesis
Absorbance of long-wave (thermal infrared) radiation	High level	High absorbance contributes to internal heating. High level of re-radiation as thermal radiation
Aerodynamic roughness (z_o)	0.01–0.7 mm	Aerodynamic smoothness leads to a low turbulent transfer of heat between the atmosphere and the snow surface

Radiation shield

The radiation reflective properties of snow are an important aspect of the global climate system. Most short-wave radiation is reflected by cold snow, but long-wave radiation is absorbed and re-emitted. As noted earlier, the reflective properties of snow can be markedly reduced by snow biota such as algae.

The short-wave radiation which is not reflected by snow cover is absorbed largely within the top 30 cm of the snow pack. Radiation penetration is greater the shorter the wave length, and has important implications in terms of energy availability (for photosynthesis) and UV-induced photo-inhibition. Algae and other microorganisms occurring at the snow surface are frequently exposed to excessive levels of damaging radiation (Section 3.12).

Insulation

The low thermal conductivity of snow arises due to its high porosity and air content and gives it a high thermal insulation capacity. This is important in protecting microorganisms, animals, and plants from the chilling effects of wind and severe winter temperatures.

Long-term stability of the snow system

Loss of snow cover occurs due to sublimation (conversion to vapour in the frozen state) and melting, both of which are driven by energy absorption. Most of the physical properties of snow tend to reduce these losses, including the high values for latent heat of vaporization and the high latent heat of fusion. The high reflectance of short-wave radiation, low thermal conductivity, and low aerodynamic roughness also reduce heat uptake and temperature increase.

The relative stability of the snow ecosystem is particularly important during melt, allowing algae to migrate and bloom prior to the formation of resistant spores and completion of the life cycle (Figure 3.26).

Water reservoir

Under appropriate weather conditions, snow is a reservoir for water that provides habitat and food sources for microbes, invertebrates and small mammals. Almost all of the physical properties listed in Table 2.15 have some effect on the energy content and temperature of snow, which determines the water content.

The liquid water content of snow rises rapidly as the temperature increases above 0°C. In addition to ambient temperature, other factors such as rainfall and liquid retention capacity of snow are also important in determining the water content.

Transport medium

During melt, snow acts as an important transport medium, with flagellate algae and other biota exhibiting vertical movement within the water film surrounding the snow crystals. Migration within the snow column is an important aspect of the life cycle of snow algae (Figure 3.26).

2.17.5 Snow and ice microorganisms

In the last three decades, considerable advances have been made to our understanding of the taxonomy, ecology, physiology, and interrelationships of the microorganisms present in snow and ice environments (Hoham and Duval, 2001). The biology of these microorganisms may be considered in relation to general biodiversity, adaptations and identity of characteristic species, population levels, and interactions between organisms.

Biodiversity in frozen environments

The range of organisms that have been identified in different snow and ice environments are summarized in Table 2.16. Although some of these studies concentrate on particular groups of organisms, the general picture is one of considerable microbial diversity in snow environments.

Within each of the these major groups of organisms, certain species appear to be particularly adapted to snow environments. Many of these are psychrophiles, only able to grow at low temperature (1–4°C). Organisms that only occur in snow environments include species of algae, fungi and various invertebrates. Within the fungi, psychrophilic yeasts appear to be particularly widespread in snow environments. Filamentous fungi include species in the genera *Chionaster*, *Selenotila*, and *Phacidium*. Further details of snow algae are given in Section 3.12, and on other biota in the section on snow microbe associations.

Table 2.16 Microbial diversity in different ice/snow environments (information taken from Hoham and Duval, 2001)

Environment	Microorganisms identified	Reference
Agassiz ice cap, Canada	(1) Isolated at 1 and 4°C 4 Gram$^+$ bacteria, 12 yeasts (2) Isolated at 25°C bacteria (Gram$^+$ and Gram$^-$ rods, cocci) fungi, actinomycetes	Handfield *et al.*, 1992
Snow cover in Tioga Pass, USA	Diverse, correlated populations of bacteria and algae	Thomas, 1994
Snow	466 species of microorganisms reported – including fungi, bacteria, and algae	Kol, 1968
Polar ice, snow, and soil	Psychrophilic yeasts	Sinclair and Stokes, 1965
Snow ski area, USA	Ciliated protozoa, snow algae, and fungi	Bidigare *et al.*, 1993
Alpine snow	Bacteria, fungi, algae, protozoa, rotifers, tardigrades, insects	Hoham *et al.*, 1993

Populations of snow biota

Some snow environments have low populations of biota. Low microbial recovery rates from the arctic ice cap (Handfield *et al.*, 1992) for example indicated very limited microbial activity at this site.

In other situations, particularly where there is melting snow, populations of biota such as algae can become very high. High populations of green algae are particularly well seen from the green, red and orange snow blooms that have been observed in the now fields of North America, Europe, and Antarctica. Studies by Thomas (1994) showed algal populations reaching levels of 4×10^4 cells ml^{-1} in coloured snow, dropping to values of 0.1– 5 per cent of this value in white snow. High algal levels appeared to be correlated with high bacterial populations, which reached highest levels of 3.2×10^5 cells ml^{-1} in red snow, falling to about 10–30 per cent of this figure in white snow.

Snow microbe interactions

Interactions between microbes are an important aspect of snow ecosystems and can be identified at three levels – observations on diversity within food webs, correlations between populations of organisms, and direct observations of associations between individual organisms.

Foodwebs   The wide range of biota reported in various studies on snow ecosystems (Table 2.16) are indicative of complex food webs. In many cases, these food webs are active for only a limited period of time, while free water is available.

Population correlations   Correlations between population peaks are indicative of a trophic or spatial (e.g., epiphytic) association between biota and have been reported particularly in relation to algal populations. These correlations include the following.

- *Algal and bacterial populations.* Thomas (1994) reported a direct correlation between bacterial

and algal populations in mountain snow in California, USA. The highest populations of bacteria (3.2×10^5 cells ml^{-1}) were found with large populations of algae (4.9×10^4 cells ml^{-1}) in red-coloured snow.

- *Algae and filamentous fungi.* Both parasitic (Kol, 1968) and non-parasitic fungi show clear populations with algal populations. The latter include snow fungi in the genera *Chionaster* and *Selenotila* (Hoham *et al.*, 1993).

- *Algae and yeasts.* It is clear that many fungi occurring in snow environments are present as yeasts. Hoham and Duval (2001) report direct correlations between high yeast and algal populations in areas of mountain snow.

The above correlations of populations of heterotrophic bacteria, filamentous fungi, and yeasts with algae suggests that these organisms are simply growing on dissolved organic carbon (DOC) secreted by algal cells. This indicates that a microbial loop is operating in these ecosystems, much as it does in lake and river environments.

Direct associations   Direct associations between snow biota have been observed at the microscopical level by various authors, including bacterial and fungal associations with algae. Hoham *et al.* (1993), for example, reported an association between bacteria and fungi with *Chloromonas* snow algae from Eastern USA. These organisms were observed adhering to the outer gelatinous coat of resting spores of the alga. Another example is provided by the snow fungus *Phacidiun infestans*, which grows through melting snowpacks in some parts of the Pacific Northwest of the USA. Snow algae can be seen adhering passively to the fungus, but physical connections between them do not occur as in lichen symbiosis.

In these and other instances, observed associations between snow biota are difficult to interpret in functional terms. Spatial association does not necessarily imply symbiosis or any degree of metabolic cooperation between the participants.

3

Algae: the major microbial biomass in freshwater systems

In the freshwater environment, light energy conversion and related synthesis of carbon compounds is carried out by three major groups of organisms (primary producers) – higher plants (macrophytes), algae, and photosynthetic bacteria. Algae are the main microorganisms involved in this process and may be defined as simple plants (lacking roots, stems, and leaves) that have chlorophyll-a as their primary photosynthetic pigment and lack a sterile covering of cells around the reproductive cells. Although algae include prokaryote organisms (blue-green algae), the closely-related photosynthetic bacteria differ in terms of cell size, pigmentation, and physiology (strict anaerobes, not evolving oxygen) and are generally placed in a distinct category. Photosynthetic bacteria will be considered separately in Chapters 4 and 6.

In this chapter, and elsewhere in the book, the term 'algae' includes many different types of organism, and is really a term of convenience. Although commonly used, the term becomes problematic in the light of modern knowledge on the diverse origins of the group, with blue-greens fundamentally different from eukaryotic algae – and within these, heterokont algae quite distinct from other eukaryote organisms (Section 3.1.2, Table 3.3).

Algae as autotrophs and heterotrophs. In addition to requiring light (phototrophs) the majority of algae are also autotrophs – using carbon dioxide as their sole source of carbon. Algae also typically use inorganic sources of nitrogen, phosphorus, and sulphur as sole sources of these elements for biosynthesis. Although most algae are autotrophic, some do have the ability to assimilate organic compounds in certain environmental situations. Furthermore, although most algae are photoautotrophs in relation to bulk supplies of energy and inorganic nutrients, they frequently require small quantities of organic growth factors, including one or more of cobalamin, biotin, and thiamine (Swift, 1980). The development of heterotrophy is considered further in Section 3.11.

Algae as freshwater microorganisms. Freshwater algae range in size from microscopic organisms (unicellular and colonial) to macroscopic forms which are visible to the naked eye and appear plant-like. Planktonic algae are typically microscopic (micro-algae), and are part of the microbial community which is the main theme of this book. In contrast, various benthic or attached algae are macroscopic, and do not fit into the broad area of aquatic microorganisms. The green algae in

Freshwater Microbiology: Biodiversity and Dynamic Interactions of Microorganisms in the Aquatic Environment David C. Sigee
© 2005 John Wiley & Sons, Ltd ISBNs: 0-471-48529-2 (pbk) 0-471-48528-4 (hbk)

Table 3.1 Major divisions of freshwater algae: microscopical appearance, motility, and typical habitat (data from Lee, 1997, Van den Hoek et al., 1995, and Wehr and Sheath, 2003)

Algal division (class)	Index of biodiversity*	Typical colour	Typical morphology of freshwater species	Motility (vegetative cells/colonies)	Aquatic habitats	Typical examples
1. Blue-green algae *Cyanophyta*	124	Blue-green	Microscopic or visible – usually colonial	Buoyancy regulation Some can glide	Lakes and streams Planktonic or attached	*Synechocystis Microcystis*
2. Green algae *Chlorophyta*	302	Grass-green	Microscopic or visible – unicellular or filamentous colonial	Some unicells and colonies with flagella	Lakes, rivers, estuaries Planktonic or attached	*Chlamydomonas Cladophora*
3. Euglenoids *Euglenophyta*	10	Various colours	Microscopic – unicellular	Mostly with flagella	Lakes and ponds Planktonic	*Euglena Colacium*
4. Yellow-green algae: **Eustigmatophyta Raphidiophyta Tribophyta**	90	Yellow-green	Microscopic – unicellular or filamentous	Flagellate and non-flagellate forms	Planktonic, benthic and epiphytic Wide habitat range	*Chlorobotrys Vischeria*
5. Dinoflagellates **Dinophyta**	37	Red-brown	Microscopic – unicellular	All with flagella	Lakes and estuaries Planktonic	*Ceratium Peridinium*
6. Cryptomonads *Cryptophyta*	12	Various colours	Microscopic – unicellular	Mostly with flagella	Lakes Planktonic	*Rhodomonas Cryptomonas*
7. Chrysophytes† *Chrysophyta*	72	Golden brown	Microscopic – unicellular or colonial	Some with flagella	Lakes and streams Planktonic	*Mallomonas Dinobryon*
8. Diatoms **Bacillariophyta**	118	Golden brown	Microscopic – unicellular or filamentous colonies	Gliding movement on substrate	Lakes, rivers, estuaries Planktonic or attached	*Stephanodiscus Aulacoseira*
9. Red algae **Rhodophyta**	25	Red	Microscopic or visible – unicellular or colonial	Non-motile	Mainly streams, some lakes Attached	*Batrachospermum Bangia*
10. Brown algae **Phaeophyta**	4	Brown	Visible – multicellular cushions and crustose thalli	Non-motile	Lakes and streams Attached	*Pleurocladia Heribaudiella*

*Biodiversity: number of genera within the group in the USA (Wehr and Sheath, 2003)
†Including haptophyte and synurophyte algae

particular include large filamentous forms such as *Cladophora* and *Chara*, and description of these is limited to their ecological role as attached algae or periphyton.

Freshwater algae constitute a diverse group of biota and occupy a wide range of aquatic habitats. In this chapter we will consider this group of organisms in terms their taxonomic grouping, molecular characterization, variation in size and shape, activities within the freshwater environment, strategies for survival, and general biodiversity. Other aspects such as photosynthesis (Chapter 4) and competition for nutrients (Chapter 5) are considered elsewhere.

A.   TAXONOMIC AND MOLECULAR CHARACTERIZATION

3.1   Major taxonomic divisions of freshwater algae

Freshwater algae do not occur as a formal taxonomic group of organisms, but represent a loose and diverse collection of divisions with members that are united by possession of the general features outlined previously. The heterogeneity of freshwater algae is emphasized by the fact that they are split between two of the major domains of living organisms (Table 1.2) – the Bacteria (a prokaryote group that includes blue-green algae) and the Eukarya (including all eukaryote algae). Although there is little consensus among phycologists in terms of exact groupings, they are separated here into 10 principle divisions or classes. Information on the major characteristics of these divisions is summarized in Table 3.1 (microscopical appearance, motility, and habitat) and Table 3.3 (biochemical and cytological characteristics), with a description of each group in Section 3.1.3. Further information on algal classification can be obtained from a range of standard texts including Lee (1997), Van den Hoek *et al.* (1995), Wehr and Sheath (2003), and John *et al.* (2002). Molecular characterization of freshwater algae is of increasing taxonomic and diagnostic importance, and is discussed in Section 3.3.

3.1.1   Microscopical appearance, motility and ecological features (Table 3.1)

Examination of environmental samples under the light microscope reveals a wide range in algal morphology and size (Figure 3.1 and Table 3.2), with variation from unicellular to colonial forms and (in fresh samples) differences in colour, presence or absence of an outer layer of mucilage, and motility.

Motility occurs both on solid surfaces (benthic algae) and within the water column (planktonic forms) and involves both active (mucilage extrusion, cilia and flagella) and inactive (buoyancy mechanisms) processes. Planktonic diatoms are non-motile, depending on water movement to maintain their position within the water column. Within the eukaryote algae, three groups – the euglenoids, dinoflagellates, and cryptomonads are entirely unicellular and are actively motile by flagella.

Algae are present in all freshwater environments including lotic and lentic systems, plus snowfields, aerosols, and a range of extreme aquatic situations (Section 2.15). Within lotic and lentic systems, certain algal groups (euglenoids, dinoflagellates, cryptomonads, and chrysophytes) show a preference for planktonic conditions, while others (blue-green algae, green algae, and diatoms) are equally planktonic or benthic. Although the great majority of freshwater algal species have a widespread geographic distribution (cosmopolitan), there are some species of chrysophytes, green algae, red algae, and diatoms (Figure 2.8) which are restricted (endemic) to certain geographic regions or particular water bodies. Some freshwater species are also able to survive in brackish (partly saline) water, which highlights the freshwater or marine affinities of the different major groups.

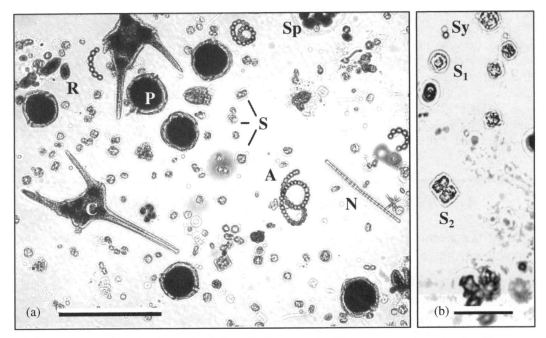

Figure 3.1 Taxonomic diversity in a mixed phytoplankton sample (a) low power view, showing blue-green algae (*Anabaena*, A; *Aphanizomenon*, N), green algae (*Sphaerocystis*, Sp), dinoflagellates (*Ceratium*, C; *Peridinium*, P), cryptomonads (*Cryptomonas*, R) and diatoms (*Stephanodiscus*, S) (scale–100 µm); (b) high power view, showing details of some of the smaller phytoplankton including *Stephanodiscus* in face (S1) and side view (S2-pair of cells), and the unicellular blue-green alga *Synechococcus* (Sy, pair of cells) (scale–25 µm). The cells have been fixed in iodine in preparation for cell/colony counts (see Table 3.2). In these fields of view, cell size (greatest axial dimension) ranges from ~180 µm (*Ceratium*) to ~2 µm (*Synechococcus*). Cell/colony shape ranges from spherical (*Peridinium*) to oval (*Cryptomonas*) to the extended forms of *Ceratium* (unicell) and *Aphanizomenon* (colony)

Table 3.2 Estimating major algal species populations in a mixed phytoplankton sample as organism counts and biovolumes (unpublished data from Andrew Dean, with permission)

Species	Count (cells or colonies ml^{-1})	% total phytoplankton count	Unit species biovolume (µm^3)	Total species biovolume × 10^5 (µm^3 ml^{-1})	% total phytoplankton biovolume
Stephanodiscus minutula	12000	78	380	46	13
Cryptomonas ovata	775	5	1050	8.1	2
Rhodomonas minuta	680	4	145	1.0	<1
Anabaena flos-aquae[*]	120	1	2165	2.6	<1
Synechococcus aeruginosa	30	<1	20	0.005	<1
Aphanizomenon flos-aquae[*]	480	3	1520	7.3	2
Ceratium hirundinella	30	<1	40,000	12	4
Peridinium cinctum	578	4	48,000	280	77

*Colonial species.
Phytoplankton population determinations are from the sample illustrated in Figure 3.1, and are presented as direct count and biovolumes of particular species:
- *unit biovolume*: volume of a single cell or colony, estimated from linear measurements;
- *total biovolume*: volume of the entire population of a particular species, determined as the product of until biovolume and cell count;
- *percentage contributions* of particular algae (e.g., *Stephanodiscus*, *Peridinium*) to the total phytoplankton count and total phytoplankton biovolume (shaded columns) differ markedly in relation to unit (cell or colony) size.

Table 3.3 Major divisions of freshwater algae: biochemical and cytological characteristics (data from Lee, 1997, Van den Hoek et al., 1995, and Wehr and Sheath 2003)

Algal division (class)	Pigmentation[*]			Starch-like reserve	External covering	Chloroplast Fine-structure		Flagella (vegetative cells and gametes)
	Chlorophylls	Carotenes	Phycobilins			Outer membranes	Thylakoid groups	
1. Blue-green algae *Cyanophyta*	**a**	β	+	Cyano-phycean starch$^\alpha$	Peptidoglycan matrices or walls	0	0	0
2. Green algae **Chlorophyta**	**a,b**	α,β,γ		True starch$^\alpha$	Cellulose walls, scales	2	2–6	0–many
3. Euglenoids *Euglenophyta*	**a,b**	β,γ		Paramylon$^\beta$	Protein pellicle	3	3	1–2 emergent
4. Yellow-green algae: **Eustigmatophyta Raphidiophyta Tribophyta**	**a,c**	α,β		Chryso-laminarin$^\beta$	Cellulose in some	4	3	2 unequal (heterokont)
5. Dinoflagellates *Dinophyta*	**a,c$_2$**	β		True starch$^\alpha$	Cellulose theca (or naked)	3	3	2 unequal (heterokont)
6. Cryptomonads *Cryptophyta*	**a,c$_2$**	α,β	+	True starch$^\alpha$	Cellulose periplast	4	2	2 equal
7. Chrysophytes[†] *Chrysophyta*	**a,c$_1$, c$_2$,c$_3$,**	α,β,ε		Chryso-laminarin$^\beta$	Pectin, plus minerals and silica	4	3	2 unequal (heterokont)
8. Diatoms[†] *Bacillariophyta*	**a,c$_1$c$_2$c$_3$**	β,ε		Chryso-laminarin$^\beta$	Opaline silica frustule	4	3	1, reproductive cells only
9. Red algae *Rhodophyta*	**a,d**	α,β	+	Floridean starch$^\alpha$	Walls with galactose polymer matrix	2	0	0
10. Brown algae[†] *Phaeophyta*	**a,c$_1$c$_2$c$_3$**	β,ε		Laminarin$^\beta$	Walls with alginate matrix	4	3	2 unequal (heterokont) reproductive cells only

[*] Major pigments are shown in bold type Starch-like reserves $^\alpha$:α – 1,4 glucan $^\beta$:β – 1,3 glucan
[†] Chrysophytes, diatoms and brown algae are sometimes grouped as classes within the phylum Heterokontophyta.

Major algal groups: biodiversity in freshwater and marine environments

Comparison of marine and freshwater systems The different divisions of freshwater algae differ in their relation to freshwater and marine environments. The degree of adaptation and evolution within these separate aquatic systems is indicated by the relative abundance and extent of species diversity that occurs for particular groups. None of the major algal divisions listed in Table 3.1 is exclusively freshwater, but certain groups exhibit greater abundance and diversity in freshwater systems – including blue-green and green algae. Other groups such as diatoms and chrysophytes are well-represented in both situations, while dinoflagellates, red algae, and brown algae show greater biodiversity in marine systems. Red algae and brown algae, in particular, have relatively few freshwater representatives.

Biodiversity in freshwater systems Biodiversity within the major algal divisions in freshwater systems is indicated by the range of habitats that are colonized and by the diversity of genetic, physiological, biochemical, and structural characteristics that occur within the group. The number of genera within divisions provides an index of phenotypic biodiversity and in the freshwater algae of North America (a well-characterized group occurring over a wide geographic area) ranges from >300 (Chlorophyta) to <5 (Phaeophyta) (Table 3.1).

3.1.2 Biochemical and cytological characteristics (Table 3.3)

Although biochemical and cytological features are important markers in distinguishing different algal groups (e.g., Descy *et al.*, 2000), they have limited use as aids to the identification of field samples since they typically require detailed laboratory (chemical and microscopical) analysis.

Biochemical features

Major biochemical features include pigmentation, storage products, external covering (cell wall) com-

position, and identity of osmotically-active low MW organic solutes (Table 4.5).

Pigmentation The pigmentation of algae is derived from three main groups of molecules – chlorophylls, carotenoids, and phycobilins (Van den Hoek *et al.*, 1995). The predominantly green colour of these organisms (from chlorophyll) is frequently modified by the presence of the other pigments. The golden-brown colour of diatoms, for example, arises due to the masking of chlorophyll by the accessory pigment fucoxanthin and blue-green algae have a bluish tinge due to phycocyanin.

Algal pigments are localized within the algal cell in association with the photosynthetic or thylakoid membranes. Chlorophylls are composed of a porphyrin ring system with a central magnesium atom, and occur as four main types – chlorophyll-a, -b, -c (c_1, c_2, and c_3) and -d. Chlorophyll-a (Figure 4.10) occurs in all photosynthetic algae as the primary photosynthetic pigment (the light receptor of photosystem I of the light reaction) and varies from 0.3–3 percent of algal dry weight. The other chlorophylls function as accessory pigments and have a limited but distinctive distribution within the different algal groups (Table 3.3).

Carotenoids are long-chain molecules that can be divided into two main groups: carotenes – oxygen-free hydrocarbons, and xanthophylls, their oxygenated derivatives. Of the four carotenes present in algae, β-carotene, occurs in all the algal groups while α-, γ- and ε-carotenes have a more restricted occurrence (Table 3.3). Xanthophylls occur as a wide range of molecules, with approximately 30 different types being recognized and forming a distinctive pattern of distribution within the different algal groups.

Phycobilins are water-soluble red or blue pigments located on (blue-green algae, red algae) or inside (cryptophytes) the photosynthetic membranes. The pigment molecule or chromophore is a tetrapyrrole and occurs in a combination with non-pigmented protein (the apoprotein) to form the phycobiliprotein. The blue chromophore is phycocyanobilin and the red chromophore phycoerythrobilin.

Storage products Algae contain a range of high and low molecular weight carbohydrate storage products. The high MW starch-like compounds are either α-1,4 linked or β-1,3 linked glucans, and are diagnostic for particular algal groups (Table 3.3). Low MW storage carbohydrates include the sugars sucrose (important reserve in green algae and euglenoids) and trehalose (blue-green algae), with various glycosides and polyols (e.g., mannitol) being important in red and brown algae respectively.

Protein reserves include cyanophycin, present in blue-green algae as an important storage product of fixed nitrogen (Figure 5.15). Lipid reserves are also widely present in algae, and appear to be particularly prominent in dinoflagellates and diatoms. Polyphosphates are also a major storage product throughout the algal groups and are important in the luxury consumption of phosphate (Section 5.8).

Cell wall composition   The outer covering of algae typically forms a continuous discrete structure (cell wall), which is variously referred to as a pellicle (euglenoids), theca (dinoflagellates), periplast (cryptomonads), and frustule (diatoms). The pellicle and periplast occur within the plasmalemma, the rest are external to it.

In general, algal cell walls are composed of two constituents – a skeletal or fibrillar component plus an amorphous matrix. The most common skeletal component is cellulose (a polymer of 1,4 linked β-D-glucose), but other macromolecules – including pectin, peptidoglycan (mucopeptide), and protein – may also be involved (Table 3.3). Amorphous mucilaginous components are an important part of cell wall structure in red and brown algae, and form a separate (mucilage) layer in many algal groups. Diatoms are unique in having a cell wall made of amorphous, hydrated silica that is associated with proteins, polysaccharides, and lipids (Fischer *et al.*, 1999). Details of cell wall formation in diatoms are given in Section 5.10.

Cytological characteristics

Cytological features are of fundamental importance in distinguishing the different algal groups.

The most fundamental division within the algal assemblage, into prokaryotes (blue-green algae) and

eukaryotes (other groups), is based on cell structure – with blue-green algae lacking a cell nucleus and the distinctive cytoplasmic organelles typical of eukaryote cells (Figure 3.2). The various fine structural and metabolic features that separate prokaryotes from eukaryotes, and distinguish blue-greens from other (eukaryotic) algae are shown in Table 1.2. The relatively 'primitive' state of blue-green algae in no way detracts from their ecological and evolutionary success within the freshwater environment.

Within the eukaryotic algae, other important cytological features include the fine-structure of the chloroplast, the number and appearance of

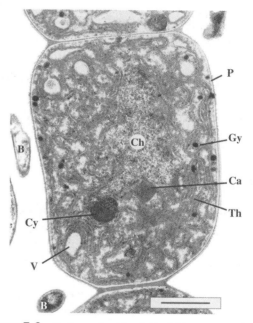

Figure 3.2   Prokaryotic identity of blue-green algae: Transmission electron micrograph of a section of *Anabaena*, showing:
Ch – central region of diffuse chromatin (no limiting membrane),
Cy – cyanophycin granule (nitrogen storage),
Ca – carboxysome (polyhedral body),
Gy – glycogen granule (cyanophycean starch),
Th – peripheral thylakoid membranes,
V – vacuole,
P – thin peptidoglycan cell wall
Epiphytic bacteria (**B**) are embedded in a layer of surface mucilage surrounding the *Anabaena* cell (not visible in this preparation) (scale – 1 μm)

flagella, and the fine structure of nuclear chromatin. Chloroplasts differ in terms of the number of outer membranes (two envelope membranes plus additional endoplasmic reticular membranes) and the grouping of thylakoids – ranging from single membranes (not grouped) to aggregates of up to six (Table 3.3). Considerable diversity occurs in relation to flagellar structure. Several groups of algae (Table 3.3) are united by the presence of unequal (heterokont) flagella, with a long forward-directed tinsel flagellum and a backward-directed smooth flagellum. The fine structure of nuclear chromatin is diagnostic for one particular group of algae, the dinoflagellates, which have distinctive permanently-condensed chromosomes (Figure 3.23). The reader is referred to standard texts on phycology (e.g., Van den Hoek *et al.*, 1995) for a comprehensive review of algal cytology.

3.1.3   *General summary of the different groups*

General features of the different algal groups, with links to their reference in subsequent chapters in the book, are given below.

Blue-green algae (Cyanophyta)

Also referred to as cyanobacteria, this is the only prokaryote group of algae. *Cyanophyta* are important constituents of periphyton and phytoplankton communities, where they are present both as unicellular (picoplankton) and colonial forms. Blue-green algae are thought to have evolved during the early Precambrian era, when they were exposed to a reducing atmosphere and high levels of irradiation (particularly UV). During their long existence, they have colonized nearly all freshwater, marine, and terrestrial habitats, including such extreme environments as hotsprings (up to 70°C), hypersaline lakes, high arctic and alpine lakes, and hot and cold deserts. The general success of blue-green algae in aquatic environments has been attributed to the following

- Efficient light harvesting mechanisms, with ability to adjust to spectral differences by variations in accessory pigments.

- Continued photosynthesis at low concentrations of CO_2 and high pH. These conditions are particularly apparent during bloom formation in eutrophic waters, which eukaryotic algae are not able to tolerate.

- Resistance to damaging radiation, by producing a range of compounds that act as photoprotectants by absorbing short-wavelengths (Section 4.9.3).

- Temperature adaptations – different species can grow in extreme hot or cold environments. In temperate and tropical lakes, blue-green algae are able to maintain growth rates at high summer temperatures where other algae show temperature inhibition.

- Buoyancy mechanism for positioning in the water column. This energetically-efficient process allows blue-green algae to carry out diurnal migration within the water column between surface waters (high light and predation) and lower waters (high phosphate and nitrate concentrations, avoidance of photoinhibition).

- Efficiency in nutrient uptake. Blue-green algae show a whole range of features related to nutrient uptake:

  - they are able to dominate both oligotrophic (as picopoplankton) and eutrophic (large colonial greens) environments; in eutrophic conditions, dense algal blooms out-compete eutrophic algae;

  - some blue-greens can fix gaseous nitrogen, allowing them to grow at low N/P ratios;

  - blue-greens produce siderophores under conditions of iron stress, allowing them to scavenge Fe(III) at limiting environmental levels;

  - blue-greens do not require exogenous sources for vitamin requirements.

- They have chemical (toxins) and physical (large colonies) mechanisms to resist removal by filter-feeding zooplankton.

- They have evolved specialized symbiotic bacterial associations, which are particularly important in relation to heterocyst functions.

Some of these characteristics have specific importance in the ability of blue-green algae to form dominant blooms in eutrophic waters, and are discussed further in Section 10.6.

Green algae (Chlorophyta)

These comprise the most diverse class of algae. They are important as both planktonic and attached organisms, with morphologies ranging from simple unicells to complex colonial forms. Some macroscopic members of the green algae (*Chara, Cladophora*) have a higher-plant like appearance and are important members of the periphyton. Certain groups within the green algae have specific ecological requirements, including flagellated chlorophytes (nutrient-rich standing waters) and coccoid unicells and colonies (high light, nutrient, and temperature, standing waters). Desmids are more common in ponds and ditches that have low conductance and low to moderate nutrient levels.

Euglenoids (Euglenophyta)

In terms of general abundance and species diversity, this is a relatively minor group of freshwater algae. These organisms may become particularly abundant, however, in the phytoplankton of standing waters rich in nutrients where they may be readily identified under the microscope by their unicellular spindle-shaped morphology and active motility.

Yellow-green algae (Eustigmatophyta, Raphidiophyta and Tribophyta)

Yellow-green algae are a diverse group of freshwater organisms, occurring in a wide range of habitats, and with a large number of reported genera (at least 90 in North America). Although widespread, these algae are not very prominent members of the freshwater flora, since many are small coccoid forms that occur only in small numbers.

Dinoflagellates (Dinophyta)

Dinoflagellates are often relatively minor components of of lake phytoplankton in relation to species counts. The large size of these organisms, however, means that their biovolume contribution to phytoplankton is much more significant – and they often make a major contribution to overall algal biomass (Figure 3.1).

Along with colonial blue-greens, these algae are prime examples of *K*-selected organisms (Section 3.10.2), and tend to form blooms in temperate lakes towards the end of summer when stratification is stable and epilimnion nutrients are in decline. Under such conditions, these highly motile organisms are able to migrate into the nutrient-rich hypolimnion to obtain their supplies of nitrate and phosphate (Figure 3.22).

Cryptomonads (Cryptophyta)

These unicellular organisms are particularly diverse in temperate regions, where they typically occur as phytoplankton in lakes and ponds. The short cell cycle and ability for active growth of cryptomonads means that they are particularly common during the clear-water phase of the annual cycle in temperate lakes, along with other *r*-selected organisms.

Chrysophytes (Chrysophyta)

The majority of chrysophytes are unicellular or simple colonial forms, and are typically planktonic although attached forms do exist. They are typically associated with standing waters which have low to moderate nutrients, alkalinity, and conductance (pH slightly acid to neutral).

Diatoms (Bacillariophyta)

Diatoms initially appeared in the fossil record about 185 million years ago, and have been abundant in surface waters for the past 115–110 million years. The major biomass of diatoms occurs in marine systems, where they are the most important microbial primary producers and the major contributors to global carbon fixation.

These micro-algae are also of major importance in freshwaters, where they occur as both planktonic and attached (biofilm) organisms in lakes, streams, and estuaries. Diatoms may be unicellular or colonial, and constitute one of the largest classes of freshwater micro-algae (Fischer *et al.*, 1999). The cell wall (frustule) is unique among living organisms in being almost entirely composed of silica (Section 5.10), and these organisms are well known for the intriguing species-specific design and ornamentation of this rigid and very dense structure.

Diatoms are divided into two main groups – centric diatoms (radial symmetry, typically planktonic) and pennate (bilateral symmetry, many benthic species), with further taxonomic subdivisions in relation to frustule morphology (Lee, 1997). The taxonomic composition of diatom communities provides a useful indicator of environmental characteristics, and has been widely used to monitor changes in the salinity (Section 2.15.3), nutrient status (Section 10.3), acidity (Section 3.7.3), and general hydrological disturbance of lakes.

Red algae (Rhodophyta)

Red algae are predominantly marine in distribution, with only 3 per cent of over 5000 species worldwide occurring in true freshwater habitats (Wehr and Sheath, 2003). Although freshwater red algae (such as the large filamentous alga *Batrachospermum*) are largely found in streams and rivers, these organisms may also occur as marine invaders of lakes and brackish environments.

Certain freshwater red algae in the littoral zones of the Great lakes Basin (USA), for example, appear to be originally marine and to have lost the capacity for sexual reproduction. These include the filamentous red alga *Bangia atropurpurea* (Lin and Blum, 1977), which reproduces only by asexual monospores – in contrast to marine species which undergo alternation of generations and carry out sexual reproduction. Attached red algae (e.g., *Chroodactylon ramosum*) also contribute to the epiphytic flora of lake periphyton.

Brown algae (Phaeophyta)

As with red algae, brown algae are almost entirely marine – with less than 1 per cent of species present in freshwater habitats (Wehr and Sheath, 2003). These species are entirely benthic, either in lakes or rivers, and have a very scattered distribution.

Freshwater brown algae include genera such as *Pleurocladia* and *Heribaudiella*, and are the least diverse of all freshwater algae. Their morphologies are based on a relatively simple filamentous structure, and they lack the complex macro-morphology typical of the brown seaweeds.

3.2 Algal species: taxonomy and intraspecific variation

Although algal species are typically well-defined units, based principally on morphological criteria, there is often a wide variation within these taxa in terms of morphology, size and shape, molecular genetics, biochemistry, and microscopical analysis. This intra-speciic variation is important for two main reasons.

- *Algal taxonomy.* Intraspecific variation has relevance to the definition of individual species, and is also used in some cases to define subspecific taxonomic units (subspecies or strains).

- *Biodiversity.* Although biodiversity is normally considered in relation to the variety of species (see Section 3.13), variation within species is also important as a major source of diversity within the environment.

3.2.1 Taxonomy of algal species

Definition of algal species on morphological criteria may present problems where the criteria are not well-defined or where there is intra-specific variation. Morphological polymorphism and geographical variation have been noted for species of the blue-green algae *Microcystis* (Komarek, 1991) and *Anabaena* (Baker, 1991, 1992). Individual species often appear to be a continuous transitional sequence or 'cline' of strains, with a gradation of taxonomic characteristics (Baker, 1991). Variation in environmental factors may also create a wide variation in cell and filament morphology (Gorham *et al.*, 1964; Doers and Parker, 1988; Zirbel *et al.*, 2000).

3.2.2   Chemical diversity within species – enzyme analysis, molecular groups, and elemental composition

In addition to key biochemical differences between the major algal groups (Table 3.3), variations in chemical composition also occur at genus to sub-species level. Chemical and biochemical analysis of intra-specific variation can be carried out in a variety of ways, providing information on such diverse aspects as enzyme composition, lipid composition (Hayakawa *et al.*, 2002), isotope ratios, molecular groups, and elemental composition. All of these approaches combine to emphasize chemical diversity at the species and intra-specific level.

Enzyme analysis

In the blue-green algae, electrophoretic analysis of polymorphic enzymes (allozymes) has revealed distinct patterns of occurrence within *Anabaena* (Stulp and Stam, 1984) and *Microcystis* (Kato *et al.*, 1991). Allozyme analysis of blue-green algae is relatively simple since individual organisms typically have a single allozyme for each polymorphic enzyme, compared to a pair of allozymes (cytosol and plastid) in higher (eukaryote) algae. The studies carried out on *Microcystis* (Kato *et al.*, 1991) involved analysis of four polymorphic enzymes and demonstrated clear and consistent allozyme combinations for two morphologically-defined species, but marked heterogeneity in a third. The latter showed well-defined differences between strains, indicating clear genotypic variation at subspecies level.

Stable carbon and nitrogen isotope ratios

Isotope ratios within cells provide a very sensitive index of biochemical function, depending on both extrinsic (environmental) and intrinsic (metabolic) characteristics (Yoshii *et al.*, 1999).

Stable carbon ($\delta^{13}C$) and nitrogen ($\delta^{15}N$) ratios of phytoplankton vary spatially and temporally according to environmental conditions. The ($\delta^{13}C$) value of microalgae is determined largely by the organism's metabolism and growth rate, with enhancement of the phytoplankton ($\delta^{13}C$) level in temperate and tropical climates when growth is high. The $\delta^{15}N$ of phytoplankton is closely correlated with the form of nitrogen used and with algal growth rate. Within the ecosystem as a whole, $\delta^{13}C$ and $\delta^{15}N$ values differ markedly between biota, and show a close correlation with trophic status.

Molecular groups and elemental composition

Information on intraspecific biodiversity in relation to the chemistry of individual cells can only be obtained by microscopical techniques. These include the use of chemiluminescence to determine variation in nitrate concentrations (Villareal and Lipschultz, 1995), analysis of molecular species using Fourier-transform infra-red (FTIR) analysis (Sigee *et al.*, 2002) and the determination of elemental composition using proton-probe analysis (Brook *et al.*, 1988) and X-ray microanalysis (Sigee *et al.*, 1998). All of these approaches indicate wide variability in the chemical composition of cells within particular species in environmental samples.

Infrared analysis   FTIR analysis of individual algal cells gives information on the vibrational states of molecular groups present within major macromolecules (Sigee *et al.*, 2002). Analysis of single colonies of the green algal *Pediastrum*, for example, gave very clear spectra with 12 distinct bands corresponding to a range of molecular groups (Figure 3.3). Significant differences between spectra occurred within a single environmental

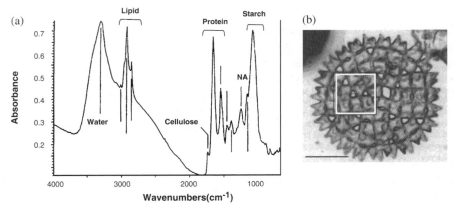

Figure 3.3   Molecular composition of algal cells – FTIR spectrum of the colonial green alga *Pediastrum* (figure taken from Sigee *et al.*, 2002 and reproduced with permission from the European Journal of Phycology). (a) Fourier Transform Infra-red (FTIR) spectrum from colony (b). Some of the major bands, with vibrational modes, are: **Water**: v(O—H) stretching **Lipid**: (v_{as}(CH$_2$) stretching of methylene **Protein**: amide I & II **Nucleic acid** (**NA**) v_{as}(>P=O) stretching of phosphodiesters **Starch**: v(C—O) stretching (b) Single colony of *Pediastrum* – the spectrum was taken from the area marked by a white square (scale = 25 μm)

micro-population of this organism, demonstrating intra-specific heterogeneity at the cellular level.

X-ray microanalysis   XRMA provides information on the elemental composition of biological micro-samples (Sigee *et al.*, 1993), such as single cells and colonies. A detailed account of the application of this technique to cells of the dinoflagellate *Ceratium* (Sigee *et al.*, 1999a) is given in Section 5.1.4 (see Figure 5.2). Significant differences in the elemental composition of *Ceratium* cells were noted both within and between micro-populations (in mixed phytoplankton samples) isolated from the lake environment.

Similar results have been obtained with other freshwater algae, including the demonstration of distinct cell sub-populations of *Anabaena* and *Microcystis* (Sigee and Levado, 2000) in relation to silicon content.

3.3   Molecular analysis

Extraction and sequencing of DNA has been carried out on laboratory and environmental samples of a wide range of freshwater algae. In the case of prokaryote (blue-green algae) organisms this has

involved analysis of nucleoid DNA (Kaneko *et al.*, 1996), while in eukaryote algae gene sequences present in nuclear, mitochondrial (Chesnick *et al.*, 1996; Laflamme and Lee, 2003), and chloroplast (Reith, 1999) DNA have been examined.

DNA studies on freshwater (and marine) algae have provided novel approaches to:

• understanding phylogenetic relationships,

• molecular characterization and identification of major taxonomic groups, species and subspecies in environmental samples,

• identification and characterization of gene activity relevant to environmental and biotic interactions.

This volume is concerned particularly with the last two of these aspects, dealing respectively with the occurrence and activities of algae within the aquatic environment.

3.3.1   *Molecular characterization and identification of algae*

Reliable characterization of taxonomic groups is clearly important in environmental studies, since the definition of particular units needs to be clear and to be established on a firm biological basis.

The use of molecular techniques has been particularly useful for blue-green algae, giving clearer resolution of species and sub-species (strains). The delineation of species within this group is currently based largely on morphological features such as the arrangement (e.g., filamentous, colonial, or single cells), size, and shape of cells. In many cases, however, closely related species have overlapping morphology and individual species show geographic or environmental variation. As a result, characterization and identifying species of blue-green algae in environmental samples can be confusing and difficult.

Molecular characterization of species or subspecies makes use of a particular DNA sequence to define the taxonomic unit. Use of this approach with a specific algal sample involves amplification of the particular sequence to produce enough DNA for analysis, followed by the analysis itself – using restriction fragment length polymorphism (RFLP) or direct sequencing. The choice of the DNA sequence to be amplified, and hence the appropriate PCR primer, is important for two reasons – prevention of amplification of contaminant DNA, and level of taxonomic resolution.

Avoidance of contamination

One of the potential problems with the PCR process is that non-target DNA may also be amplified in addition to target DNA. Amplification of bacterial DNA in blue-green algal cultures may be avoided by the use of bacteria-free (axenic) samples, or by designing blue-green algal-specific PCR primers that can discriminate between the two types of DNA. The latter may be achieved either by the use of well-characterized genes common to both organisms (e.g., 16s ribosomal RNA genes (rDNA)) where the gene sequence is different in target and contaminant situations, or by using gene sequences in the target organism that are not present in contaminants (e.g., genes that code for phycobilin pigments in blue-green algae).

Taxonomic resolution

Blue-green algae contain a number of well-characterized genes which are not present in bacteria and have potential for taxonomic differentiation. These include the genes encoding the main light-harvesting accessory proteins phycocyanin (*cpc*), allophycocyanin (*apc*) and phycoerythrin (*cpe*). The problem with using functional gene sequences for taxonomic purposes is that they are typically highly conserved and have sufficient sequence divergence for comparisons only at genus level and above. More variable gene sequences for use at species and sub-species level can be obtained by amplifying the non-coding DNA region between genes, referred to as the intergenic spacer (IGS) or internal transcribed spacer (ITS) sequences.

Detection of particular gene sequences in environmental samples can be carried out in two main ways – direct sequence analysis of the sample, or use of molecular probes (Table 3.4).

Direct sequence analysis Sequence analysis has been used to provide general information on

Table 3.4 Molecular identification of algal species in aquatic environmental samples

Environmental sample	Technique	Reference
Picoplankton in Lake Baikal	Direct sequencing	Semenova and Kuznedelov (1998)
Colonial blue-greens: *Anabaena*, *Microcystis*, *Nodularia*	PCR-RFLP analysis of the *cpcB-A* intergenic spacer and flanking regions	Bolch *et al.* (1996)
Flagellate nanoplankton	Small subunit rRNA probes for *Paraphysomonas* (chrysophyte)	Caron *et al.* (1999)
Mixed diatom populations and laboratory cultures	Large subunit rRNA probe for *Pseudo-Nitzschia* (diatom)	Scholin *et al.* (1997)
Estuarine river samples	Use of internal transcribed spacer-specific PCR assays for *Pfiesteria* (dinoflagellate)	Litaker *et al.* (2003)

species composition of poorly-characterized environmental samples and also to reveal variations on genotype composition within species.

Analysis of poorly-characterized samples. Molecular analysis has been particularly useful in the case of picoplankton, where traditional methods of identification are difficult due to the minute dimensions of these organisms. The use of culture techniques as part of the identification process also present problems since many of these organisms have stringent nutritional requirements.

This approach was used to study the species composition of picoplankton in Lake Baikal (Russia), where a mixture of autotrophic and heterotrophic organisms become prominent in summer (Semenova and Kuznedelov, 1998). These authors carried out isolation, PCR amplification, sequencing, and phylogenetic analysis of the picoplankton DNA. Comparative analysis of rRNA sequences coding for the $5'$-terminal region of 16s rRNA genes led to the construction of a phylogenetic tree from the combined data set. This indicated the presence of seven autotrophic (algal) species, 15 heterotrophic species of picoplankton plus some unidentied organisms. Named autotrophic species included five blue-green algae and two green algae, while the heterotrophs comprised Actinomycetes and a range of bacteria.

Genotypic variations within species. Molecular techniques are increasingly being used to study genotypic variations within species. Studies by Bolch *et al.* (1996), for example, used the intergenic spacer between genes encoding the β- and α-phycocyanin subunits to study genetic variation within 19 isolates of blue-green algae belonging to three morphologically-defined species (morphospecies). Restriction enzyme digestion of the amplification products revealed clear fragment patterns that placed the strains within three genetically-defined species – *Anabaena circinalis*, *Microcystis aeruginosa*, and *Nodularia spumigena*. Within each species, differences in the fragment pattern (restriction fragment length polymorphism, RFLP) revealed further subdivision (polymorphism) into genetic

strains. *M. aeruginosa* was particularly polymorphic, while the other two species were less so. Denaturing gradient gel electrophoresis of sections of the internal transcribed spacer between rRNA genes (Janse *et al.*, 2003) has also proved useful in detecting intrageneric and intraspecific variation in *Microcystis*.

Recent studies by Kurmayer and Kutzenberger (2003) have used real-time PCR to investigate the frequency of microcystin genes in natural populations of *Microcystis* sp. sampled from Lake Wannsee (Germany). Two DNA sequences were monitored – the *mcyB* gene (as an index of microcystin production) and the intergenic region within the phycocyanin (PC) operon (to quantify total population). Phytoplankton microsamples from the lake showed that *mcy* genotypes made up only a small proportion (1–38 per cent) of the total *Microcystis* population. The mean proportion of *mcy* genotypes was relatively stable over time, indicating that the straight *Microcystis* cell count could be directly used to infer the number of microcystin-producing cells.

Identification of algae – molecular probes

The development of species-specific oligonucleotide probes from DNA sequence data, followed by *in situ* hybridization, has considerable potential for the identification and counting of algae in environmental samples (Table 3.4).

As with direct sequencing, this technique has particular advantages with small unicellular algae, where there are often relatively few morphological features available for identification. Caron *et al.* (1999) sequenced the small-subunit ribosomal genes of four species of the colourless chrysophyte genus *Paraphysomonas*, leading to the development of oligonucleotide probes for *P. imperforata* and *P. bandaiensis*.

Molecular probes have major potential for the detection of nuisance algae, particularly those that produce toxins (Pomati *et al.*, 2000). They have been used, for example, to distinguish toxic from non-toxic diatom species (Scholin *et al.*, 1997), where differentiation would otherwise require the time-consuming application of scanning and transmission electron microscopy. They have also been used for the rapid identification of *Pfiesteria piscicida*, a potentially toxic dinoflagellate that has been the cause of extensive fish mortalities in coastal rivers of the eastern USA (Section 10.1.4). Litaker *et al.* (2003) used unique sequences in the internal transcribed spacer (ITS) regions ITS1 and ITS2 to develop PCR assays capable of detecting *Pfiesteria* in natural river assemblages. These have been successfully used to detect the potentially harmful organism in the St. Johns River system, Florida (USA).

3.3.2 Investigation of gene function in freshwater algae

Molecular techniques have been used to sequence and characterize a wide selection of genes in freshwater algae. The genome of *Synechocystis* (blue-green alga), for example, has been completely sequenced (Kaneko *et al.*, 1996) and analysis of open reading frames (ORFs) has lead to the identification of over 3000 putative genes. These include genes for biosynthesis (amino acids, cofactors), cell processes (e.g., chemotaxis, cell division), metabolism (energy processes, fatty acids), nucleic acid activities (synthesis, transcription, translation), and energy transduction (photosynthesis and respiration). In the eukaryote algae, *chlamydomonas* has proved a model organism for the molecular genetic dissection of photosynthesis (Dent *et al.*, 2001).

Molecular responses to environmental change

Many examples of gene expression in algae involve a response to environmental factors (Table 3.5), and molecular biology has particular potential in exploring the ways in which environmental change mediates these effects. These studies have been particularly intensive in blue-green algae, where the comparatively small size of the prokaryote genome makes the genetic system of these organisms more amenable to study compared to eukaryotes. Molecular responses to environmental change are considered in relation to light modulation of algal activities, algal responses to limiting nutrients, environmental control of differentiation, and responses to stress.

Light modulation of genetic activity Light activation of algal genes involves the response of specific photoreceptors and a signal transduction process leading to either transcriptional control of gene activity or post-transcriptional control of the gene products (Reith, 1999; Johnson and Golden, 1999).

Light responsive genes are particularly important for optimizing photosynthesis at low light levels, and encode proteins involved in pigment synthesis, chlorophyll-a binding proteins, and phycobiliproteins. Induction of these genes by different wavelengths of light suggests a complex modulation of gene activity within the constantly changing light conditions of the aquatic environment. This is considered further in Section 4.6.2 in relation to photoadaptation.

Light activation of gene activity is also important in other biological activities (Table 3.5) such as induction of DNA repair processes (Section 4.9.1), general modulation of the diurnal clock (Section 4.10.3) and aspects of nitrogen metabolism, including nitrogen fixation (Section 5.6.4) and assimilation (Section 5.5.3). Other light-mediated activities include the synthesis of microcystin-related proteins and the synthesis of gas vesicle proteins, and are discussed below.

Light modulated synthesis of microcystin and microcystin-related protein. Microcystin is an important secondary metabolite that is released by the blue-green alga *Microcystis*, and is a major toxin associated with bloom formation (Section 10.7.3). Synthesis of microcystin is carried out by a group of proteins, the microcystin synthetase complex, which is encoded by a cluster of genes including *mcy*A, *mcy*B, and *mcy*D (see Figure 3.4 and Dittmann *et al.*, 2001). The microcystin gene cluster

Table 3.5 Some examples of environmentally-controlled genetic activity in algae, summarized in relation to four inter-related aspects – the response to light, uptake of nutrients, differentiation, and stress responses

Gene activity	Algae	Induction
Light-responsive genes	Blue-green algae	Light
Synthesis of photosynthetic proteins and pigments[a]		
DNA repair – photoreactivation of dimers[b]	*Chlamydomonas* (green alga)	UV-A and blue light
Synthesis of microcystin-related proteins[c]	*Microcystis* (blue-green)	Blue light
Gas vesicle formation[d]	*Pseudanabaena* (blue-green)	Low light intensities
Modulation of diurnal clock[e]	Blue-green algae	Light
Nutrient-related gene activity		
Nitrogen fixation: synthesis of nitrogenase and cyanophycin synthetase[f]	Blue-green algae	Low nitrogen levels
Induction of nitrate reductase[g]	General	Light, nitrate and carbohydrate concentration
Silicic acid transporter (SIT) proteins[h]	Diatoms	Silicic acid concentrations Factors controlling the cell cycle
Cell wall protein formation (frustulins)[i]	Diatoms	Factors controlling the cell cycle
Synthesis of flavodoxin[j]	*Synechococcus* and other blue-green algae	Iron depletion
Gene control of differentiation		
Heterocyst differentiation[k]	*Calothrix* (blue-green)	Red and green light
Differentiation of hormogonia[l]	*Nostoc* (blue-green alga)	Light intensity, culture medium
Conversion of vegetative cells to gametes[m,n]	*Chlamydomonas, Closterium* (green algae)	Two-stage process promoted by nitrogen starvation, followed by blue light signal
Response to environmental stress		
Switch to alternative transcription (σ) factors[o]	*Synechocystis* (blue-green alga)	Salt stress, heat shock, acute nutrient starvation

References: [a]Golden (1995); [b]Petersen and Small (2001); [c]Dittmann *et al.* (2001); [d]Damerval *et al.* (1991); [e]Johnson and Golden (1999); [f]Schneegurt *et al.* (2000); [g]Heldt (1997); [h]Martin-Jézéquel *et al.* (2000); [i]Fischer *et al.* (1999); [j]Porta *et al.* (2003); [k]Campbell *et al.* (1993); [l]Waaland *et al.* (1971); [m]Musgrave (1993); [n]Fukumoto *et al.* (2003); [o]Huckauf *et al.*, 2000

also contains a gene that encodes an ABC transporter protein, suggesting that the microcystin is actively secreted by the algal cell. Cells that are active in microcystin synthesis also produce two microcystin-related proteins, MrpA and MrpB, encoded by genes *mrp*A and *mrp*B. Synthesis of these proteins:

- is closely coupled to microcystin synthesis – Mcy mutants fail to synthesize both microcystin and the microcystin-related proteins;

- is modulated by light – synthesis of MrpA is strongly promoted by blue light, and is inhibited by high levels of white light;

- is probably promoted by close proximity of other *Microcystis* cells (quorum sensing) – evidence for this comes from the close homology of MrpA with RhiA protein of *Rhizobium leguminosarum*, which is subject to quorum-sensing regulation, and has been shown to control the expression of a large number of genes – including peptide

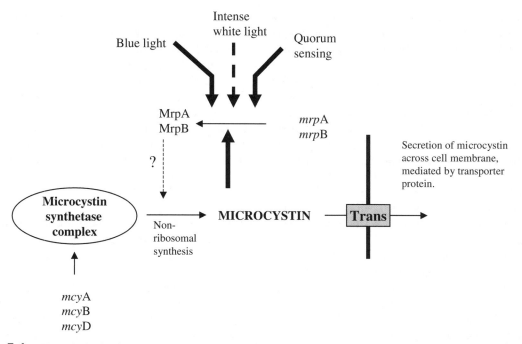

Figure 3.4. Hypothetical scheme to show possible feedback interactions between the synthesis of microcystin and microcystin-related proteins (Mrps): expression of Mrps is promoted (⟶) or inhibited (– – ⟶) by external and internal factors (including microcystin synthesis) and may in turn modulate the synthesis of microcystin (based on Dittmann *et al.*, 2001)

synthetase genes, exopolysaccharide genes and virulence factors.

The different responses of *mrp*AB transcript accumulation by mutant (no response) and wildtype (response) cells show that *mrp*AB transcription is microcystin-dependent, and that the response of the *mrp* genes to blue light is also microcystin dependent. Although the role of Mrp protein has yet to be clarified, the studies of Dittmann *et al.* (2001) suggest that it is part of a quorum sensing, light dependent system which controls the production and possibly the export of microcystin.

Induction of gas vesicle formation in blue-green algae. Formation of gas vesicles in these organisms is important in their migration and depth regulation within the water column (Section 3.8), and may be light-regulated. The induction of gas vesicle protein

(GvpA) in the blue-green algae *Pseudanabaena* (Damerval *et al.*, 1991) is particularly interesting since it is promoted by low but not high light intensity. Algae cultured at low light (5 µmol photons m^{-2} s^{-1}) produced 0.4kb transcripts of the gvpA gene, which were lost within 12 hours after transfer of the cells to high light (50 µmol m^{-2} s^{-1}) conditions. Transcription was resumed on transfer back to a low-light regime. The potential role of light-induced GvpA production in the regulation of buoyancy is summarized in Figure 3.20.

Algal responses to limiting nutrients The induction of genes which promote inorganic nutrient uptake is a key aspect in the ability of algae to respond to fluctuations in nutrient availability. This is seen, for example, in the induction of nitrate reductase by threshold levels of nitrate (Section 5.5.2) and by the Si-induction of silicic acid transporter (*SIT*) genes (Martin-Jézéquel *et al.*, 2000;

Section 5.10.2). Inhibition of nitrogenase synthesis by threshold nitrate levels (Section 5.6) represents the reverse situation of nutrient suppression of gene activity.

Exposure to conditions in which nutrients are limiting may also trigger a molecular response. Limitation in Fe availability, for example, leads to the replacement (in both eukaryote and blue-green algae) of the electron transfer catalyst ferredoxin by the iron-free but functionally-equivalent protein flavodoxin (Porta *et al.*, 2003). Environmental populations of planktonic micro-algae exhibit flavodoxin accumulation as a biochemical marker for conditions of iron depletion. In blue-green algae, flavodoxin is encoded by the gene *isiB*, which combines with *isiA* to form an operon under the control of the Fe-dependent repressor *Fur*. Decrease in external Fe concentration to sub-critical levels inactivates *Fur* and leads to expression of the *isiAB* operon. This operon is widely spread throughout blue-green algae, and in most cases is only responsive to the environmental parameter of Fe deficiency. The occurrence of flavodoxin accumulation and *isiAB* transcription as diagnostic markers for iron deficiency has led to the construction of a blue-green algal (*Synechococcus* sp. PCC 7942) reporter strain for freshwater environments (Porta *et al.*, 2003). This has *luxAB* reporter genes fused to the *isiAB* promoter, and provides a means of assessing low Fe availability, as perceived by the test organism.

Environmental control of differentiation

Environmental factors such as the concentration of inorganic and organic nutrients may act on their own or in combination with light to control differentiation processes. These include the differentiation of heterocysts and hormogonia in blue-green algae, and the conversion of vegetative to gamete cells in *Chlamydomonas* (Musgrave, 1993) The transition to gamete cells in this organism can be mediated by transference of a culture to a low-nitrate medium, and is an interesting example of differentiation within a haploid organism. Sexual differentiation in the green alga *Closterium*, is also controlled by the environmental nutrient and light regime (Fukumoto *et al.*, 2003). In this organism, transcription of a sexual cell division-inducing

pheromone occurs in mating type minus (mt-) cells only, and is suppressed by continuous dark and supplementation of a nitrogen source.

Control of transcription in prokaryotes is typically modulated by transcription proteins or sigma (σ) factors (Sections 3.4, 1.4.1, 4.10, 6.6.4). Sigma factors have also been identified in some eukaryote algae. In the red alga *Cyanidium*, for example, nuclear-encoded σ-factors have been shown to transcribe subsets of plastid genes for photosynthesis during chloroplast development (Liu *et al.*, 1995).

Responses to stress   Stress factors include all those environmental changes which have an adverse effect on biological function, and have been considered previously in relation to ecosystems (Section 1.6). They also operate at the cellular and molecular level, and freshwater algae are able to adapt to sudden but moderate changes such as salt stress, heat shock, acute nutrient starvation, and high light levels (Section 4.9) by a range of adaptive molecular processes.

Processes of microorganism acclimation to environmental stress are mainly regulated at the level of transcriptional activation or repression, where the change is acting on single genes. An example of this is the light induction of the DNA repair enzyme photolyase as a response to damaging light levels (Figure 4.23).

In addition to stress responses involving single genes, groups of genes may also be regulated in stressed cells by the activity of alternative sigma factors which replace the primary sigma factors that normally operate under favourable conditions (Huckauf *et al.*, 2000). In blue-green algae, these alternative sigma factors have been identified mainly in group 3 of the σ^{70} family. The role of alternative sigma factors in stress responses has been studied in the single-celled blue-green alga *Synechocystis* sp. by examining the effect of mutagenesis on specific sigma genes (Huckauf *et al.*, 2000). These studies identified three gene products (group 3 σ factors) as important in stress responses – regulatory protein RsbU (important in regenerating growth after nitrogen- and sulphur-starvation), SigF protein (required for the induction of salt stress proteins), and SigH (induction of heat-shock proteins).

B. SIZE, SHAPE, AND SURFACE MUCILAGE

3.4 Phytoplankton size and shape

Freshwater phytoplankton, composed of photo-synthetic bacteria and algae, shows considerable variation in the size and shape of individual organisms (cells or colonies), as illustrated in Figure 3.1. Phytoplankton dimensions are important in relation to enumeration and assessment of biovolume, designation of size category (picoplankton to macroplankton), and biological activity.

3.4.1 Cell counts and biovolume

Enumeration of species within mixed phytoplankton samples is normally carried out after fixing the sample in iodine, allowing the cells and colonies to settle in a tall container, then transferring a concentrated aliquot of suspension to a counting chamber (e.g., Sedgwick rafter slide). Counts of individual species are usually made using a light microscope with inverted objectives, ensuring that adequate numbers of organisms are counted to achieve statistical validity.

Although counts of individual species provide useful information of the total numbers of organisms present, and can be used to study aspects such as population change, fungal infection rate, and biodiversity – they give little indication of contribution to the total phytoplankton biomass. In the phytoplankton sample shown in Figure 3.1, for example, the small diatom *Stephanodiscus minutula* was the most common alga present. On a direct count basis (Table 3.2), this organism accounted for 78 per cent of the total phytoplankton population, compared with only 4 per cent for the dinoflagellate *Peridinium*. If the volume (biovolume) of individual cells/colonies is taken into account, however, *Stephanodiscus* now records only 13 per cent of the total phytoplankton population, while *Peridinium* is 77 per cent. These biovolume determinations give a much better estimate of relative biomass and provide a baseline in terms of contribution to overall productivity.

3.4.2 From picoplankton to macroplankton

Freshwater phytoplankton can be divided into four main groups (picoplankton to macroplankton, Table 3.6) on the basis of size. The range of linear size seen in phototrophic planktonic organisms is from photosynthetic bacteria (0.2μm cell diameter) to large colonial algae such as *Microcystis* (2000μm colony diameter), and represents variation over four orders of magnitude. The smallest planktonic blue-green alga, *Prochlorococcus marinus*, is found in marine rather than freshwater systems and has a diameter ranging from 0.5 to 0.7μm, reaching concentrations of up to 4×10^5 cells ml^{-1} (Palinska *et al.*, 2000).

Table 3.6 Size range of phytoplankton

Category	Linear size (cell or colony diameter)	Biovolume[*] (μm^3)	Unicellular organisms	Colonial organisms
Picoplankton	0.2 to 2 μm	4.2×10^{-3} to 4.2	Photosynthetic bacteria Blue-green algae (*Synechococcus, Synechocystis*)	–
Nanoplankton	2 to 20 μm	4.2 to 4.2×10^3	Blue-green algae Cryptophytes (*Cryptomonas Rhodomonas*)	–
Microplankton	20–200 μm	4.2×10^3 to 4.2×10^6	Dinoflagellates (*Ceratium, Peridinium*)	Diatoms (*Asterionella*)
Macroplankton	>200 μm	$> 4.2 \times 10^6$	–	Blue-green algae (*Anabaena, Microcystis*)

[*]Biovolume values are based on a sphere (volume $= {}^4/_3 \prod r^3$)

Although freshwater planktonic algae might be regarded simply as a relatively uniform group of microscopic organisms, the linear size range of algae encountered in the lake environment is equivalent to the range of plant sizes seen in terrestrial environments such as a tropical rain forest. If biovolumes are considered, the size range from picoplankton to macroplankton extends over nine orders of magnitude (Table 3.6), and *Microcystis* is greater than the bacterial cell by a factor of 10^{12}.

The smallest organisms (picoplankton and nanoplankton – see Figure 3.5) have often been difficult in the past to detect and enumerate by conventional microscopy, and their role in the freshwater environment has been underestimated. The more recent use of fluorescent microscopy and the development of species-specific oligonucleotide probes (Caron *et al.*, 1999) has improved this dramatically. A variety of micro-algae are now known to be widespread in the freshwater environment and in some cases to make a major contribution to the overall phytoplankton biomass. Studies by Happey-Wood (1988), for example, have shown that micro-chlorophytes account for more than 90 per cent of the algal cell count in some oligotrophic lakes in North Wales (UK), where they comprise more than 75 per cent of the algal biomass expressed as cell volumes. In other lakes such as Lake Baikal (Russia), the nanoplankton community is dominated by unicellular blue-green algae (Nagata *et al.*, 1994) such as *Synechocystis limnetica*. In this oligotrophic lake, these organisms generate a major bloom in late summer (Figure 2.9) – accounting for about 60 per cent of the total annual primary production (Semovski *et al.*, 2000).

3.4.3 Biological significance of size and shape

The major distinction between prokaryotic phytoplankton (blue-green algae, photosynthetic bacteria) and eukaryotic phytoplankton noted earlier has important implications for cell size, with most prokaryote cells having diameters <5 µm. Some eukaryote planktonic organisms, such as heterotrophic nanoflagellates also approach this size range – with diameters typically <10 µm – but are ultimately limited by the need to accommodate a nucleus and various eukaryote organelles within the cell. In addition to size, the range of cell and colony shape is also remarkable, with many organisms having an elongate shape, further extended by spines or cellular processes (Figure 3.1).

The size and shape of phytoplankton has important implications for a range of biological functions, including physiological processes (surface exchange of materials, light absorption, ability for rapid growth), distribution in the water column (passive movement, sedimentation, motility) and resistance to ingestion by zooplankton.

Exchange of materials at the cell surface

The passive diffusion of solutes and dissolved gases across the surface of phytoplankton cells is a two-way process and is governed by Fick's Laws of Diffusion. For each solute, the rate of diffusion for the whole cell depends primarily on surface area, concentration gradient, and coefficient of molecular diffusion (see Section 3.7.1).

Cell size is an important aspect of this exchange process since the volume of the organism determines total metabolic demand (solute uptake) or excess (solute loss) – which drive the concentration gradient. With a particular concentration gradient and diffusion coefficient, the size-related surface area determines the total passive exchange that can take place. These two aspects of demand and exchange capacity are brought together in the surface area to volume (S/V) ratio. Organisms with a low S/V ratio have potential problems since the area over which exchange can take place is low in relation to requirements. Spherical organisms have the smallest possible ratio, since:

$$S/V = \frac{4\pi r^2}{4/3\pi r^3} = \frac{3}{r} \qquad (3.1)$$

where r = radius of sphere. For single-celled spherical algae, an increase in size will reduce the S/V ratio. This has important evolutionary implications, where the maintenance of a high S/V ratio during the evolution of large-sized organisms is achieved by the development of elongate shapes and by

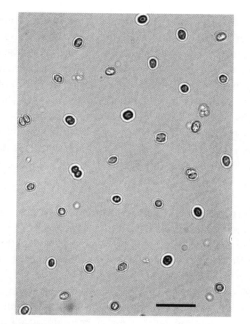

Figure 3.5 *Synechococcus aeruginosa* (Cyanophyceae): this single-celled blue-green alga is a common member of the nanoplankton, with a cell diameter of approximately 2–5 μm. This sample was obtained from a mid-summer *Synechococcus* bloom which developed in a stratified eutrophic lake at a time when epilimnion nutrient concentrations were still high (soluble nitrogen: $1\,mg\,l^{-1}$, soluble phosphate: $50\,\mu g\,l^{-1}$, scale $-20\,\mu m$)

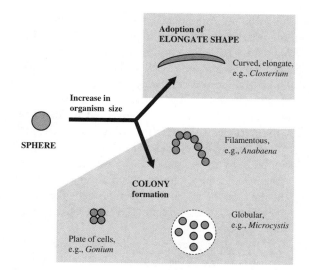

Figure 3.6 Hypothetical scheme to show maintenance of a high S/V ratio during the evolution of larger-sized phytoplankton. Algae avoid an evolutionary decrease in surface area to volume (S/V) ratio that would occur by simple increase in spherical size by adopting elongate shapes or forming colonies. This is particularly the case for K-selected organisms, which include the dinoflagellate *Ceratium* (with elongate spines) and the colonial blue-green algae *Anabaena* (filamentous) and *Microcystis* (globular). In colonial algae, the key S/V ratio is that of individual cells rather than the whole colony since diffusion of metabolites occurs directly to and from the separate cells

colony formation (Figure 3.6). Many large-sized K-selected planktonic algae that inhabit stable, crowded environments (Section 3.10.2) have a colonial form (*Anabaena, Microcystis*) or have elongate extensions (*Ceratium*).

The ecological importance of organism size in relation to nutrient transfer is also indicated by a comparison of algal species which dominate low-nutrient and high-nutrient environments. Within major algal groups such as Chlorophyceae and Cyanophyceae, the smaller single-cell organisms tend to dominate oligotrophic sites, while larger colonial forms tend to predominate in eutrophic systems. The higher surface/volume ratio of microplanktonic algae leads to a more efficient uptake of nitrate and phosphate, giving these organisms a competitive advantage in nutrient-limiting conditions. The smallest free living lake biota, bacteria,

have an even greater competitive advantage over algal cells in low-nutrient conditions (Section 6.12). Although unicellular picoplankton and nanoplankton are particularly characteristic of oligotrophic systems, they also occur in eutrophic lakes where they may form dominant populations. This is seen in Figure 3.5, with a mid-summer bloom of *Synechococcus*, and is also recognized in Reynolds' (1990) classification of algal succession in relation to lake nutrient levels – where hypereutrophic lakes are characterized by a sequence of unicellular algae (Table 10.4).

Light absorption and cell size

Absorption of light by pigments contained within a suspension of cells is different from the situation where the same amount of pigment is dispersed throughout the whole liquid medium. This packaging effect has implications for the absorption of light by phytoplankton populations and the attenuation of irradiance within water columns.

Light absorption by individual particles (e.g., algal cells) in suspension is governed by the size of the particles (d) and the absorption coefficient (a), and may be considered either in terms of absorption by the whole cell or absorption per unit pigment molecule (pigment-specific absorption).

In the case of an entire spherical cell, the 'efficiency factor for absorption (Q_a)' may be defined:

$$Q_a = \frac{\text{energy absorbed within the sphere}}{\text{energy incident on its geometrical cross section}}$$
(3.2)

Q_a directly varies with particle size (d) and absorption (a) coefficient and can be used to determine the light absorption of an entire population of cells in suspension (Morel and Bricaud, 1981).

In contrast to whole cells, pigment-specific light absorption increases with decreasing cell size and decreasing internal molecular shading. Under conditions of equal pigment concentration, small cells would have higher light absorption per unit pigment with implications for the efficiency of photosynthesis. Cell size is also important in the absorption of UV radiation and the damage to pigment and other macromolecules during the process of photoinhibition. In general, smaller phytoplankton cells have higher rates of UV absorption and are thus more prone to photoinhibition, as discussed in Section 4.9.4.

Small size as an ecological response

Small-celled unicellular algae have a low biomass, short cell cycle and rapid growth rate. These organisms are adapted to the exploitation of transient new environments, (where conditions for growth are time-limited) and situations of environmental stress. Such conditions favour organisms that follow an opportunistic or r-selection strategy (see Section 1.2.6).

Conditions of time-limited growth Small-celled (pico- and nanoplanktonic) algae tend to dominate the phytoplankton of sub-polar lakes, where the period of summer growth is restricted. In Lake Baikal (Russia), for example, there is only a short period (July–September) of warm (10–15°C) surface water, during which autotrophic nanoplankton predominate (Figure 2.9). These organisms reach counts of up to 2×10^6 cells ml^{-1} within the euphotic zone (about 15m), with the blue-green alga *Synechocystis* as the major organism.

Conditions of environmental stress The inverse relationship between the size of freshwater biota and the effects of both internal and external environmental stress factors has been noted previously in Section 1.6 (Figure 1.13).

Zooplankton grazing pressure operates as a major internal stress factor during the seasonal cycle of many temperate lakes, with a pronounced shift to small (r-selected) algae during the clear-water phase (Figure 3.14). External stress, such as heavy-metal pollution, may also have a dramatic effect. Studies by Cattaneo *et al.* (1998), for example, demonstrated a marked decline in the size of algae (diatoms), protozoa (thecamoebians), and zooplankton (cladocerans) in Lago d'Orta (Italy) over a 50-year period. This change in the mean size of individual organisms began with the onset of copper (and other metal) pollution, and was primarily due to a shift in taxonomic composition within the different groups. Reduction in body size also occurred within a single taxon, the diatom *Achnanthes minutissima*.

Passive movement of cells and colonies

Phytoplankton units (single cells or colonies) are exposed to continuous random water movement within the water column. This ranges from oscillations at the molecular level, causing Brownian motion of the smallest phytoplankton cells (along with bacteria and viruses), to larger-scale water movement which occurs as turbulence.

In conditions of turbulence, the fixed flow to which small phytoplankters are exposed is a laminar shear, causing the organism to move (translate) and

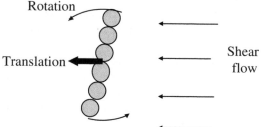

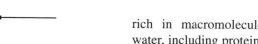

Figure 3.7 Displacement of phytoplankton in flowing water; linear colonies and single cells undergo forward movement (translation) and rotation under conditions of linear flow

rotate (Figure 3.7). Such passive rotation is important in increasing the nutrient flux, and needs to be taken into account in considering the uptake of nutrients in turbulent conditions. It is also important in other planktonic processes such as aggregate formation and predator–prey interactions, since it determines the volume swept by the phytoplankter and hence the likelihood of encounter with other particles or with the appendages of herbivores.

Laboratory studies on the passive movement of chain-forming diatoms such as *Skeletonema* (Karp-Boss and Jumars, 1998) under conditions of steady shear flow demonstrated periodic rotation. In diatoms where the chain forms a rigid structure, the period of rotation depended on the magnitude of the shear rate and the axis ratio (major axis/minor axis) of the colony.

Sedimentation within the water column

Sedimentation is the downward movement of non-motile plankton (typically with a rigid cell wall, lacking cilia or flagella) relative to the adjacent water medium, in response to gravitational forces ($g = 9.8081$ m s^{-2}).

Sedimentation occurs where the density of a particle exceeds that of the surrounding liquid medium, which is 1000 kg m^{-3} for pure water. In the case of phytoplankton cells, the protoplasm is

rich in macromolecules which are denser than water, including proteins ($\sim$1300 kg m^{-3}), carbohydrates ($\sim$1500 kg m^{-3}), and nucleic acids ($\sim$1700 kg m^{-3}). These molecules are contained within an aqueous cytosol, and the overall density of phytoplankton cells or colonies is typically $>$1050 kg m^{-3}. This may be considerably increased by the presence of dense inorganic inclusions such as polyphosphate bodies ($\sim$2500 kg m^{-3}) and opaline silica cell wall material ($\sim$2600 kg m^{-3}). Conversely, cell density can be reduced by the internal presence of low-density lipids (860 kg m^{-3}), gas vacuoles, and the occurrence of external mucilage.

The passive rate of sedimentation of a spherical body under conditions of laminar flow, is given by the Stokes equation:

$$\nu_s = 2gr^2(\rho - \sigma)(9\eta)^{-1} \qquad (3.3)$$

where ν_s is the terminal velocity of the particle (m s^{-1}), g is gravitational acceleration (m s^{-2}), η is the coefficient of viscosity of the liquid medium (kg m s^{-1}), ρ and σ are the density of the particle and liquid medium respectively (kg m^{-3}), and r is the particle radius (m).

According to this equation, for spherical cells, size and density are key factors affecting the rate of sinking. Major factors that can cause deviation from this equation are non-laminar flow and non-spherical shape. Most algae are not exactly spherical, and their deviation from a sphere (ϕ_r) will result in a larger surface area and a greater frictional resistance, thus reducing the sinking rate. This can be incorporated into a modified Stokes equation:

$$\nu_t = 2gr^2(\rho' - \rho)/(9\eta \cdot \phi_r) \qquad (3.4)$$

where ν_t is the terminal velocity of the particle (m s^{-1}) and ϕ_r is the coefficient of form resistance (deviation from sphericity).

The variety of shapes seen in phytoplankton cells and colonies will thus have a major effect on sinking rates, and the development of attenuated shapes such as elongate colonies and the presence of spines may be interpreted in part as an evolutionary development to reduce passive sinking and maintain cells within the zone of light availability. Although many phytoplankton cells/colonies have developed attenuate morphology, the extent to which this occurs (as measured by surface area/volume ratio) remains within strict limits – see Section 4.3.1.

For buoyant particles such as gas-vacuolate blue-green algal cells (where the density is less than water) the value for ν_t is negative, representing a rate of upward flotation.

Size and motility

The size range of flagellate algae is limited by the propulsive force of the flagella (Fenchel, 1991). These organisms typically have a regular flagellar beat frequency of about 50 Hz, with a flagellar length not exceeding 50 µm. The swimming velocity of cells that propel themselves by one or a few flagella is typically about 0.1–0.2 mm s^{-1}.

In order to swim, the swimming velocity (forward propulsion) must exceed the sinking velocity. With spherical cells of density 1.1, Stoke's law predicts sinking velocities of 0.1–0.2 mm s^{-1} for cells with a diameter in the range 25–50 µm. This sets the upper size limit for flagellates. The size range of typical flagellates that use flagella for locomotion is about 4–20 µm, with many (particularly the heterotrophic nanoflagellates) having diameters <10 µm. A decrease in size to this level allows for greater motility due to a decrease in the gravitational vector. An alternative strategy to decreasing size in relation to flagellar motility is to adopt other forms of movement. Many of the larger flagellates (e.g., large euglenoids) do not swim, but are capable of gliding over surfaces. Another possibility for circumventing the problem of cell size in flagellates is to regulate buoyancy, and reduce sinking velocity. This is seen in the dinoflagellate *Noctiluca*.

Resistance to ingestion by zooplankton

Grazing of phytoplankton cells by zooplankton is the major factor that limits algal growth in the freshwater environment. Ingestion of algae by zooplankton can be limited by a range of algal characteristics including size and shape, surface chemistry, presence of mucilage, and algal toxins (Section 9.8.3).

Many algae have evolved a strategy to reduce zooplankton grazing by maximizing size (greatest axial linear dimension), including single cells (e.g., dinoflagellates *Ceratium*, *Peridinium*) and colonies (e.g., blue-green *Anabaena*, *Microcystis*).

3.4.4 Variation in size and shape within phytoplankton populations

The size and shape of planktonic algae clearly have considerable adaptive value since they have direct influence on both growth (compare *r*- and *K*-selected cells) and loss (sedimentation, zooplankton ingestion) rates. These parameters are thus important in competition between algae, determining the abundance of particular species within the aquatic environment and their individual contribution to community biomass (Duarte *et al.*, 1990). Differences in size and shape will also lead to differences in surface area/volume (*S/V*) ratios of cells and colonies – with resulting differences in physiological activities (see previous section).

Most studies on size/shape variation in phytoplankton populations have concentrated on differences between species (mixed populations), but size plasticity within species may also be important.

Variations in size and shape within mixed populations

Phytoplankton populations sampled from a particular site typically contain a mixture of different species with an apparently unrestricted range of morphologies, varying from spherical unicellular and colonial forms to single cells and colonies with highly extended and irregular shapes (Figure 3.1).

In such a situation, it might be expected that derived parameters, such as *S/V* ratios would also extend over a wide and unrestricted range. This was

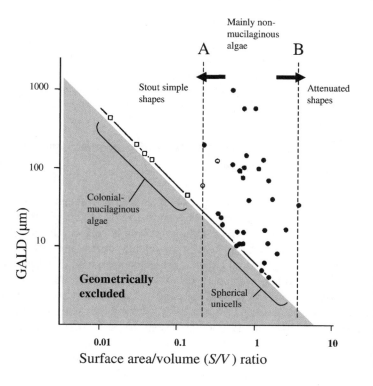

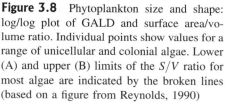

Figure 3.8 Phytoplankton size and shape: log/log plot of GALD and surface area/volume ratio. Individual points show values for a range of unicellular and colonial algae. Lower (A) and upper (B) limits of the S/V ratio for most algae are indicated by the broken lines (based on a figure from Reynolds, 1990)

initially investigated by Lewis (1976) who examined the relationship between unit size (expressed as the greatest axial linear dimension – GALD) and the surface/volume ratio of the 27 most abundant phytoplankton species in a tropical lake. Values were plotted for various theoretical shapes and for phytoplankton data (similar to Figure 3.8). The major conclusion was that phytoplankton S/V ratios were conserved within a range much narrower than expected by random choice of shapes. Unlike the theoretical shapes, phytoplankton organisms occupied quite a narrow band, with particular sizes ranging from spherical (low S/V ratio) to attenuate (high S/V ratio) shapes. Extending the Lewis plot to a wider range of planktonic algae (Reynolds, 1990; Figure 3.8) indicates that conservation of S/V ratios holds particularly for non-mucilaginous algae, but not for mucilaginous colonial forms.

In non-mucilaginous phytoplankton, a lower limit to the S/V ratio is to be expected in terms of algal physiology, since a point is reached at which the

surface is no longer able to provide metabolite exchange levels appropriate to volume requirements. The presence of an upper S/V boundary may seem surprising, however, since it is not immediately clear in terms of cell physiology why there should be a limitation to the development of attenuate shape and high S/V ratios. The upper S/V limit may relate to the need for a critical level of sedimentation to ensure continuous replacement of the microenvironment adjacent to the cell surface. This replacement is required for efficient nutrient uptake and for the loss of unwanted metabolites.

The occurrence of mucilaginous colonial algae such as *Microcystis* below the lower S/V limit emphasizes the separateness of cells within the colony. These aggregations are not limited by size since metabolites are able to diffuse to and from individual cells through the mucilage matrix. The physiologically relevant S/V ratio in these organisms is that of individual cells rather than the whole colony (Figure 3.6).

Size variation within species

Considerable size plasticity occurs within species, with both single cell and colonial algae showing great variation in overall dimensions. Although this plasticity partly reflects the variation expected within a normal distribution, evidence suggests that algae are also able to modulate size in relation to functional parameters such as absorption of light and general competitive ability. These functional parameters may in turn relate to environmental features such as phytoplankton community biomass, water turbulence, and nutrient availability.

Light absorption   Control of colony size in blue-green algae, for example, may be important in controlling sedimentation and optimizing light absorption. In studies on a nutrient-rich lake, Robarts and Zohary (1984) noted that light absorption by the dominant alga *Microcystis aeruginosa* increased with reduction in colony size (at comparable lake water chlorophyll-a concentrations), and that the euphotic zone accordingly increased with colony size.

General competitive ability   Algal size may also be important in relation to general competitive ability. Comparative studies (Duarte *et al.*, 1990) of different lakes have suggested that the average size of phyoplankton tends to increase with increasing community biomass. In the case of Florida lakes, for example, these studies have shown that the average size of individuals within genera tended to increase as the ½ power of their biomass within the community. These observations have been interpreted as evidence for size-dependent differences in competitive ability, with small algae being better competitors in sparse communities and large algae competing more effectively in dense communities. The adaptive value of algal size is emphasized by the fact that genera capable of large size variations tend to thrive in both sparse and dense communities, and are more ubiquitous than genera with a limited size plasticity.

Water turbulence (Steinberg and Geller, 1990) may have a direct physical effect on cell and colony size. Conditions of high turbulence, occurring in systems such as dammed rivers, lead to algal populations of small mean size. This was particularly apparent in algae such as *Microcystis aeruginosa* and *Coelosphaerium kuetzingianum*, where the reduction in size under turbulent conditions resulted in these species occurring as small single cells within the picoplankton.

3.5 Mucilaginous and non-mucilaginous algae

In addition to differences in size and shape, freshwater algae also vary considerably in the presence of extracellular mucilage. This is seen particularly clearly in the lake environment, where some species of planktonic algae have cells embedded in a large volume of mucilage – while others have no apparently detectable mucilage at all. The presence of mucilage is not readily observed by bright-field light microscopy, but can be seen under phase contrast (Figure 3.9) and by special preparation techniques such as negative staining and fluorescence microscopy of lectin-stained preparations. Atomic force microscopy (AFM) can also be used to provide high-resolution information on the topography and material properties of the mucilage layer of living algae. Recent AFM studies on three diatoms, for example, have revealed differences in mucilage surface nanostructure and in the dynamic properties (adhesion and stretching) of surface polymer chains (Higgins *et al.*, 2003).

An outer layer of mucilage increases overall size and confers a distinctive surface chemistry to the algal cells. Mucilage has an important role in the biology of these organisms and has ecological implications in terms of biogeochemical cycles. Even algae that are regarded as 'non-mucilaginous' typically have a surface layer of polysaccharide material. Combined carbohydrate analysis and atomic force microscopy demonstrated a thin sugar-rich layer of surface mucilage in the diatom *Pinnularia* (Chiovitti *et al.*, 2003), for example, which would be difficult to see under the light microscope. This thin mucilage layer in diatoms has particular importance in separating the cell wall from the external aquatic medium and reducing silica solubilization in living organisms (Figure 5.24).

3.5.1   Chemical composition of mucilage

Mucilage is composed of a complex macromolecular network enclosing a water matrix. Although water is the major component (>95 per cent) by weight, the chemistry of mucilage is dominated by the macromolecular component and the exposed sugar and charged groups which are associated with this. Relatively little is known about the detailed chemistry of algal mucilage, although chemical analyses have been carried out on the surface slime of a number of blue-green algae. In *Microcystis flos-aquae*, for example, where individual cells are embedded in a globular mass of surface slime, the macromolecular component is almost entirely polysaccharide, with no detectable protein (Plude *et al.*, 1991). Gas chromatographic analysis of the polysaccharide indicates a composition closely similar to higher plant pectin, with galacturonic acid as the main sugar plus minor quantities of neutral sugars (galactose, glucose, xylose, mannose, and rhamnose). The surface mucilage of other blue-green algae is also polysaccharide-based, but differences occur in sugar composition and detectable levels of protein are present in some cases. Variations in chemical composition may account for differences in mucilage appearance and consistency in different algae.

Surface sugars

Surface sugars are an important biochemical feature of both mucilaginous algae (where they are exposed throughout the mucilage layer) and non-mucilaginous algae, and have been studied using lectin probes. These multivalent carbohydrate-binding proteins are highly specific to particular sugar groups and include molecules such as concanavalin A (ConA – binds to mannose, glucose, and N-acetylglucosamine) and peanut agglutinin (PNA – binds to galactose and N-acetylgalactosamine).

Using a panel of seven fluorescent-labelled lectins, Sengbusch and Muller (1983) demonstrated great variety in the occurrence of surface sugars in selected species from the major algal groups. The results showed clear patterns of lectin-binding which appeared to be species specific, but not specific to the major algal group. Within the aquatic environment, surface biochemistry is clearly an important aspect of biodiversity – with each individual species expressing its own particular combination of sugar molecules at the environmental interface. The functional role of these surface sugars has not been elucidated, but they have clear potential as receptor molecules for communication (within and between algal species), species-pathogen recognition, and the attachment of a wide range of organisms (including protozoa, fungi, bacteria, and viruses).

3.5.2   Role of mucilage in phytoplankton activities

The presence of a layer of surface mucilage can be seen as an ecological strategy that has evolved in all the major algal groups and affects a number of characteristics – including an increase in unit (single cell or colony) size, approximation to a spherical shape, decrease in overall density, and the acquisition of a characteristic surface chemistry. This mucilage is important for both planktonic and benthic micro-algae.

Phytoplankton

For planktonic algae, surface mucilage influences a number of biological activities.

Sinking rate   The effect of surface mucilage on the sinking rate within the water column is defined by the Stokes Equation (3.3). The speed of sedimentation will depend on the balance between increase in unit size (increasing the sedimentation rate) and the decrease in density (decreasing the rate).

Grazing and digestion by zooplankton   Surface mucilage is an important factor in the selection of food particles by zooplankton (Section 9.8.3) – influencing the size and shape of the algal cell or colony and also its surface chemistry. Once the alga has been ingested, the presence of a surface layer of mucilage may also prevent digestion within the zooplankton alimentary canal, allowing the algal cells to be voided in a viable state.

Association of epiphytic biota

Mucilage provides an increased area for attachment and a solid substratum for a range of epiphytic organisms – including bacteria, protozoa, fungi, and other algae. In addition to providing a site of attachment for these organisms mucilage may also provide a medium within which motile epiphytic organisms can move, and in some cases provides an major source of nutrients.

Surface mucilage is particularly important in large colonial algae, where an extensive phycosphere community can develop forming a self- contained micro-ecosystem (Section 6.13, Figure 6.23).

Adsorption of anions and cations

Although physical binding (adsorption) of anions and cations occurs at the surface of both mucilaginous and non-mucilaginous algae, recent studies have shown that algae with a large amount of mucilage tend to have a higher capacity for ion adsorption. This has potential significance for the uptake of nutrients from the aquatic environment and the ability of algal cells to resist heavy-metal toxicity.

Laboratory studies (Figure 3.9) on Cu adsorption by cultured algal cells demonstrate a linear relationship between adsorption (expressed as log Cu per unit dry mass) and Cu concentration in the medium (log values). This indicates a close fit to the Freundlich adsorption isotherm, with the mucilage behaving kinetically as a monolayer containing heterogeneous binding sites. These adsorption characteristics are not restricted to algal cells, but are also typical of other microorganisms such as fungi (Sag *et al.*, 1998).

The heterogeneous binding sites in algal mucilage include positively and negatively charged groups, which promote the adsorption of both anions and cations. The degree of adsorption and the balance between anion and cation binding depends on the composition of the mucilage. In the case of *Microcystis flos-aquae*, for example, the mucilage is composed largely of galacturonic acid, and charge attraction of cations to carboxyl groups is the main factor in ion adsorption (Plude *et al.*, 1991). In addition to carboxyl groups, sugars with particular configurations of hydroxyl groups can also complex some metals (Angyal, 1972). The ability of mucilage to chelate Fe may be mediated by the presence of specific iron-binding compounds (siderophores) such as hydroxamic acid, catechols and phenols.

Benthic algae

Surface mucilage is important for benthic microalgae such as diatoms, where it is involved in attachment and locomotion.

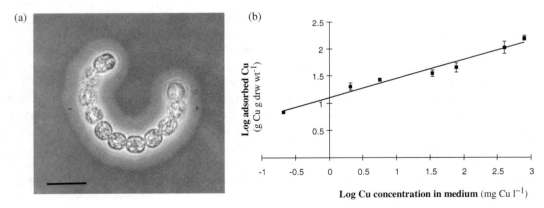

Figure 3.9   Adsorption of cations by surface mucilage of *Anabaena*. (a) Phase contrast image of a single filament of *Anabaena spiroides*, showing a 5 μm-thick mucilage sheath (scale = 10 μm). (b) Freundlich isotherm plot for Cu adsorption by cultured *Anabaena* cells, showing a close fit ($p < 0.01$) to the model. The Freundlich adsorption capacity constant (K_f) was 12.6, indicating a much greater capacity for adsorption compared with non- mucilaginous algal species (K_f values 0.4–3.6) (Unpublished data from Christina Tien, with permission)

Some benthic diatoms are able to glide over solid surfaces by the extrusion of mucopolysaccharide filaments through special 'mucilage pores,' leaving a mucilaginous trail over the substratum. Motility in these organisms is restricted to two major groups – the pennate diatoms (where the mucilage pores occur along the central ridge, or raphe), and centric diatoms with special surface projections (labiate processes). Studies on the pennate diatom *Navicula* (Edgar and Zavortink, 1983) have shown that the mucus filaments are produced by secretory vesicles along the length of the raphe slit, and that control of secretion is mediated via actin filaments. Mucilage secretion is a periodic event and, in response to an appropriate stimulus, occurs via mucilage pores which typically occur at the ends of the raphe. The mucilage is released into the raphe, streaming in one direction until it hits the fixed substratum – when it forces the diatom to move in the opposite dirdection. In pennate diatoms, the direction of movement follows a curved or straight trajectory (depending on the shape of the raphe), and observed speeds vary from $2–14\,\mu ms^{-1}$ at room temperature (Lee, 1997). Speed of movement can vary with the type of substrate, ranging in *Nitzschia* from $2.7\,\mu ms^{-1}$ on glass to only $0.8\,\mu ms^{-1}$ on soft agar.

3.5.3 Environmental impact and biogeochemical cycles

Many mucilaginous algae, including blue-green algae such as *Microcystis* and *Anabaena*, have worldwide occurrence throughout a range of freshwater systems. These organisms often form large populations in natural waters, with the high levels of surface mucilage influencing the chemical and ecological properties of the ecosystem. Studies on a temperate eutrophic lake (Tien *et al.*, 2002) showed that cell-associated mucilage occupied $0.1–7 \times 10^{-3}$ per cent of lake water volume within the epilimnion over the annual cycle, reaching the highest value during the autumn bloom of *Microcystis*. Seasonal changes in the total volume of associated mucilage reflected the succession of mucilage-forming species, but did not correlate with the concentration of soluble extracellular polysaccharide (EPS) in the lake water. Soluble EPS ranged in concentration from $2.5–60\,mg\,l^{-1}$, with peaks in the spring diatom bloom ('non-mucilaginous') and the late clear-water phase – but not the *Microcystis* (mucilaginous) bloom. This suggests that surface mucilage (cell-associated polysaccharide) does not simply diffuse into the surrounding water, and that the soluble lake water EPS concentration relates more to other factors such as algal secretion (Section 4.7.2), and bacterial and zooplankton activity.

Whatever the derivation of soluble EPS, it has a combined role with surface mucilage in the adsorption of cations from lake water – influencing the availability of trace nutrients, sequestering toxic metals, affecting pH, and playing a role in several biogeochemical cycles. EPS has been implicated, for example, in the manganese cycle where it is involved in the oxidative precipitation of manganese nodules in certain lakes (Richardson *et al.*, 1988).

C.   ACTIVITIES WITHIN THE FRESHWATER ENVIRONMENT

3.6   Benthic algae: interactions with planktonic algae and ecological significance

3.6.1   Planktonic and benthic algae

Benthic algae are present in the bottom sediments of almost all aquatic systems, where they require adequate light to carry out photosynthesis and growth. In terms of primary productivity, they form a distinct community (periphyton) from pelagic algae (phytoplankton), which photosynthesize, grow, and spend their vegetative phase in the water column. The two communities are ecologically related, however, since individual species contribute to both populations and dynamic interactions occur between the two groups.

Differences between benthic and planktonic algae

Some general distinctions separate these major groups of algae (Sterenson *et al.*, 1996). Phytoplankton is typically composed of micro-algae (<200 μm diameter), while periphyton is a mixture of micro- and macro-algal (>200 μm) species. Particular taxonomic groups, such as filamentous green algae and pennate diatoms are typical of benthic conditions, while volvocales and centric diatoms are characteristic of the pelagic environment. Various morphological and physiological features are also diagnostic. The attachment structures of filamentous algae and the raphe, mucilage pads, and stalks of pennate diatoms are clear adaptations to the benthic environment. Conversely, the development of motile mechanisms (flagella, gas vacuoles) is typical of planktonic forms. Other differences that have been suggested include a higher specific gravity (higher settling speed) in benthic species, and adaptation to greater nutrient diversity in the benthic environment.

Differences between planktonic and benthic algae also occur in relation to the structure and location of the community. In contrast to the dispersed state of phytoplankton, benthic algae are typically closely associated within a densely packed periphyton mat. This influences the growth of cells and their rates of primary production and nutrient uptake. Internal shading results in a wide variation in photosynthetic activities within the dense periphyton biomass, leading to major differences in the photosynthesis-irradiance (*P–I*) response curves between phytoplankton and periphyton communities (Section 4.3). The location and close packing of benthic algae also influences the way in which these biota are grazed. In contrast to the dispersion feeding of phytoplankton by (largely crustacean) zooplankton, benthic algae are largely removed by the scraping and gathering activities of benthic molluscs and insect larvae (Section 9.13).

Benthic and planktonic phases

Freshwater algae share one important characteristic with other lake microorganisms (such as bacteria,

fungi, and protozoa) – individual species typically have both pelagic (present in the main water body) and benthic (associated with sediments) phases. In species which are mainly planktonic, the benthic phase may simply be a metabolically inactive and resistant spore which over-winters in the sediments. In species that are primarily benthic, the planktonic phase may originate by detachment of single cells or colony fragments, or by the formation of gametes and zoospores as part of the life cycle. Pelagic phases are particularly important in benthic organisms for dispersal and colonization of new environments.

The occurrence and interactions of planktonic and benthic algae depend primarily on light availability to the sediments (depth of water column) and the displacement of unattached algae by the water current, both of which relate to the hydrology of the water body. Some of the interactions between planktonic and attached phases are illustrated in Figure 3.10, which shows two contrasting environmental situations. In deep lakes, where lack of light penetration prevents growth of benthic algae, sediment populations are mainly resistant spores – which are in seasonal equilibrium with planktonic populations. Although there are no attached benthic algae, attached algae do occur as epiphytes which are in balance with free planktonic populations. These epiphytic algae are associated with larger colonial algae as part of the algal microcosm (Figure 6.23) and also with other lake organisms. The unicellular blue-green algae *Cyanothece* and *Synechococcus* are occasionally observed as epiphytes within colonies of *Gomphosphaeria*, and the euglenoid alga *Colacium* occurs as a green epibiont on the surface cuticle of microcrustaceans and rotifers (Wehr and Sheath, 2003). In contrast to deep lakes, shallow river systems typically have extensive populations of actively growing algae on sediments and as epiphytic growths on macrophytes, both of which relate to planktonic phases in terms of detachment and recruitment.

In lakes, periphyton communities are associated particularly with the littoral zone, and are spatially separated from the main water body which is dominated by phytoplankton. Although lake periphyton are subject to some physical disturbance

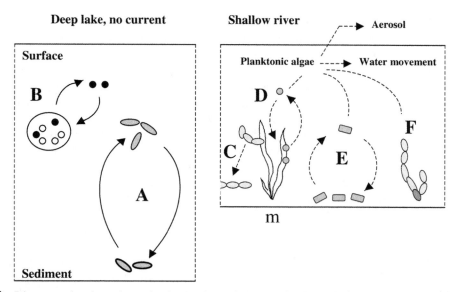

Figure 3.10 Diagram showing dynamic interactions between freely pelagic and non-pelagic (benthic or attached) phases of algae in contrasting lake and river situations (not to scale)

Deep lake: sediments are not within the photic zone, so there are no photosynthetic benthic algae. **A.** Benthic resistant spores give rise to planktonic populations, and vice versa. **B.** Some free planktonic algae occur in equilibrium with attached (epiphytic) populations associated with large colonial algae and other organisms.

Shallow river. High light levels support vigorous growths of attached epiphytic and benthic algal communities, which are detached by the current to form phytoplankton:

- *Epiphytic algae* present on macrophytes (m) include filamentous and unicellular forms: **C.** filamentous algae may detach to form filament masses (metaphyton); **D.** unicellular algae are in equilibrium with planktonic populations.
- *Benthic algae* (periphyton): **E.** attached micro-algae (e.g., diatoms) interchange with planktonic cells by detachment and settlement; **F.** macroscopic filamentous algae can also detach (giving rise to planktonic fragments) and settle

(wave action) they do not experience the flow characteristics of river systems, with the continuous displacement and detachment of microorganisms which is important in lotic dispersal.

Competition between planktonic and benthic algae

Competition between planktonic and benthic algae for light and nutrients shows close similarities to the interactions that occur between phytoplankton and higher plants (macrophytes) – Section 4.8. In both cases, benthic and pelagic primary producers exhibit various competitive interactions, which in the case of phytoplankton and periphyton may involve (Vadeboncoeur *et al.*, 2003):

- more rapid uptake of nutrients from the water column by phytoplankton due to the smaller surface/volume ratio of pelagic algae, and the fact that nutrient uptake by periphyton is constrained by boundary layer kinetics;

- closer access of periphyton to sediment nutrients, allowing them to regulate the availability of these growth resources to phytoplankton;

- attenuation of light by phytoplankton, limiting periphyton productivity.

These interactions may lead to different responses by pelagic and benthic algae under conditions of changing light and nutrients, resulting in inverse relationships between population development in the two groups of organism. Eutrophication resulting from fertilization experiments, for example, may lead to increased phytoplankton biomass – with a compensatory decline in the level of periphyton. This transition was observed in shallow, oligotrophic Greenland lakes (Vadebonceur et al., 2003), with a eutrophication switch from benthic to pelagic dominance of primary productivity. The change in productivity was accompanied by a corresponding shift from periphyton to phytoplankton in the diets of zoobenthos, as demonstrated by carbon stable isotope analysis. Although benthic and pelagic habitats in these lakes were energetically linked through food web connections, eutrophication depleted and finally eliminated the benthic primary production pathway.

3.6.2 Lake Periphyton

Although the algal population of lakes is normally considered in terms of phytoplankton, benthic algae also make a major contribution to lake ecology and to primary productivity. This is particularly the case for shallow lakes, where a large part of the bottom receives enough light to support photosynthesis and periphyton (Section 1.2.4) can dominate carbon fixation. Even in deep lakes, where the area of periphyton is restricted to the shoreline (littoral zone), the role of periphyton in the whole-lake carbon budget can be substantial (Lowe, 1996). Lakes with steep sides, such as Lake Tahoe (USA) have relatively narrow littoral zones with a correspondingly lower proportion of periphyton carbon fixation.

The development of periphyton communities in lakes is strongly influenced by environmental variables such as light, turbulence, water chemistry, macrophyte dominance, and grazing pressure. Climate is also important, and the seasonal changes which determine periphyton development in a dimictic lake of temperate zones are clearly quite different from those operating in a permanently frozen (amictic) lake in Antarctica. In temperate lakes, the periphyton community receives light every day, but experiences great seasonal fluctuations. Periphyton in polar lakes experiences alternating 6 month periods of light and dark, and is permanently under ice.

Lentic periphyton communities are often complex and highly structured, and are composed of both autotrophic (algae) and heterotrophic (fungi and bacteria) components. The algae occur as two major growth forms:

- Large filamentous algae (mainly green algae and blue-green), which are firmly attached to the sediment or to macrophytes, and extend into the water medium. Filamentous algae with coarse cellulose cells walls, such as Cladophora, can themselves support extensive populations of epiphytes. Cladophora creates an important microhabitat for epiphytic microbes and communities of invertebrates, increasing the surface area of the eulittoral zone by an estimated factor of >2000.

- Single-celled and small colonial algae, present on rocks (epilithon) and large plants or algae (epiphyton). These algae include a wide range of diatoms, plus blue-green and some red algae.

Although these different growth forms and taxonomic groups are intermixed within the benthic community, periphyton often shows a clear shoreline zonation – with an upper eulittoral zone dominated by green algae and diatoms and a lower sublittoral zone dominated by blue-green algae and diatoms. These two zones are determined by differences in light availability, water turbulence, seasonal variation, and macrophyte dominance (Figure 3.11).

The eulittoral zone roughly corresponds in depth to the epilimnion of the water column, and shows similar seasonal variations in light and temperature. Algae within this zone show clear seasonal variation, and are characterized by species which have firm attachment to the substratum. Many of these

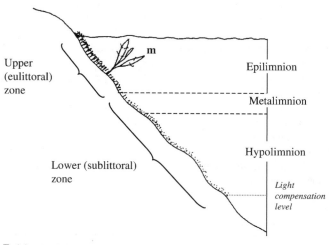

Eulittoral zone: high sunlight, wave action and temperature variation, with macrophyte (m) domination in some lakes. Sediments composed of rocks and coarse material. Periphyton dominated by green filamentous algae and epiphytes

Sublittoral zone: thermally-stable region, with low light and turbulence. Little macrophyte development. Sediments consist of fine silts and detritus. Dominated by epipelic-epipsammic diatoms and blue-green algae.

Figure 3.11 Periphyton distribution and habitat variation in a thermally-stratified lake (based on a figure from Lowe, 1996)

Although the littoral periphyton community can be grouped into two broad zones, changes in periphyton and environmental parameters also form an interrelated continuum with depth. In addition to the above factors, variation in grazing also has an important impact on community diversity.

algae grow best under conditions of water movement (rheophilous), and the eulittoral community is closely similar to the periphyton of well-lit streams. The species composition of the eulittoral zone varies with water quality (Table 3.7). In high nutrient (eutrophic) hard-water lakes, dominant filamentous genera include green algae (*Cladophora*, *Ulothrix*, *Oedogonium*) and occasionally the red alga, *Bangia*. The epilithon and epiphyton of these lakes are dominated particularly by diatoms such as *Navicula* and *Cymbella*. Oligotrophic softwater lakes tend to have a shift from filamentous to unicellular algae, with a preponderance of diatoms and desmids (*Cosmarium*, *Staurastrum*). Diatoms are particularly important in these environments, and in oligotrophic Lake Tahoe (USA), for example, the eulittoral zone is dominated by stalked diatoms such as *Gomphoneis*. Filamentous algae which do occur in these conditions are represented by members of the Zygnemataceae (green algae) such as *Mougeotia*.

Below the eulittoral zone, where wave motion, light energy and temperature variation are reduced, a marked change occurs in the periphyton community. The distribution and abundance of periphyton populations in the sublittoral zone is also strongly influenced by the fine silts and detritus that occur below the zone of wave activity. These deep, thermally stable, lentic habitats support a rich assemblage of algal species. Although relatively few studies have been carried out on these environments, various workers have reported communities of diatoms and blue-green algae. Caljon and Cocquyt (1992), for example, reported 227 diatom taxa from the surface sediments of Lake Tanganyika, and Roberts and Boylen (1988) discovered extensive communities of epipelic filamentous and colonial blue-green algae in the deep littoral zone of an Adirondack lake (USA).

This deep-water algal periphyton can only occur where the euphotic zone extends down to the lake sediment, and is typically not present in eutrophic

Table 3.7 Comparison of algal periphyton in eutrophic and oligotrophic lakes (adapted from Lowe, 1996)

	Oligotrophic, soft-water lake	Eutrophic, hard-water lake
Eulittoral Zone	**Unicellular and colonial epilithic algae**	**Filamentous and epiphytic algae**
Filamentous forms	**Green algae**: (Zygnemataceae) *Mougeotia*	**Green algae**: *Cladophora, Ulothrix, Oedogonium* **Red algae**: *Bangia*
Unicellular and simple colonial	**Green algae**: (Desmids) *Cosmarium, Stuarastrum* **Diatoms**: *Eunotia, Tabellaria*	**Blue-greens**: *Chamaesiphon Lyngbya* **Diatoms**: *Cocconeis, Cymbella*
Sublittoral Zone	**Extensive growth of diatoms and blue-green algae**	**Few taxa, below the photic zone**
Filamentous and colonial	**Blue-greens**: *Hapalosiphon, Calothrix*	
Unicellular	**Diatoms**: *Eunotia Navicula*	

lakes. Changes in the light regime of lake sediments and the periphyton community may arise secondarily due to changes in nutrient levels and other chemical effects. The deep-water periphyton habitat has been eliminated in many lakes, for example, where increased phytoplankton levels resulting from eutrophication have lead to a reduction in the depth of light penetration. Conversely, acidification of poorly-buffered lakes tends to have a direct impact on phytoplankton (more than periphyton), reducing phytoplankton populations, increasing light penetration, and secondarily increasing periphyton biomass (Lowe, 1996).

3.6.3 Benthic algae in flowing waters

In rivers, benthic algae become the main algal component in the ecosystem due to the displacement of phytoplankton by the water current, and its subsequent loss from the system. These algae are frequently attached to the substratum and include both macroscopic filamentous forms (*Cladophora, Spirogyra, Chara, Vaucheria*) and microscopic forms (particularly diatoms) which are part of the periphyton assemblage. Attached algae may be grouped in terms of the substratum, with some present on higher plants and macro-algal surfaces (epiphytes), while others are associated with inorganic substrates

such as rocks (epilithic algae), sand (epipsammic algae), and fine sediments (epipelic forms). Dispersal of algae within freshwater systems occurs primarily by water movement of detached single cells and fragments of filaments, which contribute to the limited plankton occurring in this environment (Fig. 2.14). It seems likely that some algae may also enter the atmosphere in aerosols and undergo aerial dispersal. A wide selection of Chlorophyta (including *Chlorella, Chlorococcum*, and *Chlamydomonas*) have been recovered from aerial samples, with a much smaller range of chrysophytes and blue-green algae also identified (Brown *et al.*, 1964).

3.6.4 Ecological role of benthic algae

Benthic algae play a major ecological role in streams, wetlands, estuaries, and shallow lakes, where they are important in primary production, chemical transformation, retention of nutrients by sediments, physical stabilization of sediments and the provision of habitats for other organisms (Stevenson, 1996).

Primary production by benthic algae

Although the main source of energy in streams was once thought to be detritus from terrestrial origin, it is now realized that primary production by benthic algae is equally important in many mid-sized streams (Minshall, 1978). Benthic algae are also significant primary producers in shallow lakes, ponds, and wetlands. In wetlands, photosynthetic activity and growth of epiphytic algae (with phytoplankton) occurs at those times of year when higher plants can be out-competed due to low macrophyte growth and high algal turnover rates (Section 4.8).

Chemical transformation

In many aquatic ecosystems, benthic algae have an important role as chemical modulators, transforming inorganic chemicals into organic forms. In low-nitrogen habitats, for example, benthic blue-green algae play a key part in the nitrogen cycle – converting N_2 to NH_3 and amino acids. In medium to high nitrogen habitats benthic algae are also involved in the daytime uptake of nitrate, leading to a diurnal variation in the concentrations of this nutrient in some streams. Benthic algae are also primary harvesters of inorganic phosphate in stream nutrient spiralling (Case Study 5.1) and in the removal of phosphate from inflow steams at the edge (littoral zone) of lakes and wetlands.

Nutrient retention by sediments

Benthic algae on surface sediments intercept the passage of inorganic nutrients into the water body, acting as important sinks prior to release into the water column. Epiphytic algae also have a particularly important role to play on wetlands, where macrophytes pump nutrients out of the sediments into the main water body. Epiphytic algae present on the macrophyte surfaces trap nutrients before they reach the water column, returning them to the sediments when the algae settle to the bottom.

Physical stabilization of sediments

Substrata in a wide variety of aquatic habitats are stabilized by benthic algae, which prevent displacement of the sediment at times of aquatic disturbance. This stabilization ranges from the presence of epipelic diatoms and other unicellular algae on estuarine mudflats to the extensive overgrowth of filamentous forms (blue-greens, *Vaucheria*) on sands and other coarse sediments. In addition to this stabilization effect, benthic filamentous algae can also trap inorganic materials, leading to sediment build-up in regions of algal colonization.

Habitats for other organisms

Communities of benthic algae are important habitats for many other aquatic organisms. These communities provide refuges from high current flow, protection from predators, and access to other organisms (potential food supply) within the localized area. Hummocks of *Chara*, for example, support a considerable density and diversity of aquatic invertebrates in streams where sand particles provide conditions which are otherwise too unstable and low in nutrient to support invertebrate populations. Cladophora and other filamentous algae support considerable numbers of epiphytes, along with populations of small invertebrates such as chironomids, amphipods, and small meiofauna. Even algal biofilms, dominated by diatoms, can support substantial numbers of chironomids and meiofauna.

3.7 Temporal activities of freshwater algae

Freshwater environments are relatively unstable and subject to continuous change, varying from transient alterations in environmental factors (e.g., light intensity) to longer-term changes in hydrology and climate. The ability of algae to respond to changes in their microenvironment, and in the ecosystem as a whole, is clearly important to their biological success.

Temporal activities of freshwater algae (Figure 3.12) vary with the time interval that is being considered (fractions of a second to hundreds of years) and include short-term cellular events,

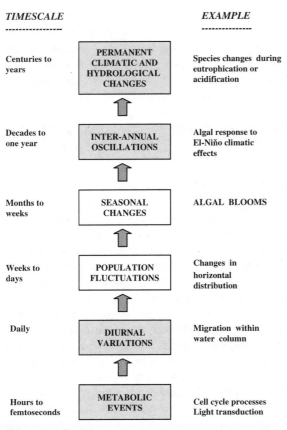

TIMESCALE

EXAMPLE

| Centuries to years | PERMANENT CLIMATIC AND HYDROLOGICAL CHANGES | Species changes during eutrophication or acidification |

| Decades to one year | INTER-ANNUAL OSCILLATIONS | Algal response to El-Niño climatic effects |

| Months to weeks | SEASONAL CHANGES | ALGAL BLOOMS |

| Weeks to days | POPULATION FLUCTUATIONS | Changes in horizontal distribution |

| Daily | DIURNAL VARIATIONS | Migration within water column |

| Hours to femtoseconds | METABOLIC EVENTS | Cell cycle processes Light transduction |

Figure 3.12 Temporal activities of freshwater phytoplankton. The time scale may be arbitrarily divided into short-term (bottom shaded) boxes, medium-term, and long-term (top boxes) periods

medium-term changes in algal succession, and long-term alterations to the ecosystem. Some of these aspects are considered below, while others are discussed elsewhere – including diurnal activities (Section 4.10) and algal bloom formation (Section 10.6).

3.7.1 Short-term changes: molecular and cellular processes

Physiological processes within algal cells occur over periods ranging from femtoseconds (10^{-15}s, e.g., light harvesting and electron transfer processes) to minutes and hours (cell cycle events).

Rapid molecular events: light harvesting and electron transfer in photosystem II

In eukaryote algae, the initial harvesting of light and subsequent generation of high-energy electrons provides a good example of rapid molecular processes at the short end of the timescale shown in Figure 3.12. These processes occur in the 400kD photosystem II light-harvesting complex (LHC II) present in plastid thylakoid membranes and involve the activities of two separate molecular structures – the antenna complex and photochemical reaction centre (Figure 3.13).

Light photons are collected by chlorophyll molecules within the antenna complex. The energy of excitation from a single photon is then transferred to chlorophyll P680 within the photochemical reaction centre, raising the energy of an electron in the reduced molecule from its ground state to a highly excited state. This high-energy electron is then captured by a pheophytin (Pheo) acceptor molecule, converting chlorophyll P680 to P680$^+$ in one of the most powerful biological oxidation reactions. The reduced state of Pheo$^-$ then stimulates electron capture by the quinone acceptor Q_A, with subsequent serial electron transfer to Q_B, plastoquinones, cytochrome b6/f, and on to Photosystem I. At times of inactivity, the reaction centre is in an 'open' condition, with P680 in a reduced state and the electron acceptors Pheo, Q_A, and Q_B in an oxidized form. Photon excitation of P680 initiates the flow of electrons within the PSII reaction centre and on to the plastoquinone pool.

The time course of these molecular events is extremely rapid (Kolber and Falkowski, 1993). Following the initial impact of a photon of light, excitation of P680 occurs within 100fs, reoxidation of Q_A- occurs within 0.6 ms, and the onward passage of electrons from the plastoquinone pool occurs within 2–15 ms, depending upon water temperature. A single linked Photosystem II pathway can process up to 66 reactions per second at near freezing temperatures and up to 500 at about 50°C, provided that the PSII reaction centre continues to receive the next electron the instant that it reopens. Under such conditions, the pathway would be operating at full capacity and the rate of photosynthesis

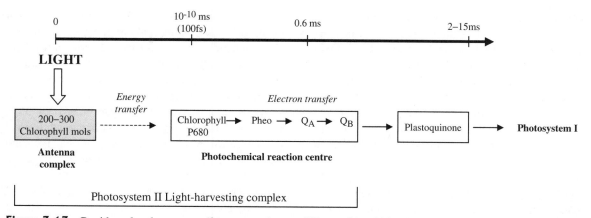

Figure 3.13 Rapid molecular events (femtoseconds to milliseconds) within the Photosystem II Light-Harvesting Complex (LHCII) of eukaryote algae (see also Figures 4.6 and 4.24)

would be at the point of light saturation by the incident photon flux density.

Light harvesting and electron transfer in the PSII light-harvesting complex initiate the process of photosynthesis in eukaryote algae (Section 4.2) and are a key part of the light-dependent reactions occurring in the thylakoid membranes (Figure 4.4). Further information on the PSII photochemical reaction centre is given in relation to photoinhibition and the effects of UV-B radiation on algal cells (Section 4.9, Figure 4.24).

Cell cycle processes

Cell cycle processes provide an example of longer-term physiological processes, operating over periods of minutes to hours. These processes include aspects such as new cell-wall formation, which is initiated once cytokinesis has occurred. In the case of diatoms, this involves the formation of a silicon deposition vesicle and the activities of special cell-wall forming proteins (Figures 5.28 and 5.29).

The duration of the cell cycle, from daughter cell, through growth and division, to the next daughter cells, varies considerably between algal species. Under conditions of optimum temperature, nutrient supply, and light, duration depends primarily on cell size – with the shortest cell cycles being recorded for picoplankton such as *Synechococcus* (where

duration may be as low as 2 hours). Minimum cycle length for nanoplanktonic algae such as *Chlamydomonas* and *Chlorella* are about 6 hours, while larger single-celled organisms (e.g., dinoflagellates) and colonial forms typically have much longer cell cycles.

The overall duration of the cell cycle depends on the time required for doubling of the biomass, which in turn depends on the uptake of sufficient quantities of nutrients such as carbon, nitrogen, and phosphorus. Where this uptake occurs as a simple diffusion process (e.g., CO_2), the minimum time taken for completion of the activity can be calculated in terms of physical parameters and cell size.

Time required for diffusion of CO_2 in *Chlorella*

The number of moles of a solute (n) which can diffuse across an area of cell membrane (a) per unit time (t) depends on the gradient in solute concentration (C_o) at the cell surface (dC_o/dx) and the coefficient of molecular diffusion (m) of the substance. The value for 'n' can be determined (Reynolds, 1997) from the equation:

$$n = a\,m(dC_o/dx)t \qquad (3.5)$$

Using the calculations of Reynolds (1997), a single spherical cell of *Chlorella* with a diameter of about 4×10^{-6} m would have values for surface area (a) and volume (v) of 50.3×10^{-12} m^2 and 33.5×10^{-18} m^3 respectively, and a carbon content of 0.63×10^{-12} mol. Aquatic concentrations of

CO_2 in air-equilibrated conditions range from 11–23 µmol CO_2 l^{-1}. With an external CO_2 concentration of 11 µmol CO_2 l^{-1} and a molecular diffusion coefficient (m) of about 10^{-9} m^{-2} s^{-1}, the rate of solute diffusion (n) would be 275×10^{-18} mol s^{-1}. This assumes a constant diffusion gradient, with depletion of CO_2 from a water thickness equal to the cell radius. At this rate, the time taken to deliver the entire doubling requirement of carbon to the cell by simple diffusion would be about 2300 s or just over 38 minutes. Once carbon has entered the cell, the time requirements (per cell cycle) of natural daylight for photosynthesis and the duration of carbon fixation have been estimated at about 25 minutes and 4.4 hours respectively.

With other nutrients, temporal values for the doubling requirement uptake by *Chlorella* of nitrogen and phosphorus (by simple diffusion) have been calculated at 540 s (9 minutes) and 2360 s (just over 39 minutes) respectively (Reynolds, 1997).

Responses to environmental change may involve a gradation of temporal processes. The up-regulation of nitrate reductase (NR) during light-mediated nitrate assimilation, for example. involves both short-term (within minutes) regulation by an activation/inhibition system and longer-term (within hours) light-induced gene activation and NR induction (Figure 5.13).

3.7.2 Medium-term changes: algal succession

Algal succession involves temporal changes in the biomass and species composition of natural populations. These changes are an important aspect of aquatic ecosystems, since they typically define the major microbial biomass (micro-algae) within the environment and have a major impact on other fresh water biota.

Algal succession will be considered in relation to two particular examples – (a) seasonal changes in lake phytoplankton (1 year period), and (b) algal succession in the development of stream biofilm and periphyton communities (a period of weeks to months). Although these transitions provide good examples of medium-term changes in the time scale sequence (Figure 3.12), long-term changes in phytoplankton succession are also an important aspect

of lake development (see next section). Seasonal changes in lake phytoplankton and biofilm succession are similar in some respects. In both cases there is a clear algal transition, with early successional species being replaced by late successional species in a defined and predictable sequence. Algal succession in phytoplankton differs from that of biofilms in being relatively long-term, and being determined to a large extent by abiotic factors (e.g., light and temperature). Unlike biofilms, seasonal changes in phytoplankton populations do not involve permanent change either in the biota or in the physical environment.

The changes in species composition which occur during pelagic and benthic algal succession often involve interactions between algae. These interactions may be direct or indirect and are of four main types – induced environmental change, species competition, species antagonism, and non-competitive adaptation (Tuji, 2000; McCormick and Stevenson, 1991).

- *Induced environmental change.* Early successional species bring about changes in the local environment which either facilitate or inhibit the establishment and growth of subsequent species.

- *Species competition.* Competition arises when organisms depend on the same consumable resource and when that resource is present in limited supply. Consumption of this resource by one species limits access by other species and reduces their growth and development. Competition occurs in relation to three major resources – light, space, and nutrients.

- *Species antagonism.* Where direct inhibition of other species is mediated via anti-microbial compounds such as antibiotics.

- *Non-competitive adaptation.* Species change is determined largely by adaptations to particular environmental conditions which occur at different times in the sequence rather than by direct interactions between species.

Species succession in phytoplankton populations are driven mainly by non-competitive adaptation (e.g., early spring growth of diatoms under conditions which other algae cannot tolerate) and species competition. In contrast to this, community changes in biofilms are more equally determined by all of these factors.

Seasonal changes in phytoplankton populations

In temperate climates, well-marked changes in the total biomass and species composition of lake phytoplankton occur over the annual cycle.

Individual species show bursts of population growth, which typically last for 30–100 generations, and can react rapidly to shifts in environmental conditions (Heinonen *et al.*, 2000). Seasonal changes in a typical temperate eutrophic lake are shown in Figure 3.14 and can be separated into four major phases: spring diatom bloom, clear-water phase, mixed summer/autumn bloom, and over-wintering phase (Steinberg and Geller, 1993).

Spring diatom bloom The massive increase in population of diatoms seen in spring is triggered by

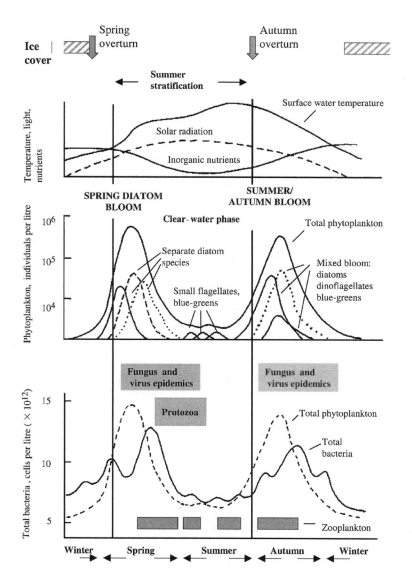

Figure 3.14 Effects of seasonal changes in the physical environment on microbial populations in the pelagic zone of a temperate lake (based on a figure from Horne and Goldman, 1994) Increases in solar radiation lead to the development of a spring diatom bloom, which reaches a peak after stratification. The bloom becomes limited due to depletion of inorganic nutrients in the epilimnion, and is followed by a clear-water phase (low algal biomass) and then a late summer/autumn bloom. Epidemics of fungi and viruses that infect the phytoplankton contribute to the termination of the blooms. Zooplankton grazing is particularly intense in the clear-water phase and involves a sequence of crustacean and rotifer populations throughout the summer and autumn period. Heterotrophic bacteria feed on the excretory products (dissolved organic carbon) of the phytoplankton and have major peaks which follow the algal blooms (Straskrabova and Komarkova, 1979). Populations of ciliate and flagellate protozoa, which graze the bacteria, follow on from the bacterial peaks

a rise in water temperature and an increase in light intensity and daylength. These changes initially occur in the unstratified lake, where diatoms are well suited to the turbulent state of the water and are able to grow under conditions of relatively low light intensity.

Early in the spring diatom bloom the lake becomes stratified, leading to the isolation of the surface layer (epilimnion). Nutrients within this layer – particularly Si and P – become depleted, ultimately falling to levels which may limit diatom growth. Limitation may also occur due to fungal infection and zooplankton grazing, leading to a crash in the diatom population.

Clear-water phase   The end of the diatom bloom is characterized by a period of low algal biomass and high zooplankton population. Under conditions of high zooplankton grazing pressure, algae are restricted to species with rapid growth and short cell cycles (*r*-selected species, Section 3.10.2). Zooplankton populations occur as irregular and patchy groups within the water column, and development of algal populations is transient within the rapidly changing biotic environment.

During this phase, phytoplankton is dominated by small unicellular algae, particularly cryptomonads and green algae. Towards the end of the clear-water phase the development of significant numbers of inedible algal species leads to a decline of zooplankton and the start of the summer bloom.

Mixed summer/autumn bloom   The mixed summer/autumn algal bloom develops under conditions of increased temperature, day length and light intensity, and decreasing epilimnion nutrients (particularly soluble N and P). This phase is characterized by a succession of blue-green algal populations and dinoflagellates (all of which are inedible to zooplankton) and diatoms. The occurrence of high populations of inedible algae means that the zooplankton population is maintained at a low level at a time when the algal biomass reaches maximum seasonal levels. The final part of the summer bloom occurs with the autumn overturn. Destratification of the lake results in a large-scale increase in inorganic nutrients (N, P, Si) which promotes the growth of a wide range of algae – including diatoms – at the end of the growing season.

Over-wintering phase   Progressive reduction in temperature and light leads to reduction in algal growth. Circulation of algae within the water column also leads to phytoplankton populations, with many algae sinking to the bottom of the lake. Throughout winter, phytoplankton biomass is limited and is dominated by diatoms and blue-green algae.

The four major phases seen in the above sequence may vary in magnitude within an individual lake from one year to the next. This can be seen in Figure 9.12, where the annual cycle for Rostherne Mere (UK) in the year 2000 was unusual in having the spring diatom bloom repressed by intense calanoid copepod grazing (Section 9.8). This led to a 'clear-water' phase which was dominated by the expected *r*-selected algae (particularly cryptomonads), but these developed to bloom proportions in this normally low-biomass phase.

The seasonal sequence also varies considerably with lake trophic status and water circulation characteristics. Oligotrophic and continuously mixed (polymictic) lakes tend to have diatom dominance throughout the entire summer period. In stratified lakes, increasing nutrient level causes a progressive decrease in the extent of the early seasonal diatom bloom (Table 10.4).

Algal succession in the development of stream biofilm and periphyton communities

Benthic colonization of stream environments frequently leads to the development of a periphyton community. This is dominated by attached algae, but also contains a wide range of associated microorganisms and invertebrates. The development and continuation of the periphyton community involves four major stages (Figure 3.15):

- initial colonization of the exposed surface,

- development of a diatom biofilm,

- colonization of the biofilm by filamentous algae to form a mature periphyton community,

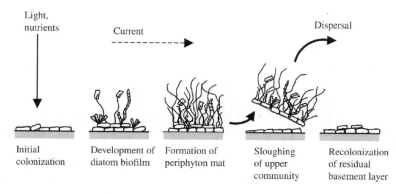

Figure 3.15 Accumulation of benthic algal cells on river substrata and the process of periphyton mat development (based on a figure from Tuchman, 1996)

- loss of fragments of the community due to the stream current, leading to exposure of earlier surfaces and regeneration of a new community.

Colonization and biofilm development Diatoms are particularly important in the initial colonization of exposed surfaces and the formation of an algal biofilm. Various authors (Tuji, 2000;

McCormick and Stevenson, 1991) have demonstrated a clear diatom succession, with transition from early colonizing (phase I) to mid-successional (phase II) and late-successional (phase III) species. The position of a particular species within this time sequence depends on its growth form, growth dynamics, and adaptation to a high- or low-light environment (Table 3.8).

Table 3.8 Diatom succession in algal biofilms (information from McCormick and Stevenson, 1991, and Tuji, 2000)

Species	Growth form	Growth Dynamics	Light Adaptation
Early colonizers (phase I)	*Attached diatoms*		High irradiance species High G_{max} and I_s values
Gomphonema angustatum	Short chains of cells	Settlement from high population level in water column	
Meridion circulare	Attached rosettes	Fast immigration due to rapid reproduction after attachment	
Surirella ovata	Single cells or pairs	Fast immigration due to rapid attachment	
Mid-successional species (phase II)	*Attached diatoms*		Low irradiance species Low G_{max} and I_s values
Cymbella sp., *Gomphonema olivaceum*	Vertical growth via long mucilaginous stalks	High growth rate maintained in dense algal conditions	
Late-successional species (phase III)	*Unattached, tangled diatoms*		Low irradiance species Moderate G_{max} and very low I_s values
Fragilaria vaucheriae	Unicellular	High growth rate maintained in dense algal conditions	
Aulacoseira varians	Chain forming		
Cocconeis placentula	Prostrate		

Light adaptation parameters: G_{max} – estimated maximum growth rate, I_s – estimated light intensity at half G_{max}

The rate of colonization of the newly exposed surface (r), which occurs by settlement from planktonic populations, depends on the summation of several processes (McCormick and Stevenson, 1991):

$$r = (R - D) + (I - E) \qquad (3.6)$$

where r = rate of accumulation of benthic diatoms, R = reproductive rate, D = death rate, I = immigration rate, and E = emigration rate.

In the above study, five diatom species dominated the colonization sequence – with three phase I and two phase II species. Early successional species were adapted in different ways for rapid colonization of the substrate by having high population counts in the water column (high immigration rate), high probabilites of attachment (low emigration rate), or high levels of division (high reproductive rate) on the exposed surface. Early colonizers had a limited growth form, which restricted their development to the immediate surface of the substratum. Mid-succession species were slow to establish, but then had high growth rates. The ability of these phase II diatoms to develop mucilaginous stalks gave them a vertically-extended growth which allowed them to displace earlier species by out-competing them for light, nutrients, and space. The final phase of diatom development is characterized by the entry of diatom species which cannot attach to the substratum, but become entangled within the vertical community of stalked diatoms. These phase III diatoms include both unicellular organisms and chain-forming species.

With development of the biofilm, there is a transition from high irradiance (early colonizers) to low irradiance (subsequent colonizers) species (Tuji, 2000) in line with the development of a vertical structure to the community. Early colonizers have high values for maximum growth rate (G_{max}) and half-saturation coefficient (I_s) – (Table 3.8). Phase II diatoms produced stalks at low-light intensities, increasing their elevation within the biofilm, and showing clear low-light adaptations. Phase III species were highly adapted to low irradiance, with very low I_s values. These diatoms are able to survive the very low-light levels within the middle of the dense diatom community, where they show substantial growth (moderate G_{max} values). Comparison of the irradiation responses of

these three groups of biofilm diatoms shows that the efficiency of light is closely linked to successional strategy.

In disturbance-prone ecosystems such as rapid-flowing head-water streams, algal succession may not progress beyond the formation of a diatom biofilm. In less stressful environments an extensive periphyton community may develop.

Maturation of the periphyton community

Further development of the biofilm involves the formation of a mixed algal community of stalked diatoms, filamentous green and blue-green algae, forming a loosely associated upper canopy. Resources become limiting at the base of the mat as cell densities increase, forming a vertical resource gradient. Cells in the upper canopy experience conditions of high light and access to inorganic nutrients in the water column, while cells in the lower part of the periphyton mat experience low light and low nutrient levels (Tuchman, 1996). Photoadaptation to low light involves an increase in photopigment concentrations to maximize photon collection. Heterotrophy may also be important as a survival mechanism for basement cells.

As the periphyton mat grows and extends into the water medium, periodic sloughing due to current shearing forces results in exposure of the original substratum or regions of the lower canopy. Recolonization of this exposed surface results in a continuation in the cycle of biofilm development.

In biofilm succession, where community change is occurring over a short (2–3 week) period, seasonal changes in the environment are not important. Algal succession is determined initially by non-competitive adaptation (establishment mechanisms) followed by induced environmental change (substrate formation), species competition, and possibly species antagonism.

3.7.3 Long-term changes: variations over a number of years

Inter-annual oscillations

Although lakes and reservoirs of temperate regions typically show a repeated annual pattern of seasonal change, these freshwater systems also show variations from one year to the next (Hoyos and Comin, 1999). Such inter-annual variability is an intrinsic feature of many water bodies and is caused by climatic fluctuations and the effects of human activities. Regional and local climatic variations are responsible for differences in a wide range of parameters which have a direct influence on phytoplankton populations, including physical (timing of ice break-up, mixing depth, water residence time), chemical (concentrations of inorganic nutrients), and biological (over-wintering survival of fish and zooplankton populations) characteristics. As a result of these variations, phytoplankton populations show marked annual differences in overall productivity and the precise sequence of species succession. Lake Sanabria (Spain) provides a good example of this, where climatic variations in rainfall have a marked effect on water residence time, affecting the accumulation of nutrients and the development of blue-green algal blooms (Hoyos and Comin, 1999).

In many cases, local conditions relate to broader climatic trends, such as the El-Niño Southern Oscillation. This low frequency (within a decade) variation has been shown (Anderson *et al.*, 1996) to correlate well with observed inter-annual variability in temperate lakes such as those of the Wisconsin area (USA). Distances over which these climatic influences extend may be considerable, as shown by the sensitivity of lakes in southern Europe to climatic variability caused by ocean–atmosphere interactions in remote areas of the South Pacific (Rodo *et al.*, 1997).

Long-term permanent changes in lakes

Long-term changes to lakes also involve more permanent alterations than the inter-annual oscillations seen previously, extending beyond a 10 year period and in some cases to hundreds of years. These chronic changes relate to internal factors (such as the build-up of organic material in the lake, leading to long-term eutrophication) or to external changes such as heavy-metal pollution, prolonged inflow of nutrients (eutrophication), and the effects of acid rain (acidification).

Confirmation of long-term changes in general ecology and resident phytoplankton requires an extended data series, which can be obtained by analysis of lake sediments. In many lakes, uninterrupted deposition of both phytoplankton and benthic algae occurs over a prolonged time period, leading to the formation of well-preserved layers in chronological sequence. Analysis of diatoms within these sediment cores is particularly useful, since these organisms can be readily identified from their preserved frustules and are useful indicators of particular pH and nutrient conditions.

Long-term changes in algal populations have been demonstrated from sediment analyses in relation to the following.

- *Heavy-metal pollution.* Studies by Cattaneo *et al.* (1998) on the sediments of Lago d'Orta (Italy) showed taxonomic shifts in the populations of algae (diatoms) and other lake biota (thecamoebians, cladocerans) over a 50 year period, as a response to industrial effluent contamination (primarily copper pollution). These studies have also demonstrated the long-term inverse relationship between size and environmental stress (Sections 1.6 and 3.4.3).

- *Eutrophication.* Various indices of algal species composition can be employed in the assessment of trophic status (Section 10.3.1), including the proportion of araphid pennate/centric diatoms (A/C ratio). This has been used by Byron and Eloranta (1984) to study long-term changes in Lake Tahoe (USA), where sediment analyses demonstrated a dramatic shift in the A/C ratio during the 1950s (Figure 10.3). These diatom changes indicate a sudden transition from ultra-oligotrophic to oligotrophic status at a time of increased lakeshore human habitation.

- *Acidification.* Studies on the sediments of Loch Dee (UK) have demonstrated a long-term acidification of the lake (pH 6.3 to pH 5.6) over a 130 year period (Flower *et al.*, 1987). During this time, diatoms typical of near-neutral conditions (*Cyclotella comensis*, *Asterionella formosa*) have been replaced by diatoms more adapted

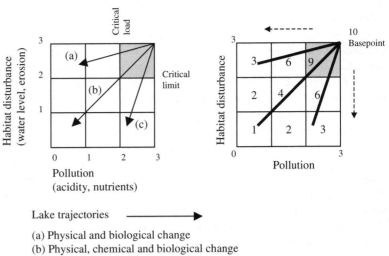

Figure 3.16 Diagrammatic representation of long-term biological changes in lakes in relation to pollution and habitat disturbance. *Left*: hypothetical changes over time for three lakes (a), (b) and (c). *Right*: quantitative expression of these changes in relation to the original (or natural) state, with 1 = most changed, 9 = least changed, and 10 = pristine. Restoration of the lakes, with changes in phytoplankton and other biota back to the original state, would involve a return to the 'target box' (shaded square) (based on a figure from Battarbee, 1999)

to a lower pH (*Eunotia veneris*, *Frustulia rhomboides*).

With many lakes, long-term effects not only involve changes in water quality but also physical disturbance, including changes in water level and erosion. These different effects are largely independent and can be visualized in terms of a 'naturalness index', representing the degree to which a lake has changed in relation to its natural baseline. This concept has been developed in relation to lake rehabilitation, where the direction of restoration would be towards a target ecosystem of minimal habitat disturbance and pollution, represented by the 'target box' in Figure 3.16 (Battarbee, 1999).

3.8 Phytoplankton distribution within the water column

The vertical distribution of phytoplankton in a stratified lake is highly dynamic and is closely related to environmental parameters and their seasonal/diurnal changes within the water column. It has major importance for the growth and survival of the algae concerned, and also has implications for the biology of other organisms such as bacteria, protozoa, and zooplankton. The vertical distribution of phytoplankton is considered initially in relation to a particular lake (case study), with subsequent discussion of diurnal migration and vertical positioning.

Case study 3.1   Vertical zonation of Phytoplankton in a stratified lake

Distribution of algae in the water column of a mixed (unstratified) lake may be relatively uniform, but the development of stratification imposes major vertical differences in the physical–chemical environment and in the occurrence of algae. A specific example of algal zonation is illustrated in Figure 3.17, which shows the mid-day vertical pattern of different phytoplankton groups during bloom formation in a stratified eutrophic lake (Rostherne Mere, UK). At this time of year, the lake was dominated by dinoflagellates (mainly *Ceratium hirundinella*) and blue-green algae (mainly *Anabaena minima*), populations of which showed high levels in the epilimnion (0-4 m), low levels in the metalimnion (4-8m) and moderate levels in the hypolimnion (10-22m). Other algal groups–diatoms, green algae, chrysophytes and cryptophytes occurred at much lower biomass levels, but had a broadly similar distribution.

These patterns of algal distribution relate to zonation in the physico-chemical environment (Figure 3.18), with highest algal populations occurring under conditions of high light intensity, temperature, and oxygen concentration (epilimnion). Algal populations were at a minimum under conditions of severe oxygen depletion (metalimnion), rising to moderate levels in the lower part of the water column – where oxygen levels were restored and concentrations of major nutrients (P, N, Si) were at a maximum.

Although the distributions of particular groups were broadly comparable, differences in detail did occur. Dinoflagellates and blue-green algae had their highest surface concentrations within the top 2 m of the surface, for example, while diatoms peaked at a depth of 2-4 m, and green algae at 4-6 m. Individual species also showed major differences in their vertical distribution, suggesting that different algae have different strategies determining their position in the water column.

In general, the distribution of algae at any one point in time within the water column of a lake will depend on three major factors – active migration, passive movement, and pre-existing population levels (Figure 3.19).

3.8.1   Active Migration of Algae

Active migration involves the ability of algae to move themselves independently within the water body and is mediated by the intracellular production of gas vacuoles (buoyancy mechanism) or by some type of active motility (flagella or streaming movements). Algae (such as diatoms and non-motile chlorophytes) which do not possess either of these mechanisms are not capable of independent movement within the water column. During the period of major growth within the water column, migration of phytoplankton involves two distinct but interrelated activities – diurnal migration (large-scale movement) and depth regulation (small-scale vertical positioning) in relation to external stimuli. The entry of algal cells into the water column from the benthic sediments at the beginning of the growth phase is a further major migratory activity of these organisms (Section 3.10.1).

Individual algae and algal groups differ in the extent to which they exhibit diurnal migration or vertical positioning (Table 3.9). For three of the groups (chrysophytes, cryptophytes, and green algae) diurnal migration is not well documented, but these algae do show strong vertical positioning within lake strata – in some cases over long periods of time (Sandgren, 1988b). Other groups, such as dinoflagellates and blue-green algae, are more defined in their patterns of diurnal migration, but also show vertical positioning within their diurnal cycle.

Diurnal migration of algae

Many freshwater algae have the ability for diurnal migration between the lake surface (epilimnion) and deeper layers (metalimnion or hypolimnion), moving up the water column during the morning and down later in the day. This migration allows these organisms to combine day-time photosynthesis in the euphotic zone with night-time uptake of nitrates and phosphates in the nutrient-rich hypolimnion, and has been investigated particularly in relation to dinoflagellates and blue-green algae.

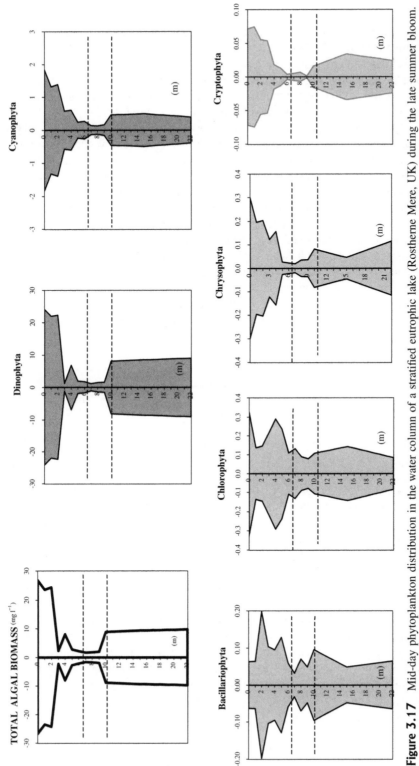

Figure 3.17 Mid-day phytoplankton distribution in the water column of a stratified eutrophic lake (Rostherne Mere, UK) during the late summer bloom. The biomass of total hytoplankton, and algae within different taxonomic groups, is expressed as mg l^{-1} (total width of each graph) and shows considerable variation with depth. The main thermocline is indicated by broken lines. The vertical distribution in total algal biomass (top left) may show major differences from the profile of chlorophyll concentration (Section 4.6.3) (taken, with permission, from Levado, 2001)

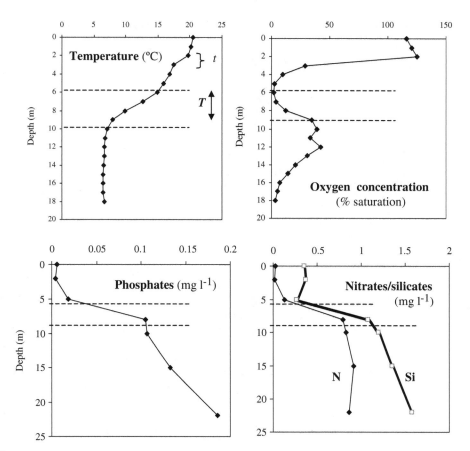

Figure 3.18 Physico-chemical characteristics of the water column during the late summer (August) bloom of a eutrophic lake (Rostherne Mere, UK). The water column shows clear thermal stratification, with a well marked thermocline (T) at 6–10 m, and a hypolimnion below 10 m which is stable at 6°C. The epilimnion is biphasic, with a temporary thermocline (t) at 2–3 m. Depth changes in oxygen concentration are highly complex, with supersaturation (reaching 130 per cent) at the top of the epilimnion – typical of an intense surface bloom, dropping to minimal levels in the lower epilimnion and thermocline. The upper hypolimnion has a broad peak of oxygen concentration (reaching 50 per cent saturation) corresponding to a region of high algal biomass (see Figure 3.17). Inorganic nutrients (phosphates, nitrates, silicates) are almost completely depleted in the epilimnion, but reach high levels in the hypolimnion (taken, with permission, from Levado, 2001)

Diurnal migration of dinoflagellates As with blue-green algae, the diurnal movement of dinoflagellates relates to their strong migratory mechanism and their ability to dominate stratified lakes in late summer. At this time of year, these algae are only able to maintain their growth by nutrient uptake from the lower parts of the water column. Some evidence for hypolimnetic uptake has been provided by studies of Mitchell and Galland (1981), where

sampling of *Gymnodinium* at Lake Mahinerangi (New Zealand) in the early morning showed that cells located in the hypolimnion (12 m depth) had higher concentrations of phosphorus compared with cells present at the lake surface.

Dinoflagellates, such as *Peridinium* and *Ceratium*, are strongly motile and positively phototactic, with clear patterns of diurnal migration – rising to the lake surface during the light period and descending

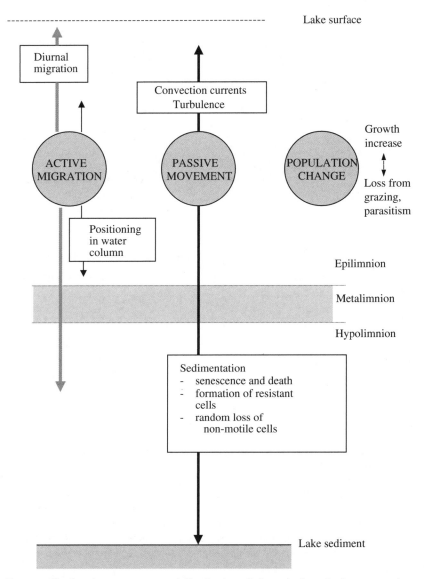

Figure 3.19   Factors affecting the occurrence and distribution of phytoplankton in the water column of a stratified lake

or dispersing at night. The diurnal migration of these organisms is closely tied to a circadian rhythm in their phototactic response, which is high in the first half of the day and subsequently declines. Patterns of dinoflagellate diurnal migration vary with light gradients, and are also influenced by temperature variations, nutrient limitation, and population age (Pollingher, 1988).

Buoyancy migration and gas vacuole formation in blue-green algae   Buoyancy migration in blue-green algae is promoted by the formation of gas vesicles (Waaland and Branton, 1969), which are present within larger membrane-bound gas vacuoles. The only other group to possess these gas vesicles are bacteria, and they are almost exclusively restricted to microorganisms living in

Table 3.9 Migratory characteristics of motile algae within the water column of standing waters

Algal group	Diurnal migration	Positioning in water column
Blue-green algae[a]	Effective buoyancy mechanism with well-defined diurnal migration	Unicellular algae often form stratified populations in oligotrophic lakes
		Colonial forms show rapid positioning
Green algae[b]	Demonstrated in *Pandorina*, but not well documented for other motile greens	Various motile greens reported to occur in restricted depth zones
Dinoflagellates[c]	Typically strongly motile and phototactic with well-defined diurnal migration	Vertical positioning in relation to light and oxygen (avoidance of anoxia)
Cryptomonads[d]	Demonstrated for some species	Many occur as stably stratified populations
Chrysophytes[e]	Demonstrated for some species in special (very stable, shallow epilimnion) habitats, but not a general characteristic	Many have ability for phototactic positioning in defined strata, e.g., metalimnion, stable population under ice

References: [a]Paerl (1988); [b]Happey-Wood (1988); [c]Pollingher (1988); [d]Klaveness (1988); [e]Sandgren (1988a)

aquatic habitats. They have been found in over 150 species of prokaryotes present in 11 phyla within the Kingdom Eubacteria (including blue-green algae) and two phyla in the Kingdom Archaebacteria (Table 1.2).

Gas vesicle structure and composition. In all the organisms studied, gas vesicles have a similar cylindrical morphology and are constructed from homologous proteins. These features, together with fundamental gene sequence homologies, suggest that the ability to migrate within the water column via gas vesicle/buoyancy regulation has evolved once in an ancestral prokaryote, and is now an important feature of several major groups of aquatic prokaryotes.

Gas vesicles are cylindrical structures, occurring as stacks within gas vacuoles. These reduce the specific gravity of cells below that of the surrounding aquatic medium, allowing the organism to rise and become positioned at a particular depth within the water column. A full account of the structure, synthesis, and role of gas vesicles in regulating buoyancy has been given by Walsby (1994). These inert, gas-filled structures are produced by the synthesis of two key proteins – GvpA and GvpC. GvpA forms the main framework of the vesicle, and is a small MW (7.4 kDa) hydrophobic protein that is

arranged in a linear crystalline array. GvpC is a larger (22 kDa) protein which stabilizes the structure by adhering to the inside of the GvpA framework. GvpC contains a highly conserved motif of 33 amino acid residues, and differs markedly from GvpA in having s preponderance of hydrophilic residues. The molecular genetics of vesicle synthesis has now been well characterized, with two genes (*gvpA* and *gvpC*) encoding the two major proteins.

Gas vesicle formation and collapse. The buoyancy of blue-green algal cells is lowered by loss of gas vesicles (due to collapse at high turgor pressure) and is raised by vesicle increase. Formation of new gas vesicles occurs by a process of self-assembly, following the expression of GvpA and GvpC by activation of *gvpA* and *gvpC* genes. In some blue-green algae, activation of *gvp* genes is constitutive (occurring after existing vesicles have been destroyed by pressure) while in others (e.g., *Pseudanabaena*) it is induced at low light intensity. Dissolved gases subsequently enter the gas vacuole by diffusion, and the buoyant cell rises in the water column.

Role of vesicles in diurnal migration. Diurnal migration results from alternating phases of decreasing (daytime) and increasing (nighttime)

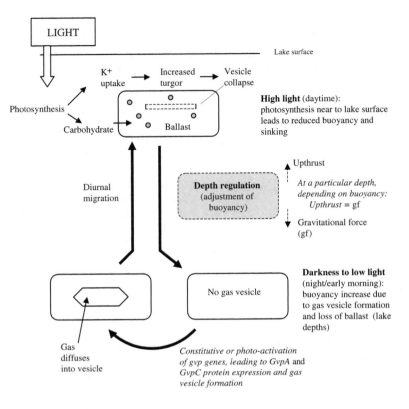

Figure 3.20 General scheme for buoyancy control of diurnal migration and regulation of depth in blue-green algae, based on data from a variety of species (Walsby, 1994). Periodic diurnal migration (⟶) results from an alternating decrease in buoyancy during daytime, and an increase at night/early morning. For simplicity, a single gas vesicle is shown (not drawn to scale, and with surrounding gas vacuole omitted). Carbohydrate ballast – ○

buoyancy (Figure 3.20). During daytime, high rates of photosynthesis at the lake surface result in decreased buoyancy, causing the organism to sink in the water column at the end of the light period. The decreased buoyancy can arise in three ways:

- formation of photosynthetic product, resulting in the formation of carbohydrate storage material (mainly glycogen), which acts as ballast – increasing the specific gravity of cells;

- photosynthetically-mediated uptake of K^+ leads to an increase in turgor pressure, which may lead to collapse of the gas vesicles and a drop in buoyancy.

- inhibition of *gvpA* and *gvpC* gene expression may occur at high light intensity, preventing the formation of new gas vesicles.

Different blue-green algae appear to differ in the relative extent to which these three processes occur.

Reverse processes occur at night, with removal of ballast, loss of turgor (K^+) and formation of new vesicles all contributing to an increased buoyancy and a rise in algae within the water column at the beginning of the next light period.

In conditions of bright light, the algae may position themselves at a particular depth (or depths) within the water column (see next section). At this point, the light intensity promotes sufficient ballast formation to counteract the upward thrust from the gas vesicles, which are continuously formed during the division process.

In some situations, environmental conditions may prevent ballast formation, leading to a loss of buoyancy control. This has been reported in conditions of iron depletion, where lack of uptake of Fe (III) results in a block of photosynthesis. Under such conditions, the blue-green algae are trapped at the surface of the lake, unable to carry out either diurnal migration and or depth regulation. Inability to escape from the damaging radiation causes death to a large part of the population within a relatively short (~2 hour) period.

Diurnal migration and dissolved organic carbon accumulation Diurnal migration of phytoplankton occurs as a direct response to changes in the abiotic lake environment (light regime), and can in turn lead to further abiotic changes, particularly in relation to water chemistry. Various studies have demonstrated diurnal changes in the concentration of dissolved organic carbon (DOC) in lakes (Burney, 1994), which in some cases can be linked to migration of algae within the water column and the daytime release of DOC in the epilimnion.

Analysis of surface water in Lake Havgardsjoen (Sweden), for example, demonstrated a huge mid-day pulse of DOC (>129 mg l^{-1}), which appeared to relate to the presence of migratory populations of blue-green algae and dinoflagellates at this time (Søndergren, 1984). In subsurface water, DOC increased during the day to reach a maximum of 15 mg l^{-1}, subsequently decreasing to less than 10 mg l^{-1} at night. Elevated daytime concentrations of DOC have also been demonstrated in other lakes, including Lake Eyrie, USA (diurnal range 2.3–4.2 mg l^{-1}, Nalewajko *et al.*, 1980) and Lake Oern, Denmark (diurnal range 2.7–3.5 mg l^{-1}, Søndergaard, 1984). In other cases, where diurnal fluctuations of DOC have not been demonstrated, the lack of correlation to algal migration has been variously attributed to high levels of externally-derived DOC (obscuring the internal fluctuations), close coupling of DOC utilization to its release, and lakes where *in situ* biological activity is minimal.

More recent studies (Burney, 1994) have concentrated on specific groups of compounds rather than total DOC, demonstrating diurnal fluctuations in the lake water concentrations of a range of algal exudates – including dissolved free amino acids (DFAA), dissolved mono- and disaccharides, and the metabolic regulator cyclic adenosine $3':5'$-monophosphate (cAMP).

Vertical positioning in the water column

Many motile algae are able to position themselves by active movement (taxis) at particular depths in the water column in response to external stimuli. These include gravitational field (gravitaxis), light direction and intensity (phototaxis), temperature and chemical (chemotaxis) gradients – including variations in oxygen concentration. Recent studies on phototaxis in phytoplanktonic flagellates (Clegg *et al.*, 2003) have shown that the preferred light level matches the optimal irradiance for growth in some organisms, but is sub-optimal in others, suggesting that light may also be used as an ecological indicator of other environmental conditions.

Blue-green algae carry out depth regulation by adjustment of buoyancy (Figure 3.20). This involves control of gas vesicle gene expression, collapse of vesicles by alteration of turgor pressure, and formation of carbohydrate ballast by photosynthesis. Stabilization at a particular depth occurs when the opposing forces of upthrust and gravitation are equal.

Depth regulation may involve the long-term establishment of stable populations at fixed depths in the lake (no diurnal migration) or temporary positioning in relation to ambient conditions (combined with diurnal migration).

Long-term stratified algal populations In thermally stratified lakes, chrysophytes, cryptophytes, flagellate green algae, and unicellular blue-green algae (picoplankton) often occur as long-term stable populations at fixed depths within the water column. In some situations, these algae are restricted to the metalimnion layer where they persist throughout summer at reduced temperature and very low light. These adverse environmental conditions are compensated for by reduced zooplankton grazing, close proximity to nutrient rich hypolimnion and the ability of many algae to supplement reduced carbon fixation by heterotrophic nutrition. Such metalimnetic algae are actively growing cells, forming a substantial biomass (chlorophyll maximum) and are not simply sinking out of the epilimnion layer. Lakes that develop metalimnetic maxima are often of low nutrient status, with low productivity, clear water, and a euphotic zone extending into the metalimnion (Klaveness, 1988). The importance of motility in establishing algal zonation is emphasized by depth analysis of green algae, with motile greens such as *Pandorina* and *Volvox* typically forming narrower depth zones than non-motile greens such as *Sphaerocystis* and *Oocystis* (Happey-Wood, 1988).

Temporary positioning of algae   Many motile algae (including flagellate and buoyant species) are able to make rapid and short-term depth adjustments within the water column in relation to changing growth-related environmental factors. In some cases, the responses of individual algal species to particular stimuli have been studied under controlled laboratory conditions. These laboratory studies often employ continuous monitoring of single cells and whole populations of algae exposed to different stimuli, and are particularly useful since they reveal:

- the type and precision of the response,

- changes in motility,

- changes in short-term responses within the diurnal cycle,

- adverse effects of high light intensity (photoinhibition) on algal orientation and motility (see Section 4.9).

Results obtained from laboratory studies suggest that the upward movement of flagellate algae within the water column at low light intensities is typically mediated by positive phototaxis, supported by negative gravitaxis (Hader, 1995). At higher levels of illumination the upward movement of cells is normally reversed with a change to negative phototaxis. In contrast to this, some organisms show a change from positive phototaxis to diaphototaxis (movement perpendicular to the incident light beam) at higher light intensity. This has been demonstrated particularly in dinoflagellates where it provides an effective mechanism for cells and populations to stay at a selected level within the water column.

The wide variation in vertical positioning of micropopulations or individual cells within the water column at any point in time suggests considerable heterogeneity within the overall population in terms of environmental responses. Although algae such as *Ceratium* and *Peridinium*, for example, may show large-scale lake-surface accumulation in lakes, individual cells are distributed throughout the water column. These cells are in stable equilibrium with their microenvironment and have higher chlorophyll contents compared with cells higher present at a higher light intensity (Sandgren, 1988a). The vertical spread of these cells is limited, however, to aerated regions of the water column. Dinoflagellates are very sensitive to oxygen concentration, and their vertical migrations do not extend into anoxic hypolimnia.

Algae also differ markedly in the speed of their positioning. Blue-green algae in particular are capable of very rapid vertical movements, with upward speeds of about 3m hour^{-1} being recorded for *Microcystis* and *Aphanizomenon* (Paerl, 1988). Such migration rates greatly exceed the flagellar swimming speeds of motile eukaryote algae such as dinoflagellates and green algae. The vertical distribution of these organisms may be quite complex, with persistent buoyancy leading to the accumulation of surface scums overlying populations of migrating cells.

3.8.2   Passive movement of algae within the water column

Passive movement of algae within the water column is brought about by water movement (convection currents, wind turbulence) and sedimentation.

Wind disturbance of the epilimnion can have a major effect on the distribution of algae in lakes, breaking stratification and disrupting the migration and positioning patterns noted earlier. Surface accumulation and bloom formation of blue-green algae, for example, requires a well-stratified and static water column for full development. In certain situations, convection currents may also be important for water movement, and have been implicated in algal mixing within the epilimnion (Happey-Wood, 1988).

Passive sinking of algae (sedimentation) within the water column potentially affects all algae, since all objects are subject to gravitational force. The tendency for sedimentation is counteracted by algal motility, specific buoyancy mechanisms, and water turbulence, and is influenced by intrinsic features of cell size, shape, and density as defined by the Stokes equation (see previously). Sedimentation of algae within the water column is a continuous process resulting from senescence and cell death (loss of motility mechanism), formation of resistant cells (increased density), and random loss of non-motile cells from the euphotic zone. Sedimentation of diatoms, which are maintained in suspension by water turbulence, is increased by the development of a static water column at the onset and progression of stratification.

Sedimentation is of particular significance to diatoms because of their high density (silica cell wall) and lack of planktonic motility. These organisms have evolved either in the direction of reducing the rate of sedimentation or by adapting their life cycle to periodic sinking. The nanoplanktonic diatom *Stephanodiscus hantzschii* provides an example of the former, where the evolution of small cell size reduces sinking rate to <0.05 m day^{-1}, and promotes planktonic existence. At the other end of the scale, the filamentous diatom *Aulacoseira granulata* has a high density (1.13 g ml^{-1}) and sedimentation rate (>1 m.day^{-1}) and has evolved a life cycle where planktonic existence is restricted to a turbulent part of the annual cycle (Sommer, 1988).

3.9 Freshwater algae and nutrient status of the environment

The development of phytoplankton populations in lakes and rivers is related to a range of environmental factors, including temperature, light, soluble gases, and levels of dissolved inorganic nutrients in the water body. Concentrations of silicate, nitrate, and phosphate, in particular, influence the growth of algae and the species present within the ecosystem. The quantitative effects of inorganic nutrients in

promoting algal growth are considered in Chapter 5, and the role of algae in the biological effects of environmental nutrient enrichment (eutrophication) is considered in Chapter 10.

3.9.1 Phytoplankton species composition and lake nutrient status

The trophic state of freshwater lakes has a major impact on the type of algae that dominate the ecosystem and in the seasonal phytoplankton succession (Table 10.4).

Ecological preferences in lakes

General ecological preferences of different algal groups within the lake environment are summarized in Table 3.10. It should be emphasized, however, that individual species and groups can often exist in a wide range of environments, so this summary must be regarded as a very broad overview. The presence or absence of particular algal groups and species has considerable diagnostic value in the assessment of lake trophic status, as discussed in Section 10.3. Although nutrient status (particularly the concentrations of phosphates and nitrates) is clearly important, other environmental aspects such as temperature, pH, salinity, and water turbulence are also relevant. Some of these aspects, such as oligotrophic status and acidity, tend to be correlated, with ologotrophic algae such as chrysophytes being tolerant of both conditions.

In green algae, blue-green algae, and to a lesser extend the chrysophytes, the broad spectrum of morphology is paralleled by a related range of ecological preferences. In each of these phylla, the small unicellular subgroups tend to be particularly characteristic of oligotrophic waters, while the large colonial forms dominate eutrophic lakes. The ability of small, single-celled algae to exist and out-compete larger colonial forms at low nutrient level can be related to their higher surface/volume ratio and resulting greater efficiency in nutrient uptake (see previously).

The major environmental requirements of planktonic diatoms are water turbulence and high silicon

Table 3.10 General ecological preferences (standing waters) for algae within the major taxonomic groups (information taken mainly from Sandgren, 1988b, and Wehr and Sheath, 2003)

Group	Typical member of group	Preferred nutrient status	Physico-chemical preferences	General occurrence
Blue-green algae				
• Small coccoid forms	*Synechococcus, Aphanothece*	Range from oligotrophic to eutrophic waters	Some adapted to low light regimes, altered spectral conditions, high salinity	Particularly important in large oligotrophic and mesotrophic lakes
• Large colonial algae	*Anabaena, Microcystis*	Mesotrophic–eutrophic Many can grow at low N/P ratios	Establish massive blooms under conditions of low turbulence and high irradiation Can tolerate low CO_2, high O_2 and high pH	Common in productive lakes throughout temperate and tropical regions Frequently occur as late summer dominants
Green algae				
• Small flagellates	*Chlamydomonas*	Common in oligotrophic lakes	Some species of *Chlamydomonas* typical of acid conditions	Occur in a wide range of standing waters,
• Colonial flagellates	*Eudorina, Volvox*	Mesotrophic to eutrophic waters	Often found in temporary conditions – ponds and rain pools	Do not usually form dominant populations.
• Desmids	*Staurastrum, Cosmarium*	Oligotrophic to mesotrophic lakes (low conductance and low Ca content)	Greatest diversity seen in waters with pH 4–7	Greatest diversity occurs on the sediment surface in shallow areas. Relatively few species are truly planktonic

Group	Examples	Nutrient status	Ecological characteristics	Distribution
• Non-motile coccoid and colonial	*Chlorella, Sphaerocystis*	Oligotrophic to eutrophic lakes	Suited to early stratification when light is not limiting and turbidity/turbulence are limited	Ubiquitous and widely-distributed
Dinoflagellates	*Ceratium, Peridinium*	Typically dominant in mesotrophic to eutrophic lakes	Prefer abundant light. Avoid salinity (halophobic), anoxic conditions. Growth and migration disturbed by turbulent conditions	Frequently occur as late summer dominants in temperate mesotrophic and eutrophic lakes
Cryptomonads	*Rhodomonas, Cryptomonas*	Found over a wide range of nutrient status	Can tolerate variable salinity and deviant chemical composition	Wide range of spatial and temporal niche occupation in temporal lakes
Chrysophytes	*Ochromonas* (oligotrophic) *Dinobryon* (meso/eutrophic)	Typically dominant in oligotrophic lakes	Low summer temperatures. Low conductivity. Prefer neutral–acid lakes	Common in north temperate oligotrophic lakes. Also seen in alpine, arctic, and some tropical lakes
Diatoms	*Cyclotella*[1], *Asterionella*[2] *Stephanodiscus*[3]	Preferences range from oligotrophic[1] to mesotrophic[2] and eutrophic[3] conditions	Non-motile plankton, so need water turbulence to stay in suspension. Many adapted to low light, low temperature and survival on sediments	Occur throughout all standing water systems as planktonic and benthic populations Dominate most temperate lakes in winter and spring

level. Although this group does not show any over-all preference in terms of nutrient status, individual members have strict trophic preferences over the whole spectrum of nutrient levels.

Eutrophication and phytoplankton change

Increased inorganic nutrient status (particularly in relation to nitrate and phosphate availability) leads to changes in phytoplankton composition and bio-mass. Although such nutrient changes are normally considered in relation to human activities (anthro-pogenic eutrophication – Chapter 10), they may also occur as natural processes.

The term 'eutrophication' was originally used in reference to the natural ageing process of lakes, and for a long time it was assumed that oligotrophic lakes were 'primitive', 'evolving' to eutrophic water bodies as part of a natural progression. Although this has certainly occurred in some cases, the long-term continuation of oligotrophic lakes such as Lake Baikal (Russia) indicates it is not universal.

In lakes where 'natural eutrophication' occurs, long-term inflow of water carries soluble nutrients into the lake where they are taken up by phyto-plankton, sequestered within biomass, and depos-ited onto sediments in a process of continued accumulation. This leads to increased 'internal loading' and an overall increase in nutrient status. The degree to which such accumulated uptake can occur depends partly on the mean annual concen-tration of nutrient which occurs in the system. In the case of phosphorus, for example, this may be expressed (Vollenweider and Kerekes, 1980):

$$[P_\lambda] = [P]_q (\tau_p/\tau_w) \qquad (3.7)$$

where $[P_\lambda]$ is the mean annual phosphorus concen-tration in the water body, τ_p and τ_w are the respec-tive mean retention times of phosphorus and water in the system, and $[P]_q$ is the average phosphorus concentration in the inflow. The values τ_p and τ_w depend on the internal hydrology (area loading, water circulation, mean depth) of the lake, and $[P]_q$ depends on the external hydrology and catch-ment area. In sub-polar lakes such as Lake Baikal,

the limited phytoplankton growth season will also impose a restriction on nutrient uptake and biomass accumulation.

3.9.2 Nutrient status of river environments – effect on benthic algal biofilms

Benthic algal biofilms dominate the river environ-ment and contain a complex mixture of algae including green algae, blue-greens, and diatoms. The algal composition of biofilms is influenced by a range of interrelated environmental factors includ-ing water flow, water depth, light regime, chemistry, nutrient status (eutrophication), pollution, and graz-ing. The effects of these parameters have been investigated particularly in relation to diatoms (Lock et al., 1984; Round, 1993) since these organ-isms:

- occur throughout all rivers – diatoms comprise the largest and most prevalent group of algae in river systems,

- can be rapidly and easily sampled,

- have species that are very sensitive to water quality (chemistry), eutrophication, and pollution – the ecological requirements of species are well documented,

- have a rapid growth cycle and respond rapidly to perturbation of the environment.

Rivers form a chemical and biological continuum, with increases in nutrient content from the upper reaches to the estuary. Longitudinal changes in physical and chemical parameters are paralleled by qualitative and quantitative changes in the algal biofilm. These have been summarized in some detail for diatoms by Round (1993), who has sug-gested longitudinal subdivision of increasingly eutrophic rivers into five major zones (Table 3.11).

The increasing state of eutrophication seen in Table 3.11 would be typical of a river that com-mences with clean water at the source (Zone 1), followed by increasing input of nutrients (agriculture,

Table 3.11 Diatom transition along a river course with increasing nutrient pollution – division into five major zones (information taken from Round, 1993)

Zone	Physical Parameters and Water Quality	Dominant Diatom Species	General Aspects of Biofilm
ZONE 1 Clean water in uppermost reaches	High water flow pH 3.6–4.1 Clear, well-aerated water	*Eunotia exigua* *Achnanthes microcephala* Small-celled species Directly attached to the stone surface	Little development of an associated mucilage community
ZONE 2 Nutrient richer, higher pH	pH 5.6 – 7.1 Clear, well-aerated water	*Hannaea arcus* *Fragilaria capucina* *Achnanthes minutissima*	Development of a mucilaginous biofilm with complex community
ZONE 3 Nutrient rich	pH 6.5 – 7.3	*Achnanthes minutissima* *Cymbella minuta* *Amphora pedunculus* *Cocconeis placentula*	Wide range of diatom species Complex, well-structured community
ZONE 4 Eutrophic with restricted diatom flora	Moderate organic pollution Decline in water quality – high nitrogen levels	*Gomphonema parvulum*	Decline in water quality is indicated by loss of the *Amphora, Cymbella, Cocconeis* complex
ZONE 5 Diatom flora grossly restricted by organic pollutant influx	Highly eutrophic, organically polluted water Extremely poor water quality	*Navicula atomus* *Nitzschia palea* Small-celled diatoms	Limited diatom community

sewage) lower down (Zone 5). Individual rivers vary in their development of these zones. Most eutrophic rivers that occur in the UK, for example, are typical of Zone 3 over most of their length.

As with the variation in lake nutrient status seen earlier, changes in river trophic status are also accompanied by other parameters such as pH and oxygenation. Different trophic states (Zones 1–5) are characterized by particular dominant species and also by groups or assemblages of diatoms. The *Cymbella, Amphora, Cocconeis* complex seen in Zone 3, for example, is typical of moderately rich waters, but is lost with the pollution input seen in Zones 4 and 5.

D. STRATEGIES FOR SURVIVAL

3.10 Strategies for survival: the planktonic environment

Complex interactions and constraints within the freshwater environment have resulted in the evolution of a wide range of strategies for survival. This is particularly the case for planktonic algae, which have to contend with a number of interrelated growth-limiting factors, including:

- competition with other algae for light, space, and nutrients,

- limiting light conditions – this includes elimination from the euphotic zone, caused by gravitation and water movement,

- adverse winter conditions of light and temperature (seasonal variation in temperate lakes),

- fungal and viral attack,

- grazing activities of zooplankton, protozoa, and fish.

Planktonic algae have evolved different ways of dealing with these constraints. These include various strategies for maintaining their position in the water column (Section 3.8), avoidance of grazing (Section 9.8), and adaptations to low light (Section 4.6) and nutrient (Section 5.8) levels – all of which lead to major differences between algae in structure, physiology, and life style. Two further aspects will be considered here, the ability of algae to exploit environmental conditions in a particular part of the seasonal cycle by adopting a meroplanktonic life style, and adaptations to unstable and stable environments by *r*- and *K*-selection. The development of heterotrophic nutrition in relation to light limitation (and high nutrients) is considered in Section 3.11, and the adoption of strategies to survive the limitations of snow environments in Section 3.12.

3.10.1   Meroplanktonic algae

The majority of lake algae are holoplanktonic, being present within the water column over much of the annual cycle, and competing with other algae for light and nutrients during this period. In contrast to this, meroplanktonic algae have evolved a relatively limited planktonic existence, with a large part of the annual cycle being spent as a dormant phase on the lake sediments. Meroplanktonic algae include the diatom *Melosira* and the blue-green alga *Microcystis*, which have their planktonic phase at different times in the annual cycle of temperate lakes.

The evolution of the meroplanktonic state allows particular phytoplankton species to grow over a limited phase of the seasonal cycle to which they are adapted. This may either be a time of low

competition from other algae (*Melosira*) or when competition is intense (*Microcystis*).

Melosira

Many diatoms in freshwater environments are able to escape the effects of intense competition in the top part of the water column by mechanisms such as subthermocline growth or prolonged vegetative survival on sediments (Wehr and Sheath, 2003). *Melosira* presents a good example of the latter, being present on the sediments for most of the year, but moving into the water column at a time when the populations of other algae are generally limited.

Melosira is a filamentous diatom with thick cell walls (high cell silicon content) and a high rate of sedimentation, requiring water turbulence to keep in suspension. Studies by Lund (1954) have shown that populations of this alga rise from the lake sediment in autumn, persisting over winter as planktonic cells (Figure 3.21). Although water inorganic nutrients are present at elevated concentrations over winter, algal growth does not commence until spring, when increases in light and temperature trigger nutrient uptake and a brief population increase. This is brought to a close by stratification, which results in a reduction in turbulence and sedimentation of

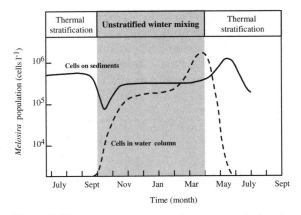

Figure 3.21 Seasonal cycle of the meroplanktonic alga, *Melosira italica* in Blelham Tarn (English Lake District) (figure adapted and redrawn from Horne and Goldman, 1994, based on the studies of Lund, 1954)

cells to the lake bottom. Cells with stored nutrients are able to survive on the lake sediment until the next planktonic phase. The early population increase in spring makes use of the brief improvement in growth conditions that occurs prior to stratification, and precedes the growth of other phytoplankton species – thus avoiding competition.

Melosira is closely related to another filamentous diatom, *Aulacoseira*, which has been mentioned previously in relation to the endemic flora of Lake Baikal, Russia (Figure 2.8) and the formation of early-season blooms below ice (Figure 2.3).

Microcystis

This blue-green alga (Figure 6.24) overwinters as vegetative colonies on the lake sediment, where it persists until early summer, when some of the colonies rise into the upper water column in response to increased light intensity (Figure 3.22). Successful recruitment of benthic cells to the water column requires a preceding phase of high water clarity (Reynolds, 1997) and light penetration, possibly involving the photo-activation of gas vesicle formation (Section 3.8.1). Other seasonal changes that trigger activation include increasing temperature and the onset of anoxic conditions. As the

colonies enter the water column, passive dispersion may also be promoted by wind-induced turbulence.

Recent studies (Brunberg and Blomqvist, 2003) have shown that recruitment from shallow areas of the lake is particularly important. Initial (early summer) abundance of *Microcystis* on surface sediments in shallow and deep areas of Lake Limmaren (central Sweden) ranged from $25–28 \times 10^6$ colonies m^{-2}. Subsequent recruitment to the water column (over a 17 week period) amounted to 12.9×10^6 colonies m^{-2} for shallow water (i.e., 50 per cent of original abundance) and 2.3 colonies m^{-2} for deep sites (8 per cent of original abundance). These data indicate a clear link between planktonic populations of this alga throughout the lake and benthic populations in shallow regions, which in the case of deep lakes means the peripheral littoral zone.

Initial inoculum levels in the water column are quite low, but continued growth leads to planktonic populations which dominate the lake towards the end of the stratification period – at a time of low nutrient concentrations. Populations of *Microcystis* are able to survive nutrient-limiting conditions in the top part of the water column by the initial use of stored materials from their dormant phase and by diurnal migration into the nutrient-rich hypolimnion. The ability to migrate into nutrient-rich regions of the water column give *Microcystis* a

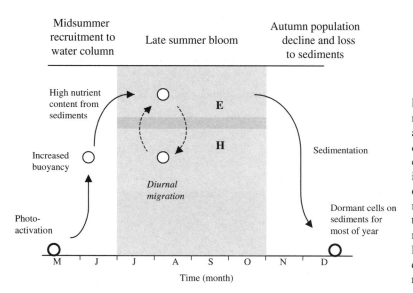

Figure 3.22 Annual cycle of the meroplanktonic colonial blue-green alga *Microcystis*. Colonies of *Microcystis* (O) are present in the water column over a limited period, spending most of the year in a dormant state on lake sediments. During the summer bloom, a period of intense phytoplankton competition, diurnal migration into the nutrient-rich hypolimnion (H) makes the organism independent of declining epilimnion (E) nutrient concentrations

competitive advantage over less motile algae at a time when algal populations in the epilimnion are dense and competition is high. *Microcystis* does, however, have to compete with other algae such as dinoflagellates, which are also able to migrate into the nutrient-rich hypolimnion, and the seasonal progression of many eutrophic lakes involves summer dominance by either *Microcystis* or *Ceratium*. Whether a particular season favours *Microcystis* or *Ceratium* dominance appears to depend on which population develops first. 'Microcystis years' have been linked to early recruitment of colonies into the water column, correlating with early conditions of light penetration and water clarity (Reynolds, 1997).

Populations of *Microcystis* come to the end of their bloom phase in autumn, falling to the bottom of the lake as resistant cells where they may remain viable for a number of years.

3.10.2   Strategies for unstable and stable environments: r-selected and K-selected algae

R-selection and *K*-selection represent adaptive strategies to exploit two contrasting environmental situations (Section 1.2.6). *R*-selected species are characterized by small size, high growth rate and short cell cycle – and are able to increase in population rapidly under conditions of low population density and low species competition. These organisms are particularly prominent in temperate lakes during the clear-water phase. During this time, grazing pressure by herbivores is patchy and intense, and algal growth is limited to short periods in those parts of the lake with lower densities of zooplankton. *K*-selected species, conversely, typically have large size, low growth rate, and long cell cycles and are adapted to conditions of high population density and high competition.

Examples of *r*-selected and *K*-selected phytoplankton are given in Table 3.12, with data for growth rate and cell/colony size. Growth rate can be expressed as the generation time (the time taken for the population to double) or the exponential growth constant (k') – which is derived from the exponential growth equation:

$$N_t = N_0 e^{k'} \tag{3.8}$$

where N_0 = biomass at zero time, N_t = biomass at time t, k' = the exponential growth constant.

The value k' can be used as a measure of growth rate, and in the case of *r*-selected algae is typically

Table 3.12   Growth rate and size differences between various *r*-selected and *K*-selected algae. Maximum exponential growth rates, k', are shown for laboratory cultures grown under continuous saturating illumination, at temperatures in the range 20–25°C (taken partly from Reynolds, 1990)

Species	Exponential growth constant (k') (ln units day^{-1})	Approximate size of single cells or colonies[*]
r-selected species		
Chlamydomonas spp	2.29–2.91	upto 40 μm
Chlorella pyrenoidosa	2.15	3–5 μm
Synechococcus sp.	2.01	4–15 μm (depending on species)
Scenedesmus obliquus	1.52	
Cryptomonas erosa	0.83	32 μm
K-selected species		
Aphanizomenon flos-aquae[*]	0.98	upto 2 mm
Anabaena flos-aquae[*]	0.78	upto 2 mm
Microcystis aeruginosa[*]	0.48	upto 2 mm
Ceratium hirundinella	0.21	upto 400 μm

>1, while *K*-selected algae typically have values <1. Values for k' tend to decrease with increase in organism (single cell or colony) size.

Dinoflagellates as K-selected organisms

Of all the algal groups, dinoflagellates provide some of the best examples of *K*-selected species – as indicated by the very low k' value for *Ceratium hirundinella* (Table 3.12). The evolution of these organisms has been dominated by the development of huge amounts of nuclear DNA, reaching levels of 200 pg cell^{-1} in *Gonyaulax* – compared to levels of 0.05–3 pg cell^{-1} for other eukaryote cells. The evolution of these very high levels of nuclear DNA (most of which is genetically redundant) has paralleled unique methods of packaging (Sigee, 1984), with dinoflagellate DNA being permanently condensed as chromosomes (Figure 3.23) which are largely stabilized by divalent cations and not conventional eukaryote histones. The high levels of dinoflagellate DNA directly correlate with large cell size and result in the long cell cycle – due to the length of time required for DNA replication (interphase) and chromosome separation (karyokinesis).

The dinoflagellates *Ceratium* and *Peridinium* are common examples of *K*-selected organisms in moderate to high nutrient waters, often dominating the rather static lake environment in late summer when the water column is stably stratified and grazing is limited. Alternative domination of late summer standing waters by *Ceratium* or *Microcystis* has been discussed in the previous section.

3.11 Heterotrophic nutrition in freshwater algae

Although the majority of freshwater algae are photoautotrophic, carrying out photosynthesis and forming complex organic compounds from inorganic precursors, some also have the facility for heterotrophic nutrition. These organisms are able to use external organic compounds for energy, metabolism, and growth (Sanders, 1991; Tuchman, 1996).

The ability of some algae to supplement their autotrophic life style by the uptake of complex organic carbon from the environment occurs in two separate ways:

- *organotrophy*: direct uptake of soluble organic molecules by absorption over the cell surface, and

- *phagotrophy*: ingestion of particulate organic matter.

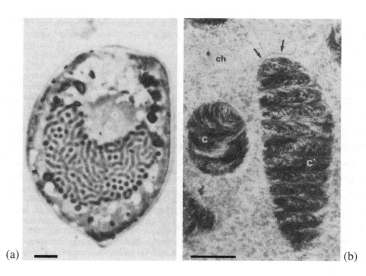

(a)　(b)

Figure 3.23 Distinctive nuclear fine-structure in dinoflagellates:
(a) Light micrograph of thin section of fixed interphase cell of *Prorocentrum*, showing the large central nucleus with numerous condensed chromosomes (scale = 2 μm)
(b) Electron micrograph-detail of transverse (c) and longitudinally (c′) sectioned chromosomes, showing the distinct fibrillar appearance. One of the chromosomes is attached (arrows) to an intranuclear cytoplasmic channel (scale = 0.5 μm). Figure taken from Sigee (1984) and reproduced with permission of the Linnean Society of London.)

Table 3.13 Transition from autotrophy to heterotrophy in flagellate algae (vertical arrows emphasize the transition from facultative to obligate heterotrophy)

Mode of Nutrition	Characteristics
Autotrophy	Synthesis of all required organic compounds from inorganic sources. Algae use light energy (phototrophy) for this
Heterotrophy	Requirement for external organic compounds – either as specific metabolites or as a general carbon source
Organotrophy	Uptake of soluble organic molecules at cell surface
Auxotrophy	Metabolic requirement for specific external metabolites (vitamins, growth factors)
Photo-Organotrophy	Facultative organotrophs, able to obtain carbon from CO_2 or soluble organic compounds Energy from light
Chemo-Organotrophy	Facultative or obligate organotrophs obtain carbon from organic compounds Energy from organic compounds
Phagotrophy	Uptake of particulate matter by phagocytosis
Mixotrophy	Facultative phagotrophs, able to carry out photosynthesis and phagotrophy
Phagotrophy	Obligate phagotrophs Feeding via food vacuoles for the uptake of particulate matter

In each case, the transition from autotrophy to heterotrophy occurs along a gradient of nutritive states (Table 3.13), ranging from facultative organotrophy and phagotrophy (where requirement for external organic compounds may be minimal) to obligate states (where dependence is total). These facultative states include auxotrophy, photo-organotrophy and mixotrophy (Table 3.13). The development of heterotrophy has arisen in response to two quite different sets of environmental situations as follows.

- *High external nutrients*. Nutrient enrichment of the environment, with high external concentrations of soluble organic compounds may occur in various ways. These include large-scale enrichment of aquatic systems due to agricultural pollution, as well as micro-enrichment of epiphytic and benthic algae from vascular plants or via exudates/faeces of molluscs, insects, crustaceans, and other organisms crawling over the benthic flora. The presence of soluble organics presents the opportunity for nutrient absorption at the cell surface and favours growth of organotrophic algae.

- *Limitations to photosynthesis*. Conditions of low light intensity and/or limiting inorganic nutrients, may cause severe impairment of photosynthesis and thus the facility for autotrophic nutrition. This situation has lead to the development of mixotrophy and phagotrophy (Wilcox and Wedemeyer, 1991; Raven, 1997; Li *et al.*, 2000).

There are many situations of low light intensity where facultative heterotrophy would provide advantages over obligate autotrophy. On a large spatial and temporal scale, conditions of light limitation result from seasonal, geographic, and hydrological factors. Seasonal effects include snow-covered ice on lakes and rivers, turbid spring floodwaters, and autogenic (self) shading within late-summer algal blooms. The geographic influence of latitude is particularly severe in polar and subarctic regions, and hydrological influences include the development of algal communities in deep lentic and lotic ecosystems. On a more local scale, seen for example in forest streams, shading from surrounding terrestrial vegetation may impose severe light limitation, and at the microlevel – shading

Table 3.14   Auxotrophic requirements for vitamin B_{12} (cyanocobalamin)

Algal group	General requirement	Specific requirement
Euglenoids	All require B_{12} (photoauxotrophic)	Absolute requirement for *Euglena gracilis* – 4900 to 22 000 molecules of B_{12} needed for cell division[a]
Chrysophytes	Many phagotrophs, also require B_{12}	*Oochromonas danica* can synthesize all vitamins except B_{12} and several fat-soluble vitamins[b]
Prymnesiophytes	Generally require either thiamine or vitamiun B_{12}	*Coccolithus huxleyi* needs only thiamine[c]
Diatoms	Most diatoms need B_{12}, except some estuarine species	Estimated at 5–13.8 molecules μm^{-3} for cultures of *Skeletonema costatum*[d]

References: [a]Carrell (1969); [b]Stoltze *et al.* (1969); [c]Lee (1997); [d]Guillard and Cassie (1963).

within benthic algal communities (periphyton) reduces light penetration to the basal layer.

3.11.1   Organotrophy

Organotrophy (also referred to as saprotrophy or osmotrophy) may either involve the specific uptake of particular metabolites or the bulk non-specific uptake of organic materials as a simple carbon source. This uptake takes place over the whole cell surface and occurs either via active transport at the plasmalemma or by pinocytosis.

Auxotrophy: specific uptake of particular metabolites

With some algae, organotrophy involves a specific requirement for small amounts of particular metabolites. This is referred to as auxotrophy, and differs markedly from the more general situation where there is bulk uptake of organic carbon at the surface, with no molecular specificity. Many photosynthetic algae, when cultured in the laboratory, show a metabolic requirement for small quantities of particular organic molecules in their surrounding medium. These organisms may lack the ability to synthesize the complete suite of macromolecules required for growth, and require small levels of organic supplements such as vitamins (particularly vitamin B_{12}; thiamine, and biotin) and organic growth factors.

Requirement for vitamin B_{12} has been demonstrated in various groups of algae (Table 3.14), where it occurs irrespective of both photosynthetic (euglenoids) and phagocytic (chrysophytes) activity. The use of defined media for laboratory cultures permits precise estimation of the required level of B_{12} supplementation, which in the case of *Skeletonema* may be as low as 5 molecules μm^{-3} (Table 3.14). The need for specific vitamins may have important ecological implications. This has been demonstrated for the marine prymnesiophyte *Coccolithus huxleyi*, which requires the vitamin thiamine but not B_{12}, it is thus able to grow regularly and out-compete other algae in the vitamin-deficient offshore waters of the Sargasso sea (Lee, 1997).

Photo-organotrophy and chemo-organotrophy: non-specific uptake of organic materials as a simple carbon source

Many laboratory studies have shown the ability of planktonic and benthic algal species to be cultured heterotrophically in complete darkness, with the addition of organic supplements (Tuchman, 1996). The carbon supplements typically involve the addition of one to three simple organic substrates to the normal growth medium, including sugars (e.g., glucose, fructose), organic acids (acetate, lactate), amino acids (leucine, glycine), glycerol, and urea. The algae include blue-greens, chrysophytes, dinoflagellates, diatoms, and green algae.

Organotrophy may be particularly important to planktonic diatoms, where short-term periods of low turbulence can cause the organisms to drop out of the photic zone. The ability of diatoms such as *Cyclotella cryptica* (Hellebust, 1971) to grow heterotrophically in the dark allows the alga to survive on sediments prior to resuspension in the water column when turbulence returns. This organism is able to use glucose as the sole carbon source, requiring a 24 hour lag period for metabolic adjustment to the heterotrophic mode of nutrition.

The ability of many freshwater algae to grow under controlled laboratory conditions in the dark in the presence of a carbon source provides strong experimental evidence for the wide occurrence of organotrophy. In some cases, direct uptake has been demonstrated using radiolabelled compounds such as [^{14}C]-glucose or [^{14}C]-leucine. Although organic carbon uptake under conditions of prolonged darkness provides critical evidence for organotrophy, many algae carry out this process in the light. The energy sources required for carbon uptake in these two situations are respectively organic compounds (chemo-organotrophy) and solar radiation (photo-organotrophy).

Photo-organotrophy   In autotrophic algae, the dark-fixation of CO_2 is coupled to the generation of ATP and NADPH by light-dependent processes as part of the normal process of photosynthesis. Recent studies (reported in Tuchman, 1996) have demonstrated that this coupling may become disrupted in the presence of external organic compounds in the light. Illumination of photosynthetic reaction centres leads to automatic activation of light-dependent reactions, with no switch-off until the quality/quantity of irradiance decreases below the threshold for pigment stimulation. Associated light-independent reactions can be deactivated in the light if adequate concentrations of organic carbon substrates are available in the surrounding environment. If this occurs, the Calvin cycle becomes blocked and carbon fixation ceases. Light dependent reactions continue to function, providing energy which can be used to actively transport exogenous carbon substrates into the cell, assemble energy-rich storage compounds or simply be channelled into other cellular metabolic pathways.

Photo-organotrophy can be regarded as a facultative activity in which the balance between autotrophy and heterotroophy depends on substrate availability, and possibly other environmental factors as well.

Chemo-organotrophy   Chemo-organotrophs obtain energy from organic compounds and include both facultative and obligate species.

Many facultative organotrophs normally exist by phagotrophy, but are able to take up soluble organics in appropriate circumstances. In these organisms, soluble organic nutrient may be taken up with particulate material and be absorbed across the phagocytic vacuole membrane. Several of these algae have been cultured heterotrophically in the laboratory on high nutrient liquid media, including various chrysophytes (species of *Ochromonas* and *Poterioochromonas*) and dinoflagellates (species of *Crypthecodinium*, *Oxyrrhis*, and *Gyrodinium*). In all cases the required concentration of organic solutes in the culture medium was much higher than normally encountered in environments where the algae are found, suggesting that organotrophy is not the normal mode of nutrition of these organisms under environmental conditions.

Algae with an absolute dependency on organotrophic nutrition typically live in organically polluted environments such as low pH sites enriched with animal waste or sewage works. Such obligate heterotrophs include several colourless euglenoid algae (*Astasia*, *Cyclidiopsis*) and other organisms normally regarded as protozoa (*Chilomonas*, *Polytomella*). Flagellates living in such environments tend to be out-competed by bacteria for nutrients due to their lower surface to volume ratio. For this reason, nutrient levels need to be very high for the flagellates to survive, and in most environments saprotrophy alone is insufficient to support the growth of a flagellate population.

3.11.2   Phagotrophy

The evolutionary development of phagotrophy is largely a feature of flagellate algae, and has been observed in three major groups of organisms – chrysophytes, dinoflagellates, and prymnesiophytes. The trend towards phagotrophy in algae may have given rise to a wide range of flagellates that are now completely heterotrophic and are of indeterminate taxonomic status. Many of these are very small (<10 μm diameter) and are part of a mixed assemblage referred to as heterotrophic nanoflagellates (HNFs). The ecological importance of HNFs in planktonic food webs has been emphasized in Chapter 2, and is considered further in Chapter 9.

The capture and ingestion of particulate food by flagellate algae involves a wide range of prey – including bacteria, blue-green algae, eukaryote algae, protozoa, and metazoan gametes. Algal phagotrophs are raptorial feeders, obtaining prey by direct interception. This is in contrast to filter and diffusion feeding, which has been adopted by some non-algal phagotrophs such as protozoa and some indeterminate flagellates. Once the prey has been caught, various mechanisms have evolved for the ingestion process. This range of activities is seen particularly well in the dinoflagellates, where a large number of species are known to feed phagotrophically (Wilcox and Wedemayer, 1991). Some of these dinoflagellates (e.g., *Amphidinium cryophilum*) are able to ingest food through a cytoplasmic extension (peduncle), drawing whole cells or parts of cells into a nascent food vacuole prior to digestion. Other dinoflagellates envelop their prey in a membranous structure, carry out external digestion, then take up the nutrients. *Noctiluca* uses a flexible adhesive tentacle to bring food material to a point of ingestion – the cytostome.

Other algae simply ingest particulate food material at the cell surface, with the formation of a phagocytic vesicle. Where this occurs, the food source appears to relate mainly to the size of the phagotroph, with (large) dinoflagellates ingesting small eukaryotes such as cryptophytes. Smaller chrysophytes and prymnesiophytes ingest mainly bacteria (Table 3.15). Environmental and laboratory studies have demonstrated a range of factors which may promote phagotrophic activity – including prey density, light intensity, and inorganic nutrient (particularly N and P) concentrations. Recent studies (Skovgaard and Hansen, 2003) have suggested that toxin production by Prymnesiophytes may be important in motile prey immobilization and ingestion.

Mixotrophy: the ability to carry out phagotrophy and photosynthesis

Various algae are able to carry out both photosynthesis and phagotrophy. With increasing development of the phagotrophic habit, the algal cell moves from dependence on photosynthesis (obligate phototroph) to only limited dependence (facultative phototroph). The balance between phototrophic and heterotrophic nutrition has shifted even further with some algae, where the photosynthetic capability of the algal cell has been lost (non-phototrophic) but the cell is dependent on ingested algae for intermittent photosynthesis (facultative symbiosis) and heterotrophic nutrition via phagocytosis. These three states are illustrated within the dinoflagellates (Table 3.15), where there is a wide range of heterotrophic nutrition.

Obligate phototrophs   In obligate phototrophs, such as the estuarine dinoflagellate *Gyrodinium galatheanum* (Li *et al.*, 2000), photosynthesis is an essential source of carbon at all times. The rate of phagotrophy increases with light intensity up to a particular saturation level and this organism is not able to grow in the dark – even in the presence of food. In *Gyrodinium galatheanum,* where there is a light-dependent feeding pattern, phagotrophy is clearly not a response to low light conditions. The major role of phagotrophy in obligate phototrophs is probably to supply additional nutrients (N and P) which are needed for photosynthetic carbon assimilation. Phagotrophy may also be important in supplying trace organic growth factors, since feeding (at a reduced rate) also occurs in nutrient-rich conditions.

Facultative phototrophs   Photosynthesis is not essential for the growth of these algae. The dinoflagellates *Fragilidium subglobosum* (Skovgaard, 1996), and *Amphidinium cryophilum* (Wilcox and Wedemeyer, 1991) are facultative phototrophs and

Table 3.15 Mixotrophy in dinoflagellates and chrysophytes: the balance between phototrophy and heterotrophy in selected species (obligate phototrophs are indicated by shading)

| Organisms | Phototrophy | Heterotrophic Nutrition | | |
		Mode of nutrition	Stimulated by	Food Source
Dinoflagellates				
Gyrodinium galatheanum[a]	Obligate	Facultative phagotrophy	High light Low inorganic nutrients	Cryptophytes
Fragilidium subglobosum[a]	Facultative	Facultative phagotrophy (obligate at low light levels)	Low light	
Amphidinium cryophilum[b]	Facultative	Facultative phagotrophy (obligate at low light levels)		Other dino-flagellates
Gymnodinium acidotum *Gymnodinium aeruginosum*[c]	Not phototrophic (rely on ingested organism)	Facultative Symbiosis Obligate periodic phagotrophy		Cryptophytes
Chrysophytes				
Dinobryon cylindricum[d]	Obligate phototroph	Facultative phagotrophy		Bacteria
Poterioochromonas malhamensis[d]	Facultative phototroph	Facultative phagotrophy and saprotrophy		Bacteria Glucose uptake

References: [a]Li *et al.* (2000); [b]Wilcox and Wedemayer (1991); [c]Fields and Rhodes (1991); [d]Raven (1997).

phagotrophs – able to grow equally well phototrophically or completely phagotrophically in the dark. The cold-water dinoflagellate *Amphidinium cryophilum* has been described as a photosynthetic winter species, and grows in freshwater ponds which are frequently covered in ice and heavy snow. The phagotrophic habit is a clear environmental adaptation, supplementing carbon and other nutrient input overwinter and allowing the organism to grow at a time when irradiance (and photosynthesis) is reduced by short days, low sun angle, and ice/snow cover. At low light levels, cells feed phagotrophically and are almost colourless, while at high light levels they feed much less frequently (if at all) and are brightly pigmented (Wilcox and Wedemayer, 1991).

Non-phototrophic algae Two species of algae, *Gymnodinium acidotum* and *Gymnodinium aerugi-*

nosum can be regarded as non-phototrophic, in that they appear to have completely lost their own chloroplasts and their ability to carry out autonomous photosynthesis. Instead, they rely on ingested cryptophyte cells for short periods of imported photosynthetic activity (symbiosis) followed by periodic phagocytosis (Figure 3.24). Resistant cysts of these algae germinate in the laboratory to release colourless motile cells, which remain colourless and die within a few days unless they are mixed with cryptophytes (Fields and Rhodes, 1991). Ingestion of cryptophytes leads to digestion of the cryptophyte nucleus, but chloroplasts (and other cytoplasmic organelles) are retained intact and functional over a period of about 12 days, supplying the host cell with photosynthetic products. Within this time period, the dinoflagellate has to ingest more cryptophytes to maintain the symbiotic association and stay alive. The colourless cell released

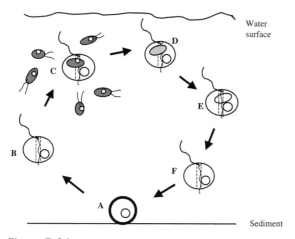

Figure 3.24 Environmental cycle of *Gymnodinium acidotum*, (based on the laboratory studies of Fields and Rhodes, 1991).
(A) Resistant cyst, resting on lake sediment, (B) Released colourless, motile cell, migrates up to the surface water (positively phototactic) and populations of cryptophytes. (C) Ingestion of cryptophyte cells. (D) Symbiotic phase – within a few days, the cryptophyte nucleus has been digested but chloroplasts remain functional. (E) Completion of phagotrophy – chloroplasts and other cryptophyte organelles undergo digestion. (F) Colourless dinoflagellate, needs to ingest more cryptophyte cells, or will encyst

from the cyst is positively phototactic, and is presumably carried into the surface waters of its lake environment to make contact with populations of cryptophyte cells. Growth and division of the dinoflagellate will continue during the period of the cryptophyte bloom, but will cease when this terminates resulting in cyst formation and dormancy until the next opportunity for symbiosis occurs. This association between dinoflagellate and cryptophyte is of interest in relation to the ecology of the lake, but also has evolutionary relevance as a possible early stage in the permanent incorporation of actively functional foreign plastids in recipient host cells.

Obligate and facultative phototrophs have also been observed in phagotrophic chrysophytes (Raven, 1997). *Dinobryon cylindricum* (obligate phototroph) cannot survive in continuous darkness, but ingestion of bacteria (bacterivory) is needed for rapid sustained growth in light. Bacterivory supplies

up to 25 per cent of the carbon in *Dinobryon* cells, but probably has a more important role in supplying major inorganic nutrients and growth factors. Studies by Auclair (1995), for example, have suggested that ingestion of blue-green algae is a major source of Fe for phagotrophic chrysophytes in Canadian Lakes.

3.12   Survival in snow and ice: adaptations of cryophilic algae

A wide range of algae are adapted to grow and survive in low temperature (snow and ice) environments. As noted earlier (Section 2.17), such environments occur over a wide area of the Earth's surface and support very distinctive ecosystems with organisms exposed to a range of adverse environmental conditions. The most notable feature in such environments is the absence of free water for most of the year, so algae spend most of their time in a frozen and inert physiological state.

The biological features of snow algae have been looked at particularly in snow packs, where a burst of physiological activity occurs during the annual melt process (Hoham and Duval, 2001). The colonization of melting snow by these organisms and their subsequent growth to high population levels within the snow matrix represents one of the more remarkable sets of adaptations to a particular freshwater system.

3.12.1   Major groups of cryophilic algae

Cryophilic (snow and ice) algae include representatives from all of the major groups noted earlier in this chapter (Tables 3.1 and 3.3). Some of the typical species routinely encountered in low temperature situations are summarized in Table 3.16, together with their typical locations and other ecological characteristics. As with other freshwater systems, true red algae (class Rhodophyta) are poorly represented. Reports of blooms of 'red algae' in snow environments typically refer to high populations of green algae with carotenoid pigmentation or – less commonly – to high levels of other algae such as dinoflagellates.

Table 3.16 Taxonomic diversity and habitats of snow and ice algae

Algal Group	Representative species	Typical snow and ice habitat
Blue-green algae	*Phormidium frigidum*	Aerobic or anaerobic mats under ice
	Lyngbya martensiana	
	Anabaena cylindrica	Ponds and ice surface
	Nostoc commune	
Green algae	*Chlamydomonas nivalis*	Melting snow-packs and ponds
	Chloromonas brevispina	Can form red blooms, with populations
	Chloromonas nivalis	of 10^5–10^6 cells ml^{-1} of snow water
Euglenoids	*Notosolenus sp.*	Melting snow-packs and ponds
	Euglena sp.	
Dinoflagellates	*Gyrodinium sp.*	Snow surfaces, occasionally forming red
	Gymnodinium pascheri	blooms
Cryptomonads	*Cryptomonas frigoris*	Melting snow-packs and ponds
Chrysophytes	*Chromulina chionophila*	Melting snow-packs and ponds
	Ochromonas sp.	
Diatoms	*Navicula sp.*	Widely present in aerobic conditions of
	Nitzschia sp.	polar lakes, particularly in algal mats
	Pinnularia sp.	

Different algae tend to occupy their own specific environments within snow and ice systems. Some areas of melting snow are very rich in algae, particularly flagellates that belong to the order Volvocales of the green algae, including *Chlamydomonas* and *Chloromonas*. Other algal flagellates in snow include chrysophytes, euglenoids, cryptomonads, and dinoflagellates, some of which are illustrated in Figure 3.25. Individual organisms are found both as actively motile cells and non-motile resistant spores, depending on snow conditions.

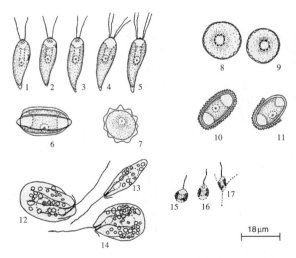

Figure 3.25 Typical examples of snow algae found in southwestern United States.

Chloromonas nivalis, a green alga. 1–3: vegetative cells with two flagella, papilla, two apical contractile vacuoles, centrally located nucleus and parietal cup-shaped chloroplast. 4–5: motile zygotes (planozygotes) with four flagella, otherwise similar to vegetative cells. 6–7: zygospores with central nucleus surrounded by parietal chloroplast. (Note two large lipid bodies, one at each pole of the cell, 6.)
Chlamydomonas nivalis, a green alga. 8–9: resting spores with smooth cell wall, red protoplast, and central pyrenoid surrounded by starch plates. *Chloromonas brevispina*, a green alga. 10–11: zygospores with central nucleus surrounded by parietal chloroplast. (Note two large lipid bodies, one at each pole of the cell and shedding of primary wall with spines and exposure of inner smooth secondary wall, 11.)
Notosolenus sp., a colourless euglenoid. 12–14: cells with ridges on pellicle, central to posterior nucleus, numerous paramylon bodies, one contractile vacuole at anterior end of cell, and two flagella, the longer projected anteriorly, the shorter posteriorly. Cells in narrow diameter view (13) and wide diameter view (12–14).
Chromulina chionophila, a golden alga. 15–17: cells with a parietal, band chloroplast, one contractile vacuole, a single flagellum, and 8–25 chrysolaminarin bodies. (Note three rhizopodia extending from the plasmalemma, 17.)
Figure taken from Hoham and Duval (2001) and reproduced with permission of Cambridge University Press

Algae without an actively motile stage, such as blue-green algae and diatoms, are not normally found in snow packs, but do occur in other low-temperature locations. In Antarctic dry valley lakes, for example, algal mats present under the ice are dominated by blue-green algae in the genera *Lyngbya* and *Phormidium* (Wharton *et al.*, 1983). Benthic (pennate) diatoms in the genera *Navicula*, *Nitzschia* and *Pinnularia* are common in such mats under aerobic conditions, where they can be seen gliding between filaments of the algae. Green algae are rare in below-ice samples, but are present in surface ice and snow, along with blue-green algae such as *Anabaena* and *Nostoc*. Many of the cryophilic algae noted in Table 3.16 are restricted to ice and snow environments and are not found in temperate lakes or other water bodies.

3.12.2   Life cycles of snow algae

Studies on the life cycles of snow algae have been carried out mainly in relation to green algal flagellates – the dominant algae in the snow environment.

The active phase of the life cycle occurs in spring or summer when water is released as the snow melts, nutrients and gases become available and light penetrates into the snowpack. The life cycle of *Chloromonas* (Figure 3.26) is typical of snow algae and involves four main phases – zygospore germination, colonization of the snowpack, population increase, and formation of zygotes. These phases correlate with physical and chemical factors at the time of snowmelt.

- The active phase begins with germination of resting zygospores at the bottom of the snowpack. These spores are present either on the snow/soil interface, or on the snow/snow interface in persistent snowfields. Germination of these spores is triggered by the persistence of air temperature above freezing over several days and by increased light levels, and involves division by meiosis to produce biflagellate zoospores.

SNOW-AIR INTERFACE

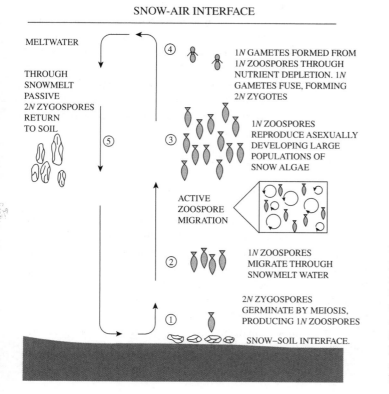

MELTWATER

THROUGH SNOWMELT PASSIVE 2*N* ZYGOSPORES RETURN TO SOIL

④   1*N* GAMETES FORMED FROM 1*N* ZOOSPORES THROUGH NUTRIENT DEPLETION. 1*N* GAMETES FUSE, FORMING 2*N* ZYGOTES

⑤

③   1*N* ZOOSPORES REPRODUCE ASEXUALLY DEVELOPING LARGE POPULATIONS OF SNOW ALGAE

ACTIVE ZOOSPORE MIGRATION

②   1*N* ZOOSPORES MIGRATE THROUGH SNOWMELT WATER

①   2*N* ZYGOSPORES GERMINATE BY MEIOSIS, PRODUCING 1*N* ZOOSPORES

SNOW–SOIL INTERFACE.

Figure 3.26   Life cycle of the snow algal flagellate *Chloromonas* in a melting snowpack. The cycle commences with zygospore germination (1), followed by active zoospore migration and positioning in the snowpack (2), population increase (3), and formation of zygospores (4), (5). Algal cells are present in the haploid (1N) or diploid (2N) state. Figure taken from Hoham and Duval, 2001, and reproduced with permission of Cambridge University Press.

- The zoospores swim in the liquid meltwater around the snow particles towards the upper part of the snowpack. This is the dispersive or colonization phase of the life cycle and leads to vertical positioning of the zoospores within the snowpack in relation to irradiance levels and spectral composition.

- Zoospores reproduce asexually, generating large populations of snow algae. Visible blooms of snow algae often occur a few days after germination.

- Nutrient depletion results in the differentiation of zoospores to form gametes. These fuse to form resting zygotes, which eventually deposit on the soil surface when the snowpack has melted – or remain on old snow in persistent snowfields. Populations of snow algae stay in the same locality from year to year. In conditions where the snow completely melts, the spores remain dormant through summer and may undergo meiosis during the first freezes in autumn. They are subsequently covered in new snow and germinatewhen conditions ameliorate the following year.

3.12.3   Physiological adaptations of snow algae

Survival and active growth in the snow environment requires the ability to adapt to a relatively brief growth period and to withstand high levels of solar irradiation, low nutrients and temperature, high acidity, and extended periods of desiccation.

Brief growth period

Many snow algae are motile and single-celled. These characteristics reflect the need for free water (for movement and growth) and the relatively brief time period (snow melt) when this may be available. During this brief growth phase, motility, unicellularity, and reproductive strategy are all important in making optimal use of the environment.

Rapid colonization and motility   Motility is important for colonization of the snowpack when free water (melt water) becomes available. Having flagella allows the cells to move within the water film that surrounds snow crystals at the time of snow melt. Other snow microbes that are not flagellates (non-motile algae, fungi, and bacteria) are passively moved within the snow, melt water.

Rapid population growth: *r*-selected species
Snow algae are typically good examples of *r*-selected species, single-celled with small size and short cell cycle. Such organisms show maximum increase in population under appropriate (i.e., free water) conditions, and are able to dominate the new environment within a short period of time.

Reproductive strategy   The production of resting spores is an essential part of the life cycle of snow algae, ensuring survival between the brief periods of growth. The fastest route to producing these resting spores would be by asexual reproduction, since more time is needed to complete the sexual phase of the snow algal life cycle. Field observations suggest that in the most severe environments, where the snowpack is inconsistent from year to year, asexual reproduction may be favoured by natural selection (Hoham and Duval, 2001). This ecological strategy, however, would lead to a decrease in genetic diversity and algae living in such environments may be headed for evolutionary extinction.

Protection from harmful irradiation

The exposed, often high-altitude habitats of snow algae means that these organisms frequently have to tolerate excessive levels of irradiation. Shortwave radiation values as high as 86 000 lux, for example, have been measured in exposed alpine snow containing *Chlamydomonas nivalis* (Mosser *et al.*, 1977). Ultraviolet radiation is particularly high in snow environments due to reduced atmospheric shielding (alpine regions) or where the stratospheric ozone layer is relatively thin (polar regions). The damaging effects of solar radiation, particularly UV light, are discussed in Section 4.9. Specific effects on snow algae have been recorded for *Chlamydomonas nivalis*, where 90 minutes of UV-B irradiation impaired photomovement, motility, and velocity (Hader and Hader, 1989). Exposure to UV radiation has also been shown to reduce photosynthesis by as much as 25 per cent in 'red snow' and 85 per cent in 'green snow'.

As with other algae in exposed environments (Section 4.9.6) snow algae have demonstrated a wide range of high-light and UV-protection responses, including the intracellular production of the following.

- *Photoprotectants*. These include secondary carotenoid pigments such as astaxanthin (see below) and shikimate pathway metabolites, acting as a passive radiation shield.

- *Antioxidants*. In *Chlamydomonas nivalis*, these are known to arrest photoinhibitory damage by the reduction of free radical production in the thylakoid membrane.

- *Repair systems*. A photoreactivation enzyme in *Chromulina chionophilia* is known to repair UV damage to chlorophyll and other photosynthetic pigments.

Astaxanthin: the major pigment of 'red' snow algae   One of the most conspicuous features of snow algae is the development of red pigmentation. In various green algae, including the widespread snow alga *Chlamydomonas nivalis*, the major red pigment has been identified as the xanthophyll astaxanthin. Large accumulations of this pigment occur within the cell in extra-chloroplastic lipid droplets, where they are considered to reduce photoinhibition and cell damage.

The development of red pigmentation occurs at high levels of irradiation, but also depends on nutrient levels, particularly nitrogen availability (Figure 3.27, Bidigare *et al.*, 1993). Cells of *Chlamydomonas* collected from exposed snow slopes remained green when nutrient levels supported a high growth rate. Such cells maintain a high turnover of the Q_b protein that reduces the cells, susceptibility to photoinhibition from damaging wavelengths of light. The effects of nutrient limitation depends on the level of incident radiation. Under conditions of nitrogen depletion but continued high light level, growth rate is reduced and chlorophyll breakdown is paralleled by the synthesis of astaxanthin. At low light levels, chlorophyll is not degraded and the cells remain green.

In addition to red astaxanthin pigment, orange and yellow-orange carotenoids also accumulate in vegetative cells and zygotes of some algae. Orange carotenoid accumulation, for example, occurs in the polar lipid bodies of zygotes of *Chloromonas* (Figure 3.25). The orange zygotes are located in the upper snow layers where irradiation levels are highest, while zygotes collected from the lower snow layers lack carotenoid accumulation and are usually green (Hoham and Duval, 2001).

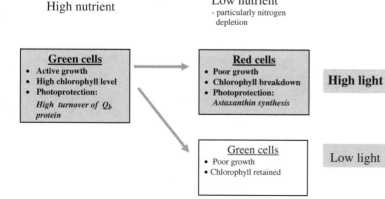

Figure 3.27 showing:

High nutrient | Low nutrient - particularly nitrogen depletion

Green cells
- Active growth
- High chlorophyll level
- Photoprotection: *High turnover of Q_b protein*

Red cells
- Poor growth
- Chlorophyll breakdown
- Photoprotection: *Astaxanthin synthesis*

High light

Green cells
- Poor growth
- Chlorophyll retained

Low light

Figure 3.27   Influence of light and nutrient on pigmentation of snow algae. High light conditions occur in exposed environments, low light in areas covered by forest canopy (based on information in Hoham and Duval, 2001)

Nutrient limitation

Nutrient levels are frequently very limited in snow environments (see Chapter 2). Adaptations of snow algae to low-nutrient conditions include the requirement for mineral but not organic molecules, their ability to remain inert over long periods as resistant zygotes, and the ability of motile stages to migrate towards nutrient sources (chemotaxis). Snow algae rapidly colonize habitats where nutrients are high, including snow surrounding sea bird colonies in Antarctica.

The small size of (unicellular) snow algae, with high surface area/volume ratios, also optimizes nutrient absorption under limiting conditions.

Temperature

Various workers have queried whether snow algae are adapted specifically to low temperatures or are simply surviving at the edge of their temperature range. In order to answer this question, algae have been isolated and their growth characteristics studied under controlled laboratory conditions. Hoham (1975) has suggested that true snow algae are obligate cryophiles (or psychrophiles), growing optimally below 5°C, abnormally at 10°C, and not surviving over this limit. Obligate cryophiles include species of *Chloromonas* and *Chlainomonas* and strains of *Chlamydomonas nivalis*. Other snow algae, such as *Raphidonema nivale*, are able to grow at temperatures up to 15°C in the laboratory, and qualify as facultative cryophiles.

Specific adaptations to life in the cold include the ability to photosynthesize at low temperatures and the occurrence of a variable fatty acid ratio.

Fatty acid ratio   Various studies have indicated higher ratios of monounsaturated/saturated fatty acids in snow algae compared with their temperate relatives. Such ratios are considered to have a cryoprotective function since high levels of unsaturated fatty acids cause increased membrane fluidity, which is important in maintaining membrane transport at low temperatures. It has been suggested that the growth temperature range of an organism depends primarily on the ability to regulate its membrane fluidity, and the ability to maintain this fluidity at low temperature. In cold-adapted organisms, the lower temperature limit is thus not determined by membrane constraints, but by the freezing properties of aqueous solutions inside and outside the cell.

Photosynthesis

Photosynthetic optima for snow algae show some variation between species. Many snow algae show maximum photosynthetic activity at −3 to 4°C (Hoham, 1975; Mosser *et al.*, 1977) while others, such as *Chlamydomonas nivalis*, photosynthesize optimally at 10–20°C, but retain substantial activity at −3 to 4°C. The ability of snow algae to carry out photosynthesis at freezing temperatures is a key factor in their survival.

Acidity

The pH of snow which contains algae has been widely sampled in different continents and has indicated universally acidic conditions, with pH values typically in the range 4–6.2. Laboratory studies on *Chloromonas pichinchae*, which normally grows at pH 4.9–5.2 in nature, have demonstrated a growth optimum at pH 6. Other isolates of *Chloromonas* indicate adaptation to more acid conditions. Little is known about the biochemical and physiological mechanisms which underlie this low pH tolerance.

Desiccation

Desiccation presents the final challenge for many snow algae, which have to survive long periods of potential water loss. This is particularly the case for resistant spores. In many of these cells, the presence of thick complex walls, with primary and secondary layers, may reduce desiccation.

E. BIODIVERSITY IN THE ALGAL COMMUNITY

Preceding sections of this chapter have emphasized the huge diversity of living organisms (biodiversity) that occurs within this major group of freshwater biota. This has been noted in relation to taxonomic features, molecular characteristics, size and shape, environmental preferences, interactions with other biota, and strategies for survival. This section considers diversity in relation to species composition.

3.13 Variety of freshwater algae: indices of species diversity

The assessment of diversity in freshwater biota is complex, both in terms of the criteria used (see above) and the target organisms being assessed. These may be particular taxonomic groups (e.g., blue-green algae) or broad taxonomic assemblages (all algal species), and may be assessed in relation to a single point in time, an extended time period (e.g., a phases of the seasonal cycle), or a particular micro-habitat.

The biodiversity of freshwater algae within a specific environment is normally considered in reference to the range of species present. Assessment of biodiversity in aquatic habitats is important for a number of reasons, including:

- comparison of different natural habitats (e.g., different lake systems) in relation to geographic, hydrological, and specific environmental parameters;

- assessment of the effects of human activities (or their effects) such as chemical pollution, eutrophication, and global warming;

- understanding fundamental aspects of community structure dynamics.

In this section, algal species diversity will be considered specifically in relation to one particular situation – the lake planktonic environment.

3.13.1 The Paradox of Phytoplankton Diversity

One of the remarkable aspects of the lake environment is the large number of phytoplankton species that is present at any one time. Between 30 and 40 algal species may be routinely identified in mid-summer samples from the water column of a temperate eutrophic lake, the final total depending on the skills of the identifier and the number of samples examined. Such species diversity appears as a paradox (Hutchinson, 1961), since it might be expected that a single group of organisms (algae) competing for resources within a uniform and relatively unstructured environment would have low diversity due to competitive elimination. Explanations for the high diversity of freshwater phytoplankton are based on the fact that interactions in the planktonic environment are highly complex, and fall into three main areas as follows

- *Environmental complexity.* Although the freshwater environment might appear to be relatively uniform, particularly within the epilimnion, there are some important variations. This is particularly the case for the biotic environment, with very patchy distributions of algae and zooplankton resulting in micro-environmental variation in nutrient availability, algal competition, and grazing pressure.

- *Diversity in adaptive strategies.* As noted earlier, algae have responded to a range of environmental constraints in a number of ways with different strategies for survival. Competition between algae will not, therefore, involve a single characteristic relating to a single environmental feature (e.g., relative ability to take up phosphate ions from a phosphorus-limited epilimnion), but a whole range of adaptations relating to multiple environmental characteristics.

- *Non-adaptive interactions.* This approach to phytoplankton diversity postulates that there is insufficient time for competitive interactions and adaptations to completely and fully operate, since interactions in the freshwater environment are essentially short-term. Community composition at any one time is considered to be determined more by the pre-existing situation than by current species adaptations. This situation is dominated by historical factors and the pre-existing community (referred to as 'founder-effects').

3.13.2   Biodiversity indices

Biodiversity within a group of organisms may be determined numerically in relation to a variety of parameters (referred to as importance values) including biomass, productivity, and organism count (Odum, 1997). Most estimates of species diversity are determined from species counts, which can be used to generate bio-indices of three main types – species richness, species evenness/dominance, and a combined index of biodiversity.

Species richness

This index relates to the total number of species present in the population sample – the greater the number of species the greater the measure of biodiversity. Species richness may be assessed in two main ways as follows.

Total number of species (S)   Species richness is often determined simply in relation to the total number of species. The major problem with this index is that the value (S) may depend on sample size – the bigger the sample the more species there are likely to be.

Margalef index (d)   This index avoids the complication of sample size, by incorporating the number of individuals (N) in the sample. The index thus represents a measure of the number of species relative to the total number of individuals present in the sample, where

$$d = (S - 1)/\log N \qquad (3.9)$$

Species evenness/dominance

The evenness of species occurrence is also an important measure of biodiversity. Considering two hypothetical samples (I and II), each composed of four species (A, B, C and D):

Sample I. Total 100 individuals; species A (25), B(25), C (25), and D (25).
Sample II. Total 100 individuals; species A (94), B (2), C(2), and D(2).

Sample I would be regarded as more diverse than sample II, though species richness is the same. Sample I has greater evenness of species occurrence, but a lower dominance of any species. Sample II is the converse.

Species evenness may be determined as *Pielou's evenness index* (J'), where:

$$J' = H'(\text{observed})/H' \max \qquad (3.10)$$

and $H'_{\max}$ is the maximum possible diversity that would occur if all species were equally abundant.

Dominance is the converse of evenness, and can be determined using the Simpson index (D), which relates the number of individuals in each species (n_1) to the total number of individuals in the sample (N), where:

$$D = \sum (n_1/N)^2 \qquad (3.11)$$

Comparison of indices for species evenness and dominance over a number of population samples shows a clear inverse relationship.

Combined index of biodiversity

Clearly, species richness and evenness (or dominance) both contribute towards mixed population diversity, and a most useful approach would be to bring these together as a single value. The Shannon–Wiener Diversity Index is the most common combined measure of diversity, taking into account richness, evenness, and abundance of the community structure and assumes that individuals are randomly sampled from an infinitely large population

(Mason, 2002). The equation below is used to calculate the Shannon–Wiener Index:

$$H' = \sum P_i(\ln P_i) \qquad (3.12)$$

where H' is diversity and P_i the proportion of individuals found in the ith species (estimated as n_i/N (total number)).

3.13.3   Numerical comparison of phytoplankton populations

Lake nutrient status

In general, lakes with a very high nutrient status have relatively low species diversity (Margalef, 1958). Studies by Reynolds (1990), for example, compared UK lake sites of decreasing nutrient status – Blelham enclosures (fertiliser-enriched, never P-limiting), Crose Mere (rarely P-limiting), and Windermere (chronically P-limiting). Both mean and maximum seasonal diversity (Margalef index) increased along this sequence, with maximum levels in Lake Windermere. Further reduction in lake nutrient status towards oligotrophy results in a decrease in phytoplankton diversity. This was demonstrated by Debacon and McIntyre (1991), who obtained seasonal Shannon index values in the range 0.5–2.5 for the ultra-oligotrophic Crater lake in Oregon (USA).

Seasonal changes in phytoplankton biodiversity

The major seasonal changes that occur in phytoplankton biomass, with the succession of spring diatom bloom, clear-water phase, and mixed summer algal bloom (Figure 3.14), are also accompanied by changes in biodiversity. An example of this is presented in Figure 3.28, which shows changes in algal biomass, with related indices of species richness, species dominance, and general biodiversity in a typical temperate eutrophic lake (Levado and Sigee, unpublished observations).

During this annual sequence, the spring diatom bloom was a period of moderate species richness, with low dominance – since the diatom population

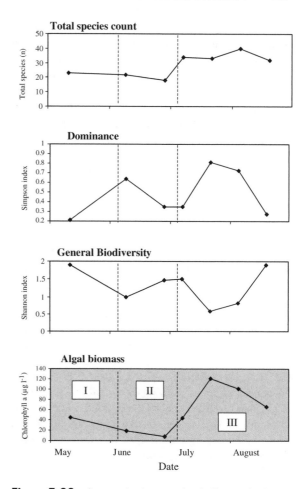

Figure 3.28   Seasonal changes in indices of phytoplankton biodiversity in a temperate eutrophic lake. Vertical broken lines demarcate the end of the diatom bloom (I), the clear-water phase (II), and the late summer mixed algal bloom (III). (Taken with permission, from Levado, 2001.)

was composed of a balanced mixture of species. Increased dominance and low biodiversity (Shannon index) occurred in the early clear-water phase. This was characterised by a relatively small number of r-selected species, with dominance by green algae such as *Ankyra, Chlamydomonas* and *Sphaerocystis*.

Species richness reached highest levels during the development of the mixed summer algal bloom, emphasising the wide range of species that grew towards the end of lake stratification (see

Figure 3.17). This build up in species richness towards the end of the growth period is in line with the laboratory studies of Odum (1962), who studied the transition of a fresh culture of algal species to a mature steady-state situation. Initially the culture was dominated by a few rapidly-growing species, but eventually a steady state mixed population of high biomass and biodiversity was reached. Park (1980) has suggested, in relation to terrestrial ecosystems, that the increase in species richness occurring during ecological succession is due to an increased availability of habitat niches. In the case of the aquatic environment, we can consider 'habitat niches' as the ability of different phytoplankton species to compete and exploit the environment in different ways, with the diversity of competitive strategies noted previously.

The maximum values for species diversity that occurred during the summer bloom were matched by minimum indices for the combined diversity index. This was because of developing high population levels of colonial blue-green algae and dinoflagellates, resulting in a high index of dominance. Increased summer dominance in this lake (Figure 3.28) was associated with the conditions of intense competition that relate to very high phytoplankton biomass ($>100 \,\mu$g chlorophyll 1^{-1}) and signals the tendency of highly-adapted K-selected species to out-compete other algae.

3.13.4 Biodiversity and ecosystem function

Various aspects of ecosystem function relate to biodiversity within the biological community, particularly the diversity of primary producers (see Chapter 1). Seasonal changes in phytoplankton biodiversity, for example, relate to aspects of lake community dynamics such as algal productivity and ecosystem stability.

Biodiversity and algal productivity

Biodiversity is at a minimum during the period of minimum productivity (clear-water phase), with increased biodiversity typically being associated with periods of higher algal growth (spring diatom bloom and mixed summer bloom). The relationship between biodiversity and productivity breaks down, however, under conditions of intense summer bloom formation – when dinoflagellates or blue-green algae may completely dominate the lake and cause a marked reduction in species richness.

Biodiversity and ecosystem stability

Ecosystem stability may be defined in various ways (see Chapter 1), including the degree of community change and species diversity.

On both of these criteria, the clear-water phase can be seen as a time of high instability within the annual cycle. During this time there are major changes in the biomass and species content of phytoplankton populations, with zooplankton populations also undergoing rapid changes and maintaining a high grazing pressure (high stress level) on phytoplankton. Biodiversity (measured as species richness or Shannon index) typically reaches low levels during this phase.

4

Competition for Light

Sunlight (solar radiation) is important to aquatic microorganisms in four major and inter-related ways.

- *Determination of their physico-chemical environment*. Light is a major determinant of the dynamics and structure of aquatic environments. Energy from the sun provides the heat that generates the Earth's wind patterns, resulting in mixing of the surface layers of lakes and oceans. Transformation of light to heat energy within surface waters results in a localized temperature increase which combines with wind energy to produce thermal stratification (Figure 2.10). This in turn limits the distribution of nutrients within the water column and thus has a secondary effect on aquatic chemistry. Light also has direct effects on chemical characteristics and is important, for example, in the degradation of humic acids in peaty lakes to more biologically available compounds (DeHaan, 1993).

- *Production of biomass*. Light penetration of surface waters drives the photosynthetic activity of the primary producers. These include both planktonic and benthic organisms, and comprise three major groups – higher plants, algae, and photosynthetic bacteria. The photosynthetic reducing power which is generated by light is important for both carbon (CO_2) and nitrogen (NO_3, NO_2, NH_4)

assimilation. Nitrogen fixation (in colonial blue-green algae) and phosphorus metabolism are also closely linked to photosynthesis, but the cellular energy required for uptake and deposition of silicon in diatoms is derived solely from aerobic respiration without any involvement of photosynthetic energy (Martin-Jézéquel *et al.*, 2000).

- *Damage to cell processes*. In extreme conditions, where general light intensity is high, or where there is a high level of ultraviolet radiation, damage to cell metabolic and genetic processes (photoinhibition) can seriously impair biological activities.

- *Induction of periodic seasonal and diurnal activities*. Temporal changes in the occurrence and activities of freshwater biota are considerably influenced by the periodicity of light. This affects both seasonal activities (where day length is important) and diurnal activities (where alternation of light and dark entrains circadian rhythms).

This chapter initially considers aspects of light intensity and wavelength within aquatic systems. (Section 1). The major part of the chapter then deals with light as a resource for the production of biomass (Sections 2–8), followed by a consideration of the damaging effects of light (Section 9) and the importance of light periodicity (Section 10).

Freshwater Microbiology: Biodiversity and Dynamic Interactions of Microorganisms in the Aquatic Environment   David C. Sigee
© 2005 John Wiley & Sons, Ltd   ISBNs: 0-471-48529-2 (pbk) 0-471-48528-4 (hbk)

4.1 The light environment

4.1.1 Physical properties of light: terms and units of measurement

Solar radiation can be considered either as a continuous wave of energy or as discrete packets (photons) of excitation. These two concepts give rise to two fundamental measurements of light:

- wavelength, which describes the quality of light and the type of effect it will have on living organisms, and

- photon intensity, which describes the quantity of light to which freshwater microorganisms are exposed.

Solar energy has a wavelength range of about 100–3000 nm, with most of the energy being present between 300–2000 nm. The spectral range can be separated into three main regions (Figure 4.1) – ultraviolet radiation (100–400 nm), visible light (approximately 400–700 nm), and infrared radiation (700–300 nm). These three groups comprise 3 per cent, 46 per cent and 51 per cent respectively of solar radiation arriving at the Earth's outer atmosphere. The visible radiation also coincidentally corresponds to the 'photosynthetically available radiation' (PAR) which drives the photosynthetic activities of most phototrophic organisms (algae and higher plants). Infrared radiation appears to have little specific metabolic effect on living organisms, although photosynthetic bacteria do absorb some of it for photosynthesis (Figure 4.11). Ultraviolet light is of major importance in causing damage to microorganisms and is discussed in Section 4.9.

The measurement of light is based on the physical properties of electromagnetic radiation and is carried out using a 'quantum sensor'. Numbers of photons are usually expressed directly in moles, where 6.02×10^{23} photons equal 1 mol. Commonly used terms and units for different light parameters are given in Table 4.1 and include the number of incident photons (photon flux, photon flux density),

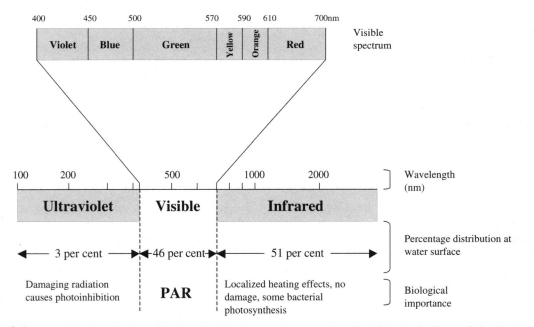

Figure 4.1 Wavelengths of solar radiation, showing the occurrence and major biological effects of the three major bands – ultraviolet, visible, and infrared

Table 4.1   Radiometric measurement of light

Term	Property	Unit of measurement
Photon flux	The number of photons arriving per unit time	μmol^* photons s^{-1}
Photon flux density (PFD)	The number of incident photons, per unit area and time interval	μmol photons m^{-2} s^{-1}
Irradiance	The amount of incident energy, per unit area and time interval	W m^{-2} or J m^{-2} s^{-1}
Photosynthetically active radiation (PAR)	Wavelength of radiation within the waveband 400–700 nm	nm
Spectral composition	The distribution of light intensity at different wavelengths	Photon flux or irradiance at specific bands (nm)

$^*\mu$moles have now replaced μEinsteins as the unit for the number of photons.

energy content (irradiance), and wavelength (range of wavelength, spectral composition).

Light intensity in both environmental (e.g., Figure 4.22) and laboratory (e.g., Figure 4.8) systems is typically expressed as photon flux density (PFD) and is an important parameter is assessing the light-response of living organisms in relation to photosynthesis, growth, and photoinhibition. The terms 'photosynthetic PFD' and 'photosynthetic irradiance' are also sometimes used to denote the amount of incident light within the restricted range of photosynthetically active radiation.

4.1.2   Light thresholds for biological activities

In the freshwater environment, photon flux density varies with time (seasonal, diurnal), water quality (content of dissolved substances, presence of biota), and depth in the water column (see Section 4.1.3). Light intensity is quantitatively important for the photosynthesis and growth of autotrophic microorganisms (Section 4.2.4), and is also qualitatively important in determining the presence or absence of particular biological activities. This qualitative effect operates through threshold levels, some of which are shown in Figure 4.2.

Threshold levels of light are particularly relevant to the activities of photosynthetic organisms. At the top end of the PFD scale, these include light saturation of photosynthesis in periphyton and phytoplankton. With decreasing light intensity levels, a sequence of thresholds operate for the growth and survival of macrophytes, algae, and photosynthetic bacteria.

- *Macrophytes.* The threshold PFD level for macrophyte survival (45–90 μmol m^{-1} s^{-1}) is ecologically important and is directly linked to microbial activity, since in eutrophic waters shading by algal blooms reduces PFD below the critical level and causes eradication of the macrophyte flora (Chapter 10).

- *Algae.* In temperate and polar lakes, critical levels of light intensity are required at the beginning of the growing season to trigger the onset of the spring diatom bloom. Studies on laboratory cultures (*PI* curves, Section 4.2.4) indicate the onset of net photosynthesis and growth at PFD values of 0.1–1 μmol m^{-1} s^{-1}.

- *Photosynthetic bacteria.* These organismns are restricted to low-light anaerobic regions of the water column, and the onset of growth is triggered by very low-light levels (<1 μmol m^{-1} s^{-1})

In addition to the direct effects of light on photosynthetic microorganisms, indirect effects on freshwater microbes also occur via the influence of light on zooplankton and fish. The onset of maximal vertical migration of *Daphnia hyalina* requires

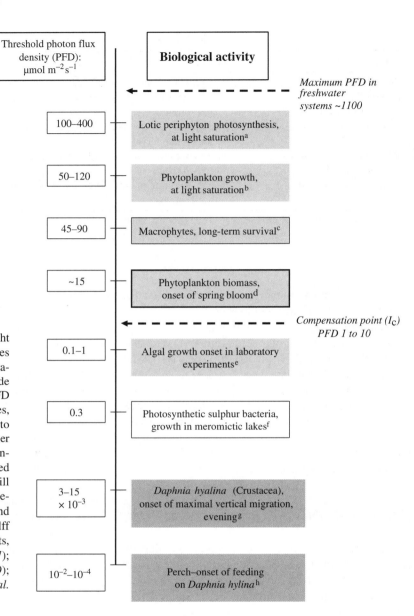

Figure 4.2 Threshold levels of light intensity affecting biological activities in freshwater environmental and laboratory situations. The order of magnitude is given for threshold levels of PFD that directly affect microbial activities, or influence other biota that relate to microorganisms. References and further information: [a]laboratory studies on intact periphyton communities obtained from different streams, Boston and Hill (1991); [b]Harris (1980); [c]north temperate macrophytes, Sand-Jenson and Borum (1991); [d]polar lake, Kalff et al. (1972); [e]laboratory experiments, Middelboe and Markager (1997); [f]meromictic lakes, Pfennig (1989); [g/h]temperate lake, Ringelberg et al. (1991) (data from Kalff, 2002)

threshold light levels of 10^{-3} μmol m^{-1} s^{-1}, and has implications for the grazing of microorganisms in surface waters. This light level is equivalent to the irradiance from a full moon, which also allows fish such as perch to feed visually on zooplankton in surface waters at night – indirectly influencing populations of algae and other microorganisms via the food web.

4.1.3 Light under water: refraction, absorption, and scattering

Both the overall intensity of light and the spectral composition are modified by its passage through water, largely due the effects of refraction, absorption, and scattering. Refraction of light, with separation into different wavelengths, occurs as the radiation passes through regions of different refractive index – including the air/water interface and strata of different density in the water column. Absorption (conversion of light to heat energy) and scattering (reflection) are mediated by water molecules, dissolved substances, and particulate matter.

Decrease in light intensity

The intensity of light at the surface of a body of water does not normally exceed 2000 μmol photons m^{-2} s^{-1}, equivalent to about 400 Wm^{-2} in energy units. Loss of light due to absorption and scattering results in an exponential decrease in intensity with depth. The fraction of light lost per metre of water is expressed mathematically as the extinction coefficient (ε_λ). For parallel beams of monochromatic (single wavelength) light, the intensity of light at a particular depth in the water column is given by the Lambert–Beer law, which can be expressed as:

$$I_z = I_0 e^{-\varepsilon_\lambda z} \qquad (4.1)$$

where: I_z = light intensity at depth z (mol photon m^{-2} s^{-1}), I_0 = light intensity penetrating the water surface (mol photon m^{-2} s^{-1}), ε_λ = extinction coefficient for the particular wavelength (m^{-1}), and z = depth (m).

The extinction coefficient (ε_λ) is very wavelength-dependent and can be separated into separate components which relate to light absorption by water molecules (ε_w), dissolved substances (ε_d) and particulate matter (ε_p). For pure water, where ε_d and ε_p are zero, $\varepsilon_\lambda = \varepsilon_w$.

Light zonation of water column   The top part of the water column has a distinct zonation in relation to light availability (Figure 4.3). The depth at

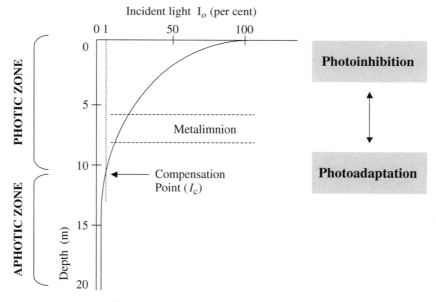

Figure 4.3   Vertical distribution of light in a stratified lake.

Light intensity at the compensation point (I_c) is typically in the range 1–10 μmol photons m^{-2} s^{-1}, compared with maximum irradiances at the surface of the water column of about 1100 μmol photons m^{-2} s^{-1} (Hill, 1996). The penetration of light in relation to lake stratification is shown in Figure 2.11. Extension of the photic zone (region of net photosynthesis) below the metalimnion is typical of low nutrient (oligotrophic) lakes. In eutrophic lakes the presence of dense algal populations limits the photic zone to the epilimnion. The range of light intensities within the photic zone leads to the contrasting processes of photoinhibition (Section 4.9) and photoadaptation (Section 4.6). The depths quoted in this figure vary with time, and from lake to lake which light intensity reaches the point where O_2 evolution by photosynthesis equals O_2 uptake by respiration is referred to as the compensation point. Light intensity at the compensation point, I_c, varies with environmental (e.g., temperature) and physiological characteristics, but is normally about 0.1–1 per cent of the value at the lake surface. The vertical distance between the water surface and compensation point is the photic zone, and is the region within which net anabolic (synthetic) processes occur in phototrophic organisms. Below this, in the aphotic zone, respiration exceeds photosynthesis and metabolism is predominantly katabolic.

The depth of the photic zone varies widely between water bodies. In eutrophic lakes, light attenuation with depth is pronounced due to high phytoplankton biomass and the photic zone does not extend below the thermocline. In contrast, oligotrophic lakes have little particulate matter and the photic zone may extend below the thermocline.

In the case of shallow lakes, rivers, and wetlands, the photic zone may extend down to the sediments or bed of the body of water. This promotes growth of microbenthic algae, which can become major primary producers within the particular ecosystem.

Changes in the wavelength of light

The spectral composition of underwater radiation is important to lake biota in various ways, including photosynthesis, light-responsive activities (phototaxis, diurnal variation), and radiation-induced damage to cells and molecules.

Underwater separation of light into different wavelengths is complex, with the end result depending on the balance between the processes of refraction, absorption, and scattering. The bending of light by refraction as the radiation passes through water of different densities is higher for short wavelengths (blue light) than for long wavelengths (red light). Absorption of light is also wavelength-selective, being most pronounced at both ends of the spectrum. As light passes through the water column, the ultraviolet and infrared ends of the spectrum are absorbed first, leaving a progressively narrower band of light, restricted to the blue-green part of

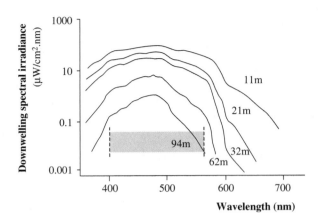

Figure 4.4 Changes in spectral composition of light with depth in an oligotrophic lake (Lake Tahoe, USA). At depths below 30 m, most of the transmitted light is in the violet to green part of the spectrum (grey block). Depth values are given to the nearest metre (based on a figure from Smith *et al.*, 1973)

the spectrum (Figure 4.4). Reflection of light by particulate or molecular material may also cause changes in spectral composition. The predominant scattering of blue light by water molecules gives clear-water (oligotrophic lakes) a distinctive blue colour, and results in a clear shift of transmitted light towards the red end of the spectrum with depth.

Changes in the spectral composition of light with depth differs considerably between lakes, varying particularly with the content of dissolved substances (e.g., humic acids), phytoplankton content, and the amount of nonliving organic and inorganic material. Depth changes in light intensity and composition are least in clear-water lakes, but may become very pronounced in high nutrient lakes with extensive phytoplankton populations. Photons in the blue and green portions of the spectrum are rapidly attenuated in turbid or coloured waters, while photons in the red part of the spectrum show least alteration under such conditions.

4.1.4 Light energy conversion: from water surface to algal biomass

Only a small proportion of the light energy incident at the water surface is converted to chemical energy

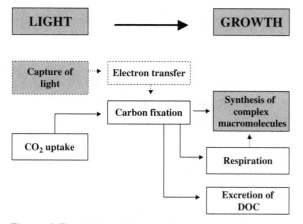

Figure 4.5 Linkage between the capture of light and the synthesis of complex macromolecules for growth; energy losses occur during the process of photosynthesis and in subsequent respiration and excretion of dissolved organic carbon (DOC)

in the form of algal biomass. The loss of energy is partly due to depletion within the water medium and partly due to losses during light transformation within the algal cell (Figure 4.5).

Factors affecting light depletion may be very different when comparing phytoplankton and benthic algal communities.

Phytoplankton

Approximately 1–2 per cent of incident light is lost by reflectance at the water surface (varying with the angle of incidence). Subsequent transmission within the water medium leads to a decrease in light intensity and changes in spectral composition, as described previously. The proportion of light that is captured by phytoplankton cells within the water column will depend on:

● light removal and alteration by physical parameters – particularly water molecules, dissolved matter, and non-living particulate matter;

● light removal and alteration by biological factors such as other phytoplankton cells (self shading) and macrophytes, and

● how much algal biomass is available for light uptake.

Calculations of the proportion of light quanta absorbed at the water surface which are subsequently captured by phytoplankton range from 4–80 per cent (Table 4.2) varying with chlorophyll-a concentration in individual water bodies. The effect of water turbidity is indicated by comparison of the clear-water of Lake Constance (70 per cent light capture at 30 mg chl-a m^{-3}) with the turbid

Table 4.2 Capture of light energy by phytoplankton populations

Lake	Properties	Algal biomass (mg chl-a m^{-3})	Light capture by phytoplankton
Lake Kinneret[a] (Sea of Galilee)	Eutrophic (dominated by *Ceratium*)	5–100	4–60%
Lake Minnetonka[b] (USA)	Eutrophic (dominated by blue-green algae)	3–100	8–80%
Lough Neagh[c] (Ireland)	Eutrophic turbid	27–92	20–50%
Lake Windermere[d] (UK)	Mesotrophic (dominated by *Asterionella*)	1–7	5–25%
Lake Constance[e] (Germany)	Mesotrophic Clear water	1–30	4–70%

References: [a]Dubinsky and Berman (1981); [b]Megard *et al.* (1979); [c]Jewson (1977); [d]Talling (1960); [e]Tilzer (1983)

water of Lough Neagh (50 per cent light capture at 92 mg chl-a m^{-3}).

Benthic algae

Although physical light absorption and alteration within the water column is also important for benthic algae, interception and absorption by other lake organisms may impose extreme conditions. Benthic algae may experience particularly severe light attenuation from the water surface due to:

- their existence in a physically compressed community (biofilms, algal mats) that promotes self-shading,

- light attenuation by the whole water column, including the entire mass of phytoplankton,

- shading by macrophytes – benthic algal communities can only occur in shallow water (due to the need for light penetration) which also favours the growth of macrophytes,

- benthic algae are fixed in position, while planktonic algae are able to migrate in the water column to optimal light conditions.

After the removal of light by the environment, further losses are involved during the conversion of the light (captured by phytoplankton cells) to the chemical energy of the primary photosynthetic product (phosphoglyceric acid) and subsequently to proteins, lipids, and carbohydrates (Figure 4.5). These losses include inefficiencies during the electron transfer and carbon fixation parts of photosynthesis, and are calculated by Kirk (1994) to allow a maximum transformation of light energy (absorbed in the reaction centres) to carbohydrate as 25 per cent. Subsequent conversion of carbohydrate to algal biomass involves further energy depletion (Figure 4.5), including loss of newly-synthesized material by respiration and leakage of small molecular weight dissolved organic carbon (DOC) into the environment. All of these factors result in an estimated maximum conversion of light to biomass in algal cells of about 18 per cent.

4.2 Photosynthetic processes in the freshwater environment

Photosynthetic processes form the basis for phytoplankton productivity and the net assimilation of carbon in aquatic ecosystems (Williams *et al.*, 2002).

4.2.1 Light and dark reactions

Photosynthesis takes place in the chloroplasts of eukaryotic algae and within the entire cells of blue-green algae and photosynthetic bacteria, and can be separated into two processes: light-dependent reactions and dark reactions (Figure 4.6).

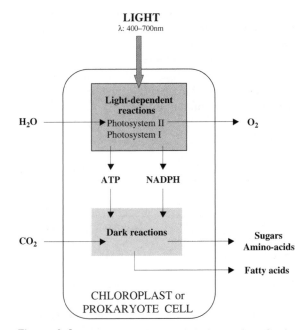

Figure 4.6 Light-dependent and dark reactions in the process of photosynthesis (eukaryotic and blue-green algae); details of the photosystem II light-harvesting complex are given in Figures 3.13 and 4.24

Light reactions

These occur on specialized photosynthetic (thylakoid) membranes and involve the capture of light energy by photosynthetic pigments and the transfer of this energy to adenosine triphosphate (ATP) and nicotinamide adenine dinucleotide phosphtate hydride (NADPH). In organisms that carry out oxygenic photosynthesis (eukaryotic and prokaryotic algae), water is split as part of the reaction, releasing oxygen and making available electrons which convert NADP to NADPH. Light reactions are temperature-independent since they are photochemical in nature and are not driven by enzymatic processes.

Light reactions are of two main sorts – photosystem I (PS I) which is associated with the conversion of NADP to NADPH, and photosystem II (PS II) responsible for the splitting of water. Each system involves a specific macromolecular assembly, the light-harvesting complex, which is divided into two separate regions – a surface antennal complex and an internal reaction centre. Absorption of light energy is carried out by surface pigments within the antennal complex, with subsequent funnelling of the energy to the reaction centre. This contains a single molecule of chlorophyll-a (P680) to mediate electron transfer to electron chains. Antennal pigments include chlorophyll-a, plus a range of accessory pigments such as chlorophyll-b, chlorophyll-c, carotenoids, phycocyanins, and phycoerythrins. The time course of photon excitation and electron transfer in the photosystem II light-harvesting complex (LHCII) are given in Section 3.7 (Figure 3.13) and a diagram of molecular interactions within LHCII in Figure 4.24.

Although both photosystems are required for complete photosynthesis, they are physically distinct, have independent requirements for photons, and may have different types of antennal pigments. In the red and blue-green algae, for example, chlorophyll-a is associated particularly with PS I, while phycoerythrins are present mainly in PS II harvesting complexes.

Dark reactions

These are associated with the fixation of CO_2, and although they do not require the presence of light, they follow light reactions by only a few hundredths of a second. These reactions occur in the spaces between thylakoid membranes and utilize the energy and reducing power of ATP and NADPH to reduce CO_2 to hexose, ultimately generating a variety of sugars, amino- and fatty acids.

4.2.2   Photosynthetic microorganisms

Two distinct groups of photosynthetic microorganisms occur in the freshwater environment – algae (including blue-green algae) and photosynthetic bacteria.

Algae are the most prominent of the two, occurring in almost all freshwater environments and typically having greater biomass and greater productivity. Algae are also more dispersed within individual environments, occurring both as pelagic (present throughout the water column) and benthic forms, whereas photosynthetic bacteria are typically restricted to limited microhabitats (anaerobic zones). Finally, algae are more biologically diverse, present as both eukaryotes and prokaryotes, with a wide range of structural and physiological features.

Most of the work on light and photosynthesis has been carried out in relation to algae, and much of the subsequent description relates specifically to them. Photosynthetic bacteria and their contribution to ecosystem productivity are considered separately in Section 4.5.

Light is important to photosynthetic organisms both as a general growth resource (generating ATP and carbohydrates) and also as a more specific factor in cell processes such as gamete differentiation. Light has a key role, for example, in promoting sexual reproduction (gamete formation) in green algae. In *Scenedesmus*, illumination is important at two critical periods in the laboratory induction of gamete formation – shortly after nutrient (N) withdrawal, and when somatic cells are becoming irreversibly committed to gamete differentiation (Cain and Trainor, 1976).

4.2.3   Measurement of photosynthesis

Photosynthesis, involving the first two of the above processes (light capture to carbon fixation), is typically measured either as O_2 evolution or CO_2 uptake. Measurement via O_2 evolution (by chemical analysis, an oxygen electrode, or manometry) is normally carried out where photosynthesis is particularly active, as in many laboratory studies. For field measurements, where photosynthesis is often at a low level, a more sensitive technique may be required and the process is normally determined as fixation of $^{14}CO_2$. This is carried out by the addition of small amounts of [^{14}C]-bicarbonate to bottles containing phytoplankton samples, then suspending the sealed bottles at points in the aquatic environment for a defined incubation period (normally a few hours) where photosynthesis is being determined. The algae are then deposited on a membrane filter, treated with acid, and assayed for the amount of radioactivity incorporated.

Although measurement of photosynthesis by $^{14}CO_2$ uptake has been widely criticized (see below), its strength lies in its sensitivity – not for slow rates but for low biomass concentrations.

The rate of photosynthesis may be expressed either as *gross photosynthesis* P_g (where O_2 uptake and CO_2 evolution during respiration are not taken into account) or *net photosynthesis* P_n (where corrections are made for respiration). Measuring photosynthesis simply as overall CO_2 uptake would thus give a measure of net photosynthesis, to which CO_2 generated during respiration would have to be added to obtain a gross rate of photosynthesis. The converse is true where photosynthesis is measured by O_2 evolution. Although the terms gross and net photosynthesis are clear in terms of definition, it is not always apparent exactly what is being measured. The use of $^{14}CO_2$ fixation to measure CO_2 uptake introduces a complication in terms of the fate of the labelled carbon atom, and there is some controversy as to whether this method measures gross or net photosynthesis, or gives an intermediate value.

In addition to gas exchange methods, *in situ* photosynthesis can also be determined by the use of active fluorescence (Kolber and Falkowski, 1993). This involves estimating photosynthetic rates from light-stimulated changes in the quantum yield of chlorophyll fluorescence, and has the advantage that artefacts associated with isolating phytoplankton assemblages in bottles are avoided and that there are no problems in differentiating between net and gross procedures.

Whatever procedure is used for measurement, the rate of photosynthesis can be expressed either in relation to environmental dimensions (per unit area or per unit volume of water) or per unit phytoplankton biomass (specific photosynthetic rate).

4.2.4 Photosynthetic response to varying light intensity

The relationship between rate of photosynthesis (P) and light intensity or irradiance (I) can be investigated in the laboratory or under environmental conditions. Bottles containing algae can be suspended at different points in the water column of a lake, for example, where the attenuation of light with depth provides the necessary range of irradiance values.

Photosynthesis – irradiance curves of particular algae are measured for two main reasons: to evaluate ecophysiological responses to light, and to predict *in situ* photosynthesis over a range of irradiance conditions (Hill, 1996).

Photosynthesis – irradiance curves

In freshwater algae (and higher plants) the rate of photosynthesis typically varies in a non-linear relationship with light intensity (Figure 4.7). As light intensity increases from zero, the photosynthetic response undergoes three main phases – light limited increase (zone A), light saturation (zone B), and light-inhibited decrease (zone C).

Phase 1: light limitation In the dark, photosynthesis does not occur, so exchanges of O_2 and CO_2 are solely due to respiration. As light intensity is increased from zero, there is an initial low rate of photosynthesis, which increases with light intensity until the gas exchange of photosynthesis exactly compensates for that of respiration. The irradiance value (I_c) at this point is the *light compensation point*. Further increase in light intensity results in

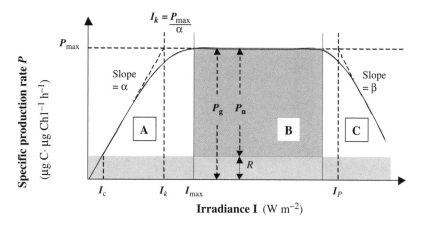

Figure 4.7 A generalized photosynthesis-irradiance $(P - I)$ curve for a discrete algal population, growing under particular conditions of temperature and nutrient availability. At any point on the horizontal axis, net photosynthesis (P_n) equals gross photosynthesis (P_g) minus respiration (R). The parameter I_p provides a measure of the onset of photoinhibition. Based on a figure from Kalff, 2002.

net photosynthesis, with an initial linear increase in the rate of photosynthesis with light intensity. The slope of the graph during the linear phase (α) represents the highest rate of increase of photosynthesis with time, and has a value which is determined by the rate of the light-dependent reactions. At these low irradiances the rate of photosynthesis is limited primarily by the number of photons captured by photosynthetic pigments, and the value 'α' provides a measure of the efficiency with which this occurs.

Phase 2: light saturation   With further increase in irradiance the graph becomes non-linear as light becomes saturating. Photosynthesis over this non-linear part of the PI graph is limited primarily by dark reactions (Figure 4.6) such as those controlled by ribulose 1,5-bisphosphate carboxylase (RUBISCO) activity. Saturation of photosynthesis might also occur with light-dependent reactions if the light-harvesting complexes of photosystem II were no longer able to accept all photons that are incident on the antenna complexes (Section 3.7). The maximum rate of photosynthesis (P_{max}) is reached when the system becomes fully saturated at light intensity I_{max} and is sometimes referred to as the *photosynthetic capacity*.

Phase 3: photoinhibition   In addition to limitation of photosynthesis by light dependent (phase 1) and dark-dependent (phase 2) reactions, photoinhibition may also be important (phase 3).

This is the damaging effect of light on a range of cell activities, including photosynthesis. The effects of photoinhibition are initially seen at mid-level irradiances, resulting in an extended non-linear part of the curve and higher values for I_{max}. Photosynthesis at irradiances beyond I_{max} also shows a sharp decrease, with the slope of the curve (β) providing a measure of the degree of photoinhibition under these conditions. The processes involved in photoinhibition, and mechanisms for avoidance, are discussed in Section 4.9.

Variation in PI curves

The onset of light saturation (transition from straight line to curve) is an important measure of algal light response, but is difficult to define from the graph (Figure 4.7). In view of this, an arbitrary measure can be obtained from the intersection of the line (slope α) with the line at P_{max}. The irradiance value (I_k) at this point is a marker for the onset of light saturated photosynthesis, and can be used as an index of species acclimation to different light levels.

All photosynthetic rate parameters vary with cell size and morphology, taxonomy and environmental factors such as previous light regime, CO_2 concentration, and temperature. The light intensity

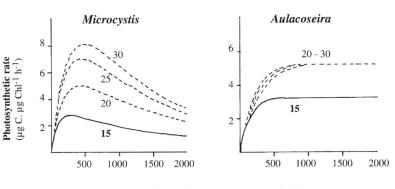

Figure 4.8 Photosynthesis – irradiance (*PI*) curves for a blue-green alga (*Microcystis*) and a diatom (*Aulacoseira*): laboratory studies on the two algae showed marked differences in their photosynthetic response to temperature increase (15–30°C), and in their photoinhibition over the proscribed photon flux range (figure redrawn from Coles and Jones, 2000)

required to reach light compensation point (I_c), saturate photosynthesis (I_k) and achieve maximum photosynthesis level (I_{max}) varies markedly from one algal species to another (Figure 4.8), as also does the degree of photoinhibition.

4.3   Light as a growth resource

Light is an environmental growth resource, leading to the synthesis of macromolecules, increased production of biomass, and potential increase in population. The role of light as a growth resource may be compared with that of other environmental resources such as inorganic nutrients (phosphates, nitrates, silicates) and CO_2 concentration (Raven, 1984).

- The potential supply of light to algal cells (mol photons m^{-2} s^{-1}) can be expressed in similar units to the potential supply of inorganic nutrients (mol nutrient m^{-2} s^{-1}).

- In both cases, acquisition of the resource occurs as a two-step process – physical uptake and metabolic assimilation. In the case of light, this involves absorption by the light-harvesting complex followed by primary photochemistry, while nutrient acquisition involves transport across the plasmalemma followed by primary assimilation via respective enzyme complexes.

- At low resource levels, acquisition of light and nutrients both show limiting concentrations,

below which net growth does not occur. These limiting concentrations are determined partly by the efficiency of resource collection (light absorption, nutrient uptake) at low levels, but also by the balance of metabolic activities. In the case of light, the limiting light intensity (compensation point) at the bottom of the euphotic zone has been defined as the point at which the energetic gains of photosynthesis are just balanced by respiratory losses. An analogous situation occurs with inorganic nutrients, where the initial uptake and assimilation may be counterbalanced by subsequent leakage and regeneration at subsequent points in metabolism.

4.3.1   Strategies for light uptake and utilization

As with inorganic nutrient resources, competition between algae has lead to the evolution of strategies for maximum uptake under different environmental conditions and in the optimum use of growth materials derived from it. In the case of light, these strategies involve adaptations and evolutionary commitments within different algal groups at all levels of uptake and utilization (Figure 4.5):

- capture of light, by photosynthetic pigments within a light harvesting complex,

- conversion of kinetic (light) to potential (molecular) energy – this involves electron transfer and carbon fixation,

Table 4.3 Typical light energy requirements for different algal groups. Values for I_c, I_{max} and I_p (see Figure 4.7) are very approximate, showing considerable variation within algal groups. Information from Horne and Goldman, 1994

| Algal group | Light intensity (μmol m^{-2} s^{-1}) | | | Seasonal adaptations |
	Light compensation point (I_c)	Maximum photosynthesis (I_{max})	Onset of photoinhibition (I_p)	
Blue–green	5	39	>200	Adapted to late summer bloom formation – high I_p, low I_{max}
Dinoflagellates	7	47	233	Adapted to late summer bloom formation – high I_p, low I_{max}
Diatoms	6	84	86	Adapted to early seasonal growth, moderate I_{max}, low I_p
Green	21	211	>250	Adapted to rapid growth during early to mid summer

• synthesis of low MW compounds for physiological activities (e.g., osmoregulation) and complex macromolecules for growth (anabolic processes).

These aspects have evolved in various ways within the different groups of algae, leading to clear distinctions in terms of their ability to grow at different minimum and maximum light levels (Table 4.3), their energy requirements in relation to cell wall synthesis and motility (Table 4.4) and their use of different small MW compounds in osmoregulation (Table 4.5).

4.3.2 Light–photosynthetic response in different algae

Differences in light-response between different algae, and the importance of environmental conditions, can be seen by comparison of the photosynthesis – irradiance ($P - I$) curves for blue-green algae and diatoms at different temperatures. This is illustrated in Figure 4.8, where *PI* curves are shown for two commonly-occurring members of the lake phytoplankton community – *Microcystis* (blue-green alga) and *Aulacoseira* (diatom). *Microcystis* (and other blue-greens) showed a general increase in the rate of photosynthesis with temperature (up to 30°C), but the diatom did not respond to increases

above 20°C. In both cases, the rate of photosynthesis increased with photon flux density up to levels of 210–550 μmol photons m^{-2} s^{-1} (about 10–25 per cent full sunlight), above which photoinhibition was observed in *Microcystis* but not *Aulacoseira* (Coles and Jones, 2000).

Athough considerable variation exists in $P–I$ curves between individual algae and in relation to environmental parameters, each major group of algae has its own characteristics. The advantage displayed by *Microcystis* over *Aulacoseira* at high temperatures, for example, by having higher values for P_{max} and α under these conditions, is typical of other members within the blue-greens and diatoms.

The general response of different algal groups (blue-green algae, dinoflagellates, diatoms, and green algae) have been summarized by Horne and Goldman (1994) in relation to light requirements for minimum photosynthesis (I_c), maximum photosynthesis (I_{max}) and the onset of photoinhibition (I_p) – as shown in Table 4.3. The data indicate that:

• the range of light intensities over which algae grow varies considerably;

• for most algae, photosynthesis commences at light intensities of around 5–7 μmol m^{-2} s^{-1} – the threshold for green algae is substantially higher at 21 µmol m^{-2} s^{-1};

- the light intensity at which maximum photosynthesis occurs is also higher for green algae compared with blue-greens, dinoflagellates, and diatoms;

- photoinhibition occurs at much lower light levels in diatoms (86 µmol m^{-2} s^{-1}) compared with other algal groups (>200 µmol m^{-2} s^{-1}).

The differences in light response exhibited by the different algal groups represent varying strategies in their adaptation to environmental light conditions. This is shown in the seasonal succession of algae, where periods of major dominance broadly coincide with appropriate light conditions (Table 4.3). In the case of diatoms, for example, the range of light that is most readily used fits the spring bloom period where light levels are generally low due to circulation of water prior to stratification and to ambient seasonal conditions. In polymictic lakes of temperate regions, continued mixing and circulation of water may lead to reduced average light exposure and domination by diatoms over much of the year. Green algae tend to dominate in the summer, particularly during the clear-water phase, when light intensities are high. Although these algae can use up to 211 µmol m^{-2} s^{-1}, this exceeds the light intensity normally available in the epilimnion.

4.3.3 Conservation of energy

Algae show adaptations in terms of conservation or optimal use of the energy that has been transformed by photosynthesis. A major part of this energy is used in the synthesis of macromolecules for growth, of which cell wall materials represent a major investment. Comparison of different algae shows that the energy required for cell wall construction varies considerably (Table 4.4).

The silica cell wall of diatoms requires about 12 times less energy for construction compared with the cellulose and peptidoglycan cell walls of other algae, giving these organisms the potential to

Table 4.4 Energy requirements for molecular synthesis and cell processes by freshwater algae (modified from Raven, 1984; Werner, 1977)

Process	Energy required (pW)
Synthesis of cell wall material, proteins and lipids (per atom C or Si)	
Cellulose (most algae)	2.5
Peptidoglycan (blue-greens)	3.4
Silica (diatoms)	0.11
Protein (gas vacuoles)	4.8
Lipid (cell membrane)	6.0
Activities in the freshwater environment	
Contractile vacuole activity	3.4
Cell motion by flagella (50 µm s^{-1})	0.00024

out-compete other algae on an energy basis. This may contribute to the evolutionary success of these organisms and the fact that diatoms are by far the most common algae in both fresh and saline waters. The saving of energy by diatoms does have a trade-off, however, in terms of silicon requirement and high mass. The need for silicon means that growth of diatom populations may become suddenly limited when supplies of the element become depleted. The high mass of diatom cell walls confers a high specific gravity and sedimentation rate, so that many species require turbulent water conditions to stay in suspension, and may be restricted to particular types of lake or particular times of year. Complete absence of a cell wall saves on energy of construction, but the naked cell requires considerable expenditure of energy in using a contractile vacuole to maintain its osmotic balance so this strategy saves little energy in the freshwater environment.

Optimal use of energy is also important in movement within the water column. Active swimming of motile organisms such as cryptophytes and dinoflagellates involves continuous expenditure of energy. Organisms such as blue-green algae that use buoyancy to regulate their position within the water column require energy to synthesize the gas vacuole proteins, but once these have become established regulation is by the deposition and removal of carbohydrate ballast (Section 3.8.1, Figure 3.20). This is a normal part of the diurnal activities of these organisms, so the diurnal migration in the water column requires little extra energy. The formation of ballast at the lake surface under conditions of high light will also sink the cells out of the zone of photoinhibition, so depth regulation in relation to short-term changes in light intensity will also require little energy. The lack of any buoyancy or active motility mechanism in diatoms means that they have no direct energy expenditure in terms of maintaining their position within the water column, but does impose reliance on water turbulence.

In temperate lakes, the relative success of these different strategies during the annual cycle depends on the physico/chemical environment and on competition with other algae. Diatoms are clearly adapted to turbulent water conditions from autumn to spring, but tend to be replaced by organisms with buoyancy regulation or active motility during the more static conditions of the stratified water column in summer. Although actively motile organisms such as dinoflagellates require high energy expenditure to migrate in the water column, their ability for nocturnal movement from a nutrient-depleted epilimnion down to lower parts of the hypolimnion for nutrient uptake gives them a competitive advantage in late summer. Blue-green algae are also able to migrate into the hypolimnion, but their buoyancy

mechanism requires active photosynthesis for ballast formation. This may become limited in autumn due to conditions of reduced light, leading to later dominance by dinoflagellates.

4.3.4 Diversity in small molecular weight solutes and osmoregulation

Photosynthesis is closely linked to the generation of low molecular weight organic solutes (Figure 4.6) that are involved in osmoregulation. The formation of simple sugars, in particular, results in the formation of a range of osmotically-active compounds which are diagnostic for particular algal groups (Table 4.5). These compounds are important in the osmotic balance within freshwater environments, where algal cells are surrounded by a hypotonic medium. In such conditions the tendency for water to enter by endosmosis is either counterbalanced by cell wall pressure, or (in naked algae) requires continuous expulsion of water by contractile vacuole activity. In conditions of increasing salinity, where the external medium becomes hypertonic, higher intracellular concentrations of osmotically-active solutes are required to balance the elevated external molarity.

The importance of light in the osmoregulatory process is illustrated by the diatom *Cyclindrotheca fusiformis*, where photosynthesis is responsible for maintaining the free internal mannose concentration at an adequate level to balance outside osmolarity (Paul, 1979). In the dark, free mannose can be

Table 4.5 Some examples of osmotically-active low molecular weight organic solutes in different algal groups

Algal group	Environmental groups	Low MW solute
Blue-green algae[a]	Freshwater environments	Glucosyl-glycerol
	Estuaries	Sucrose or trehalose
	Hypersaline lakes	Quaternary ammonium compounds
Green algae[b]	Euryhaline species, e.g., *Dunaliella*	Glycerol
Diatoms[c]	Euryhaline species, e.g., *Cylindrotheca*	Mannose
Red algae[d]	Members present in ion-rich waters, e.g., *Bangia atropurpurea*	Floridoside
Brown algae[e]	Some species present in varying salinity (estuaries and saltmarshes)	D-mannitol

References: [a]Warr *et al.* (1987); [b]Ginzburg and Ginzburg (1981); [c]Paul (1979); [d]Reed (1985); [e]Reed *et al.* (1985)

generated from its polymer, polymannose. Under light conditions, a decrease in external molarity (hypotonic conditions) results in a conversion of free mannose to polymannose. If the external molarity is raised, polymannose synthesis is inhibited and free mannose is synthesized directly through CO_2 fixation.

Some algae such as *Dunaliella* are euryhaline, able to grow over a wide range of salt concentrations. This alga exists as two varieties able to grow at NaCl concentrations of >0.5 M (halotolerant cells)) and >2 M (halophilic cells) respectively (Ginzburg and Ginzburg, 1981). The halotolerant strain has large red cells which contain high concentrations of glycerol, and is able to grow in environments of high salinity such as the Great Salt Lake of Utah (USA), where it forms dense populations colouring the water red or green.

Blue-green algae form a particularly interesting group in terms of osmoregulation, with different small MW molecules being involved in different environmental situations (Warr *et al.*, 1987) as follows.

- In environments with little variation in salinity, blue-green algae respond to increases in salinity by the formation of glucosyl-glycerol. This is formed relatively slowly, but is adequate to adjust to small changes in salinity.

- In estuarine environments, where salinity changes occur rapidly as the tide goes in and out, blue-green algae produce sucrose or trehalose as the major osmoregulant. This production occurs over a short time period, allowing a rapid response to environmental change.

- In hypersaline environments, such as the Great Salt Lake of Utah (USA), blue-green algae make long-term adjustment to high salinity by the continuous production of osmotically-active quaternary ammonium compounds glycine, betaine, and glutamate betaine.

In addition to the production of small MW organic compounds, internal molarity may also be controlled via the internal concentration of inorganic ions such as K^+. This is particularly important in blue-green algae, where diurnal fluctuations in photosynthetically-mediated K^+ uptake result in fluctuations in internal pressure (turgor) and the periodic formation and loss of gas vesicles (Section 3.8.2, Figure 3.20).

4.4 Algal growth and productivity

The photosynthetic conversion of light (kinetic) energy into chemical (potential) energy, with the formation of reduced carbon compounds, is the key process which generates biomass within aquatic systems. The rate of synthesis of biological material (primary production) is important in relation to the net increase in biomass and in the development of algal populations.

4.4.1 Primary production: concepts and terms

Algal growth requires the synthesis of a wide range of cellular material and photosynthetic production of organic carbon compounds is clearly only part of this process. The rate of synthesis of algal biomass is an important parameter since it determines the increase in algal population, and is referred to as the 'gross primary production'. Although this term is widely used, there is no generally accepted definition. Different workers have variously defined gross primary production as follows.

- The rate of conversion of light energy into chemical energy (Platt *et al.*, 1980). This reflects the view held by theoretical ecologists and plant biophysicists, who consider primary production in terms of energy transformations during the initial steps of photosynthesis.

- The rate of organic carbon production resulting from photosynthetic activity (Williams, 1993). This definition considers production in terms of carbon flow and reflects the viewpoint of the community ecologist and plant physiologist.

- The rate of assimilation of inorganic carbon and nutrients into organic and inorganic matter by autotrophs (modified from Underwood and Kromkamp, 1999). This definition considers primary production in terms of the overall production of biomass (including inorganic material such as silica cell walls), and follows the approach of workers such as biogeochemists who are interested in the full range of nutrients within the environment.

4.4.2   Primary production and algal biomass

Gross primary production does not directly translate into algal biomass since the synthetic increase in biomaterials is countered by a range of loss process, including respiration. Net primary production is normally defined as gross primary production minus autotrophic respiration. Other loss processes are also important, including excretion of materials (e.g., enzymes and dissolved organic carbon), cell lysis, and activities of other freshwater biota (e.g., zooplankton grazing, fungal and viral parasitism). In the case of phytoplankton, loss of dead cells from the euphotic zone by sedimentation, and exchange of living cells between the water column and sediments are also important. The growth of phytoplankton can be expressed mathematically:

$$\Delta B / \Delta t = (P - R - E)B - G_P - S \pm B_E \quad (4.2)$$

where: $\Delta B / \Delta t$ = increase in biomass with time, P = rate of gross primary production, R = rate of respiration, E = excretion and cell lysis, B = existing biomass, G_P = loss by grazing or parasitism, S = loss by sedimention of dead cells, and B_E = exchange of biomass between sediments and water column.

Seasonal relationship between growth and photosynthesis

Investigations of the factors that determine seasonal changes in phytoplankton populations often focus on photosynthetic and growth responses of phytoplankton. Because growth of a population of algal cells primarily depends on the increase in protein

biomass, which directly relates to carbon assimilation and photosynthesis, the relationship between specific growth rate (μ) and environmental conditions (light, temperature) might be expected to reflect the photosynthetic rate P. The increase in protein and carbon biomass (growth rate) depends on a balance of factors, however (Figure 4.5), and although photosynthesis, growth rate, and population increase are often closely coupled (Coles and Jones, 2000), this is not always so. The relationship between photosynthesis and population increase may vary because of a range of imbalances in cellular processes as follows.

- Respiration rate does not change in direct proportion to photosynthesis at different light and temperature levels.

- Cell division rates may peak at lower temperatures than photosynthetic rates.

- Extracellular release of DOC (loss of biomass) may be less than 5 per cent at optimum light and temperature but can increase to nearly 40 per cent under conditions of low light and high temperature. DOC release under conditions of high light (causing photoinhibition) may also be excessive.

- Net photosynthesis and carbon assimilation are restricted to the illuminated part of the day, while protein synthesis may continue in the dark.

4.4.3   Field measurements of primary productivity

Most measurements of algal productivity involve determination of photosynthetic activity, either in terms of CO_2 fixation or O_2 evolution. The $^{14}CO_2$ uptake method developed by Steemann Nielson (1952) has been extensively used to estimate primary production, which is expressed as fixed carbon mass per unit area per unit time. Productivity may also be considered in terms of long-term development of biomass (algal production). This may be expressed as units of chlorophyll-a

Table 4.6 Algal production in wetland communities (data taken from Goldsborough and Robinson, 1996)

Algal type	Wetland	Biomass (chlorophyll-a concentration)	Carbon fixation rate
Epipelon	Freshwater marsh (sediment surface)[a]	0–5 mg m^{-2}	0.2–0.5 µg cm^{-2} h^{-1}
	Freshwater marsh (core sample)[b]	10–400 mg m^{-2}	
	Salt marsh (core sample)	20–150 mg m^{-2}	1.5–6.0 µg cm^{-2} h^{-1}
Epiphyton	Freshwater marsh (attachment surface)[c]	0–4 mg cm^{-2}	0.2–0.5 µg cm^{-2} h^{-1}
	Freshwater marsh (sediment surface)[d]	7–650 mg m^{-2}	
	Peat bog	2–240 mg m^{-2}	
Metaphyton	Freshwater marsh[e]	0–80 mg m^{-2} (dry weight)	1.4–114 mg cm^{-2} y^{-1}
Phytoplankton	Freshwater marsh[f]	5–100 µg l^{-2}	3–16 µg l^{-2} h^{-1}
		(5–100 mg m^{-2})	(3–16 mg m^{-2} h^{-1})

This table indicates typical values for the range of algal production (expressed as total biomass and rate of carbon fixation) that different workers have measured for epipelon, epiphyton, metaphyton, and phytoplankton in different wetland environments. These data should be interpreted cautiously due to variation arising from geographic and seasonal factors, differences in analytical methods, and differences in the way that data are expressed.

[a]Biomass taken from sediment surface; [b]biomass in core sample; [c]biomass per unit surface to which the algae are attached; [d]biomass per wetland surface area; [e]biomass expressed as dry weight rather than chlorophyll-a mass; [f]biomass expressed per litre and per wetland surface area, assuming a depth of 1 m

per unit area (benthic algae) or per unit volume of water (phytoplankton). Measurements of algal production as biomass are particularly useful in complex aquatic environments (e.g., wetlands, Table 4.6) where carbon fixation by different algal groups may be difficult to determine.

Assessment of primary productivity varies with complexity of the ecosystem, ranging from relatively simple planktonic populations in lakes to more complex communities in wetlands.

Primary productivity of lake phytoplankton

Primary productivity varies considerably in relation to the nutrient status of the lake, environmental conditions and type of algae present. In some cases, productivity occurs mainly via the picoplankton. Measurement of photosynthetic uptake of ^{14}C-bicarbonate during the summer growth phase of Lake Baikal (Russia), for example, indicated total productivity values of 36 µg C l^{-1} day^{-1}, with over 80 per cent being carried out by unicellular blue-green algal picoplankton and nanoplankton (Nagata *et al.*, 1994).

Primary productivity in wetland communities

Measurement of algal productivity in wetland systems is complicated in two main ways.

- The algae being monitored occur as a diverse assemblage of attached and free-floating forms, and can be divided into four main types on the basis of their life style and microhabitat – epipelic algae (present on mud surfaces), epiphytic algae (attached to higher plant and macroalgal surfaces), metaphyton (surface or benthic floating masses of filamentous green algae), and phytoplankton (free-floating cells and colonies entrained in the water column).

- Because of their different microhabitats, measurements of biomass and carbon fixation tend to use different units in terms of the environmental parameter, making comparability difficult. Thus the biomass of epipelic algae is typically expressed per unit area of wetland, epiphytic algae per unit area of attachment substrate (e.g., higher plant surface), and phytoplankton per unit volume of water.

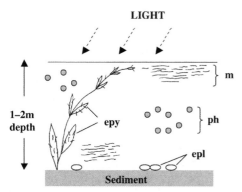

Figure 4.9 Light and algal productivity in a wetland system: diagram showing complexity of algal productivity in a wetland system, with four main groups of algae contributing to carbon fixation – epiphytic algae (epy) on macrophytes, bundles of filamentous metaphyton (m), phytoplankton (ph), and epipelic algae (epl) on fine sediments. Competition for light affects the population growth and productivity of these different groups (not drawn to scale)

The complexities of light availability, algal types, and productivity are illustrated in Figure 4.9. Epipelon biomass can be measured in two main ways. Most studies of freshwater systems involve the collection of an algal sample by placement of lens paper onto the exposed sediment surface. Estimates of epipelic algae by this method tend to give quite low biomass values (<10 mg m^{-2} chlorophyll-a). These low values to some extent reflect the fact that growth of wetland epipelon is limited by shading from submerged and emergent macrophytes, their associated epiphytes, floating metaphyton mats, and phytoplankton blooms – all of which reduce light levels at the sediment surface over much of the growth season. Low biomass values may also reflect limited sample pick-up by this technique, which relies on the migration of algal cells into the lens paper. The other method of measuring epipelic biomass is from a core sample. This has the advantage that all epipelic algae are measured, but suffers from the drawback that degraded pigments from sedimented phytoplankton may also be present, resulting in an overestimate of living biomass in some situations. Chlorophyll-a analysis from core samples give values that are much higher and are more comparable to salt marsh epipelon (Table 4.6). Higher epipelon biomass values might be expected in salt marshes, since the algae in these systems are part of a much more open community. Although salt marsh determinations are also taken from core samples, there is less contamination from phytoplankton in this situation.

Epiphyton biomass is not usually expressed per unit area of wetland because of relating plant surface area to this. Data are normally based on algae that are attached to morphologically simple artificial substrata, with biomass values in freshwater marshes typically <5 µg cm^{-2} of chlorophyll-a. In those studies where epiphyton biomass has been estimated per unit wetland area (Table 4.6), epiphyton biomass is equivalent to the level of algae present in other shallow waters.

The relatively little quantitative information available for metaphyton is normally expressed in terms of dry weight, making comparability with other algae difficult. Biomass of metaphyton is generally <10 mg cm^{-2}, rising to much higher levels when metaphyton is undergoing rapid growth and forming extensive surface mats.

Levels of phytoplankton in wetland systems are highly variable, but normally exceed 50 µg l^{-1} chlorophyll-a. Assuming a homogeneous distribution of phytoplankton within the water column and a depth of 1 m, this value is equivalent to 50 mg m^{-2} of chlorophyll-a per unit wetland area. Small eutrophic wetlands often have chlorophyll-a levels well in excess of this value (>200 mg m^{-2} chlorophyll-a). These wetland phytoplankton levels indicate biomass levels as high as those of epiphyton and metaphyton in less turbid areas, and are as high as phytoplankton levels that develop in eutrophic lakes.

4.5 Photosynthetic bacteria

Although algae are normally the major microbial primary producers in freshwater environments in terms of overall biomass and productivity, photosynthetic bacteria are also frequently present and contribute to carbon fixation within the ecosystem.

General taxonomic and ecological aspects of photosynthetic bacteria are discussed in Section 6.7.

This section focuses on photosynthetic activities which differ from those of algae in lacking oxygen evolution and only occurring under anaerobic conditions. Bacterial photosynthesis requires external electron donors, such as reduced sulphur compounds or organic compounds in the photosynthetic reduction of CO_2. Assimilation of CO_2 via the Calvin Cycle occurs in all photosynthetic bacteria. Some strains in the purple sulphur and non-sulphur bacteria are also able (mixotrophic) to use simple organic molecules such as fatty acids as their carbon source, separately or in combination with CO_2 uptake.

4.5.1 Major groups

Three major groups of photosynthetic bacteria have been recognized – the green sulphur bacteria (Chlorobacteriaceae), purple sulphur bacteria (Thiorhodaceae), and the purple and brown non-sulphur bacteria (Athiorhodaceae). The Chlorobacteriaceae are almost entirely obligate phototrophs, only able to grow under conditions of light and CO_2 availability. Thiorhodaceae are also predominantly photoautotrophs, but many species are potentially mixotrophic, able to photo-assimilate simple organic compounds. Athiorhodaceae photometabolize simple organic substances and are inhibited by H_2S. Several species have the potential for aerobic heterotrophy.

4.5.2 Photosynthetic pigments

The absorption characteristics of bacterial photosynthetic pigments have considerable ecological significance. These pigments are of two main types – bacteriochlorophylls and carotenoids – matching the light requirements of surface and deep water habitats respectively.

Bacteriochlorophylls

Four major bacteriochlorophylls (a–d) have been isolated from photosynthetic bacteria, differing from chlorophyll-a (present in plants and algae) in details of various side groups (Figure 4.10). Although there is no simple correlation between taxonomic groups and type of bacteriochlorophyll, bacteriochlorophyll-a does tend to predominate in the purple sulphur and non-sulphur bacteria. Bacteriochlorophyll-b appears to be restricted to these two groups, while bacteriochlorophylls-c and -d are typically found in the green sulphur bacteria.

Bacteriochlorophylls differ from chlorophyll-a in having absorption maxima in the red and near infrared part of the spectrum (Figure 4.11). Since this part of the spectrum is strongly absorbed by water, these pigments will function primarily in exposed situations such as shallow ponds and estuarine mud flats.

Carotenoids

Cultures of purple sulphur and non-sulphur bacteria have a variety of colours, ranging from peach to brown, pink, and purple red – varying with population density, sulphur content, and age of culture. These colours are mainly due to the presence of carotenoids, which mask the bacteriochlorophylls and in the absence of which the purple bacteria would have the coloration of blue-green algae.

More than 25 carotenoids have been characterized in photosynthetic bacteria. These have been divided into five major groups in terms of chemical structure and biosynthesis (Pfennig, 1967), with a diversity of carotenoids (Groups I–IV) being typical of purple bacteria, and Group V typical of the green sulphur bacteria. Taxonomic distinctions in terms of carotenoid content are shown by the presence of characteristic carotenoids (Table 6.5) for green sulphur bacteria (Chlorobactene), purple sulphur bacteria (Okenone), and purple non-sulphur bacteria (OH-Spheroidenone). Purple sulphur and non-sulphur bacteria show similarity in the mutual possession of Lycopene and Spirilloxanthin, but with none of the major carotenoids found in green sulphur bacteria.

Carotenoids are physiologically and ecologically important in absorbing a different range of wavelengths compared with bacteriochlorophylls. Red

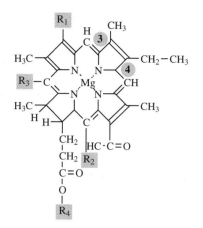

	R$_1$	3	4	R$_2$	R$_3$	R$_4$
Chlorophyll-a	—CH=CH$_2$	—		—C$\stackrel{O}{\underset{O \cdot CH_3}{}}$	phytyl ester (C$_{20}$H$_{11}$O —)	—H
Bacteriochlorophyll-a	—C$\stackrel{O}{\underset{CH_3}{}}$	dihydro		—C$\stackrel{O}{\underset{O-CH_3}{}}$	phytyl ester	—H
Bacteriochlorophyll-c	—C$\underset{H}{\overset{OH}{-}}CH_3$	—		—H	farnesyl ester (C$_{16}$H$_{25}$O —)	—CH$_3$
Bacteriochlorophyll-d	—C$\underset{H}{\overset{OH}{-}}CH_3$	—		—H	farnesyl ester	—H

Figure 4.10 Relationship between chlorophyll-a and some of the principal bacteriochlorophylls. Chlorophyll-a (top figure), present throughout the algae, is a cyclic tetrapyrole with a central magnesium atom. Bacterial chlorophylls differ from this in the identity of four side chains (R1–R4) and two chemical bonds (3,4) (figure adapted and redrawn from Pfennig, 1967)

and brown bacterial carotenoids selectively absorb light in the blue-green part of the spectrum, giving an absorption maximum at 450–550 mμ, which is the prominent wavelength in deep lakes. The presence of these pigments in deep-water purple and green bacteria (giving the latter a distinct brownish tinge) is clearly important for light absorption in this part of the water column.

4.5.3 Bacterial primary productivity

Photosynthetic bacteria are able to assimilate CO_2 in the presence of light and are thus primary producers. Although much is known about their physiology and occurrence in freshwater systems,

relatively little is known about their contribution to overall productivity in aquatic environments.

Studies on primary productivity of photosynthetic bacteria in lakes have tended to occur over a limited part of the annual cycle, at a time when bacterial populations were temporarily high. Work by Czeczuga (1968a,b) in Poland, for example, has demonstrated high rates of bacterial primary productivity during successive years of *Chlorobium* and *Thiopedia* blooms, reaching carbon fixation levels of 55 and 157 mg C m^{-2} day^{-1} respectively (Table 4.7). These values represent a substantial part of the overall primary productivity, in one case exceeding the algal contribution.

In spite of temporary maxima in CO_2 fixation by photosynthetic bacteria, their contribution to

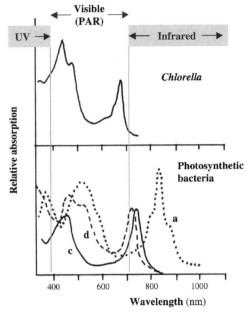

Figure 4.11 Absorption spectra of intact cells of photosynthetic bacteria and the green alga *Chlorella*, recorded in saturated sucrose solution. The bacteria were: **a**: *Chromatium okenii* (purple bacterium containing bacteriochlorophyll-a); **c**: *Chlorobium thiosulfatophilum* (green bacterium, bacteriochlorophyll-c); **d**: *Chlorobium phaeobacteroiodes* (green bacterium, bacteriochlorophyll-d). In the bacteria, a major portion of the solar radiation used for photosynthesis comes from the infrared part of the spectrum. In algae, photosynthetically available radiation (PAR) coincides with the visible light range (figure adapted and redrawn from Pfennig, 1967)

productivity over the whole annual cycle is considered to be generally quite low. Exceptions to this may occur where algal production is depressed due to circumstances such as input of allochthonous materials. This is the case at Smith Hole Lake, USA (Wetzel, 1973) where the mean annual productivity of photosynthetic bacteria was about half that of the algae (Table 4.7).

4.6 Photoadaptation: responses of aquatic algae to limited supplies of light energy

In most aquatic environments, the supply of light may vary considerably (both spatially and temporally) and may become limited in relation to algal optimal photosynthetic requirements.

In the water column of a typical temperate lake or river, for example, midday light intensities during the summer vary from high levels at the water surface (over 800 μmol m^{-2} s^{-1}, on a bright day) to zero, just below the euphotic zone. These light levels have potentially damaging effects on algal cells at the water surface (photoinhibition – Section 4.9), while the provision of energy for photosynthesis becomes extremely limited in the lower part of the water column.

The ability of algal cells to continue photosynthesis and to survive in environmental conditions of

Table 4.7 Comparison of bacterial and algal productivity[*] in a variety of lakes

Lake	Algal productivity	Bacterial productivity	Bacteria
Wadolek, Poland[a]			
June 1966	32.8	55.3	*Chlorobium* in
July 1966	144.5	19.4	metalimnion
Sept 1966	24.6	19.1	
Muliczne, Poland[b]			
Aug 1967	478	157	*Thiopedia* in
Sept 1967	258	136	hypolimnion
Oct 1967	281	28	
Smith Hole, USA[c]			
Annual period	194	91	Late summer population of *Chromatium*

References: [a]Czeczuga (1968a); [b]Czeczuga (1968b); [c]Wetzel (1973)
[*]Productivity measured as mg C m^{-2} day^{-1}

low light availability is referred to as photoadaptation. Both planktonic and benthic algae are able to adjust their molecular and physiological activities in relation to different levels of light availability (Geider and Platt, 1986).

4.6.1 Different aspects of light limitation

During daylight hours, exposure of algal cells to the sun's energy may become limited in three main ways:

- Temporal changes in light intensity, where alternating levels of high and low irradiance affect the continuity of physiological processes and the degree of adaptation at any one point in time. In the planktonic environment, variations in light intensity may be caused by factors such as moving cloud cover and changes in shading by drifting populations of algae and zooplankton. Circulation of algae within the water column due to water turbulence also limits overall light availability, and is important during periods of seasonal non-stratification, in polymictic lakes, and in water bodies subject to artificial disturbance (hypolimnetic aeration). Benthic communities are also subject to similar time limitations. In forest streams, for example, sun beams (sunflecks) penetrating the leaf canopy can contribute 10–85 per cent of total daily irradiance (Hill, 1996), but are very transient, moving across the streambed as the sun moves across the sky.

- Continuous exposure to low light intensity, where algal cells have a long-term location within a particular light-limiting microenvironment. In the case of lake phytoplankton, this may involve the permanent location of populations within lower parts of the euphotic zone during the summer growth period. During winter, low ambient light levels combine with water mixing in unfrozen lakes to limit photosynthesis. Frozen Arctic and Antarctic lakes are particularly limited in relation to light penetration.
  Benthic algal communities (periphyton) frequently experience conditions of external and

internal light limitation. External limitation involves a reduction in light levels reaching the community, while internal limitation occurs due to self-shading within the algal mat.

External light limitation is important in both stream and lake environments. The shading effects of terrestrial vegetation are particularly pronounced in streams, where leaf canopies can intercept 95 per cent or more of incident sunlight, reducing maximum photon flux densities (PFDs) to less than $40\,\mu\text{mol m}^{-2}\,\text{s}^{-1}$. Benthic algae in lakes experience low light levels due to attenuation within the water column. This affects benthic algal communities at the edge of lakes (where light-mediated differences in community structure occur with depth) and also deep-water periphyton communities in oligotrophic clear-water lakes. In Lake Tahoe (USA), for example, 1 per cent of the light penetrates to sediments 60 m below the surface, allowing the development of a benthic community at the bottom of the water column (Lowe, 1996).

- Limited range of wavelengths. Both planktonic and benthic algae may become exposed to a limited range of wavelengths due to differential absorption of light above the water surface and within the water column. Differential absorption prior to water entry is particularly important in relation to forest streams, where the spectral composition of light is altered as it filters through terrestrial vegetation. Leaf absorption of red and blue wavelengths leads to a light environment which is weighted towards green wavelengths. Passage of light through the water column of lakes and streams results in progressive loss of wavelengths at both ends of the visible spectrum with depth, leading to a predominantly green/blue spectrum.

The ability of algal cells to condition themselves to limiting light conditions is achieved by a range of molecular and physiological modifications. These vary in time from short-term changes in light-responsive gene activity to long-term alterations in cell compartmentation and pigment content.

4.6.2 Variable light intensity: light-responsive gene expression

Phytoplankton cells are able to react to transient increases in light intensity by activation of specific light-responsive genes. This activation has been investigated particularly in relation to blue-green algae and involves three main processes – stimulation of wavelength-specific receptors, light-responsive signal transduction pathways, and activation of a range of genes.

Receptor–signal transduction pathways

In addition to light, blue-green algae are also able to sense and respond to changes in nutrient and oxygen concentrations.

Cyclic nucleotides, including cyclic adenosine monophosphate (cAMP) and cyclic guanosine monophosphate (cGMP), appear to play a key role in the signal transduction process. In *Anabaena* and other blue-green algae, cAMP concentration changes rapidly in response to environmental changes, with an increase in concentration on transition from light to dark and vice versa. These results suggest that cAMP acts as a second messenger of light signal transduction in blue-green algae, moderating light-responsive activity of a variety of processes. These include cell motility, as demonstrated in the unicellular blue-green alga *Synechocystis* by Terauchi and Ohmori (1999). Mutants of this alga which lack the enzyme adenylate cyclase lose their motility. This enzyme is required for cAMP synthesis, and motility can be restored by exogenous addition of cAMP.

Under standard growth conditions, the two second messengers cAMP and cGMP normally occur at similar levels in blue-green algae but changes in light, nutrients, and oxygen lead to changes in the level of both nucleotides. The intracellular levels of both molecules are regulated by cyclic nucleotide phosphodiesterases, which play a pivotal role in cAMP and cGMP signal transduction. Ochoa de Alda and Houmard (2000) have identified two putative phosphodiesterases by screening the complete genome sequence of the unicellular blue-green alga *Synechocystis*, using bioinformatic tools and applying protein sequence-function studies. These phosphodiesterase genes are surrounded by a cluster of regulatory genes, with all the genes being transcribed as a single unit. The regulatory genes encode polypeptides that act as two component systems, containing both signal transmitter (sensory histidine kinase) and receiver (response regulator) domains. Two of these genes contain signalling motifs known to regulate light-stimulated signalling pathways in other organisms, particularly involving blue light. In *Synechocystis*, blue light is known to induce changes in the intracellular level of cAMP, suggesting that the putative cAMP phosphodiesterase is part of the blue light signal transduction pathway.

Light-responsive genes

Signal transduction of changes in light intensity allows cells to respond to environmental change by modulating the expression of light-responsive genes, thus producing an integrated response. Light-modulated gene expression affects a wide range of metabolic activities in freshwater algae (Section 3.3.2, Table 3.5), various examples of which are considered in different sections in this book.

Some specific examples of light-responsive gene activity are summarized in Table 4.8. Details of light induction of microcystin-related proteins and gas vesicle formation are considered elsewhere (Section 3.3), but the light-responsive formation of photosystem proteins and pigments is considered here since it is particularly relevant to photoadaptation. This light-modulated activity means that algae can make responses to light fluctuations within the highly variable aquatic environment, optimizing their photosynthetic activity over relatively short time periods. This occurs by promoting the synthesis of photosystem II centre proteins, chlorophyll-a binding proteins and phycobilin pigments (phycocyanin and phycerythrin). Wavelength-specificity for light-responsive gene expression varies considerably between genes (Table 4.8), suggesting the involvement of multiple receptors and multiple transduction pathways. With some genes, the activation wavelength relates directly to the function of the end-product. The synthesis of phycocyanin in

Table 4.8   Some examples of light-responsive gene expression in blue-green algae

Organism	Genes	Gene product	Induction
Light Reception and Signal Transduction[a]			
Synechocystis	*slr2098*	Proteins containing light-sensing	Blue light receptor activity
	slr2104	*domain*	
	Sll1624	cGMP phosphodiesterase	Regulation of cGMP and
	Slr2100	cAMP phosphodiesterase	cAMP levels
Photosystem Protein and Pigment Synthesis[b]			
Synechococcus	*psbA*	Photosystem II centre proteins	Light intensity or spectral quality
	psbB		(blue/red ratio)
Synechococcus	*hliA*	Chlorophyll-a binding protein	White or blue light
Calothrix	**cpc genes**	Phycocyanin	
	cpcB1A1 operon	–	Constitutive
	cpcB2A2 operon	–	Red light
	cpe genes	Phycoerythrin	
	cpeBA operon	Phycoerythrin apoproteins	Green light
Microcystin-Related Proteins[c]			
Microcystis	*mrpA*	MrpA protein	Induction of MrpA by blue light
	mrpB	MrpB protein	Inhibition by high intensity white light
Gas Vesicle Protein Expression[d]			
Pseudanabaena	*gvpA*	GvpA protein	Gene activation and protein transcription at low but not high irradiance

Shaded area indicates putative designation, derived from genome sequence
References: [a]Ochoa de Alda and Houmard (2000); [b]Golden (1995); [c]Dittmann *et al.* (2001); [d]Damerval *et al.* (1991)

the blue-green alga *Calothrix* (*cpc*B2 operon), for example, is promoted by red light while the synthesis of phyocerythrin (*cpe* operon) is activated by green light.

With some light-dependent physiological processes, light-responsive gene expression and protein induction are not sufficiently rapid to provide adequate response to short-term environmental changes. This is the case for nitrate assimilation, for example, where a more responsive reversible enzyme inhibition/activation system has also developed (Figure 5.13).

4.6.3 Photosynthetic responses to low light intensity

Phytoplankton and benthic algal cells can adapt to different light levels by adjustments to their physiology, as indicated by changes in the characteristics of their photosynthesis – irradiance (*P–I*)

curves. These changes reflect alterations in the balance of biochemical processes within photosynthesis and by adjustments in the allocation of photosynthetic products to different metabolic pools. Short-term changes in photosynthetic parameters are triggered by activation of the light-responsive genes discussed previously.

Physiological adaptations to low light levels: characteristics of P–I curves

Quantitative parameters in the *P–I* relationship (Figure 4.7) vary with ambient light conditions in both planktonic and benthic algae. Adaptations to low light intensities are usually characterized by:

- lower photosynthetic rates at saturation irradiances (P_{max}),

- a low compensation point (I_c),

- increased photosynthetic efficiency at low irradiances (increased α), and by

- decreased saturation parameters (lower I_k or I_{max}).

Relatively little information is available on I_c levels, though some species of phytoplankton and benthic algae are known to have values less than $<2\mu mol\ m^{-2}\ s^{-1}$, and are clearly adapted to habitats where photons are scarce (Hill, 1996).

Differences in photosynthetic efficiency and saturation levels have been demonstrated in samples from different light environments for both planktonic and algae.

Planktonic algae Physiological adaptations to low light levels have been demonstrated particularly clearly in cultures of planktonic algae (Figure 4.12). Cells of the green alga *Scenedesmus*, for example,

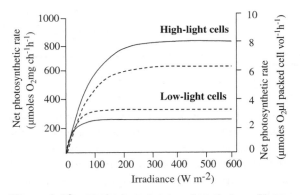

Figure 4.12 Photoadaptation in cultured algae. Photosynthesis – irradiance (*P–I*) curves are shown for laboratory populations of the green alga, *Scenedesmus obliquus*, grown in continuous culture at high ($28\ W\ m^{-2}$) and low ($5\ W\ m^{-2}$) irradiance levels. High-light cultures contained a lower chlorophyll concentration compared with low-light cultures (7.8 mg and 12.8 mg chlorophyll ml^{-1} packed cell volume, respectively). The photosynthetic rate is expressed either as O_2 evolved per mg chlorophyll (——) or O_2 per unit cellular biomass (packed volume, -----). Photosynthesis in low-light cells saturated at much lower irradiance levels compared with high-light cultures (I_{max} value of ~150 compared with ~300 W m^{-2}) and had lower saturated rates of photosynthesis (P_{max} value of ~260 compared with ~800 μmoles O_2 mg chl^{-1} h^{-1}) (plotted from data of Senger and Fleischhacker, 1978)

showed markedly different *P–I* responses when cultured at high and low light levels (Senger and Fleischhacker, 1978). Continuous cultures of low-light grown cells saturated at much lower irradiance levels (I_{max}) compared to high-light grown cells, with a lower maximum photosynthesis rate (P_{max}) per unit chlorophyll.

One of the most extreme cases of light adaptation in the planktonic environment is provided by micro-algae living under ice in Arctic and Antarctic lakes. These organisms exhibit light saturation at less than $10\ \mu mol\ m^{-2}\ s^{-1}$ (Cota, 1985).

Benthic algae Analyses of *P–I* curves from benthic algae (Boston and Hill, 1991; Hill, 1996) provide an interesting contrast to the phytoplankton data discussed previously. These laboratory studies on intact stream periphyton communities obtained from shaded (maximum light 50 μmol m^{-2} s^{-1}) and open sites (maximum 1100 μmol m^{-2} s^{-1}) demonstrated that *P–I* responses of benthic algae were not well related to the ambient light environment, but showed better correlation to algal biomass. The key difference between phytoplankton and periphyton in terms of *P–I* response is that the phytoplankton population occurs in suspension, while periphyton occurs as dense mats – with self-shading and other aspects of development in the matrix affecting the light response. Even under bright light, due to the thickness of the periphyton mat, the overall community *P–I* response may be more typical of low light conditions. This is because the *P–I* curve is an integration of the responses of cells that are high-light adapted (upper layer) and shade-adapted (lower layer). For periphyton mats, biomass and related thickness are an important and, in some cases, dominant influence on the *P–I* responses of the community.

Despite the varied light regimes of benthic algae, most *P–I* studies indicate I_k values within a narrow range (100–400 μmol m^{-2} s^{-1}). These saturation irradiances are generally much higher than ambient irradiance values occurring *in situ*. In heavily shaded forest streams, for example, photosynthesis by attached algae typically saturates at between 100–200 μmol m^{-2} s^{-1}, yet maximum irradiances during summer (time of greatest shade) rarely exceed 30 μmol m^{-2} s^{-1} (Hill, 1996). This suggests

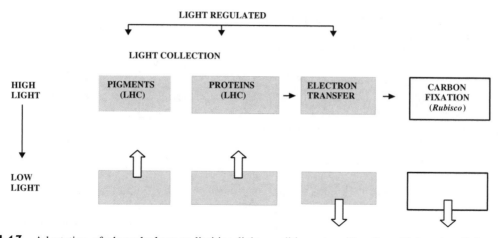

Figure 4.13 Adaptation of phytoplankton to limiting light conditions: transition from high to low light conditions leads to up-regulation (⇑) and down-regulation (⇓) of specific metabolic activities; levels of pigments and proteins in the light-harvesting complex (LHC) are light-regulated, as also is the rate of electron transfer

that light saturation irradiance levels for benthic algae are not normally reached in the lotic environment, and that photosynthesis in streams is typically light-limited. The relatively high light saturation values for stream benthic algae is surprising and may reflect the fact that the photosynthetic machinery of these organisms is geared to harness the higher irradiances that occur infrequently (e.g., sunflecks) or seasonally (e.g., winter and spring maxima for streams shaded by deciduous trees).

Biochemical adaptations

For phytoplankton populations, differences in *P–I* responses reflect adaptations to low light at the biochemical level. These adaptations occur in relation to the machinery of photosynthesis (balance of chemical activities, chlorophyll concentration) and the allocation of photosynthetic products.

Balance of chemical activities Within photosynthesis, algal cells need to balance the processes of electron generation (by the light-harvesting complex), electron transfer, and carbon fixation. Reduction in light intensity leads to an immediate reduction in the generation of electrons, and requires a rapid cell response. Transfer from high

to low light intensity involves an up-regulation in the synthesis of pigments and light-harvesting complex proteins, and a down-regulation of electron transfer and carbon fixation (Figure 4.13).

Studies on the green alga *Dunaliella tertiolecta* (Escoubas *et al.*, 1995) have shown that the signal regulating these photoadaptive changes is provided by the redox state of the plastoquinone pool within the electron transfer chain of photosystem II. A shift from high to low light intensity results in a decrease in electron flow from the photosystem II light-harvesting complex, conversion of the plastoquinone pool to a highly oxidized state, and increased synthesis of mRNA encoding the light-harvesting complex proteins (*cab* mRNA). The rate of synthesis of these proteins is enhanced after a lag phase during which *cab* mRNA accumulates. A shift from low to high light intensity has the converse effects.

Chlorophyll concentration Up-regulation of pigment synthesis at lower light intensities, leading to increased pigment content, has been well-documented for both prokaryote and eukaryote algae (Falkowski and La Roche, 1991). These changes in chlorophyll content can be directly measured in laboratory (Figure 4.12) and also in environmental phytoplankton samples as an increase in the concentration of chlorophyll per unit cell biomass. This may be expressed as the chlorophyll-a/carbon ratio (θ). The changes also lead to a decrease in the rate of carbon fixation per unit mass of chlorophyll (chlorophyll-specific photosynthesis rate (P^{Chl}), though the rate of carbon fixation per unit mass ($P^{Biomass}$) may remain unchanged.

Photoadaptation of phytoplankton under environmental conditions is demonstrated by physiological observations of depth samples and also by the occurrence of deep chlorophyll maxima as follows.

Physiology of depth samples. Photoadaptation within the water column is well illustrated by the studies of Tilzer and Goldman (1978) on depth variations in algal physiology in Lake Tahoe. These authors sampled phytoplankton from three depths (surface, 50 m, and 105 m) and incubated these samples with $^{14}CO_2$ at a series of depths throughout the water column (samples re-suspended down to 105 m). The results show that photosynthetic rates expressed per unit chlorophyll (P^{Chl}) were much higher in samples originally derived from the middle and top of the water column. Values for carbon fixation per unit biomass ($P^{Biomass}$) showed more similarity in the three samples (Figure 4.14), and showed clear variation with original sample depth. Phytoplankton originally sampled at the lake surface showed a peak of $P^{Biomass}$ at a much shallower level (about 20 m) compared with phytoplankton originally derived from 105 m, indicating clear algal adaptations to local light regimes within the water column. The growth rate of phytoplankton at different depths (μ_d) can be related to the parameters P^{Chl} and θ, where:

$$\mu_d = P^{Chl}\theta \qquad (4.3)$$

Deep chlorophyll maxima. The increased chlorophyll/biomass ratio that occurs during photoadaptation also provides an explanation for sub-surface chlorophyll maxima which are frequently observed in the clear waters of oligotrophic lakes (Fennel and Boss, 2003). These chlorophyll maxima do not correspond to peaks in phytoplankton biomass, but are localized regions of high chlorophyll concentration (per unit carbon algal biomass) at the bottom of the euphotic zone, where there is an increased

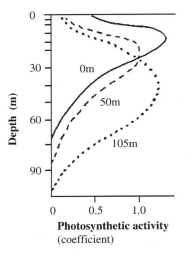

Figure 4.14 Photodaptation at different depths in a clear lake (Lake Tahoe, USA): vertical profiles of photosynthetic activity are shown for phytoplankton samples collected from three depths – 0 m (solid line), 50 m (broken line) and 105 m (dotted line), and then incubated between lake surface and 105 m depth; photosynthetic activity is expressed as a coefficient (photosynthesis rate per unit carbon mass). Figure adapted and redrawn from Tilzer and Goldman, 1978.

nutrient supply with depth and conditions are oxygenated. Studies by Fennel and Boss (2003) on algal distribution in the water column of oligotrophic Crater Lake (USA) showed that biomass and chlorophyll maxima are generated by fundamentally different processes. Under steady-state conditions, maxima in algal biomass occur primarily in relation to positioning mechanisms (Section 3.8) and the local balance between growth increase and losses (respiration and grazing), while the vertical distribution of chlorophyll is determined mainly by photoadaptation.

Allocation of photosynthetic resources
Changes in the above balance of activities within photosynthesis occur as a short-term response (minutes to hours) to changes in light intensity. Continuation of high or low light conditions leads to more long-term (hours to days) changes in pigment content and in the fate of the photosynthetic products.

Geider and MacIntyre (1996) have modelled these changes with a view to establishing long-term quantitative parameters of photoadaptation, and have identified three main destinations for fixed carbon – light-harvesting apparatus, biosynthetic apparatus (for growth and division), and energy storage reserves. In their dynamic model of phytoplankton growth and light adaptation (Geider and MacIntyre, 1996), the relative distribution of photosynthate to these different pools varied considerably with light intensity (Figure 4.15). At high irradiance, the proportion of fixed carbon being cycled back to the light-harvesting apparatus was minimal, with substantial allocation to energy storage reserves. At low levels of irradiance, the balance between these pools was reversed. At all levels of irradiance, the relative flux of carbon to the biosynthetic apparatus was substantial, emphasizing the need for continued cell growth and population increase.

4.6.4 Spectral composition of light: changes in pigment composition

The spectral composition of light in the freshwater environment varies with depth and with water quality. These changes in the light spectrum impose limitations on light absorption by algal cells, which may respond by making compensatory adjustments to the balance of their photosynthetic pigments.

The effect of changes in spectral composition have been investigated by Pick (1991), who examined the abundance and composition of small-celled blue-green algae (picocyanobacteria) in the surface waters of 38 freshwater lakes of varying trophic status and dissolved inorganic carbon (DIC) content. Two major groups of these organisms could be distinguished on the basis of the presence of the accessory pigments phycocyanin and phycoerythrin, with absorption peaks in the orange-red (620–650 nm) and green (peak 550 nm) parts of the spectrum respectively. One group of the algae (PE cells) had both pigments, while the other (non-PE cells) completely lacked phycoerythrin. The proportion of PE cells within the picocyanobacterial assemblage declined markedly in lakes which

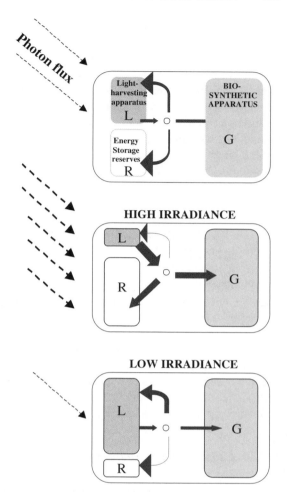

Figure 4.15 Photoadaptation: allocation of resources at different light levels.

Top panel: Dynamic model of phytoplankton growth and photoadaptation, identifying three intracellular pools (L,R,G). Solid arrows illustrate the flux of excitation energy into a control point (O), and of photosynthate out of the control point.

Middle panel: Cells adapted to high irradiance have small L and high R. Solid arrows indicate the high energy flux into the control point and allocation to the different pools.

Lower panel: Cell adapted to low light level, with much reduced energy flux to control point. The proportion of photosynthate allocated to synthesis of L is relatively greater under low-light conditions, although the absolute amount may be smaller due to the light limitation of photosynthesis. At all light levels a substantial allocation goes to pool G to maintain growth and population increase (based on a figure from Geider and MacIntyre, 1996) were nutrient enriched or were coloured by humic material. These lakes showed high light absorption and a shift in predominant spectral composition towards the yellow part of the spectrum. The change in algal composition indicated a change in pigmentation to optimize longer wavelength absorption under these conditions. Conversely, PE cells were particularly abundant in oligotrophic and mesotrophic lakes, where conditions of high light penetration and substantial levels of blue/green wavelengths prevail.

4.7 Carbon uptake and excretion by algal cells

Uptake of CO_2 by freshwater algae, the major microbial primary producer, has been considered so far in terms of energy input (solar radiation), carbon fixation, and growth of the organisms involved. Carbon incorporation and processing during photosynthesis also has major effects on the chemical composition of the aquatic environment, both in terms of CO_2 removal and excretion of dissolved organic substances (Figure 4.17).

4.7.1 Changes in environmental CO_2 and pH

Carbon dioxide is one of the major inorganic nutrient requirements for phototrophic organisms, occurring as the most oxidized and abundant form of inorganic carbon available in the biosphere. It is present in solution as dissolved CO_2, carbonic acid (H_2CO_3), and as bicarbonate (HCO_3^-) and carbonate (CO_3^{2-}) ions. (Stumm and Morgan, 1970). Between pH4 and pH11, these species are linked by the following equations:

$$CO_2 + H_2O = H_2CO_3$$
$$H_2CO_3 = H^+ + HCO_3^- \qquad (4.4)$$
$$HCO_3^- = H^+ + CO_3^{2-}$$

Total inorganic carbon (C_T)

$$C_T = [CO_2] + [H_2CO_3] + [HCO_3^-] + [CO_3^{2-}]$$

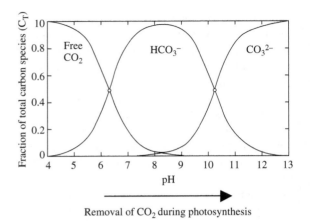

Removal of CO_2 during photosynthesis

Figure 4.16   Effect of pH on the distribution of inorganic carbon species (based on a figure from Shapiro, 1990)

The relative proportions of these chemical species is governed by the equilibrium shown in Figure 4.16, the state of which depends to a large extent on pH. Free CO_2 is present in significant amounts between pH4 and pH7, decreasing rapidly by pH8 and occurring only at about 0.003 per cent of the total carbon species by pH9. This environmental equilibrium is influenced by various factors, including the net removal of CO_2 by autotrophs (photosynthesis), CO_2 generation by heterotrophs (respiration), and addition of lime as part of management activities.

Autotrophic activity

Active photosynthesis of macrophytes and algae (phytoplankton, benthic algae) results in the removal of free CO_2 from solution. This leads in turn to the reaction of bicarbonate ions with water to generate more CO_2 and CO_3^{2-} ions. The carbonate ions react with water to form hydroxyl (OH^-) ions, giving rise to an increase in pH. The extent of the pH rise depends on the buffering capacity of the water, which has a pronounced maximum of pH8.1–8.3 in fresh waters (Shapiro, 1990). Because of this, a small increase in pH brought about by photosynthesis results in a large fall in the concentration of free CO_2. Accumulations of algae (particularly blue-green algae) at the lake surface thus tend to increase pH and make CO_2 less available under conditions of bright light and resulting high rates of photosynthesis.

Heterotrophic activity

Addition of CO_2 to lake and river water occurs from both external (atmospheric) and internal (heterotrophic) sources. Many aquatic systems are heterotrophic, or contain heterotrophic compartments, and are supersaturated with CO_2. This occurs particularly in anaerobic regions of high bacterial population, such as the hypolimnion and sediments of eutrophic lakes (Figure 5.9). Direct addition of any of the carbon species in Equation (4.4) will have an effect, increasing the total amount of inorganic carbon present and affecting the concentration of all components. Addition of bicarbonate for example will increase the absolute levels of CO_2 and H_2CO_3.

Addition of lime to aquatic systems

Addition of elements other than carbon may also be important in affecting the total concentrations and balance of carbon species in lake water. This is particularly the case for calcium, which is often applied to commercial fishponds (Kvet *et al.*, 2002) in the form of limestone ($CaCO_3$) or lime (a mixture of calcium hydroxide and calcium bicarbonate). Addition of lime to water of moderate to low alkalinity results in an increase in pH due to the removal of free CO_2 by the calcium hydroxide and the formation of soluble calcium bicarbonate. This rise in pH increases the buffering capacity of the water and accompanies a rise in the level of inorganic carbon pool in the aquatic system. These changes in water chemistry increase the primary productivity of both macrophytes and algae, ultimately resulting in increased levels of fish production.

The total concentration of inorganic carbon in aquatic environments, and its distribution between the various chemical forms, is relatively constant in the surface waters of oceans, but varies considerably in freshwater systems. In freshwater bodies, pH can range from 1.0–11.0, with total inorganic carbon concentrations varying from 10 mmol m^{-3} (acidic waters) to 100 mol m^{-3} (bicarbonate-rich waters at high pH). The uptake of CO_2 by algae can be severely limited at the low end of this concentration spectrum, particularly as inorganic carbon species have low diffusion coefficients and low rates for conversion between chemical species at low pH values. Variation in algal species composition between waters of different pH can be partly attributed to the adaptations of different organisms to differences in CO_2 availability.

4.7.2 Excretion of dissolved organic carbon by phytoplankton cells

Phytoplankton cells release a wide range of soluble organic compounds into the surrounding water medium. These make an important contribution to the overall dissolved material present in the ecosystem and are referred to as dissolved organic carbon (DOC) or dissolved organic matter (DOM). Many of these compounds are derived directly from the process of photosynthesis (see later), hence their consideration in this part of the book.

DOC released by phytoplankton cells includes carbohydrates (Münster, 1984), polypeptides (Walsby 1974), free amino-acids, organic acids (particularly glycollic acid – Pant and Fogg, 1976) and cyclic AMP (Francko and Wetzel, 1982). Considerable molecular diversity occurs within each of these major groups. Analysis of soluble carbohydrates in the Plussee (Germany), for example, involved the determination of 26 monosaccharides and six disaccharides (Münster, 1984). The major derivation of lake water concentrations of these organic compounds from phytoplankton cells is indicated by environmental observations of seasonal correlations of DOC with algal biomass, diurnal correlations with phytoplankton activities (Chapter 3.7), and laboratory studies on photosynthesis and DOC release (see below).

Range of activities involved in DOC release

Release of DOC occurs from both healthy and degenerating (lytic) algal cells (Figure 4.17). In healthy cells, DOC exudation takes place due to passive diffusion (leakage of photosynthetic products) and active secretion (e.g., production of extracellular enzymes, siderophores). The passive

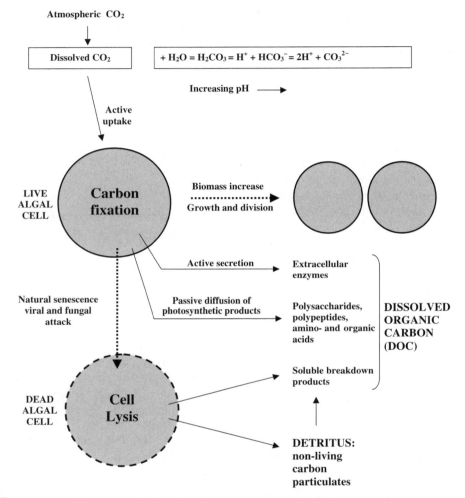

Figure 4.17   Uptake of CO_2 and production of carbon compounds in the freshwater environment.

CO_2 availability for algal uptake and photosynthesis depends on the ionic equilibrium within the surrounding microenvironment. Carbon compounds produced during photosynthesis persist in the environment in various ways – becoming incorporated into algal biomass (growth and division), released as dissolved organic carbon (DOC), and sinking to the bottom of the water body as detritus. The three types of material subsequently enter the food chain due to consumption by herbivores (living biomass), uptake by bacteria (DOC – microbial loop) and ingestion by detritivores (sedimenting or benthic detritus). DOC is derived via active secretion, passive efflux of small MW photosynthetic products, and by loss of soluble organic materials during cell lysis. Release of DOC by active secretion results in a continuous food supply to planktonic bacteria (see Section 6.10), with positive correlation between bacterial growth and algal production (Cole *et al.*, 1988). Major DOC release by cell lysis occurs during periods of algal decline, with peak bacterial populations following phytoplankton population maxima by periods of days or weeks (Straskrabova and Komarkova, 1979) diffusion of these substances out of algal cells depends on their intracellular and extracellular concentrations, and the membrane permeability of the cell and is directly related to the level of photosynthesis (see below). Measurements of algal extracellular DOC in natural phytoplankton assemblages have generally indicated higher rates of release compared with those recorded in the laboratory using axenic algal cultures.

Although algae are probably the major source of DOC in many standing freshwater systems, these compounds are also derived by secretion and lysis of other biota, including bacteria and zooplankton. Viruses, with sizes $<0.2\,\mu m$, are also part of the defined soluble fraction. In addition to autochthonous DOC (derived from algae and other biota contained within the water body), DOC may also enter the system by inflow from external sources. Such allochthonous DOC includes humic acids and other material derived from the catchment area, and is particularly important in flowing (lotic) systems (Chapter 1.7).

The derivation of DOC from different sources is indicated by infra-red analysis of filtered water samples. Infra-red (FTIR) absorption spectra (Figure 4.18). show clear bands of carbohydrate, polypeptide, and lipid material, consistent with origins via passive release, active secretion, and lytic activities of phytoplankton cells and other biota.

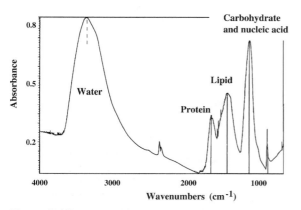

Figure 4.18 Molecular composition of DOC from a eutrophic lake (unpublished data from Sigee, Dean and White)

Ecological significance of DOC release by phytoplankton

Although the release of DOC represents only a small fraction (usually <5 per cent) of primary production, the dominance of phytoplankton over most of the Earth's surface makes this release a significant global carbon flow, with important implications for both marine and freshwater systems. DOC exudation is ecologically significant in a number of ways.

- *Decrease in net primary productivity.* DOC release represents a loss of carbon which had been fixed by algal cells during photosynthesis, and thus decreases the net primary productivity. Most of this loss can be regarded as 'accidental' in the sense that the majority of these compounds are simply leaking out of the algal cells, and their release does not bring any apparent benefit to the algal cells. The loss of carbon is accompanied by a loss of nitrogen and phosphorus compounds.

- *Photosynthetic balance.* Although the loss of photosynthetic products is essentially a wasteful process, it may be important in enabling phytoplankton to dispose of excess carbon products accumulation of which may overwhelm the cell's ability to process fixed carbon (Murray, 1995).

- *Nutrient uptake and retrieval.* Some of the compounds released by algae are actively secreted and are important in nutrient uptake and retrieval. The enzyme alkaline phosphatase, for example, is excreted by many phytoplankton species under conditions of low phosphorus availability (Section 5.8.3). This acts by releasing inorganic phosphate from organic phosphate complexes within the aquatic medium and is important in phosphorus recycling. Siderophores, released by blue-green algae, are important under conditions of iron limitation for the scavenging of Fe^{3+} ions from the surrounding environment (Section 5.12.1).

- *Microbial loop.* DOC represents a major carbon source for heterotrophic organisms, including bacteria, protozoa, and some algae. The release of DOC by algal cells is the main driving force for the microbial loop, promoting the growth of these microorganisms and the flux of carbon throughout the whole food web. It is a major aspect of the carbon cycle within the aquatic environment.

- *Indirect interactions between phytoplankton and bacteria.* The excretion of DOC by phytoplankton has direct effects on bacterial productivity, which may in turn have reverse adverse or beneficial effects back on the algal population. As an example of the former, competition between algae and bacteria for inorganic nutrients will be increased with greater bacterial populations, reducing the levels of nitrate and phosphate available to the algae. Increased DOC levels will also increase the population of antagonistic bacteria.

Specific beneficial effects include the promotion of surface symbiotic bacteria, some of which are important in algal nitrogen fixation. Murray (1995) has suggested that promotion of the microbial loop by phytoplankton is also an important defence against viruses. The increased bacterial population will result in increased levels of random algal virus adsorption to bacterial cells, with rapid destruction of non-infecting viruses at the bacterial surface.

Photosynthesis and DOC release

Early studies on primary productivity in algae indicated that, during photosynthesis, algae release a substantial quantity of their newly-formed photosynthetic products into the surrounding medium as dissolved organic carbon. The quantitative relationship between carbon fixation during photosynthesis and release as DOC has been studied by supplying axenic algal cultures with radiolabelled carbon (^{14}C) and following the course of transfer. The amount of ^{14}C which is taken up into organic carbon in cells (primary productivity) and excreted into culture medium (DOC) can be determined by scintillation counting after removal of inorganic carbon and by making appropriate background corrections.

Fallowfield and Daft (1988) used this approach (summarized in Figure 4.19) to investigate the relationship between photosynthesis and DOC release in a range of blue-green and green algae, and to determine the effect of one major external factor – the presence of lytic bacteria – on this relationship.

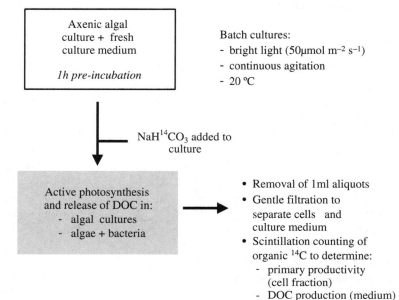

Figure 4.19 Carbon fixation and DOC release in cultured algae, with and without added bacteria (experiment of Fallowfield and Daft, 1988)

Under the conditions used in these experiments:

- all algae showed active photosynthesis, with higher rates of carbon fixation in blue-green (5–9 μgC μgChl-a^{-1} h^{-1}) compared with green algae (1–5 μgC μgChl-a^{-1} h^{-1}),

- in all cases, active photosynthesis was associated with the release of significant amounts of DOC.

The rate of DOC release was related to the rate of carbon fixation. The high carbon assimilation rates of blue-green algae were associated with particularly high rates (0.2–0.4 μgC μgChl-a^{-1} h^{-1}) of DOC release in these organisms. In individual cultures, the linear increase in carbon fixation was matched by a linear increase in DOC production, indicating a close coupling between the two processes. In general, the loss of carbon as DOC ranged from 1–7 per cent of productivity.

Inoculation of the cultures with the bacterium *Lysobacter*, an organism closely associated with bloom-forming blue-green algae, reduced the level of carbon fixation and increased the amount of DOC released by blue-green but not green algae. The promotion of increased blue-green algal DOC release by this organism suggests a specific uncoupling of this process from photosynthesis. The excretion of algal organic molecules has clear benefits for the bacterium, since *Lysobacter* showed active growth in algal culture filtrates, taking up a range of amino-acids present within the DOC. This has potential environmental importance for increasing the availability of algal-derived DOC to associated bacteria such as *Lysobacter*, and to heterotrophic organisms generally within the plankton.

Location of DOC release and assimilation within the water column

Conventional theory postulates that photosynthetically derived DOC is secreted by algae in the euphotic zone, where it is locally recycled via the microbial loop. Soluble carbon is thus assimilated by heterotrophic bacteria in the top part of the water column, while particulate carbon sediments into the hypolimnion and is broken down in the lower water column and benthic zone.

Although this model holds for some lakes (e.g., Crater Lake, USA), it does not apply to others. Recent studies by McManus *et al.* (2003) on Lake Superior (USA), for example, have shown a substantial amount of DOC passing into hypolimnion where it is broken down and contributes to deep lake oxygen consumption. Measured rates of oxygen consumption in this lake substantially exceed those that would occur simply from particulate carbon.

4.8 Competition for light and carbon dioxide between algae and higher plants

In most aquatic habitats, environmental conditions support the growth of higher plants (macrophytes) as well as algae. Both groups of biota are actively photosynthetic (primary producers) and compete for carbon dioxide, light, space, and inorganic nutrients.

4.8.1 *The balance between algae and macrophytes in different aquatic environments*

The balance between algae and macrophytes varies with the type of aquatic system and the water quality.

Lakes

In temperate climates, lake macrophytes are largely restricted to the edge of the lake (littoral zone) or other shallow regions, and there is relatively little overall competition for light with phytoplankton. The development of macrophytes in the littoral zone increases with progression from oligotrophic to eutrophic status, but very high nutrient levels lead to intense algal blooms and repression of higher plants (Sections 10.2 and 10.7).

Tropical and subtropical lakes often show extensive development of floating macrophytes such as the water hyacinth (*Eichornia crassipes*) and water lettuce (*Pistia stratoites*). These may cover large areas of lake water and totally dominate primary production in the lake.

Wetlands

Wetland sites such as shallow pools and water meadows typically have extensive developments of macrophytes, which can dominate primary production and out-compete algae. The interception of light by both rooted and floating macrophytes can be very high (Goldsborough and Robinson, 1996). During periods of maximum *Phragmites* growth, up to 95 per cent of incident radiation can be removed by the developing canopy. Similarly, transmisison of light through floating *Lemna* mats may be only 0.1 per cent of surface irradiance.

Macrophyte dominance is particularly well seen in seasonal wetlands. In North West Australia, for example, the wet season (October to April) brings high humidity, thunderstorms, and torrential rain, with extensive flooding of large tracts of land. These seasonal wetlands support a wide range of rapidly-growing hydrophytes (Figure 4.20), which die or become dormant with the onset of the dry season.

More permanent wetlands, such as those of the Třeboň Biosphere Reserve (Czech Republic), also show marked seasonal variation and have been the subject of intensive research (Kvet *et al.*, 2002, Section 2.8).

Figure 4.20 Macrophyte domination of seasonal wetlands at the edge of Fogg dam floodplain, North West Australia.

This photograph was taken at the end of the rainy season, with waterlilies (*Nymphaea violacea*) in the foreground and stands of the lotus lily (*Nelumbo nucifera*) in the distance amongst paperbark trees (*Melaleuca* sp.). Other water plants common in this wetland include the water lettuce (*Pistia stratoites*), yellow bladderwort (*Utricularia gibba*), and Monochoria (*Monochoria cyanea*). Within a short period of time this wetland will be largely dry, with sunny areas covered by grassland

---

Case study 4.1   Competition between algae and macrophytes in shallow lakes of the Třeboň wetlands

Competition between algae and higher plants has been investigated in particular detail (Pokorny *et al.*, 2002b) in the small shallow lakes of the Třeboň Biosphere Wetlands. These lakes are used as fishponds and, as with other Central European sites, are intensively managed with high fish stocks and application of lime and organic fertilizers (Section 2.8). Their shallow hydrology and high nutrient status promote active growth of both algae and higher plants, competition between which is complicated by the diversity and seasonal changes of the two groups of organisms.

Macrophyte diversity and seasonal changes

Diversity   Macrophytes can be divided into two main groups (Figure 4.21) as follows.

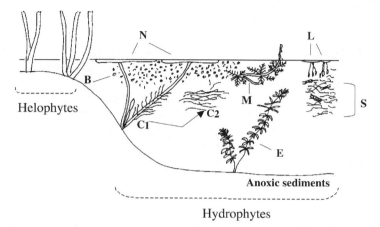

Figure 4.21   Diversity and competition between algae and higher plants in a shallow eutrophic pond.

Higher plants comprise helophytes (littoral zone) and hydrophytes (main water body), some of which are rooted (*Elodea* – E, Water lily – N) while others are free floating (*Myriophyllum* – M). *Lemna* (L) is a member of the free-floating surface community (pleuston). Algae include early bloom-forming plankton (*Chlorococcales* – B) and macrophyte-attached periphyton such as *Cladophora* (C1). Later in the season, rafts of free-floating filamentous algae may dominate the water body, some of which were originally attached (*Cladophora* – C2) while others (*Spirogyra* – S) were not. Diagram constructed from a description of fishponds in the Třeboň wetlands (Pokorny *et al.*, 2002b)

- *Hydrophytes* are plants living in water which are either completely submerged or semi-submerged. Completely submerged hydrophytes include rooted pondweeds, attached to the substratum (e.g., *Elodea*), and free-oating plants (*Ceratophyllum, Myriophyllum*). Semi-submerged plants include plants with leaves that oat on the water surface and are either attached (rooted) to the sediment (e.g., waterlilies – *Nymphoea* and *Nuphar*) or free-oating (e.g., duckweed – *Lemna*). Duckweeds are of particular interest in terms of algal competition. They are highly simplied hydrophytes, and as members of the free-oating surface biota of the lake (pleuston) are sometimes referred to as pleustophytes. Their short life cycle, coupled with small size and rapid vegetative propagation, permit a rapid population increase when growth conditions improve and make their behaviour and role in water bodies similar to the algae.

- *Helophytes* are plants rooted in flooded or waterlogged soil, whose photosynthetic organs are predominantly or completely emerged. These plants are typical of the edge of lakes and ponds (littoral zone) and include plants such as bulrush (*Typha*), the common reed (*Phragmites*), and *Glyceria*.

Seasonal changes   Perennial hydrophytes and helophytes grow rapidly during the early part of the growing season, using storage products contained within their rhizomes and other underground organs. This characteristic enables them to out-compete algae for light, CO_2, and nutrients and have a substantial growth phase prior to the development of the summer algal bloom. Annual or quasi-annual hydrophytes such as *Elodea canadensis* and small *Potamogeton* species have a major growth period during the clear-water phase, using the high irradiation input during this window of opportunity between algal blooms.

Algal diversity and seasonal development

Diversity Algae that compete with macrophytes are present in the fishponds either as phytoplankton or attached to the plant surfaces (periphyton).

In most of the Třeboň fishponds, the range and development of phytoplankton is typical of other eutrophic water bodies, with an initial diatom bloom followed by a clear-water phase, then a mixed summer bloom of green algae, diatoms, and some blue-greens. Early application of manure or other fertilizer, however, can bring forward the summer bloom, promoting intense early development of phytoplankton which dominates the water body throughout the season. This rapidly-growing algal bloom is composed mainly of single celled or simple colonial forms, with green algae (order Chlorococcales) being particularly prominent.

In addition to these simple forms, filamentous algae are also prominent in some of the more highly eutrophic fishponds, particularly those with low fish populations. Some of the filamentous algae (species of *Oedogonium* and *Cladophora*) are initially observed as epiphytes on submerged macrophytes (periphyton). These periphytic algae have seasonal cycles of colonization, growth as attached colonies, and subsequent detachment to form free-oating mats. Other lamentous algae, such as *Spirogyra* and *Hydrodictyon*, do not form part of the periphyton community and are always planktonic.

Seasonal changes Seasonal changes in algae have important implications for competition with macrophytes. Under conditions of poor to moderate nutrient levels, the seasonal progression of algal development will allow early macrophyte growth, particularly in the clear-water phase. In hypereutrophic water bodies, however, the early and intense bloom of green algae (particularly Chlorococcales) promoted by high nutrient input retards the development of macrophytes from the start of the growing season by shading the bottom of the pond and taking up CO_2. The early development of a green algal bloom also reduces or eliminates the clear-water phase, depriving annual and other hydrophytes of an important period of illumination.

Seasonal development of filamentous epiphytic algae on hydrophyte surfaces has a more localized effect, but also inhibits macrophyte growth by shading and CO_2 removal.

4.8.2 Physiological and environmental adaptations in the competition between algae and macrophytes

The high nutrient status of the Třeboň fishponds favours algal domination, and macrophytes need to employ a range of ecological and physiological adaptative strategies to survive in such eutrophic and hypereutrophic conditions. These strategies include seasonal growth patterns, which are important in relation to light and CO_2 availability, and are part of a broad range of environmental adaptations (Pokorny *et al.*, 2002b) which affect competition between these biota (Table 4.9).

Assimilation of CO_2 and bicarbonate ions

The ability of vascular plants and algae to assimilate bicarbonate ions in addition to free CO_2 is one of the most important adaptations to life in the aquatic environment. This physiological adaptation allows both sets of organisms to tolerate high pH and overcome frequent depletion of free CO_2 resulting from intense photosynthesis.

In general, algae are more efficient than higher plants in the assimilation of bicarbonate ions, which gives them a competitive advantage when pH rises to high values. In addition to this, not all higher plants show this activity. The hydrophyte

Table 4.9 Competition between algae and macrophytes – summary of adaptations (information taken from Pokorny et al., 2002b)

	Macrophytes	Algae
Environmental conditions	Favoured by moderate to low nutrient conditions	Out-compete macrophytes for light and CO_2 in high-nutrient shallow conditions
Major types	Hydrophytes Helophytes	Phytoplankton Periphyton
Seasonal competition for light	Early seasonal growth Major growth during clear-water phase	Phytoplankton – dense spring/summer bloom Periphyton – early colonization of plant surfaces
Uptake of bicarbonate	Limited to certain species/varieties Relatively inefficient	All algae can assimilate bicarbonate Highly efficient
Biomass position within the water column	Photosynthetic growth concentrated near to water surface	Algae position themselves in euphotic zone
Anoxic sediments	Free-floating macrophytes not affected Rooted macrophytes tolerate anoxic stress	Not directly affected
Mineral nutrient uptake	Uptake from sediments (roots) and water column	Water column only, but highly efficient
Low light levels	Low light saturation levels and compensation point	Low light saturation levels and compensation point

Ceratophyllum demersum, for example, has light-adapted and shade-adapted varieties which differ in bicarbonate uptake. The light-adapted varieties are able to exist in competition with algae under high light levels, and can overcome the marked fluctuations in pH by bicarbonate as well as free CO_2 assimilation. Shade-adapted varieties grow in water where the pH remains relatively low (close to neutral) and are only able to assimilate free CO_2.

Under extreme conditions, intense uptake of CO_2 by higher plants results in pH levels rising to values of 10–10.5. At this level, bicarbonate and carbonate ions are present but free CO_2 is absent, and only bicarbonate users such as *Elodea canadensis*, small species of *Potamogeton*, and the varieties of *Ceratophyllum* noted above are able to survive. Lack of free CO_2 is a major limitation for many higher plant species in eutrophic and hypereutrophic environments, and the photosynthesis of macrophytes that rely on free CO_2 ceases at pH 8.5–9.

Growth and life cycles

The importance of rapid growth early in the season is an important aspect of algal – macrophyte competition and has been emphasized in the previous section.

As well as timing of growth, algae and macrophytes are also able to concentrate their photosynthetic biomass in the top part of the water column, where light and CO_2 availability are optimal. Reduction in water transparency caused by algal growth results in the accumulation of vascular plant photosynthetic organs at or near the water surface, and enables them to compete with or even suppress phytoplankton. The evolutionary development of surface-floating leaves or partial shoot emergence is the ultimate expression of this trend, and confers substantial advantage over algae in the competition for light and CO_2.

Adaptations to anoxic sediments

One important effect of intense algal blooms is the imposition of a steep oxygen gradient within the water column, with high levels at the water surface and low levels at the bottom. This gradient arises due to high algal photosynthetic activity in the euphotic zone, coupled with bacterial break-down of the increased algal biomass which sinks to the lake sediments.

Anoxic sediments have no direct effect on phytoplankton, but they do cause physiological stress to the root and rhizome systems of attached macrophytes. These plants show a number of adaptations to the anoxic conditions, including the presence of high levels of reserve materials, allowing them to consume large amounts of carbohydrate in inefficient anaerobic metabolism. Macrophytes also typically possess air-conducting aerenchymatous tissues, allowing them to transfer oxygen into anaerobic regions. These adaptations enable vascular plants living in aquatic environments to survive long periods of anoxic stress, and are particularly well developed in emergent macrophytes (helophytes) which are rooted in low-oxygen sediments close to the water surface.

Mineral nutrient uptake

Uptake of inorganic nutrients is discussed later in relation to nitrates (Section 5.5) and phosphates (Section 5.8), and is considered here specifically in relation to competition between algae and higher plants. In oligotrophic and mesotrophic environments, shortage of mineral nutrients is the main limiting factor for both groups of organism. Although oligotrophic water bodies tend to show little development of macrophytes (Figure 4.25), rooted vascular plants are present with increased nutrient levels and do have some advantages over both planktonic and attached filamentous algae in limiting nutrient conditions. These plants are able to take up scarce mineral nutrients via their root systems and recycle them internally, while algae have little chance to grow within the low-nutrient water column.

In eutrophic conditions, both sediments and the water column are rich in nutrients, allowing algae to grow abundantly and out-compete macrophytes. Algal growth in such conditions is accentuated by increased release of phosphorus from anoxic sediments, which is then transported upwards through the water column by thermal mixing during the night. Although this increased phosphorus supply enhances the growth of all aquatic plants, there is higher relative uptake by algae (compared to macrophytes) due to their greater surface area/volume ratio.

Both algae and macrophytes are able to take up and store mineral nutrients at times of high availability, as noted elsewhere in relation to phosphorus (Section 5.8.3). This luxury uptake effectively uncouples the organism from limiting external conditions, and is important in their competitive interactions.

Shading and competition for light

Interception of light is an important aspect of competition between algae and higher plants. Excessive growth of filamentous algae at the water surface can cause a dramatic decrease in the penetration of photosynthetically-active radiation (PAR), with 98 per cent being removed within the top 1 cm of the water column.

Submerged macrophytes are shaded by both phytoplankton and periphyton, and can in turn deprive algae of light. Quite apart from competition between these organisms, dense growth of algae and higher plants may also result in self-shading. Self-shading by hydrophytes is one of the major factors in limiting their growth in eutrophic and hypereutrophic waters.

Hydrophyte adaptations to avoid or minimize the effects of shading by planktonic and periphytic algae include the following.

• Early growth prior to development of algal blooms (see previously).

• Development of aerial foliage and floating leaves, both of which intercept light before it reaches algae.

• Production of numerous growth tips early in the season, developing new shoot surface area at a higher rate than that of colonization by periphytic algae. In some vascular hydrophytes (e.g., *Elodea*), the smooth cuticle of this new growth reduces surface colonization by periphytic algae which can, however, form dense communities on older leaves and on the surfaces of other macrophytes.

- Physiological adaptation to low irradiance. This is typical of submerged hydrophyte leaves, particularly those shaded by dense phytoplankton or periphyton. Photo-adapted leaves of hydrophytes have low photosynthesis saturation levels and low light compensation points, with respective values of 20–40 Wm^{-2} and 1–2 Wm^{-2} being reported by Pokorny *et al.* (2002b). In situations where phytoplankton levels are low, it is the periphyton which may take over as the major algal competitor for light. These biota have a lower CO_2 compensation point than macrophytes, and their higher efficiency of bicarbonate uptake means that they not only deprive the vascular plant of light but also CO_2.

Although ability to grow at low light levels is important, filamentous algae do not show any gradation in adaptation within the water column. The light compensation point for algae deep in the water column is the same as those at the surface. These organisms appear to rely simply on rapid growth rather than physiological adaptations to out-compete hydrophytes.

The existence of hydrophytes in these conditions is variable, and although a wide range are able to survive algal competition in high nutrient conditions, others cannot. Pokorny *et al.* (2002b), for example, report that the water lily *Nymphaea candida* and other sensitive vascular plants are rapidly disappearing from the increasingly hyper-eutrophic fishponds of the Třeboň wetlands.

In addition to interactions within the water column, competition for light is also an important factor at the edge of ponds and lakes, where shading of littoral algae by helophytes is ecologically significant.

4.9 Damaging effects of high levels of solar radiation: photoinhibition

Aquatic organisms are frequently exposed to high light intensities, with irradiance values at the top of the water column in lakes and other water bodies exceeding 800 μmol m^{-2} s^{-1} on bright summer days (Hill, 1996). Although these aquatic light levels are considerably less than values of up to 1800 μmol m^{-2} s^{-1} encountered in terrestrial environments (Horne and Goldman, 1994) the radiation may still have an adverse effect on the metabolism and activities of algae and other biota close to the water surface. High light intensity is an example of an environmental stress factor which can perturb life processes over a range of biological levels, from ecosystems (Section 1.6) to cell physiology (this section) and molecular control mechanisms (Section 3.3.2).

Photoinhibition may be defined as the destructive effects of light on cell processes leading to impairment of metabolic activities, reduced growth, and in some situations cell death. The most obvious examples of photoinhibition come from depth measurements of photosynthetic activity close to the lake or river surface, where there is a clear depression in carbon fixation without any decrease in algal biomass, as measured by chlorophyll-a concentration (Figure 4.22).

The adverse effects of light irradiation are particularly important in relation to phytoplankton and algal communities in extreme high-light environments. The biological implications of photoinhibition will be considered in relation to the mechanisms of metabolic destruction, algal strategies for avoiding or minimizing photoinhibition, and the importance of this process in different environments.

4.9.1 Specific mechanisms of photoinhibition

The destructive effects of light are increased with a reduction in the wavelength of the irradiation. The inhibitory effects seen with photosynthetically active radiation (PAR 400–700 nm) are thus much

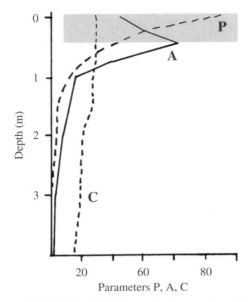

Figure 4.22 Photoinhibition of the spring diatom bloom in the Neuse River (North Carolina, USA).

Inhibition of photosynthesis in the top 0.5 m of the water column (■) is indicated by decreased productivity (A) at high chlorophyll level (C). The phytoplankton community was mainly composed of *Aulacoseira*, *Cyclotella*, and *Navicula*.

P: photosynthetically active radiation (µmol photons m^{-2} s^{-1} × 10); **A:** algal productivity (mg C m^{-3} h^{-1}); **C:** chlorophyll-a content (mg Chl-a m^{-3}) (figure adapted and redrawn from Paerl *et al.*, 1983)

enhanced with a shift to the ultraviolet end of the spectrum, comprising UV-A (315–400 nm) and UV-B (280–315 nm) radiation (Wellburn, 1998). Light incident at the water surface has a variable content of UV-B, depending on the solar angle and the thickness of the ozone layer. Penetration of the water column results in a rapid loss of UV radiation, with 99 per cent absorption of UV-A in clear-water occurring within 2–5 m of the surface and equivalent absorption of UV-B occurring within 0.5 m. In spite of the rapid attenuation of UV-B in the water column, this radiation will have major effects on microorganisms in surface waters and many workers have used this wavelength band in studies on the molecular mechanism of photoinhibition. Incident

levels of UV-B, with associated microbiological effects, are expected to increase in line with predicted reductions in the stratospheric ozone layer (Madronich *et al.*, 1998).

The destructive effects of visible and ultraviolet light are mediated both by direct action on molecular bonds and by the generation of active (superoxide) oxygen species, which leads to photo-oxidation of a range of molecular species. These effects are similar for both aquatic algae and higher plants (Jansen *et al.*, 1998) and involve direct effects on macro-molecular targets (DNA, photosynthetic machinery, photoreceptor, and motor organelle proteins) as well as indirect effects on a range of cellular processes.

DNA damage and repair

Absorbance of UV-B irradiation by DNA triggers the dimerization of thymine pairs (Figure 4.23), with the formation largely of cyclobutane-pyrimidine dimers (CPDs). These DNA modifications are mutagenic. They also cause disruptions to cellular metabolism because of blockages in gene transcription and DNA replication arising from the inability of RNA- and DNA-polymerase to read through unrepaired dimers. Such blockages in protein synthesis and DNA replication inhibit cell growth and can limit increases in algal population.

Repair of UV-B damaged DNA occurs via light-dependent photoreactivation, using photolyase enzymes which restore the bases to their original state (Figure 4.23). Photolyases carry light-absorbing components (chromophores) which have absorption maxima between 350 and 400 nm and are activated primarily by UV-A, and to a lesser extent, blue light. Since UV-A and blue light invariably accompany UV-B irradiation under natural conditions, the photoreactive role of these specific wavelengths is environmentally highly effective. Light-dependent repair of CPDs has been demonstrated in both eukaryote (Petersen *et al.*, 1999; Petersen and Small, 2001) and prokaryote (Takao *et al.*, 1989) algae, and is closely similar to the system operating in higher plants (Jansen *et al.*, 1998).

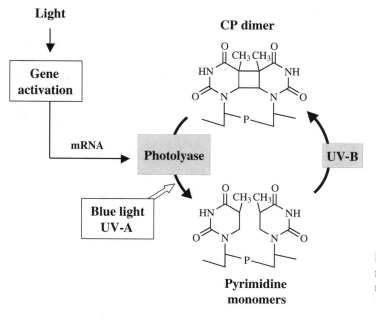

Figure 4.23 General scheme for the formation and photoreactivation of DNA dimers (figure derived from Jansen *et al.*, 1998)

Photosynthetic machinery

Although there is a clear advantage in terms of competition for light and promotion of photosynthesis for algae to be at the top of the euphotic zone, extensive residence (for more than a few hours) can expose these organisms to harmful radiation. Many investigators (e.g., Horne, 1975) have linked impairment of photosynthesis to the high levels of light in surface waters. As noted previously, this effect is seen in Figure 4.22, where there is a sharp depression in phytoplankton carbon fixation close to the water surface, even though light intensity and chlorophyll-a concentration increase in this region.

Mechanisms underlying this environmental effect can be investigated in laboratory cultures, where the damaging effect of light radiation is revealed by plotting photosynthesis – irradiation curves. As noted previously (Figure 4.7), these show a levelling off of photosynthesis, with a maximum value (P_{max}) at irradiance (I_{max}), followed by a decline due to photoinhibition. The onset of this decline marks the point at which light radiation damage causes a net reduction in the rate of photosynthesis, and it is this light intensity which is quoted in Table 4.3. Modelling of light saturation curves by Platt *et al.* (1980) indicates, however, that the process of photoinhibition commences at irradiation levels lower than I_{max}, before the net decrease in photosynthesis. Theoretical curves without photoinhibition reach a higher maximum rate of photosynthesis (P_s) at an irradiance value higher than I_{max}, and there is no subsequent decline once the system has become saturated (e.g., *Aulacoseira*, Figure 4.8).

Visible and UV light have a range of effects on the photosynthetic machinery, including bleaching of photosynthetic pigments, damage to photosynthetic (thylakoid) membranes and reduced activity of the enzyme ribulose bisphosphate carboxylase (Rubisco) (Jansen *et al.*, 1998). Bleaching of photosynthetic pigments has been demonstrated (Häder, 1995) by comparing the absorption spectra of intact algal cells, with and without exposure to solar radiation, and has demonstrated pigment loss specifically in the wavelength absorption range of the carotenoids and chlorophylls.

In addition to these general effects, UV-B causes direct and specific damage to the PSII complex, which catalyses the transfer of electrons from water to plastoquinone (see Section 3.7, Figure 3.13). The degradation of PSII is maximal at 300 nm, with shorter wavelengths having much less effect. Within the highly-structured protein pigment complex (Figure 4.24), UV-B radiation causes rapid light-driven degradation of the central chlorophyll-a molecule (P680), two key proteins – D1/D2 and the electron acceptors quinone A (Q_A) and quinone B (Q_B). This leads to an accumulation of inactive PSII complexes. Repair of the accessory proteins in PSII complexes is thought to be mediated by rapid turnover of D1 protein (driven by photosynthetically active radiation) and also combined turnover of D1-D2 proteins (mediated by UV) – Jansen *et al.* (1998).

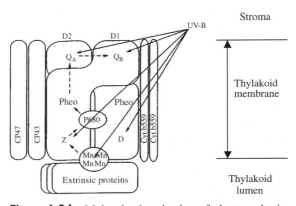

Figure 4.24 Molecular inactivation of photosynthesis: effects of UV-B radiation on photosystem II. Schematic representation of the photochemical reaction centre of photosystem II, showing some of the major proteins and electron acceptors grouped around the chlorophyll-a molecule (P680) with its cluster of Mn atoms. The broken arrows indicate the migration of electrons through P680, the primary electron donor. Proposed molecular targets for photoinhibition (solid arrows) are: (1) the chlorophyll molecule, with the cluster of four Mn atoms which catalyse the splitting of water and the generation of electrons – extrinsic proteins are involved in stabilizing this reaction; (2) redox-active tyrosines (Z and D), located on proteins D1 and D2 – Z normally serves as the electron donor to P680; (3) plastoquinones QA and QB – electrons pass to these secondary acceptors from pheophytin (Pheo), which acts as a primary electron acceptor (figure taken with permission from Jansen *et al.*, 1998), see also Figures 3.13 and 4.5

Photoreceptor and motor organelle proteins

Orientation and diurnal migration of planktonic organisms within the water column involve both receptor activity to external stimuli (light, gravity) and motility. Recent laboratory studies ((Häder, 1995) have shown that high levels of solar radiation can impair both of these activities in a wide range of cultured algae (including euglenoids, cryptomonads, and dinoflagellates), acting by denaturation of photoreceptor and motor organelle proteins. These photoinhibitory effects lead to an inability of the organisms to adapt to constantly changing conditions in the environment, and may affect diurnal migration.

Laboratory studies on photoinhibition of light and gravitational responses have been carried out mainly on dinoflagellates, which show pronounced diurnal migrations and active short-term responses in relation to these stimuli. Häder (1995) used continuous CCD camera recording and computer analysis of individual cell movements under controlled laboratory conditions to assess the directional precision of the dinoflagellate response and the general effects on motility. Control cultures of a test dinoflagellate (tentatively identified as *Gymnodinium*) showed a high precision of orientation under lateral illumination. Exposure to high levels of unfiltered solar radiation lead to a sharp decline in the directional precision of the photoresponse after 40 minutes. The percentage of motile cells also showed a marked decline after a threshold period of time, though in these experiments the average swimming motility of the motile fraction remained constant at about 250 μm s^{-1}.

4.9.2 General effects of photoinhibition

The specific effects of photoinhibition on particular target molecules and particular processes lead to a wide range of secondary effects. UV-B irradiation in particular leads to general inhibition of cell division and growth, decreased levels of ATP, reduced rates of nutrient (ammonia, nitrate, phosphate) uptake, and loss of motility (Hessen *et al.*, 1995).

Some of these general effects may be quite complex, varying with time, and also with radiation intensity and physiological state of the cells. Exposure of cells of *Chlamydomonas* to UV-B radiation (Hessen *et al.*, 1995), for example, leads to an initial increase in growth at low radiation levels (1.2 kJ m^{-2}) with a subsequent decline. Marked inhibition of growth occurs throughout the time course at high radiation levels (16 kJ m^{-2}). Stimulation and inhibition of growth in *Chlamydomonas* is closely correlated with phosphorus uptake and loss of flagella, demonstrating a close interlinking of the physiological effects of photoinhibition.

The immediate effects of UV-B irradiation on cell processes, and the recovery of cells from this photoinhibition, vary with the physiological state of cells. In *Chlamydomonas* the degree of phosphorus limitation appears to be particularly important, with a higher recovery of flagella in P-limited cells (87 per cent within 72 hours after termination of radiation) compared with only 34 per cent of non P-limited cells.

Different algae have varying susceptibilities to UV-B radiation. Larger species of diatoms are reported (Hessen *et al.*, 1995) to be more susceptible than smaller ones, and flagellated algae more susceptible than non-flagellated. These differential effects of photoinhibition will potentially lead to changes in community composition under natural conditions.

4.9.3 Strategies for the avoidance of photoinhibition

Primary producers face the challenge that they depend on solar energy in the middle range (>400 nm) for photosynthesis, but are damaged in high intensity visible light and by UV irradiation. Of the two groups of primary producers, the potential risk of radiation damage is particularly high for algae (which are frequently found close to the water surface), but is not significant for photosynthetic bacteria which are limited to the lower part of the euphotic zone.

Algae are able to minimize photoinhibition in two main ways – by avoidance (migration) and by reducing the harmful effects of the radiation *in situ*.

Avoidance of radiation

Algae are able to migrate from regions of high light intensity in both planktonic and benthic environments.

Planktonic environments In planktonic environments, avoidance by actively motile (flagellated) algae and by algae with a buoyancy mechanism, involves migration away from the water surface to lower regions of the water column. Such avoidance mechanisms are typically achieved by phototaxis, supported by other receptor processes (see previously). In the case of blue-green algae, sinking under conditions of high surface illumination may also depend on the formation of ballast by photosynthesis. Any interference with this process may lead to the colonies being stranded at the water surface, resulting in large-scale destruction of the algal population.

Benthic environments Various groups of benthic algae, including blue-green algae, diatoms, and desmids (green algae) are able to carry out light-related active movement on substratum.

Blue-green algae. Many blue-green algae, including both single-celled and filamentous forms, are capable of gliding movements. Single cells tend to make irregular jerky movements while filamentous forms such as *Oscillatoria* glide on substratum in a more controlled way, achieving speeds of 2–11 μm s^{-1} (Van den Hoek, 1995). Recent studies have suggested that movement is achieved by numerous microfibrils in the cell wall, which are spirally wound around the cell. Waves propagated in rapid succession along these fibrils are thought to produce a rotating forward movement of the whole organism, through friction between microfibrils and substratum. As the algae creep around like slugs, they leave behind them a trail of mucilage. The production of this mucilage was originally thought to be the means of locomotion, but is now considered to provide a solid substratum over which the organisms can move.

The movement of benthic blue-green algae in response to high light intensity involves three types of response, each mediated by its own set of photoreceptor pigments and its own physiological mechanism:

- *phototaxis*, where the light source determines the direction of movement – blue-green algae exhibit positive phototaxis to dim light and negative phototaxis to bright light;

- *photokinesis,* where the speed rather than the direction of movement is influenced by light intensity;

- *photophobic response,* where a rapid reversal of movement is induced by a sudden increase or decrease in light intensity.

As a result of these mechanisms, populations of benthic algae tend to accumulate in regions of moderate light intensity, avoiding high illumination. This may either involve lateral movement on benthic surfaces, or vertical movement into the substratum of environments such as mudflats (Figure 2.19) where there are marked diurnal migrations. *Diatoms.* Some benthic diatoms, including various pennate and centric forms, are able to migrate over substrata by the extrusion of mucilage (Section 3.5). In some cases, (reported in Lee, 1997) movement is influenced by light. Cell clumps of *Nitzschia palea*, obtained from liquid culture, tend to disperse on transference to a glass slide in the presence of light but do not disperse in darkness.

Reduction of harmful effects

High levels of solar radiation do not always lead to inhibition of photosynthesis. This is seen in such diverse freshwater environments as snowfields and lake pelagic systems, where high levels of irradiation can lead to protective physiological changes in the micro-algae.

Photoprotection in different freshwater environments In snow environments, physiological responses to harmful irradiation have been studied particularly in relation to green algae, such as *Chlamydomonas nivalis* (Section 3.12.3). High levels of visible light and UV radiation induce the formation of photoprotectants, anti-oxidants, and photoreactive repair systems. Formation of the xanthophyll pigment astaxanthin (photoprotectant) is particularly intense under conditions of high light and low nutrient (Figure 3.27), leading to the formation of red blooms of snow algae referred to as 'red snow'. In lake pelagic environments, irradiation-induced physiological adaptations in phytoplankton have been inferred from studies such as those by Paerl *et al.* (1983) on the summer bloom of the blue-green alga *Microcystis aeruginosa*. The rate of carbon fixation in high-light adapted populations of this organism increased right up to the water surface, compared with surface inhibition in non-adapted populations seen at other times of the year (e.g., spring diatom bloom, Figure 4.22). This ability to withstand high levels of solar radiation can be largely attributed to the production of UV-absorbing molecules which essentially block-out the radiation, limiting the damaging effects on other molecules in the cell.

Photoprotection in blue-green algae Blue-green algae are particularly effective in carrying this out. The development of such protection mechanisms in these organisms may relate to the fact that they arose in the early preCambrian-period, when UV levels were much higher than at present due to the absence of absorbing gases (O_2 and O_3) in the atmosphere (Dillon and Castenholz, 1999). The evolutionary development of this strategy was also important in ecological terms, since fossil evidence has indicated that many of these pre-Cambrian algae were present in shallow pools where an avoidance mechanism would not have been relevant.

Blue-green algae produce various compounds that absorb short-wavelength radiation, particularly UV light, and act as photoprotectants. These UV-absorbing compounds include mycosporin-like amino-acids (MAAs), a blue-green algal-specific compound (scytonemin), and pigments that are associated with the photosynthetic machinery (carotenoids). MAAs are generally present within blue-green and other algae, fungi, aquatic invertebrates, and fish, and clearly have a widespread role for photoprotection within aquatic biota (Karentz, 1999).

Scytonemin is an inducible pigment which is specific to blue-green algae. It is contained in the extracellular sheath and is synthesized in response to irradiation by wavelengths at the UV-blue end of the spectrum. Scytonemin absorbs strongly in the blue, UV-A, UV-B and UV-C wavelengths and has been shown to protect metabolically active and inactive blue-green algal cells against the deleterious effects of UV radiation. The ability to synthesize scytonemin is known only in blue-green algae, and is found in all five subgroups of the phylum (Dillon and Castenholz, 1999). The widespread occurrence of scytonemin in this phylum, particularly in different evolutionary lines that are believed to have diversified early in the history of the group, suggests that the protection mechanism arose early in the evolution of these algae at a time when UV radiation was an important stress factor. The strong absorption of UV-C radiation (190–280 nm) has little ecological relevance under present conditions, but would have been important in pre-Cambrian times.

Reduction in the harmful effects of radiation can also be achieved by adjustments to photosynthetic pigment composition, increasing the proportion of carotenoid pigments that act in a similar way to scytonemin by absorbing the radiation and limiting its penetration to other parts of the cell. Increased carotenoid/chlorophyll-a ratios have been observed (Paerl *et al.*, 1983) for developing bloom populations of *Microcystis*, leading to a high resistance to photoinhibition and greater photosynthetic efficiency.

Role of carotenoids in photoprotection Carotenoids exhibit strong *in vitro* absorbance in the UV and near-UV parts of the spectrum, and increased proportions of these pigments increase light utilization in the low and middle ranges of the photosynthetically active radiation (PAR) as well as providing protection from UV damage.

The protective effect of carotenoids involves direct removal of UV light by absorption and also alleviation of damage caused by the formation of highly reactive chlorophyll-a and oxygen species. Under UV irradiation, Chlorophyll-a can be converted to a reactive triplet state (Chl-a^T), leading in turn to the formation of triplet oxygen (O_2^T):

$$Chl\text{-}a \xrightarrow{h\nu} Chl\text{-}a^T \qquad (4.5)$$
$$Chl\text{-}a^T + O_2 \longrightarrow Chl\text{-}a + O_2^T$$

O_2^T can lead to photooxidative destruction of a range of organic compounds, including chlorophyll-a.

Carotenoids act by transferring the triplet state from Chl-a^T and O_2^T in the following ways:

$$Chl\text{-}a^T + Carot. \longrightarrow Chl\text{-}a \ \ (\text{ground state}) + Carot^T$$
$$Carot.^T \longrightarrow Carot. \ \ (\text{ground state}) + heat$$
$$(4.6)$$

$$Carot. + O_2^T \longrightarrow Carot.^T + O_2 \ (\text{ground state})$$
$$Carot.^T \longrightarrow Carot. \ (\text{ground state}) + heat$$
$$(4.7)$$

Individual carotenoid pigments differ in their biochemical roles. Zeaxanthin is an important photoprotectant, with a limited role in photosynthesis, while β-carotene is the converse.

4.9.4 Photoinhibition and cell size

A variety of evidence in the literature suggests that cell size is a key factor in determining sensitivity to ultraviolet radiation. This evidence comes from theoretical analysis and modelling studies, experimental results from cultured algae, and studies on environmental assemblages (Laurion and Vincent, 1998).

In theoretical terms, two related aspects suggest that small cells should be more sensitive than large ones.

- Pigment-specific light absorption increases with decreasing cell size and decreasing internal molecular shading (referred to as the 'package effect' – see Section 3.4.3). This leads to higher UV exposure per unit pigment and per unit cell volume.

- The protective efficiency of UV-absorbing molecules such as mycosporine-like amino-acids is a function of cell concentration, but also of cell size through its effect on molecular self-shading. Model calculations suggest that while nano- and microplankton cells contain sufficient UV-screening pigments to give protection, these are not sufficient within the small cell volume of picoplankton.

Experimental and environmental evidence for the importance of cell size in relation to UV damage comes mainly from oceanographic studies. Studies by Karentz et al., 1991, for example, showed that the UV sensitivity of 12 species of Antarctic diatoms was highest in smaller cells, which sustained more damage per unit DNA and were killed by a lower UV flux, compared with larger cells. Other studies have suggested that factors other than size may take precedence. In the sub-Arctic lake communities studies by Laurion and Vincent (1998), cell size was not a good index of UV sensitivity. The blue-green algal dominated picoplankton in these lakes was less sensitive to the impact of UV irradiation than would have been predicted simply on the basis of cell size.

Although cell size is important for UV sensitivity, picoplankton cells may have their own adaptations to minimize the effect. Blue-green algal picoplankton in particular dominate high-UV freshwater and marine environments (tropical oceans) and have a wide range of effective defenses against UV damage. These include multiple copies of their genome, protective pigments and enzymes against reactive oxygen species, and repair mechanisms for UV-damaged DNA and photosystems.

4.9.5 Lack of photoinhibition in benthic communities

Benthic light environments are highly variable, with maximum light intensities to which algae are exposed varying from near zero in deep, turbid waters to full sunlight in shallow, clear conditions.

Laboratory studies on stream periphyton communities (Hill, 1996) suggest that benthic algae taken from shaded sites are capable of showing photoinibition, while algae from exposed sites are not. These two groups have photosynthesis – irradiation (P–I) curves corresponding to Figure 4.8 (left and right respectively), with laboratory irradiance levels

up to 1200 μmol m^{-2} s^{-1}. Although shaded benthic algae are capable of showing photoinhibition, their low-light environment would not normally reach levels to cause this.

Although laboratory P–I curves of exposed benthic algae do not give absolute proof of absence of photoinhibition under exposed environmental conditions, they do suggest that it is much less important in benthic than in planktonic organisms. Lack of photoinhibition in benthic algae may be partly attributed to the permanent development of protective pigments in these fixed communities. The typical occurrence of benthic algae in dense mats will also be important (Dodds et al., 1999), since high irradiance will be rapidly attenuated due to self-shading. Even if photoinhibition occurs at the mat surface, this will be masked by lack of photoinhibition through the rest of the community.

4.9.6 Photoinhibition in extreme high-light environments

The degree of exposure of phytoplankton to high-intensity visible and UV light varies with the type of water body and with climate, altitude, and water quality. In many situations, photoinhibition has only a limited effect on algal productivity. In temperate lakes, for example, although high levels of solar radiation occur on sunny days at the water surface, the algae are able to migrate down the water column to a position of optimum light. In other lakes, this avoidance of photoinhibition is not possible. Many polar and some alpine lakes are very shallow. These lakes also have very high light exposure with clear oligotrophic water, so that photoinhibition occurs throughout a major part of the water column. The overall decrease in pelagic productivity in these lakes is balanced by the development of a community of benthic algae and mosses, in some cases forming an extensive growth on the lake bed.

The effects of UV irradiation (UVR) on aquatic systems in extreme environments is particularly relevant at the present time, since these systems are currently experiencing increased levels of underwater UVR due to stratospheric ozone depletion and to climate-related changes in spectral

Table 4.10 Ecological sites of high photoinhibition

Site	Light characteristics	Pelagic community	Benthic community
Alpine lakes			
Finstertaler See (European alps)[a]	High photoinhibition – very intense light in summer	Phytoplankton dominated by dinoflagellate Limited zooplankton	Dominated by blue-green algae and mosses – varied microfauna
Polar lakes			
Algal lake[b]	Maximum photoinhibition at midday in summer	Low productivity	Blue-green algae and mosses – limited microfauna
Lac L'eau Claire. Lac Kayouk[c] Canada Oligotrophic		Dominated by blue-green algal picoplankton (*Synechococcus*)	

References: [a]Pechlaner *et al.* (1972); [b]Goldman *et al.* (1972); [c]Laurion and Vincent (1998)

UVR attenuation in the water column (Laurion and Vincent, 1998). The effects of increased UVR on size (see previous section) may be particularly important, since any shift in size would have major implications for pelagic food web processes and for interactions with the benthos via sedimentation.

Some examples of freshwater environments with high levels of photoinhibition are summarized in Table 4.10, and include Antarctic lakes and Alpine lakes.

Antarctic lakes

The periphery of the Antarctic continent is characterized by the presence of many lakes which are small, shallow, and are much influenced by the surrounding ocean (maritime lakes). Although these lakes do not have a high level of precipitation, they do not dry up because of the low temperature ($<10°C$ throughout the year) and high humidity. These lakes typically have a low nutrient input and a low productivity. Many of the algae present are washed in from surrounding terrain or from the bottom of the lake.

In summer, at a time of long daylight hours and high insolation, photosynthetic rates are generally low – with minimum rates at midday and highest rates at midnight. This pattern arises due to severe limitation of photosynthesis for most of the day during Antarctic summer. In these lakes where photoinhibition is greatest at noon, photosynthesis is inversely related to light intensity. In autumn a more normal pattern of productivity occurs, since overall light levels are much less than summer and maximum productivity returns to the middle of the day.

Benthic communities of blue-green algae dominate the bottoms of shallow lakes, providing a major habitat for Antarctic invertebrates such as protozoa, rotifers, crustacea, copepods, and Cladocera. These populations typically have little species diversity due to the isolation of Antarctica from the world's land masses, resulting in limited colonization of this remote region.

Alpine lakes

These are found in most mountainous regions of the world. They are typically cold, deep lakes that are ice-free for most of the year. The light regime of these lakes at high altitude differs from lowland lakes in an increased level of solar radiation, rich in UV light. The oligotrophic nature of many of these lakes (e.g., Lake Tahoe, USA – Figure 4.25) gives the water a high degree of clarity, enhancing UV penetration and increasing further the adverse effects of solar radiation.

Figure 4.25 Lake Tahoe (USA): the largest Alpine lake in North America, showing intense irradiation of phytoplankton over part of the year.

The clear water of this oligotrophic water body (see foreground) permits a high level of light penetration, with resulting photoinhibition. The lake occurs at an altitude of 1920 m, and receives most of its precipitation as snow rather than rain. The phytoplankton of the lake is dominated by diatoms and chrysophytes, which remain dominant seasonally throughout the summer. Algal populations are regulated primarily by nutrient- and light-limited growth, coupled with cell death in the surface waters (high insolation) and natural waste due to random sedimentation. The lake shows an absence of shoreline macrophyte development typical of oligotrophic systems (photograph by Rosemary Sigee)

In many cases the low algal productivity is matched by a relatively simple ecosystem. Verderer Finstertaler See in Austria (Pechlaner, 1972) remains below 10°C over most of the summer, and has dinoflagellates as the only common group of phytoplankton. The zooplankton population is composed of two rotifers and one copepod. Although the planktonic flora and fauna are limited, extensive benthic flora develops on the lake beds where sufficient light penetrates. These plants provide a microenvironment for an extensive community of nematodes, oligochaetes, ostracods, and chironomid larvae.

Light penetration and photoinhibition vary with the seasonal cycle. In winter months, populations of dinoflagellates accumulate under the ice, where they adapt to low light levels. Melting of the ice and increased insolation in spring lead to greater penetration of the water column, with downward migration of the algae to depths of 5–15 m to avoid photoinhibition.

4.10 Periodic effects of light on seasonal and diurnal activities of freshwater biota

Periodic changes in the intensity and quality pf light have major effects on the biology of aquatic organisms. These effects are enhanced by the low light absorption (a) and scattering (b) coefficients of water (Table 2.2) which maximize the depths to which these fluctuations exert an influence. Periodic changes operate over two main time scales – seasonal and diurnal.

4.10.1 Seasonal periodicity

The importance of seasonal changes in light intensity on the annual succession of phytoplankton has been previously noted (Figure 3.14), and can be explained in terms of general environmental adaptations and competitive ability of different algal groups.

In addition to this rather broad effect of light on microbial interactions, annual changes in light

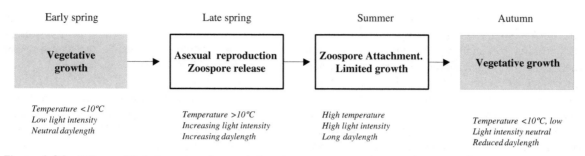

Figure 4.26 Effects of light intensity, photoperiod, and temperature on the growth pattern of *Ulothrix zonata*.
Light and temperature affect the balance between vegetative growth and asexual reproduction. The limited growth of attached zoospores in summer is partly because this balance is against growth, and also because of intense competition from another littoral alga – *Cladophora glomerata*

intensity, wavelength, and daylength may also have a more specific effect on the succession and growth patterns of particular algae and other microorganisms.

The filamentous green alga *Ulothrix zonata* provides a good example of such specific and direct effects of light on growth pattern (Graham *et al.*, 1985). This alga is common in many northern lakes of the USA and in Canada, where it grows abundantly in spring and autumn in shallow waters. At these times of year the increase in biomass occurs due to high vegetative growth. The decline in growth rate and total biomass in late spring is due to a switch from vegetative growth to asexual reproduction (Figure 4.26) – with the formation of zoospores and the disruption of vegetative filaments. Experiments on laboratory cultures have shown that this transition is favoured by high temperatures (>10°C), high light levels (about 520 μE m^{-2} s^{-1}) and photoperiods of either short day (8:16 light – dark) or long day (16:8 light – dark) cycles. Conversely, zoospore formation is minimal under conditions of low temperature (<10°C), low light levels, and neutral daylengths (12:12 light – dark). Seasonal changes in the growth pattern of *Ulothrix* can largely be explained in terms of these environmental parameters. In spring, the switch to zoospore production results from the combined effects of rising water temperatures and increasing light intensity and daylength. During

summer, when conditions limit growth, *Ulothrix* survives as short filaments on rocks or as epiphytes on other algae. Growth is resumed in autumn when temperatures and light levels decline, and daylength shifts towards neutral.

4.10.2 Diurnal changes

The response of freshwater organisms to periodic changes in light becomes particularly marked in relation to the daily alternation of day and night. This well-defined environmental regime leads to biological periodicities that are typically both 'diel' (occurring at 24 hour intervals) and 'diurnal' (occurring every day, as a sequence of activity within a 24 hour time frame). 'Circadian rhythms', are an example of diurnal change, and are metabolic or behavioural cycles with a periodicity of about 24 hour.

General aspects

Diurnal periodicities in freshwater biota include patterns of growth and death (Lee and Rhee, 1999), metabolism (photosynthesis, nitrogen fixation) and behaviour (general motility and migration). Some of these responses (e.g., photosynthesis) are a direct physical result of the presence or absence of light, while others (e.g., vertical migration) relate more to an underlying, ultimately cellular, timing mechanism. This intracellular clock operates via diurnal variations in gene transcription and related molecular activity (see later).

Organisms in the pelagic environment have greater light exposure compared to benthic biota, and show particularly strong periodic behaviour. This includes the diurnal migrations of phytoplankton and zooplankton (Figure 5.19), and the related movements of fish predators. These often show a similar light-regulated pattern to migrating zooplankton as they follow their prey. Benthic organisms on lake sediments and in streams also show diurnal activity. Epibenthic animals, living on the surface of sediments, avoid light and predators by hiding under stones during the day and only emerging at night. Crayfish and other predators also emerge from under rocks and plants at night to begin foraging, and deep water fish may move into shallower regions at dawn or dusk for their major feeding activities.

Circadian rhythms: endogenous activity and entrainment

Although these diurnal activities relate to external changes in the environment, such as changes in light, temperature, oxygen concentration, and food availability, they may be driven by innate rhythms in the organism that continue irrespective of environmental alteration. Such daily or circadian rhythms operate as endogenous biological programs which time metabolic and/or behavioural events to occur at optimum phases of the daily cycle. These rhythms have three diagnostic characteristics:

- in constant conditions the programmes free-run with a period of approximately 24 hours duration;

- in an environmental cycle involving periodic alternation of characteristics such as light–dark, or high–low temperature, the rhythm will take on the period of the environmental cycle – this is referred to as entrainment;

- the period of the free-running rhythm varies little with temperature. Within the physiological range circadian rhythms are temperature-compensated.

The mechanisms that control circadian rhythms and their biological role in the freshwater environment have been investigated particularly in blue-green algae and dinoflagellates.

4.10.3 Circadian rhythms in blue-green algae

The occurrence of circadian rhythms in blue-green algae was first demonstrated by Grobbelaar *et al.* (1986), who studied the rhythm of nitrogen fixation and amino-acid uptake in the unicellular freshwater blue-green alga *Synechococcus*. These authors were the first to report the isolation of mutants affecting these processes, establishing a genetical molecular basis for control of the diurnal cycle. Circadian rhythms are now known to be expressed in other genera of blue-green algae as well, including *Anabaena*, *Cyanothece*, and *Triochodesmium* (Johnson and Golden, 1999).

Molecular studies on the control of the diurnal cycle in these algae were advanced considerably by the creation of a reporter strain of *Synechococcus*. This was achieved by inserting a DNA construct in which the luciferase gene set *luxAB* was expressed under the control of the promoter for a *Synechococcus* photosystem II gene, *psbAI*. Luminescence in this genetically-modified organism follows a circadian cycle, rising during the day and falling at night, and conforms to all three criteria for circadian rhythms noted previously (Liu *et al.*, 1995).

The control sequence

Further studies have now revealed a chain of molecular events controlling the circadian cycle (Figure 4.27), which can be considered in three parts – the central clock, light modulation, and control of circadian gene expression (Johnson and Golden, 1999).

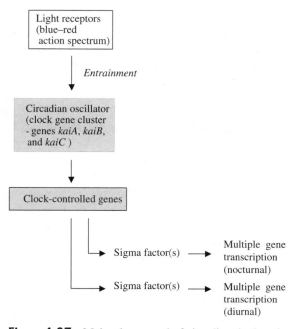

Figure 4.27 Molecular control of circadian rhythms in blue-green algae (information from Johnson and Golden, 1999)

- *The central clock.* Central control of the circadian rhythm resides in a molecular oscillator (clock) composed of a cluster of three genes – *kaiA*, *kaiB*, and *kaiC*. The clock gene products interact with each other, forming an autoregulatory feedback loop which has rhythmic periodicity. Deletion or inactivation of any of the *kai* genes causes disruption of the diurnal cycle, with changes in the cycle length, but has no effect on cell viability.

- *Light modulation.* Light/dark signals are considered to be the primary environmental events which set the phase of circadian clocks. Spectral analysis of circadian activation indicates that blue and red light are most effective in blue-green algae, while green and far-red light are ineffective. The pigments involved in light reception for circadian rhythms in blue-green algae have not yet been determined, and may be specific for this activity. They do not appear to include either phytochromes (phasing effect of red light not reversed by far red) or photosynthetic pigments (action spectrum for circadian entrainment does not coincide with that for photosynthesis).

- *Circadian gene expression.* Liu *et al.* (1995), investigated the extent of circadian control of gene expression in *Synechococcus* by inserting a luciferase gene set (*luxAB*) into the genome to achieve random insertion throughout the chromosome. Screening the luminescence patterns from the large number (~ 800) of clones indicated that circadian expression was widespread throughout the genome, and could be separated into two major categories, day-time and night-time expression. Expression of day-time genes builds up throughout the day, with a trough at dawn and a peak at dusk, and includes the *psbAI* gene for photosystem II. In contrast to this, night-time genes build up their expression during the night, with a minimum at dusk and a maximum transcription at dawn. Night-time genes include *purF* which encodes glutamine PRPP amidotransferase (involved in the purine synthetic pathway)

Because of the large number of genes which are controlled by the circadian oscillator, it seems unlikely that each is controlled by a specific transcription factor. It seems most probable (Johnson and Golden, 1999) that global transcription (sigma) factors are involved, produced by genes that are activated by the oscillator (clock-controlled genes), and promoting expression of groups of circadian genes which encode specific nocturnal or daytime biological activities.

Circadian rhythms and biological processes

The molecular control of circadian rhythms in blue-green algae is of prime importance to the biology of these organisms, where it is involved in the diurnal timing of processes such as photosynthesis, carbohydrate deposition, amino-acid uptake, and nitrogenase activity. These activities are seen particularly well in the process of nitrogen fixation by unicellular blue-greens such as *Cyanothece* (Section 5.6.4,

Figure 5.16), where the diurnal separation of photosynthesis and nitrogenase activity requires clockwork precision.

Maintenance and timing of circadian rhythms may also be of general importance in competition between algae. Comparison of wild type and mutant cells (with different clock periods) suggests that the presence or absence of a fully operational 24 hour clock does not affect viability since growth of monocultures was not affected by mutation. Laboratory competition experiments, however, involving mixtures of strains showed that cycle duration is of key importance in cell interactions. Using light – dark regimens of different cycle length, Ouyang *et al.* (1998) showed that organisms whose endogenous cycle most closely matched the external regimen invariably out-competed those algae who were less well matched – irrespective of whether they were wild type or mutant.

These competition experiments are consistent with the idea that the circadian programme orders cellular processes to match environmental cycles optimally. When this correspondence is lost, fitness and competitive ability are reduced. Circadian rhythms are clearly a major aspect of interactions between blue-green algae and their environment. The extent and importance of such rhythms in other freshwater biota remains to be seen.

4.10.4 Circadian rhythms in dinoflagellates

Although nothing is yet known of the central clock in dinoflagellates, recent studies on the expression of clock-related genes indicates many similarities to blue-green algae. Okamoto and Hastings (2003) carried out a genome-wide study of circadian gene expression in *Pyrocystis lunula* using microarray technology, and demonstrated that 3 per cent of the genes screened were circadian-controlled. These included transcription factors, proteases, light-harvesting proteins, transporters, and metabolic enzymes. As with the blue-green algae, circadian regulation operated mainly at transcription level, with most gene expression occurring in early day and late night circadian time (CT).

5

Inorganic nutrients: uptake and cycling in freshwater systems

This chapter deals specifically with inorganic constituents in freshwater systems, focussing on those compounds (containing nitrogen, phosphorus, and silicon) which are major nutrients for freshwater microorganisms. Other important ionic components are considered in Section 5.1, and micronutrients (trace elements) in Sections 5.11 and 5.12.

5.1 Chemical composition of natural waters

Natural waters are composed of particulate material within an aquatic matrix (Section 2.1). The chemical components can thus be considered as four major groups, with soluble organic and inorganic material dissolved in the matrix and insoluble organic and inorganic material present as the particulate fraction (Figure 5.1). Organic material is composed largely of carbon-based molecules, but does also typically have associated inorganic (mainly ionic) components. Insoluble components are operationally defined as those which will not pass through a 0.2 μm filter membrane and, in the case of organic material, comprise living and dead biomass (with the exception of most virus particles).

Soluble organic and inorganic components arise partly as a result of biological transformations within the aquatic system (Figure 5.1), and partly from outside sources (external loading) and from sediments (internal loading). Insoluble inorganic material is largely composed of dense non-biological particulate matter that is kept in suspension by water turbulence, the requirement for which means that high levels of these particulates are more typical of lotic compared to lentic systems. The most notable exception to the non-biological origin of inorganic particulates are diatom cell-wall (frustule) fragments, which are generated by death and break-up of diatom populations. Apart from this, these inorganic particulates are not directly involved in biological transformations within the aquatic system.

In this section, the chemical composition of freshwater systems is considered in relation to the general inorganic components of lakes and rivers, followed by the derivation of these materials via aerial deposition and water inflow. The chemical requirements and composition of freshwater biota are then discussed, together with the role of freshwater organisms in the cycling of inorganic components in the aquatic environment.

5.1.1 Soluble inorganic matter in lakes and rivers

Freshwaters contain soluble inorganic matter in ionic form and as dissolved gases. Typical major ions present in natural waters include (in order of decreasing concentration) the cations Ca^{2+}, Na^+,

Freshwater Microbiology: Biodiversity and Dynamic Interactions of Microorganisms in the Aquatic Environment   David C. Sigee
© 2005 John Wiley & Sons, Ltd   ISBNs: 0-471-48529-2 (pbk) 0-471-48528-4 (hbk)

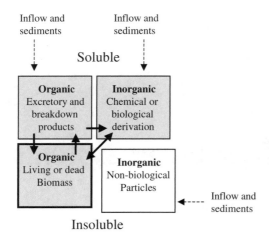

Figure 5.1 Four major states of chemical components in freshwater systems – shaded quadrants show the three states involved in biological transformations. Soluble inorganic material has either a chemical (e.g., dissolved from rocks) or a biological origin, while insoluble inorganic components are mainly mineral particulates. Soluble organic compounds are derived either as secretory (e.g., algal exudates) or breakdown products of aquatic biomass. These may be taken up by aquatic organisms (heterotrophic bacteria) or converted to soluble inorganic material in the process of mineralization. The different chemical components are derived either from biological interchange (→) or from external sources (---→)

centrations and form part of the dissolved organic carbon (DOC). Of the dissolved gases, free N_2 is important within the nitrogen cycle, and concentrations of CO_2/O_2 fluctuate with photosynthetic and respiratory activity. The equilibrium between gaseous CO_2 and carbonate/bicarbonate ions is a key aspect of the pH-buffering system of natural waters (Section 4.7.1).

Aquatic concentrations of major cations and anions

The overall concentration of soluble inorganic components is normally expressed as total dissolved solids (TDS), which is defined as the sum of the concentrations of the dissolved major ions. The world average TDS for river systems is about 110 mg l^{-1}, which corrects to approximately 100 mg l^{-1} for natural TDS once pollution has been taken into account (Berner and Berner, 1987). Mean values for TDS and major cations vary considerably between individual rivers and between continents (Table 5.1), reflecting differences in natural and anthropogenic inputs. The term 'salinity' is sometimes used interchangeably with total dissolved solids, but is more strictly applied to the sum of concentrations of all anions and cations dissolved in the water (Allan, 1995). The concentration of total dissolved ions can be most easily measured as conductivity, with units expressed as micro-Siemens (μS) cm^{-1}. The ability to use conductivity for this measurement arises from

Mg^{2+} and K^+, and the anions HCO_3^-, SO_4^{2-}, Cl^- (Table 5.1). The ionic balance of the water medium is also influenced by the presence of organic cations and anions which are present at much smaller con-

Table 5.1 Ionic composition of river water (mg l^{-1}) (adapted from Berner and Berner, 1987)

	Total dissolved solids	Ca^{2+}	Mg^{2+}	Na^+	K^+	Cl^-	SO_4^{2-}	HCO_3^-	SiO_2	Discharge (km^3 y^{-1})	Run-off ratio[‡]
World average[*]											
Actual	110.1	14.7	3.7	7.4	1.4	8.3	11.5	53.0	10.4	37.4	0.46
Natural[†]	99.6	13.4	3.4	5.2	1.3	5.8	8.3	52.0	10.4		
Europe	212.8	31.7	6.7	16.5	1.8	20.0	35.5	86.0	6.9	2.6	0.42
South America	54.6	6.3	1.4	3.3	1.0	4.1	3.8	24.4	10.3	11.0	0.41

[*]World average is derived from data obtained from North America, South America, Europe, Africa, Asia, and Oceania.
[†]Actual values are corrected to exclude pollution.
[‡]Run-off ratio = average run-off per unit area/average rainfall.

the fact that distilled water has a very high resistance to electron flow, and the presence of ions in the water reduces that resistance.

5.1.2 Aerial deposition of nutrients

Although external loading of freshwater systems occurs mainly via terrestrial inflow (see below), there is increasing realization that direct aerial deposition onto the water surface can also be a significant factor in promoting nutrient availability and stimulating microbial productivity.

Aerial deposition of nitrogen and phosphorus via rain-water, aerosols, and dust particles is particularly important for oligotrophic waters such as the eastern basin of the Mediterranean (Markaki et al., 2003) and freshwater lakes in the Sierra Nevada, USA (Sickman et al., 2003). These atmospherically-derived nutrients are particularly important for stratified waters, since they are directly delivered to the upper regions of the water column – where nutrients are most depleted and where light availability promotes rapid uptake into photosynthetic organisms. Atmospheric input is particularly high during summer and autumn, when stratification limits the contribution of nutrients from the lower part of the water column.

In the eastern Mediterranean (Markaki et al., 2003), aerial deposition of dissolved organic nitrogen (DIN) amounts to about 43 mmol N m^{-2} year, with dissolved inorganic phosphorus (DIP) at about 193 μmol P m^{-2} year. Observed DIN/DIP ratios are in the range 63–349, which is higher than the elemental ratio of 16 within the biomass (Redfield ratio, Section 5.1.4). The atmospheric input of bio-available N and P thus presents an unbalanced contribution to microbial uptake and productivity.

Recent increases in nitrate concentration in some Sierra Nevada and Rocky Mountain (USA) lakes suggests that N loading to aquatic systems in that part of the world is intensifying (Sickman et al., 2003). In Lake Tahoe (California/Nevada, USA), increased atmospheric input of nitrogen has contributed to the decline in water quality, and a shift from co-limitation of phytoplankton by N and P, to persistent P-limitation.

5.1.3 Nutrient inflow from terrestrial sources

The major external loading of freshwater systems, and source of nutrients for microbial productivity, comes from terrestrial deposition of rain and inflow from the surrounding catchment area. The ionic composition of river water (and derived lake and wetland systems) depends on the composition of rainfall, with further contributions from weathering of rocks and human activities (pollution).

Indirect contribution from rainfall

The concentrations of major ions in rain-water (Table 5.2) are considerably less than river water, with total dissolved solute (TDS) values of a few milligrams per litre (Berner and Berner, 1987). Some of the ions in rain-water (Na^+, K^+, Ca^{2+}, Mg^{2+}, and Cl^-) are derived mainly from particles in the air, while others (So_4^{2-}, NH_4^+, and NO_3^-) come primarily from atmospheric gases. Marine and coastal rain has particularly high concentrations of Mg^{2+}, Na^+, and Cl^-, reflecting the importance of marine salts. In contrast, continental rain collects elements from land-based particulate material, which has higher levels of Ca^{2+}, NH_4^+, and NO_3^-. The relative importance of rain input to the

Table 5.2 Typical concentrations of major ions in rainfall (mg l^{-1}) (adapted from Berner and Berner, 1987)

Ion	Ca^{2+}	Mg^{2+}	Na^+	K^+	NH_4^+	H^+	Cl^-	So_4^{2-}	NO_3^-
Continental rain	0.2–4	0.05–0.5	0.2–1	0.1–0.5	0.1–0.5	pH 4–6	0.2–2	1–3	0.4–1.3
Marine and coastal rain	0.2–1.5	0.4–1.5	1–5	0.2–0.6	0.01–0.05	pH 5–6	1–10	1–3	0.1–0.5

chemical composition stream- and lake-water varies considerably with time (climatic and seasonal change) and locality.

Weathering of rocks

The higher concentration of ions in river water compared with rain-water partly results from evaporation, and is partly due to derivation from bedrock and soil sources. The relative importance of evaporation on ion concentration can be estimated from the run-off ratio, which is the average run off per unit area divided by the annual rainfall. On a global scale, the average run-off ratio is 0.46, which means that 46 per cent of precipitation becomes run off. If the 54 per cent loss is due to evaporation, ion concentration from rain to river water would be $\times 2.2$ (the concentration factor). The actual value is roughly $\times 20$, emphasizing the role of other sources such as rock weathering and anthropogenic input.

The importance of evaporation for individual ions can be estimated by comparing rain-water concentrations ($\times$ concentration factor) with river-water values. Mean values for North America (Allan, 1995) suggest that about 10–15 per cent of Ca, Na, and Cl in USA rivers comes from rain, compared with a quarter of K and almost half the sulphate. In contrast, almost none of the HCO_3^- or SiO_2 comes from rain.

In natural and unpolluted systems, weathering of rocks is the major source of inorganic chemicals in river and lake water. The great majority of the world's rivers have TDS that contain more than 50 per cent HCO_3^- plus 10–30 per cent (Cl^- and So_4^{2-}) – reflecting derivation from sedimentary rocks and particularly carbonate minerals (Berner and Berner, 1987).

Anthropogenic sources

Human activities are leading to a broad increase in inorganic concentrations in river and lake systems. Although this is perceived mainly in relation to eutrophication and the input of nitrates and phosphates (see Chapter 10), increases occur across the board. This is seen in Table 5.1, where the world average river water concentrations of almost all ions appear to be increased by human activities. Pollution from domestic sewage, fertilizers, and road salt is particularly important in relation to freshwater concentrations of Na, Cl, and SO_4. Berner and Berner (1987) estimate that about 28 per cent of the sodium in river water is anthropogenic.

5.1.4 Chemical requirements and composition of freshwater biota

Many of the dissolved inorganic constituents present in water are required for the production of biomass by both heterotrophic (e.g., bacteria) and autotrophic (e.g., phytoplankton) microorganisms – and qualify as 'inorganic nutrients'. The bulk of these inorganic nutrients enter the food chain via phytoplankton assimilation and are taken up in relation to algal requirements and nutrient availability. Information on nutrient uptake are given in Sections 5.5 (nitrogen), 5.8 (phosphorus), and 5.10 (silicon).

Nutrient requirements and biological roles

Inorganic nutrients can be broadly divided into two main groups: macronutrients required in substantial amounts, and micronutrients required in trace quantities.

Some of the aquatic source compounds and biological roles of these nutrients are summarized in Table 5.3. Macronutrients (including compounds of nitrogen, phosphorus, silicon, and sulphur) are major components of macromolecules and inorganic cell wall material (Si) and have a key structural role in the cell. Si differs from the other macronutrients in having a very limited biological role. This element appears to have mainly structural (non-metabolic) significance, and is limited as a required nutrient to just one major taxonomic group of microorganisms – the diatoms. Micronutrients often have specific metabolic roles in the activities of freshwater microorganisms, and are discussed in Section 5.11. In addition to macro- and

Table 5.3 Occurrence and biological roles of major elements present in the freshwater environment

Element	Compounds present in water	Major biological role
Carbon	Gas: CO_2 Anions: HCO_3^-, CO_3^{2-}	Photosynthesis pH of water
Oxygen	Present as gas and oxidized compounds	Aerobic respiration Oxidation reactions
Nitrogen[*]	Gas: N_2 Cation: NH_4^+ Anions: NO_2^- NO_3^-	Component of amino-acids and proteins Ammonia and nitrite occur as transient breakdown products – oxidized to nitrate
Phosphorus[*]	Anions: ortho- and polyphosphates	Component of: nucleotides (DNA, RNA, ATP) Polyphosphate storage
Silicon[*]	Anions: silicates	Wall component of diatoms and other algae
Sulphur	Anions: So_4^{2-} S^{2-}	Component of amino-acids cysteine and methionine
Chlorine	Anion: Cl^-	Ionic balance in cell, osmoregulation, enzyme reactivity
Sodium Potassium	Cations: Na^+, K^+	Ionic balance in cell
Magnesium Calcium	Cations: Mg^{2+}, Ca^{2+} (free and bound state)	Ionic balance in cell Structural component of cell walls and exoskeleton Metabolic importance (Ca) as second messenger

[*]Elements that may occur in limiting amounts.

micro-nutrients, which have clear structural or biochemical roles, other inorganic constituents have no apparent biological function. This is the case for aluminium, which is taken up or adsorbed by phytoplankton cells as simple Al^{3+} ions, or related ionic complexes.

The ecological importance of biologically-available compounds of N, P, and Si is emphasized by their role in determining primary and secondary productivity. This is recognized in the use of nitrate and phosphate concentrations to define the trophic status of lakes (Section 10.1.1, Table 10.1) and rivers (Section 10.1.4, Table 10.2). Competition for nutrients between freshwater microbiota is an important factor in determining the dominance of particular species and in the trophic interactions that occur in lakes and rivers.

Nutrient uptake into algae

The chemical composition of algae (and bacteria) is directly determined by uptake of inorganic material (largely as anions and cations) from the surrounding water medium. Other aquatic organisms that feed on algae and bacteria derive their inorganic components mainly from their food supply.

The uptake of inorganic nutrients by algae in lakes and other freshwater systems is important for various reasons:

- As the major microbial biomass during the period of seasonal growth, these organisms represent the principal site of nutrient uptake from the aquatic environment. In stratified lakes, progressive loss of phosphates, nitrates, silicates, and other inorganic components from the epilimnion during spring and summer is principally due to removal by phytoplankton.

- Entry of inorganic chemicals into algal cells represents the major route of entry into the freshwater food chain, with subsequent passage into the biomass of herbivores and carnivores. If little alteration occurs during biomass transformation, the chemical composition of organisms along the food chain would be expected to have fundamental similarities.

- Algae that are not ingested sink to the bottom of the water body, where their breakdown contributes to the high concentrations of inorganic nutrients in the hypolimnion and sediments. Nutrient uptake and sedimentation of surface phytoplankton is thus a major contributor to the chemical heterogeneity that develops within the water column, and leads to cycling of elements in the aquatic system.

Elemental composition of freshwater microorganisms

The inorganic chemical composition of freshwater biota can be evaluated in various ways, including ionic concentrations and elemental composition. The latter gives an overview of all the elemental constituents present within biomass, irrespective of ionic/non-ionic state and elemental species, and can be determined in two main ways – analysis of bulk samples and microscopical analysis of single cells or other micro-samples.

- Bulk analysis, using techniques such as atomic absorption spectrophotometry (AAS), has the advantage of high sensitivity (<1 ppm for most elements) and the fact that elemental composition is averaged over a broad category of biota (e.g., phytoplankton, zooplankton).

- Microscopical techniques such as X-ray microanalysis (XRMA) have less sensitivity (0.1 per cent or 10^3 ppm) compared with bulk analysis but have the advantage of high spatial resolution, allowing the study of single cells and individual species within mixed samples. In XRMA, the electron probe penetrates the cell to a depth of 1–15 μm (depending on accelerating voltage) generating X-rays for analysis. At low voltages and large cells we are looking particularly at the elemental composition of the cytoplasm periphery, cell wall, and outer surface.

XRMA has been widely used in biology for looking at the elemental composition of micro-samples (Sigee et al., 1993) and is particularly useful in the aquatic environment for studying individual species of phytoplankton and other biota within mixed environmental populations. This technique allows us to examine biodiversity at the intraspecific level and has been applied to the detailed examination of a variety of freshwater algae including diatoms, green algae, blue-greens, and dinoflagellates. Analysis of the commonly occurring freshwater dinoflagellate Ceratium hirundinella provides a typical example and illustrates the range of information that can be obtained with this technique.

Case study 5.1   Elemental composition of Ceratium hirundinella

X-ray microanalysis was carried out (Sigee et al., 1999a) on individual Ceratium cells within mixed phytoplankton samples obtained from separate depths (0–8 m) in a eutrophic lake. X-ray emission spectra typically showed clear peaks of Si, P, S, and Cl, plus elements occurring as monovalent (Na, K) and divalent (Mg, Ca) cations (Figure 5.2). Other elements were not detected using this technique due to their low atomic number (C, H, O, and N) or low concentration (e.g., most transition metals). Quantitative analysis of micro-populations of Ceratium provided information on the following:

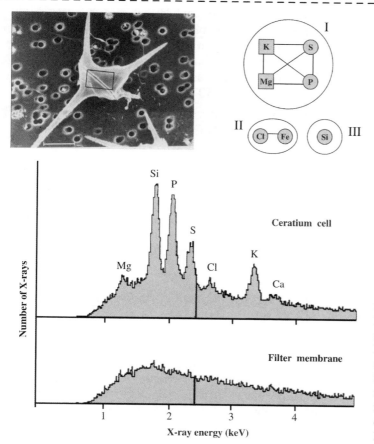

Figure 5.2 Elemental composition of the dinoflagellate *Ceratium hirundinella* (X-ray microanalysis)

Top left: scanning electron micrograph of single *Ceratium* cell – the position of the microprobe for X-ray analysis is indicated by the rectangle. Background circles are pores in the filter membrane (Scale = 30 μm)

Top right: major groups of detectable elements (summary of different micro-populations, based on correlation and Factor analysis) – Na and Ca did not show consistent correlations, and Si occurred in isolation

Bottom: X-ray emission spectra from *Ceratium* cell (clear peaks of Mg, Si, P, S, Cl, K, and Ca) and filter membrane (control – no clear peaks). Peaks of C, O, H, and N were not detectable in this analysis and the high Si peak represents surface accumulation of this element (figure taken from Sigee *et al.*, 1999a) and reproduced with permission of Inter-Research

- *Mean elemental concentrations.* Potassium was the main detectable cation present within the cells (mean concentration about 170 mmol kg^{-1} dry weight), while P was the main anionic element (320 mmol). The high level of Si recorded in these analyses was interpreted as due to localized accumulation in the outer region of the cell, possibly by surface adsorption.

- *Differences between micropopulations.* Comparison of samples ($n = 40$) of *Ceratium* at different depths (0–8 m) within a stratified lake suggested that significant differences do occur, suggesting some patchiness in the elemental composition of micropopulations. Differences were not great, however, and the overall results suggest an underlying homogeneity in the population of *Ceratium* throughout the sampled water column – consistent with the known high mobility and rapid migration of these cells within the water body .

- *Variation within the micropopulation.* All of the detectable elements showed considerable variation between cells. In the 2 m depth sample, for example, the individual cell concentration of K ranged from 0–248 mmol kg^{-1} dry weight (mean 157) and corresponding values for P were 16–458 (mean 294). Other elements showed similar variation between the single cell analyses, suggesting that micropopulations of *Ceratium* are very heterogeneous in terms of their elemental composition.

- *Elemental correlations.* Although elemental concentrations vary considerably between analyses, this variation is not random but occurs within a tight framework of elemental correlations – with particularly significant correlations between Mg, P, K, and S, as summarized in Figure 5.2. These correlations were further supported by defined elemental ratios, with particular constancy in concentrations of Mg/K, P/K, and Mg/P within analyses. Factor analysis of elemental concentrations within micro-populations also revealed Mg, P, K, S as a major elemental group, accounting for 30–40 per cent of the sample variance.

X-ray microanalysis of other aquatic microorganisms

The elemental composition seen in *Ceratium* is typical of other freshwater microorganisms, including different phytoplankton species, bacteria (Booth *et al.*, 1987), and protozoa (Abraham Peskir *et al.*, 1997). In all, XRMA identified K as the major cation and P as the major anionic element – with Na, Mg, Si, S, Cl, and Ca also being routinely detected.

As noted previously, the intracellular level of these elements represents a considerable increase in concentration compared to surrounding aquatic values. The concentration of P in cells of *Ceratium*, for example, was estimated at about 10^4 times greater than surrounding water (total soluble P) compared to values of 10^2–10^3 for K (Sigee *et al.*, 1998). The frequent occurrence of Si in XRMA emission spectra from phytoplankton cells is of particular interest, suggesting that this element is widely associated with algal surfaces and that it is not restricted to the cell walls of diatoms and chrysophytes.

Other algae also show the wide range of concentrations seen in *Ceratium*, suggesting that elemental heterogeneity is an important aspect of environmental biodiversity within phytoplankton at the intra-specific level.

One of the most interesting aspects to have emerged from the XRMA studies is the central group of inter-correlated elements – Mg, P, K, S, referred to as the major correlation matrix. This statistical matrix probably reflects real elemental associations in relation to macromolecules – proteins and nucleic acids, that make up the main bulk of the algal biomass. Other routinely-detected elements such as Na and Cl are not normally part of the matrix, suggesting their separate location principally in vacuoles or cytosol. Si was also not part of the correlation matrix in any of the algae analysed, consistent with its presence as a discrete component at the cell surface.

Correlation matrix and Redfield ratio

Future XRMA studies including quantitation of N, C, and other elements may well extend the major correlation matrix to a much broader chemical assemblage. The existing XRMA studies themselves follow on from the early work of Ketchum and Redfield (1949), who established that algal biomass contains C:N:P in a fixed atomic ratio of 106:16:1 (about 42:7:1 by weight) – known as the Redfield ratio. Within certain limitations (Falkowski, 2000) the Redfield ratio is a constant feature of both photoplankton and zooplankton (freshwater and marine). One important aspect of the correlation matrix (including the Redfield associations) is that it relates algal chemical composition to environmental nutrient availability. Studies by Sigee *et al.* (1998) on seasonal changes in *Ceratium* showed marked alterations in elemental composition over a 3 month period. A sharp decrease in the concentration of intracellular P occurred during the early part of the sampling sequence, coinciding with environmental decreases in phosphorus availability and a rapid increase in cell population. Due to constraints imposed by the correlation matrix, the fall in internal P concentration was matched with falls in the concentration of Mg, K, and S, even though the environmental availability of these other elements was not limiting.

5.1.5 Nutrient availability and cycling in aquatic systems

Key nutrient elements such as carbon, nitrogen, and phosphorus are present in the aquatic environment in the four major states identified in Figure 5.1. Nutrients enter the food chain mainly in dissolved inorganic form by direct uptake into algae and bacteria. The availability of nutrients thus depends primarily on soluble inorganic concentrations within the ecosystem, which in turn depends on entry of nutrients into the ecosystem from outside (external loading, allochthonous derivation), entry into the water medium from the sediments (internal loading, autochthonous derivation), and nutrient cycling within the ecosystem.

The assessment of nutrient availability for freshwater microorganisms is usually made in relation to measured aquatic concentrations. New molecular approaches using reporter-gene technology (Porta *et al.*, 2003), however, open up the possibility of direct sensing by the organisms concerned (Section 3.3.2).

Nutrient cycling and recycling processes

'Nutrient cycling' refers to the sequence of nutrient transfers which occur between different chemical states (Figure 5.1) and between different ecological compartments (e.g., different groups of biota, detritus, water medium) in the freshwater system. Following initial uptake into biomass, the chemical state of the nutrient undergoes a sequence of changes, finally being released back into the water and reversion to the soluble inorganic (biologically-available) form with completion of the nutrient cycle. The regeneration of biologically-available nutrient is referred to as 'nutrient recycling' and is part of the general cycling process.

The cycling of particular nutrients (e.g., phosphorus) involves three main processes:

- uptake of soluble inorganic nutrients (e.g., phosphate) by freshwater algae and bacteria, with conversion to insoluble organic biomass;

- transfer of the nutrient from one organism to another through the food chain – during these transformations the nutrient is associated with a succession of macromolecules, which in the case of phosphorus includes nucleic acids and P-rich proteins;

- release back to the environment in a soluble organic or inorganic form. The release to water, and transformation of nutrients to biologically available forms, is referred to as remineralization and completes the nutrient cycle. As part of this process, some of the soluble organic nutrient is converted to soluble inorganic form (e.g., soluble organic phosphate to phosphate anions), the most readily assimilable nutrient state.

The conversion of insoluble organic material to soluble organic or inorganic molecules can occur in various ways:

- passive leakage or excretion of nutrients from algal cells – the release of soluble organic nutrients, also referred to as dissolved organic carbon (DOC) and dissolved organic material (DOM), is considered separately in relation to algal photosynthesis (Section 4.7.2) and bacterial productivity (Section 6.10);

- cell lysis by physical/chemical agents or viruses;

- breakdown of insoluble biomass and further hydrolysis of soluble organic compounds by algal or bacterial extracellular enzymes (proteases, phosphatases); and

- consumption of algae by consumers (protozoa, zooplankton, fish) and nutrient release by sloppy feeding, excretion, or leaching of faeces.

Nutrient cycling is important for the regeneration of nutrients in freshwater systems, and the continued growth of phytoplankton and benthic algae. In temperate climates the contribution of nutrient recycling to nutrient availability shows strong seasonal variation, as shown particularly by lake ecosystems.

Rate of nutrient recycling

With any aquatic system, the rate of recycling of a particular nutrient can be quantified as the turnover rate (k) within an ecosystem compartment (e.g., water, algae, detritus) per unit time. Nutrient recycling may also be quantified as the flux or organic nutrient remineralized per unit time (expressed as mass time^{-1}).

The rate of nutrient recycling depends on the rates of each of the key processes noted above – uptake, biomass transfer, and remineralization. This rate has important implications for primary productivity since one inorganic nutrient (CO_2) is directly required for carbon fixation, and other inorganic nutrients (e.g., nitrates, phosphates) are also required for biomass formation. In the hypothetical case of closed ecosystems, where fresh nutrients are not derived from inflow (external loading) or sediment (internal loading) sources, primary production would be a direct function of nutrient recycling. In practice, most ecosystems are clearly not closed, and nutrients for primary production are obtained from both external (allochthonous) and internal (autochthonous) sources. As a broad generalization, the greater the external supply of nutrients the less important is internal nutrient recycling in meeting biological requirements.

Nutrient cycling in streams

Streams are open ecosystems that receive the major input of nutrients from their catchment areas. This would suggest that internal cycling is not of major importance in these systems and that the supply of nutrients for primary productivity would not be dependent on internal recycling processes. A range of studies (summarized in Mulholland, 1996) suggest, however, that algal productivity is often nutrient-limited, indicating that nutrient recycling may be important after all. The apparent paradox may be explained in terms of the spatial parameters of the stream environment, where nutrient influx spreads throughout the whole body of water, but the primary producers (benthic algae) are limited to the substratum. These biota are able to assimilate nutrients only from their immediate microenvironment, leading to local limitation. Local nutrient recycling becomes

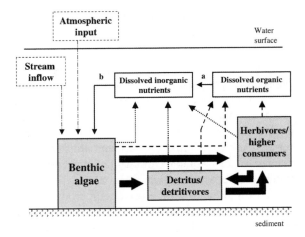

Figure 5.3 General scheme for nutrient cycling in streams (based on a figure from Mulholland, 1996)

–·–·► External derivation, from atmosphere and stream inflow

➡ Biomass change or transfer

– – –► Excretion or conversion to dissolved organic matter (DOM)

············► Excretion or conversion to dissolved inorganic nutrients (DIN)

———► a. Conversion of organic to inorganic nutrients (mineralization)

b. Uptake of soluble inorganic nutrients

especially important in this situation, particularly as remineralization occurs largely on the stream bottom – in close proximity to the algal population.

A general scheme for nutrient cycling in streams is shown in Figure 5.3, and involves all major groups of microorganisms, plus invertebrates and other benthic biota. Major uptake of dissolved inorganic nutrients (e.g., nitrates, phosphates) is carried out by benthic algae. Nutrient transfer to herbivores (including protozoa) and then higher consumers occurs via the food chain, and death of all types of biota leads to the accumulation of dead biomass (detritus). Insoluble biomass is converted to soluble organic material via secretory activity (algae), excretion (herbivores/ carnivores), and breakdown of detritus (by saprophytic fungi, bacteria, and protozoa). Viruses and parasitic fungi enhance the release of organic material during infection and ultimate host cell death. Dissolved inorganic nutrients are released by all groups of biota, and are also derived from dissolved organic nutrients by the activity of bacterial and algal extracellular enzymes.

The unidirectional flow of water adds a further dimension to nutrient availability in streams, displacing the nutrients downstream as they complete their cycle through the biota. The effect of river flow has been incorporated into quantitative aspects of the cycling process by introducing the notion of nutrient spiralling. In this concept, the distance travelled by a nutrient atom, while completing one cycle through the ecosystem, is called the spiralling length (S). The inverse of S (distance^{-1}) is a measure of the rate of nutrient cycling within the spatial dimension of the ecosystem; S^{-1} can be related to the nutrient turnover rate (k) and the downstream velocity of nutrient flow (v) by the equation:

$$S^{-1} = k \cdot v^{-1} \qquad (5.1)$$

Although water flow has an important general role in the dynamics of river systems by displacing both nutrients and freshwater biota downstream, there are localized regions within the general environment where this is minimal. These low-flow regions arise in various ways, and include the general benthic boundary layer occurring at the base of the water column (Figure 2.14) plus supplementary boundary layers which may be directly associated with benthic organisms. These are formed by the interaction of river flow with the biomass surface (Figure 5.7), and result in regions of relatively static water both inside and at the edge of benthic communities (Mulholland, 1996).

Boundary zones are important in localized nutrient cycling associated with benthic communities since they support the wide range of unattached organisms (algae, protozoa, invertebrates) that are involved in the cycling process – and would otherwise be swept away by the current. They are also important sites of nutrient retention, since both soluble and particulate nutrients remain within the locality, rather than being lost downstream. As noted earlier, localized nutrient cycling and nutrient regeneration are particularly important in river systems, where they may be a major factor in maintaining high levels of primary productivity.

Nutrient cycling in lakes

Cycling of nutrients in lake environments involves similar transformations to those seen in streams. The major differences are that initial uptake occurs into phytoplankton rather than benthic algal biomass, and water flow is so low that nutrient spiralling is not significant. The algal release of soluble photosynthetic products may also be more significant in lake systems, particularly under bloom conditions.

Seasonal variations in external loading, internal loading, water circulation, and nutrient cycling all contribute to changes in nutrient availability in temperate and mediterranean lakes (Figure 5.4). Major inflow via streams, ground water, and rain occur in winter, which is also a time of maximum distribution of nutrients throughout the water column due to vertical mixing. Little transfer of nutrients through biota (cycling) occurs at this time of year because biomass is low, and biological activity is minimal.

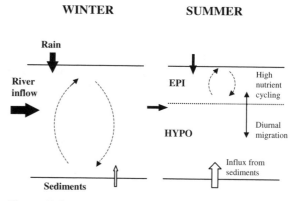

Figure 5.4 Seasonal variation in nutrient input, water circulation, and biological cycling in a eutrophic lake that has differences in winter/summer rainfall. *In winter*, nutrient inflow (external loading) exceeds biomass uptake and soluble nutrients accumulate. Nutrients are dispersed throughout the water column by circulation (---▶).

In summer, there is reduced inflow, but major influx occurs from sediments (internal loading). Circulation is restricted to the epilimnion (EPI). Nutrient removal by phytoplankton causes depletion in the epilimnion, but biological cycling leads to some regeneration. Some input also occurs from the hypolimnion (HYPO) as a result of diurnal migration

The reverse situation occurs in summer, where nutrient inflow is reduced, vertical mixing is limited due to thermal stratification but cycling is at a maximum. The process of nutrient cycling is enhanced by the vertical migrations of phytoplankton and zooplankton, with high nutrient uptake in the hypolimnion (nutrient source) and maximum transfer and release in the epilimnion (nutrient sink). In tropical lakes, nutrient concentrations show little seasonal variation and nutrient availability is more dependent on the biological process of cycling than on physical parameters.

More specific aspects of physicochemical and biological factors on elemental cycles are considered in the sections on phosphorus, nitrogen, and silicon in the freshwater environment.

5.2   Nutrient uptake and growth kinetics

Direct uptake of inorganic nutrients such as nitrate and phosphate occurs in relation to two main groups of freshwater biota – algae and bacteria. Other organisms obtain their inorganic quota almost entirely via the food chain. This section deals primarily with inorganic nutrient uptake in algae, though the quantitative empirical relationships apply equally to bacteria.

5.2.1   Empirical models for algal nutrient kinetics

Under appropriate experimental conditions, the uptake of inorganic nutrients by algae and the resulting growth of these organisms closely relate to external (surrounding water) concentrations. This can be demonstrated in the laboratory context, where nutrient uptake and growth rate have been widely studied in relation to single algal species (monoculture), and single nutrients such as nitrate or phosphate. These studies have led to the development of three empirical models (Figure 5.5) for the nutrient-growth

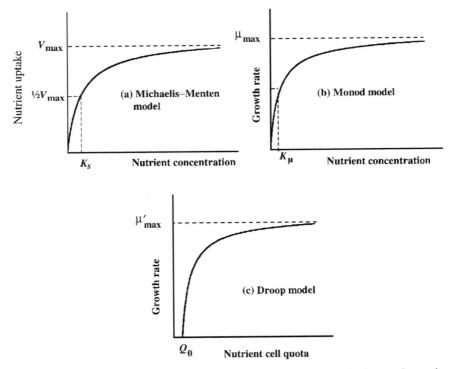

Figure 5.5 Graphical representation of algal nutrient uptake and nutrient-limited growth as described by the Michaelis–Menten, Monod, and Droop models

relationships of algae (Borchardt, 1996). These models relate nutrient uptake to external nutrient concentrations (Michaelis–Menten model), cell growth to external nutrient concentrations (Monod model), and cell growth to internal nutrient concentrations (Droop model).

Michaelis–Menten model for nutrient uptake

The initial model (Figure 5.5(a)) was developed by Dugdale (1967), who proposed that uptake of nutrients by algal cells conformed to a rectangular hyperbola function which was mathematically equivalent to the Michaelis–Menten formula for enzyme kinetics:

$$V = V_{max} \frac{S}{K_S + S} \qquad (5.2)$$

where V = rate of specific nutrient uptake (time^{-1}), V_{max} = maximum specific uptake rate, S = substrate concentration (μmoles l^{-1}), and K_S = the substrate concentration that yields half the maximum uptake rate.

The use of Michaelis–Menten kinetics was particularly attractive to researchers studying nutrient relationships, since only two parameters were needed to define the nutrient acquisition and growth potential of a particular alga. The half-saturation constant (K_S) provides a measure of the ability of the alga to take up nutrient at low external concentrations, and the specific uptake rate (V) gives a value for nutrient assimilation and potential algal growth rate under particular environmental conditions. Evidence from laboratory studies quickly established that Michaelis–Mentem kinetics could be applied to the algal uptake of a variety of nutrients including nitrate, phosphate, silicate, and vitamin B$_{12}$. The possibility of using a single model for both uptake and growth was challenged, however, when it became clear that nutrient uptake and growth were coupled only under steady-state conditions such as occur in a chemostat.

Monod model for cell growth

Various workers proposed that algal growth could be more appropriately modelled by the Monod equation. This has been developed in relation to bacterial growth, and is analogous to the Michaelis–Menten equation:

$$\mu = \mu_{max} \frac{S}{K_\mu + S} \qquad (5.3)$$

where μ = specific growth rate (time^{-1}, μ_{max} = maximum growth rate, S = substrate concentration (μmoles l^{-1}), and K_μ = half saturation constant for growth for the limiting nutrient.

Graphical expression of the Monod equation is shown in Figure 5.5(b). The relationship between the half-saturation constants for nutrient uptake (K_S) and growth (K_μ) depends on the extent to which the alga can utilize or vary internal stores under conditions of nutrient limitation. When the capacity for varying intracellular stores is considerable, as with phosphorus, then K_μ is much less than K_S. When the range of storage capacity at low nutrient levels is small, as in the case of carbon, then K_μ approaches the value for K_S (Borchardt, 1996). The $K_\mu : K_S$ ratio for a particular element thus reflects the ability of an alga to vary its intracellular storage under limiting conditions, and may have adaptive significance for particular environmental situations.

Applications of the Monod model to characterize specific growth rates worked well with some elements (e.g., carbon) but not others (e.g., phosphorus), where deviations from the formula were reported at low growth rates. These deviations may have arisen in part from problems in accurately measuring very low concentrations of nutrients, and in part from multiphasic growth kinetics where the growth substrate relationship changes with substrate concentration.

In spite of these limitations, the Monod model has been successfully applied to data derived from short-term batch–culture experiments, and has been very useful in predicting competitive interactions between algae, since:

- μ_{max}, defines the maximum growth rate when the nutrient is not limiting, and predicts competitive ability at high nutrient concentration, and

- K_μ, the half-saturation coefficient, is a measure of the relative ability of a species to use low levels of nutrient, and predicts competitive ability at limiting nutrient concentrations.

Droop model for cell growth

Problems in modelling nutrient uptake and algal growth with the Michaelis–Menten and Monod models led to the development of a further model (Figure 5.5(c)) for nutrient-limited algal growth, based on the concept of the cell quota (Q). This is the total concentration of the nutrient within the cell and includes soluble nutrient that is present via short-term uptake (Michaelis–Menten equation) plus intracellular reserves which have been derived over a longer time period. Although ability to take up nutrients at limiting concentration is important, the presence of intracellular reserves is also significant and contributes to the total or cell quota (Q).

The Droop model proposes that algal growth is a function of the total intracellular concentration (Q) of the limiting element, where:

$$\mu = \mu'_{max}(1 - Q_0/Q) \qquad (5.4)$$

and Q = cell quota of the growth limiting element, Q_0 = minimum internal concentration required for growth (i.e., value for Q when $\mu = 0$), and μ'_{max} = maximum growth rate at infinite Q.

The ability of algal cells to store nutrients at high external concentration means that the maximum growth rate defined by the Droop equation (μ'_{max}) is greater than that of the Monod model (μ_{max}) by a factor characteristic of the particular element. For a particular alga, the value for the μ'_{max}/μ_{max} ratio indicates the range of storage capacity that occurs at maximum uptake. This ratio is nutrient-specific, and is analogous to the $K_\mu : K_S$ ratio at the other end of the substrate concentration scale.

By using the cell quota (Q), the Droop equation relates cell growth to both nutrient uptake (external substrate concentration) and internal reserves. The cell quota for a particular nutrient is considered later in relation to luxury consumption and storage of phosphorus (Section 5.8.2).

5.2.2   Competition and growth in the aquatic environment

One of the major benefits in studying the nutrient growth-response of algal species in the laboratory is that the results obtained can be used to assess competitive interactions and growth at the environmental level. This has proved useful with both planktonic and benthic algal communities.

Competition for nutrients in lake phytoplankton

The empirical relationships for nutrient kinetics involved responses of single algal species to changes in the external concentration of individual nutrients. The environmental situation is obviously more complex than this, with mixed populations of species competing for a range of nutrients several of which may be at low concentration. Nutrient competition in the environment can be considered initially in terms of a single nutrient at low level, then for multiple low availability.

Growth and competition in relation to a single nutrient   Comparison of μ_{max} and K_μ values (Monod equation) for algae within mixed populations gives a prediction of their relative growth rates at different nutrient concentrations and thus their ability to compete. In a mixed population of two algae, the species with the lowest K_μ value will dominate at low nutrient concentration, while the species with highest μ_{max} will dominate when the nutrient is non-limiting. An example of how this might operate is given in Figure 5.6, where two algal species are competing for phosphorus.

Growth and competition in relation to multiple nutrients   Prediction of the competition outcome with small numbers of species in relation to a single limiting nutrient is relatively straightforward. The situation becomes more complex when two or more nutrients are involved.

An important rule of algal nutrient kinetics is that growth of an algal species can only be limited by one nutrient at a time (principle of single nutrient limitation). Thus, although two or more nutrients

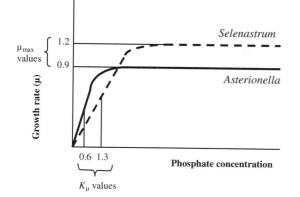

Figure 5.6 Competition between two algae for a single nutrient: phosphorus uptake and growth kinetics (Monod curves) are shown for a green alga (*Selenastrum*) and a diatom (*Asterionella*). At low phosphate concentrations *Asterionella* out-competes *Selenastrum* due to its lower K_μ value. At high phosphate concentrations, *Selenastrum* out-competes *Asterionella* because of its higher μ_{max} (figure based on data from Tilman and Kilham, 1976, and Brown and Button, 1979)

may be near to growth-limiting concentration, only one will be actually limiting. Laboratory studies on mixed algal populations exposed to 'simultaneous limitation' with different nutrient combinations showed sharp transitions in terms of limitation by one or another nutrient. These experiments indicate that algae respond to multiple depletion of nutrients in a threshold manner. The threshold occurs at the nutrient balance that matches the growth requirement of the algal species. If a particular species has optimum growth at an N:P ratio of 18:1, then the species will be P-limited at an external concentration ratio of 19:1 and N-limited at an external ration of 17:1. A change in the external balance of these elements from 19:1 to 17:1 will thus result in a switch from P-limitation to N-limitation. Optimum ratios of particular nutrient pairs vary between algal species and are a fundamental aspect of differential nutrient removal, which is considered in resource ratio partitioning theory.

The ecological significance of optimum nutrient ratios for algal competition can be seen in the seasonal succession of the diatoms *Asterionella* and *Cyclotella* (Figure 5.25), where changes in external concentrations of phosphorus and silicon lead to major temporal changes in algal dominance. In this situation, the Monod equation can be used to determine the optimum nutrient balance for each species which is required to achieve maximum growth.

Although conditions in the environment are rarely steady-state, the Monod model does appear to provide a good prediction of competition between algae and highlights the role of inter-specific competition in structuring phytoplankton communities. In addition to seasonal changes (see above), competition between *Asterionella* and *Cyclotella* along a natural gradient of silicate and phosphate in Lake Michigan also shows a close spatial fit to predictions based on the Monod model (Tilman, 1977).

Competition for resources in benthic communities

The growth of periphyton is strongly influenced by light, space, and nutrient availability (Lowe, 1996). Nutrients may be derived either from the water column or from the substratum as follows.

- *Water column*. Increases in water-column nutrients may lead to increases in benthic algal biomass, but the secondary effect of light shading due to increased phytoplankton growth may lead to an inverse relationship.

- *Substratum*. Nutrient studies on periphyton communities have generally shown that substratum or algal mat chemistry have a greater impact on growth than water-column chemistry.

The use of microelectrodes and nutrient manipulation experiments have enabled researchers to measure and manipulate nutrient availability over short distances within periphyton communities. This has led to the appreciation of the algal mat as a true community of species, with each species responding differently to resource availability.

Within periphyton communities, individual species have been identified with the specialized ability to access particular resources which are limiting to other algae. In lentic periphyton (Fairchild *et al.*, 1985), such 'resource specialists' include:

- phosphorus specialists – species of *Achnanthidium* and *Gomphonema* (diatoms),

- nitrogen specialists – species of *Epithemia*, *Rhopalodia* (diatoms), and *Anabaena* (blue-green alga), and

- space-light specialists – naviculoid diatoms and *Stigeoclonium tenue* (blue-green alga).

Resource specialists usually employ special mechanisms in the exploitation of one particular resource. Space-light specialists, for example, are able to grow rapidly to optimize space and light but require relatively high concentrations of phosphate and nitrate. Some nitrogen specialists have the specific ability to fix atmospheric nitrogen. Particular lentic microhabitats favour particular resource specialists. The epiphytic diatom *Epithemia adnata*, a nitrogen specialist, often dominates microhabitats where the ratio of available nitrogen/phosphorus is relatively low.

Although nutrient resource limitation and exploitation operate at the species level, the response of the whole community represents the sum of all the species activities. Within the periphyton community, which includes bacteria, fungi, and other microorganisms (in addition to algae), the whole-community response to nutrients may be highly complex.

5.2.3 Nutrient availability and water movement

Nutrient availability and kinetics are both influenced by water movement around algal cells. Application of the Monod model, relating algal growth to external nutrient concentrations, depends on steady-state conditions – including constant external substrate concentration and constant physical conditions of light, temperature, and water movement. Water movement around algal cells is important in both planktonic and benthic environments. Within the water column, it occurs during turbulent mixing

and also during sedimentation of single cells and colonies. In lentic environments, current flow is important in relation to benthic algae – including both filamentous forms and algal biofilms.

Under conditions of water movement, there is a gradient of water velocity around the algal cell, ranging from zero (at the surface) to the velocity of the current (further out). This external shell of decreasing water velocity is termed the 'boundary layer,' and has been defined as the region with a velocity gradient from 0–90 per cent of the free external velocity (Borchardt, 1996). The boundary layer can typically be divided (Figure 5.7) into an

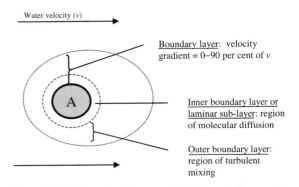

Figure 5.7   Boundary layer and diffusion shell around a stationary algal cell (A) which is exposed to a water current. In the complete absence of water movement, the laminar sub-layer simply occurs as a static diffusion shell around the algal cell. Water movement creates a boundary layer and reduces the thickness of the laminar-sublayer. This decreases the distance over which molecular diffusion has to occur and increases nutrient availability

outer layer of turbulent mixing and an inner boundary layer (or laminar sub-layer) of diminished water movement involving laminar flow.

The laminar sub-layer is a shell of relatively static water which is particularly important for nutrient nutrient kinetics. Passage of nutrients across this region is mediated by molecular diffusion and occurs along a concentration gradient generated by nutrient uptake at the cell surface. Increased water movement around the cell reduces the thickness of the laminar sub-layer (which varies from 10–100 μm thick) and thus increases the availability of nutrients for uptake at the cell surface.

5.2.4   Acute nutrient deprivation as an environmental stress factor

Acute limitation in inorganic nutrients such as nitrogen and phosphorus can act as an important environmental stress factor and lead to major physiological changes in both heterotrophic and autotrophic microorganisms.

In Gram-negative bacteria, the effect of nutrient deprivation is seen in the induction of the starvation response, with fundamental changes in gene activity and cell physiology (Section 6.6.4).

Major starvation-induced physiological changes also occur in micro-algae, with alterations in biochemical and elemental composition (Lynn *et al.*, 2000) and changes in reproduction. These organisms grow and reproduce vegetatively (by division or fragmentation) under normal conditions, but typically carry out sexual reproduction and produce resistant spores (zygospores) under situations of acute nutrient limitation. This transition is important for the generation of genetic variability (sexual reproduction), for survival during adverse conditions and for the production of zoospores (dispersal phase) at the time of zygospore germination.

Conditions of nitrogen starvation and specific light regimes have been shown to be an important environmental trigger for the induction of gene activity and related gamete formation in various unicellular algae (Table 3.5, Section 3.3.2). Similar conditions influence the growth and development of multicellular algae. The filamentous green alga *Rhizoclonium riparium*, for example, produces reproductive initials when stressed – usually by nutrient limitation or extreme temperatures. Thick-walled akinetes, containing large starch reserves, are produced as a resting stage, and are released on the death of the parent cell to germinate when conditions become favourable. Germination of the akinete normally results either in the direct formation of a vegetative filament or in the production of zoospores. Functional gametes are rarely produced in this alga (Hall and Walmesley, 1991).

A.   NITROGEN

5.3   Biological availability of nitrogen in freshwater environments

Nitrogen is an essential component of all living organisms, occurring at about 5 per cent of the dry weight. This element is incorporated into proteins, nucleic acids, and many other biomolecules, where it is present in oxidation state-III (as in NH_3). Nitrogen is found within fresh water habitats in a wide range of forms (Table 5.4), of which the soluble states are important in uptake and assimilation by microorganisms.

5.3.1   Soluble nitrogenous compounds

Nitrogen is distinguished from other key elements such as phosphorus and silicon in occurring as a gas (N_2), anion (NO_3^-, NO_2^-) and cation (NH_4^+) in the soluble inorganic state.

Table 5.4   Forms of nitrogen in water

	Organic	Inorganic
Soluble	*Dissolved organic nitrogen* Autochthonous: urea – $CO(NH_2)_2$, proteins, and nucleic acids Allochthonous: humic and fulvic acids	*Dissolved inorganic nitrogen* Anions: nitrate, nitrite Cation: ammonia dissolved gas: N_2
Insoluble	Complex biomass – living and dead biota	Largely particulate material derived from rocks and sediments

Dissolved inorganic nitrogen: growth of autotrophic and heterotrophic microorganisms

During the autotrophic growth of freshwater algae and photosynthetic bacteria, the requirement for nitrogen is met in two ways:

- assimilation of dissolved nitrate, nitrite, or ammonia,

- fixation of dissolved molecular nitrogen (N_2).

These constituents form the dissolved inorganic nitrogen (DIN) of the aquatic environment, with nitrate being the major constituent in most types of freshwaters.

Nitrogen uptake by heterotrophic bacteria occurs by assimilation of either inorganic or organic soluble nitrogen (see below).

Dissolved organic nitrogen: heterotrophic growth

Dissolved organic nitrogen (DON) is a heterogeneous group of organic compounds which ranges from well-defined constituents such as urea and amino-acids to more complex (and poorly characterized) compounds such as humic and fulvic acids (Table 5.4). Much of the nitrogen functionality in DON occurs as amide groups, though heterocyclic nitrogen compounds such as pyrroles and pyridines may also be important. Although phytoplankton uptake of nitrogen does not normally involve DON, a wide range of heterotrophic (saprotrophic) organisms, including bacteria, protozoa, and algae (Section 3.10.3) are able to assimilate low MW nitrogenous organic compounds directly from the environment.

Seasonal variation in nitrate availability

Nitrogen may enter water bodies in two main ways (Figure 5.4) – external loading (where it is derived by inflow from other water bodies, rain, or via air) and internal loading (where it comes from internal sources, particularly sediments). The entry of nitrogenous compounds, and their removal by lake biota (particularly algae) often follows a marked seasonal

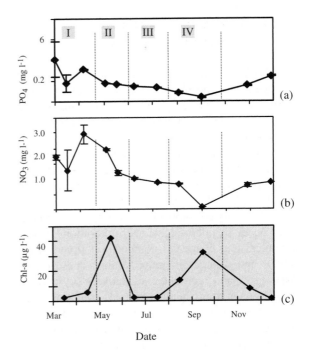

Figure 5.8 Seasonal changes in the concentrations of phosphate (a) and nitrate (b) in a eutrophic lake (Rostherne Mere, UK). The seasonal cycle (1999) is divided into four phases in relation to the phytoplankton sequence (c). Concentrations of inorganic nutrients reach a maximum value towards the end of the winter phase (I), falling rapidly with the development of the spring diatom bloom (II). Concentrations show little change during the clear-water phase (III), but become reduced to minimal levels with the growth of the late summer/autumn bloom (IV). De-stratification, with increase in soluble nutrients, occurs in October (taken, with permission, from Levado, 2001)

pattern. This is shown in Figure 5.8 for a typical temperate freshwater lake, where major external loading in winter months, coupled with low levels of uptake by biota, leads to a positive build up in biological available material, particularly nitrates. In summer, dryer weather typically leads to a reduced external loading. This is partly compensated for by an increased internal loading, arising due to the release of ammonia, nitrites, and nitrates from the sediment due to increased bacterial activity. Increased uptake by algae and other organisms frequently leads to a sharp decline in available nitrogen during summer. In stratified lakes, this occurs

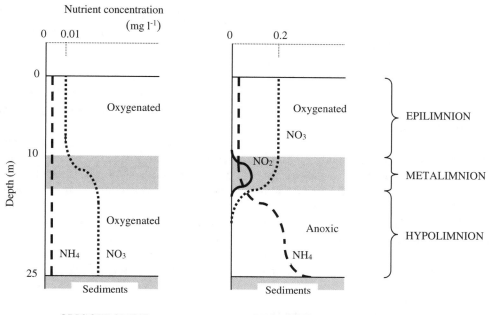

Figure 5.9   Distribution of nitrate, nitrite, and ammonia in the stratified water column of oligotrophic and eutrophic lakes in midsummer (based on a figure from Horne and Goldman, 1994). Nitrites are almost entirely absent from the water column of the oligotrophic lake. The low nitrate levels of the oligotrophic example are typical of a lake which is both nitrate and phosphate-limited (e.g., Lake Tahoe, USA). Other oligotrophic lakes (e.g., Lake Superior, USA) are only phosphate-limited and would have higher nitrate values

particularly in the epilimnion, where greatest removal via biomass in the top part of the water column is coupled with restricted vertical diffusion due to the limitations of stratification.

The vertical distribution of different nitrogen species also varies in stratified lakes with the trophic status of the system, and the degree of oxygenation which is allowed by heterotrophic bacterial activity (Figure 5.9). Once stratification has become established, eutrophic lakes typically show high levels of nitrogen throughout the water column, occurring as nitrate in the oxygenated epilimnion, but as nitrite and ammonia in the metalimnion and hypolimnion respectively due to the anoxic conditions present. In oligotrophic lakes, where often oxygenated conditions prevail throughout the water column, soluble nitrogen is present mainly as nitrate – with only low levels of ammonia and nitrites occurring.

Nitrogen limitation

As one of the major elements required for phytoplankton and benthic algal growth, nitrogen may occur at limiting concentrations in lakes and streams under particular circumstances. In standing waters, this is particularly the case for eutrophic lakes with high P concentrations and for oligotrophic lakes with low P and N. In general, nitrate levels of less than $100\,\mu g\,l^{-1}$ are limiting, with no limitation at $400\,\mu g\,l^{-1}$. Concentrations of $10\,000\,\mu g\,l^{-1}$ are typically in excess of growth requirements.

Nitrogen limitation not only has implications for general growth, but may also trigger more specific effects (Section 5.2.4). The role of nitrogen in controlling sexual processes in microscopic green algae, for example, has been widely investigated in laboratory cultures. Gene activation and related sexual processes are typically induced in these organisms by lowering the concentration, or completely eliminating, the source of available nitrogen (Section 3.3.2).

5.4   The nitrogen cycle

As with other key elements (Case Study 5.1) nitrogen undergoes a complex sequence of biologically-driven transformations in the freshwater environment. These are known collectively as the nitrogen cycle, a schematic diagram of which (with particular reference to standing waters) is shown in Figure 5.10. The cycle can be considered as five major routes of nitrogen transformation: nitrate

entry and uptake, biomass transformations, remineralization, nitrification/denitrification, and nitrogen fixation.

5.4.1   Nitrate entry and uptake (soluble inorganic to insoluble organic nitrogen)

Nitrate is the major biologically-available form of nitrogen. It enters aquatic systems via rain and soil, passing from rivers to lakes, and is then taken up by algae – which constitute the major biomass within the aquatic system. Conversion of nitrate to algal protein involves a transformation from inorganic to complex organic (combined with carbon compounds) nitrogen. Details of nitrogen uptake by algae are given in Section 5.5.

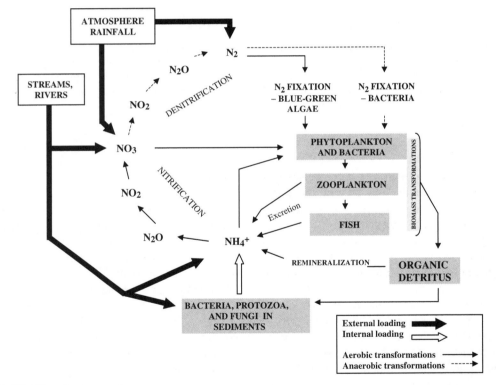

Figure 5.10   The nitrogen cycle.
Major nutrient flows of the freshwater nitrogen cycle are shown diagrammatically for a typical lake system. Various aspects such as the microbial loop and the effect of viral and fungal parasites, are omitted from the figure. External loading occurs mainly via streams and rivers. Atmospheric contribution includes nitrogen gas (microbial fixation) and aerial deposition (rain, aerosols, and dust particles)

5.4.2 Complex organic nitrogen (biomass) transformations (successive states of insoluble organic nitrogen)

Complex organic nitrogen in algal and bacterial biomass is subsequently broken down and reconverted to other forms of biomass as part of the trophic sequence which involves consumption by herbivores, and primary and secondary carnivores. The organic nitrogen thus passes along the food chain, ultimately ending up as detritus which passes to the sediments.

In addition to the classic food chain, biomass conversions and nitrogen transfer also occur via parasitic activity and the microbial loop. Parasitic activity particularly involves the effects of fungi (on phytoplankton) and viruses (on phytoplankton and bacteria). Within the microbial loop (Section 1.5.2), biomass conversions involve the loss and uptake of algal DOC, and the ingestion of bacteria by protozoon populations.

5.4.3 Remineralization (insoluble organic to soluble inorganic nitrogen)

Most organic nitrogen in aquatic ecosystems is present as plant or animal nitrogen within living organisms or dead material (organic detritus). The latter is mainly particulate and is converted to dissolved organic nitrogen (DON) by heterotrophic bacteria. The conversion of complex organic nitrogen to (inorganic) ammonia is the reverse of the previous process and is the first step known in remineralization – the transformation of an organic nutrient back to a biologically available inorganic form (Section 5.1.5). The conversion of ammonia to nitrate constitutes the final step in the remineralization process.

The overall process of oxidation of organic matter by aerobic heterotrophic microorganisms (using dissolved oxygen as the terminal electron acceptor) and return of nutrients to the inorganic form can be expressed by the equation (Kalff, 2002):

$$\overset{\text{Biomass}}{\overbrace{(CH_2O)_{106}(NH_3)_{16}(H_3PO_4)}} + 138O_2 \longrightarrow$$

$$106CO_2 + 16HNO_3 + H_3PO_4 + 122H_2O \quad (5.5)$$

(where $\Delta G_0{}^1$, the maximum amount of free energy generated by the reaction, is $-3190\,\mathrm{kJmol^{-1}}$).

In lakes, synthesis of organic nitrogen (biomass) occurs primarily in the euphotic zone while breakdown processes (leading to remineralization) take place mainly as a result of bacterial, fungal, and protozoon activity in the lower part of the water column and in the sediments.

In addition to enzymatic breakdown and degradation of biomass, remineralization may also occur by other processes (Case Study 5.1) such as excretion of nitrogen-rich metabolites by zooplankton and fish, and cell lysis resulting from parasitic infection.

5.4.4 Nitrification/denitrification (oxidation/ reduction of soluble inorganic compounds)

Major transformations of soluble inorganic compounds occur on lake and river sediments. Remains of all types of biota (detritus) sink to the bottom of the water body, decomposing to ammonia. This is converted under aerobic conditions to nitrate (nitrification). Conversion of nitrate to nitrogen occurs under anaerobic conditions, and involves a major loss of available nitrogen (denitrification) within the aquatic system.

Nitrification

The process of nitrification is largely carried out by chemosynthetic bacteria (Table 6.2) in the genera *Nitrosomonas* and *Nitrobacter*. The biological oxidation of ammonia to nitrate occurs as a two-stage process:

$$NH_4^+ \xrightarrow[\Delta G_0{}^1 = -275\,\mathrm{kJmol^{-1}}]{\text{\textit{Nitrosomonas}}} NO_2^- \xrightarrow[\Delta G_0{}^1 = -75.8\,\mathrm{kJmol^{-1}}]{\text{\textit{Nitrobacter}}} NO_3^-$$

$$(5.6)$$

The free energy released by these oxidation reactions ($\Delta G_0{}^1$) is used for the synthesis of organic matter, and is considerably greater in the first stage (nitrite formation) of nitrification.

The overall process of nitrification can be expressed:

$$NH_4^+ + 2O_2 \longrightarrow NO_3^- + H_2O + 2H^+ \qquad (5.7)$$

The consumption of two moles of dissolved oxygen (DO) for each complete conversion of NH_4^+ to NO_3^- exerts a major demand on the oxygen pool. Removal of DO by this process can lead to anaerobic hypolimnia and sediments in lakes. Severe oxygen depletion may also occur in rivers and streams, particularly downstream of sewage treatment outfalls, where large quantities of NH_4^+ and organic matter are released.

Denitrification

Denitrification is the reduction of nitrogen oxides (NO_3^-, NO_2^-) to dinitrogen gas (N_2), with gaseous oxides (NO, N_2O) as intermediate products. The process is largely carried out by facultative anaerobic bacteria in the genus *Pseudomonas*, but other bacteria (*Achromobacter*, *Bacillus*, *Micrococcus*) and some fungi may also be involved. As facultative anaerobes, all of these microorganisms have the ability to use dissolved oxygen as the terminal acceptor in aerobic conditions, but can use NO_3^- as the terminal acceptor in respiration when DO becomes

limiting. The anaerobic oxidation of organic matter during the process of denitrification, with the formation of gaseous nitrogen, can be expressed by the equation (Kalff, 2002):

$$\overbrace{(CH_2O)_{106}(NH_3)_{16}(H_3PO_4)}^{\text{Biomass}} + 94.4HNO_3 \longrightarrow$$

$$106CO_2 + 55.2N_2 + H_3PO_4 + 177.2H_2O$$

$$(\Delta G_0^1 = -3030\ kJmol^{-1}) \qquad (5.8)$$

The process of nitrification leads to the formation of a readily available source of nitrogen which can be taken up by algae and bacteria. Denitrification, on the other hand, results in the removal of this available nitrogen – with conversion to a form (N_2) which is ultimately lost from the aquatic system.

The processes of nitrification and denitrification are closely linked, both biochemically and spatially, within the aquatic system. Nitrification typically occurs on the aerobic side of the aerobic/anaerobic interface which occurs within the water column (oxycline) or on the sediments. In the sediments of oligotrophic lakes (Figure 5.11), nitrification takes place in the upper aerobic regions, where there is a high supply of NH_4 resulting from biomass breakdown. Denitrification occurs lower down, on the anaerobic side of the interface, where there is a

Figure 5.11 Role of the sediment in nitrification and denitrification in an oligotrophic lake. Sedimentation (se) of dead phytoplankton cells (detritus) leads to biomass accumulation and decomposition (de). Ammonia is formed in the anaerobic sediments, and is mainly converted to nitrate by nitrification (ni) just above the aerobic/anaerobic interface, though some ammonia diffuses into the aerobic sediments and water column. Nitrate is denitrified (dni) to gaseous nitrogen just below the aerobic/anaerobic sediment interface (diagram not to scale)

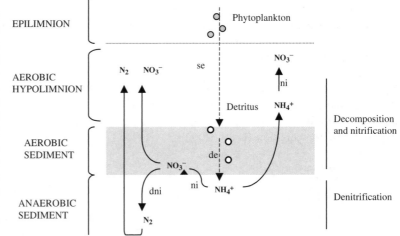

supply of nitrate from nitrification to act as the electron acceptor for respiration. Nitrate (from nitrification) and nitrogen gas (from denitrification) diffuse from the sediments into the water column.

Sediments of oligotrophic lakes, wetlands, and estuarine systems, where there is an aerobic surface and anaerobic subsurface, thus exhibit both nitrification and denitrification. In eutrophic lakes, where both sediment and hypolimnion are anaerobic, denitrification occurs throughout the lower part of the water body, and nitrification in the aerobic parts of the water column.

5.5   Uptake of nitrate and ammonium ions by algae

Uptake of readily available nitrogen (mainly as nitrate or ammonium ions) into algal cells (and aquatic macrophytes) is the first major step in the incorporation of nitrogen into the freshwater ecosystem biomass. Of the two forms of available nitrogen, ammonia is the preferred choice since energy is not required for uptake and there is no induction period of enzymes required for assimilation. Urea, which sometimes reaches appreciable concentrations in natural waters, is utilized as readily as ammonium ions.

5.5.1   Biochemical processes

Assimilation of nitrate involves the active uptake of this anion into the algal cell, followed by its conversion to ammonia using the reducing power of photosynthesis. Subsequent formation of amino-acids occurs in the plastid and is mediated by two key enzymes – glutamine synthetase and glutamate synthase (also named glutamine–oxoglutarate amino transferase, abbreviated to GOGAT).

The reduction of nitrate inside the algal cell occurs as a two-stage process as follows (Figure 5.12).

- Nitrate is converted to nitrite in the cytosol by the enzyme nitrate reductase, which requires molyb-

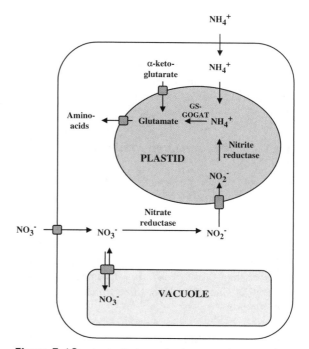

Figure 5.12   Assimilation of nitrate and ammonium ions by a eukaryote algal cell (based on a figure from Heldt, 1997).

Nitrate is taken up by the algal cell and can be temporarily stored in the vacuole or reduced in the cytosol to nitrite, which then passes into the plastid for further reduction to ammonium ions. These are subsequently converted to glutamine (enzyme glutamine synthetase, GS) then to glutamate by reaction with α-ketoglutarate (enzyme glutamate synthase, abbreviated to GOGAT). Selective transfer of metabolites across compartmental boundaries is indicated by ▬▣▶

denum as cofactor. Reducing power for NO_3^- reduction may be supplied from the plastid via reduced triose phosphates and/or dicarboxylates as demonstrated in higher plants (Larsson *et al.*, 1985).

- The NO_2 thus formed is further reduced to NH_4^+ in the plastid, using ferredoxin as electron donor – with six electrons derived directly from non-cyclic electron transport. This conversion is mediated by the enzyme nitrite reductase, which is localized to the plastid and requires Fe and Cu as cofactors.

Ammonium ions are taken up and transported directly to the plastid for conversion to amino-acids via the GS-GOGAT pathway.

5.5.2 Species variations in nitrate uptake

The rate of nitrate uptake differs considerably between algae, varying with:

- *species*: some species are genetically adapted for high rates of nitrate uptake – species of *Chlorella*, for example, are reported to be more active than those of *Microcystis* in nitrate incorporation since they have three times more nitrate reductase;

- *cell size*: smaller-sized species have a competitive advantage in nitrogen-limiting conditions due to their higher surface area/volume ratio, resulting in a more rapid diffusion per unit volume compared with larger-celled species;

- *nitrate pre-conditioning*: some algae are able to adapt to low levels of nitrate by adjusting the threshold of nitrate reductase induction.

5.5.3 Environmental regulation of nitrate assimilation

In the freshwater environment, as with the terrestrial situation, nitrate assimilation must be tightly regulated to avoid overproduction of amino-acids or build-up of toxic intracellular levels of nitrite. Studies on higher plants (Heldt, 1997) have shown that regulation of nitrate assimilation occurs through control of the level of the enzyme nitrate reductase present in the cytosol. This can be regulated in two ways – control of gene expression and reversible activation/inactivation of the enzyme molecule (Figure 5.13).

Control of gene expression

Nitrate reductase is a very short-lived protein (half-life a few hours), so that regulation of its synthesis provides an effective strategy for controlling its levels within the cell. Various environmental factors control the synthesis of the enzyme at the level of gene expression, some of which promote synthesis (nitrate concentration, light, glucose, and other carbohydrates) while others are inhibitory (glutamine and other amino-acids). These factors appear to act via receptors which monitor nitrate reductase demand in relation to the requirement for amino-acids and to the supply of carbon skeletons from CO_2 assimilation. Activation of nitrate reductase in

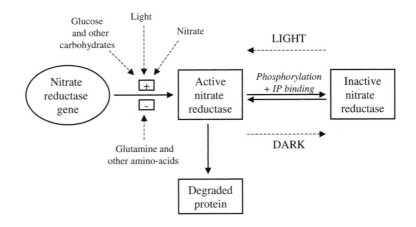

Figure 5.13 Regulation of nitrate reductase (NR) in algal cells
Induction of NR synthesis and degradation of protein both occur over a period of hours. Dark inactivation involves reversible phosphorylation and binding of an inhibitory protein (IP) and occurs within minutes (based on a figure from Heldt, 1997)

the light, for example, means that there is a direct coupling of nitrate assimilation and photosynthesis so that carbon skeletons will be immediately available for amino-acid formation in the plastid.

Activation/inactivation of nitrate reductase

Control of gene expression allows regulation of enzyme activity over a time span of hours. This would not be sufficiently rapid to prevent accumulation of nitrite during sudden shading or decrease in insolation. Rapid inactivation of nitrate reductase over a period of minutes can be achieved, however, by dark-induced phosphorylation, followed by binding of an inhibitor protein to the phosphorylated molecule. This inactivation is reversed by light and provides a sensitive fine-tuning to the environmental control of nitrate assimilation.

5.5.4 Nitrogen uptake, CO₂ assimilation, and photosynthesis

Uptake of NO_3, NO_2, and NH_4 has important implications in relation to CO_2 assimilation for two reasons – competition for photosynthetic reducing power and alteration of CO_2 availability in the surrounding microenvironment.

Competition for photosynthetic reducing power

Next to CO_2 fixation, the photosynthetic reduction of NO_3^- to NH_4^+ is quantitatively the most important reduction process in the algal cell. With both CO_2 and NO_3^- assimilation requiring photosynthetic activity, it might be expected that competition for reducing capacity would occur between these processes. This does not normally occur, however, since CO_2 assimilation and nitrate assimilation have to be matched to each other. Nitrate assimilation can only progress to completion when CO_2 assimilation provides the carbon skeletons for amino-acid formation, lack of which may lead to accumulation of toxic nitrites in the algal cell. As noted earlier, control of nitrate assimilation is closely coupled

to processes that promote carbohydrate formation in algal cells – including the external presence of glucose (and other carbohydrates) and light.

Laboratory studies on cultures of the green alga *Scenedesmus* have confirmed that competition between these processes does not occur at moderate to high light intensities, but that it is important at low light intensities (Larsson *et al.*, 1985). Competitive interactions between carbon and nitrogen assimilation were pronounced under these low light conditions, where the rate of non-cyclic electron transport becomes limiting for reductive purposes. These results may have important environmental implications where variations in light intensity occur and suggest that increasing depth in the water column, for example, will have major effects on the balance between these assimilation processes.

CO₂ availability

The form of inorganic nitrogen which is taken up by algae that are active in photosynthesis also directly affects the pH of the surrounding water medium and indirectly influences CO_2 availability. Laboratory studies on the green alga *Scenedesmus* (noted previously) demonstrated a 3 hour rise in pH to 9.7 when NO_3^- was the main source of nitrogen (due to the generation of OH^- ions), but little increase (not exceeding pH 7.5) with NH_4^+ ions. With this substrate, release of H^+ ions moderates the pH increase that normally occurs with high rates of photosynthesis and increases CO_2 availability via the pH-sensitive CO_2/carbonate equilibrium of freshwaters.

In view of the pH effect seen with nitrate ions, external conditions of high pH (low external CO_2 availability) and high nitrate concentration will lead to a situation in which nitrate reduction can outcompete carbon fixation in the algal cell, resulting in a further reduction in CO_2 uptake and photosynthetic carbon production. It is clearly advantageous for aquatic algae (and higher plants) if most inorganic nitrogen in the water occurs as ammonium ions rather than nitrate, which may be a further factor in explaining the preference that these have for ammonium ions as the main source of nitrogen.

5.6   Nitrogen fixation

The fixation of nitrogen in freshwater environments is carried out by prokaryotes – either blue-green algae or bacteria. This section considers nitrogen fixation in terms of its ecological significance, biochemistry of the fixation process, and the various strategies of nitrogen fixation which have evolved in different organisms.

5.6.1   Ecological significance of nitrogen fixation

The contribution that nitrogen fixation makes to the nitrogen budget of standing waters varies with the nutrient status of the water body – particularly the total levels of available nutrients and the balance between nitrogen and phosphorus (Table 5.5). The greatest impact of nitrogen fixation (50–80 per cent of the annual supply) occurs in eutrophic (e.g., Clear Lake, USA) or mesotrophic (e.g., Pyramid Lake, USA) systems where nitrogen is limiting. The importance of nitrogen fixation is considerably less in eutrophic lakes with high levels of N and P, and is negligible in oligotrophic water bodies.

The ecological significance of nitrogen fixation also varies with annual cycle, and may be completely absent at certain times of the year – even in nitrogen-limited lakes. In Clear Lake (USA), for example, Horne and Goldman (1972) reported two major phases of nitrogen fixation by blue-green algae. In the first phase, a spring population of *Aphanizomenon* initially grew using nitrates, but then continued growth by nitrogen fixation when these became exhausted. The growth of this alga was terminated when nitrogen fixation was inhibited by release of ammonia gas (see below) from anaerobic sediments, giving rise to a mid-summer period with no nitrogen fixation. The second phase of nitrogen fixation occurred in late summer when a decrease in the concentration of ammonia permitted growth of *Anabaena*. This was terminated at autumn overturn, when increased concentrations of nitrate ended the fixation process (see below).

5.6.2   The nitrogenase enzyme and strategies of fixation

Nitrogen fixation is mediated by a single enzyme, nitrogenase, which is composed of two subunits. The larger (300 kDa) subunit has associated Fe and Mo and is coded for by the *nifDK* gene. The smaller (35 kDa) subunit has associated Fe and is coded for by the *nifH* gene.

$$N_2 \xrightarrow[\text{ATP}]{\text{nitrogenase}} NH_4^+ \quad \text{(Anaerobic conditions)} \quad (5.9)$$

Nitrogenase activity requires a supply of energy (ATP) and will only occur under anaerobic conditions, since the enzyme is irreversibly denatured by oxygen. Activity of the enzyme is inhibited by NH_4^+ ions, and synthesis is repressed by NO_3^- ions.

Table 5.5   Significance of nitrogen fixation in lakes of different nutrient status (examples are from the USA, Horne and Goldman, 1994)

Type of lake	Nutrient status	Nitrogen fixation
Eutrophic (Lake Mendota, Wisconsin)	High levels of phosphorus and nitrates	Nitrogen fixation contributes 5–10 per cent of annual supply
Eutrophic (Clear Lake, California)	High levels of phosphorus, low nitrates	N₂ fixation: 50–80 per cent of annual supply; allows algal bloom to persist in spring
Mesotrophic (Pyramid Lake, Nevada)		
Oligotrophic (Lake Tahoe, California)	Low levels of phosphorus and nitrates	Negligible

Table 5.6 Strategies of nitrogen fixation in the freshwater environment (shaded boxes – organisms occurring mainly in sediments or as benthic communities)

Prokaryote	Examples	Strategy
Blue-green algae		
(a) colonial, with heterocysts	*Anabaena*	Specialized cell (heterocyst)
	Aphanizomenon	associated bacteria
	Gleotrichia	
	Nodularia	
	Nostoc	
(b) unicellular	*Gloeothece*	Diurnal separation of photosynthesis
	Synechococcus	and N fixation
	Cyanothece	
N$_2$-fixing bacteria	*Azotobacter*	Restricted to anaerobic environment
	Chlorobium	
	Clostridium	

The occurrence of one type of enzyme, the activity of which is inhibited by oxygen, suggests that the biochemical mechanism of nitrogen fixation has evolved just once and at a time when reducing conditions prevailed in the Earth's atmosphere. The requirement for anaerobic conditions imposes major limitations on nitrogen fixation under the present conditions of an oxygen-rich atmosphere, and is particularly severe for photosynthetic oxygen-generating organisms such as blue-green algae. These limitations have led to three main strategies of nitrogen fixation in the freshwater environment – the development of specialized anaerobic cells (heterocysts), diurnal separation of photosynthesis, and nitrogen fixation and restriction to an anaerobic environment (Table 5.6).

5.6.3 Heterocysts: nitrogen fixation by colonial blue-green algae

Various filamentous blue-green algae have solved the problem of carrying out photosynthesis and nitrogen fixation at the same time by spatial separation of these processes in vegetative cells and specialized anaerobic cells (heterocysts) respectively. These nitrogen fixing blue-greens include both planktonic (*Aphanizomenon*, *Anabaena*) and attached (*Gleotrichia*, *Nodularia*, *Cylindrospermum*, *Nostoc*) forms. The latter form part of the periphyton assemblage typical of streams and rivers. Other colonial blue-green algae such as *Microcystis* and

Gomphosphaeria, which do not have heterocysts, are not able to carry out nitrogen fixation.

Heterocysts typically appear as spherical, thick-walled cells and are quite distinct from vegetative cells and resistant spores (akinetes) – Figure 5.14. Nitrogenase is able to exist and function within heterocysts due to the internal anaerobic environment of these cells. This is maintained by:

- removal of oxygen due to high rates of respiration,

- restriction of oxygen inflow into the cell due to a thick cell wall,

- lack of photosynthetic generation of oxygen due to an absence of photosystem II,

- maintenance of an anaerobic microenvironment immediately around heterocysts due to the present of associated bacteria which remove oxygen by respiration.

Heterocysts are involved both in the fixation of nitrogen and in the formation of a derived nitrogenous compound, cyanophycin (Sherman *et al.*, 2000; Figure 5.15). This nitrogen storage and transport compound is produced non-ribosomally from equimolar amounts of the amino-acids asparagine and arginine, themselves synthesized via the ammonia generated by nitrogen fixation. Cyanophycin is very heterogeneous in size, ranging in molecular weight from 15–100 kDa. It occurs in a very dynamic state within heterocysts, where the level is

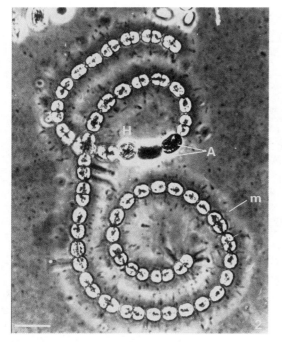

Figure 5.14 Nitrogen fixation in filamentous blue-green algae

Light microscope (phase contrast) image of a filament of *Anabaena spiroides* showing a single spherical heterocyst (H) adjacent to two elongate akinetes (A – resistant spores). The presence of a mucilage sheath (m) containing epiphytic bacteria can be clearly seen (scale = 10 μm; figure taken from Clay *et al.* (1991) and reproduced with permission from Scanning Microscopy Inc.)

controlled via the balance of synthesis (by cyanophycin synthetase), degradation (depolymerized by cyanophycinase), and transport out of the heterocyst into surrounding vegetative cells. This transport is part of a two-way traffic, with photosynthetic products moving from vegetative cells into the non-photosynthetic heterocysts. Heterocysts can thus be regarded as specialized cells, active in nitrogen fixation, with a dynamic reservoir of cyanophycin which is partly controlled by the enzymes cyanophycin synthetase and cyanophycinase.

The fixation of nitrogen by these filamentous blue-green algae is dependent on photosynthetic activity and is thus light-requiring. It is also metabolically expensive, since the radiant energy used for fixation will not be available for growth. As a consequence of this, the fixation process normally stops when NH_4^+ and NO_3^- become readily available or when irradiance is low. The direct dependence of nitrogen fixation on photosynthesis limits the process in planktonic algae to the euphotic zone (Figure 5.17), and also to that part of the diurnal cycle when light is at maximum intensity (Ashton, 1979).

5.6.4   Diurnal control of nitrogen fixation: unicellular blue-green algae

Various unicellular blue-green algae, including *Gleothece*, *Synechococcus*, and *Cyanothece* (Figure 5.16) have separated the processes of photosynthesis and nitrogen fixation in time by evolving a very defined metabolic sequence of molecular

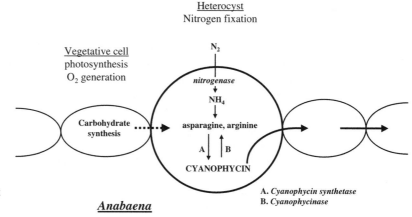

Figure 5.15 Role of heterocyst in nitrogen fixation

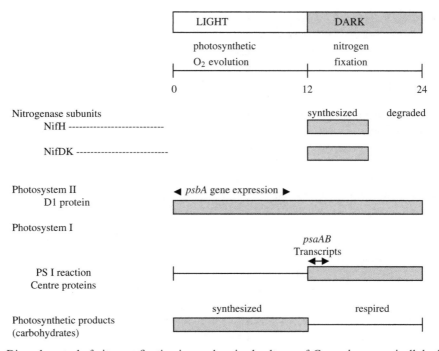

Figure 5.16   Diurnal control of nitrogen fixation in synchronized cultures of *Cyanothece*, a unicellular blue-green alga (figure based on information in Schneegurt *et al.*, 2000)

control during the diurnal cycle. Studies on laboratory cultures of cells exposed to alternate light-dark cycles (Schneegurt *et al.*, 2000) have shown that photosynthetic evolution of oxygen is restricted to the light period (0–12 hours), while nitrogen fixation only occurs in the dark phase (12–24 hours).

The underlying circadian regulatory mechanism is independent of light stimuli, since diurnal oscillations still continue in conditions of continuous darkness. The circadian oscillator operates as described previously, controlling photosynthesis and nitrogen fixation in a precise way (Figure 5.16). During the diurnal cycle, photosynthetic activity (with associated evolution of oxygen) is initiated at the onset of the light period by activation of photosystem II genes. Carbohydrate synthesis is promoted during the light period (0–12 hours), followed by a phase of active respiration (12–14 hours) which establishes an anaerobic intracellular atmosphere. These reducing conditions are maintained over the dark period by down-regulation of oxygen evolution – even in continuous light.

During the dark period, control of nitrogen fixation is mediated by tight regulation of the nitrogenase complex at both the transcriptional (gene activation) and post-transcriptional (protein degradation) level. *NifH* and *NifDK* genes are activated at the beginning of the dark period, leading to a burst of nitrogenase synthesis and the rapid onset of fixation activity. This is rapidly terminated at the end of the dark period by protease digestion of the nitrogenase subunits. The period of nitrogen fixation coincides with a phase of active photosystem I but not photosystem II activity, suggesting that cyclic photophosphorylation is important in ATP generation for nitrogenase activity over this time.

5.6.5 Anaerobic environment: nitrogen-fixing bacteria

The third major group of prokaryotes that are able to convert gaseous to organic nitrogen, the nitrogen–fixing bacteria, have not evolved a strategy for carrying out this process under oxygenated conditions and are strictly limited to anaerobic environments.

These bacteria comprise a mixed assemblage of organisms, including heterotrophic genera such as *Azotobacter* and *Clostridium*, methane-oxidizing bacteria and photosynthetic bacteria. *Azotobacter* is present as an epiphyte on submerged macrophytes and in planktonic form within the hypolimnion, with little population variation in relation to the trophic status of the water body. In contrast to this, *Clostridium* is mainly present in the lower part of the water column and on sediments, where its population is considerably enhanced by the anaerobic conditions which feature in eutrophic and some mesotrophic lakes (Table 5.7). Methane oxidizing bacteria are also present in the water column and sediment interface.

Photosynthetic bacteria such as *Chlorobium* and *Pelodictyon* carry out nitrogen fixation in conjunction with photosynthesis. Oxygen is not evolved during this process, and these organisms are

Table 5.7 Populations of *Clostridium* in different lakes

	Water column (numbers ml^{-1})	Sediments (numbers g^{-1} wet weight)
Oligotrophic	0	0–10
Mesotrophic	0–10	0–10^3
Eutrophic	1–20	10^2–10^4

typically found at the top of the hypolimnion in mesotrophic or eutrophic lakes where the water column is static, anaerobic, and has adequate light available.

Different strategies of nitrogen fixation by aquatic microorganisms lead to major differences in the location of these activities within the freshwater environment. This is seen particularly in freshwater lakes, where nitrogen fixation by different organisms can take place throughout the whole of the water column (Figure 5.17).

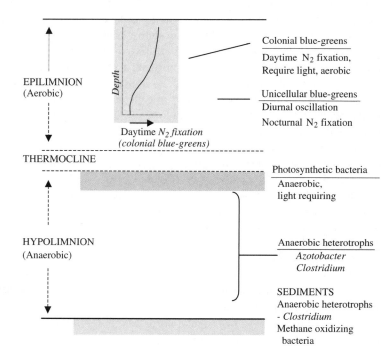

Figure 5.17 Nitrogen fixation in the water column and sediments of a eutrophic lake: the nitrogen-fixing activities of blue-green algae, photosynthetic bacteria and anaerobic heterotrophs occur at different positions within the column and at different times in the diurnal cycle

B.   PHOSPHORUS

5.7   Occurrence and biological availability of phosphorus

Phosphorus is an essential element in all living systems, where it is normally present at levels of about 0.3–0.5 per cent dry weight. It is important in cells as a structural molecule (present in phospholipids and nucleic acids), as a major storage component (particularly polyphosphates), and is involved in energy transformations (ATP).

5.7.1   *Phosphorus availability and limitation*

Phosphorus is present in the aquatic environment as soluble organic (DOM), insoluble organic (lake biota and detritus), and soluble inorganic (phosphate) forms. In most freshwaters, phosphorus occurs mainly within the lake biota – as insoluble organic P. Assimilation of phosphorus by freshwater algae is restricted to uptake of phosphate ions (PO_4^{3-}), the concentration of which is referred to as 'total soluble phosphate' or TSP. Dissolved organic P, the other soluble component, is readily converted to phosphate by phosphatase enzymes, and together with TSP, constitutes the 'biologically available phosphorus' (BAP) within the water. BAP cannot be determined chemically, but must be measured by algal assay. In eutrophic lakes, BAP correlates well with TSP, but in oligotrophic lakes it shows a closer relationship to 'total reactive phosphorus' (TRP), which is the TSP plus the small particulate fraction.

Phosphorus limitation

Although the requirement for P by freshwater organisms is considerably less than N (Redfield ratio), it is normally P which is growth-limiting in freshwater systems. This limitation depends on the balance between P and N availability, and arises for two main reasons – lower levels of P supply, and the ability of phytoplankton to carry out nitrogen fixation.

Low P availability   This arises due to the relatively low inflow levels of P into aquatic systems, due to poor release from rocks and root interception in feeder streams and terrestrial supply routes. Those phosphate ions that enter the aquatic system are liable to further depletion by adsorption onto suspended particulate matter and sediments.

No P-fixation   In many lakes, nitrogen fixation by blue-green algae is able to compensate for existing deficiencies in nitrogen concentration, leaving P as the limiting nutrient. Unlike nitrogen, there is no gas phase for phosphorus, with no 'P fixation' – either inorganic (within the atmosphere) or biotic (within the aquatic system). There is equally no loss of P to the atmosphere, so the absence of a P gaseous phase is of debatable significance.

Phosphorus limitation within a particular water body is indicated by a number of parameters, including:

- low concentrations of available P. As noted earlier, P limitation within a particular system depends on the relative availability of P (phosphate concentration) and N (nitrate concentration) in relation to the Redfield ratio. Phosphorus limitation is particularly likely in water with high nitrate levels, where the N/P ratio is >10:1;

- low concentrations of P within the lake biota, where the content of P is less than expected on the Redfield ratio – this may be determined in terms of particulate carbon (PC), phosphorus (PP), and nitrogen (PN), with phosphorus limitation indicated by ratios of PC/PN >106, and PN/PP>16;

- enhanced dynamics of phosphorus uptake. In P limited conditions, metabolic adaptations by phytoplankton to maximize P uptake are indicated by high levels of alkaline phosphatase secretion (Section 5.8.3) and rapid uptake of $^{32}PO_4$ from the aquatic medium.

These criteria were used by Evans and Prepas (1997) to investigate factors restricting the growth of phytoplankton biomass in prairie saline lakes in Alberta (USA), demonstrating two particular water bodies as being P-limited.

In a phosphorus-limited lake, the growth of algal biomass is ultimately determined by the amount of available P. The total amount of biomass that grows during the period of maximum productivity (summer) is directly proportional to the total level of phosphorus (soluble and insoluble) which accumulates in winter prior to the onset of phytoplankton growth. During the summer growth phase, the standing crop of phytoplankton at any point in time is determined by the amount of biologically-available phosphorus present throughout the water column.

5.7.2 The phosphorus cycle

As the part of biologically-available phosphorus (BAP) that is most readily assimilated by algae, phosphate occupies a central role in the chemical transformations of this element in aquatic systems (Figure 5.18). In a lake, the concentration of phos-

phate in the water directly controls the mass of phytoplankton and other organisms that develop during the major growth phase, and is a key determinant of the trophic status of the water body. As with the other major inorganic nutrients, the concentration of phosphorus in lake water is determined by three major factors – external loading (entry of phosphorus into the lake from outside, via water inflow), internal loading (entry from sediments), and nutrient cycling (phosphorus release from lake biota). External loading of temperate lakes with phosphorus occurs mainly in the winter months, while internal loading and cycling are most important in summer (Figure 5.4).

Phosphorus that is taken up by phytoplankton has three main fates:

- passage to the bottom of the lake within the dead algal cells (detritus) that continuously sediment within the water column – although most of this phosphorus ultimately passes back into the water column via internal loading from the sediments, it is lost for phytoplankton growth in the short term;

- some of it enters the food chain, making a long-term contribution to the zooplankton and fish biomass;

- phosphorus recycling: some of it passes back into the water column, mainly due to short-term release from lake biota.

External loading

This involves entry of phosphorus into lakes and other water bodies via streams, rivers, ground water, and run-off from land. These carry both particulate and soluble phosphorus, varying with the hydrology and the terrain – and also with human activities. Input of agricultural fertilizers, and industrial and human effluent, may cause a major increase in phosphorus loading, particularly soluble organic and inorganic P. External loading is a major cause of eutrophication with many types of water body (Table 10.8) including lakes (Section 10.8.1) and wetlands (Case Study 10.4).

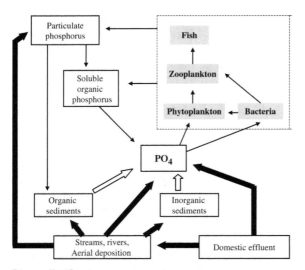

Figure 5.18 Phosphorus cycle in lakes.
External loading (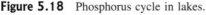) Internal loading (⇨).
Inflow of phosphorus into lake systems occurs mainly via streams and rivers, but aerial deposition is also important; this involves rain, aerosol, and dust precipitation

Internal loading

Continuous sedimentation of phosphorus-rich detritus within the water column leads to the build-up of a layer of organic material, with the continuous release of phosphate by bacterial decay. The passage of this phosphate into the water column depends on the state of oxygenation of the sediment/water interface. Under anoxic conditions, there is free diffusion of phosphate ions from sediment to water column along a concentration gradient. Release of phosphate is considerably reduced ($\times 10^3$ slower) in the presence of oxygen, due to adsorption of the phosphate anion to particulate matter under these conditions.

$$\text{Particle} + PO_4 \xrightarrow{\;O_2\;} PO_4\text{-Particle} \qquad (5.10)$$
$$\textit{free} \longleftarrow \textit{adsorbed state}$$

The adsorption process is highly complex, with an equilibrium that varies with pH, redox potential (E_h), and oxygen concentration.

Release of phosphate from sediments can be very important in lakes with a long turnover time, since summer P release will affect the spring concentration, and hence the maximum attainable biomass in the following season. Phosphate release is much less significant in unproductive lakes, where accumulation of detritus is minimal and conditions at the sediment surface are aerobic.

Phosphorus recycling

Phosphorus recycling involves the short-term re-entry of soluble nutrient into the water column due to release from lake biota (Case Study 5.1). This includes:

- direct release from phytoplankton cells – by leakage of metabolites or death and lysis within the water column: radiotracer studies on algal cultures have shown that in exceptional circumstances algae can lose up to 10–20 per cent of their cell phosphorus per hour, with normal losses of <3 per cent under sustained conditions (Mulholland, 1996);

- excretion from zooplankton, fish, and other components of the food chain.

Recycling of phosphorus is environmentally important since absorbed nutrient becomes available again in the short term for phytoplankton and bacterial growth.

Role of zooplankton in phosphorus recycling   The activities of zooplankton are particularly relevant to phosphorus recycling since they constitute the major biomass involved in the excretory process. These organisms excrete approximately 10 per cent of their total body phosphorus on a daily basis.

The rate of excretion varies with feeding rate, water temperature, and type of food and is closely tied to the diurnal cycle (Figure 5.19). Zooplankton rise to the lake surface at night, where they become active in phytoplankton grazing at maximum temperature within the water column. Under these conditions, processes of digestion and excretion are at an optimum. Migration of zooplankton to the lower part of the water column during the day has the converse effect of lower rates of grazing, digestion, and excretion. The zooplankton diurnal cycle thus operates to achieve maximum release of excreted phosphorus in the upper part of the water column, where it is most directly available for uptake by phytoplankton.

The soluble excretory products of zooplankton are approximately 50 per cent inorganic and 50 per cent organic phosphorus. These excretory products are particularly important in supporting phytoplankton growth in phosphorus-limited water bodies. Since zooplankton distribution is typically very patchy, the distribution of recycled phosphorus will also be irregular across the lake surface.

The continuous removal of phytoplankton by zooplankton, and cycling the element back to a bio-available form, has important implications for the progression of phytoplankton populations. In the hypothetical absence of zooplankton (Figure 5.20), phytoplankton populations would be expected to follow a succession of maxima, with intervening periods of phytoplankton breakdown and phosphorus release. This situation does occur under conditions of nutrient pollution (Section 10.2.3), when zooplankton populations are severely depressed by the growth of colonial blue-green algal blooms. In contrast to this, where substantial zooplankton populations are present, the resulting

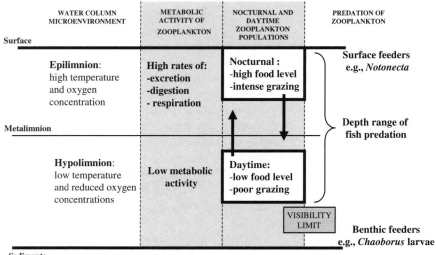

Figure 5.19   Diurnal variation in grazing rates and metabolic activity of zooplankton.

The diagram illustrates diurnal changes in the activity of zooplankton within the water column of a stratified lake during the seasonal period of diel vertical migration (DVM). Shaded columns emphasize the importance of this activity to phosphorus recycling, with high rates of grazing and nutrient excretion by nocturnal populations within the epilimnion. As an example, the zooplankton populations indicated are typical of Lake Maarsseveen (Holland), where the main nocturnal (midnight) distribution of *Daphnia hyalina* ranged from 0–4 m, and the main daytime (noon) distribution was 8–12 m (Ringelberg *et al.*, 1996). The daytime distribution of *Daphnia* in this lake was just above the light threshold depths for visual predation by juvenile perch (*Perca fluviatilis*), which ranged from 18–21 m (visibility limit) (aspects of fish/zooplankton interactions are considered in Section 9.9.3, Figure 9.15)

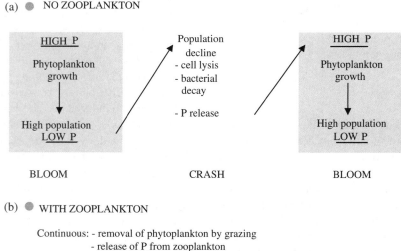

Figure 5.20   Postulated effect of zooplankton grazing on the growth of lake phytoplankton.

In the hypothetical absence of zooplankton (a) reduction of soluble lakewater phosphorus (P) by phytoplankton uptake leads to nutrient limitation and the population crashes. Subsequent release of P during cell lysis promotes a further algal bloom. The presence of zooplankton (b) leads to continuity of P cycling and algal growth, without the extremes of population growth and decline seen in (a) continuous phytoplankton grazing and phosphorus release promote a more regular progression of phytoplankton growth – without the bloom-crash sequence which would otherwise occur. The grazing (and phosphorus-cycling) activities of zooplankton in standing waters may thus paradoxically promote the long-term presence and growth of phytoplankton populations.

5.8 Adaptations of freshwater microorganisms to low phosphorus concentrations

As the major limiting nutrient, phosphorus is often present as a scarce resource within the freshwater environment. Ability of lake biota such as phytoplankton and bacteria to survive at low phosphate levels is important for their continued presence in low nutrient environments and for the ability of individual species to compete under P-limiting conditions. These organisms have adopted three main strategies to survive at low phosphate levels – low K_μ values in the kinetics of phosphorus uptake (and low absolute values of P uptake saturation), storage of phosphorus during times of non-limitation (luxury uptake), and secretion of alkaline phosphatase to utilize soluble organic phosphorus.

5.8.1 Kinetics of phosphorus uptake

As with other nutrients, the relationship between growth rate and substrate concentration (Monod curve) varies considerably between species. This can be illustrated (Figure 5.6) by comparison of a diatom (*Asterionella*) and a green alga (*Selenastrum*). *Asterionella* has a lower half-saturation or K_μ value (0.6) compared with *Selenastrum* (1.3) and is able to take up phosphate and use it more effectively for growth at lower concentrations. At low phosphate concentrations, if there are no other limiting elements or other environmental constraints, *Asterionella* will out-compete *Selenastrum* and dominate a mixed culture of the two algae. The reverse is true at high phosphate concentrations, where *Selenastrum* is able to use phosphate more

effectively than *Asterionella* in promoting growth and has a higher value for the maximum growth rate (μ_{max} 1.2 compared with 0.9).

Although differences in phosphate K_μ values between algal species may relate particularly to differences in physiology (e.g., active uptake of specific ions), other factors such as cell size are also probably important. This is illustrated by the general ability of bacteria to out-compete eukaryote algae for phosphate ions (Bratbak, 1987), where the higher surface area/volume ratio of the prokaryote gives a potentially more efficient relationship between ion uptake (surface area) and cell growth (volume). As noted in Section 6.12.1, the greater efficiency of nutrient uptake per unit volume in bacteria is only translated into higher growth rates when the bacteria are not carbon-limited.

5.8.2 Luxury consumption of phosphate

The uptake of phosphorus by microorganisms does not only occur when the rate of growth is nutrient-limited (Figure 5.21).

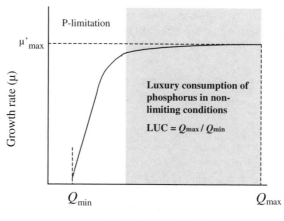

Phosphorus cell quota (Q_P)

Figure 5.21 Luxury consumption of phosphorus (P) under non-limiting conditions (Droop model). The cell P content (quota) ranges from a minimum level (Q_{min}) when external P just supports growth, to a maximum value (Q_{max}) when uptake is at its highest level. The luxury uptake coefficient (LUC) provides a measure of the ability to take up P over a wide range of concentrations and store P under non-limiting conditions

Under P-limiting conditions, when the rate of algal growth depends on external phosphate concentration, most of the anion taken up into algal cells is used directly for growth by incorporation into a range of macromolecules, including nucleic acids, phospholipids and phosphorylated proteins. At higher phosphate concentrations, when growth is not nutrient-limited, excess phosphate may be taken up into intracellular storage compounds such as insoluble polyphosphate. Uptake and storage of P under such conditions is referred to as 'luxury consumption' and provides an important reserve for future conditions of low P-availability. Luxury consumption may occur where phosphate levels are particularly high (Figure 5.21) or where other key nutrients are in short supply and are themselves limiting.

Polyphosphate granules

Phosphorus is stored as insoluble polyphosphate inclusions in a wide range of phytoplankton cells. In blue-green algae, they may appear as very distinctive electron-dense deposits in ultrathin section. Polyphosphates consist of orthophosphate groups linked by high-energy bonds, and are generated by the activity of polyphosphate kinase.

$$P + P \xrightleftharpoons[\substack{ATP \quad ADP}]{\substack{\text{Polyphosphate} \\ \text{kinase}}} P{\sim}P \quad \text{Polyphosphate} \tag{5.11}$$

Polyphosphate deposition has a primary role in providing medium- to long-term phosphorus storage capacity, which occurs without any increase in the osmotic potential of the cell due to the high insolubility of the end-product. Polyphosphate deposition has also been suggested as being important in other ways (Kornberg, 1995), including:

- Physiological adjustment to varying external phosphate levels: diurnal migration in the water column exposes cells to a range of external phosphate concentrations. Temporary build-up and break-down of polyphosphate effectively integrates phosphate assimilation over the diurnal period.

- Disposal of pollutant phosphates: at very high phosphate levels (occurring for example in some domestic or industrial effluents) formation of polyphosphates sequesters phosphorus within the cell in an inert form.

- Energy reservoir: polyphosphates not only store phosphorus but also energy in the form of high-energy bonds. Break-down of these molecules leads to the formation of ATP, which thus becomes available for use by other cell processes.

- Metal chelation: strongly electro-negative polyphosphate deposits act as powerful chelators of metal cations, providing an important sink and reservoir for metals within the cell.

- Buffer against alkali: polyphosphates occur in complex equilibrium with phosphate and hydrogen ions, acting as a buffer for alkalis within the cell.

- Buoyancy regulation: along with other dense deposits in the cell, polyphosphates contribute to the ballast, a key component of the buoyancy regulation process (Figure 3.20).

Luxury consumption and the P quota

Increase in phosphate availability and uptake promotes cell growth and increases the mean total amount of P per cell (the phosphorus quota, Q_P) within the algal population. Values for Q_P obtained under different conditions of P availability provide a useful measure of the capacity of different algal species for luxury uptake. The luxury uptake coefficient (LUC) is defined as the ratio of $Q_{\max}/Q_{\min}$ – where $Q_{\max}$ and $Q_{\min}$ are values for Q_P at maximum and minimum cumulative phosphorus uptake respectively (Figure 5.21). Values for the luxury uptake coefficient vary considerably between algae, ranging from values of <10 (e.g., *Cyclotella*) to >100 (colonial blue-green algae). Algae with a high LUC are able to continue growth, with up to 20 cell divisions, in protracted conditions of acute P limitation.

Growth conditions	Poly-P	Sugar-P	Nucleic acid	Other
N$_2$ fixation High P	3	22	55	20
Low P	6	-	75	19
NO$_3$ uptake High P	15	-	60	25

Figure 5.22 Storage of phosphorus in cultured *Anabaena* under different growth conditions (data from Thomson *et al.*, 1994, expressed as a percentage of the total intracellular phosphorus content)

Phosphorus storage in blue-green algae

Colonial blue-green algae tend to have high luxury uptake coefficients, suggesting that P storage is particularly important in this group. This may be due in part to the meroplanktonic nature of algae such as *Microcystis*, with the build-up of high Q_P values on nutrient-rich sediments during the benthic phase (Section 3.10.1, Figure 3.22). These algae are able to use the stored phosphorus for prolonged growth after vertical migration into the P-depleted epilimnion, and are able to persist at the lake surface due to diurnal migrations into nutrient-rich hypolimnion. The ability of these algae to store phosphate plays a major role in their colonization and domination of lakes in late summer, at a time when phosphate and nitrate levels in many water bodies are falling to very low levels. The stored phosphorus has considerable environmental significance, since it effectively uncouples cell growth at this time of year from the constraints of low external P concentrations.

The proportion of phosphorus in blue-green algae which can be stored as polyphosphate varies considerably, ranging from 15 per cent (*Anabaena*) to 60 per cent (*Oscillatoria*). Although polyphosphate is generally considered to be the main P storage compound, this is not always the case. In *Anabaena*, the pathway for P storage is closely dependent on nitrogen metabolism (Thomson *et al.*, 1994). Under nutrient-limiting conditions, phosphorus occurs mainly in nucleic acids (Figure 5.22), with minimal luxury storage. Transfer from low to high phosphorus levels, with nitrogen supplied as nitrate, leads to substantial storage of P as polyphosphate. Transfer from low to high phosphate levels without a nitrate supply promotes nitrogen fixation, with storage of P as sugar-P complexes rather than polyphosphate.

5.8.3 Secretion of alkaline phosphatase

Alkaline phosphatase (APA) is produced by both phytoplankton and benthic algal cells as a response to low phosphate levels (Mulholland, 1996). It acts by breaking down soluble aquatic organic phosphates and releasing phosphate ions (soluble inorganic phosphorus) – which can be directly assimilated by algal cells (Figure 5.23). Secretion of the enzyme is induced by low, and repressed by high, aquatic PO$_4^{3-}$ levels. It is also induced under conditions of internal phosphorus deficiency, when the cell quota

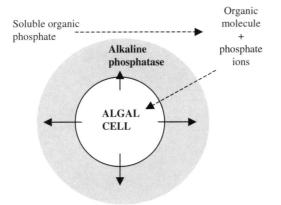

Figure 5.23 Secretion of alkaline phosphatase by algal cells in nutrient-limiting conditions

(Q_P) falls to critically low levels. Hydrolysis of organic phosphorus compounds by phosphatases leads to increased phosphorus uptake by algal cells and increased rates of nutrient cycling. The 'phosphatase-hydrolysable phosphorus may be present in the general aquatic medium or adsorbed to sediments, where it may form a particularly important source of phosphorus for benthic algae in stream environments.

The ability of algal cells to release phosphate ions from soluble organic phosphates in the aquatic environment is particularly important under P-limiting conditions, where differences between species in terms of induction and amount of enzyme produced may be an important factor in inter-specific competition. Although the phosphate released into the environment would be potentially available to all algal cells, most benefit would come to the APA-producing organism. This is because distribution of secreted enzyme and release of the cleaved PO_4 will tend to be restricted to its immediate microenvironment (Figure 5.23). This localized effect will also be enhanced by the relatively short half-life (hours) of the enzyme.

The level of APA in the aquatic medium can be determined colorimetrically by release of dye (3-*o*-methylfluorescein) from artificial substrate:

$$3\text{-}o\text{-methylfluorescein phosphate} \xrightarrow{\text{alkaline phosphatase}}$$
$$3\text{-}o\text{-methylfluorescein} + \text{inorganic phosphate}$$

$$(5.12)$$

Enzyme concentration can be expressed relative to phytoplankton biomass (nmol methylfluorescein $(\mu g \text{ chl-a})^{-1} h^{-1}$), and serves as a useful environmental indicator of P-limiting conditions (Evans and Prepas, 1997).

C. SILICON: A WIDELY-AVAILABLE ELEMENT OF LIMITED METABOLIC IMPORTANCE

Silicon is the second most common element (after oxygen) at the Earth's surface, where it is present mainly as silica (SiO_2). This occurs in soil and rocks both as free SiO_2 and as complex silicates (e.g., sodium felspar – $NaAlSi_3O_8$), and is frequently associated with aluminium.

Weathering of rocks leads to the release of Si as soluble silicic acid (H_2SiO_4), plus colloidal and particulate (clays) forms, of which only silicic acid can be taken up by freshwater biota. Colloidal and particulate forms of Si have an important physicochemical role in the freshwater environment for the adsorption of major nutrients such as phosphate and ammonium ions. Weathering of silicon-rich rocks is also important as a source of associated cations (e.g., sodium ions) and as a contributor to the buffering system of freshwaters by the generation of bicarbonate ions.

$$2NaAlSi_3O_8 \quad + \quad 2CO_2 + 3H_2O =$$
$$\text{sodium} \qquad\qquad \text{rainwater,}$$
$$\text{springs, streams}$$

$$4SiO_2 + Al_2Si_2O_5(OH)_4 + 2Na^+ + 2HCO_3^-$$
$$\text{soluble} \qquad \text{kaolinite} \qquad\qquad\qquad (5.13)$$
$$\text{silica}$$

5.9 The silicon cycle

Chemical transformations of silicon in the freshwater environment are relatively simple compared

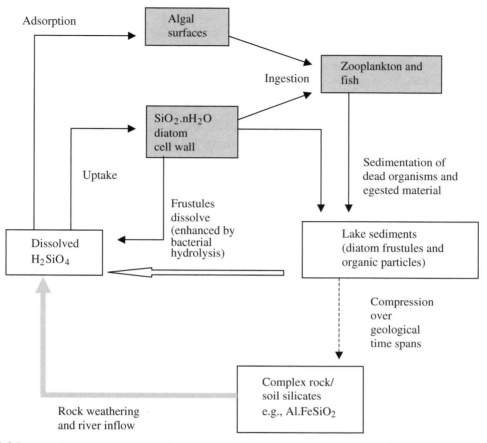

Figure 5.24 The Silicon Cycle in lakes, emphasizing the role of diatoms.
External loading (⟹) Internal loading (⟹)

to nitrogen and phosphorus, with one-way flow from rocks to sediments (Figure 5.24). Unlike nitrogen and phosphorus, silicon is available to biota only as a single form (silicic acid, also referred to as 'reactive silica'), and its mass uptake is restricted to particular groups of organisms, including phytoplankton (diatoms, chrysophytes) and sponges. Soluble silicon is not secreted by zooplankton and fish, and its cycling (conversion of insoluble to soluble silicon) within the water column is largely restricted to the dissolution of diatom cell walls (frustules).

The mass uptake of silicon by diatoms for cell-wall formation represents the major biological transformation in the aquatic environment. The high silicon content of diatom cell walls can be readily demonstrated by X-ray microanalysis (Krivtsov *et al.*, 2002; Section 5.1.4), with the occurrence of a massive Si peak in X-ray emission spectra. X-ray analysis of the surfaces of other phytoplankton cells directly sampled from lake environments (e.g., Figure 5.2) also reveals the routine presence of silicon, suggesting that this element is associated (possibly by adsorption) with the surfaces of a wide range of phytoplankton cells. Analysis of lake samples of the blue-green algae *Anabaena* (Sigee *et al.*, 1999c) and *Microcystis* (Sigee and Levado, 2000) have indicated considerable heterogeneity in the presence of surface silicon, with the occurrence of distinct subpopulations of high-Si and low-Si cells. Phytoplankton-associated silicon (present as cell-wall material or adsorbed at the cell surface) enters the food chain via ingestion by zooplankton, and ultimately passes to the lake sediment, where it combines with phytoplankton debris to promote the internal loading for the water body. Various X-ray studies (see above) have demonstrated a close correlation in the detection of phytoplankton-associated silicon and aluminium, suggesting that the latter element enters the food chain in parallel with the Si.

Solubilization of silicon from the cell walls of dead diatoms is important in the regeneration of silicic acid, and occurs from sedimenting cells and benthic detritus. Studies by Patrick and Holding (1985) have shown that natural populations of freshwater bacteria are able to increase the rate of solubilization, apparently by the production of hydrolytic enzymes.

Apart from its importance as a major cell-wall constituent for diatoms (and to a lesser extent other biota), silicon appears to have little biological significance for freshwater microorganisms. Its role in diatom cell walls is structural, and this element appears to have little (if any) metabolic function – which is surprising in view of the widespread occurrence of silicon in the freshwater environment.

5.10   Silicon and diatoms

The uptake of soluble silica and its deposition as insoluble cell-wall material is unique to diatoms and has major implications for the evolutionary success of this algal group. Diatoms are widely present in almost all freshwater environments – both as benthic and pelagic organisms. This success must be attributed – at least in part – to their strategy of having a silica cell wall, with the energy savings in wall synthesis and the resulting growth efficiency that this brings (see Section 4.3). The silica cell walls of diatoms are rigid, with a defined architecture (Vrieling et al., 2000), and once they have been formed no change in size and shape can take place. Such a cell wall is unique within living organisms.

The uptake and deposition of silica by diatoms also has two major environmental implications. First, the kinetics of Si uptake and its relationship to cell growth determines the ability of different diatom species to out-compete others at particular Si concentrations. Second, the uptake of soluble Si and its deposition as insoluble Si represents a major geochemical process in aquatic systems. The effect of diatom populations in carrying out large-scale conversion of soluble to insoluble Si (referred to as mineralization or biosilification) is a good example of the ability of freshwater organisms to make major chemical changes to their environment. Diatom growth in the Amazon estuary, for example, results in the production of an estimated 15×10^6 tons of diatom cell-wall material per annum, removing 25 per cent of the river's dissolved silica (quoted in Horne and Goldman, 1994).

5.10.1   Si uptake and phytoplankton succession

Phytoplankton succession in temperate lakes typically involves a substantial spring diatom bloom (Figure 3.14) – with stripping of Si from lake water, and extensive removal also of other nutrients such as nitrate and phosphate. Rapid decline of the diatom bloom (prior to the clear-water phase) can result from a range of factors (see Chapter 3), including a decrease in the concentration of silicate below a critical level. The dependency of diatom growth on silicon concentration is particularly acute because over 95 per cent of the cell wall is composed of Si, there is no storage mechanism to allow luxury consumption when Si is in excess, and there is no recycling of Si from zooplankton within the water column.

Close examination of diatom counts within the spring bloom reveals a sequence of separate populations, with genera such as Asterionella typically having an early peak in the sequence and Cyclotella a late peak. These population changes take place under rapidly changing concentrations of silicon and phosphorus (Figure 5.25). Competition between Asterionella and Cyclotella depends on the kinetics of nutrient uptake in relation to the prevailing concentrations (Tilman and Kilham, 1976). The growth and competitive ability of each alga will be at a maximum when the uptake and use of both silicon and phosphorus most closely fits the specific physio-

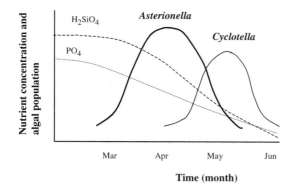

Figure 5.25 Diagram showing typical changes in populations of the diatoms *Asterionella* and *Cyclotella* during the spring bloom of temperate lakes: the growth of *Asterionella* correlates with a sharp fall in silicon and phosphorus lake water concentrations (nutrient concentrations on different scales)

logical processes of that organism. This can be determined using key parameters for each nutrient from the Monod equation (Section 5.2, Figure 5.5), including specific growth rate (μ), half saturation coefficient (K_μ) and substrate concentration (S), where:

$$\mu = \mu_{max} \frac{S}{K_\mu + S} \qquad (5.14)$$

For a particular alga, such as *Asterionella*, growth rates in $P(\mu_P)$ and silicon (μ_{Si}) will be similar when:

$$\mu_P = \mu_{Si} \qquad (5.15)$$

or

$$\left[\mu_{maxSi} \frac{S_{Si}}{K_{Si} + S_{Si}} \right] = \left[\mu_{maxP} \frac{S_P}{K_P + S_P} \right] \qquad (5.16)$$

Thus, nutrients will be in balanced supply for growth when their concentration ratio is similar to that for the half-saturation coefficients, and:

$$\frac{S_{Si}}{S_P} = \frac{K_{Si}}{K_P} \qquad (5.17)$$

Optimum conditions for the two diatoms are shown in Figure 5.26. *Asterionella* will be most competitive when the Si/P substrate concentration ratio is 97, and *Cyclotella* when the ratio is 5.6, which correspond to the situation in early and late spring respectively. On either side of these values, the algae will be P-limited or Si-limited respectively.

5.10.2 Si uptake and cell-wall formation

Diatom cells are physiologically adapted for the bulk assimilation and incorporation of silicon into newly-forming cell-wall material, leading to the occurrence of an Si-rich external frustule (Figure 5.27). Silicon uptake and cell processing are dynamic activities which require energy from aerobic respiration, without any involvement of photosynthetic energy, and can take place in the dark. Silicon metabolism in these organisms thus differs markedly from the uptake and biochemical conversions of nitrogen and phosphorus – which are closely linked to photophosphorylation, and are light-requiring.

Recent studies (Martin-Jézéquel *et al.*, 2000), have revealed new insights into the molecular processes involved in silicification. These occur over a discrete period within the cell cycle (immediately preceding cell division to daughter cell separation) and can be considered in relation to Si uptake, intracellular transport, and internal deposition (Figure 5.27).

Silicon uptake

Studies on cultures of the diatom *Navicula* (Sullivan, 1976) have demonstrated that the first step in silicon mineralization is the active uptake of silicic acid. Using radioactive germanium (combined with silicic acid, $^{68}Ge\text{-}Si(OH)_4$) as tracer, Sullivan demonstrated that the uptake of Si was dependent on external pH and temperature, and was promoted by some monovalent cations (Na^+, K^+) but not others (Li^+, NH_4^+). This energy-requiring process was partially inhibited by uncouplers and inhibitors of oxidative phosphorylation. The trans-membrane transport of silicic acid was sensitive to sulphydryl blocking agents, and varied with stage of growth cycle, reaching a maximum in early stationary phase cultures. Active uptake of silicon led to an estimated 250-fold concentration gradient of silicic acid compared with the external medium. Specific rates of Si

P + Si : the two major elements that determine spring growth

Diatom	K_P	K_{Si}	$\mu_{max}Si$
Asterionella	0.04	3.9	1.1
Cyclotella	0.25	1.4	1.3

K_P = half saturation constant (K_μ) for P (µmoles)

K_{Si} = half saturation constant (K_μ) for Si (µmoles)

$$\textit{Asterionella.}\ \frac{K_{Si}}{K_P} = \frac{3.9}{0.04} = 97 \qquad \textit{Cyclotella.}\ \frac{K_{Si}}{K_P} = \frac{1.4}{0.25} = 5.6$$

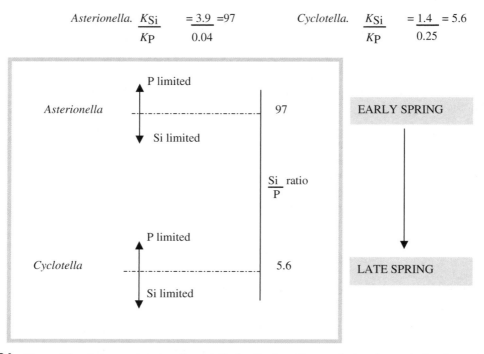

Figure 5.26 Competition between *Asterionella* and *Cyclotella* for silicon and phosphorus during the spring diatom bloom

uptake (V) and Si-dependent cell division rate (μ) followed Michaelis–Menten and Monod kinetics.

The identification of a saturable uptake system indicates that Si uptake is carrier-mediated (Martin-Jézéquel *et al.*, 2000). Molecular characterization of the silicic transport system commenced with the initial cloning and functional identification of cDNAs encoding a silicic acid transporter (SIT) in the diatom *Cylindrotheca* (Hildebrand *et al.*, 1997). Further studies by Hildebrand *et al.* (1998) have identified five different types of *SIT* genes, maxi-

mum transcriptive activity of which corresponds to the main phase of Si uptake (just before the period of silica deposition). The pattern of *SIT* mRNA synthesis suggests that different transporters have different roles in Si uptake. Some of these are at the cell surface and are involved in Si transport before (low-affinity transporters) and during (high-affinity transporters) the major phase of bulk uptake. Other transporters are thought to occur as components of organelle membranes, promoting uptake and subsequent transport within the cytoplasm.

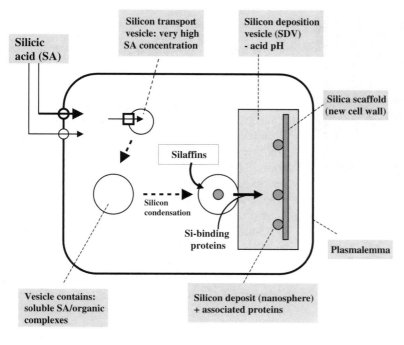

Figure 5.27 Proposed sequence of silicon uptake, transport, and deposition during cell wall formation in a diatom. These processes may be separated in time, with a major phase of Si uptake preceding cell division and deposition occurring between mitosis and daughter cell separation. Cellular transport of Si occurs across membranes (⟶) and by movement of silicon transport vesicles within the cytoplasm (⁃⁃➤). Three different transmembrane silicic acid transporter (SIT) complexes are shown here: *plasmalemma complexes*: low affinity, low capacity ◯, high affinity, high capacity **◯**, *intracellular complex*: ☐ (based on information from Martin-Jézéquel *et al.*, 2000)

Intracellular transport

Relatively little is known about the mechanism of intracellular Si transport, though it seems likely that specialized vesicles are involved (Figure 5.27). Various investigators have demonstrated intracellular pools of soluble Si at concentrations far in excess of the saturation level for silicic acid. This suggests that the silicic acid is complexed with low MW organic material, possibly within the membrane-bound vesicles.

Close coupling of Si transport and silica deposition are indicated in some diatom species by comparison of maximum uptake rates (V_{max}) with the quota of cell wall silicon (QW_{Si}). In synchronized cultures of *Navicula pelliculosa*, for example, respective values were 27.9 (V_{max}) and 42–75 fmol cell^{-1} (QW_{Si}) – suggesting a transport period of 1.5–2.7 hours under conditions of maximum uptake. This is consistent with the measured rate of population increase (Sullivan, 1977), suggesting that all Si taken up into the cell is directly transported and deposited without any diversion to internal storage.

This close coupling implies a tight control of internal transport processes.

Silicon deposition

The chemical industrial synthesis of silica-based materials such as catalysts, resins, and molecular sieves requires extreme pressures, temperatures, and pH. In contrast, silica condensation by diatoms occurs at ambient temperatures and pressures, resulting in a great morphological diversity of cell-wall structure and ornamentation. Individual diatom species have their own distinctive pattern of silica deposition (Fig. 5.28).

Silicon is deposited at the site of cell-wall formation as a hydrated amorphous derivative, with the general formula $[Si_nO_{2.n-(nx12)}(OH)_{nx}]$, where $x < 4$ (Martin-Jézéquel *et al.*, 2000). The massive frustule structure may account for half the cell's dry weight. Frustules are highly perforated, and occur as two overlapping and separate halves – the epitheca and hypotheca (Figure 5.29). Division of the diatom cell

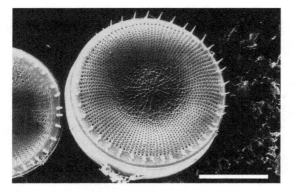

Figure 5.28 Scanning electron micrograph of the diatom *Stephanodiscus minutula*, showing complex ornamentation of the silica cell wall.

This centric diatom is disk-shaped, with cell-wall valves (frustules) which have radiating rows of pores (punctae). The valve has a central depressed area, with marginal spines at the edge. This particular species is common in the phytoplankton of eutrophic lakes (scale = 15 μm)

involves separation of the two halves, which then become the epitheca of each daughter cell.

The silica frustules of diatoms have a thin outer covering or pellicle (composed primarily of proteins and carbohydrates, with some lipid), which is important in preventing loss of silica by dissolution into the aquatic medium.

Molecular control of biosilification   Deposition of silica (biosilification) occurs within a specialized vacuole, the 'silica deposition vesicle' (SDV) contained within the daughter cell (Figure 5.27). Soluble silicic acid taken up at the cell surface becomes concentrated, condensed (as nanospheres), transported into the SDV, then deposited onto the Si framework of the new cell wall (Si scaffold) – resulting in a mass of insoluble silica.

Recent studies (Fischer *et al.*, 1999, Kroger *et al.*, 1999) have suggested that condensation and deposition of the silica is mediated by two major groups of cell-wall proteins (Figure 5.28). This may be summarized:

$$\text{Soluble silica} \xrightarrow{\substack{Condensation \\ proteins \\ (silaffins)}} \text{silica deposit} \xrightarrow{\substack{Deposition \\ (frustulin\text{-}like) \\ proteins}} \text{attachment}$$
(silicic acid)   to cell wall

Condensation appears to be mediated by one particular group of proteins, the 'silaffins' which have been identified within the hydrofluoric acid extractable protein (HEP) fraction from diatom cell walls. These proteins are composed of polycationic peptides, with lysine residues conjugated to polyamine and dimethyl side chains. This molecular configuration triggers the condensation of silica to small deposits (nanospheres) of insoluble material. *In vitro* polymerization of Si by silaffins parallels the biological situation in two key respects – the requirement for acid conditions (Vrieling *et al.*, 1999), and the generation of nanospheres.

The final stage of Si incorporation into the new cell probably involves a protein-mediated deposition, in which the protein attaches the Si to sites on the cell wall scaffold. A class of diatom cell-wall proteins, the 'frustulins', have recently been isolated and characterized (Kroger *et al.*, 1999) which give some indication of how this might occur. These proteins bind calcium rather than silicon, but have cation-binding sites, possible cell-wall recognition sites and an N-terminal-recognition sequence for entry in to the SDV.

Cell division and cell-wall formation

Cell-wall formation occurs within each daughter cell on completion of cytoplasmic separation (cytokinesis). New cell walls are formed in relation to each of the two parental valves (frustules) – the epitheca and hypotheca (Figure 5.29). During cell division, each valve produces a smaller complementary valve, resulting in a progressive decrease in cell size. This eventually leads to sexual reproduction, where the zygote (auxospore) expands in size to form the next vegetative cell. Size is thus restored to the original value, and the next division sequence commences.

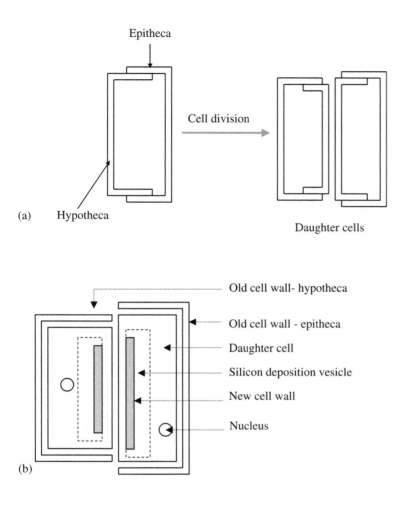

Epitheca

Cell division

(a)   Hypotheca

Daughter cells

Old cell wall- hypotheca

Old cell wall - epitheca

Daughter cell

Silicon deposition vesicle

New cell wall

Nucleus

(b)

Figure 5.29   Cell division and cell-wall formation in diatoms: (a) Formation of new cell walls from the epitheca and hypotheca, resulting in two daughter cells. Cell division leads to a progressive decrease in size, since the daughter formed from the hypotheca is smaller than the mother cell. (b) Formation of new cell walls after cytokinesis, within the silicon deposition vesicles of the daughter cells

D.   TRACE ELEMENTS

In addition to macronutrients (inorganic compounds containing carbon, nitrogen, phosphorus, and silicon) natural waters also contain a variety of dissolved chemicals containing elements which are essential but are required in smaller quantities. These micronutrients include some of the anions and cations noted in Tables 5.1 and 5.2, plus a range of (trace) elements occurring at much smaller concentrations.

Trace elements include the metals manganese, iron, cobalt, copper, zinc, selenium, molybdenum,

and barium, plus non-metals such as fluorine. Other elements, such as vanadium and cadmium are also frequently included on the list, but their primary functions and absolute requirements are not clear. The concentrations of trace elements in natural waters vary with water chemistry and temporal (seasonal and diurnal) factors, but average concentrations (measured as μgl^{-1}, Table 5.8) range from 1–500 for Fe and Mn, 1–15 for the transition metals Cu and Zn, and are typically <1 for Co, Mo, Cd and V.

Table 5.8   Range in average concentrations ($\mu g \ l^{-1}$) of soluble trace elements in surface waters of lakes and rivers; (data derived from a range of lakes and rivers, summarized from Wetzel, 2001)

Element	Fe	Mn	Cu	Zn	Co	Mo	V
Range	1–540	0.3–140	1–10	1–13	0.05–0.9	0.4–30	0.03–0.1

5.11   Biological role of trace elements

Evidence for the role of these elements as required micronutrients comes from several sources, including their environmental uptake by freshwater microorganisms, stimulation of growth and metabolic activities by addition to natural waters, their requirement for *in vitro* culture of algae and other microorganisms, and their known biochemical functions.

5.11.1   Environmental uptake of trace elements

Uptake of trace elements by phytoplankton has been clearly shown by the studies of Groth (1971), who determined the concentrations of a range of elements in the epilimnion, hypolimnion, phytoplankton, and sediments of Schöhsee, northern Germany.

The average concentration of elements dissolved in epilimnion and hypolimnion (Table 5.9) indicated considerably higher levels below the thermocline in the case of Fe and Mn, significantly higher levels in the case of C, Mo, and Zn, but no significant difference with Cu. Differences in dissolved elemental concentration that develop between the epilimnion

and hypolimnion of a stratified lake can be partly attributed to removal (uptake by biota) from the epilimnion, coupled with increases (release from sedimenting detritus and from sediments) in the hypolimnion. The other major factor is the difference in redox potential (E_h) between the two parts of the water column. This is particularly important in the case of Fe, where oxygenated conditions of the epilimnion promote the conversion of Fe^{2+} to Fe^{3+}, a much less soluble form of iron.

Comparison of metal concentrations in biomass (phytoplankton) and aquatic medium (epilimnion water) indicate a very high concentration of trace elements by lake biota (Table 5.9). Enrichment ratios (concentration factors) were of the order of $\times 10^3$ for each metal, due either to metabolic uptake and retention by cells, or to adsorption at the cell surface, or both.

Phytoplankton uptake of trace metals in the epilimnion, followed by sedimentation and deposition of biomass on the sediments leads to very high accumulation of these elements at the bottom of the hypolimnion. Sediment concentration factors (compared with the epilimnion) were invariably a factor of $\times 10^3$, and included higher values 10^5 (Mn) and 10^6 (Fe) in two cases.

Table 5.9   Average concentrations of trace elements in the epilimnion, hypolimnion, phytoplankton, and sediments of Schöhsee, Northern Germany (data from Groth, 1971)

	Mn	Fe	Cu	Zn	Co	Mo
Epilimnion (E) $\mu g \ l^{-1}$	4.5	15	1.0	1.8	0.03	0.21
Hypolimnion (H) $\mu g \ l^{-1}$	590	425	0.9	1.9	0.07	0.30
Enrichment ratio H/E	130	28	0.9	1.1	2.3	1.4
Phytoplankton (P) $\mu g \ g^{-1}$	130	950	60	110	1.1	4.2
Enrichment ratio P/E	29×10^3	63×10^3	60×10^3	61×10^3	37×10^3	20×10^3
Sediments (S) $\mu g \ g^{-1}$	1600	58 000	95	350	8.3	1.4
Enrichment ratio S/E	355×10^3	3900×10^3	95×10^3	195×10^3	280×10^3	6.7×10^3

5.11.2 Stimulation of growth in aquatic environments

Various workers have shown that induced increases in the bioavailability of trace elements – by alteration of pH or direct addition of metal ions – can stimulate primary productivity and specific metabolic activities. These effects can be demonstrated on environmental samples in the laboratory (bottle bioassays) or by direct addition to the environments. Environmental results suggest that trace metals can become limiting at certain times of year, with important ecological effects on planktonic algae and other microorganisms.

pH effects are particularly important in the bioavailability of Fe. Bottle bioassays carried out by Evans and Prepas (1997) demonstrated that phytoplankton growth in Fe-limited saline lakes could be promoted either by the addition of Fe or by partial acidification of the sample (titrating 25 per cent or 50 per cent of the total alkalinity).

Limitations in Fe may become particularly acute during blue-green algal blooms, when concentrations of free ions are reduced by high pH and algal uptake, and the highly oxygenated lake water results in all the ionic iron being in the ferric state. Wurtsbaugh and Horne (1983) demonstrated dramatic increases in nitrogen fixation when iron was added to lake samples taken during blooms of the blue-green algae *Anabaena* and *Aphanizomenon*. Under Fe-limiting conditions, addition of iron to lake samples has a dual effect on the metabolic activity of heterocysts (Table 5.5) – promoting the synthesis of nitrogenase and respiratory electron transfer. Stimulation of respiratory activity generates the reducing atmosphere in the heterocyst needed to protect the nitrogenase enzyme.

Addition of molybdenum, the other trace metal required for nitrogen fixation (and uptake) may have a similar effect. Bottle bioassays carried out by Goldman (1960) showed that phytoplankton in Castle Lake, California (USA) was molybdenum-limited at certain times of year. Addition of molybdenum to the whole lake, increasing the aquatic concentration from $<0.2 \, \mu g \, l^{-1}$ to $7.7 \, \mu g \, l^{-1}$, increased both primary and secondary (zooplankton) productivity. Increases in the populations of cladocerans and copepods were particularly prominent, with effects lasting for several years.

The promotion of phytoplankton growth by Fe and Mo, without any toxic effects, are well established. With other trace metals, such as copper, the toxic effects may be more prominent. Although copper is metabolically important for photosynthesis and nitrate uptake, and occurs at very low concentration in natural waters, deficiencies in availability have not been clearly established. Addition of chelated copper to eutrophic Clear Lake, California (USA) caused a slight stimulation of photosynthesis and nitrogen fixation at very low supplementation, but a rapid toxic effect soon became apparent (Horne, 1975).

5.11.3 Importance of trace metals in the culture of aquatic algae

The importance of trace metals such as Fe, Mn, Mo, Cu, and Zn for algal growth is indicated by their requirement as constituents of laboratory growth media (Huntsman and Sunda, 1980). Other elements such as boron (B) and vanadium (V) may also be necessary, but the requirement is so low that their presence as impurities is sufficient to support growth.

Chelating agents such as citrate and ethylenediamine tetra-acetic acid (EDTA, usually added as the sodium salt) are also added to the medium. These agents have a high affinity to metal ions, binding with them in a reversible equilibrium where there is always a proportion of free metal ions. The chelating agent (with metal) is too large a molecule to be taken up by the cell, but the presence of free metal ions ensures that there is a continuous low level of availability for metal uptake by algal cells. The addition of these compounds thus provides exploitable soluble reserves of trace metals without problems of toxicity. The chelating agents used in laboratory cultures have the same effect as various organic degradation products (e.g., humic and fulvic acids) and algal secretory products (siderophores) that are present in natural waters.

5.11.4 Biochemical roles of trace elements

The biochemical roles of trace metals are based on one important property – their ability to occur in two or more ionic states. Conversion between the states requires or releases only small amounts of energy. This gives them the ability readily to accept or donate electrons at particular redox levels, and to take part in oxidation/reduction reactions. In the case of iron, for example (Kalff, 2002):

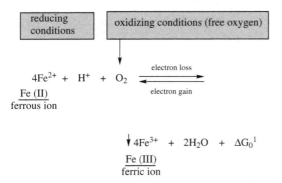

$$(5.18)$$

(maximum amount of free energy $(\Delta G_0^1) = -4\,kJmol^{-1}$).

Conversion of ferrous to ferric ions occurs in the presence of free oxygen, with the release of only $4\,kJmol^{-1}$ free energy (ΔG_0^1). Ferric ions are highly insoluble, so this conversion also results in precipitation and loss of soluble iron from the water column.

Trace metals are important to freshwater organisms as enzyme cofactors, micro-substrates and structural components (Table 5.10).

Enzyme cofactors

Copper and iron have key roles as enzyme cofactors in the electron transfer complexes present on the inner mitochondrial membranes of all aerobic microorganisms. Within these complexes, iron is combined with sulphur as an Fe-S centre, which is involved in the acceptance and donation of electrons along a gradient of reduction/oxidation (redox) potential. Iron and copper have an identical role in the photosynthetic electron transfer chain occurring

on the thylakoid membranes of blue-green and eukaryote algae. In addition to their role in energy transduction, trace metals are also important as enzyme cofactors in detoxication of superoxide radicals (Kitayama and Kitayama, 1999), nitrogen fixation, and nitrate assimilation.

Trace metals as micro-substrates

Iron and manganese are important as bacterial substrates in anaerobic environments, where they replace oxygen as the terminal electron acceptor in the oxidation of organic substances (see Table 6.2). These electron acceptors are part of a sequence (Mn^{4+}, NO_3^-, Fe^{3+}, SO_4^{2-}) that runs with decreasing oxygen concentrations (Figure 6.6). Bacteria that use these metals as electron acceptors are facultative anaerobes, able to use either oxygen or the trace metal as electron acceptor. This ability is particularly important for bacteria living on aquatic sediments – giving them the ability to switch their metabolism in conditions of fluctuating oxygen tension, or to migrate into subsurface anaerobic sediments to access sources of substrate that are not available to obligate aerobes.

The oxidation of organic matter, with the associated reduction of manganese and iron, may be summarized as follows.

Reduction of insoluble Mn(IV) to soluble Mn(II)

$(CH_2O)_{106}(NH_3)_{16}(H_3PO_4) + 236MnO_2 + 472H^+$

$\longrightarrow 236Mn^{2+} + 106CO_2 + 8N_2 + H_3PO_4$

$+ 366H_2O$ (5.19)

maximum amount of free energy $(\Delta G_0^1) = -3050\,kJmol^{-1}$.

Reduction of insoluble Fe(III) to soluble Fe(II)

$(CH_2O)_{106}(NH_3)_{16}(H_3PO_4)$

$+ 212Fe_2O_3$ (or $424FeOOH$) $+ 848H^+$

$\longrightarrow 424Fe^+ + 106CO_2 + 16NH_3 + H_3PO_4$

$+ 530H_2O$ (or $742H_2O$) (5.20)

$(\Delta G_0^1) = -1410\,kJmol^{-1}$ (Fe_2O_3)

$(\Delta G_0^1) = -1330\,kJmol^{-1}$ ($FeOOH$)

Table 5.10 Some important biological roles of trace elements (micronutrients): trace metals quoted are manganese (Mn), iron (Fe), cobalt (Co), copper (Cu), zinc (Zn) molybdenum (Mo), and barium (Ba) – magnesium (Mg) is not a trace element but is included as a specific biochemical microconstituent

General biological role	Enzyme or structural component	Trace metal	Biological characteristics
(a) Enzyme cofactors			
Detoxification of superoxide radicals	Superoxide dismutase (SOD)	**Fe, Mn**	Identified in various algae, including *Chlamydomonas*
Nitrogen fixation[a]	Nitrogenase large subunit	**Fe**	Blue-green algae and bacteria
	Small subunit	**Fe, Mo**	Coded by nif*DK* and nif*H genes*
Nitrate assimilation[b]	Nitrate reductase	**Mo**	Widely present in algae and bacteria
	Nitrite reductase	**Fe, Cu**	
Respiratory electron transfer	NADH dehydrogenase complex	**Fe**	22 polypeptide chains. Contains at least five Fe-S centres
	b-c$_1$ complex	**Fe**	Dimer. Each monomer has an Fe-S centre
	Cytochrome oxidase complex	**Cu**	Dimer. Each monomer contains two Cu atoms
Photosynthetic electron transfer	Chlorophyll	**Mg**	Light-harvesting and energy transfer
	Plastocyanin	**Cu**	Small protein, mobile electron carrier
	Ferredoxin	**Fe**	Small protein, with single Fe-S centre
Photosynthetic carbon assimilation	Carbonic anhydrase	**Zn**	Catalysis of critical rate-limiting step in carbon uptake
(b) Structural component[c]	Protozoon statoliths. Ba sulphate crystals in desmid algae[e]	**Ba**	Geo-receptors (statoliths) Unknown function
(c) Micro-substrate			
Oxidation/reduction reactions[d]	Co-substrate in oxidation of organic matter	**Mn, Fe**	Used by facultative anaerobic bacteria.
Vitamin	Vitamin B12	**Co**	Required by algae

References: [a]Section 5.6.2; [b]Section 5.5.1; [c]Section 9.4.1; [d]Section 6.6.3; [e]John *et al.*, 2002

Oxidation of organic matter via Fe(III) can involve either Fe_2O_3 or FeOOH, and in both cases generates considerably less energy (for cell synthetic activities) than oxidation via Mn(IV) or via oxygen as the terminal acceptor ($\Delta G_0{}^1 = -3190\,\mathrm{kJmol^{-1}}$).

Trace metals may also be important microsubstrates for algae, where they are a structural component of key organic molecules. The presence of cobalt in vitamin B12 is an example of this. In this molecule, a single Co atom lies at the centre of four pyrrole rings in a similar way to the positioning of Mg in chlorophyll.

5.12 Cycling of iron and other trace metals in the aquatic environment

5.12.1 The iron cycle

Availability of Fe in natural waters

Iron is present in natural waters as free cation, cation chelated to soluble organic material, and fine particulate material (inorganic precipitates). In lakes and running waters, the total (soluble and insoluble) iron content may reach values of 10 mg l^{-1}. These values are considerably higher than the solubility values of about 10 μg l^{-1} for ferric ions in well-oxygenated, neutral waters which are predicted from laboratory measurements of iron solubility (Mill, 1980), and are largely due to adsorption or complexation of ferric and ferrous ions dissolved organic matter (DOC).

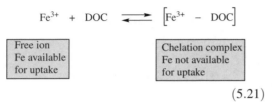

$$(5.21)$$

Complexed Fe is not available for direct uptake by algal cells or other microorganisms, and the very low solubility of ferric ions in aerated waters provides major problems for microorganisms. Problems of low iron availability are partly compensated by internal biochemical changes (Section 3.3.2) and the secretion of siderophores.

Some organisms, such as blue-green algae, have increased their ability to scavenge Fe by the production of extracellular ferric-specific chelating agents or 'siderophores' (Wilhelm *et al.*, 1996). These low MW compounds function as extracellular ligands which aid in the solubilization and uptake of Fe^{3+} in environments where the low availability of Fe^{3+} serves to limit growth. The production of siderophores by terrestrial microorganisms and their role in iron-acquisition have been well documented for bacteria and fungi (Winkelmann, 1991), but are less clear for blue-green algae and their part in aquatic iron cycles.

Siderophores of blue-green algae appear to function in a two-stage process (Wilhelm *et al.*, 1996):

- release of hydrophilic hydroxamate-type siderophores into the immediate microenvironment, resulting in a Fe^{3+}-binding and causing solubilization of Fe^{3+};

- exchange of the hydroxamate-bound Fe^{3+} with a surface-associated hydrophobic catechol-type siderophore which transports the iron to the cell.

It has been suggested that this sequence, which differs from the one-step process seen in terrestrial systems, may be an advantage to aquatic microorganisms by enhancing the re-encounter efficiency of the cell with its ligands (Wilhelm *et al.*, 1996).

The secretion and retrieval of siderophores to obtain Fe from the aquatic environment operates at the micro level, and shows some similarity to the secretion of alkaline phosphatase to retrieve soluble phosphate (Section 5.8, Figure 5.23). Both systems operate most effectively under nutrient-limiting conditions, and in both cases the secretion of retrieval compounds has been documented for both prokaryote and eukaryote algae. The release of siderophores by blue-green algae may be an important aspect of microbial competition in freshwater systems, paralleling the role of these chemicals in the process of iron retrieval that occurs in terrestrial microhabitats such at soil and leaf surfaces (Sigee, 1993).

Seasonal changes in iron availability

In temperate climates, the occurrence and availability of iron in natural waters varies with the seasonal cycle. Seasonal variations in iron availability involve interrelated aquatic changes in both total soluble Fe and ionic species (ferric/ferrous), and involve complex interactions between the water column and the sediments (Figure 5.30). Much of our understanding of the cycling of iron in low DOC, non-acidified waters comes from the classic work of Mortimer (1941), who studied ionic interchanges in Esthwaite Water (UK).

Seasonal changes in iron availability are particularly prominent in eutrophic lakes, where the range of oxygenation in the hypolimnion and sediments ranges from aerobic to very reducing (Figure 5.9 and 5.30). During the period of winter mixing, iron is present largely as the oxidized state in the oxygenated water column and on aerobic surface sediments. At this time of year, iron is present as soluble Fe^{3+} in the water column (chelated to DOC) and as insoluble precipitates of hydroxy and phosphate anions on the surface of sediments. Ferric iron forms a range of insoluble oxides and oxyhydroxides

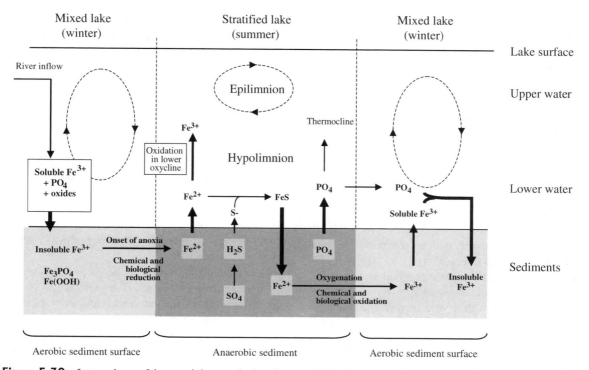

Figure 5.30 Interactions of iron, sulphur, and phosphorus within the water column and sediments of a temperate eutrophic lake (arrows denote chemical transitions ($\longrightarrow$) and water movement ($\dashrightarrow$))

(FeOOH), which flocculate in the water column and settle on sediments – often forming a rusty brown layer of $Fe(OH)_3$.

As the lake becomes stratified and the oxygen level in the lower part of the water column declines, ferric ions are converted to the ferrous state by biological and chemical reduction. Similar processes occur in the lower (hyporheic) zone of rivers and in wetlands. The biological process involves bacterial reduction of iron as part of the respiratory oxidation of organic matter. Ferric iron becomes the terminal electron acceptor once supplies of oxygen, manganese ions (Mn^{4+}), and nitrate (NO_3^-) have become exhausted. The chemical oxidation/reduction of iron depends on the redox state of the system and also pH, with a 100 per cent rise on the oxidation rate of Fe^{2+} for every unit rise in pH.

The conversion of insoluble Fe^{3+} to soluble Fe^{2+} in anaerobic sediments leads to a marked release of iron from sediments. The internal loading of Fe^{2+} from anaerobic sediments can be very high, yielding concentrations in excess of $1000\,\mu g\,l^{-1}$ in the anoxic hypolimnia of low carbonate (noncalcareous) lakes. Ferrous ions that have diffused into the hypolimnion are rapidly removed by freshly-released sulphide (S-)ions, forming insoluble ferric sulphide. The ferrous ions are also chemically and biologically oxidized when they come into contact with dissolved oxygen. The biological oxidation is mediated by bacteria in the genus *Siderocapsa* and occurs in the lower part of the oxycline (during stratification) and in the water column and sediments on return of oxygenated conditions at lake overturn. The seasonal phase of re-oxygenation is accompanied by a dramatic fall in the concentration of soluble iron in the water column. This results mainly from chemical (non-biological) causes, including the large-scale oxidation of soluble Fe^{2+} to insoluble Fe^{3+} and the formation of insoluble ferric phosphate from high phosphate concentrations previously released from anaerobic sediments.

The cycling of iron between the sediments and water column of aquatic systems (Figure 5.30) also affects the cycling of other elements such as phosphorus and sulphur. Changes in Fe concentrations and speciation influence the availability of these other elements to freshwater biota, influencing their concentrations and flux within the water column.

5.12.2 The manganese cycle

Seasonal changes in Mn within the water column of stratified lakes and other water bodies closely resemble the iron cycle, with an alternation between insoluble oxidized states (Mn[III], Mn [IV]) in aerobic conditions and a soluble reduced state (Mn[II]) when oxygen becomes depleted.

The two metals differ, however, in the redox potential (E_h) at which these transitions take place. As hypolimnia establish and conditions become reducing, oxidized states of Mn ([III] and [IV]) become transformed at a higher E_h value than Fe^{3+}, resulting in the presence of substantial levels of soluble Mn[II] at a time when Fe is still largely in the insoluble form. Relatively little reduced sulphur is formed at this high E_h, so newly-formed Mn[II] that is released from sediments precipitates mainly as oxyhydroxide flocs and not MnS.

When conditions in the lower part of the water column become oxidized at the other end of the stratification period, the oxidation of Mn[II] commences before Fe[II]. The chemical oxidation of Mn[II], however, is much slower than Fe[II] at pH 6–7, and in the case of Mn is mainly carried out by bacteria – particularly those in the genus *Metallogenium*.

A further important difference between Fe and Mn that affects metal availability in the soluble state is that, unlike Fe, little oxidized Mn is chelated with DOC (Urban *et al.*, 1990). Mn forms complex insoluble flocs, containing large amounts of Ca, Mg, Si, P, S, Cl, K, and Ba, which sediment to the bottom of the water column – removing Mn and other micronutrients from the planktonic environment.

Iron, manganese, and other geochemical cycles

In addition to the impact of the Fe cycle on transitions and fluxes of P and S (see previously), Fe and Mn also influence the circulation and availability of trace metals in aquatic systems.

Circulation of Fe[III] and Mn [III, IV] to the upper part of the water column leads to the formation of FeOOH and MnOOH aggregates. These scavenge redox-sensitive trace metals from neutral oxygenated surface waters before sedimenting to the lower part of the water column. Accumulation of these complexes on aerobic sediments leads to the retention of trace metals (with oxidized Fe and Mn) in an insoluble state. When oxygen becomes depleted, trace elements are released with reduction and solubilization of Fe and Mn, and may return to the surface layer where they become available again to planktonic organisms.

6

Bacteria: the main heterotrophic microorganisms in freshwater systems

Bacteria occur as one of three major groups of prokaryotes in the freshwater environment, differing from blue-green algae in their heterotrophic mode of nutrition but showing many physiological similarities to the actinomycetes (Section 8.2). The majority of bacterial cells have a maximum linear dimension in the picoplankton (0.2–2 μm) range, though some freshwater bacteria fall into the femtoplankton (<0.2 μm) and nanoplankton (2–20 μm) categories.

These organisms can be readily observed in water samples by light (dark field, phase-contrast) microscopy or by transmission and scanning electron microscopy. Figure 6.1 shows the typical scanning electron microscope appearance of freshwater bacteria that have been grown in liquid nutrient medium. The environmental scanning electron microscope (ESEM) image visualizes the preparation in the wet state, showing the cells in their native state and revealing the presence of copious mucilage. This is secreted by the bacteria and accumulates in the culture medium (Figure 6.1(a)). Chemical processing removes mucilage, giving a much 'cleaner' preparation, which may be more useful for making bacterial counts. The relatively uniform appearance of the laboratory monoculture differs markedly from environmental samples (Figure 6.11), where considerable variation occurs in relation to size and morphology.

In this chapter, freshwater bacteria are discussed in terms of their general diversity and ecological significance, genetic interactions (particularly gene transfer), metabolic diversity, productivity, and interactions with phytoplankton.

A. GENERAL DIVERSITY WITHIN THE ENVIRONMENT

6.1 General diversity, habitat preferences, and ecological significance of freshwater bacteria

Freshwater bacteria are a very diverse assemblage of prokaryote organisms, varying in their morphology, physiology, and ecological preferences.

6.1.1 General diversity

Bacteria may be conveniently grouped into a number of natural assemblages (Holt and Krieg, 1994), based on characteristics such as cell shape, spore-forming capabilities, and whether they are aerobic/anaerobic or Gram-positive/Gram-negative (Table 6.1). Some of these groups are uniform and

Freshwater Microbiology: Biodiversity and Dynamic Interactions of Microorganisms in the Aquatic Environment   David C. Sigee
© 2005 John Wiley & Sons, Ltd   ISBNs: 0-471-48529-2 (pbk) 0-471-48528-4 (hbk)

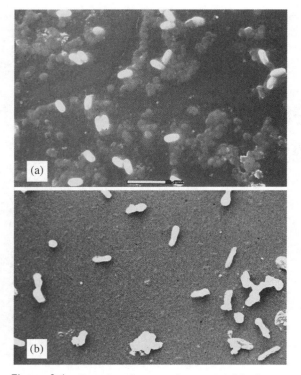

Figure 6.1 Scanning electron microscopy of freshwater bacteria. Freshwater isolates, cultured in liquid nutrient medium, are imaged:

(a) in the wet state, by environmental scanning electron microscopy (ESEM), and (b) after chemical fixation, by conventional microscopy (CSEM). The rod shape of these bacteria is seen in both images, but only ESEM gives a clear visualization of cells in their natural state and the presence of the copious mucilage (grey floccules). EM counts of total bacteria are normally carried out from CSEM preparations, where cells are not obscured by mucilage (scale = 10 μm)

distinctive (e.g., Group 1, Spirochaetes), while others (e.g., Group 7, sulphur bacteria) are very heterogeneous. The wide range of phenotypic characteristics which contribute to this diversity reflect the underlying genetic diversity discussed in Section 6.3.

Most aquatic bacteria are Gram-negative, with relatively few (e.g., *Bacillus pituitans*) Gram-positive representatives. Bacterial size varies with both taxonomic and nutritive state, and the various shapes include rod, helical, vibrioid (curved), and coccoid (spherical) forms (Figure 6.11). Other dis-

tinguishing features relate to the mode of division (e.g., conspicuous budding morphology) and the development of the extracellular sheath, which is very prominent in bacteria such as *Sphaerotilus* but poorly developed in pseudomonads. Motility provides another distinguishing feature, with some bacteria (e.g., *Thiopedia*) being completely non-motile, while others are motile via flagella which may occur over the whole cell's surface or be present as polar groups at one (e.g., *Pseudomonas*) or both (e.g., *Spirillum*) ends of the bacterium. Under light microscope bright-field illumination, freshwater bacteria typically appear refractive (Figure 6.24(a)) and colourless, though some species have bright green or purple coloration (photosynthetic organisms) or other types of pigmentation.

Differences in life-style are also a prominent feature of freshwater bacteria. Most bacteria are free living, obtaining nutrient supplies from substrates present in the aquatic medium. Other bacteria show clear trophic associations with other organisms, such as symbiosis (Section 6.13.3) and parasitism. *Bdellovibrio* is an important bacterial parasite of various microorganisms, with a life cycle which includes both free-living (non-nutritive) and parasitic phases.

6.1.2 Habitat preferences

Bacteria are widespread throughout the freshwater environment, forming extensive pelagic and benthic populations in a wide range of habitats including mudflats, bogs, sulphur springs, lakes, and rivers (Table 6.1). Although some bacteria, such as *Escherichia coli*, are present as accidental contaminants, most freshwater bacteria show close physiological adaptations to their environment. Strict anaerobes, for example, are confined to anoxic sediments and hypolimnia. In some cases, particular organisms (e.g., *Bacillus pituitans*) have a very restricted habitat range, while others such as *Pseudomonas aeruginosa* are very widespread – being routinely found in freshwater, soil, and aerial samples.

Diversity of microenvironments within individual habitats leads to a range of adaptations for both

Table 6.1   Some commonly-occurring aquatic bacteria, illustrating the variety of phenotypic and habitat diversity in freshwater systems. The variety of bacteria commonly found in standing and flowing waters is shown by a range of examples, grouped according to their phenotypic and ecological characteristics. Individual species are arranged within selected numbered groups, which represent a relatively small fraction of the thirty five phenotypic assemblages enumerated in Bergeys Manual of Systematic Bacteriology (Holt et al., 1994).

This table does not include some of the more specialised parasitic forms (such as Mycoplasmas) or bacteria present in extreme environments (particularly Archaebacteria). Other aspects of freshwater bacterial diversity are shown in Tables 2.10 (temperature range), 2.13 (Archaebacteria), 6.2 (metabolism), 6.5 (photosynthetic bacteria), 6.8 (antagonistic interactions) and Fig. 2.20 (pH range)

Bacterium	Some key phenotypic features	Habitat
Group 1: helically shaped, motile bacteria (Spirochaetes); single celled, contain periplasmic flagella enclosed within an outer sheath, Gram-negative		
Spirochaeta plicatilis	Helical cells Capable of locomotion both in suspension and on solid surfaces Anaerobic or microaerophilic	Common in brackish and marine freshwaters rich in H_2S
Group 2: aerobic or microaerophilic; motile helical or vibrioid (comma-shaped) cells, Gram-negative		
Spirillum volutans	Rigid helical rods Bipolar tufts of flagella Breakdown of organic material	Microaerophile. Eutrophic, stagnant fresh or saline waters
Bdellovibrio bacteriovorus	Small, curved cells Motile via single polar flagella at one or both ends of cells Parasite of bacteria and algae Biphasic life cycle	Widely present throughout freshwaters and soil
Group 4: aerobic rod- and coccoid-shaped cells, Gram-negative		
Azotobacter chroococcum	Typically occurs as large ovoid cells, but varies in shape (pleomorphic) Motile (with peritrichous flagella) or non-motile Important N_2 fixer	Range of soil and freshwater systems
Methylomonas methanica	Straight, curved or branched rods Motile via a single polar flagellum Methane oxidising bacterium	Aerobic conditions close to anaerobic sediments
Pseudomonas aeruginosa	Motile (polar flagella), rod-shaped General breakdown of organic material	Very common throughout aerobic freshwater environments
Group 5: facultative anaerobes, rod shaped, Gram-negative		
Escherichia coli	General breakdown of organic material	Faecal contaminant of freshwaters
Group 7: strictly anaerobic, morphologically diverse bacteria; sulphate- or sulphur-reducing, Gram-negative		
Desulfovibrio desulfuricans	Spiral to vibrioid-shaped cells Motile by single or tufts of polar flagella Important sulphur reducing bacterium	Anaerobic sediments and hypolimnia
Group 10: photosynthetic bacteria, not generating oxygen; wide range of shape, Gram-negative		
Thiopedia rosea	Spherical to ovoid cells Non-motile Obligately anaerobic and phototrophic Bright purple-red colour	Anaerobic sediments and top of hypolimnia

Table 6.1   (*Continued*)

Bacterium	Some key phenotypic features	Habitat
Rhodospirillum rubrum	Photosynthetic purple non-sulphur bacterium Spiral cells, motile by polar flagella Anaerobic to aerobic growth	Lakes and mudflats
Group 13: budding and/or appendaged bacteria; a very diverse group, Gram-negative		
Caulobacter vibrioides	Rod shaped or vibrioid cells, with a single flagellum at one end of the cell and a stalk at the other Non-budding Motile and attached phases in life cycle Predatory	Widely present in a range of freshwater systems
Group 14: bacteria contained within a sheath of extracellular material, growing as chains of cells in filaments, Gram-negative		
Sphaerotilus natans	Occurs as chains of rod-shaped cells, with prominent sheath and holdfasts for attachment Important in breakdown of organic material	Attached to submerged plants and stones, in high nutrient flowing waters
Group 15: gliding motility (on solid surfaces); Gram-negative, diverse morphology		
Beggiatoa alba	Long filamentous bacterium, occurring as single cells or in filaments Cells often contain distinct sulphur inclusions Oxidises H_2S to S	Occurs in sediments at interface between underlying anoxic high sulphide zone and the overlying oxic zone
Cytophaga hutchinsonii	Rod-shaped, with gliding motility Important in breakdown of polysaccharides (cellulose and chitin) and proteins	Lake and river sediments, associated with decomposing organic matter
Lysobacter enzymogenes	Thin rods Able to lyse a variety of microorganisms, including blue-green algae	Widely present in freshwaters
Group 18: endospore-forming Gram-positive rods and cocci		
Bacillus pituitans	Rod-shaped Motile by peritrichous flagella Decomposition of protein, forming H_2S	Bog lakes of high organic content

pelagic and benthic organisms. In the lake environment, for example, different species of pelagic bacteria position themselves in the water column in relation to local conditions of light intensity, oxygen level, and nutrient concentrations. Attached bacteria also occupy specific niches within the lake or river environment, being associated with particular pelagic and benthic algae, macrophytes, and sediments. Bacterial biofilms form important microbial communities in lakes (air–water interface at lake surface, sediments) and in streams (air–water interface, stone and plant surfaces). The physiology of attached bacteria is different in many respects

from pelagic organisms (e.g., growth characteristics, resistance to anti-microbial compounds), and the close proximity of bacterial cells within dense attached communities is important for activities such as gene transfer.

6.1.3   Environmental significance of freshwater bacteria

With extensive benthic populations in lakes and rivers, and pelagic population levels of 10^2–10^8 cells ml^{-1}, bacteria are by far the most abundant of all the free-living freshwater biota, and are ecologically important in a number of ways. Within the freshwater environment, bacteria are:

- taxonomically very varied, and make a major contribution to the phenotypic, genetic and molecular biodiversity,

- the main heterotrophic microorganisms in different aquatic systems,

- a very diverse assemblage in terms of metabolic activities, occupying key roles in geochemical cycles,

- particularly important in anaerobic environments, where algae and other free-living biota are much less metabolically active,

- a key component of the microbial food web, within which they are particularly important in recycling algal secretory products via the microbial loop,

- able to outcompete algae for nitrates and phosphates under nutrient-limiting conditions,

- able to form key associations with other biota, particularly algae, and

- the primary colonizers of many benthic habitats, forming permanent bacterial biofilms on many different types of substrate.

Bacteria are the most opportunistic of all the free-living freshwater biota and are the ultimate *r*-strategists (see Section 1.2.6). In this respect they surpass almost all algal groups in terms of small size and short cell-cycle duration, and in terms of high rates of growth and nutrient absorption. Within the water column, bacteria differ from pelagic algae in showing one further adaptation towards *r*-selection – the majority of organisms within the population are metabolically inactive, awaiting transient favourable environmental conditions to trigger growth and division. These inactive forms appear as tiny micro-cells within environmental samples (Figure 6.11).

6.2 Taxonomic, biochemical, and molecular characterization of freshwater bacteria

In many instances, studies on freshwater bacteria in environmental samples require some knowledge of the identity of the organisms concerned. This may involve a formal identification of individual organisms at species level, or a less rigorous characterization of individual bacteria (or mixed populations) in relation to biochemical and molecular aspects. This range of analysis is considered in relation to species identification, detection of marker strains, and biochemical characterization of whole communities.

6.2.1 Species identification

A comprehensive analysis of species diversity in aquatic bacterial communities would typically involve a combination of classical taxonomic and molecular techniques.

Classical taxonomic analysis

Classical identification of individual bacterial species in environmental samples typically involves isolation, laboratory culture, then taxonomic characterization. Within the broad assemblages listed in Table 6.1, classification of bacteria into families, genera and species is based on a wide range of phenotypic characteristics (Holt and Krieg, 1994). These include culture conditions, colony morphology, biochemical characteristics, and detailed (high-resolution) morphology.

The use of *in vitro* culture as part of the classical analysis and identification may be problematic. Only a minute proportion (about 1 per cent) of environmental organisms can be cultured in the laboratory, and even where culture is possible, the bacteria may be very slow growing (e.g., *Desulphivibrio*). Where bacteria can be cultured, the use of a standard identification procedure such as the MICROLOG BIOLOG™ system (developed by

Don Whitley Scientific™) can prove useful. This works by measuring an organism's ability to utilize a number of different carbon sources in order to create an individual fingerprint for a particular isolate – which can then be compared to reference species within a data base. Apart from this, direct application of classical techniques to environmental sample is largely restricted to the use of light microscopy, but there are limitations with this approach due to similarities in the appearance of closely-related species. Because of problems in establishing clear taxonomic characteristics, freshwater biologists often tend to emphasize broad metabolic groups of bacteria (Table 6.2) rather than strict taxonomic units. These metabolic groups tend to use the same energy, carbon, and electron sources plus the same electron donor and form important ecological units (Pedros-Alio, 1989).

Molecular analysis

Molecular analysis has the advantage that it can be equally applied to both cultivated and environmental samples. A typical example of the comprehensive use of molecular techniques is given by the work of Brinkmeyer *et al.* (2003), who investigated the diversity and structure of bacterial communities in Arctic and Antarctic ice. These investigators sequenced 16S rRNA from over 200 pure cultures of bacteria, demonstrating that α- and γ-proteobacteria, plus organisms within the *Cytophaga-Flavobacterium* assemblage, were the dominant taxonomic bacterial groups at both poles. Direct identification and quantification of particular taxa in environmental samples was carried out using *in situ* fluorescence hybridization with new oligonucleotide probes which were developed to target *Octadecabacter*, *Glaciecola*, *Psychrobacter*, *Polaribacter*, and other genera.

A similar molecular analysis of the bacterial samples from the surface waters of a humic lake (Burkert *et al.*, 2003) provided comprehensive information on taxonomic diversity within the community and also demonstrated shifts in community composition in relation to the availability of exogenous dissolved organic carbon (DOC). The pelagic bacterial community of this lake contained two major groups of bacteria with fundamentally different growth strategies – β-proteobacteria showed a marked population increase with short-term DOC enrichment, while members of the class Actinobacteria did not respond.

The use of 16S rRNA analysis for taxonomic identification requires the existence of extensive clone libraries of 16S rRNA genes, which are currently limited for certain key groups of bacteria. The *Cytophaga–Flavobacterium* group of bacteria, for example, is known to have a potentially unique role in the utilization of organic material in freshwaters and to be abundant in aquatic ecosystems. Under-representation of this assemblage in clone libraries means that relatively little is currently known about the diversity and abundance of uncultured members in environmental populations (Kirchman *et al.*, 2003).

6.2.2   Genetic markers: detection of particular strains in the aquatic environment

The growth of molecular biology has proved useful in developing the use of genetic markers to tag and enumerate particular strains within the environment. This becomes particularly relevant for monitoring the release of genetically-engineered organisms which may find their way into freshwater systems. Not only is the spread and survival of released organisms important, but the survival and transfer of genetically-modified (recombinant) plasmids which they contain is also relevant. The use of standard culture techniques (where possible) to monitor such organisms is clearly not appropriate in this context, since these techniques are not genetically specific to the modified organism.

Bacteria that have been labelled with an introduced novel gene (the genetic marker) can be identified either on the basis of a specific phenotypic characteristic (e.g., antibiotic resistance) in culture, or directly in the environmental sample using a specific molecular probe. Studies attempting to assess the survival and transfer of recombinant organisms in aquatic environments have typically relied on plasmids which contain antibiotic resistance markers. Attempts to follow the transfer and survival of such plasmids in aquatic systems can be severely limited by the presence of such genes in the natural population, resulting in a high background resistance level. The development of marker systems which avoid the use of antibiotic resistance circumvents this problem, and has been used by Winstanley *et al.* (1992) to detect pseudomonads released into lake water. The system developed by these workers involves the detection of a marker gene (xylE) and its product catechol 2,3 dioxygenase. The criteria for such genetic markers – low background level in natural populations, stability of the marker plasmid in different host environments, and adequate expression of the marker gene – are broadly satisfied in this case.

6.2.3 Biochemical characterization of bacterial communities

Biochemical characterization of bacterial communities has considerable potential for environmental studies, and was used by Findlay *et al.* (2003), in their investigation of stream bacteria. This approach was used in conjunction with molecular analysis (see previously).

Case study 6.1 Changes in bacterial community function and composition as a response to variations in the supply of dissolved organic material (DOM)

Findlay *et al.* (2003) studied substrate-mediated shifts in the biological properties of stream sediment bacteria in relation to both function (extracellular enzyme activities) and molecular composition (RAPD analysis).

Extracellular enzyme activities

Extracellular enzymes were assayed in bacterial suspensions obtained by vortexing gravel particles coated with bacterial biofilm. Fluorescent-labelled substrates were used to measure potential activities of a range of extracellular enzymes – including β-glucosidase, α-glucosidase, N-acetylglucosaminidase, β-xylosidase, phosphatase, leucine aminopeptidase, esterase, and endopeptidase. Addition of both labile (readily assimilable) and recalcitrant (poorly assimilable) dissolved organic material promoted increases in extracellular enzyme level, with greatest stimulation from the labile substrate. Different types of DOM led to different patterns of enzyme activity, indicating major shifts in community function as a substrate response.

RAPD analysis

Banding patterns of randomly amplified DNA (RAPD) provide a sensitive characteristic for analysing alterations in community composition. Community DNA was extracted from gravel samples containing biofilm bacteria, then DNA sequences amplified using RAPD analysis. This involved the use of six arbitrary primers amplifying DNA sequences of varying length, followed by separation of the amplified regions (amplicons) by gel electrophoresis. Because the primers differ in sequence, each produces a different distribution of amplicons, resulting in a size pattern which represents a molecular 'community signature.'

RAPD analysis of stream bacteria showed clear distinctions between communities in relation to original environmental location (with strong clustering of samples according to stream type) and also in relation to DOM. Marked divergence of communities occurred in relation to labile versus recalcitrant DOM, indicating a major shift in community composition as a response to organic substrate composition.

B. GENETIC INTERACTIONS

The ecological significance of genetic activity in aquatic bacteria has already been emphasized in relation to biofilm communities (Section 1.4.1), where physiological differences between biofilm and planktonic organisms relate to quorum sensing and the expression of stationary-phase genes (Figure 1.7). This section considers aspects of genetic diversity in aquatic bacteria, mechanisms of gene transfer and evidence for gene transfer in freshwater systems.

6.3   Genetic diversity

Genetic diversity involves variation in genetic composition and gene expression, and is a key factor in:

• the variety of bacterial metabolic activities in the freshwater environment, and

• ecological adaptations of bacteria to particular habitats.

The genetic composition of particular bacteria and bacterial groups within the microbial community may be considered in relation to the relative contributions of chromosomal and accessory DNA, the ecological importance of gene transfer in freshwater systems, and the complexity of the 'community genome.'

Expression of individual genes is regulated by both external (e.g., specific substrate availability) and internal (e.g., stage of cell cycle) factors (Prescott *et al.*, 2002). Major environmental changes may induce the expression of a whole array of genes, which in prokaryotes (bacteria and blue-green algae) may be triggered by the expression of a controlling transcription factor. In bacteria,

this is seen in the response of biofilm communities to high population density (quorum sensing – Section 1.4.1) and the response of bacterioplankton to acute nutrient deprivation (starvation response – Section 6.6.4). In blue-green algae, multiple induction of genes via sigma factors has been demonstrated during exposure to physico-chemical stress (heat, salinity, nutrient deprivation – Section 3.3.2) and in the diurnal oscillation of gene expression in (Section 4.10).

6.3.1 Chromosomal and accessory DNA

Genetic diversity in freshwater bacteria, as with bacteria generally, is characterized by the presence of two separate genetic systems, composed of chromosomal and extra-chromosomal DNA. Chromosomal DNA carries the main gene bank of the bacterial cell, contains all the genes that are essential for growth and division of the bacterium, and replicates in strict synchrony with the timing of the cell cycle. Extra-chromosomal DNA is present in the bacterial cell as separate, relatively short-sequence fragments, and includes phages, plasmids, transposons, and insertion sequences. This DNA can be regarded as accessory to chromosomal DNA, since it typically encodes for non-essential characteristics such as UVC resistance, catabolism of unusual carbon sources, resistance to antibiotics and heavy metals, and metabolites involved in secondary metabolism. These accessory elements are present in variable amounts and are able to replicate independently of the cell cycle.

Genetic diversity in bacteria is promoted by the active transfer of genes between organisms, which involves the transport and incorporation of accessory elements (containing the transferring gene)

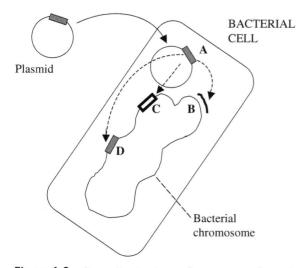

Plasmid

BACTERIAL CELL

Bacterial chromosome

Figure 6.2 Generalized scheme for gene transfer and expression in a recipient bacterial cell. Entry of accessory DNA (e.g., plasmid) by transduction, transformation, or conjugation leads to internal expression of a gene (▬) on the plasmid (A), or transfer to the chromosomal DNA, resulting in: (B) recombination (replacement of a region of DNA in the chromosome by homologous, introduced DNA) or (C) mutation (insertion of the DNA into a genome sequence, causing loss of change of gene function in that sequence) or (D) transposition (insertion of a novel or existing active gene sequence into a new part of the chromosome)

from one cell to another (Figure 6.2). Transferred genes may have long-term and short-term effects as follows.

• Many of the transferred genes are retained on the transferred DNA (e.g., plasmid) and are expressed in the host bacterial cell. Plasmid-encoded genes often confer short-term environmental advantage, such as resistance to heavy metals in a polluted habitat and resistance to ultra-violet (UVC) damage to DNA under conditions of high irradiation.

Many bacteria (e.g., *Escherichia coli*) are able to repair DNA damage caused by UVC irradiation via a chromosomally encoded UVC-inducible, mutagenic repair system, which is mediated via a key macromolecule – RecA protein. The common freshwater bacterium *Pseudomonas aer-* *uginosa* has little resistance to the damaging effects of UVC radiation since it does not contain this system, but it does contain RecA protein (Miller *et al.*, 1990). Many naturally-occurring plasmids of *Pseudomonas* species encode for UVC resistance and are found abundantly in aquatic habitats. Introduction of such plasmids into RecA$^+$ strains of *Pseudomonas aeruginosa* allowed UVC-inducible repair, allowing these organisms to grow and survive in exposed environments. The UVC-repair system is also mutagenic, so introduction of UVC plasmids also resulted in an increased mutation rate.

• Accessory elements may also carry and introduce fragments of DNA which become incorporated into the main chromosome. The chromosomal alterations are of three main types (Figure 6.2) – recombination, mutation, and transposition. Recombination involves replacement of a region of chromosomal DNA by homologous DNA introduced into the bacterial cell, and is an important aspect of DNA repair. Direct insertion of the introduced DNA into the bacterial chromosome can cause loss or alteration of existing gene function (mutation) or expression of the gene in a new region of the chromosome (transposition). Alteration of gene expression on the chromosome has long-term significance, since the modified chromosomal DNA replicates and passes to daughter cells in a stable manner (Miller *et al.*, 1990).

The incorporation and expression of plasmid genes in different bacteria is important for phenotypic diversity within the bacterial community. This includes the expression of novel genes introduced via human activity as well as natural mutations arising within the ecosystem (see next section).

6.3.2 The ecological importance of gene transfer in freshwater systems

The potential for gene transfer between bacteria in aquatic systems has been the subject of much research (Wellington and van Elsas, 1992) and is of interest for two main reasons – introduction and spread of novel genes arising from human activities, and the natural process of bacterial evolution in freshwater environments.

Introduction of novel genes

This has received much attention in recent years because of potential problems with:

- the discharge of bacteria with plasmids containing antibiotic resistance genes (originating from clinical and agricultural sources) from sewage and waste-treatment facilities. The possible transfer of these plasmids to the indigenous microbial flora and to human pathogens imposes a potential threat to human health;

- the potential release of genetically-engineered microorganisms (GEMs) into the terrestrial environment, with aquatic contamination and transfer of the novel genes to other freshwater biota. Genetically engineered bacteria which might be released into the environment include a wide range of organisms that have been modified for agricultural purposes – other possibilities include the use of GEMs which have been modified for the detoxification of polluted soil, water, and land-fill sites by the manipulation of normal catabolic activities (Winstanley *et al.*, 1992).

In both cases, transfer and spread of the novel genes within the aquatic environment has potential implications for human health. Introduction of novel genes may also be important within the natural ecological framework, affecting the short-term survival and long-term evolution of bacterial species.

Gene transfer within the ecosystem

The transfer and spread of genes (including natural mutations) within aquatic bacterial populations is important for a number of key processes that affect the ability of bacteria to adapt to changes in the freshwater environment and compete with other organisms (Young, 1992).

- Gene transfer speeds up the process of bacterial evolution by bringing together new genes that originally arose as advantageous mutations in different bacterial lines. The origin of these mutations can thus be exploited by the entire bacterial community, allowing these organisms to adapt more rapidly to new environments.

- Gene transfer gives accessory genetic elements a major role in contributing to genetic diversity. As noted previously, these elements are of various types and code for phenotypic characteristics such as antibiotic resistance, heavy metal resistance, and degradation of complex organic compounds – all of which may give short-term ecological advantage. The role of accessory elements as agents for rapid transfer across bacterial lineages is indicated by the fact that the genetic control of conjugation is plasmid-mediated.

- The introduction of new DNA into the bacterial cell, particularly into the bacterial chromosome, not only generates genetic novelty but is also important for maintaining the *status quo*. Gene transfer by transformation is thought to be particularly important in DNA repair, replacing DNA that has become non-functional due to adverse mutation restoring the original gene function.

6.3.3   Total genetic diversity: the 'community genome'

Total genetic diversity can be estimated from the molecular complexity of the collective DNA sample that is isolated from the entire community (Torsvik *et al.*, 2000). The DNA complexity can be determined by measuring the re-annealing (reassociation) rate for single-stranded DNA in solution under defined conditions. If the microbial community can be regarded for the sake of the calculation as a 'single species', the reassociation data provide an estimate of the total genome size of this species. The total number of putative species can then be derived by dividing this value by the standard genome size of a 'typical species' such as *Escherichia coli*.

As an example of this, Ovreas *et al.* (2003) quote values for the total DNA complexities of bacterial communities in pristine sediments in the range 2.7×10^{10} to 4.8×10^{10} bp. Taking the genome size of *E. coli* as 4.1×10^6 bp gives a DNA complexity within the sediments that is equivalent to 6500–11 500 genomes. The number of bacterial species that actually contribute to this diversity must be a matter of speculation.

6.4 Mechanisms for gene transfer in freshwater systems

Three main mechanisms for bacterial gene transfer have been demonstrated in the freshwater environment (Pickup, 1992; Van Elsas, 1992; Ehlers, 2000) – transformation (uptake of naked DNA), transduction (mediated by bacteriophages), and conjugation (requiring cell-to-cell contact). The transferred genes may either by carried on intact plasmids (frequently the case with transduction and conjugation) or on other types of DNA fragment (transformation). Once initial transfer has occurred, long-term survival of the introduced gene will depend either on persistence of the DNA fragment within the bacterial population or on transposition of the gene to permanent genomic or extra-chromosomal DNA within the bacterial cell (Figure 6.2).

Movement of genes within bacterial cells includes the activities of insertion elements and transposons. Insertion elements are DNA sequences which encode the movement of the DNA sequence from one location to another. These elements carry no selectable markers and are typically less than 1 kb in size. Transposons, in comparison, are larger and carry other genes, often encoding antibiotic resistance, flanked by insertion sequences. Insertion of transposons into the bacterial genome has a variety of effects including inactivation of the gene or operon into which they transpose, delivery of novel

genes contained within their sequence, and rearrangement of other DNA sequences within the bacterial DNA. The transfer of transposons from host to host can be mediated by any of the three transfer mechanisms noted below.

6.4.1 Transformation: uptake of exogenous DNA

The ability of planktonic or attached (e.g., biofilm) bacteria to take up external (exogenous) DNA from the water medium and incorporate it into their genome is referred to as transformation. In this process (Figure 6.3), genes present on exogenous DNA originally derived from a donor bacterium are transferred to a new cellular environment, where they may become active as novel genes or replace existing genes (DNA repair).

Requirements for transformation

The two major requirements for transformation in the aquatic environment are free availability of extra-cellular DNA and the physiological capability of recipient cells (referred to as 'transformable' or 'competent'). The conversion of non-competent to competent bacterial cells is prompted by environmental factors and is considered in the following section.

Exogenous DNA appears to be universally available in freshwater systems, occurring over a wide range of concentrations (1–200 ng ml^{-1}). This compares with a standard level of approximately 27 ng ml^{-1} in sea water (Paul *et al.*, 1988 and 1991). Although this DNA occurs as insoluble macromolecular fragments within the aquatic medium, it is technically part of the dissolved organic material (DOM) or dissolved organic carbon (DOC) since it passes through a 0.2 μm filter membrane. Unlike most of the DOC derived from lake biota, which is secreted by living algal cells, exogenous DNA is produced by breakdown or lysis which results from cell death (Paul, 1989). Within the freshwater environment, the exogenous DNA is heterogeneous both in terms of origin (derived from bacteria, algae,

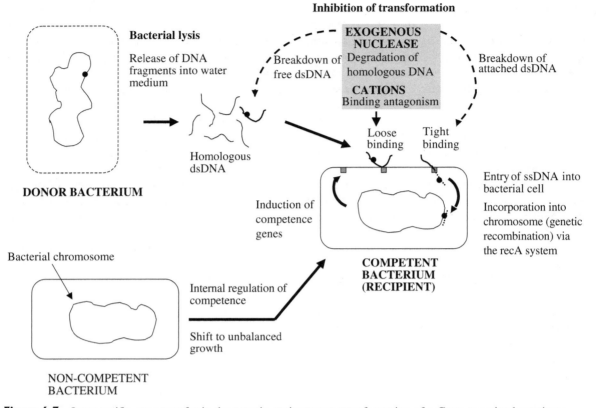

Figure 6.3 Intraspecific gene transfer in the aquatic environment: transformation of a Gram-negative bacterium
▪ – DNA receptor, ● – transferred gene, DNA: double-stranded (——), single-stranded (- - - -)

invertebrates) and state – comprising both single-stranded (ssDNA) and double-stranded (dsDNA) fragments. This DNA is in a continuous dynamic flux and has a rapid turnover, with persistence times in the range 4–24 hours. DNA is continuously being released from lysed cells into the freshwater system, and is continuously being removed by processes of adsorption (to cell surfaces and other particulate matter) and enzymatic breakdown (by exogenous nuclease enzymes).

Bacterial transformation only involves homologous (i.e., from the same strain or species) dsDNA. DNA from other organisms, ssDNA, and RNA are not taken up and may actually interfere with the transformation process by competing for adsorption sites on the recipient bacterial surface.

The transformation process

The natural process of transformation can be thought of as a sequence of three distinct events – (1) development of competence, (2) DNA binding, and (3) DNA uptake and incorporation into the host cell (Stewart, 1992; Figure 6.3).

Development of competence   Competence is a regulated state in transformable bacteria. In Gram-negative organisms, this is internally controlled and appears to be correlated with a state of limited or unbalanced growth (Figure 6.3). Competence may be promoted where environments are inherently nutrient-limited or where high bacterial densities in niche habitats (e.g., biofilms) have a localized growth-limiting effect. In Gram-positive bacteria, the induction of competence depends on the external accumulation of secreted polypeptide competence factors to a critical level, and is largely determined by population density (Stewart, 1992).

The development of bacterial competence involves the induction of various competence genes, resulting in the synthesis of surface DNA receptors and other key proteins.

DNA binding   The binding of exogenous DNA to the surface of competent bacterial cells is closely dependent on the chemical composition of the water medium – including concentration of homologous DNA, presence of heterologous DNA, concentration of nuclease enzymes, and ionic composition.

The binding process involves a sequence of events, with initial bacterial/DNA contact, followed by loose (electrostatic) attachment to the protein receptor then firmer (covalent) association. The aquatic concentration of homologous dsDNA is clearly important for initial contact and binding, and in some bacteria the presence of specific nucleotide recognition sequences are also necessary. Any factors which reduce the level of homologous dsDNA in the environment (e.g., presence of nuclease enzymes) or interfere with the binding process (e.g., presence of heterologous DNAs) will inhibit transformation. In addition to these effects, loose binding of DNA is antagonized by high cation levels, and at each stage of binding the surface DNA is vulnerable to attack by nuclease enzymes.

The opposing effects of ion antagonism, interference by other nucleic acids, and enzyme degradation on bacterial surface receptor binding can be viewed as a complex framework of competitive interactions in the freshwater environment.

DNA uptake and incorporation into the bacterial cell   DNA uptake and incorporation represents the final stage of bacterial transformation. The dsDNA attached to the surface receptor is nicked by a surface nuclease, which sequentially degrades one of the strands allowing the other strand to enter in isolation. The entry of a single-stranded intermediate into the cytoplasm allows the key bacterial recombination protein (recA protein) to bind

to the transformed strand. This protein is then activated for binding to the double-stranded chromosomal DNA, subsequently positioning and inserting the transformed strand at the homologous region of the main chromosome (genetic recombination). More detailed information on the molecular events taking place during transformation can be found in standard texts on microbiology (e.g., Prescott *et al.*, 2002).

Significance of bacterial transformation in the aquatic environment

Although the extent to which transformation occurs in aquatic environments is not known, external DNA availability and the presence of favourable environmental (nutrient limitation, localized cell concentrations) conditions suggest that it may be significant. In terms of environmental significance, however, it is important to distinguish between gene transfer at the interspecific (between species) and intraspecific (within species) levels.

Transformation is generally considered to have only limited ecological potential (Van Elsas, 1992) for gene transfer at the interspecific level, since:

- only a limited number of bacterial species are known to be transformable, so there is a limit on host bacteria,

- base-sequence homology is often required for stable maintenance of introduced DNA, so there is a limit on donor DNA,

- there are restrictions to the size and type of DNA that can be successfully taken up by transformable cells, placing further limitations on the range of DNA uptake.

These limitations are reduced when considering the uptake of small fragments of homologous DNA within species, and transformation may have more significance for DNA repair (see previously) than gene transfer. The relevance of transformation to DNA repair is emphasized by the fact that most transformable bacteria use their primary recombination (recA) system for the final step in the transformation process.

6.4.2 Transduction: gene transfer between bacteria via bacteriophages

The potential for transductive gene transfer within aquatic bacterial communities has been shown by experimental studies on *Pseudomonas aeruginosa* model systems. These have demonstrated that stable phage-mediated transfer of both chromosomal (Morrison *et al.*, 1979) and plasmid (Saye *et al.*, 1987) genes can occur between strains of the same species at significant frequencies within freshwater habitats.

Although the potential for this process is clearly present in aquatic environments, the general importance of transduction for gene transfer may be limited (Van Elsas, 1992), since:

- bacteriophages often have restricted host ranges, reducing the variety of species available for gene exchange,

- there is a limit to the size of bacterial DNA that can be packaged inside phage heads.

The mechanism and significance of transduction within aquatic bacterial communities is discussed in Section 7.8.6.

6.4.3 Conjugation: transfer of plasmid DNA by direct cell contact

Gene transfer between bacterial cells by conjugation typically involves the transfer of plasmids and is probably the genetic exchange system of greatest importance within freshwater bacterial communities. Plasmids (and other extra-chromosomal elements) have evolved to achieve maximum gene expression within bacterial populations, and have developed capabilities for:

- optimal transfer between conjugating bacterial cells,

- post-conjugation replication (i.e., survival) in the new (transconjugant) host cells – plasmids typically have 'wide-range' replication functions,

- post-conjugation gene expression in the new host cells.

The process of conjugation which leads to plasmid transfer is itself determined almost entirely by plasmid rather than chromosomal DNA. This encodes specific transfer functions allowing the DNA to pass from one bacterial cell to another. These conjugative plasmids, which promote their own transfer during conjugation, are also able to promote the transfer of non-conjugative plasmids. This is referred to as mobilization, and requires the presence of a *mob* site on the non-conjugative plasmid. Numerous genetically distinct and non-interacting systems have evolved to effect plasmid transfer. Some systems (broad range) have the capacity to encode plasmid transfer to a wide range of phyllogenetically-unrelated organisms, while other (narrow-range) systems only permit plasmid transfer to a small group of closely-related bacteria.

The importance of conjugative gene transfer in the freshwater environment is underscored by the role of plasmids in this process. Much of the evidence for gene transfer in the aquatic environment comes from studies on plasmids and their movement within bacterial populations.

6.5 Evidence for gene transfer in the aquatic environment

Evidence for gene transfer between bacteria in the freshwater environment comes from three major sources – retrospective analysis, laboratory (*in vitro*), and field (*in situ*) experimental studies (Pickup, 1992).

6.5.1 Retrospective analysis

Epidemiological analysis of plasmid populations in natural communities of bacteria has provided much useful information on gene transfer. Retrospective analysis of plasmid transfer involves three stages:

- isolation of specific groups of bacteria such as pseudomonads, antibiotic-resistant bacteria, and mercury-resistant bacteria from the environment,

- size and molecular analysis of the plasmids,

- inference of transfer events from molecular similarities found between plasmids isolated from related and unrelated bacteria.

Various published reports provide evidence for gene transfer in different aquatic environments, including rivers, estuarine and river sediments and lakes.

Case study 6.2   Plasmid-borne resistance in aquatic bacteria

In one of the earliest studies on plasmid diversity in aquatic bacteria Jobling *et al.* (1988) isolated bacteria from the River Mersey (UK) and analysed them for their tolerance to mercury ($HgCl_2$). Approximately 40 per cent of the bacterial population exhibited mercury resistance (HG^R), and in 13 out of 52 HG^R isolates the tolerance was encoded on a single plasmid which could be transferred to *Escherichia coli* in conjugal matings. These 13 HG^R plasmids:

- ranged in size from 75 kb to >250 kb,

- encoded the same HG^R phenotype, all expressing a narrow spectrum mercury resistance,

- could be classified by restriction mapping into three distinct groups,

- included one plasmid group that was present in eight of the 13 isolates, comprising a range of Gram-negative species (in the genera *Alcaligenes*, *Pseudomonas*, *Klebsiella*, *Enterobacter*, and *Acinobacter*). This plasmid was exactly similar in terms of size and restriction mapping in all these organisms.

These results indicate that plasmids play an important role in the response of bacteria to contaminating mercury in this aquatic environment. The presence of a defined single HG^R conjugative plasmid in different taxonomic groups suggests that there has been widespread plasmid transfer within the environment. The occurrence of the HG^R gene in three different groups of plasmid also implies that there has been considerable genetic rearrangement.

6.5.2   Laboratory (in vitro) studies on plasmid transfer

The basis of laboratory experiments on plasmid transfer in bacteria is to mix donor cells (plasmid-containing strains) with plasmid-free recipient cells, promote mating between the bacteria, then determine the characteristics and frequency of transconjugant formation. Laboratory (*in vitro*) studies on freshwater bacteria have the advantage over field (*in situ*) studies in that gene transfer can be studied between defined strains of organisms under controlled physical, chemical, and biological conditions. In order to match the field situation, however, laboratory studies normally involve the use of bacteria that have originally been isolated from the freshwater environment, with mating being carried out in conditions which closely approximate to the river or lake situation.

Case study 6.3   Plasmid transfer in Pseudomonas aeruginosa

Laboratory studies by O'Morchoe *et al.* (1988) have demonstrated plasmid transfer between strains of the freshwater bacterium *Pseudomonas aeruginosa* (Figure 6.4). Separate cultures of donor and recipient bacteria were grown to high population level (10^8 CFU ml^{-1}), then mixed together. Mating was carried out

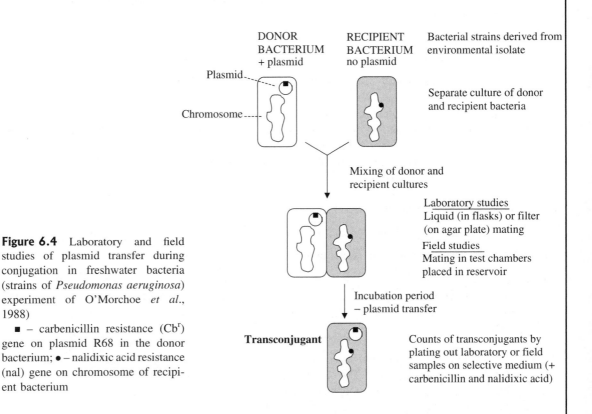

Figure 6.4 Laboratory and field studies of plasmid transfer during conjugation in freshwater bacteria (strains of *Pseudomonas aeruginosa*) experiment of O'Morchoe *et al.*, 1988)

■ – carbenicillin resistance (Cb^r) gene on plasmid R68 in the donor bacterium; ● – nalidixic acid resistance (nal) gene on chromosome of recipient bacterium

either in suspension (liquid mating) or on filter membranes. After a 2 hour incubation period, transconjugants were identified in-and counted by-plating out on medium selective for particular antibiotic markers. The plasmid transfer frequency, expressed as the number of transconjugants per donor cell, ranged from 10^{-3} to 10^{-6}, and was influenced by the following.

- *The ratio of donors to recipients.* An excess of recipient cells in the mating mix increased the frequency of transconjugant cells recovered by as much as 100-fold.

- *The presence of a natural microbial community.* Mating was tested in both sterilized and natural lake water. When the natural microbial community was present (unsterilized water) the frequency of transfer was typically lower than in comparable matings carried out in sterilized water.

- *The mating substratum.* Mating carried out on a solid surface (filter mating) generated a higher level of transconjugants compared to mating in suspension (liquid mating).

These results are similar to those obtained by other workers (Pickup, 1992) and suggest that plasmid transfer between bacteria will be particularly high in biofilms (compared to planktonic bacteria), where population counts are high and frequent contacts occur within the microenvironment. The addition of a natural microbial community to the laboratory system introduces other bacteria and microorganisms such as protozoa. The presence of other bacteria appears to exert an effect on the recorded level of gene transfer, either affecting the transfer of plasmids or the detectability of transconjugants. Protozoa probably influence the end result by ingesting experimental (donor, recipient, and transconjugant) bacteria along with other bacterial cells.

6.5.3 Field (in situ) studies on bacterial gene transfer

Early experiments on gene transfer focussed on R-factor transmission between *Escherichia coli* and other coliform bacteria and were laboratory-orientated (Pickup, 1992). These studies then developed a more environmental approach, attempting to create lake or river conditions in terms of the organisms involved (field isolates) and the laboratory microenvironment for plasmid transfer (see previous section).

This approach was extended into the field, with enclosed systems (dialysis tubing, Teflon bags, diffusion chambers) containing mating mixtures of bacteria being placed directly into the freshwater environment. In their studies on gene transfer between strains of *Pseudomonas aeruginosa*, O'Morchoe *et al.* (1988) also carried out test matings within enclosed chambers placed in a reservoir environment (Figure 6.4). A significant number of transconjugants was recovered from these field trials, demonstrating gene transfer under semi-natural conditions, but the frequency of detectable transfers was considerably lower than had been obtained from the laboratory matings involving pure cultures. The presence of other microorganisms in the field probably has the same effects as the addition of natural microbial populations to laboratory systems (see previously).

The use of enclosed systems may be further extended to studies on unenclosed systems. Bale *et al.* (1988) studied plasmid transfer between pseudomonad species in river epilithon by attaching filters containing donor and recipient cells to the surface of stones and submerging them in river water. In such situations, transfer of marker plasmids can be investigated between donor and recipient bacteria (within the filter) or between donor bacteria and the indigenous epilithic bacteria which can be isolated as a mixed natural suspension from the stone surface.

Plasmid transfer studies carried out in a variety of running- and standing-water environments (Pickup, 1992), show that gene transfer frequency is highly variable, ranging from high (10^{-1}) to almost undetectable (10^{-9}) transconjugants per donor cell. In all reported studies, however, some plasmid transfer was detectable, and was typically in the range 10^{-3} to 10^{-6} transconjugants per donor cell. These studies indicate that plasmid transfer between bacteria is widely significant within the freshwater environment.

Relative importance of environmental conditions on gene transfer

Relatively little direct information is available on the importance of environmental factors in gene transfer, and much of the speculation on this is derived from laboratory studies. In general, this process would be expected to be stimulated by conditions which promote:

- enhanced survival and activity of the host cells (favourable temperatures, high nutrients), and

- the specific processes of transformation, transduction, or conjugation.

Environmental factors influencing the different processes of gene transfer are highly varied and are clearly important in maintaining genetic diversity within the bacterial community. These include aspects such as concentrations of soluble DNA, exogenous nuclease and cations (transformation), nutrient availability (induction of competence in transformation), presence of bacteriophage virions (transduction), and proximity for cell contact (conjugation, transformation).

The relative importance of the three processes of gene transfer may also vary between microenvironments, with conjugation possibly predominating in high population density biofilms – while transformation and transduction may be relatively more important in dispersed planktonic populations.

C. METABOLIC ACTIVITIES

6.6 Metabolic diversity of freshwater bacteria

Variations in the metabolic activities of bacteria are an important aspect of their diversity within the freshwater environment, and reflect their different roles and locations within the ecosystem. In this section metabolic diversity is considered in relation to key metabolic parameters, variations in CO_2 fixation, aerobic and anaerobic decomposition of organic substrates and the metabolic transition which occurs from high- to low-nutrient availability.

6.6.1   Key metabolic parameters

Four key features define the metabolic status of individual freshwater bacteria.

- *The source of energy.* Either from light (photo-trophs) by photosynthesis or from chemical energy (chemotrophs). Chemotrophs use energy obtained from energy-yielding (exergonic) reactions to oxidize organic matter.

- *The source of electrons for growth (electron donor).* These are obtained either from organic (organotrophs), or from chemical compounds (lithotrophs) such as sulphide, hydrogen, andwater.

- *The source of carbon, required for synthesis of bacterial biomass.* This is obtained either by reducing CO_2 during photosynthesis (auto-trophs), or from complex organic compounds (heterotrophs).

- *The terminal electron acceptor.* The final electron acceptor in the process of respiration involves either oxygen (aerobic respiration) or other molecules (e.g., sulphate, nitrate) in anaerobic respiration.

The main role of bacteria in the freshwater environment is the breakdown of organic biomass and the recycling of various key elements (nitrogen, phosphorus, sulphur) which are present within the various organic compounds. In line with this, the majority of freshwater bacteria are heterotrophic, living on organic carbon compounds present in the aquatic medium or in the sediments. Using the above terminology, the typical bacterium in the water column or substratum of a freshwater system is a chemo-organo-heterotroph, while the typical algal cell is a photo-litho-autotroph (Figure 6.5).

Classification of freshwater bacteria in relation to their metabolic activities (Table 6.2) provides a set of ecologically relevant groups, some of which (chemosynthetic autotrophs, most non-photosynthetic heterotrophs) occur in aerobic environments, while others (e.g., photosynthetic autotrophs, photosynthetic heterotrophs, fermentation bacteria) are present in anaerobic hypolimnia, sediments and other anoxic sites.

6.6.2   CO₂ fixation

Although CO_2 fixation in freshwater environments is usually associated with the photosynthetic requirements of algae and macrophytes, some bacteria also have this activity. Bacteria which are able

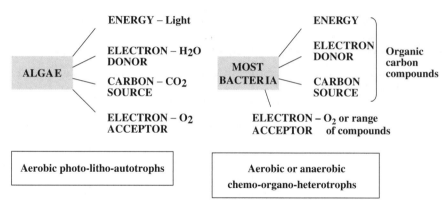

Figure 6.5 General metabolic comparison of freshwater algae and bacteria: algae are normally simply referred to as autotrophs and bacteria as heterotrophs

to assimilate CO_2 directly are of three types, as follows.

- *Photosynthetic bacteria (photoautotrophs)*. Using light energy to mediate CO_2 uptake. Major characteristics of these organisms are described in Section 6.7.

- *Chemosynthetic bacteria (chemoautotrophs)*. These organisms typically occur within the water column at the boundary layer between aerobic and anaerobic zones, using reduced inorganic compounds as energy substrates. The reduced compounds are largely derived by decomposition of organic matter within the anaerobic hypolimnion. Chemosynthetic fixation of CO_2 is particularly high in conditions of steep redox potential gradient which occur at the top of the anaerobic zone, but is low in other parts of the water column. The requirement for an anaerobic hypolimnion means that chemosynthesis tends to be prominent in eutrophic rather than oligotrophic lakes, and in the stable long-term redox gradients that develop in meromictic lakes.

- *Heterotrophic bacteria*. Most non-photosynthetic fixation of CO_2 is carried out by heterotrophic bacteria. This dark CO_2 fixation provides a useful parameter for measuring heterotrophic productivity.

6.6.3 Breakdown of organic matter in aerobic and anaerobic environments

The breakdown of biological material by bacteria ultimately involves an oxidation/reduction reaction, with the transfer of electrons from the organic substrate (oxidation) to an electron acceptor (reduction). This may be summarized:

$$(Substrate + \bar{e}) + Acceptor \longrightarrow Substrate$$
$$+ (Acceptor + \bar{e}) + free\ energy \qquad (6.1)$$

Bacteria act as catalysts in mediating this reaction, which can only take place in conditions which are thermodynamically suitable. The energy that is released by this process can be quantified as the free energy per mole of substrate (ΔG_0^1). Only a fraction of this energy (1–50 per cent) becomes directly available to the bacterial cell, and is used for catabolic and anabolic (synthetic) processes, both of which promote bacterial productivity and population increase. The free energy that is released in the oxidation process varies with substrate and electron acceptor, which depend on environmental conditions – including the presence (aerobic environment) and absence (anaerobic environment) of oxygen.

Table 6.2 Classification of major groups of freshwater bacteria according to their metabolism; aerobic bacteria (shaded boxes) occur in the epilimnion of stratified eutrophic lakes, and throughout the oxygenated water column of oligotrophic lakes and running waters; anaerobic bacteria occur in the hypolimnion and sediments of eutrophic lakes and other anoxic environments

Major metabolic types	Energy source and electron donor	Electron acceptor	Carbon source	End products (other than carbon)	Organism
Photosynthetic autotrophs	Light, H_2S, S, $S_2O_3^{2-}$, H_2	H_2O	CO_2	S, SO_4^{2-}, H_2O	Green and purple sulphur bacteria
Chemosynthetic autotrophs	NH_4^+, NO_2^-	O_2	CO_2	NO_3^-	Nitrifying bacteria
	H_2S	O_2	CO_2	S°, H_2O, SO_4^{2-},	Colourless sulphur bacteria
	H_2	CO_2	CO_2	CH_4	Methanogens
Photosynthetic heterotrophs	Light, organic substances (sugars, alcohols, acids)	H_2O	Organic substances	H_2O	Non-sulphur purple bacteria
Non-photosynthetic heterotrophs (selected types only)					
(a) Majority of freshwater bacteria	Organic substances	O_2	Organic substances	Organic acids, alcohols, CO_2 etc.	Aerobic heterotrophic bacteria
(b) Denitrifers	Organic substances	NO_3^-	Organic substances	N_2, NO_2^-, NH_3	Denitrifying bacteria
(c) Sulphate reducers	Primarily organic substances	SO_3^{2-}, $S_2O_3^{2-}$ ($S_4O_6^{2-}$), NO_3^-	Organic substances	$H_2S(S_2O_3^{2-})$, N_2	Sulphate reducing bacteria
(d) Mn and Fe reducers	Organic substances	Fe^{3+}, Mn^{4+}	Organic substances	Fe^{2+}, Mn^{2+} CO_2, N_2, NH_3	Facultative anaerobes
(e) Fermenters	Organic substances	Organic substances	Organic substances	H_2, CO_2, organic acids, NH_3, CH_4, H_2S	Fermentation bacteria

Aerobic conditions

Aerobic conditions may be defined as environments where oxygen is freely available, and is used in the oxidation of organic (and some inorganic) substrates. Aerobic environments contain obligate aerobic microorganisms (which are restricted to oxygen as the secondary electron acceptor) and facultative anaerobes (which are able to use other secondary acceptors in addition to oxygen).

In a lake, the concentration of dissolved oxygen (DO) ranges from supersaturation (lake surface) to very low levels in the hypolimnion and sediments. This range of oxygen concentration correlates with a range of oxidizing ability or oxidation/reduction potential (redox potential). Redox potential (E_h) can be measured in reference to a standard electrode, and is normally expressed as millivolts (mV). In well-oxygenated environments, the redox potential is normally in excess of $+360$mV, falling to -500 mV in highly reducing conditions.

Because of the connection between free energy released during substrate oxidation and bacterial growth, bacteria that are involved in high-energy reactions have the potential for higher growth rates and can out-compete those that mediate lower-energy reactions. In aerobic environments, oxygen which generates most free energy, will become the main electron acceptor and bacteria that use this to oxidize their substrates will predominate. The surface waters of rivers and lakes are thus populated by obligate aerobes which use complex organic compounds as a source of energy, carbon, and electrons. The oxidation–reduction reaction can be summarized (Kalff, 2002):

Organic biomass

$$(CH_2O)_{106}(NH_3)_{16}(H_3PO_4) + 138O_2 \longrightarrow$$
$$106CO_2 + 16HNO_3 + H_3PO_4 + 122H_2O \quad (6.2)$$

Free energy (ΔG_0^1) = -3190 kJmol^{-1}

The organic compounds which are degraded by bacteria in lakes include material from dead biota (particularly algae that have sedimented out of the epilimnion) and extracellular products (dissolved organic carbon, DOC). In oligotrophic lakes, both of these sources of carbon are minimal, limiting the activity of aerobic heterotrophs and the removal of oxygen, so that the whole water column remains aerobic. In eutrophic lakes, high substrate levels lead to much higher oxygen uptake, resulting in anaerobic conditions. This is particularly acute in the hypolimnion, where there is greatest accumulation of dead organic matter and no photosynthetic generation of oxygen by algae. In streams and rivers, where the entire water column is typically well-oxygenated, much aerobic degradation by bacteria occurs on the sediments. The role of bacteria in the decomposition of leaf litter is discussed in Section 8.5.2 (Figure 8.4), and typically follows a phase of fungal activity.

Although most bacteria in aerobic environments are involved in the oxidation of organic material, some are able to use inorganic compounds. Such chemolithotrophic bacteria are active in many aerobic situations and are responsible for the oxidation of a range of inorganic substrates including methane, hydrogen sulphide, ammonium ions, nitrite ions, and ferrous ions (Table 6.3). The maximum amount of free energy (ΔG_0^1) released per mole of reactants ranges from -4 to -810 kJ in these reactions, and is considerable less than the value of -3190 kJmol^{-1} (see above) generated by oxidation of biomass. Many of the bacteria involved in these reactions

Table 6.3 Oxidation of selected reduced inorganic compounds in aerobic conditions by chemolithotrophic bacteria

Substrate	Reaction	Free energy released[*]
Methane	$CH_4 + 2O_2 \longrightarrow CO_2 + 2H_2O$	-810
Hydrogen sulphide	$HS^- + 2O_2 \longrightarrow SO_4^{2-} + H^+$	-797
Ammonium ions	$2NH_4^+ + 3O_2 \longrightarrow 2NO_2^- + 4H^+ + 2H_2O$	-275
Nitrite ions	$2NO_2^- + O_2 \longrightarrow 2NO_3^-$	-75.8
Ferrous ions	$4Fe^{2+} + O_2 + 4H^+ \longrightarrow 4Fe^{3+} + 2H_2O$	-4

[*]The maximum free energy released (ΔG_0^1) is expressed as kJ mol^{-1}, and represents the change in standard free energy when 1 mole of reactants is converted to 1 mole of product (at 25°C, 1 atmosphere pressure and at pH7, information from Kalff, 2002)

are present in the oxygenated microzone of mud–water interfaces, where there is a ready supply of substrate by diffusion from the lower sediments (internal loading).

Anaerobic decomposition of organic matter

Anaerobic environments are those where the concentration of oxygen is too low for it to be used as an electron acceptor. These environments may be divided into two main groups – low oxygen environments, and anoxic environments, where oxygen is completely absent.

- *Low oxygen environments.* Although oxygen is the electron acceptor in fully-oxygenated environments, removal of DO by metabolic processes may reduce the availability to such an extent that other electron acceptors (Mn^{4+}, NO_3^-, Fe^{3+}, and SO_4^{2-}) become used instead.

  The free oxidation energy released per molecule of organic matter via each of these acceptors varies considerably, from -380 to $-3050 \, kJmol^{-1}$ (Figure 6.6). Since oxidation of organic matter follows a sequence in which acceptors generating most energy take precedence, each of these acceptors will be used in turn until depleted levels lead on to the next one. This process continues until all oxidizable substrates or all electron acceptors are removed from the system. This chemical sequence (Figure 6.6) is paralleled by an ecological succession, in which whole communities of bacteria change with the chemical environment. Bacteria that use Mn^{4+}, NO_3^-, Fe^{3+} are facultative anaerobes, able to use either oxygen or an inorganic electron acceptor – depending on prevailing conditions.

- *Anoxic environments.* These environments, where oxygen is completely absent, contain populations of obligate anaerobic bacteria. Oxidation/reduction reactions using inorganic electron acceptors are severely restricted, though sulphate-reducing bacteria are able to use SO_4^{2-} when oxygen is completely exhausted (E_h <200 mV). Most

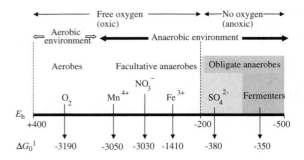

Figure 6.6 Sequence of electron acceptors during the transition from aerobic to anaerobic conditions: the sequence of electron acceptors follows the quantitative sequence of free energy (ΔG_0^1) released from one mol of organic matter during the oxidation or fermentation process (data from Kalff, 2002)

breakdown of organic material occurs by fermentation processes.

In anaerobic conditions, organic matter is metabolized by a variety of heterotrophic bacteria, which obtain energy by substrate phosphorylations. In this situation, where oxygen no longer acts as the universal hydrogen acceptor, the situation becomes complex, with a range of organic and inorganic compounds taking over this role. In some cases, the same compound can act as hydrogen acceptor or donor, depending on environmental conditions.

Large quantities of organic detritus are degraded under the anaerobic conditions which occur in the hypolimnion and sediments of eutrophic lakes, and in the sediments of ponds, rivers, and waste-treatment plants (septic tanks, anaerobic lagoons). The process of anaerobic degradation converts biomass to CO_2, CH_4, and NH_3 and can be summarized (Kalff, 2002):

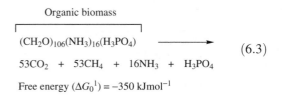

$$(CH_2O)_{106}(NH_3)_{16}(H_3PO_4) \longrightarrow \quad (6.3)$$

$$53CO_2 \; + \; 53CH_4 \; + \; 16NH_3 \; + \; H_3PO_4$$

Free energy (ΔG_0^1) = $-350 \, Jmol^{-1}$

Comparison with the oxidation process (using oxygen as electron acceptor) shows that bacteria in the

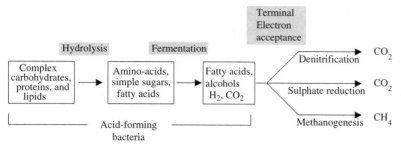

Figure 6.7 Stages of anaerobic metabolism of complex organic compounds (based on a figure from Wetzel, 1983)

oxygenated part of the water column obtain much more energy from aerobic breakdown of organic material compared with those carrying out anaerobic fermentation on the sediments.

The process of fermentation occurs as two distinct stages (Figure 6.7) as follows.

1. *Initial hydrolysis and fermentation.* Hydrolytic and fermentative conversion of proteins, carbohydrates, and fats to a range of breakdown products (primarily fatty acids) is carried out by a heterogeneous group of facultative and obligate anaerobic bacteria. These bacteria generate large amounts of organic acids and are collectively referred to as acid formers. The degradative activity of these organisms results in the formation of CO_2 and various reduced end products, including H_2, H_2S, acetic, proprionic, lactic, and butyric acids, ethanol, and amines. These compounds would accumulate in the anaerobic environment if they were not metabolized in various ways.

2. *Removal of breakdown products.* Removal of oxidizable intermediate and end products is carried out by obligate anaerobes which are able to use a range of hydrogen acceptors such as sulphates, nitrates, and CO_2. The different groups of organisms that carry out this terminal oxidation include the following.

- *Denitrifying and sulphate-reducing bacteria* which use nitrate and sulphate as the ultimate

electron acceptor. The activity of these organisms is limited by sulphate and nitrate availability, which has to diffuse from the epilimnion into the hypolimnion in eutrophic lakes. In aerobic sediments, bacteria such as *Thioploca* obtain their nitrate at the surface, then migrate into anaerobic regions to use the oxygen in nitrate for sulphur oxidation (Zemskya, 2001).

- *Methane-producing bacteria*, operate under strict anaerobic conditions, and include two rod-shaped (*Methanobacterium*, *Methanobacillus*) and two coccoid (*Methanococcus*, *Methanosarcina*) genera. Methane is generated by one of two processes.

In the first situation, bacteria use CO_2 as the acceptor of hydrogen derived from the organic acids:

$$CO_2 + 8H \longrightarrow CH_4 + 2H_2O \qquad (6.4)$$

In the second process, acetic acid is directly converted to CO_2 and methane:

$$CH_3COOH \longrightarrow CH_4 + CO_2 \qquad (6.5)$$

The methane generated by reduction of CO_2 escapes to aerobic regions where it is readily oxidized.

Below the top 1–2 mm, deep (profundal) sediments of stratified lakes are permanently anoxic, and methane production is a major activity as a key terminal process in the anaerobic decomposition of organic matter (Nüsslein et al., 2003). Fermentation becomes the dominant breakdown process in such conditions because the only alternative, sulphate reduction (Figure 6.6), is limited in most freshwater lakes by low sulphate availability. In North American lakes, more than half of the total carbon output to the sediment was found to be converted to CH_4, with H_2/CO_2 and acetate as the immediate substrates (see above). Studies on Lake Kinneret (Israel) demonstrated a direct coupling in deep sediments between the oxidation of acetate and the generation of methane (Nüsslein et al., 2003).

- *Photosynthetic bacteria*. Reduced substrates such as H_2S are released from decaying biomass as part of the fermentation process. Subsequent oxidation of reduced substrates by photosynthetic bacteria involves the removal of hydrogen and electrons to drive the reduction of CO_2 in the process of photophosphorylation. The uptake of CO_2 by these organisms as part of an assimilatory sulphate-reduction system is thus important for both carbon compound formation and the disposal of excess hydrogen and electrons from the reduced substrates.

Other end products of anaerobic fermentations, including H_2, CH_4, H_2S, and N_2 can also be metabolized by photosynthetic bacteria, which are able to use hydrogen as an electron acceptor simultaneously with the sulphate assimilatory reduction system. Unlike the other two major groups of anaerobic bacteria involved in the terminal degradation of biomass, photosynthetic bacteria are located at the top of the anaerobic hypolimnion rather than in the lower regions and sediments.

6.6.4 Bacterial adaptations to low-nutrient environments

Many bacterial species have evolved special mechanisms to survive the adverse environmental conditions (Section 2.14.2) of low-nutrient availability.

Gram-positive bacteria tend to form dormant spores, while Gram-negative bacteria have molecular and physiological mechanisms which enable them to persist at low metabolic (but not dormant) activity until adequate nutrient levels return; they then exploit the improved growth conditions and undergo a burst of synthetic activity and population increase (Zambrano and Kolter, 1996). The ability of bacteria to live through conditions of acute nutrient deprivation is referred to as 'starvation-survival' and may be defined as 'the process of survival in the absence of energy-yielding substrates' (Morita, 1982).

Low-nutrient aquatic environments

Starvation survival has been investigated particularly in relation to the planktonic bacteria of marine environments, where they may be carried passively within water masses for many years (Menzel and Ryther, 1970) under conditions of extremely low organic carbon availability (Morita, 1997). Although many of the environmental starvation adaptations shown by aquatic bacteria (Table 6.4) have been studied specifically in marine organisms, the low-nutrient responses of freshwater organisms are expected to be closely similar.

Organic nutrients are normally available to freshwater bacteria from exogenous sources or from algae via the microbial loop (Section 1.5.2). Carbon limitation of bacterial populations become particularly acute where there is limited exogenous supply and where algal populations are low due to a lack of available inorganic nutrients (e.g., oligotrophic standing waters) or absence of light (e.g., aquifers – Section 2.14).

The starvation response

The starvation response in aquatic bacteria is a classic example of environmentally-induced molecular activity, and involves the activation of a cohort of 'starvation genes' followed by associated changes in biochemical activity, cell size and shape, and bacterial populations (Table 6.4).

Table 6.4 The starvation response in aquatic bacteria

Cell characteristic	Starvation response	Biological implications
1. Genetic changes		
– Formation of starvation-specific transcription factor (RpoS)	Activation of the rpoS gene leads to formation of RpoS	Stimulation of rpoS-controlled genes
– Stimulation of RpoS-controlled genes	Activation of 30–50 starvation genes	Genes encode starvation phenotype
2. Biochemical composition		
– Surface membrane	Decrease in membrane-specific lipid content of cell	Related to decrease in cell size
	Changes in fatty-acid composition and fluidity	Ability to utilize a greater range of external substrates
– ATP content (energy state)	Reduction of adenylate energy charge (AEC)* from 1 to value of 0.5–0.75.	Reduced AEC permits metabolic maintenance but not cell growth
– Internal storage compounds	Utilization of specific storage compounds (e.g., glycogen & PHB†)	Integration of carbon availability over time
– Internal non-storage molecules	Loss of free amino-acids Degradation of proteins and RNA	Internal metabolites and structural molecules become substrates
3. Cell size and shape	Decrease in size. Elongate bacteria become spherical	Increased S/V ratio, increasing potential for nutrient uptake
4. Bacterial populations		
– Total count	In most cases an initial increase followed by decline	Reduced food availability for bacterial consumers
– Viable count	Major decrease in viable count, with survival of a few dormant cells	Long-term survival allows future nutrient exploitation

*AEC = [ATP] + ½[ADP]/[ATP] + [ADP] + [AMP].
†PHB – polyhydroxybutyrate.

Molecular activation Transition from high- to low-nutrient environment promotes a fundamental change in cell physiology, with a switch from growth to maintenance. Energy reserves need to be mobilized and the cell has to survive in the absence of multiplication (Reeve *et al.*, 1984). These changes result from the activities of a large set of genes which are induced at the onset of starvation. This induction is the result of the initial activation of the *rpo*S gene and the formation of a starvation-specific transcription factor (RpoS or σ^s) which confers new promoter recognition sites on the RNA polymerase (Loewen and Hengge-Aronis, 1994). The role of *Rpo*S in the induction of starvation-induced dormancy (Figure 6.8) parallels its role in the transition of bacterial populations to a stationary phase during batch culture (Bohringer

et al., 1995), and the induction of stationary phase characteristics during quorum sensing in biofilms (Figure 1.7).

Regulation of RpoS by environmental changes in nutrient concentration is mediated by changes in the concentration of internal metabolites (Zambrano and Kolter, 1996), including cAMP and ppGpp (which promote transcription) and UDP-glucose (which represses translation). Factor RpoS is also regulated by post-transcriptional control of its molecular stability (Figure 6.8).

Activity of RpoS-controlled genes About 30–50 proteins are thought to be induced *via* the RpoS-controlled genes, including enzymes which are involved in hydrolysis, protein and carbohydrate synthesis, and protection of DNA (Loewen and

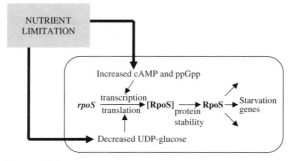

Figure 6.8 Molecular response of Gram-negative bacteria to nutrient limitation (based on a figure from Zambrano and Kolter, 1996)

Hengge-Aronis, 1994). The phenotypic effects and biological implications of this genetic activity are summarized in Table 6.4. Biochemical changes include a decrease in surface membrane (with changes in biochemical composition), decreases in ATP content, utilization of internal storage compounds and also conversion of internal non-storage molecules for energy formation. Decrease in the surface membrane, with utilization of cell contents for energy formation, results in a decrease in size and a change to spherical shape (for elongate bacteria). The small size of bacteria in the starvation state is one of the most obvious features of this metabolic condition, and may result in the formation of very tiny 'ultramicrocells'.

Reduction in biomass formation and cell division leads to a marked decrease in the bacterial population. Amy and Morita (1983) noted four distinct patterns of change during the starvation survival process, the most common of which involved an initial increase in total count followed by a prolonged decline. Viable counts also showed sharp decline, with reductions of over 99 per cent over a 4-week starvation period being recorded for some aquatic bacteria (Kurath and Morita, 1983). Resulting populations include a high proportion of dead cells, cells with minimal metabolic activity (surviving but unable to divide in the short-term), and cells with low metabolic activity that are able to respond rapidly to high nutrient and form colonies on nutrient agar.

6.7   Photosynthetic bacteria

As noted earlier (Section 4.5.1), photosynthetic bacteria can be divided into three major groups – the green sulphur bacteria (Chlorobacteriaceae), purple sulphur bacteria (Thiorhodaceae), and the purple non-sulphur bacteria (Athiorhodaceae).

6.7.1   General characteristics

Although these groups are defined primarily in terms of colour and metabolic substrate, they also show differences in terms of cell size, fine structure, photosynthetic activities (Section 4.5), involvement in the sulphur cycle (Section 6.8), general ecology, and cell motility. The main characteristics of these groups are summarized in Table 6.5.

Within each group, cell shape varies greatly between species, including spherical, elliptical, rod-shaped, and vibrioid forms. Differences in size occur between species and also between major groups. The overall size of green sulphur bacteria tends to be less than purple sulphur bacteria, which reach maximum diameters of 5–10 μm. The location of photosynthetic pigments in green sulphur bacteria within distinctive 'chlorobium vesicles' distinguishes the fine structure of these organisms from the other two major groups. Gas vacuoles are found in those organisms (purple and green sulphur bacteria) which form discrete layers in stratified lakes, but not the less ecologically-defined purple non-sulphur bacteria (Walsby, 1994). This distinction highlights the importance of gas vacuoles in the depth-regulation of photosynthetic bacteria.

6.7.2   Motility

Motility of photosynthetic bacteria has been investigated particularly in laboratory cultures, where there is considerable variation within species and within major taxonomic groups. Individual photosynthetic bacteria, like algae, come in one of three categories – non-motile, actively motile (by flagella), and passively motile (by gas vacuoles). Unlike algae, these characteristics are to some extent

Table 6.5 Major groups of anoxygenic (non O_2-evolving) photosynthetic bacteria; information from Pfenning (1967) and Holt *et al.* (1994). In addition to these three major groups of photosynthetic bacteria, Holt *et al.* (1994) also describe four other groups based on phyllogenetic characteristics

	Green sulphur bacteria	Purple sulphur bacteria	Purple non-sulphur bacteria
Freshwater genera	Chlorobacteriaceae *Chlorobium Pelodictyon*	Thiorhodaceae *Thiopedia Thiospirillum*	Athiorhodaceae *Rhodopseudomonas Rhodospirillum*
Energy source	Light	Light	Light or chemical
Characteristic pigments: chlorophyll	Bacteriochlorophylls -c and -d or -e	Bacteriochlorophyll-a or -b	Bacteriochlorophyll-a or -b
typical carotenoids	Chlorobactene β-isorenieratene	Lycopene Spirilloxanthin Okenone	Lycopene Spirilloxanthin OH-Spheroidenone
Carbon source	Almost all obligate autotrophs (CO_2)	Mostly autotrophs (CO_2), some also mixotrophic	Mixotrophs (take up simple organic compounds)
Electron donor	Sulphide and sulphur	Sulphide and sulphur	sulphide
Environmental conditions	Obligately anaerobic, high H_2S tolerance	Anaerobic, low H_2S tolerance, optimum pH9.5	Aerobic or anaerobic H_2S inhibits growth
Sulphur deposition	Extracellular	Intracellular	Limited (extracellular) or no sulphur deposition.
Specialized photosynthetic organelles	Distinctive sub-surface 'chlorosome' vesicles	Specialized internal membranes	Specialized internal membranes
Motility	Non-flagellate Some have gas vacuoles	Some have flagella Some have gas vacuoles	Some have flagella No gas vacuoles
Typical size range	Small: 0.5–2 µm diameter	Large: 2–10 µm diameter	Small: 0.5–2 µm diameter

interchangeable, with environmental factors having an important effect. Cultures of Thiorhodaceae exposed to high sulphide concentrations and light intensity undergo a conversion of all motile to non-motile cells, which sink to the bottom of the vessel and produce copious slime. Differences in motility within this group are shown in Table 6.6, where particular genera may fit into one of four categories depending on their capabilites.

The ecological significance of active and passive motility is particularly apparent in situations such as the lake water column, where both processes are important in the exact positioning that occurs at the top of the hypolimnion. Many of these organisms are killed by exposure to oxygen in the pre-

sence of light, so it is important that they do not stray up into the aerated waters of the epilimnion. Migration of purple sulphur and green sulphur bacteria has been observed during summer stratification in holomictic lakes, where upward movement of bacterial populations follows the extension of the anaerobic high-sulphide zone that accompanies the rise in the thermocline. Flagellate forms are able to migrate by their own flagellar activity. *In vitro* studies have suggested that vertical movements of these organisms occur in response to gravity (negatively geotactic) and oxygen (negatively aerotactic). The response to light is non-directional in terms of the stimulus and involves a reversal in direction of flagellar movements

Table 6.6 Active motility and gas vesicles in purple sulphur bacteria

	Genera with gas vacuoles	Genera without gas vacuoles
Genera with motile stage	*Lamprocystis*	*Thiospirillium Chromatium Thiocystis*
Genera with no motile stage	*Rhodothece Thiodictyon*	*Thiococcus Thiocapsa*

followed by a reversal in the direction of cell movement.

Gas-vacuolate photosynthetic bacteria include green sulphur (e.g., *Pelodictyon*) and purple sulphur (e.g., *Amoebobacter*) organisms, which may use their gas vacuoles in a similar way to blue-green algae (Figure 3.20) for depth regulation (Walsby, 1994). Observations by Eichler and Pfennig (1990) showed that populations of gas-vacuolate purple bacteria over-wintered in flocs of organic material in the sediments of Schleinsee (Southern Germany). At stratification, these bacteria disappeared from the sediments, becoming dispersed throughout the hypolimnion and later forming a discrete layer at the top of the hypolimnion. At autumn overturn the bacteria developed deposits of oxidized iron and manganese which acted as ballast, reducing buoyancy and depositing the cells on the sediments prior to the next over-wintering phase.

6.7.3 Ecology

The metabolic activities and requirements of photosynthetic bacteria determine their ecological niche in the aquatic environment. These organisms are often found in narrowly-defined micro-environments within relatively heterogeneous aquatic systems including ponds, ditches, marshes, estuarine mud flats, rivers, and lakes. In contrast to the Athiorhodaceae, which occur widely but never at high population density, the Thiorhodaceae and Chlorobiaceae are often present at high abundance – appearing as red or greenish layers on mud or forming substantial blooms in lakes.

Hidden blooms of photosynthetic sulphur bacteria are often found below the surface of well-stratified productive lakes, where they are restricted to the top of the hypolimnion (immediately below the thermocline) as a discrete layer. Within this layer, green sulphur bacteria are often localized below an overlaying population of purple sulphur bacteria, in accordance with the higher H_2S tolerance of the Chlorobacteriaceae.

At this interface, photosynthetic bacteria are just within the euphotic zone (light level typically <1 per cent lake surface value), with adequate supplies of substrate (reduced sulphur compounds) and under anaerobic conditions. In lakes which completely mix during the annual cycle (holomictic), the bacterial bloom is limited to the period of summer stratification when the thermocline has risen to its highest point and light intensity is maximal. Lakes which do not completely mix (meromictic) have a thermocline (with an associated anaerobic high-sulphide zone) which may be stable over many years and have associated bacterial blooms which are constant over long periods of time. Pfennig *et al.* (1966), for example, report the long-term presence of a pink bacterial layer 19.5m below the surface of meromictic lake Blankvann in Norway, from which they were able to isolate a range of purple and green sulphur bacteria. Although photosynthetic bacteria tend to accumulate at the thermocline boundary, they are not restricted to this part of the lake and studies on European lakes have shown that purple and green sulphur bacteria can be found throughout the hypolimnion to depths of 35 m (Pfennig *et al.*, 1966).

As noted previously, purple non-sulphur bacteria differ from other photosynthetic bacteria in not building up bloom populations under natural conditions. This is even true for conditions which might be expected to suit their mixotrophic metabolism, such as the presence of organic wastes (e.g., sewage ponds). This lack of bloom formation can be attributed to the presence of sulphate, which is normally present at high levels in such environments and is converted by sulphate reducing bacteria to H_2S, directly inhibiting growth of purple non-sulphur organisms. Green and purple sulphur bacteria become dominant under such conditions, eventually reducing the level of sulphide to a point at which purple non-sulphur bacteria are able to co-exist.

6.8 Bacteria and inorganic cycles

Freshwater bacteria have a key role in geochemical transitions within the aquatic environment, and are important in the cycling of metabolically-important elements such as nitrogen (Section 5.4), Fe (Section 5.12) and sulphur. The cycling of other elements, such as silicon (Section 5.9, Figure 5.24), may also involve bacterial activity. Studies by Patrick and Holding (1985) have shown that solubilization of diatom frustules (regenerating silicic acid) is increased in the presence of natural populations of freshwater bacteria.

6.8.1 Bacterial metabolism and the sulphur cycle

The cycling of sulphur within the freshwater environment involves alternate phases of anabolic and catabolic activity. Sulphate ions are taken up by all lake biota and converted to sulphydryl (−SH) groups in the synthesis of proteins. Breakdown of proteins on death of the organism results in the release and conversion of simple sulphur compounds, leading ultimately to the regeneration of sulphate ions. Four main types of microbial metabolic activity are involved in the sulphur cycle − protein decomposition, sulphate reduction, aerobic and anaerobic sulphide oxidation (Figure 6.9).

Microbial interactions involved in the cycling of sulphur within the aquatic environment are clearly localised in eutrophic lakes, where the occurrence of distinct aerobic and anaerobic zones within the water column leads to clear separation of microbial activities (Figure 6.10). The aerobic epilimnion is the main region of incorporation of inorganic sulphur compounds into lake biomass (trophogenic zone), while the anaerobic hypolimnion and sediments are the primary sites of conversion from organic to inorganic sulphur (tropholytic zones).

Dissolved inorganic sulphur occurs primarily as sulphate ions (SO_4^{2-}) within the oxygenated lake water of the epilimnion. These ions are taken up by algae and other biota and are subsequently reduced to sulphydryl (−SH) groups during protein synthesis. Death and sedimentation of lake biota leads to cell breakdown in the hypolimnion and lake sediment, with further reduction of sulphydryl groups to H_2S during the process of protein decomposition. This anaerobic process is carried out by a wide range of bacteria, including *Pseudomonas liquefaciens* and *Bacterium delicatum* (Kuznetsov, 1970). Protein decomposition occurs mainly in the surface sediments, where bacterial population counts are up to three times greater than in lake water. Reduction of sulphate generated from mineral sources also occurs in the sediment. In these anaerobic conditions, sulphate provides a source of oxygen for the oxidation of molecular hydrogen or carbon

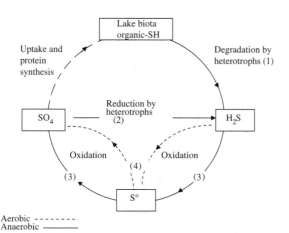

Figure 6.9 Sulphur cycle in a Eutrophic lake: (1) *Pseudomonas liquefaciens, Bacterium delicatum,* (2) *Desulfovibrio, Desulfotomaculum,* (3) *Chlorobium, Chromatium,* (4) *Beggiatoa* (H_2S to S), *Thiobacillus* (S to SO_4)

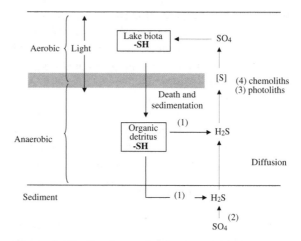

Figure 6.10 Sulphur cycle transformations in the water column of a eutrophic lake: microbial processes involve − protein decomposition (1), sulphate reduction (2), anaerobic oxidation of H_2S (3), and aerobic oxidation of H_2S (4)

compounds by bacteria such as *Desulfovibrio* and *Desulfotomaculum*.

$$H_2SO_4 + 4H_2 \longrightarrow 4H_2O + H_2S$$

$$[\Delta G_0{}^1 = -14\,\text{kJ mol}^{-1}] \quad (6.6)$$

$$H_2SO_4 + 2[CH_2O] \longrightarrow 2CO_2 + 2H_2O + H_2S$$

$$(6.7)$$

H_2S, generated in the sediments by protein decomposition and reduction of sulphate, diffuses vertically through the hypolimnion and is oxidized immediately prior to entry, or during entry, into aerobic conditions.

Anaerobic sulphur oxidizing bacteria occur at the top of the hypolimnion, and can be divided into two main groups – the green sulphur bacteria and purple sulphur bacteria. In both cases, sulphide is oxidized to sulphur or sulphate as part of a light-mediated reaction.

$$CO_2 + 2H_2S \xrightarrow{\text{light}} [CH_2O] + H_2O + S \quad (6.8)$$

$$2CO_2 + 2H_2O + H_2S \xrightarrow{\text{light}} 2[CH_2O] + H_2SO_4$$

$$(6.9)$$

The other major group of sulphur-oxidizing bacteria are colourless chemosynthetic microbes which are mostly aerobic. These organisms oxidize sulphide to sulphate via elemental sulphur, which is deposited either inside (*Beggiatoa* and *Thiothrix*) or outside the cell (*Thiobacillus*) as an intermediate.

Sulphur deposition in *Beggiatoa* and *Thiothrix*, two organisms typical of sites where H_2S is being produced, continues for as long as sulphide is available:

$$H_2S + \tfrac{1}{2}O_2 \longrightarrow S + H_2O [\Delta G_0{}^1 = -10\,\text{kJ mol}^{-1}]$$

$$(6.10)$$

Depletion of sulphide ultimately results in metabolism of internal stores of sulphur, with release of sulphate into the surrounding water.

$$S + 1\tfrac{1}{2}O_2 + H_2O \longrightarrow H_2SO_4 [\Delta G_0{}^1 = -28\,\text{kJ mol}^{-1}]$$

$$(6.11)$$

D. BACTERIAL POPULATIONS AND PRODUCTIVITY

Within the freshwater environment, the underlying growth rate of bacteria (productivity) is closely linked to population increase. Sections 6.9 and 6.10 concentrate particularly on planktonic bacteria, considering populations mainly in relation to cell numbers and total biomass. Aspects of diversity within populations are discussed in Section 6.1 and a procedure to make a theoretical estimate of total species content in Section 6.3.3.

6.9 Bacterial populations

In most freshwater environments bacteria form the largest population of all free-living biota, and are only exceeded by viruses in terms of total organisms present. The population ecology of freshwater bacteria is thus characterized by high cell counts (both planktonic and biofilm communities) and the capacity for rapid rates of reproduction. Bacterial popu-

lations tend to show marked fluctuation in response to environmental factors that promote (e.g., inorganic and organic pollutants, increase DOC levels from phytoplankton blooms) or deplete (e.g., increased levels of zooplankton or protozoon grazing) the increase in biomass.

6.9.1 Techniques for counting bacterial populations

Populations of bacteria in suspension can be enumerated either as total or viable counts. Direct counting of bacteria (total counts) is regarded as the most reliable method for evaluation of community dynamics since all cells are included. Measurements of bacterial dimensions can be used to convert counts of cell numbers to estimates of total bacterial biomass, which in turn can be combined with measures of bacterial productivity to provide useful information on the dynamics of bacterial populations.

Total counts

Although direct counts of aquatic bacteria in environmental samples are routinely carried out by light microscopy, the small size (normally 0.2–5 μm diameter) of these organisms may create problems since they are close to the resolution of the light microscope and are also within the size range of organic/inorganic particulate material typically present in lakes and rivers. Both of these problems may be largely overcome using fluorescent stains which enhance the detection of bacteria and do not normally label particulate non-living material. Various stains are available. Acridine orange forms green and red fluorescing complexes with DNA and RNA respectively when excited with light (wavelengths 436 or 490 nm), while DAPI (4'6-diamidino-2-phenylindole) forms a blue fluorescing complex with DNA at or above 390 nm.

The procedure for direct counting involves addition of the stain to the water sample, filtration of a known volume through a polycarbonate filter then examination and enumeration using epifluorescence microscopy. Preparations from lakes and rivers typically reveal a very high population of bacteria, with individual cells ranging in size and shape (Figure 6.11). Total counts can also be made under the scanning electron microscope after passing a known volume of chemically-fixed water sample through a 0.2 μm (pore diameter) filter membrane (Figure 6.1).

Viable counts

Counts of viable (metabolically-active) heterotrophic bacteria can be readily carried out by plating water samples onto nutrient agar plates and counting the number of colonies that develop. Bacterial counts, expressed as colony forming units (CFU) ml^{-1}, record those organisms that are able to grow and multiply on the nutrient medium. Although this approach can potentially give information on the total number of metabolically active heterotrophic bacteria (total heterotrophic viable count) present in

Figure 6.11   Total bacterial count
Fluorescence microscopy of DAPI-stained lake water sample reveals a high bacterial count (10^7 cells ml^{-1}) and a diverse population. Bacteria include vibrioid (v), rod-shaped (r) and coccoid (c) forms. Some of the bacteria are large and undergoing division (r), and appear to be metabolically active. Other bacteria (arrows) are under 1 μm in diameter, and are probably metabolically inert (scale = 10 μm)

the sample, there are a number of problems. Most importantly, all plating media are highly selective, and many viable organisms with complex nutrient requirements (fastidious organisms) will be excluded. Bacteria requiring specific physicochemical parameters (e.g., particular conditions of pH or oxygen concentration) may also be excluded from the count.

6.9.2   Biological significance of total and viable counts

Total bacterial counts from freshwater sites are invariably higher than viable counts. This is seen, for example, in Figure 6.12, where total bacterial counts in two freshwater lakes were 10^6–10^7 cells ml^{-1}, compared with viable counts of 10^3–10^4 cells ml^{-1}. In this situation, only 1 in 10^3 total bacteria were being recorded as viable cells.

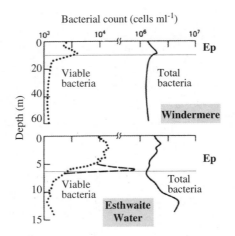

Figure 6.12 Total and viable bacterial counts in the water column of a deep mesotrophic lake (Windermere, UK) and a shallow eutrophic lake (Esthwaite Water, UK) during summer stratification:

Dashed line represents a short-term maximum in the epilimnion (Ep) (adapted and redrawn from Horne and Goldman, 1994, originally from Jones, 1971)

Differences between total and viable counts reflect the high degree of heterogeneity within natural microbial communities, and arise for two main reasons:

- most bacteria are metabolically inactive – although these organisms are able to become metabolically active when environmental conditions improve, they do not form colonies when plated out on nutrient agar and are referred to as 'non-viable';

- many metabolically active bacteria require particular growth conditions which are not satisfied in routine laboratory cultures (see above).

Values for total and viable counts of planktonic bacteria vary with the trophic status of the water body and also the position within the water column. Comparison of mesotrophic Lake Windermere and eutrophic Esthwaite Water (UK) indicated higher levels of total bacteria (approaching 10^7 cells ml^{-1}, and much higher levels of viable bacteria (over 10^4 cells ml^{-1}) under eutrophic conditions (Figure 6.12; Jones, 1977). In the nutrient-rich lake, total counts of bacteria are particularly high in the hypolimnion,

where accumulation of algal detritus provides an organically-rich environment. The contrasting presence of highest viable counts in the epilimnion of the eutrophic lake partly reflects the high levels of dissolved organic carbon (DOC) released by the standing populations of phytoplankton, but may also reflect problems in culturing strictly anaerobic and fastidious organisms from the hypolimnion.

6.10 Bacterial productivity

Bacterial productivity is a dynamic aspect of community function and an important component of biomass formation and transfer in freshwater microbial food webs. As with other lake biota, productivity refers to the intrinsic rate of increase in biomass and may be considered as 'net' and 'gross' terms, where:

$$\text{Net productivity} = \text{Gross productivity} - \text{Internal mass loss} \quad (6.12)$$

Gross productivity of planktonic bacteria is the underlying rate of increase in biomass, and is the sum total of all the anabolic (synthetic) processes that promote growth. Internal mass loss includes catabolic (breakdown) processes such as respiration, and excretory/secretory processes such as exoenzyme production.

Net productivity gives a measure of the fundamental growth rate of the population before external factors such as grazing, parasitism, and sedimentation are taken into account. Net productivity and population increase are thus quite different concepts, though productivity is clearly an important factor in population increase. Although productivity is typically considered, and frequently measured, as the rate of increase in dry weight, these values are normally converted to carbon content and expressed as the increase in mass of carbon per unit volume of water per day (mg C m^{-3} d^{-1}).

6.10.1 Measurement of productivity

Bacterial productivity can be determined with reference to a number of parameters, all of which are directly related to growth rate. These include:

- *Rate of cell division.* Determination of the number of dividing cells within the population.

- *Assessment of heterotrophic activity.* Determination of rates of decomposition of organic matter, count of viable heterotrophic bacteria. Non-light requiring fixation of CO_2 is carried out almost exclusively by heterotrophic bacteria (Wetzel, 1983) and provides a useful index of heterotrophic activity. The level of dark CO_2 fixation is low in oligotrophic lakes, both in absolute terms and as a proportion of photosynthetic CO_2 fixation. Mean 14 year values for oligotrophic Lawrence Lake, USA (Wetzel, 1983) were $4.7\,g\,C\,m^{-2}\,y^{-2}$ and 13.3 per cent respectively. Both of these parameters increase with increase in lake nutrient status.

- *Uptake of radioactive precursors.* Including general metabolites such as C^{14}-glucose and S^{35}-SO_4 (which are taken up into a range of molecules, e.g., Jones *et al.*, 1983) or specific precursors such as H^3-thymidine and H^3-leucine (which are taken up into nascent DNA and proteins respectively).

Measurement of bacterial productivity using any of the above parameters requires the subsequent use of appropriate conversion factors to express productivity as units of carbon increase.

In practice, bacterial productivity is normally determined via uptake of radioactive precursors, measuring either DNA or protein synthesis. Measurement of DNA synthesis involves *in situ* incubation of bacterial suspensions with H^3-thymidine over a brief period of time (<45 minutes), with determination of the DNA increase from radioactive uptake. Measured changes in DNA are then converted to rates of bacterial productivity in carbon units. The application of H^3-thymidine uptake is a particularly useful method to determine bacterial productivity, since the radioactive precursor is only taken up into newly-synthesized macro-molecules and is not involved in DNA turnover. Thymidine is rapidly transported across the bacterial membrane and converted to thymidine monophosphate by the enzyme thymidine kinase. Uptake of thymidine and

incorporation into replicating DNA is much less rapid in blue-green algae, eukaryotic algae, and fungi, so the rapid incubation time used in this procedure effectively ensures that only bacterial productivity is being monitored.

6.10.2 Regulation of bacterial populations and biomass

A wide range of environmental factors affect bacterial biomass in freshwater systems, including predation (mainly protozoa and rotifers), parasitic (viral) attack, and nutrient (inorganic nutrients, dissolved organic carbon) availability. In terms of the microbial food web, these different factors operate by top-down or bottom-up control (Figure 6.13). In addition to food web factors, changes in the chemical environment (e.g., seasonal hypolimnetic oxygen depletion) also affect bacterial abundances (Ricciardi-Rigault *et al.*, 2000).

The relative importance of top-down and bottom-up control in determining planktonic bacterial populations has been a matter of some discussion. Analysis of ranges of aquatic systems by different workers have supported both sides of the debate. In some cases, such cross-system investigations have indicated that bacterial abundance relates primarily to predation or lysis, while in other cases the bacterial populations relate more to available nutrient resources. Sanders *et al.*, (1992) and Berninger *et al.* (1991a), for example, have demonstrated a strong inverse correlation between bacterial abundance and predation by protozoa (heterotrophic nanoflagellates). In other studies (Gasol and Vaque, 1993), this coupling was not observed, possibly due to reduction of the protozoon populations by large zooplankton (particularly daphnids). The importance of bottom-up control is indicated by the increase in bacterial populations with progression from oligotrophic to eutrophic lakes (Billen *et al.*, 1990), and by the clear correlation between algal (primary) and bacterial (secondary) productivity and the importance of dissolved organic carbon (see below).

As a general overview, it seems clear that both types of control may operate, but the balance between the two varies with time and ecological

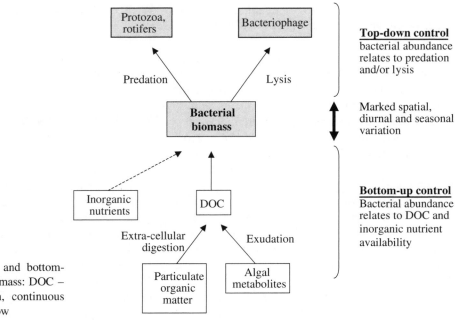

Figure 6.13 Top-down and bottom-up control of bacterial biomass: DOC – dissolved organic carbon, continuous arrows indicate carbon flow

circumstances. Studies on both marine (Ducklow and Carlson, 1992) and freshwater (Güde, 1989) systems suggest that control mechanisms vary temporally throughout the annual cycle. Marked spatial variations in predation can also occur within the water column of a freshwater lake, as demonstrated by Weinbauer and Höfle (1998).

Aspects of top-down control are considered separately in Chapter 7 (viruses) and Chapter 9 (protozoa), while one important feature of bottom-up control – the nutrient relationship between bacterial and algal populations – is considered below.

6.10.3 Primary and secondary productivity: correlation between bacterial and algal populations

Comparison of data from a range of pelagic systems indicates a high degree of correlation between bacterial and phytoplankton populations (Bird and Kalff, 1984), with close quantitative similarities between freshwater and marine environments (Cole *et al.*, 1988; Figure 6.14). Using acridine orange staining to enumerate total bacterial populations, Bird and Kalff (1984) demonstrated a close

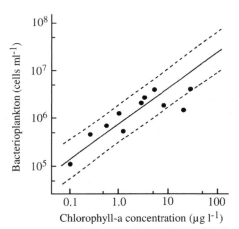

Figure 6.14 Correlation between bacterioplankton and phytoplankton populations (chlorophyll-a concentrations) in freshwater systems: broken lines are 95 per cent confidence limits of the regression line – data from marine and estuarine waters give a closely similar graph (figure taken from Bird and Kalff, 1984, and reproduced with permission of NRC Research Press)

mathematical relationship between total bacterial count (Ct) and algal biomass (chl-a), where:

$$Ct = 5.867 + 0.776 \; \log[\text{Chl-a}] \qquad (6.13)$$

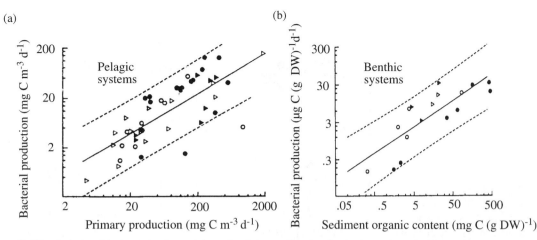

Figure 6.15 Relationship between bacterial productivity and major food source in pelagic and benthic systems: (a) bacterioplankton productivity and primary production, where the main food supply is dissolved organic carbon (DOC) released by phytoplankton (b) benthic bacterial productivity and organic content of sediment, where the main food supply is organic detritus. In both graphs bacterial productivity was measured by thymidine uptake (circles) or other methods (triangles), for freshwater (solid symbols) and marine (open symbols) systems (figure dervied from Cole *et al.*, 1988, and reproduced with permission from Inter-Research)

The correlation between populations is also matched by a close coupling between primary and secondary productivity. This is seen particularly well in the photic zone of lakes and oceans (Figure 6.15(a)), where bacterial production ranged from 0.4–150 μg C l^{-1} d^{-1} and averaged 20 per cent of primary production. The contribution of bacterial production is even greater if regions below the photic zone are also taken into account, reaching about 30 per cent of primary production throughout the entire water column. Heterotrophic bacterial productivity is clearly a major component of secondary productivity in pelagic systems, and is roughly twice the level of macrozooplankton productivity at particular levels of primary production.

The close coupling between algal and bacterial productivity is an overall property of the lake and does not extend to analyses within the water column. Depth measurements by Pace and Cole (1994) of primary and secondary productivity in a group of low-nutrient lakes showed major vertical differences between primary and secondary productivity maxima at particular points in time.

Benthic environments

In benthic systems, bacterial productivity is expressed per unit dry weight of sediment, and ranges from about 0.1–50 (g DW) d^{-1}. Most lake sediments are well below the photic zone and there is no resident photosynthetic algal population. In this situation, bacterial productivity is related more to the availability of degradable biomass rather than algal exudates. There is a general correlation between productivity, bacterial population, and sediment organic C content (Figure 6.15(b)), though some systems with high rates of benthic primary production (e.g., coral reefs) have higher rates of bacterial production than expected, simply from the organic C content.

6.10.4 Primary and secondary productivity: the role of dissolved organic carbon

The direct relationship between primary and secondary productivity in pelagic systems suggests

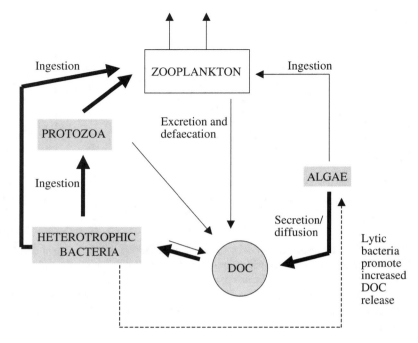

Figure 6.16 Direct and indirect links between algal and bacterial productivity in the water column of a lake: The phosphorus content of zooplankton DOC is particularly important for bacterial growth in P-limiting conditions (solid arrows indicate carbon flow, including the microbial loop (➡))

that either the growth of bacteria and phytoplankton is a separate response to common factors (e.g., inorganic nutrients, temperature) or that the growth of bacteria is directly related to particulate or soluble material derived from phytoplankton. Both of these are probably important, though the relationship between bacterial productivity and organic C content in sediments emphasizes the importance of organic substrates in promoting bacterial growth.

As noted earlier (Section 1.5.2) the flow of organic carbon from algae to bacteria via algal exudates (microbial loop) is an important part of the aquatic food web. The release of exudates as dissolved organic carbon (DOC) by algae (Section 4.7.2) involves processes of passive diffusion (photosynthetic products), active secretion (extracellular enzymes), and cell breakdown (lysis). In an actively growing and photosynthetic phytoplankton population, DOC production occurs mainly as the passive release of photosynthetic products, amounting to a maximum 10 per cent of primary productivity. This release of DOC provides a direct link between algal and bacterial productivity, with DOC occurring as a major substrate for bacterial growth. Indirect links also occur via the zooplankton population, where

part of the carbon flow from algae is released back into the environment as excreted and faecal DOC (Figure 6.16).

The ability of bacteria and other microheterotrophs to take up organic compounds released by algae was demonstrated in early studies on $^{14}CO_2$ uptake. Kinetic tracer studies by Wiebe and Smith (1977) on estuarine water samples demonstrated steady-state release of photosynthetically-derived dissolved organic carbon (PDOC) of 0.13 mg C m^{-3} h^{-1} by algae, with a rate of incorporation of PDOC into heterotrophs of 0.10–0.12 mg C m^{-3} h^{-1}. In this system, there was clearly a rapid and almost complete heterotrophic uptake of PDOC within a short time of release by the autotrophic community. Increased bacterioplankton uptake of ^{14}C-labelled extracellular phytoplankton products has been demonstrated during periods of intense algal bloom (Bell and Sakshaug, 1980).

In the case of blue-green algae, the trophic link with bacteria may be further enhanced by the activities of lytic bacteria such as *Lysobacter* (Fallowfield and Daft, 1988), the abundance of which is closely correlated with the blue-green algal population. Laboratory studies (Figure 4.19) by Fallowfield and Daft (1988) have shown that *Lysobacter* can specifically increase the release of DOC by blue-green, but not green algae, under conditions of active photosynthesis; this suggests that these organisms may have a very important role in promoting increased levels of carbon flow in the aquatic environment.

6.10.5 Bacterial productivity and aquatic food webs

Quantitative determination of pelagic and benthic bacterial productivity can give various insights into the dynamics of aquatic food webs:

Carbon flux through bacteria

The efficiency with which bacteria convert lake water DOC to bacterial biomass provides information on the total amount of carbon (carbon flux) which passes through the bacterial population. Within the photic zone of a eutrophic lake, bacterial production was about 20 per cent of primary production per unit volume of lake water. If the efficiency of conversion of substrate to bacterial biomass is 50 per cent (Cole *et al.*, 1988), then approximately 40 per cent of primary production fluxes through bacteria in this part of the water column. Within the whole water column, bacterial production averaged about 30 per cent of primary production, giving a 60 per cent flux of primary production through bacteria.

These values of 50 per cent growth efficiency are in line with the normally accepted levels of 40–60 per cent, based on observations of the uptake and efficiency of conversion of simple ^{14}C-labelled organic molecules by bacteria. Bacterial growth efficiency (BGE) not only depends on conversion of simple to complex organic molecules (bacterial productivity, BP), but also on carbon loss due to bacterial respiration BR, where:

$$BGE = BP/(BR + BP) \qquad (6.14)$$

Recent studies have shown that bacterial respiration is generally high in aquatic environments, and that growth efficiencies of bacteria should be revised down to values of <10–25 per cent in most aquatic systems. The amount of carbon flowing through the bacterial population is thus greater than originally estimated. In low-nutrient (oligotrophic) systems bacterial respiration may actually exceed phytoplankton net primary production. In this situation, where the total carbon processed by bacteria exceeds that fixed by phytoplankton, bacteria must also use other sources of carbon for respiration and the system is net heterotrophic.

Pelagic and benthic bacteria in eutrophic and oligotrophic lakes

Bacterial productivity is much greater in eutrophic compared with oligotrophic lakes, leading to the differences in population that were noted earlier (Figure 6.12). These differences in productivity can be seen in Table 6.7, where the total bacterial

Table 6.7 Bacterial productivity in the water column and on the sediments of a eutrophic and oligotrophic lake (taken from Cole *et al.*, 1988)

	Bacterial productivity in water column*	Bacterial productivity in water column	Bacterial productivity in sediments	Total productivity in column and sediments
Eutrophic lake (Lake Mendota, USA), mean depth 12.4 m	23.2	288	83.5	371.5
Oligotrophic lake (Mirror Lake, USA), mean depth 5.75 m	3.0	17.4	57.6	75

*Bacterial productivity is expressed per unit volume of water (mg C m^{-3} d^{-1}) in this column, and per unit surface area of lake (mg C m^{-2} d^{-1}) in the rest of the table (shaded columns).

productivity of water column and sediments (expressed per unit area of lake surface) in eutrophic Lake Mendota (USA) was five times greater than oligotrophic Mirror Lake (USA). This difference is much greater than expected simply in terms of depth, and reflects greater substrate availability under eutrophic conditions. The transition from eutrophic to oligotrophic lakes may signal a change from net autotrophy (where most carbon uptake is via algal photosynthesis) to net heterotrophy (where most carbon uptake is into bacteria). This switch to bacterial dominance involves a transition in carbon availability changing from readily-assimilable autochthonous (algal) DOC to poorly-assimilable allochthonous (exogenous) carbon.

Expressed on an area basis, bacterial productivity in the water column of a eutrophic lake such as Lake Mendota (USA) is much greater than in the sediments (Cole *et al.*, 1988; Table 6.7), implying that pelagic bacteria have a greater heterotrophic role than benthic ones in high nutrient conditions. The reverse appears to be true for oligotrophic lakes (Mirror Lake, Table 6.7), where benthic organisms dominate bacterial secondary productivity.

In a complex food web, where the same organic molecules may be consumed and recycled several times, secondary production by organisms such as bacteria could be nearly as large as primary production.

E. BACTERIAL COMMUNITIES IN THE LOTIC ENVIRONMENT

In freshwater systems, individual species of bacteria can occur as both planktonic (free-floating) and benthic (attached to the substratum) organisms. The equilibrium between these two states depends on environmenmtal factors such as water movement and substrate availability. In open standing waters, planktonic bacteria form the major phase, while in flowing waters displacement by water flow and greater substrate availability lead to the development of attached communities, predominant among which are biofilms.

6.11 Bacterial biofilms

A biofilm is essentially a community of microorganisms (typically algae or bacteria) which is attached to an exposed surface. In the case of bacterial biofilms, where individual cells are typically non-motile and closely associated within an extracellular matrix, the microenvironment is very different from that of the planktonic phase. Direct interactions between organisms become particularly important in biofilms, ranging from relatively simple adhesion contacts to exchange of genetic information (Section 1.4).

Biofilms occur in association with a wide range of physical interfaces in the freshwater environment (Brisou, 1995) including:

- air/water boundaries, present at the top of the water column (Section 2.16) and at the interface of submerged pockets of air and bubbles,

- water/biomass boundaries, such as the surface of algae (Section 6.13), zooplankton, and submerged higher plants,

- water/solid-inorganic surfaces, such as suspended particulate material, particulate sediments, and large rocks and stones.

Biofilms are particularly important in association with the stones and sediments of lotic communities, which are dominated by benthic organisms. They are also well-adapted to running-water environments, since the dense microbial community that develops at the solid surface is able to resist large-scale detachment by the intense shearing forces generated by water flow. The biofilm development described below is typical of the sequence that occurs on exposed rock surfaces at the bottom of a fast-flowing stream.

6.11.1 The development of biofilms

Biofilms originate by colonization of exposed surfaces and develop as a sequence of three main

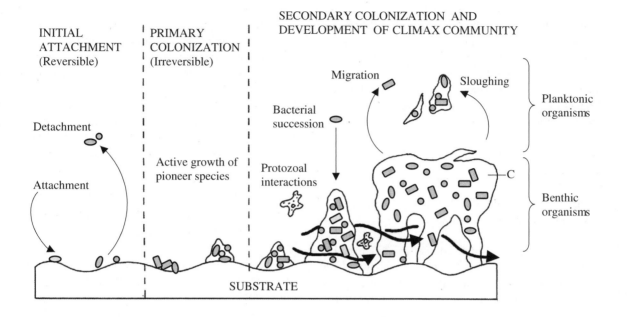

INITIAL
ATTACHMENT
(Reversible)

PRIMARY
COLONIZATION
(Irreversible)

SECONDARY COLONIZATION AND
DEVELOPMENT OF CLIMAX COMMUNITY

Detachment

Attachment

Active growth of
pioneer species

Bacterial
succession

Protozoal
interactions

Migration

Sloughing

Planktonic
organisms

C

Benthic
organisms

SUBSTRATE

Figure 6.17   Bacterial biofilm development: the flow of water (containing nutrients) within the mature community is indicated by thick arrows, C = bacterial cell cluster

phases – initial attachment, growth (primary colonization), and development of the climax community (Figure 6.17).

Initial attachment of bacteria

The site at which initial attachment occurs plays an important role in the future development and success of the biofilm. Nutrient availability will influence microbial growth, and light availability will determine future colonization by algae and possible conversion to an algal biofilm. Reversible attachment of pioneering bacteria to surfaces enables them to review the suitability of a particular microenvironment before committing to irreversible colonization of an unknown habitat.

Bacterial motility is important for initial attachment, allowing the cells to position themselves and contact the substratum without being displaced by water flow. Various studies have noted that removal of flagella from planktonic bacteria leads to a decrease in attachment to solid surfaces.

Bacterial attachment also requires the presence of a thin layer of organic material (referred to as a 'conditioning film') on the surface of the substratum. This surface conditioning film arises by adsorption of both polar and non-polar macromolecules to the solid surface. The film often has a net negative electrostatic charge, and the attraction of counterions towards this net charge typically results in a diffuse double layer which organisms must penetrate in order to attach to surface molecules. Bacterial adhesion to the substratum conditioning film is mediated via cell surface polymers, including special adhesion molecules (adhesins). These molecules are frequently arranged on the tip of surface extensions (fimbriae) which allow the organism to penetrate the double ion layer noted previously.

Primary colonization and early growth of biofilms

Primary colonization occurs once the initial (pioneer) bacteria have become irreversibly attached to the substratum and firmly established as a resident population. Growth of the biofilm involves an increase in microbial biomass plus the production of extracellular matrix. Increase in microbial biomass occurs by growth and division of existing organisms to form microcolonies.

Secondary colonization and development of a climax community

Further growth and development of the biofilm involves two major activities:

- continued growth and division of existing bacteria, forming discrete colonies – these tend to form large cell clusters, separated by water channels, within which convective flow has been demonstrated,

- secondary colonization, with the addition of new bacterial and other microbial species – addition of new bacteria occurs by coaggregation and leads to a succession of bacterial species. General changes in biological succession within biofilms are summarized in Figure 2.16, and involve a transition from physicochemical to biological interactions.

The climax community Mature biofilms, occurring on the surfaces of rocks and stones, are highly complex – both in terms of microbial diversity and three-dimensional structure. The latter is determined largely by the architecture of the polysaccharide matrix, with interconnecting channels and free water flow, as revealed by confocal scanning laser microscopy. This water flow is important in carrying nutrients and oxygen to different parts of the biofilm. Micro-electrode studies have revealed wide variations in oxygen concentrations within biofilms, with oxygen contours and local gradients following the distribution of biomass and access to water current.

Mature biofilms are in equilibrium with their aquatic surroundings and show no further increase in biomass. Continued loss of bacteria by sloughing and migration into the water column is balanced by entry of new bacteria as part of the process of bacterial succession. Other microorganisms also become incorporated into biofilms as the community develops and approaches maturity, including algae, protozoa, fungi, viruses, and invertebrates.

6.11.2 Dynamic interactions in the establishment of biofilms: the role of bacterial co-aggregation

As bacterial biofilms are composed of a number of different species, each entering the community at a particular point in time, there must be some mechanism to establish a distinct hierarchy in association.

Bacterial co-aggregation, involving recognition and specific binding between different species, provides such a mechanism for this succession. The high degree of association specificity means that organisms have a preference for particular partners, and can only enter the biofilm when these partners are in place. Early secondary colonizers have specificity for primary colonizers, and later secondary colonizers successively associate in sequence.

Bacterial co-aggregation in freshwater biofilms has been documented by various workers (see case study below). The key recognition process appears to involve lectin-sugar interactions, since co-aggregation may be abolished or reversed by the addition of simple sugars to mixed cultures where association normally occurs. Lectin–sugar interactions are known to be reversible and highly specific, and have been implicated in many biological recognition processes.

Case study 6.4 Specific recognition and adhesion amongst aquatic biofilm bacteria

Co-aggregation studies have been carried out by Rickard *et al.* (2000, 2002) to investigate specific recognition and adhesion between genetically distinct bacteria isolated from aquatic biofilms.

In the initial investigation (Rickard *et al.*, 2000), five strains of biofilm bacteria were characterized on the basis of biochemical tests, morphology, and molecular analysis (sequencing approximately 650 bases of the 16s rRNA gene). Four strains were identified as *Blastomonas natatoria* (strains 2.1, 2.3, 2.6, and 2.8) and one strain as *Micrococcus luteus*. Ability of pairs of strains to co-aggregate was determined by mixing suspensions of different bacteria under carefully controlled conditions and scoring the ability of cells to flocculate (co-aggregate) and settle out. Of the 10 possible pairwise combinations, six pairs showed strong co-aggregation. The ability of these pairs to co-aggregate showed strong dependence on their phase of growth cycle, with no expression of co-aggregation early in batch culture but strong (maximum) expression in stationary phase.

Co-aggregation between strains was mediated by the presence of protein (adhesin) molecules on the surface of one partner and saccharide (receptor) molecules on the surface of the other. The adhesin molecules were shown to be heat- and protease-sensitive lectins, while the receptors were heat- and protease-insensitive sugars with specific saccharides (galactose, galactosamine, or lactose) or saccharide combinations (galactose and galactosamine) as the recognized component. Ability to co-aggregate was abolished by the addition of particular sugars, and for all six co-aggregating pairs at least one sugar reversed co-aggregation. The pattern of co-aggregation and sugar inhibition allowed Rickard and co-workers to construct a diagrammatic representation of cell associations (Figure 6.18).

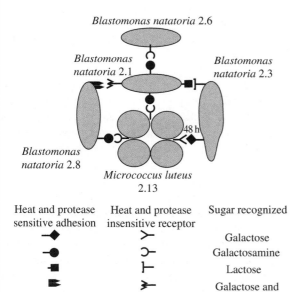

Heat and protease sensitive adhesion

Heat and protease insensitive receptor

Sugar recognized

Galactose
Galactosamine
Lactose
Galactose and galactosamine

Figure 6.18   Specific co-aggregation between biofilm bacteria.

Interactions are shown between five isolates: *M. luteus* 2.13 (grows as tetrads), *B. natatoria* 2.3 (club-shaped rod) and *B. natatoria* 2.1, 2.6, and 2.8 (small symmetrical rods). Each interaction is shown as complementary symbols representing a protein adhesin (lectin) or sugar (saccharide) receptor (figure taken from Rickard *et al.*, 2000, and reproduced with permission of the American Society for Microbiology)

These studies indicate that biofilm bacteria may carry multiple adhesions or receptors (or combinations of both) and suggest that these are important in the association and succession of cells in intact biofilms. The occurrence of maximum co-aggregation ability in stationary phase organisms is consistent with this, since the majority of biofilm bacteria are thought to be at this phase of growth.

Further studies by Rickard *et al.* (2002), involving a wider panel of biofilm isolates, showed that co-aggregation occurred between both distantly-related and closely-related organisms, operating at inter-generic, interspecific, and intraspecific levels. One particular isolate, *Blastomonas natatoria* strain 2.1, co-aggregated with a particularly wide range of bacteria, and may have a major role as a bridging organism in biofilm development.

F. BACTERIAL INTERACTIONS WITH PHYTOPLANKTON

In the water column of lakes and other standing waters, phytoplankton interact with both planktonic (free-living) and attached (epiphytic) bacteria. In the first case, the interactions are to some extent remote, while epiphytic associations involve a much closer interchange. In some cases, epiphytic associations develop into a clear symbiotic relationship, with close metabolic coupling – to the mutual benefit of both organisms.

6.12 Interactions between phytoplankton and planktonic bacteria

The section on bacterial productivity emphasized the close trophic relationship between planktonic algae and bacteria, with quantitative linking of algal and bacterial populations via the secretion and uptake of dissolved organic carbon. This particular interaction may be regarded as positive, since the growth of bacteria is promoted without any adverse effects on the algal population.

Other types of interaction between algae and bacteria are negative, having adverse effects on one or both groups of organism. These negative interactions arise because in the pelagic ecosystem, as with other ecosystems, nutrient and space resources are finite resulting in competition between different groups of organisms. Negative interactions include competition for inorganic nutrients, production of anti-microbial substances, and antagonistic activities of bacteria such as predation and parasitism (Figure 6.19).

6.12.1 Competition for inorganic nutrients

In the lake environment, phytoplankton populations experience loss of cells due to grazing, sedimentation, effects of parasitism, and natural cell death (lysis). The latter appears to be directly influenced by nutrient availability, since under axenic culture conditions an increase in algal lysis is promoted by conditions of nutrient deficiency. Under environ-

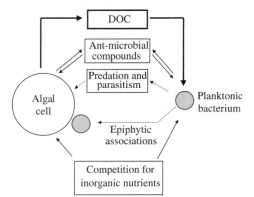

Figure 6.19 General interactions between phytoplankton and bacteria.

Planktonic algae interact with both epiphytic (close-range) and planktonic (long-distance) bacteria in a variety of ways. Many of these interactions are trophic, with the algae acting as a direct food source or competing with the bacteria. More specific interactions between phytoplankton and epiphytic bacteria are shown in Figure 6.26

mental conditions, the presence of competing bacteria may further reduce nutrient availability, leading to higher levels of algal cell death and promoting severe limitation in phytoplankton population. The impact of bacteria on phytoplankton in terms of nutrient competition has been investigated in both laboratory and field conditions.

Laboratory studies on microbial competition

Various studies have shown that bacteria are more efficient than algae in the assimilation of inorganic nutrients at low external concentrations, as indicated by the low K_s valve obtained in studies on the kinetics of nutrient uptake (Section 6.12.1). This is consistent with the smaller size of bacteria and their greater surface area/volume ratio, and allows bacteria to generally out-compete algae under nutrient-limiting conditions.

Laboratory studies by Bratbak (1987), for example, have shown that bacteria are better competitors for inorganic phosphorus compared to algae. Under inorganic nutrient limitation, however, bacteria will only out-compete algal cells when organic carbon is present in sufficient quantities to prevent C limitation. In P-limited conditions, therefore, decreasing levels of soluble P in the growth medium result in fewer algal cells and more bacteria due to differences in competitive ability. This trend continues until the bacteria become limited by a substrate other than P, probably organic carbon. The importance of DOC in inorganic nutrient competition is an important consideration in many marine and freshwater environments, where bacterial growth rates are considered to be carbon rather than mineral-nutrient limited.

Competition between algae and bacteria is complicated by the fact that metabolic activities of both competitors may enhance the growth of the other. The release of photosynthetic carbon compounds by phytoplankton may promote the competitive activities of bacteria by preventing them from becoming C–limited and thus allowing them to remove even more inorganic nutrients which would otherwise have been taken up by algae. Conversely, bacterial degradation of lysed algal cells results in the conversion of organic to inorganic P and N compounds (mineralization), thus releasing more soluble nutrients for algal uptake. These interactions are illustrated in Figure 6.20, and suggest that the effects of bacteria in out-competing algae and in limiting algal populations are not straightforward, but may vary with the physiological states of the organisms and the particular nutrient concerned. This complexity was demonstrated by the laboratory experiments of Brussaard and Riegman (1998), where addition of bacteria to diatom cultures greatly increased algal mortality under phosphate-limiting conditions, but had no effect or actually decreased mortality under conditions of N starvation.

Further information on nutrient competition between algae and bacteria, and the ways that algae are able to overcome a competitive disadvantage, are given in Section 6.12.1.

6.12.2 Antagonistic interactions between bacteria and algae

Antagonistic interactions between bacteria and algae include all those activities that directly reduce

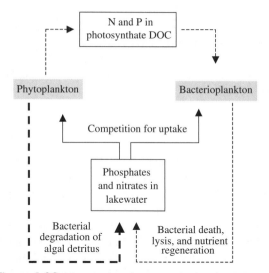

Figure 6.20 Interactions between planktonic algae and bacteria for major inorganic nutrients (nitrogen and phosphorus) in the lake water column: in addition to direct competition for nutrients, metabolic activities of phytoplankton (release of photosynthate DOC), and bacteria (degradation of algal detritus) also influence the availability and uptake of N and P

the population or growth rate of one or the other organism. In the case of algae, this involves the production of antibiotics, which may be defined as low molecular weight compounds which have a specific inhibitory effect. Bacteria act against algae via a range of mechanisms, including production of antimicrobial compounds, parasitism, predation, and epiphytic associations (Figure 6.19). Bacterial antagonists to blue-green algae have particular potential for use as biological agents in the control of nuisance algal blooms (see Chapter 10).

Antibiotic production by algae

The importance of antibiotic production by freshwater algae is based on two main lines of evidence – observations on the presence of attached bacteria in environmental samples and laboratory *in vitro* studies. Moreover, much of the evidence relating to antibiotic production by eukaryotic algae is based on marine organisms. In freshwater systems, Jones

Figure 6.21 Formation of the algal antibiotic, acrylic acid by chrysophyte algae: the antibiotic is a potent antibacterial agent

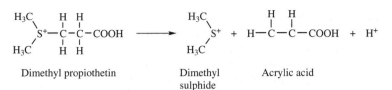

Dimethyl propiothetin   Dimethyl   Acrylic acid
                               sulphide

(1986), noted a range of bacterial attachment to algae, which diminished along the series: colonial greens – diatoms – dinoflagellates, possibly relating to an increase in antibiotic production along the sequence. *In vitro* evidence for antibiotic production is based on the bactericidal effects of algal culture filtrates or by the production of zones of inhibition around algae on agar plates. Using these techniques (Jones, 1986), antibiotic substances have been identified from various eukaryotic algae including diatoms (nucleosides, lipids), chrysophytes (acrylic acid), and green algae (chlorellin, acrylic and fatty acids). Acrylic acid (Figure 6.21) appears to be particularly widespread and to be active against a broad range of bacteria.

Environmental presence of antagonistic bacteria

The widespread occurrence and potential ecological importance of bacteria which are able to destroy freshwater algae was initially investigated by Daft *et al.* (1975), who carried out laboratory tests for the presence of antagonistic bacteria in environmental samples from freshwater lochs, reservoirs, sewage works, and soils. The presence of algal-lysing bacteria was demonstrated by spreading the water sample over algal monolayers (lawns) growing on agar, and monitoring the development of regions of lysis (plaques). Bacteria could then be isolated from the plaques, further cultured in the laboratory, and characterized. The use of algal lawns to demonstrate the presence of lytic bacteria in environmental samples closely parallels their use to demonstrate lytic viruses (Section 7.3, Figure 7.2).

The general conclusions reached from the *in vitro* study of lytic bacteria were as follows.

- The isolates of algal-lysing bacteria were all Gram-negative, non-flagellate, rod-shaped organisms and were identified as Myxobacteria. All isolates were strictly aerobic, and were able to lyse a range of bloom-forming blue-green algae, including *Anabaena*, *Aphanizomenon*, *Microcystis*, and *Oscillatoria*.

- Algal-lysing bacteria were present in all freshwater sites tested. In lakes and reservoirs, lytic bacteria were present both in the water column and on the sediment surface.

- There was a clear relationship between the population of lytic bacteria in the water column and the biomass pf phytoplankton. This was demonstrated both by comparison of different water bodies, where there was a positive correlation between the mean number of lytic bacteria and the mean annual chlorophyll concentration, and by analysis of seasonal changes in individual lochs or reservoirs. Populations of lytic bacteria were generally high in summer, where they closely followed the changes in chlorophyll concentration, and were particularly associated with blooms of blue-green algae (Figure 6.22). In winter months, lytic bacteria had virtually disappeared from the water column, but could be detected on sediment surfaces.

- The high levels of algal-lysing bacteria seen in summer months could be replenished from the lake sediment, or by inflow from the surrounding catchment. These bacteria were generally present in neutral or alkaline soil samples that were analysed from sites around the lake.

Although the widespread occurrence of these algal-lysing bacteria would suggest that they have

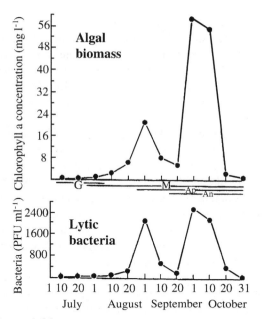

Figure 6.22 Abundance of algae and lytic bacteria in reservoir sand filters.

Total algal biomass shows a clear double peak, with individual populations of *Gomphosphaeria* (G), *Microcystis* (M), *Aphanizomenon* (Ap), and *Anabaena* (An) indicated by horizontal lines. Viable counts of lytic bacteria parallel the changes in algal biomass, indicating a close correlation in populations of the two groups of biota (figure taken from Daft *et al.*, 1975, and reproduced with permission of Blackwell Publishing Ltd)

considerable ecological significance, Daft *et al.* (1975) considered that they did not normally play a major role in the lysis of algal blooms. They appear to coexist with populations of blue-green

algae since both sets of organisms require high-nutrient conditions and attachment of the non-motile aerobic bacteria to algal cells allows them to exist as planktonic organisms under highly-oxygenated conditions. Lysis of algal cells will generate extra nutrients for the bacteria, but it is probably only when the normal equilibrium between the two organisms is altered (e.g., by inflow of high levels of bacteria in sewage effluent) that large-scale lysis and reduction of algal populations occurs.

Mode of action of antagonistic bacteria

In the above study, destruction of the algal cells typically required direct contact with the bacteria and presumably involved the activity of lytic enzymes at the bacterial surface. Although few details of the mode of action were investigated in this case, other studies (Wright, 1986) have shown that bacteria may act in various ways. In most cases these antagonistic organisms produce anti-microbial compounds, which may be secreted into the aquatic medium or retained at the bacterial surface for use in contact lysis. Bacterial antagonism may also involve more specific interactions involving parasitic relationships and predation (Table 6.8).

Anti-microbial   compounds Antimicrobial compounds that cause lysis of algae can be divided into three main groups – enzymes, antibiotics, and very low MW volatiles.

There have been several reports of algal lysis by gliding bacteria (summarized in Wright, 1986),

Table 6.8 Antagonistic interactions between bacteria and blue-green algae (information taken from Wright, 1986)

Mode of action	Processes involved	Examples
Extra-cellular compounds	Lysozyme-like enzymes	Myxobacteria
	Heat-stable, penicillin-like antibiotics	*Cellvibrio* and other bacteria
	Very low MW volatile compounds – isoamyl alcohol	*Bacillus* spp.
Contact lysis	Attachment	Myxobacteria (*Lysobacter*)
		Flexibacter sp.
Penetration	Internal parasitism	*Bdellovibrio* sp.
Predation	Entrapment and lysis	Myxobacteria
		(*Myxococcus xanthus*)

suggesting direct contact between the antagonist and the algal cell. Some of these isolates were originally considered to be *Myxobacteria*, but have since been placed in a new genus – *Lysobacter*. Lysis appears to be initiated primarily via cell-wall degradation, suggesting the activity of lysozyme-like enzymes. This group of enzymes removes the peptidoglycan cell wall of prokaryotes (including blue-green algae), producing osmotically-sensitive naked cells (spheroplasts) which are easily ruptured. A typical sequence of events appears to be as follows (Daft and Stewart, 1973).

- The bacteria move towards the photosynthesizing algal cells, possibly attracted by oxygen being evolved.

- They attach themselves to the algal cell wall (usually near to the cross septum of filamentous forms) by the polar tip.

- Activity of bacterial lysozyme-like enzymes degrades the peptidoglycan cell wall, leading to cell lysis (see above).

Heat-stable antibiotics have been isolated from culture supernatants of *Cellvibrio* and other bacteria. The substances were typically produced in stationary phase, had a low MW (1kD), and had effects similar to penicillin. These antibiotics inhibited cell-wall synthesis of vegetative cells (but not heterocysts or akinetes), leading to osmotically-sensitive spheroplasts. Vegetative cells resumed normal growth and division when sub-cultured back into normal medium. In addition to conventional antibiotics, some species of *Bacillus* have also been reported to release very low MW volatile compounds, of which isoamyl alcohol has been shown to be highly toxic to both blue-green and eukaryotic algae (Wright and Thompson, 1985; Wright *et al.*, 1991).

Parasitism   The ability of the bacterium *Bdellovibrio bacteriovorus* to infect, endoparasitize, and subsequently lyse Gram-negative bacteria is well known. These organisms have a biphasic life cycle, alternating between a non-growing free-living predatory phase and an intracellular parasitic reproductive phase. Electron microscope studies by Caiola

and Pellegrini (1984) on a lake bloom of *Microcystis aeruginosa* indicated extensive infection by a *Bdellovibrio*-like bacterium, suggesting that the parasitic activities of this organism extend to blue-green algae and may be ecologically important. After attachment and penetration of the *Microcystis* cells the bacteria were localized mainly in the periplasmic space between the host cell wall and cytoplasmic membrane. This was followed by a decrease in the amount of the blue-green algal cytoplasm, disruption of internal membrane systems (including gas vacuoles and granular inclusions), and multiplication of the endoparasite to form infective daughter cells. These would ultimately be released into the environment upon fragmentation of the cell wall and bursting of the host cell.

Predation   Predation of blue-green algae by myxobacteria has been described by Burnham *et al.* (1981) who carried out laboratory studies on the lysis of *Phormidium* by the bacterial isolate *Myxococcus xanthus*. The bacteria initially attached to the algae by formation of surface extrusions, leading to irregular clumping of bacteria around the algal filaments. Division and migration of the bacteria ultimately resulted in the complete enclosure of algae within large myxobacterial colonies, 1–6 mm in diameter. The centre of the colonies contained blue-green algae at various states of degradation, while the periphery was composed of a dense mass of myxobacterial cells. In these laboratory studies, which were carried out in an inorganic medium, growth of the blue-green algae was controlled over a long period of time – with the bacteria deriving their heterotrophic nutrition entirely from the algae and their by-products. Destruction of the algal cells was achieved even at low predator inoculum levels (down to predator/prey ratios of $1/10^3$), suggesting that these organisms may be highly effective in antagonizing phytoplankton populations in the aquatic environment.

6.13 Epiphytic associations of bacteria with phytoplankton

Bacteria are associated with a wide range of exposed surfaces in the freshwater environment, including both inorganic (rocks and fine sediments) and organic (sediment detritus, macrophytes, and planktonic biota) interfaces. The bacterial populations that develop on these surfaces may form quite complex and structured communities (e.g., biofilms on exposed rock surfaces in streams, see Section 6.11) or more simple groupings (e.g., single bacteria or small aggregates on phytoplankton). In this section we consider the epiphytic associations of bacteria with phytoplankton (particularly blue-green algae) and discuss the importance of these associations to the general ecology of lake bacteria and the physiology of phytoplankton cells. Epiphytic bacteria are considered here to include organisms within the extracellular mucilage of single cells and colonies (sometimes referred to as 'endophytes') as well as those on the outer surface.

6.13.1 Bacteria within the phycosphere

The environment immediately around the surface of algae has been referred to as the phycosphere (Bell *et al.*, 1974) and resembles other interfaces such as the root (rhizosphere) and aerial (phyllosphere) surfaces of higher plants in being a region of high nutrient concentration. These nutrients arise as extracellular products from the algae, and may support extensive populations of bacteria and other microbes (high productivity). The promotion of phycosphere bacterial populations by phytoplankton will also depend on other biochemical interactions (e.g., the production of algal antibiotics, Section 6.12.2) as well as physical properties of the algal surface (e.g., presence or absence of mucilage), both of which will vary with the physiological state and species of alga involved. The bacterial population that develops will also depend on broader aspects of the phycosphere community such as grazing activities by protozoa and the equilibrium between mucilaginous and planktonic organisms.

The phycosphere community

Although many species of phytoplankton typically have relatively low levels of epiphytic bacteria, particularly when they are in a healthy condition (Brunberg, 1999), this is not true for blue-green algae. Various morphological and physiological features of blue-green algae make them particularly suitable for forming and maintaining high populations of associated bacteria, including the tendency to aggregate as filamentous or globular colonies, and the production by many species of copious amounts of extracellular mucilage. The high nutrient phycosphere surrounding these organisms forms a microenvironment for extensive populations of bacteria, protozoa, and invertebrates.

With large colonial algae such as *Anabaena* and *Microcystis*, the phycosphere is thus a site for the development of a complex microbial community. This shows some similarity to the bacterial biofilm community discussed previously, with bacterial populations present in a polysaccharide matrix, exchanging cells with planktonic populations and subject to predation by motile and attached protozoa (Figure 6.23). These mixed epiphytic communities occur as microcosms within the water column, forming mini-ecosystems within the broader lake ecosystem, with their own dynamic interactions, local food webs and energy flows.

Microorganisms present within the mucilage include epiphytic algae (e.g., unicellular blue-greens such as *Cyanothece* and *Synechococcus*), protozoa (e.g., ciliates and choanoflagellates), and bacteria. Phycosphere microorganisms are also present at the mucilage surface (sessile stalked protozoa) and within the surrounding aquatic medium (bacteria). Dynamic interactions between microorganisms occur within and between all these regions of the algal phycosphere, resulting in a potentially complex food web in this small-scale ecosystem.

Bacteria use the mucilage as a substratum and benefit from algal secretions, but are grazed by motile and sessile protozoa. These mucilage-bound bacteria are in dynamic equilibrium with planktonic populations, and can be experimentally released for enumeration and characterization by sonication.

Ecological role of phycosphere

Within the water column the phycosphere provides the following.

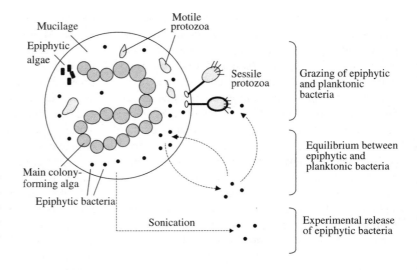

Figure 6.23 Algal microcosm: diagrammatic representation of microbial diversity and interactions in the surface microenvironment (phycosphere) of a colonial (filamentous) alga such as *Anabaena*

- A zone of attachment within an otherwise totally planktonic environment and supports the equilibrium between attached and unattached bacteria. Attached and unattached bacteria within the phycosphere probably show the same physiological differences which were exhibited by these two groups of organisms in more conventional biofilm ecosystems (Section 1.4).

- A refuge from zooplankton grazing. Zooplankton grazing free planktonic bacteria are not able to graze bacteria present on the much larger algal cells and colonies. Phycosphere bacteria are susceptible to grazing by protozoa,however, particularly ciliates and flagellates.

- In algae with surface mucilage, a substrate for heterotrophic bacteria. In the case of blue-green algae, various studies have shown that soluble as well as colloidal exudates can be readily assimilated by a range of epiphytic microbes, including heterotrophic bacteria and ciliate protozoa (Paerl, 1992).

- A site of attraction in the water column which motile organisms may locate by chemotactic response.

Although the algal phycosphere is usually considered in relation to the pelagic environment, it is also an important aspect of the benthic environment. Algae with associated epiphytes also occur in lake sediments (where living algae may be growing *in situ* or be derived by sedimentation from the water column) and as attached algae in fast-flowing water.

6.13.2 Observation and enumeration of epiphytic bacteria

Epiphytic bacteria are particularly apparent on algae with surface mucilage, including both green and blue-green algae, where they can be observed as small coccoid or rod-shaped structures using phase contrast microscopy or bright-field light microscopy (Figure 6.24). In some cases, the bacteria form a discrete layer within the mucilage occurring at a fixed distance from the algal surface, suggesting that they adopt an optimum position in relation to the algal cell and the surrounding lake water. Useful information on the location of epiphytic bacteria can also be obtained from scanning or transmission electron microscopy, where their distribution over the outer surface (SEM) or positioning within mucilage (TEM) can be seen at higher resolution.

Although populations of epiphytic bacteria are difficult to count from light or electron microscope

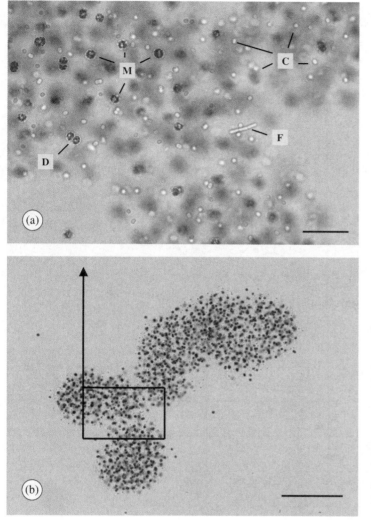

Figure 6.24 Epiphytic bacteria in *Microcystis aeruginosa*.

(a) Detailed light microscope view from colony (b) showing individual cells (M) of *Microcystis*, embedded within the mucilaginous matrix. Cells range in size from 3–5 μm, and some (D) have just completed division. Coccoid epiphytic bacteria (C) appear as bright refractive bodies, with a chain of filamentous bacteria (F) also visible (scale = 20 μm).

(b) Whole colony of *Microcystis*, composed of numerous separate cells, showing the typical irregular appearance of this colonial blue-green alga. Epiphytic bacteria are difficult to see at this magnification. The rectangle indicates the field of view of micrograph (a) (scale = 100 μm)

preparations, they can be released from phytoplankton samples in the laboratory by sonication and counted using standard techniques. Even though the completeness of bacterial release using this technique is not clear, it does provide a useful experimental approach to the enumeration of epiphytic populations.

Using this technique, Brunberg (1999) demonstrated high populations of epiphytic bacteria associated with a bloom of the blue-green alga *Microcystis* (Figure 6.24) in the water column and on the sediments of a hypereutrophic lake (Vallen-tunasjön, Sweden). Productivity of the epiphytic bacteria was also determined and compared to the free plankton population. The results showed the following.

• In the water column, *Microcystis*-associated bacteria constituted 19–40 per cent of the total bacterial abundance. These bacteria were less active (bacterial production/cell) compared with ambient water column bacteria, since their overall productivity was only 7–30 per cent of the total population.

- In the sediment, mucilage-associated bacteria constituted 1–5 per cent of the total bacterial abundance, but contributed 8–13 per cent to the total bacterial production during the summer.

In this lake, where living colonies of *Microcystis* are a significant part of the benthic biomass, the importance of epiphytic bacteria in the sediment is particularly interesting. Coupling had been previously demonstrated between *Microcystis* biomass and bacterial production in the sediment, and the

sonication data supported this by showing that *Microcystis* epiphytes were 'hot spots' of bacterial productivity compared with other sediment bacteria.

6.13.3 Specific associations between bacteria and blue-green algae

Bacteria that associate with the surface of blue-green algae occur as mixed populations and vary in

(a)

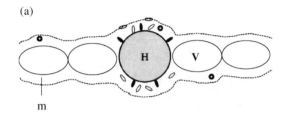

Bacterial isolate	Association specificity	*Anabaena flos-aquae*	*Anabaena cylindrica*
Flavobacterium ◎	Nonspecific – all cell types	(+/-)	(+/-)
Pseudomonas ●	Heterocysts only	(+++)	(+)
Zooglea ⬭	Heterocysts only	(+)	(+++)

Figure 6.25 Specificity and kinetics of bacterial adhesion to filaments of *Anabaena*: (a) diagram and table to illustrate adhesion specificity for three freshwater isolates – *Flavobacterium* sp, *Pseudomonas* sp, and *Zooglea* sp. H-heterocyst, V-vegetative cell, m-surface mucilage; (b) kinetics of bacterial adhesion for *Pseudomonas* sp. and *Zooglea* sp. which show increased levels of adhesion (attached cells) with increase in the concentration of planktonic (non-attached) cells (based on figures and information from Lupton and Marshall, 1981)

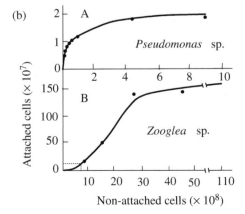

terms of the specificity of their association. This specificity differs both in terms of adhesion mechanism and metabolic co-operation.

Specificity of adhesion

Bacterial adhesion to blue-green algae varies with bacterial isolate and type of algal cell. This was investigated by Lupton and Marshall (1981), who examined the association between three bacterial isolates – *Flavobacterium* sp., *Pseudomonas* sp., and *Zooglea* sp., and filaments of *Anabaena flos-aquae* and *Anabaena cyclindrica*. The bacterial isolates had been obtained from algal cultures, and the algal filaments comprised two main cell types – heterocysts (involved in nitrogen fixation), and vegetative cells. Adhesion experiments (Figure 6.25) showed that the association of *Flavobacterium* was non-specific, while *Pseudomonas* showed high adhesion specificity to heterocysts of *A. flos-aquae*, and *Zooglea* to heterocysts of *A. cylindrica*. Differences in specificity of adhesion may reflect differences in the distribution and abundance of specific binding sites on the heterocyst surfaces.

Although both *Pseudomonas* and *Zooglea* exhibit specific association with heterocysts, they differ in the nature of their adhesion process. Electron microscopy showed that *Pseudomonas* adhered directly to the algal cell wall and was present as a monolayer of cells, while cells of *Zooglea* were largely suspended (random orientation) within the mucilage and developed as a multilayer population. These differences between the two bacterial isolates are also reflected in the experimental adhesion curves (Figure 6.25).

Both *Pseudomonas* and *Zooglea* (but not *Flavobacterium*) promoted higher rates of reduction activity by the algal cells under oxygenated conditions, suggesting that their association with algal cells may have physiological implications.

Metabolic co-operation

In the case of blue-green algae, bacterial epiphytes appear to exist in positive equilibrium with the algal cells and not as antagonists. The presence of high

populations of surface bacteria (and protozoa) is frequently observed during periods of maximum algal growth (e.g., during algal blooms), and is not a prelude to algal lysis or degradation.

The interaction between bacteria and blue-green algal cells involves a complex interchange of materials which are important both for the establishment (chemotaxis) and continued nutrition of both organisms. This symbiotic relationship is characterized by a variety of chemical links (Figure 6.26), including the secretion of chemotactic agents, photosynthates, and products of nitrogen fixation by algal cells, and the recycling of carbon, inorganic nutrients, and trace metals by bacteria. Bacteria have high efficiencies of N and P uptake under conditions of inorganic nutrient limitation (Section 6.12.1) and may provide an important route for algal uptake of N and P in these conditions. In addition to these major routes of nutrient transfer, exchange of growth factors such as vitamins, chelating agents, and other non-defined compounds may take place in this microenvironment.

Two of the above aspects, production of chemotactic agents and transfer of fixed nitrogen compounds, have been looked at in some detail.

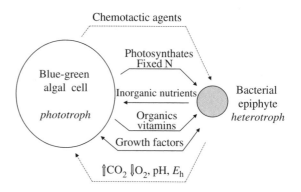

Figure 6.26 Chemical interactions in the phycosphere. Epiphytic bacteria, in direct contact or close association with algal cells, exchange a range of nutrients and metabolites with their phototroph partners. They also influence the immediate microenvironment around the algal cells, increasing the CO_2 concentration but reducing oxygen tension, pH, and redox potential (E_h). Epiphytic bacteria are themselves influenced by chemical changes (e.g., chemotactic agents) within their own microenvironment (based on a figure from Paerl, 1992)

Chemotactic agents Commonly-occurring genera of both N_2-fixing (*Anabaena*, *Aphanizomenon*) and non N_2-fixing (*Microcystis*, *Oscillatoria*) bloom-forming algae excrete compounds that attract heterotrophic bacteria (references in Paerl, 1992). The chemotactic attraction of bacteria to heterocysts (e.g., in *Anabaena* and *Aphanizomenon*) is highly specific, and involves a strong response to low molecular weight nitrogenous compounds (amino-acids, peptides) but not sugars or organic acids. In contrast, the chemotactic response to colonies which do not contain heterocysts was much broader and involved an equal response to the same array of organic compounds.

Photosynthetic and nitrogen fixation products The excretion of photosynthetic products by algal cells and the uptake of dissolved organic carbon by heterotrophic bacteria has been emphasized earlier. In the case of the blue-green algal/bacterial association, the DOC is produced by vegetative algal cells and is assimilated by bacteria in the phycosphere.

Products of nitrogen fixation (amino-acids and low molecular weight peptides) are readily assimilated by associated bacterial cells. This rapid transfer of recently fixed nitrogen was demonstrated by $^{15}N_2$ pulse labelling of non-axenic *Anabaena* colonies (Paerl, 1992). The results showed that:

- epipytic bacteria do not carry out any N_2 fixation themselves, but take up $^{15}N_2$ within 1 hour of fixation by algal cells – this indicates a close metabolic coupling between the symbionts;

- the pathways and magnitude of fixed N_2 transfer are affected by carbon metabolism. When the photosynthetic carbon demands of *Anabaena* were met, transfer of N_2 to bacteria was minimal, but under conditions of external inorganic carbon (CO_2, HCO_3^-) limitation nitrogen transfer markedly increased.

Increased transfer of fixed nitrogen under conditions of carbon limitation may represent an adaptive mechanism which promotes higher growth and respiration of the bacteria. This increased bacterial respiration changes the chemical composition of the phycosphere, increasing CO_2 concentration and decreasing O_2 concentration, pH, and redox potential (E_h). Carbon dioxide regeneration in the phycosphere is particularly important in bloom conditions, where oxygen concentrations are high and carbon dioxide may be severely limited. In this situation, epiphytic CO_2 production immediately adjacent to algal cells reduces algal carbon limitation and promotes continuation of photosynthesis. Regeneration of CO_2, with associated reduction in O_2, also has a more specific importance at the heterocyst surface, where maintenance of localized reducing conditions within a highly oxygenated general environment is essential for N_2 fixation (Section 5.6).

7

Viruses: major parasites in the freshwater environment

Although freshwater viruses have probably been the least researched of all aquatic microorganisms, their widespread occurrence and general role as parasites gives them an ecological significance equal to the other, more extensively studied, microbial groups (Wommak and Colwell, 2000).

7.1 Viruses as freshwater biota

Various ecological studies have been carried out on the role of viruses in freshwater planktonic systems (see, for example, Weinbauer and Höfle, 1998), but there is little information on viruses in benthic systems such as biofilm communities. The relative lack of ecological information on freshwater compared with marine viruses also means that information obtained from marine environments may fill useful gaps in our understanding of freshwater systems.

Viruses are small in size and are the only group of freshwater organisms which are non-cellular and belong exclusively to the femtoplankton (Section 1.2.2, Table 1.3). Although most viruses are <70 nm in diameter, a small fraction are larger than 100 nm, with some filamentous forms exceeding 1 μm in length (Middelboe *et al.*, 2003).

7.1.1 General role in the freshwater environment

Viruses are potentially important in planktonic and benthic freshwater ecosystems in a number of ways.

- They have a significant influence on the growth and productivity of the major biomass (algae) and the greatest population (bacteria) of freshwater biota.

- They cause death and lysis of freshwater microorganisms, liberating dissolved organic carbon (DOC), nitrogen, and phosphorus into the water and generating nutrients for the microbial loop. Viruses thus have a significant impact on carbon and nutrient flow in microbial food webs.

- Although viruses occur as discrete particles within the aquatic medium, their small size places them within the normally defined limits of 'soluble' organic material (<0.2 μm diameter). They thus directly contribute to the DOC as macromolecular complexes containing proteins, nucleic acid, and lipid components.

Freshwater Microbiology: Biodiversity and Dynamic Interactions of Microorganisms in the Aquatic Environment   David C. Sigee
© 2005 John Wiley & Sons, Ltd   ISBNs: 0-471-48529-2 (pbk) 0-471-48528-4 (hbk)

- They may mediate genetic exchange between their host organisms, promoting gene flow within populations of blue-green algae, eukaryotic algae, and heterotrophic bacteria respectively. Studies on virus-mediated gene transfer have been carried out particularly in relation to bacteriophages (see Section 7.8.6).

7.1.2 Major groups and taxonomy of freshwater viruses

Within a particular water body, viruses may be divided into two main groups according to whether they are naturally present (endogenous) or introduced (exogenous).

Endogenous viruses

Endogenous viruses occur as parasites of all freshwater biota, and thus influence the aquatic food web at all levels. Within the endogenous viruses, particular attention will be given in this chapter to three major groups:

- cyanophages – viruses of prokaryote (blue-green) algae,

- phycoviruses – infecting eukaryote algae,

- bacteriophages – parasites of freshwater bacteria.

Cyanophages and bacteriophages are parasites of prokaryote hosts and belong to a single taxonomic assemblage of phage viruses (Table 7.2). Phages have a head (capsid) and tail, in contrast to the phycoviruses (Table 7.3) – which constitute a single taxonomic group of viruses composed only of a capsid. Within these major groups, the taxonomy of viruses follows a nomenclature based on families, genera, and species (Tidona and Darai, 2002). Virus taxonomy changes particularly rapidly, and individual viruses are now given a Decimal Code within the universal virus database prepared for the International Committee on Taxonomy of Viruses

(ICTV). The taxonomy of freshwater viruses is considered in more detail in subsequent sections.

Two of the above groups (cyanophages and phycoviruses) are parasites of autotrophic hosts, while bacteriophages infect largely heterotrophic organisms. In addition to these three groups, parasites of other microorganisms (fungi, protozoa) and higher organisms (higher plants, invertebrates, and fish) are also clearly important in the freshwater environment. Viruses of fish such as the hematopoietic virus of salmon and trout also have commercial significance.

Exogenous viruses

Exogenous viruses are introduced into the aquatic system from an external source, such as sewage. These viruses are not a natural part of the freshwater ecosystem and do not infect the naturally-occurring freshwater biota. Examples include human enteric viruses such as coliphages, which are infective agents of gut (coliform) bacteria, and may enter the environment with their host bacteria via untreated domestic effluent. Coliphages may be used as viral indicators of the general level of faecal contamination in natural waters, and may also provide information on the source of impact. Studies by Cole *et al.* (2003) on the presence, prevalence, and concentrations of F^+ RNA and DNA coliphages in a range of municipal and agricultural wastewaters showed that F^+ RNA typing had particular potential in distinguishing faecal contamination sources.

Natural waters may also be contaminated by viruses that directly infect humans (rather than human gut bacteria) – including hepatitis A virus, human immunodeficiency virus (Chen and Suttle, 1995), and noroviruses (Lamothe *et al.*, 2003).

7.2 The virus life cycle: intracellular and free viral states

The widespread occurrence of viruses in freshwater systems (Safferman and Morris, 1967) involves their presence as two main states or subpopulations – free infective particles (virions)

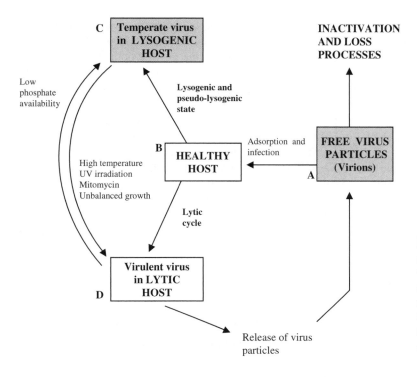

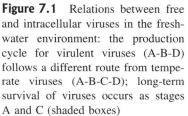

Figure 7.1 Relations between free and intracellular viruses in the freshwater environment: the production cycle for virulent viruses (A-B-D) follows a different route from temperate viruses (A-B-C-D); long-term survival of viruses occurs as stages A and C (shaded boxes)

within the environmental medium, and intracellular viruses within their specific hosts (Figure 7.1). An equilibrium exists between these two states, the balance of which varies with host availability, infectivity, and environmental conditions. At high host-population levels, the sub-populations of both intracellular and extracellular viruses increase, maintaining the balance between the two.

The extracellular population of free suspended virus particles is important for further host infection, but is also an important stage for survival of the organism. At this point in their life cycle, viruses are very exposed and are particularly vulnerable to a range of external factors which inactivate them or eliminate them from the environment (see Section 7.4.2).

The infective particles or virions are biologically inert in their free state and must infect a host cell in order to propagate. Virions identify potential host organisms by passing through the environmental medium (water) until they collide with a host cell. If this collision is with a sensitive host that has an appropriate surface receptor, the collision may result in the virus becoming firmly attached (adsorbed).

Following adsorption and infection of host cells, viruses have the potential to take over the nucleic acid and protein synthetic machinery of the host, producing more viral particles which are ultimately discharged into the environment by bursting (lysis) of the host cell. Details of viral multiplication and the host cell response during the infection process are given later in relation to cyanophages (Section 7.6.2) and phycoviruses (Section 7.7.2). In some cases, the virus is highly active in promoting the sequence of viral particle formation and host cell destruction (lytic cycle) and is referred to as 'virulent'. In other cases (lysogenic state), the virus is relatively inactive, simply being present as viral DNA (prophage) which is integrated into the host genome in an inert form, and is referred to as 'temperate'.

7.2.1 Significance of the lysogenic state

Within a population of infected cells, an equilibrium occurs between the virulent and temperate states, the point of balance depending on the particular virus/host cell relationship, and various environmental factors (Figure 7.1). Transition from lysogenic to lytic states may be promoted by factors such as high temperature, UV irradiation, and antibiotic treatment – all of which promote disturbance and irregularities in host cell growth. Growth disturbance may also occur in other ways, such as the removal of endosymbiotic algae (genus *Chlorella*) from the coelenterate *Hydra viridis*. A high proportion of the algae in *Hydra* are infected by a temperate phycovirus (*Chlorovirus*), which becomes lytic when the algae are isolated (Van Etten, 2002b).

The extent to which different aquatic viruses are temperate or virulent varies considerably, and probably varies with particular viruses and particular host-virus associations. Two closely related groups of viruses, the cyanophages and bacteriophages, both parasites of prokaryotes, differ in this respect. Cyanophages are mainly virulent (Padan and Shilo, 1973), while bacteriophages are largely temperate (Bratbak *et al.*, 1994). This difference may reflect fundamental differences in the physiology of the host cells within the freshwater environment, since the entire population of blue-green algae within a phytoplankton community is typically metabolically-active, while most of the heterotrophic bacteria are metabolically-inert. On this basis, the lysogenic state in planktonic bacteria can be regarded primarily as a dormant stage, occurring within metabolically-inert cells, with induction to the lytic condition when the host cell metabolism is activated. In accordance with this, induced conditions of low nutrient availability reduce host cell growth and promote a shift in the balance of viral activity to the lysogenic state.

The balance between the lysogenic and lytic states is clearly important in terms of growth and productivity of the host population. This is considered later in relation to planktonic bacterial populations (Section 7.8.4), where the balance between temperate/virulent bacteriophage activity has a major effect on bacterial mortality.

The state of viral infection is also important to microorganisms in terms of genetic variability. Lysogeny appears to be beneficial to host organisms in leading to increased gene transfer between cells, and thus to increased fitness in natural ecosystems.

This has been demonstrated particularly in the case of freshwater bacteria (Miller *et al.*, 1990), where lysogenic organisms served both as a source of bacteriophages, capable of mediating transduction, and as recipients of transduced plasmid and chromosomal DNA. In freshwater bacteria such as *Pseudomonas aeruginosa*, where between 45 and 100 per cent of the natural population is probably lysogenic, the potential for horizontal gene transmission is much increased, resulting in an increase in the size and flexibility of the gene pool available for natural selection. The phenotypic diversity generated by a high proportion of lysogenic bacteria, coupled with resulting high rates of phage-mediated DNA exchange (see Section 7.8.6), must be of considerable significance in a continually changing, complex aquatic environment.

7.3 Detection and quantitation of freshwater viruses

Viruses may be detected in environmental aquatic samples or infected algal cultures either within the aquatic medium itself (free particulate viruses) or within infected host cells present in suspension.

7.3.1 Free particulate viruses

Virus counts within the aquatic medium are normally carried out either by direct observation and enumeration (total count) or by their lytic effect on algal cells (viable count), involving *in vitro* assay and plaque formation. Viral populations are accordingly expressed as particle count per litre or as number of plaque-forming units (pfu) per litre, respectively. Before total or viable counts can be carried out the water sample is normally filtered and concentrated. Initial filtration through a glass filter membrane is important in removing algae, bacteria, and other extraneous particulate matter, all of which may interfere with the final count. Concentration of the filtered sample increases virus detectability and can be achieved by centrifugation ($2000 \times g$ for 5 minutes) or by dialysis against a solution of high MW such as ethylene glycol.

In addition to these standard techniques of virus detection and quantitation, soluble DNA concentration has also been used to monitor the total virus titre and the presence of particular viruses can now be detected by molecular techniques (PCR assay).

Total count

Enumeration and morphological categorization of viral particles in aquatic samples can be carried out by transmission electron microscope examination of negative-stain preparations. The appearance of free virus particles under the electron microscope is shown in Figures 7.12 (phycoviruses) and 7.14 (bacteriophages). This approach was initially adopted by Bergh *et al.* (1989) who analysed the viral content of a range of fresh and marine natural waters. These authors deposited the virus particles on electron microscope grids by centrifuging the water samples at $100\,000 \times g$ for 90 minutes prior to examination, and demonstrated much higher viral counts (up to 2.5×10^8 particles ml^{-1}) than had previously been determined by viable assay. Although this technique generates a total viral count, it gives no information on infective levels in relation to particular hosts.

Viral counts can also be carried out by epifluorescent light microscopy using stains such as the fluorochrome 4′-6′-diamidino-2-phenylindole (DAPI) and SYBR-Green 1 to detect particles smaller than 0.2 μm which contain double-stranded DNA (Noble and Fuhrman, 1998). Suttle *et al.* (1990) showed that light and electron microscope counts of virus particles in natural marine waters gave similar values (10^6–10^7 ml^{-1}).

Dissolved DNA as an index of viral concentration

Since virus particles occur within the defined soluble fraction (<0.2 μm) of natural waters, it might be expected that measurement of dissolved DNA would provide a useful index of virus content. Studies carried out by Paul *et al.* (1991) on freshwater and seawater samples, however, showed that only a small portion of the dissolved DNA was attributable to intact virus particles. Simultaneous determination of viral DNA (estimated from viral counts) and dissolved DNA indicated that nucleic acid encapsulated in viral particles averaged <5 per cent of total dissolved DNA. Not only is viral DNA a small fraction of the total, but it is difficult to measure. Viral DNA encapsulated within a protein capsid is rendered inaccessible to nucleases and fluorescent (Hoechst) stains, and is only released for detection after ethanol precipitation. The results suggest that viruses are not a major component of the soluble DNA in aquatic systems, though they may be involved in its production by lysis of bacterial and phytoplankton cells. The results also clearly indicate that measurement of total dissolved DNA does not provide a useful measure of virus concentration.

Viable count

The presence or absence of infective virus particles (for a particular microbial host) within the environmental medium can be shown by adding an aliquot of the filtered concentrate to an actively-growing microbial culture and seeing whether cell lysis occurs.

In the case of algal infection, this involves loss and degradation of chlorophyll and is most readily observed as a marked decrease in green coloration. Although algal lysis can be readily observed in liquid algal cultures, for quantitative purposes (viable counts) it is normally carried out on algal lawns, which are thin layers of concentrated algal population maintained on a gel or membrane surface (Figure 7.2). There are various ways of using algal lawns for viable counts. One simple procedure is to mix a small volume of environmental concentrate with an equal volume of cultured algal cells suspended in liquid agar, then applying it as a thin layer to a solid agar surface. After a period of time, the presence of viral particles that infect and lyse the algal cells is indicated by transparent regions of lysis (plaques) within the green algal lawn (Figure 7.2). These can be counted to determine the number of plaque-forming units (pfu) per unit volume of the original sample. In some cases, viral plaques with different size ranges occur on a single plate, indicating the presence of two or more viral strains with varying infective ability. Viable counts

Petri dish

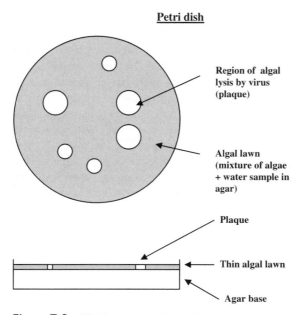

Region of algal lysis by virus (plaque)

Algal lawn (mixture of algae + water sample in agar)

Plaque

Thin algal lawn

Agar base

Figure 7.2   Viable counts of aquatic viruses using an algal lawn.

Face (top diagram) and side view of viable count assay preparation, observed after sufficient time for the development of lytic plaques. Each plaque is derived from a single infective virus particle present in the environmental water sample. Different plaque sizes relate to two separate strains of virus that are able to infect the algal host. Pure virus samples may be isolated from individual plaques for further characterization and experimentation

PCR detection of viral DNA

Detection and enumeration of aquatic viruses by total and viable count both have their problems. With total counts, electron microscope evaluation of viruses that infect particular species or groups of algae is not possible because morphological criteria for this do not exist. Even if they did, the numbers of virus particles relative to a particular algal group would probably be too low to count. Viable counts have the disadvantage that only viruses for which host algae have been isolated can be enumerated, and the development of viral plaques can take days or weeks for completion. These problems highlight the need for a rapid approach to detect, identify, and enumerate viruses which infect phytoplankton.

One possibility in terms of specificity is to apply PCR technology, using virus-specific primers. A primary consideration in developing a PCR-based method for detecting microalgal viruses is to identify a suitable viral gene for which primers can be designed. DNA polymerase (DNA *pol*) genes are ideal targets for which PCR primers of wide-ranging specificity can be designed, since although the protein transcripts contain highly conserved motifs (within exonuclease and polymerase domains) at the amino-acid level (Figure 7.3) there is some variation at the nucleotide level. In the first example

have the advantage that the assay directly relates to the host specificity of the algal cells (Padan and Shilo, 1967, 1969), but great care must be taken to exclude bacteria such as *Bdellovibrios* and *Myxococci*, which can also lyse algae and cause plaque formation. These organisms are normally removed by filtration prior to the lawn assay, but can also be destroyed by chloroform treatment (Safferman and Morris, 1964).

Individual plaques on algal lawns not only provide a viral count (each being derived from a single infective particle) but also provide a source of material for further viral analysis and characterization. Where this is the major experimental objective, the original environmental sample may be incubated with the suspension of algal cells for a number of days prior to plating, to enrich the viral titre and increase plaque formation.

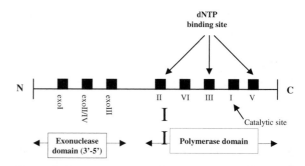

Figure 7.3   Target nucleotide sequence for phycovirus PCR primer: generic map of B-family DNA polymerases showing conserved motifs (■) and their possible functions.

The most highly conserved amino-acid sequence (I) is a catalytic site.

dNTP – deoxynucleotide triphosphate (based on a figure from Chen and Suttle, 1995) of a PCR-based technique designed to detect viruses which infect eukaryote algae, Chen and Suttle (1995) developed DNA *pol* primers to determine the presence of viruses of *Chlorella*, *Micromonas*, and *Chrysochromulina* in environmental samples. The ultimate goal will be to develop and use sensitive PCR methods to identify and quantify a whole range of viruses in freshwater and marine water samples.

7.3.2   Infected host cells

The presence of viruses within infected host cells can be shown by examination of ultrathin sections of the algal host under the transmission electron microscope (TEM). Virus particles typically appear as electron-dense structures, since their nucleic acid characteristically stains heavily with the reagents normally used with the TEM (uranyl acetate, lead citrate). These newly-formed virus particles tend to occur towards the end of the lytic cycle, and can be distinguished from electron-dense cytoplasmic structures by their diameter (about 200 nm), uniform size, polyhedral shape, and location within a discrete part of the cell (Figure 7.12 (c) and (d)). In many infected cells, virus particles occur in a distinct region of cytoplasm, the virus assembly centre.

Even where the TEM identification of intracellular viruses appears to be clear-cut, they are normally designated as 'virus-like particles' or 'VLPs', to emphasize that their identification is based on morphological and not functional criteria.

7.4   The growth and control of viral populations

7.4.1   Virus productivity

As with other freshwater microorganisms, productivity (the rate of increase in biomass) is an important ecological parameter, and can be considered as net and gross processes.

$$\text{Net productivity} = \text{Gross productivity} - \text{intrinsic loss processes} \quad (7.1)$$

In the case of viruses:

$$\text{Gross productivity} = \text{synthesis of virus biomass within host cells}$$

$$\text{Intrinsic loss processes} = \text{internal activities that reduce or delay viral release}$$

Net productivity (Equation 7.1) is the rate of increase of infective virus biomass within the water medium. The concept and measurement of productivity in freshwater viruses is complicated by the fact that these biota occur both free (virions) and in cells, and also because viral generation is closely tied to the infection and lysis of host cells (Figure 7.11).

The rate of synthesis of virus biomass within host cells (gross productivity) depends on infection cycle parameters such as rate of virion adsorption, efficiency of host cell penetration and rates of virus macromolecule formation. Intrinsic loss processes include all those internal activities that destroy virus components or limit the release (e.g. by delayed lysis) of synthesised virus particles. These are distinct from extrinsic loss factors such as sedimentation and grazing (see below). In practice, net productivity is normally considered in terms of numbers of virus particles rather than viral mass, and is expressed as the number of virus particles produced per litre of lake water per day (virus l^{-1} d^{-1}).

Net productivity can be determined by multiplying the rate of host cell lysis by the burst size. Studies by Weinbauer and Höfle (1998), for example, revealed a mean virus productivity rate of 10^9 viruses l^{-1} d^{-1} in the epilimnion of a eutrophic freshwater lake. Mean bacterial productivity in the lake was 10^9 cells l^{-1} d^{-1}, with the release of viral particles at lysis averaging 34 viruses per cell (burst size). Release of soluble cell contents at the same time as viral particles (during cell lysis) made a very small contribution (<0.1 per cent) to overall lake water DOC levels (Figure 7.4). The increased importance to bacterial mortality of viral lysis in the lower part of the water column (see Section 7.8.5) was reflected in higher burst sizes (mean value 63) and higher productivity (10^{10} viruses l^{-1} d^{-1}) in the hypolimnion. Viral productivity has also

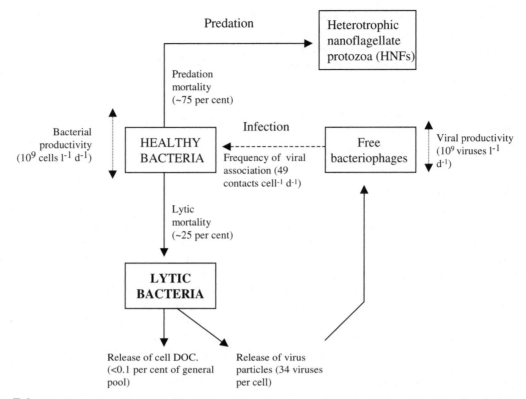

Figure 7.4   Viral and bacterial productivity in the epilimnion of eutrophic Lake Plußsee, Northern Germany. Values for lytic and predation mortalities of bacteria are expressed as a percentage of bacterial production and are highly variable, carbon flow (———▶) (based on the studies of Weinbauer and Höfle, 1998)

been monitored in terms of net change in viral abundance (total or viable water counts) over time, viral decay rates, incorporation of radioactive orthophosphate, and the fraction of host cells containing mature viral particles (Bratbak *et al.*, 1994). As yet, no particular procedure has become standard for the measurement of viral productivity in natural waters.

Based on long-term sampling (intervals of days), net changes in viral abundance tend to change (increase or decrease) at rates of about 0.9 viral units d^{-1}, irrespective of whether the data were obtained as total counts, viable counts, or as rate of incorporation of ^{32}P-labelled orthophosphate into viral DNA. Much higher rates (up to 9 viral units d^{-1}) in the change of viral abundance were observed when sampling intervals were reduced to

1–2 hours, with even higher values at shorter time intervals. These data suggest that viral productivity may vary considerably in the short term, varying with host availability and possibly diurnal variations, but is steady on a long-term basis. Long–term rates of change of viral abundance in the range 0.5–1 viral units d^{-1} correspond to the rate of many other processes affecting microbial activities in natural waters, including bacterial growth rates, grazing rates, and rates of nutrient turnover (Bratbak *et al.*, 1994).

7.4.2   Regulation of viral abundance

The abundance of viruses in natural waters depends on the release of new virus particles from infected host cells and the survival of free particles in the water medium. The release of new virus particles can occur from lytic infection of a susceptible host or by activation of a viral genome contained in a lysogenic host.

Lytic infection

Quantitative aspects of infection and viral release have been considered particularly in relation to bacteriophage systems (Bratbak *et al.*, 1994). The rate of viral infection depends on the rate of virus adsorption (dP/dt), which is proportional to the concentration of host cells (H) and virus particles (P), as shown in Equation (7.2).

$$dP/dt = kPH \qquad (7.2)$$

where k = the adsorption rate constant.

In the case of bacteriophages, experimental and theoretical (diffusion and collision theory) both give a value for k as 0.25×10^{-8} cm^3 min^{-1} (Stent, 1963).

Calculation of infection dynamics Virus adsorption and infection only occur within specific host/parasite combinations, so values for P and H refer to populations within those combinations. Approximate values for the dynamics of phage infection can be calculated for different numbers of host/parasite combinations in a particular plankton community as follows:

- assume that the abundance of bacteria and phages in a particular plankton community is, for example, 10^6 and 10^7 ml^{-1} respectively;

- if this contains 100 different phage–host systems, each containing 1 per cent of the two sets of organisms, then using Equation (7.2) for a single phage–host system:

$$dP/dt = 0.25 \times 10^{-8} \times 10^5 \times 10^4 = 2.5 \qquad (7.3)$$

The phage adsorption rate for the single system would be 2.5 min^{-1} ml^{-1} or 3.6×10^3 d^{-1} ml^{-1}. The total rate of adsorption for all phage–host systems would thus be 0.36×10^6 d^{-1} ml^{-1}.

With a total bacterial count of 10^6 cells ml^{-1}, this rate of total adsorption means that about one third of the bacterial population may experience a viral attack every day. With a burst size of 50, viral productivity would be 1.8×10^7 viruses ml^{-1} d^{-1}.

A more realistic value for the number of different phage-host systems would be about 10^5. This is based on a pfu count of individual bacterial species in natural waters of 1–100 ml^{-1}, with approximately 100 virus ml^{-1} for each system. Comparative phage adsorption values for 10^5 and 10^2 phage–host systems within the same overall bacterial and viral populations are shown in Table 7.1.

The calculations show that planktonic communities with high host species diversity (with higher numbers of phage–host systems) will have lower overall levels of adsorption and infection. During blooms, when diversity is low, the infection of

Table 7.1 Calculations of viral adsorption rates for different numbers of phage-host systems within a planktonic community (based on Bratbak *et al.*, 1994)

Number of phage-host systems	Counts per system (ml^{-1})		Adsorption rate (d^{-1} ml^{-1})	
	Bacteria	Phages	Per system	Entire bacterial population
10^2	10^4	10^5	0.36×10^4	0.36×10^6
10^5	10	10^2	0.36×10^{-4}	0.36×10^3

Total counts ml^{-1}: bacteria (10^6), phages (10^7).

individual host species and the total incidence of infections will be greater.

Induction of lysogenic host cells

Many viruses in the environment are non-virulent. This is particularly the case for the infective agents of bacteria, where more than 90 per cent of all known phages have been found to be temperate – infecting their hosts, which then enter a lysogenic state (Freifelder, 1987).

Virus production in lysogenic systems involves activation (induction) to a lytic state. The factors that bring this about in natural systems are not known, but laboratory experiments have identified various environmental factors which are effective. As noted previously (Section 7.2), these include UV irradiation, effects of antibiotics, temperature shock, and other stress factors which can lead to unbalanced growth. In any culture of lysogenic organisms, some free virus particles are always present, indicating that a small number of cells are lytic, with an equilibrium between the two states. The above environmental factors tip the balance of this equilibrium towards the lytic state, with other conditions, such as low phosphate availability, having the reverse effect.

Viral stability, inactivation, and decay

The stability of suspended virus particles (virions) is an important aspect of the freshwater biology of these organisms, since a major part of their life cycle is spent in the environment in an inert (non-parasitic) state. Stability is particularly important when host populations are at a low level and free virus particles have to survive over a long period of time if infection and continuation of the virus life cycle are to be achieved. Viral stability is also important when considering the turnover rate of the viral community. If viral particles are unstable in natural waters, the turnover time will be short and a high rate of production will be required to maintain a high particle abundance. With increasing particle stability, lower rates of productivity are required to maintain a particular level of abundance.

Although studies on viral stability, inactivation, and decay have been carried out mainly on marine systems (Bratbak et al., 1994), it seems likely that the results obtained will apply equally to freshwater environments. Studies on virus stability in natural waters have defined the term either in relation to infectivity or on the basis of particle integrity (i.e., using viable or total counts). Various interrelated environmental factors have been shown to influence viral stability.

- *Biological activity.* Virus inactivation is much greater in raw water compared to water that has had other biota removed (filtration, centrifugation) or inactivated (autoclaving, antibiotic treatment). The role of biota in the inactivation or removal of suspended virus particles is complex and probably includes processes of adsorption, secretion of extracellular hydrolytic (nuclease and peptidase) enzymes, and ingestion during filter feeding.

- *Chemical composition.* Virucidal properties of coastal seawater have been attributed to their chemical composition and particularly their content of trace metals (Kapuscinski and Michell, 1980). Reversible adsorption to particulate matter appears to protect viruses against the virucidal effects of seawater. Ionic composition of the aquatic medium may also be important. Cyanophages of filamentous blue-green algae have a requirement for divalent (particularly Mg^{2+}) cations, without which tails separate and heads burst (Martin and Benson, 1988).

- *Physical parameters.* Studies by Suttle and Chen (1992) demonstrated that sunlight has a significant effect on viral stability, increasing the rate of bacteriophage inactivation by 10–100 times in full sunlight compared with the dark. Water temperature has a variable effect, with temperatures below 20°C having no effect on exogenous coliphages T2 and T7, but with a reduced decay from 20°C to 5°C in the case of endogenous phages (Moebus, 1987).

Viral decay may simply involve disintegration of the viral particles, with release of nucleic acid from the protein capsid. The resulting particles will not be capable of host infection and will not be recorded by either viable or total count assay. Little information is available on the rate of virus inactivation in freshwater systems, but values for sea water are typically about 0.1–2 viral units d^{-1} (Bratbak *et al.*, 1994).

7.5   Control of host populations by aquatic viruses: impact on the microbial food web

In the planktonic environment, viruses have a significant effect on the growth and competitive activity of their two major hosts – phytoplankton and bacterial cells. Initially they inhibit host synthetic activities during the infection process, to be followed by complete destruction of host populations by completion of the lytic cycle. Virus infection can thus lead to a marked reduction in both primary (phytoplankton) and secondary (bacterial) productivity.

7.5.1   Metabolic effects of viruses: reduction of algal primary productivity

Destruction of phytoplankton populations by cyanophages and phycoviruses leads to a proportionate decrease in the overall level of photosynthesis and a decrease in primary productivity. In addition to this long-term effect of algal mortality, photosynthesis is also inhibited during the process of infection prior to cell death, so that productivity will also be affected in living communities which are exposed to viruses over short time periods.

The short-term effects of virus infection on algal photosynthesis can be studied in laboratory monocultures, providing useful information on variation between species. In the blue-green algae, for example, major species differences occur in terms of the timing of chlorophyll destruction and inhibition of photosynthesis during cyanophage infection (Section 7.6.2).

The effects of virus infection on mixed phytoplankton communities taken from natural waters (sea water) have been studied under laboratory

conditions at varying levels of virus concentration (Suttle, 1992, 1994). In these studies, the 2–200 nm size (virus) fraction from sea-water samples was separated out and concentrated by ultra-filtration. Addition of concentrated virus samples to the natural phytoplankton populations typically resulted in a rapid and substantial depression of CO_2 uptake, consistent with viral inhibition of photosynthesis. The time course of inhibition (Figure 7.5) suggested

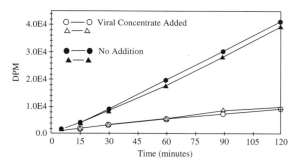

Figure 7.5   Time course of inhibition of photosynthesis by aquatic viruses: time course of ^{14}C-bicarbonate uptake into phytoplankton in natural sea water to which virus suspension has (bottom graphs) or has not (top graphs) been added (taken from Suttle, 1992, with permission from Inter-Research)

an immediate and consistent effect. The results also suggested that specific organisms were being affected within the mixed algal sample, rather than photosynthesis being reduced throughout the whole community. Variation in viral inoculum (Figure 7.6) showed that saturation was reached at a concentration factor of about 20 × ambient aquatic level. *In vivo* chlorophyll fluorescence paralleled the photosynthetic changes, suggesting a direct connection between the two (Figure 7.6).

Although, in some cases, quite high levels of photosynthetic inhibition (up to 78 per cent) were induced by relatively small additions of virus concentrate, the overall results do not imply that viral inhibition of photosynthesis and productivity were normally at this level. Assessment of the existing level of photosynthetic inhibition in particular water samples could be determined by comparison of the ambient CO_2 uptake with that predicted under zero

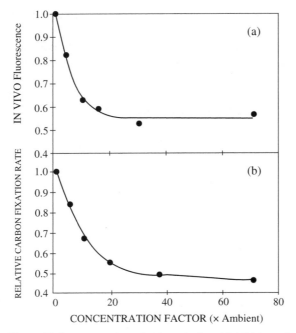

Figure 7.6 Destruction of chlorophyll and inhibition of photosynthesis in marine phytoplankton by the addition of virus concentrate: the decrease in chlorophyll fluorescence (a) and carbon fixation rate (b) are shown for a range of virus concentrations (0–80 × ambient) 8 hours after virus addition (taken from Suttle, 1992, with permission from Inter-Research)

virus concentration (Figure 7.6). In these particular samples, the existing inhibition of photosynthetic activity was quite low at about 1–4 per cent, suggesting that viruses were not having a major impact on algal productivity in the ecosystems from which they were derived.

7.5.2   Destruction of algal and bacterial populations

Viruses are a species-specific cause of cell death in host populations of both algae and bacteria, and are an important aspect of top-down control. For lytic infections, where cell death is directly related to the rate of infection, the destruction of the host population will depend on host cell and free viral population density (see previous section).

Dense (bloom) populations of algae or bacteria will be particularly susceptible to viral attack, since:

- the main species dominating the bloom (and its associated virus) will be at high population level, with resulting high adsorption rates,

- the low host species diversity means that the cumulative level of infection (and death) for the whole community will be high (see Table 7.1).

Both cyanophages and phycoviruses have been implicated in the destruction of algal blooms, and have potential as biocontrol agents. In the case of cyanophages, various workers have observed increases in virus concentrations in lake water in response to seasonal development of blue-green populations (reported in Martin and Benson, 1988). Studies on *Aphanizomenon* blooms in shallow eutrophic lakes (Coulombe and Robinson, 1987) have demonstrated evidence of virus-like particles within cells during the collapse of algal populations, suggesting cyanophage participation. The role of phycoviruses in the destruction of phytoplankton populations is discussed separately in Section 7.7.2.

Since the extent of viral infection depends on host population density, viruses are effectively competing with other freshwater biota (e.g., parasitic fungi, predatory protozoa, and zooplankton) for host organisms. An example (case study) of this is given in Case Study 7.2, where the respective parasitic and predatory interactions between bacteriophages and heterotrophic nanoflagellates for bacterial biomass are considered in a eutrophic lake.

7.5.3   Viruses and the microbial loop

In the microbial loop, carbon excreted by phytoplankton as dissolved organic carbon (DOC) is taken up by bacteria and channelled back to the grazing food web by protozoa (particularly heterotrophic nanoflagellates) and rotifers (Section 1.5.2, Figure 1.9). The effect of viruses on the microbial loop would be expected to depend on which particular group of microorganisms is being infected,

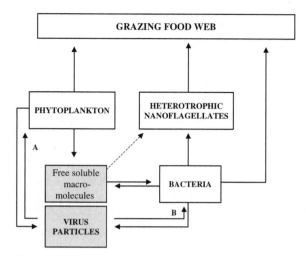

GRAZING FOOD WEB

PHYTOPLANKTON

HETEROTROPHIC
NANOFLAGELLATES

A

Free soluble
macro-
molecules

BACTERIA

B

VIRUS
PARTICLES

Figure 7.7   Viruses and the pelagic microbial food web. Arrows indicate the direction of carbon flow. Viral infection of algae (A) and bacteria (B) results in the conversion of host cell biomass to free soluble macromolecules and virus particles. Both of these, released by host cell lysis, contribute to the dissolved organic carbon (DOC) in the water medium (shaded boxes). Population growth of heterotrophic nanoflagellates (HNF) is determined mainly by bacterial availability. Productivity of phytoplankton, HNF, and bacteria determines carbon flow into zooplankton and the whole of the grazing food web. The long-term predicted effect of viral infection is to reduce carbon flow in the microbial food web

and on the time course of the infection process (Figure 7.7) since:

- Infection of phytoplankton populations by cyanophages and phycoviruses will initially increase the aquatic DOC level, and flow of carbon through the microbial loop, due to short-term cell lysis. Longer-term reduction in algal productivity will decrease DOC excretion and carbon flow.

- Reduction in the population of heterotrophic bacteria by bacteriophages would reduce the uptake of DOC by bacteria and the flow of carbon through the loop. The ability of some protozoa to take up DOC directly (organotrophy) will only partly compensate for the reduced bacterial productivity.

For both phytoplankton and bacteria, the effects of viral infection on the microbial loop are limited by host-specificity. The inhibition of carbon flow will be much greater where the algal or bacterial host is dominant, and at high population level. Microbial succession will limit the long-term effect of viruses on the microbial loop.

In addition to carbon flow, virus infections also have implications for inorganic nutrient turnover in natural waters. Lysing cells release a wide range of nitrogenous and phosphorus-containing macromolecules in addition to virus particles. In a diurnal mesocosm experiment carried out by Bratbak *et al.* (1990), 72 per cent of the bacterial population was removed by viral lysis per day. The authors calculated that the organic P released from the lysed bacteria was sufficient to support both primary (algal) and secondary (bacterial) productivity in this system. These estimates indicate that viral processes may be of quantitative significance with respect to nutrient flow in aquatic microbial systems.

7.6   Cyanophages: viruses of blue-green algae

Viruses of blue-green algae are closely similar in morphology and mode of action to bacteriophages (infecting bacteria) and are referred to as cyanophages. These viruses have a considerable impact on the ecology of blue-green algae (Suttle, 2000). They typically consist of a head and tail and contain linear double-stranded DNA (Martin and Kokjohn, 1999). Close similarities between these two groups of viruses are not surprising since both infect prokaryote hosts, with fundamental similarities in cell structure and function.

7.6.1   Classification and taxonomic characteristics

Cyanophages have a diverse morphology and are placed (along with bacteriophages) into three major families – the Myoviridae, Siphoviridae, and Podoviridae. These are distinguished in terms of tail size

Table 7.2 Examples of freshwater cyanophages and bacteriophages in the families *Myoviridae*, *Siphoviridae* and *Podoviridae*

Family		Phage Species	Host Alga or Bacterium
Myoviridae	Phages with a central tube and contractile tail, separated from the head by a neck	AS-1	*Anacystis nidulans* *Synechococcus cedrorum*
		N-1	*Nostoc muscorum*[*]
		Pseudomonas phage D312 Pseudomonas phage UT1[†]	*Pseudomonas aeruginosa* *Pseudomonas aeruginosa*
Siphoviridae	Long, noncontractile tails	SM-2	*Synechococcus elongatus* *Microcystis aeruginosa*
		S-2L	*Synechococcus sp. 698*
		Methanobacterium phage ø F3	*Methanobacterium sp.*
Podoviridae	Short tail	LPP-1	*Lyngbya* *Plectonema* *Phormidium*
		SM-1	*Synechococcus elongatus* *Microcystis aeruginosa*
		Pseudomonas phage ø PLS27	*Pseudomonas aeruginosa*

Information from Martin and Kokjohn (1999) and Tidona and Darai (2002). Each family of phages contains both cyanophages (shaded) and bacteriophages. Viruses in the different families have similar (icosahedral) capsids but are distinguished by tail morphology. Large linear viruses (family *Inoviridae*) that infect bacteria have also been detected in some freshwater environments (Middelboe *et al.*, 2003).
[*]Now *Anabaena* spp.
[†]Illustrated in Figure 7.14(d).

and contractility (Table 7.2). Cyanophages have a narrow host range, acting as parasites within specific algal groups, and are classified in relation to the blue-green algae which they infect (Table 7.2). One major group of cyanophages, for example, infect algae in the genera *Lyngbya*, *Phormidium*, and *Plectonema*, and are referred to as LPP viruses. Some genera of blue-green algae can be infected by a wide range of cyanophage viruses. This is particularly the case for *Synechococcus*, a unicellular blue-green alga which is responsible for considerable carbon fixation in the world's oceans and is also an important member of the phytoplankton in some freshwater environments.

In addition to characteristics of morphology and host specificity, cyanophages also vary in relation to their mode of infection, their dependence on host photosynthetic activity (see next section), and requirement for divalent cations (Martin and Benson, 1988). Cyanophages can be separated into two groups based on their need for monovalent or divalent cations to maintain stability in the free (virion) state. Viruses of filamentous blue-greens (e.g., LPP-1, N-1) have an absolute requirement for Mg^{2+} ions, without which they disintegrate. In contrast to this, most cyanophages of unicellular blue-greens (e.g., SM-1, AS-1) have a lesser requirement and can maintain a high level of infectivity even after prolonged dialysis in distilled water.

7.6.2 Infection of host cells

The process of infection can be studied by adding a suspension of virus particles to a liquid culture of blue-green algae, and determining the generation of new infective particles in the culture medium (pfu counts) and changes within the algal host cells (fine structure and biochemistry). The sequence of infection has been looked at particularly in the case of LPP cyanophages (Padan *et al.*, 1970) and involves a defined sequence of events.

Virus multiplication

Over a 15 hour period, viable counts in the aquatic medium follow a one-step growth curve (Figure 7.8). Virus particles are initially adsorbed on to the algal cells. After removal of unadsorbed particles, the viable count remains low for a period of about 6 hours. This latent period is the time taken for penetration, infection, and lysis of cells, liberating infective particles – which may increase up to levels of 10^6 pfu ml^{-1}.

The intracellular cyanophage growth curve can be determined by induced lysis (treatment with the enzyme lysozyme) of algal cells over a sequence of time points, followed by viable count (lawn assay) of the liberated particles (Figure 7.8). Virus levels inside algal cells remain low for about 3 hours (the eclipse period) after the start of the experiment, followed by a steep rise – reaching high internal levels just prior to the end of the latent phase.

Infection sequence

The general sequence of changes occurring in the algal cell commences with the attachment of the virus particle to the host surface. This leads to penetration of the host cell wall, injection of viral DNA, and subsequent synthesis of viral proteins and DNA. Damage to the host cell can initially be seen as invagination and destruction of the photosynthetic lamellae, to be followed by breakdown of the host DNA and finally lysis of the whole cell.

With LPP viruses, the viral proteins are restricted to the periphery of the host cell – which is where the new viral particles are formed (virogenic stroma).

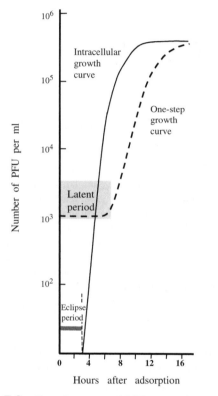

Figure 7.8 Growth curves of LPP cyanophage during infection of algal (*Plectonema*) cultures. Increases in viral particles can be monitored for both intracellular (—) and extracellular (- -) populations. Intracellular populations show a steep increase after a period of viral synthesis (eclipse period). Extracellular populations follow a typical 'one-step' curve, with increase after the latent period, within which no release by cell lysis occurs. The time sequence was considered to begin at 0 hours (1 hour after virus addition), when unadsorbed viruses were removed (based on a figure by Padan *et al.*, 1970, and reproduced with permission from The American Society for Microbiology)

Biochemical changes The biochemical changes noted earlier follow a distinct temporal sequence (Figure 7.9), with synthesis of viral proteins, host DNA breakdown, and synthesis of new viral DNA occurring at particular times. The synthesis of three main classes of viral protein can be demonstrated by immunolabelling. Inhibition of this process by the antibiotics chloramphenicol and erythromycin indicates the role of the host cell synthetic machinery in this process.

The infection cycle of LPP cyanophages is similar to bacteriophages, but slower, taking about 13 hours for completion at 26°C – compared with about 20–30 minutes for a typical bacteriophage.

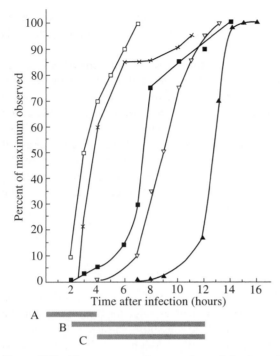

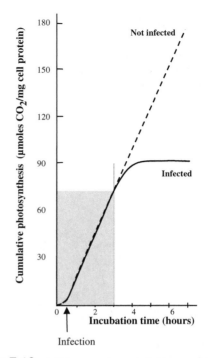

Figure 7.9   Time course of cyanophage infection of *Plectonema* cells. Showing changes in host DNA break-down (□), cells with invaginated lamellae (×), viral DNA synthesis (■), intracellular infective particles (▽) and extracellular infective particles (▲). Groups of viral proteins are shown below the graph – earliest function (A), early function proteins (B), and late structural proteins (C) (figure taken from Padan and Shilo (1973) and reproduced with permission from The American Society for Microbiology)

Figure 7.10   LPP cyanophage infection and host cell photosynthesis. Showing the cumulative rate of photosynthesis of infected (—) and uninfected (--) *Plectonema* cultures. Photosynthesis is measured as CO_2 assimilation, determined by incorporation of $NaH^{14}CO_3$ into the cells. For about 3 hours, viral infection has no effect on the rate of photosynthesis (shaded box) (based on a figure from Padan and Shilo, 1973, and reproduced with permission from The American Society for Microbiology)

Cyanophage energy requirements and effects on algal photosynthesis

All viruses need to obtain their energy from host cells. In the case of cyanophages, this may either be derived from photo- and oxidative phosphorylation (e.g., LPP and N-1 viruses) or from photophosphorylation only (e.g., SM-1 virus). Where energy is derived from both processes, algal photosynthesis is inhibited at an early stage of infection and the new virus particles tend to be formed at the cell periphery. Where energy is derived solely from photophosphorylation, photosynthesis is not inhibited until shortly before cell lysis and new viral

particles are formed in the central nucleoid – well away from the photosynthetic lamellae.

The effects of LPP cyanophage infection on host photosynthesis (Figure 7.10) indicate a 2 hour period of light requirement prior to inhibition of this process.

7.7   Phycoviruses: parasites of eukaryote algae

7.7.1   General characteristics

Viruses of eukaryotic algae (phycoviruses) have been reported for all classes of algae, infecting

Table 7.3 Phycoviruses (Family *Phycodnaviridae*): some well-characterized viruses of eukaryotic algae

Genus	Examples of species[*]	Host organism
Chlorovirus	Paramecium bursaria Chlorella virus 1 (PBCV-1)	Algal symbiont (*Chlorella*) of *Paramecium bursaria*
	Hydra viridis Chlorella virus 1 (HVCV-1)	Algal symbiont (*Chlorella*) of *Hydra viridis*
Prasinovirus	Micromonas pusilla virus SP1 (MpV-SP1)	*Micromonas pusilla*
Prymnesiovirus	Chrysochromulina brevifilum virus PW1 (CbV-PW1)	*Chrysochromulina brevifilum*
Phaeovirus	Pylaiella littoralis virus 1 (PlitV-1)	*Pylaiella littoralis*

[*]Only a few species are quoted here. In the recent virus listing by Tidona and Darai (2002), from which this table is derived, the total species so far designated were *Chlorovirus* (18 species), *Phaeovirus* (8), *Prasinovirus* (9), and *Prymnesiovirus* (2). The top three genera in the table infect unicellular (micro-) algae, with only one (shaded) operating in the freshwater environment.

both marine and freshwater organisms (Reisser, 1992, 1995; Van Etten *et al.*, 1991). Although these viruses are widespread, most of our knowledge about them comes from a limited number of virus–host systems, where the viruses concerned have been isolated and purified in the laboratory. In a recent listing of viral taxa (Tidona and Darai, 2002), phycoviruses are placed in the Family *Phycodnaviridae*, with four genera so far recognized – *Chlorovirus*, *Prasinovirus, Prymnesiovirus*, and *Phaeovirus* (Table 7.3).

Although these genera are clearly distinguished in terms of host range, electron microscope and molecular characterization show them to be closely similar at the fine-structural and genomic level. All of the viruses are icosahedral in shape (Figure 7.12(a)), with no tail, and are composed of an outer capsid with an inner core. Mean sizes for different genera range from 100–200 nm. The genome is composed of double-stranded (ds) DNA, with a linear (*Chlorococcus*) or circular (*Phaeovirus*) configuration and a size ranging from 200 to 380 kb. G-C content varies from 40–52 per cent.

Within these genera, individual species are distinguished in relation to host-specificity, and tend to be named (as with cyanophages) in relation to their host cells (Table 7.3). Groups of viruses infecting a particular host may be morphologically similar but genetically different. The group of viruses infecting the widely-occurring photosynthetic flagellate *Micromonas pusilla*, for example, are each referred to as *Micromonas pusilla* virus (MpV) and are all morphologically indistinguishable (tailless polyhedra) from each other. This group of viruses comprises seven clonal isolates, distinguishable from each other in terms of their DNA restriction analysis pattern (Cottrell and Suttle, 1991).

Within the Family *Phycodnaviridae*, chloroviruses are of particular interest as the only viruses so far characterized which infect freshwater microalgae (Van Etten, 2002a). These infective agents are unusual in having a large genome, which in the case of PBCV-1, (the most studied of all chloroviruses) is ~330 kb in size and has a predicted content of 376 protein encoding genes plus 10 transfer RNA genes. About 50 per cent of these genes encode proteins of known function that are involved in a wide range of general activities, including DNA replication and repair, nucleotide metabolism, transcription, protein synthesis and modification, cell-wall degrading enzymes, sugar and lipid metabolism, phosphorylation/dephosphorylation, DNA modification, and various miscellaneous activities. In addition to their large genome, chloroviruses such as PBCV-1 also have other unusual molecular features, including:

- the ability to encode multiple DNA transferases and DNA site-specific endonucleases,

- the ability to encode enzymes for glycosylation of their own glycoproteins,

- at least two types of introns,

- the presence of an internal lipid membrane within the capsid.

7.7.2 Host cell infection

The great majority of reports of virus infection of eukaryote algae involve random (TEM) observations of sectioned environmental material. Inference of infectivity from such observations of virus-like particles must be tentative, and there is no information on specificity and the mechanisms of infection from such studies. Laboratory studies provide a more useful approach, permitting the time course of infection to be studied under controlled conditions.

Case study 7.1   The infective life cycle of Chlorovirus

Chloroviruses are somewhat unusual in that their ability to infect algae under laboratory conditions has only been demonstrated in relation to algal endosymbionts (genus *Chlorella*) of the protozoon *Paramecium bursaria* and the coelenterate *Hydra viridis* (Van Etten and Meints, 1999). Chloroviruses have a high degree of host specificity, as indicated by the fact that distinct species infect symbiotic algae from different isolates of *Paramecium bursaria* (with no cross infection) and a single species (Hydra viridis chlorella virus – HVCV-1) infects the *Hydra* symbiont. The algal cells present in *Paramecium bursaria* (Figure 9.1) are protected from virus (PBCV-1) infection when in the host (protozoon) cell, but are susceptible once they have been isolated. Inoculation of a lawn of isolated *Chlorella* (Figure 7.2) with PBCV-1 leads to rapid algal infection, with plaques observed within 24 hours of plating. Algae from *Hydra* are more difficult to culture, since all (or a portion) of the algae are already infected, and are in a lysogenic state. The virus enters its lytic state on release of the algae, which are thus destroyed before they can be cultivated.

Figure 7.11 summarizes the time sequence of viral infection and multiplication proposed by Van Etten and Meints (1999), and can be divided into six main stages.

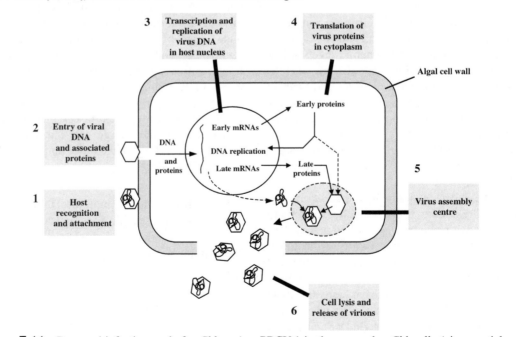

Figure 7.11 Proposed infection cycle for *Chlorovirus* PBCV-1 in the green alga *Chlorella* (virus particles not drawn to scale) (based on a figure from Van Etten and Meints, 1999)

1. Virus binding to host cell

Isolated *Chlorovirus* particles (Figure 7.12(a)) have a well-defined icosahedral symmetry and always attach to the cell wall of the algal cell via a vertex (angle) of the capsid. Mixing of viral particles with cell-wall fragments shows that viral attachment only occurs to the external side of the wall, suggesting a surface-recognition process. This recognition may involve a specific lectin-sugar type of interaction, in which virus surface proteins interact with algal surface carbohydrates. Surface recognition is further indicated by the strain-specificity of attachment, which only occurs in homologous (host-parasite) combinations. Where this occurs, algal cells may be covered by numerous attached viral particles (Figure 7.12(b)).

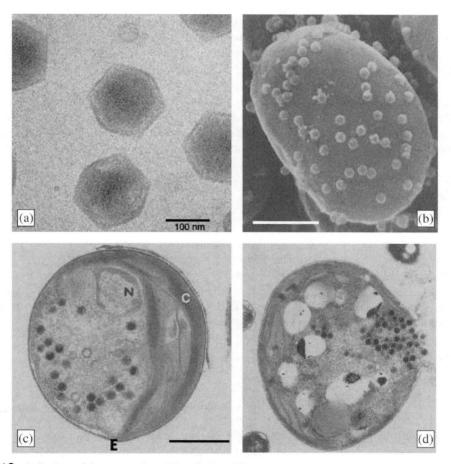

Figure 7.12 Infection of the green alga *Chlorella* by *Chlorovirus*
Transmission (TEM) and scanning (SEM) electron microscope images (by permission, from Van Etten, 2002b) (a) Cryo-preparation of isolated virus particles (virions) showing a well-defined outer capsid and a non-uniformly distributed interior mass (TEM photograph by Tim Baker); (b) pre-infective attachment of virus particles to the surface of *Chlorella* (SEM photograph by Kit Lee, scale = 1 μm); (c) 6 hour post-infection, the cells contain many hexagonal capsids, most of which are filled with DNA (dense contents), the nucleus (N) and peripheral chloroplast (C) are pushed to one side – E is cellulose cell wall (TEM photograph by Kit Lee, scale = 1 μm); (d) 20 hour post-infection, free virus particles are liberated from the bursting cell (TEM photograph by Kit Lee)

2. Entry of viral DNA

Recognition and attachment of virus particles triggers the activity of capsid-bound glycolytic enzymes, which digest the algal cell wall at the point of viral contact. Viral DNA (and associated proteins) then migrates through the hole in the wall, leaving the empty capsid on the outside. Viral DNA release only occurs after attachment to a complete cell, and does not happen when viral particles attach to cell-wall fragments.

3. Transcription and replication of viral DNA

This occurs in the host-cell nucleus. Transcription occurs in two phases, with early (beginning 5–10 minutes after infection) and late mRNAs leading to the production of two distinct groups of viral proteins. Replication of viral DNA results in the formation of numerous daughter molecules which migrate to the Virus Assembly Centre (Meints *et al.*, 1986) for incorporation into capsids.

4. Translation of virus proteins in cytoplasm

Early proteins, translated in the host cell cytoplasm, return to the nucleus to initiate DNA replication which commences 60–90 minutes after infection. A second phase of proteins are translated, many of which are targeted to the cytoplasmic Virus Assembly Centres, where virus capsids are formed. Some of the virus capsid proteins are phosphorylated, glycosylated, and myristoylated (covalent addition of myristic acid) at some site within the cell prior to capsid assembly.

5. Virus assembly

The formation of capsids and other components can be observed in the cytoplasm of the algal cell at special Virus Assembly Centres (also referred to as 'viroplasm' or 'virogenic centres') from about 2 hours after post-infection. In most algae (including *Chlorella*), nascent Phycovirus assembly occurs in the general cytosol (Figure 7.12(c)), but accumulations of virus-like particles have also been observed in the plastid or the nucleus of other algae. Some viral proteins are modified prior to assembly (see above) and lipids are also incorporated into the newly-forming viruses which have an internal lipid membrane. Viral assembly is complete with the uptake of newly-replicated viral DNA from the host nucleus.

6. Release of infective particles

The infected algal cell bursts about 2–3 hours after viral entry, with lysis of the host-cell wall and the release of 200–400 virus particles (burst size) into the surrounding medium (Figure 7.12(d)). These virions can infect other cells and be monitored as plaque-forming units in algal lawn assay.

In the *Chlorella*-virus system, virus particles are observed in ultra-thin section only during the last 20–30 per cent of the infection cycle. This would suggest that reported percentage infection data from electron microscope studies on other algae within environmental samples should be multiplied by 3–5 times to obtain true values of infection level.

Effects of PBCV-1 virus on the algal host cells include rapid inhibition of CO_2 fixation and photosynethesis. Cytological changes include displacement of the nucleus and cytoplasmic organelles, leaving one or more electron-transparent areas in the cytoplasm. These eventually become Virus Assembly Centres. Host-cell death occurs with final lysis and release of virions.

7.7.3 Ecological impact of phycoviruses

As with cyanophages, phycoviruses appear to be ecologically important, attacking a wide range of algae in the freshwater environment, reducing levels of primary productivity, and causing short-term increases in levels of dissolved organic carbon (DOC) via lysis of host cells. Phycoviruses thus have an impact on food webs via the main trophic succession and also via the microbial loop (Section 7.5.3, Figure 7.7).

Widespread occurrence

Phycoviruses are present in all types of aquatic systems. *Chlorella*-viruses are observed, for example, in a wide range of habitats of differing trophic status, from which they can be identified by plaque formation on lawns of host algae or by immuno-precipitation techniques using specific antibodies. Quantitative estimates of viruses in different habitats suggest that they can reach very high levels. Electron microscope analysis of freshwater samples indicates counts of polyhedral viruses up to levels of 10^{11} particles litre^{-1}. Whether these particles are infective and are really phycoviruses is not clear. More reliable estimates of infective particles can be obtained by plaque assay, with values of up to 10^7 PFU litre^{-1} being obtained for *Chlorella*-viruses, and 10^6 PFU litre^{-1} for *Micromonas*-viruses (Reisser, 1995).

The widespread occurrence of *Chlorella* viruses, which have so far only been identified as infective agents of endosymbiotic algae, is remarkable since the protozoa that contain the algae are not widespread and it is not known if the algae are available for infection (i.e., occur outside their protozoon hosts) in natural environments. It seems likely that other algae must also act as viral hosts in freshwater systems.

Control of phytoplankton populations

Although viruses have been widely mentioned as being potentially important in the control of freshwater and marine phytoplankton populations, criti-cal evidence for their major role in this is limited and in some cases circumstantial. Information on this comes from various directions:

- *Environmental observations.* Most studies on viruses have been carried out in marine environments, where viral activity has been implicated in the decline of blooms of several phytoplankton species (reported in Murray, 1995). These include *Emiliana huxleyi*, forming blooms in the coastal waters of Norway, and *Aureococcus anophagef-ferens*, which is responsible for recurrent 'brown tide' episodes which afflict Long Island (USA) waters, with sudden increases and collapse of these populations in late summer. The role of viruses in these situations is inferred from observations of infected algal cells, isolation of viruses that can infect the algae, and increased counts of particulate viruses in the water.

  Increased virus counts may be a corollary rather than a cause of algal death. The presence of high virus counts during a spring diatom bloom in western Norway (Bratbak *et al.*, 1990), for example, was attributed to high counts of bacteriophages (rather than phycoviruses), infecting the substantial bacterial populations that followed on from diatom lysis. These viruses were thus the result rather than the cause of algal lysis.

- *Studies in partially-controlled environments.* Studies of viral infection in small-scale enclosed volumes (mesocosms) provide a useful experimental approach to study the ecological effects of phycoviruses in semi-controlled conditions. Mesocosm studies by Bratbak *et al.* (1993), for example, demonstrated termination of blooms of the marine alga *Emiliana huxleyi* by viral activity, with succession by blooms of other algae not affected by the virus concerned. The selective effect of viral infection in directly reducing the populations of some algae but indirectly promoting the growth of others resembles the activity of parasitic fungi (Chapter 8).

- *Mathematical modelling.* In many cases, it seems likely that viruses are just one of a number of environmental aspects which cause sudden collapse of algal blooms, acting in combination with other factors such as nutrient depletion and zooplankton grazing. Modelling studies (Beltrami and Carroll, 1994) on the role of viral disease in recurrent phytoplankton blooms have indicated that even small viral infections can destabilize an otherwise stable trophic configuration between a phytoplankton species and its grazer. The ecological role of viruses in the bloom control may thus be to alter the balance of existing relationships rather than destroying the bloom by a massive infection.

Control or partial control of phytoplankton populations by viruses has important implications for the grazing food web, and the related populations of zooplankton and fish. Population control of phytoplankton by viruses acts on individual species within the mixed assembly, in contrast to other forms of control such as nutrient limitation and grazing, where the whole community is affected. The selective effect of viral attack will also tend to control the rise to dominance of particular species, providing further explanation for the paradox of phytoplankton diversity seen in aquatic systems (see Chapter 2).

7.8   Virus infection of freshwater bacteria

Bacteriophages of freshwater bacteria occur as a morphologically diverse assemblage of viruses. Most bacteriophages, including those in the families *Myoviridae*, *Siphoviridae*, and *Podoviridae*, have a head and tail (Table 7.2). Others are filamentous (family *Inoviridae*), occurring as a flexible rod which contains single-stranded DNA. Members of the *Inoviridae* include some of the largest freshwater viruses (>1 μm in length) and have been detected in both pelagic (Hofer and Sommaruga, 2001) and benthic (Middelboe *et al.*, 2003) systems. The typical appearance of a mixed bacteriophage assembly collected from a freshwater lake is shown in Figure 7.14. This TEM negative stain preparation contains examples of three morphologically-distinct types of bacteriophage.

Bacteriophages have a key role in freshwater ecosystems as parasites of a major group of heterotrophic organisms, and thus have an indirect and inhibitory influence on the removal of dissolved organic carbon via the microbial loop (pelagic foodweb) and the breakdown of sediment detritus (benthic foodweb).

7.8.1   General role of bacteriophages in the biology of freshwater bacteria

The high concentrations of phage particles noted in aquatic environments suggest that bacteriophages play a significant role in the biology of freshwater bacteria (Miller and Sayler, 1992). Recent studies further suggest that:

- bacteria are extensively infected in freshwater systems;

- in some situations, bacterial abundance is primarily controlled by viral infection rather than by grazing (heterotrophic nanoplankton);

- species composition of bacterial communities is partly controlled by differences in phage susceptibility;

- bacteriophages play a major role in gene transfer in bacterial populations, promoting the reassortment of gene pools in natural aquatic microbial communities;

- interactions between bacteriophages and bacteria are influenced by physical conditions in the aquatic environment, including the level of ultraviolet irradiation.

Information about many of these aspects is based on case studies of particular experimental organisms contained within semi-natural microenvironments, with or without the presence of a natural microbial community.

7.8.2   Bacteriophages in pelagic and benthic systems

Investigations on the interactions between viruses and bacteria have largely focused on planktonic communities, and most of the information given in the following parts of this chapter relates particularly to pelagic systems. Recent studies on benthic environments do indicate, however, that sediments are rich in viruses and that they form an important component of the benthic food web (Middelboe *et al.*, 2003).

Most studies on benthic viruses have concentrated on surface sediments, where the viruses occur at the surface of sediment particles and in the interstitial spaces – and are referred to as 'interstitial viruses'. Counts of virus-like particles indicated an abundance that exceeded levels in the water column by a factor of $10-10^3$, reaching sediment concentrations of 10^7-10^8 particles cm^{-3} (Hewson *et al.*, 2001). Values obtained for viral abundance in surface sediments along trophic gradients indicated a positive correlation with nutrient status of the water column (Maranger and Bird, 1996; Hewson *et al.*, 2001). Significant correlations also occurred between virus counts in the water column and viral abundance in the surface sediments. This connection between water column and sediments might operate in two ways:

- virus particles in the water column are carried down to the sediments by adsorption onto settling detritus material (exogenous origin), or

- bacterial populations are higher in detritus-rich sediments, and generate their own high bacteriophage populations (endogenous origin).

Although both explanations may apply, it seems likely that endogenous origin is the main factor. Direct connection between sediment bacterial and viral populations has now been demonstrated in several studies. Middelboe *et al.* (2003), for example, observed high but variable levels of both virus particles ($0.5-8 \times 10^8$ cm^{-3}) and bacteria ($0.1-4 \times 10^8$ cells cm^{-3}) in the surface layer of estuarine sediments, with positive correlations between viral abundance and bacterial populations. These authors also demonstrated a correlation between virus populations and bacterial activity (bacterial sulphate reduction and respiration), suggesting metabolic connections. This close coupling between virus and bacterial populations strongly suggests that

viruses are produced by bacteria within the sediments, and that viruses are a dynamic component of the benthic community. Examination of the interstitial viruses showed that they were dominated by large forms with a helical symmetry (family *Inoviridae*). These were morphologically quite distinct from viruses found in the water column, providing further evidence that interstitial viruses arise from within the sediments rather than by sedimentation from above.

7.8.3 Occurrence of free bacteriophages in aquatic systems

Early studies on the presence of phages in aquatic systems used viable count techniques, plating out environmental samples on bacterial lawns and counting plaques produced by host cell lysis (Figure 7.2). Within the limitations of this technique, which only records non-temperate phages infecting particular host cells, concentrations of bacteriophages in freshwater habitats (Miller and Sayler, 1992) were typically in the range $1-10^3$ plaque-forming units per ml (pfu ml^{-1}). Higher counts ($1-10^5$ pfu ml^{-1}) have been obtained from marine environments, and the frequency of enteric phages in waste water was reported to be as high as 10^8 pfu ml^{-1}.

The use of electron microscopy to determine total virus counts (Section 7.3) showed that bacteriophage abundances are much greater than originally thought. As early as 1980, Ewert and Paynter used electron microscopy to demonstrate counts of 10^9-10^{10} phage particles per ml in sewage, less than 4 per cent of which were detectable in the plaque-assay method. Total counts of phage particles in freshwater and marine systems (Bergh *et al.*, 1989; Bratbak *et al.*, 1990) demonstrated considerable variability, with values in the range 10^3-10^8 particles ml^{-1}.

7.8.4 Incidence of bacterial infection

Simultaneous counts of total phage and bacterioplankton populations typically show an excess of virus particles, with phage/bacterial ratios as high as 50:1, suggesting a high infection pressure on

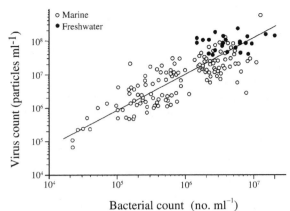

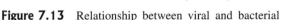

Bacterial count (no. ml^{-1})

Figure 7.13 Relationship between viral and bacterial abundance in freshwater and marine habitats.

Significant correlation occurs in marine habitats, where bacterial abundance ranges over three orders of magnitude. Lack of correlation in freshwater environments probably reflects the limited sample range. Viral counts are generally 10 times bacterial ones (figure taken from Maranger and Bird, 1995, with permission from Inter-Research)

bacterial populations. At these levels, Bergh *et al.* (1989) estimated that one third of the bacterial population would experience a phage attack every day. The difference between total virus and bacterial populations in natural waters extends over a wide range of trophic levels, with virus populations routinely being an order of magnitude greater than bacterial ones (Figure 7.13).

Evidence on the incidence of bacterial infection comes from both marine and freshwater systems and includes direct examination of environmental samples, correlation of bacterial and phage populations and experimental studies on phage-bacterial interactions.

Direct examination of aquatic bacteria

Electron microscope observations carried out by various workers (reviewed in Bratbak *et al.*, 1994) indicate that between 2 and 40 per cent of marine heterotrophic bacteria contain virus-like particles (VLPs). Since mature viral particles may be observed only during a fraction of the lytic cycle, the proportion of infected cells will be much greater than these figures indicate. In some situations, the level of infection may extend to the entire bacterial community.

Studies on freshwater systems also indicate substantial levels of bacterial infection. Weinbauer and Höfle (1998), for example, observed frequencies of visibly-infected bacteria ranging from 1.6 to 6.3 per cent at different points in the water column. As with the marine data, the actual levels of infection will be much greater than these figures suggest.

Correlation between viral and bacterioplankton abundance

Most information on the relationship between viral and bacterial abundance comes from marine habitats, where populations of both groups of organisms range over four orders of magnitude, and there is significant correlation between the two (Figure 7.13, Maranger and Bird, 1995). No clear patterns have emerged yet for freshwater systems, where studies have been largely confined to eutrophic lakes in which populations of viruses only range over two orders of magnitude.

The overall correlation between free virus (virion) concentration and bacterial populations would be expected if natural populations of aquatic bacteria were substantially infected by bacteriophages. The viral counts do not distinguish between viruses which infect algae and bacteria (or any other biota), however, and the relationship with bacterial populations could be circumstantial. In this respect, general correlations have also been noted between viral abundance and chlorophyll-a concentration, total phosphorus and bacterial production (Maranger and Bird, 1995), suggesting that populations of bacteria, algae, and viruses are all inter-correlated.

Experimental studies on bacterial/phage interactions

In natural waters, the interaction between bacteria and viruses is complicated by heterogeneity in the populations of both sets of organisms. In terms of viral diversity, Ogunseitan *et al.* (1990), for

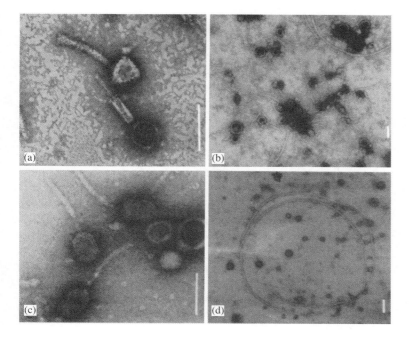

Figure 7.14 Electron micrographs of freshwater bacteriophage populations recovered from lake water by sample concentration and enrichment.

The mixed phage community from the freshwater lake (b) consisted of at least three morphologically-distinct particles: (a) phage UT1, a temperate phage (family Myoviridae), (c) a complex phage (Siphoviridae), and (d) a filamentous phage (Inoviridae) (scale = 100 nm)

Figure taken from Ogunseitan *et al.* (1990) with permission of Springer-Verlag

example, identified at least three morphologically-distinct particles within the phage community from a freshwater lake (Figure 7.14). Further diversity within these categories would be expected in relation to genomic differences and host specificity.

Experimental studies on single host–virus systems permit a more controlled approach, with clear conclusions relating to that particular combination. The interaction between particular phages and their host bacteria has been studied in enclosed experimental systems (microcosms) under controlled conditions using pure cultures of the two organisms (Figure 7.15). Ogunseitan *et al.* (1990), for example, isolated pure cultures of *Pseudomonas aeruginosa* and UT1 virus from freshwater samples, inoculated lake water microcosms and studied the level of infection over a 96 hour time period. The proportion of bacteria which contained phage DNA was determined by plating the bacteria out on agar, transferring colonies from agar plates to nylon membranes, then probing with radiolabelled DNA using standard hybridization techniques.

The ability of temperate viruses to infect large numbers of host bacteria rapidly in natural lake water was indicated by the fact that 45 per cent of the *Pseudomonas aeruginosa* population contained UT1 phage DNA within a 12 hour period. In these experiments, the process of bacterial infection was delayed in sterile compared with natural lake water, suggesting that host–phage interactions may be promoted in the presence of a viable microbial community.

7.8.5 Temperate/virulent phage equilibrium and bacterial survival

The distinction between virulent and temperate viruses results in two possible outcomes for ecosystems where bacteriophages occur in the presence of an actively-proliferating population of bacteria.

- Growth of obligately virulent phages leads to widespread destruction of bacterial cells, threatening the survival and effectiveness of the particular species or strain involved.

- Infection by temperate phages leads to a lysogenic state, in which bacteria survive. This apparent

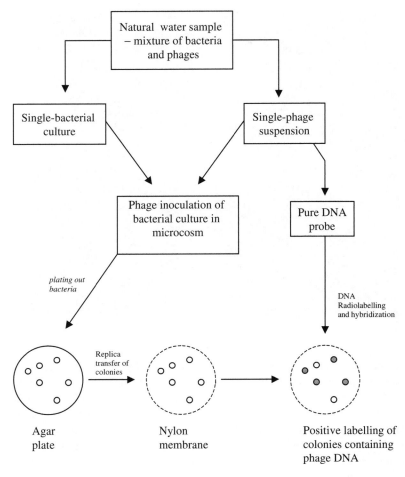

Figure 7.15 Experimental protocol to study the lysogenic infection of naturally-occurring bacterial species by compatible temperate phages. Using this approach, Ogunseitan *et al.* (1990) investigated the infection of *Pseudomonas aeruginosa* by phage UT1, using microcosms comprising 100 ml of lake water in 250 ml flasks

co-existence of bacteria and phages is important for long-term survival of the virus. This is particularly important in low-nutrient conditions – where bacterial populations are low (resulting in reduced chance of infection) and where relatively few bacteria are in an actively-metabolic state.

In practice, both lysogenic and lytic states occur within an infected bacteria population (Figure 7.1). The equilibrium between temperate and virulent states (Section 7.2) affects the productivity of both the bacterial population (incidence of mortality) and the virus assemblage (generation of phage particles).

Incidence of bacterial mortality

Where a high proportion of the bacterial population is lysogenic, any shift in the equilibrium towards the lytic state (due to environmental changes) will cause large-scale bacterial mortality and a decrease in population (Figure 7.7).

Source of phage particles

Studies by Miller *et al.* (1990) have indicated that the induction of prophages from lysogenic bacteria present in the natural microbial community provides a major source of infective phages in the freshwater environment. The concentration of *Pseudomonas aeruginosa*-specific bacteriophages in freshwater systems often reaches $10^2 - 10^3$ pfu ml^{-1}. Since half-lives in these environments are estimated at 12–24 hours, high-level production of phage particles from infected cells of *Pseudomonas aeruginosa* must continuously occur to maintain aquatic concentrations of virus particles.

The influence of environmental parameters on the equilibrium between lysogenic and lytic states has particular relevance to global warming and to increased UV irradiation resulting from depletion of the ozone layer. The effect would be particularly significant for bacterial populations, which have relatively high levels of temperate viruses and which would undergo a transition to a more lytic state. Higher temperatures and increased levels of UV radiation would thus be expected to result in increased rates of bacterial mortality (direct effects of UV, increased levels of viral lysis). The effects on viral productivity would be complex, with increased levels of free infective particles in the short term (higher rates of bacterial lysis), but reduced virion concentrations in the long term due to reduced bacterial populations and higher rates of virion degradation in the aquatic medium (direct effect of UV). Reduced levels of lysogenic bacteria would also result in a reduction in gene transmission by transduction. As more potential effects become apparent, a pattern of far-reaching phage-mediated changes emerge in possible response to continuing environmental change.

7.8.6 Bacteriophage control of planktonic bacterial populations

The role of viruses in controlling bacterial populations in natural waters is highly complex, depending on a wide range of interrelated factors. These include host/viral populations, type of interaction (primarily lytic or lysogenic), metabolic state of the bacteria, and various environmental physico-chemical parameters. Competition with other biota which also parasitize or ingest the host is additionally important, as demonstrated by Weinbauer and Höfle (1998) for bacterial populations in a eutrophic lake.

Case study 7.2   Viral lysis of bacteria in a eutrophic lake

Recent studies by Weinbauer and Höfle (1998) have demonstrated combined effects of protozoon grazing and viral lysis in reducing bacterial productivity in a eutrophic lake (Lake Plußsee, northern Germany). This work provides useful information on the quantitative characteristics of bacteriophage/bacterial interactions, and also the relative importance of lysis and predation on the control of bacterial populations at different levels within the water column.

Interactions between bacteriophage and planktonic bacteria

The parameters involved in bacteriophage/bacterial interactions relate to viral infection, bacterial mortality, and the generation of new virus particles as a result of bacterial lysis. These aspects are summarized in Figure 7.4, which includes quantitative information relating specifically to the epilimnion of Lake Plußsee (Weinbauer and Höfle, 1998). Total bacterial abundance in this part of the lake was approximately 10^9 cells l^{-1}, with viral counts an order of magnitude greater at about 10^{10} particles l^{-1}.

Viral infection   Infection of healthy bacteria by bacteriophage is preceded by viral contact. The contact rate between viruses and bacteria, estimated at 49 contacts cell^{-1} day^{-1}, emphasizes the dynamic interactions which occur prior to infection. This rate was determined from data on viral count, abundance of healthy (undamaged) bacterial cells and mean bacterial cell diameter, using the viral dynamics model of Murray and Jackson (1992).

Bacterial mortality Transmission electron microscope examination of whole bacterial cells and sectioned cells provides information on the frequency of visibly infected cells (FVIC) within the bacterial population. Bacterial mortality due to viral lysis can be determined from the FVIC using the model of Proctor *et al.* (1993), and indicated a value in the range 7–27 per cent of bacterial productivity in the epilimnion of Lake Plußsee.

Bacterial lysis Lysis of infected bacteria leads to the release of newly-formed virus particles and the remains of cell contents (DOC, soluble nitrogen, and phosphorus) into the water medium. The mean number of viral particles released from infected bacterial cells (burst size) was estimated from observation of visibly infected cells at approximately 39 virus particles per cell. The concentration of carbon released during lysis of bacterial cells was determined from the viral lysis rate of bacteria, estimated cell volumes, and a conversion factor for the carbon content of bacterial cells (350 fg of C μm^{-3}) and amounted to a mean value of 0.36 $\mu g\ l^{-1}\ d^{-1}$.

Epilimnion bacterial and virus counts were both very high. In this part of the water column, bacteriophage infection had a relatively limited effect on reducing bacterial production and the rate of virus production (0.15×10^{10} virus particles $l^{-1}\ d^{-1}$) was high. The environmental effect of bacterial lysis was minimal in terms of DOC release, amounting to less than 0.1 per cent of the general DOC pool on a daily basis.

Control of bacterial populations at different depths in the water column

One of the most interesting aspects to emerge from the Lake Plußsee study was that the importance of phage control of bacterial populations varied within the water column. Top-down control of bacterial populations in this lake involved predation by heterotrophic nanoflagellates (HNF) in addition to viral lysis. Bacterial mortality due to flagellate grazing was estimated by dividing the grazing rate by the bacterial production rate, and was added to the values obtained for bacterial mortality due to lysis to give a combined value (summed mortality) for the effect of both nanoflagellates and viruses.

The relative effects of viral lysis and predation varied within the water column (Table 7.4). At the time of sampling, Lake Plußsee was separated by a steep temperature and oxygen gradient into a warm aerobic

Table 7.4 Relative contribution of viral lysis and nanoflagellate grazing to bacterial mortality in Lake Plußsee (Weinbauer and Höfle, 1998)

Water column	Bacterial mortality due to viral lysis[*]	Summed mortality (% bacterial production)[†]	Regulation of bacterial production
Epilimnion	**25**% (8–42)	98%	Complete – mainly by flagellates
Metalimnion	**71**% (51–91)	40%	Relatively low – mainly by viruses
Hypolimnion	**92**% (88–94)	74%	Substantial – mainly by viruses

[*]Mean values (with range) indicate the viral contribution to overall bacterial mortality at different depths in the water column. Values are expressed as a percentage of the summed mortality due to lysis and grazing.
[†]Summed bacterial mortality, expressed as a percentage of bacterial productivity, gives an estimate of the combined control exerted by protozoa and viruses on bacterial growth and population increase.

epilimnion and a cold anaerobic hypolimnion. The mechanism for regulating bacterial production shifted with depth from grazing control in the epilimnion (where phages only accounted for about 25 per cent of bacterial mortality) to control by viral lysis in the hypolimnion (92 per cent phage-induced nortality).

Comparison of summed mortality rates to overall bacterial productivity provided a measure of the overall degree of control of the bacterial population (Table 7.4). Greatest control, occurred in the epilimnion, where the combined effects of protozoa and viruses accounted for 98 per cent of bacterial productivity. Least control occurred in the metalimnion, where viral lysis and HNF grazing were insufficient to control bacterial productivity. The switch from protozoon to viral control with depth may reflect the environmental preferences of protozoon populations, particularly the aerobic requirements of heterotrophic nanoflagellates.

7.8.7 Transduction: bacteriophage-mediated gene transfer between freshwater bacteria

The transfer of bacterial genes via bacteriophages (transduction) is one of the three principal modes of DNA exchange in microbial communities. General aspects of gene transfer in freshwater bacteria, including experimental approaches to demonstrate this in laboratory and field experiments, are considered separately in Sections 6.3 to 6.5.

During the formation of viral particles in the bacterial host cell, DNA replication is normally restricted to the phage DNA. Only viral DNA is incorporated into the newly-formed virions, and only viral DNA is introduced into new host cells when further infection occurs. Transduction occurs as an aberration of this process, with the incorporation of both host and viral DNA into phage capsids during the production of virus progeny (Margolin, 1987). The addition of host DNA may occur during activation of prophage, with replication of prophage plus adjacent regions of chromosomal DNA in the formation of the capsid DNA inserts (Figure 7.16). Subsequent infection of a new host cell leads to the passage of bacterial DNA (chromosomal or plasmid) in combination with the viral DNA, completing the transfer process.

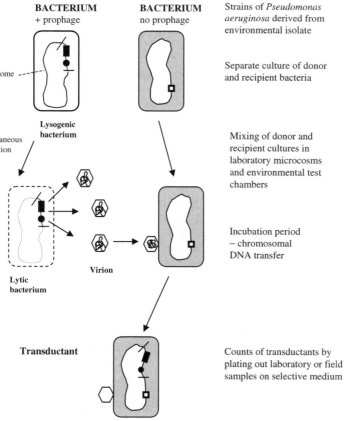

DONOR BACTERIUM
+ prophage

RECIPIENT BACTERIUM
no prophage

Chromosome

Lysogenic bacterium

Spontaneous induction

Virion

Lytic bacterium

Transductant

Strains of *Pseudomonas aeruginosa* derived from environmental isolate

Separate culture of donor and recipient bacteria

Mixing of donor and recipient cultures in laboratory microcosms and environmental test chambers

Incubation period – chromosomal DNA transfer

Counts of transductants by plating out laboratory or field samples on selective medium

Figure 7.16 Laboratory and field studies of chromosomal DNA transfer by transduction in freshwater bacteria (strains of *Pseudomonas aeruginosa*): experiment of Saye *et al.* (1990), DNA transferred by the bacteriophage is indicated diagrammatically as: ◢■■●⊦

■■ - bacteriophage DNA (prophage) in donor bacterium

● - nalidixic acid resistance (nal) gene on chromosome of donor bacterium

◻ - lysine auxotrophy locus in recipient bacterium

Transduction has generally been assumed to be a less potent mechanism for gene transfer compared with conjugation, since many bacteriophages have a relatively restricted host range and the process requires an external factor (the bacteriophage) – which may be absent or present only in limited amounts. There is also a limitation on the amount of DNA that can be packaged in the virus capsid, and the chromosomal genes transferred by this process are restricted to the region adjacent to the prophage insert. In spite of these limitations, phage-mediated gene transfer has been documented in a variety of environmentally-important bacterial species, with potential for transduction in a range of aquatic environments (Miller and Sayler, 1992).

Case study 7.3   Transduction of plasmid and chromosomal DNA in Pseudomonas aeruginosa

Miller et al. (1990) and Saye et al. (1990) studied the potential for phage-mediated gene transfer in lake water, using the commonly-occurring freshwater bacterium Pseudomonas aeruginosa and its temperate phage F116 as a model system. Transduction of plasmid and chromosomal DNA was observed under both laboratory and environmental conditions.

As with experimental studies on gene transfer during conjugation (Figure 6.4), studies on transduction involve three major aspects:

- separate culture of donor and recipient bacteria, each characterized by distinct genetic markers,

- mixing of donor and recipient bacteria, with incubation under laboratory or environmental conditions,

- identification and enumeration of bacteria containing the transferred genes (transductants in the case of phage-mediated transfer) using appropriate selective media.

This sequence is summarized in Figure 7.16 for the studies on Pseudomonas aeruginosa, illustrating the specific case of transduction of chromosomal DNA from a lysogenic donor bacterium. In these experiments, bacteriophage particles were not introduced separately into the system. Instead, virions were produced from lysogenic strains of the bacterium, containing the temperate phage F116L. Infective particles were generated by induction of the resident prophage to lytic growth, either by spontaneous induction or in response to some external stimulus during incubation. Spontaneous induction was adequate in terms of the overall levels of free virus particles produced, generating concentrations of up to 10^6 pfu ml^{-1} in mid-log-phase cultures of lysogenic strains.

Rates of gene transfer, in both laboratory microcosms and environmental test chambers, occurred at frequencies of 10^{-6} to 10^{-5} transductants per bacterial CFU. The results showed clear evidence for the potential role of transduction in freshwater microbial communities. The results also provided information on the following.

- *Comparisons of plasmid and chromosomal gene transfer.* No significant differences occurred in the frequency of plasmid and chromosomal transduction. Transduction of both single loci (as shown in Figure 7.16) and closely linked multiple loci was observed.

- *The importance of the lysogenic state.* Lysogenic bacteria can serve as both a source of the phage mediating the transduction (as in Figure 7.16), or as a recipient. Correspondingly, the transducing virus can either arise directly during the induction of a prophage from a lysogen (as in Figure 7.16) or by primary infection of a non-lysogenic cell. The greater recovery of transductants from bacterial combinations involving a lysogenized strain (either as donor or recipient), suggests that the lysogenous state is particularly important in gene transfer. By promoting higher levels of transduction, lysogeny has the potential to increase the size and flexibility of the gene pool available to natural populations of bacteria.

- *Effects of population levels.* As with conjugation-mediated gene transfer, the detectability of transductants varied with the presence or absence of the natural microbial community and the ratio of donor to recipient cells. The effect of the natural community is to reduce the recovery of both transductants and introduced donors and recipients, probably as a result of protozoan grazing.

In the transduction experiments, the phage/bacterium ratio produced in the test system was also important. The frequency of transduction was independent of donor cell concentration, suggesting that the efficiency of lytic infection and the production of phages remained constant over a relatively wide range of cell densities.

In the above experiments, gene transfer is taking place between planktonic cells. DNA transfer by transduction has also been demonstrated in biofilms, where *Pseudomonas aeruginosa* strains (with F116 phage) were introduced into the epilithon on river stones and left submerged prior to analysis and demonstration of transductants (Amin and Day, 1988).

8

Fungi and fungal-like organisms: aquatic biota with a mycelial growth form

Fungi and fungal-like organisms lack chlorophyll and have a major saprophytic (or saprotrophic) role in aquatic environments, where they are important decomposers of both plant and animal detritus. The breakdown of biomass by these organisms is important in the regeneration of soluble materials, and they play a substantial role in the carbon, nitrogen, and phosphorus cycles of lakes, rivers, and other freshwater habitats. As heterotrophic organisms, fungi are a key component of many aquatic food webs, and are in direct competition for organic material with bacteria and protozoa. Each of these groups has adopted a particular strategy to promote their saprophytic existence, which in the case of fungi involves the development of a filamentous branching growth form – the mycelial growth habit.

Fungi have been viewed traditionally as a broad group of organisms that are united in their mycelial growth form and in the associated mode of heterotrophic nutrition. This chapter considers the aquatic biology of fungi and fungal-like organisms in terms of their taxonomic and trophic diversity (Part A) and their ecological role as saprophytes and parasites (Part B). Particular emphasis is given to the role of fungi in the control of phytoplankton populations. As with other groups of microorganisms there is increasing use of molecular techniques in the analysis of biodiversity (Eggar, 1992) and functional activities.

A. ACTINOMYCETES, OOMYCETES, AND TRUE FUNGI

8.1 Fungi and fungal-like organisms: the mycelial growth habit

The mycelium is a highly successful vegetative structure which involves the production of a mass of branching tubular hyphae. These penetrate the organic substrate, secrete extracellular enzymes, and absorb soluble nutrients over their surface. The mycelial habit promotes optimal exploitation of the food source – with rapid penetration by hyphal tip extension, extensive colonization of bulk substrate and large-scale digestion within a short period of time.

The ecological success of the mycelial growth form can be gauged by the fact that it has evolved separately in at least three quite different groups of organisms – actinomycetes, oomycetes, and true fungi. These three groups have distinctive cell walls, are non-motile (though may have motile reproductive

---

Freshwater Microbiology: Biodiversity and Dynamic Interactions of Microorganisms in the Aquatic Environment David C. Sigee
© 2005 John Wiley & Sons, Ltd ISBNs: 0-471-48529-2 (pbk) 0-471-48528-4 (hbk)

Table 8.1 Three major groups of aquatic microorganisms showing a mycelial growth form

Group	Status	Possible ancestry	Key features
Actinomycetes	Prokaryote	Gram-positive Eubacteria[a]	Mucopeptide cell wall
Oomycetes[b]	Eukaryote	Heterokont Algae[b]	Cell wall: glucan/cellulose
			Main sterol: fucosterol
True fungi	Eukaryote	Choanoflagellate protozoa[b]	Cell wall: chitin
			Main sterol: ergosterol

References: [a]Ensign (1992); [b]Alexopoulos *et al.* (1996).

cells), and reproduce by means of spores. The somatic structures (mycelial hyphae) have little differentiation, with practically no division of labour (Alexopoulos *et al.*, 1996; Deacon, 1997). Differences do occur between the three groups in relation to hyphal dimensions. Hyphae of actinomycetes are normally about 1 μm in diameter, compared with 3–5 μm in most true fungi and up to 100 μm in some oomycetes.

A further assembly of organisms, the slime moulds, are also typically discussed in mycological texts, but (as with actinomycetes and oomycetes) are quite distinct from the true fungi. One particular group of slime moulds, the *Plasmodiophoromycota*, are present in fresh water environments and appear to be phylogenetically related to ciliate protozoa on the basis of small-subunit rDNA sequence homology (Castlebury, 1994).

Some of the major features of fungi and fungi-like organisms are summarized in Table 8.1. More detailed aspects of the distinguishing features and biological diversity of actinomycetes, oomycetes, and true fungi are considered in Sections 8.2 to 8.4.

8.2 Actinomycetes

Actinomycetes are a widely-distributed group of prokaryote organisms occurring in a multiplicity of natural and man-made environments, and are best known for their economic value as producers of antibiotics, vitamins, and enzymes.

The variety and ecological importance of actinomycetes in aquatic systems have often been overlooked due to problems of isolation and laboratory culture. The filamentous nature of many actinomy-

cetes typically results in strong adherence to the substratum, making them difficult to remove from organic material. Even where actinomycetes are being cultured from spores, the long incubation times that the slowly-extending colonies of these organisms require often result in overgrowth of the culture plate by bacteria and fungi. This results in culture plates being discarded before the actinomycetes even appear. In spite of these problems, actinomycetes have been widely detected (Wipat *et al.*, 1992) and isolated from aquatic systems, where they have a distinctive ecological role. The aquatic biology of these organisms will be considered in relation to their taxonomic characteristics, habitat preferences, nutritional activities, and their ability to out-compete and antagonize other microorganisms.

8.2.1 Taxonomic characteristics

Actinomycetes are prokaryotic organisms that are separated into seven major groups on the basis of oligonucleotide sequence homology. These groups comprise: actinobacteria, nocardioforms, actinoplanetes, thermomonospora, maduromycetes, streptomycetes, and a group with multilocular sprorangia (Ensign, 1992).

The fine structure and chemical composition of actinomycetes is consistent with their general prokaryote nature and specifically their Gram-positive status. They have an unusually high guanine and cytosine DNA content (in excess of 55 mol per cent), suggesting a distinct evolutionary derivation from other Gram-positive prokaryotes. Actinomycetes have a wide range of morphologies, varying from simple coccoid and rod-shaped forms to phases or

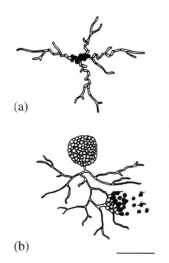

Figure 8.1 Freshwater actinomycetes: (a) *Rhodococcus coprophilus* – the fragmented mycelium has a small central group of nonmotile spores; (b) *Actinoplanes* sp. The branched mycelium bears large sporangia, one of which has burst-liberating motile zoospores (scale = 20 μm) (figure taken from Cross (1982) with permission from Blackwell Publishing)

organisms with more complex branching morphologies. Variation within single species is shown, for example, by nocardioform actinomycetes, where spherical cells grow into branched filamentous mycelia, then fragment at the onset of the stationary phase, and revert to the spherical state.

Many, but not all, actinomycetes reproduce asexually by spore formation. This ranges from simple fragmentation of unspecialized hyphal filaments (Figure 8.1(a)) to compartmentation of more specialized reproductive or aerial mycelia into chains of dormant and resistant propagules (*Streptomyces*). Reproduction is even more complex in the actinoplanetes, where hyphae enclosed in a sporangial sac fragment to form spores, which in some genera are flagellated and motile (Figure 8.1(b)).

8.2.2 Habitat

Although the majority of actinomycetes are found in soil environments, certain genera can be routinely isolated from freshwater habitats – including *Actinoplanes*, *Micromonospora*, *Nocardia*, *Strepto-*

myces, and *Thermoactinomyces*. There has been some debate as to whether such isolates are truly aquatic organisms, or whether they arise from the surrounding terrestrial environment and enter the water as wash-in or associated with organic detritus such as leaf litter. In some cases, the terrestrial origin of aquatic actinomycetes is clearly established. *Rhodococcus coprophilus* (Figure 8.1(a)), for example, is primarily associated with herbivore dung and is particularly common in farmland which is contaminated by cow manure. This actinomycete is readily washed into local watercourses, and accumulates in the sediments of associated streams and ponds. Many other actinomycetes also enter aquatic systems in spore form, sinking to the bottom of lakes and rivers where they are able to survive in a dormant condition. External derivation of these organisms in aquatic systems is indicated by their frequent isolation as spores from profundal muds and benthic sediments, where they are simply occurring in an inert and dormant state.

In other situations, where actinomycetes are found to be actively growing in organic-rich sediments, the high numbers are due to multiplication within the aquatic system and their presence is thus largely internally-derived (autochthonous). Organic debris in the littoral zone of lakes, for example, is a frequent source of actinomycetes and Cross (1982) reports high levels of *Actinoplanes* (Figure 8.1(b)) growing in leaf litter of Lake Windermere (UK). In addition to their general presence within lake and river systems, actinomycetes may be particularly frequent in various specialized aquatic habitats. These include the following.

- *Macrophyte beds*. Actinomycetes can be found actively growing in freshwater habitats in association with submerged macrophytes. The distribution and diversity of the actinomycetes depend partly on the vegetation present, with unusually high levels associated with particular plants such as *Sparganium americanum* (Wohl and McArthur, 1998).

- *Turbulent water*. Spray and foam of turbulent water provides an unusual location for actinomycetes. The high concentration (10^2–$10^3 \times$) of actinomycetes in foam samples is thought to arise as a physical phenomenon, caused by the passage of air bubbles and the concentration of hydrophobic spores at the air–water interface (Iqbal and Webster, 1973). Actinomycetes appear to provoke further foam formation in different aquatic systems, causing various industrial problems including foam formation in activated sludge plants.

- *Ground water*. Actinomycetes have also been isolated from ground water, where they may occur in extreme conditions. *Actinomyces israeli*, for example, occurs in hydrogen sulphide-rich ground waters where there are low levels of organic substrate. This anaerobic, proteolytic species is thought to occur as a commensal, obtaining its nutrients from organisms which can tolerate the low-nutrient conditions.

8.2.3 Nutrition

Actinomycetes lead an almost entirely saprophytic existence, promoting the breakdown and decay of organic detritus in lake and river sediments. These activities are important in relation to the carbon cycle and also contribute to the cycling of inorganic nutrients in aquatic ecosystems.

Genera such as *Streptomyces* and *Micromonospora* are able to secrete hydrolytic enzymes and to degrade cellulose, pentosans, chitins, and other recalcitrant organic material. Studies on the decomposition of both plant (e.g., decaying leaf matter) and animal (e.g., crustacean exoskeletons) material showed that fungi were the primary colonizers, with actinomycetes following on as secondary decomposers at a later stage. Various environmental characteristics determine the presence, distribution, and diversity of actinomycetes in relation to their role in the breakdown of organic matter. These include physico-chemical aspects (pH, turbidity, temperature, particulate size, and sediment chemistry) and biotic factors such as resident plant types and plant tissue types (healthy, necrotic, or dead). Wohl and McArthur (1998) reported a distinct distribution of actinomycetes in submerged macrophyte beds in relation to tissue types, with *Streptomyces* and *Nocardia* being present in association with necrotic and healthy plant tissue, while *Pseudonocardia* was only isolated from tissue classified as necrotic or dead.

8.2.4 Competition with other microorganisms

Although actinomycetes tend to follow fungi as part of a predictable temporal decomposition sequence, their interactions with fungi and other microorganisms are more dynamic than this procession might suggest. Interplay between the two groups of organisms involves both synergistic and antagonistic activities, including the production of antimicrobial compounds such as antibiotics and degrading enzymes.

Actinomycetes produce a wide range of antibiotics, including aminoglycosides, macrolides, and maquarimicides (Zenova and Zvyaginstev, 2002), allowing them to inhibit the growth of bacteria, fungi, viruses, and protozoa. This competitive ability may be an important factor in allowing them to enter the decomposition sequence as secondary colonizers.

Actinomycetes are also able to lyse green and blue-green algae by the production of lytic enzymes, using the algal breakdown products as substrate for their growth. Extensive lysis of blue-green algae has been demonstrated in relation to naturally-occurring algal blooms (Yamomoto *et al.*, 1998) and also in laboratory experiments (Sigee *et al.*, 1999b) and these organisms have considerable potential as blue-green algal biological control agents (see Chapter 10).

8.3 Oomycetes

Oomycetes are a common group of organisms found in both aquatic and terrestrial environments all over the world. Most of the aquatic forms are freshwater rather than marine and are commonly referred to as water moulds. These freshwater species grow mainly in well-aerated streams, rivers, ponds, and lakes where they occur most commonly in shallow waters near to the bank or shoreline.

Table 8.2 Major taxonomic groups within the oomycetes

Phylum	Characteristics	Order	Life-style
Oomycota (water moulds)	• Wide range (see text) • Zoospore biflagellate – anterior tinsel, posterior whiplash	Saprolegniales – *Saprolegnia, Aphanomyces*	Mostly saprophytes, but some important parasites of fish and crayfish
		Lagenidiales – *Olpidiopsis*	Parasites of algae, water moulds and small animals
		Leptomitales – *Apodachlya*	Saprophytes in clear, unpolluted waters
		Rhizipidiales – *Aqualinderella*	Saprophytes in anaerobic, polluted waters
		Peronosporales – *Pythium*	A few aquatic parasites of algae, fungi, and mosquito larvae

Most of the aquatic oomycetes are saprophytes (Table 8.2), living on the remains of dead animals and plants, and playing a major role in the degradation and recycling of nutrients in aquatic ecosystems. Some oomycetes are parasitic, attacking algae or animals – including rotifers, nematodes, mosquito larvae, and crayfish. A few members of this group are economically important parasites of fish such as catfish and salmon.

8.3.1 Oomycetes and true fungi

Oomycetes have traditionally been grouped with the 'true fungi' on the basis of similarities in gross morphology (Figure 8.2). These eukaryotic organisms, however, have no close phylogenetic relationship with fungi, and appear to be more closely related to heterokont algae with chlorophyll-a and -c (Alexopoulos *et al.*, 1996). Oomycetes are placed in the phylum Oomycota (Alexopoulos *et al.*, 1996), and differ from true fungi by the possession of:

- a zoospore with distinctive flagellation and fine structure,

- the production of a diploid vegetative body (thallus) in which meiosis occurs in the developing gametangia,

Figure 8.2 Detail from the edge of a colony of *Saprolegnia*, showing a profusely branched coenocytic mycelium.

Individual hyphae vary considerably in diameter and also in protoplasmic contents, ranging from clear to densely granular. Members of this genus are commonly found in clear unpolluted conditions as water moulds, living saprophytically on organic detritus. Although the mycelium gives this organism a fungal-like appearance, it belongs to the Phylum Oomycota a taxonomic group with major differences from the true fungi (see text). Scale = 20 μm

- oogamous reproduction, leading to the formation of a thick-walled oospore,

- mitochondria with tubular cristae,

- cell walls that are composed mainly of β-1,3- and β-1,6-glucans and cellulose, rather than chitin, as is the case for true fungi.

Other biochemical characteristics also separate the *O*omycota from the true fungi. Synthesis of the amino-acid lysine differs in the two groups, being produced in the oomycetes via the alpha aminoadipic pathway rather than via the diaminopimelic pathway (used by true fungi and also higher plants). Sterol metabolism also differs between the two groups. Oomycetes fall into two main groups, one of which is unable to synthesize sterols while the other can synthesize these compounds *de novo* from mevalonic acid. The most commonly formed sterol is fucosterol rather than ergosterol, which is typical of the true fungi. Other differences include the chemical identity of storage compounds (water soluble β-1,3 glucans in oomycetes, glycogen in fungi) and the presence of acyclic polyols (sugar alcohols) which are widely distributed in true fungi but appear to be absent from oomycetes.

All of these features serve to emphasize the fundamental separation of the oomycetes, indicating a separate phylogenetic origin from the true fungi – with convergent evolution of the mycelial habit.

8.3.2 Taxonomic diversity

Distinctive features of this phylum, particularly in comparison with the true fungi, have been noted previously. The somatic structures of oomycetes vary from a primitive unicellular thallus to a much-branched, filamentous mycelium. Five orders of aquatic fungi occur in this group (Table 8.2) – the *Saprolegniales*, *Lagenidiales*, *Leptomitales*, *Rhipidiales*, and *Peronosporales*.

Order Saprolegniales

These are mostly saprophytes, often forming profusely branched mycelia (Figure 8.2) around bits of decaying plant or animal material. Parasitic species do occur, however, in the genera *Saprolegnia*, *Achlya*, and *Aphanomyces*, causing major diseases of fish populations by infecting adults and their eggs. *Aphanomyces astaci* is the causative agent of 'crayfish plague', eradicating many populations of native freshwater crayfish in Europe.

Order Lagenidiales

These are a small group of aquatic fungi which are parasitic on algae, water moulds, and small animals such as rotifers, nematodes, and arthropods. The thalli of these parasites are often reduced to single cells or short, sparingly-branched filaments and typically occur inside the host cells (endobiotic). Within this group, members of the genus *Olpidiopsis* are endoparasites of water moulds belonging to the order *Saprolegniales*. Members of the genus *Lagenidium* are animal parasites and include *Lagenidium giganteaum*, a species which attacks mosquito larvae and has potential as a biological control agent of these insects (Federici, 1981).

Order Leptomitales

This order contains a few saprophytic species which resemble saprolegniales, except that the somatic hyphae are constricted at various intervals. These fungi are typically found on submerged decaying plant materials in clear unpolluted waters.

Order Rhipidiales

These fungi produce inflated thalli with a single centre of growth (monocentric) and have thin absorptive filaments (rhizoids) attaching the thallus to the substrate. As with the *Leptomitales*, these fungi are mainly saprophytes, but differ in being typical of stagnant and/or polluted waters with low levels of oxygen.

An example of this is *Aqualinderella fermentans*, which occurs in stagnant water and has unusual physiological and ecological characteristics in keeping with its anaerobic environment and which appear to be unique among free-living fungi (Natvig, 1987). Although this organism can tolerate the presence of oxygen (facultative aerobe), it has no capacity for oxidative metabolism and is obligately fermentative. In line with this, studies by Held *et al.* (1969) have revealed that *Aqualinderella* has neither cytochromes nor recognizable mitochondria.

Order Peronosporales

Members of the *Peronosporales* are the most specialized forms of the *Oomycota*, with a well-developed mycelium that consists of freely branching, slender coenocytic hyphae. Most organisms in this group are parasites, with hyphae which are intercellular or intracellular. The most specialized parasites have hyphae between the host cells and produce specialized attachments (haustoria) of various shapes and sizes.

Although *Peronosporales* are poorly represented in the freshwater environment, there are some aquatic organisms. Members of the genus *Pythium*, for example, are chiefly soil-inhabiting organisms, but include species which are parasitic on freshwater algae, fungi, and mosquito larvae (Dick, 1990).

8.4 True fungi

True fungi are a distinctive group of eukaryotic organisms that typically resemble oomycetes in the development of the mycelial habit, but differ from them in the key features noted previously. Some true fungi, such as chytrids, have lost the mycelial habit – and have adopted a unicellular morphology with a simple rhizoid system (Figure 8.7).

As with the oomycetes, they are widespread, heterotrophic organisms which obtain their nutrients by:

- saprophytic decomposition of dead material (detritus), which occurs mainly on sediments,

- parasitic interactions with the whole range of biota present in freshwater systems.

In addition to these two main modes of nutrition, some fungi are also symbiotic. Trichomycete fungi (phylum *Zygomycota*), for example, are widely present in the gut of aquatic arthropods, where there appears to be mutual benefit to host and fungus. Other fungi are predacious, capturing their prey by a range of entrapment mechanisms, then ingesting break-down products released from the digested remains. Various zygomycete fungi, for example, attack protozoa, rotifers, and nematodes.

8.4.1 Old and new terminology

Traditional taxonomy considered fungi as a very broad assemblage of organisms, dividing them into four major groups – phycomycetes, ascomycetes, *fungi imperfecti*, and basidiomycetes, all of which have representatives in the freshwater environment. Of these, phycomycetes are the only fungal group with motile spores (zoospores) and are the commonest group of fungi in freshwater systems. The *fungi imperfecti* are asexual forms derived mainly from ascomycetes and basidiomycetes, and are well represented as saprophytes and predatory fungi in freshwater situations.

The above groups are based primarily on comparative morphology and developmental patterns of the sexual reproductive structures. These relationships are now being re-examined by fine-structural, biochemical, and molecular studies, including the sequence analysis of rDNA noted above. With the emergence of new information and resulting changes in the understanding of phylogenetic relationships, there have been alterations in the arrangement of taxonomic groups and in nomenclature. A

Table 8.3 Classification of aquatic fungi: comparison of past and current names that have been applied to the major fungal groups (table adapted from Deacon, 1997, and Alexopoulos *et al.*, 1996)

Past Names		Current Name
		Phylum
Phycomycetes	Chytridiomycetes	**Chytridiomycota***
	Hyphochytridiomycetes	Hyphochytriomycota
	Oomycetes	**Oomycota**
	Plasmodiophoromycetes	Plasmodiophoromycota
	Zygomycetes	Zygomycota*
	Trichomycetes	
Ascomycetes		Ascomycota*
Fungi imperfecti		**Deuteromycota***
Basidiomycetes		Basidiomycota*

Five of the groups (*) are considered to be 'true fungi'. Groups with a substantial presence in the freshwater environment are shown in bold. Deuteromycota are a hybrid mixture of fungi without a sexual state and are mainly derived from the Ascomycota, plus a few Basidiomycota. Actinomycetes are not included in this table

comparison of past and present taxonomic names (Table 8.3) indicates that current taxonomy now places aquatic fungi in seven separate phyla plus one indeterminate assemblage (the *fungi imperfecti*). According to Alexopoulos *et al.* (1996) four of the aquatic phyla should be considered to be true fungi (Kingdom Fungi), two should be placed in a separate kingdom (Stramenopila) and the *Plasmodiophoromycetes* should be considered to be Protists (single-celled eukaryotes not belonging to any of the other major kingdoms).

Phycomycetes: a heterogeneous assemblage of important aquatic fungi

Phycomycetes dominate the fungal biota of freshwater systems and are characterized by having a mycelium which is typically composed of hyphae without cross walls and by sporangia that contain an indefinite number of spores. Motile spores (zoospores), produced by most members of the phycomycetes (excluding zygomycetes), have one or more flagella and are the major form of dispersal in this group. The body of the fungus (thallus) is composed of a vegetative (nutrient-gathering) system plus reproductive structures and varies considerably within

the group. In some orders the thallus is a well-developed much-branched mycelium, while in others, such as the chytrids, it can be reduced to a reproductive rudiment bearing a limited and simple rhizoidal system.

Although the term 'phycomycetes' provides a convenient grouping for the major assemblage of freshwater fungi, it is very heterogeneous and is no longer retained as a valid taxonomic division. This original group now encompasses three phyla which are not true fungi – the Oomycota (see previously), Hyphochytriomycota and Plasmodiophoromycota.

8.4.2 Taxonomic diversity within the true fungi

Within the aquatic environment, true fungi represent a very diverse taxonomic assemblage, with an equally diverse range of lifestyles. Each major taxonomic division (order) represents a distinct group of fungi that has evolved and adapted to freshwater conditions, adopting either a saprophytic or parasitic life style. The occurrence of four taxonomically distinct phyla (plus one heterogeneous group – the deuteromycetes) with representatives in modern freshwater systems indicates the variety of

Table 8.4 Taxonomic and trophic diversity of true fungi

Phylum	Characteristics	Order	Life-style
Chytridiomycota	• Coenocytic thallus, ranges from simple to well-developed mycelium • Only fungi that have true motile cells (zoospores and gametes); these have single posterior whiplash flagellum	Chytridiales – *Rhizophydium* – *Nowakowskiella*	Mainly parasites of algae, water moulds, animal eggs and protozoa; some saprophytes
		Blastocladiales – *Allomyces* – *Coelomomyces*	Animal and plant parasites, freeliving saprophytes
		Monoblepharidales – *Monoblepharis*	Mainly saprophytes
		Spizellomycetales – *Rozella*	Includes the endoparasite *Rozella*
Zygomycota	• Coenocytic thallus, typically well-developed mycelium • Sexual resting spores (zygospores) formed in all zygomycetes and some trichomycetes	Zygomycetes – *Zoophagus*	A few predatory aquatic species of amoebae, rotifers, and nematodes
		Trichomycetes – *Smittium*	All are obligate associates of arthropods; mostly symbiotic
Asomycota	• Septate mycelium • Production of ascospores in ascocarps	Saccharomycetales	Some saprophytic yeasts
Basidiomycota	• Septate hyphae with clamp connections and dolipore septal pore structures • Produce basidiospores on basidia	↑ *Aquatic members typically placed in Deuteromycota (have lost sexual stages)* ↓	
Deuteromycota (*Fungi imperfecti*)	• No clear sexual stages • A taxonomically mixed assemblage	Hyphomycetes	Saprophytes present in both clear and stagnant waters
		Miscellaneous group, including – *Dactylella, Arthrobotrys*	Endoparasites and predatory fungi; attack whole animals including nematodes

evolutionary routes which have led to the successful colonization and ultilization of freshwater resources by these organisms.

Summaries of the major features of these phyla are shown in Table 8.4. Full taxonomic details are not given here, and the reader is referred to a standard text on fungi (e.g., Alexopoulos *et al*., 1996) for this information.

Phylum Chytridiomycota

Members of this phylum (collectively known as 'chytrids') are widely present in both aquatic and soil habitats, and include both parasitic and saprophytic species. In many of these fungi a true mycelium is lacking, and the thallus is reduced to a single globular structure or an irregular system of rhizoids (rhizomycelium) which penetrates the host or non-living substratum. Because of their extremely small size, most parasitic chytrids can only be detected by microscopic examination. Some chytrids are endobiotic, living entirely within their hosts, while others are epibiotic – producing their reproductive organs on the surface of the living host or the dead organic material.

Four orders have important aquatic representatives within this phylum.

Order Chytridiales

This group includes many parasites of algae and water moulds, plus some saprophytes. The genus *Rhizophydium* contains species which parasitize a wide range of planktonic algae, while *Nowakoskiella* is a saprophyte that lives on decaying plant material.

Order Blastocladiales

Members of this order include parasites of plants and animals, including the genus *Coelomomyces* which is an obligate parasite of mosquito larvae. Various genera, such as *Allomyces*, are free-living saprophytes.

Order Spizellomycetales

Mostly parasites, including the genus *Rozella*, which is an obligate endoparasite of aquatic fungi belonging to the *Chytridiomycota* and the *Oomycota*.

Order Monoblepharidales

This contains only a few species and is composed mainly of saprophytes which live on submerged twigs and fruits.

Phylum Zygomycota

The phylum *Zygomycota* contains two classes, the *Zygomycetes* and *Trichomycetes*, both of which have representatives in the freshwater environment.

Class Zygomycetes

These fungi are distinguished by the production of a thick-walled sexual spore, the zygospore, and includes a number of predacious genera that attack amoebae, rotifers, and nematodes. One extensively-studied species, *Zoophagus insidians*, is a common inhabitant of freshwater pools, where it captures rotifers by the production of adhesive pegs on short lateral hyphal branches. The trapped rotifer is held until a hyphae grows through its mouth into its body, eventually penetrating its body cavity and causing disintegration of the organism.

Class Trichomycetes

Members of this group are morphologically and ecologically distinct from all other fungi. All trichomycete fungi are obligately associated with arthropods, including insects (adult and larval forms) and crustacea. The fungi typically occur in the hind-gut region, firmly attached to the chitinous gut wall by a holdfast zone.

Although the nutritional relationship between fungus and host has not been fully investigated, it appears to be symbiotic. The fungus extends into the gut lumen, where it is able to absorb nutrients from the gut contents. Evidence for host benefit is provided by the studies of Horn and Lichtwardt (1986) on mosquito larvae infected with *Smittium culisetae*. Development and metamorphosis of the larvae was restricted when completely deprived of sterols and certain B vitamins. This limitation in development was much less severe when larvae had associated fungus, compared with uninfected organisms, suggesting that the fungus may enhance fitness of its host under conditions of nutrient stress.

Phylum Ascomycota

Ascomycetes, characterized in multicellular forms by a septate mycelium and distinctive fruiting body (ascocarp) are poorly represented in the freshwater environment – apparently occurring only as aquatic yeasts in the order *Saccharonmycetales*. The absence of ascomycetes from the freshwater environment is surprising in view of their presence in marine situations, where they occur as parasites of macro-algae infecting members of the *Rhodophyta*, *Phaeophyta*, and some *Chlorophyta*.

Phylum Basidiomycota

Basidiomycetes are also atypical of freshwater environments, being restricted to organisms present on rotting wood and other organic materials in and around streams. As with the ascomycetes, these fungi only achieve ecological significance as aquatic organisms in their imperfect (asexual) state as members of the *Deuteromycota*.

Deuteromycota – asexual relatives of ascomycetes and basidiomycetes

Although sexually-defined ascomycete and basidiomycete fungi are primarily terrestrial, they do have some freshwater relatives. These occur within the large assemblage of fungi (*Deuteromycota*) which do not have sexual stages, many of which are clearly related to ascomycete and basidiomycete fungi with sexual stages. Deuteromycete fungi are important both as saprophytes and as fungi that attack small animals (parasites and predators).

One of the classic assemblages of the deuteromycetes, the hyphyomycetes, are known only from asexual spore (conidium) stages in aquatic environments. These fungi are referred to as hyphomycetes because the conidia are produced from cells which occur along the hyphae rather than being aggregated in groups. Hyphomycetes are frequently encountered in streams, and are particularly adapted to flowing water conditions (Suberkropp, 1992). Many of these organisms form conidia only when submerged in water, and rates of sporulation are directly affected by turbulence and flow rate. The tetra-radiate and sigmoid conidial shapes of these primarily saprophytic fungi (Section 8.5.2) are important for attachment to new substrates and are a major morphological adaptation to water flow.

Aquatic hyphomycetes are of particular interest in terms of convergent evolution, since the development of similar morphological characteristics and saprophytic mode of nutrition within this group appear to have arisen via separate phylogenetic lines. As noted previously, these asexual fungi have spores (conidia) that are produced by non-aggregated conidiogenous cells, and are typically branched, tetra-radiate or multi-radiate. As the sexual stages of these fungi are becoming increasingly known, it is clear (Shearer, 1993) that they are a very diverse group with connections to different major taxa within the ascomycetes – including pyrenomycetes (*Nectria*), discomycetes (*Hymenoscyphus*), and loculoascomyctes (*Massarina*). Some basidiomycetes also produce tetra-radiate conidia and basidiospores. The convergent evolution of fungi with similar conidia in the freshwater environment is probably an adaptation to the lotic habitat, since branched spores appear more able to attach to (or be trapped by) the organic substratum under flow conditions.

B. FUNGI AS SAPROPHYTES AND PARASITES

The previous section reviewed the taxonomic diversity of fungi and fungal-like organisms, and the wide range of trophic states which occur in the freshwater environment. These range from saprophytic to symbiotic, parasitic, and predatory conditions. In many cases a particular morphological characteristic or nutritive state has evolved independently in separate phylogenetic groups, indicating convergent or parallel evolution. This section considers the biological adaptations and ecological significance of fungi as saprophytes and parasites.

8.5 Saprophytic activity of fungi

Saprophytes (or saprotrophs) utilize dead organic matter for food. In their role as decomposers, fungi are particularly important in the breakdown of high molecular weight polymers such as lignin and cellulose (plant material), and chitin (insect exoskeleton) and have a key role in nutrient cycling (Moorhead and Reynolds, 1992). Degradation of high molecular weight polymers is usually the rate-limiting step in biomass degradation, with rapid removal of the monomer derivatives by the general microbial

Table 8.5 Estimated half-life times of biological polymers in natural environments (adapted from Lengeler et al., 1999)

Polymer	Half-life
Lignin	20–2000 years
Keratin (hair, wool)	1–2000 years
Humic compounds	2–200 years
Cellulose	0.01–2 months
Starch	1–10 days
Globular proteins	0.1–2 days

community. Estimated half lives of biological polymers (Table 8.5) show that lignin is particularly resistant to degradation, with keratin proteins and humic compounds also requiring long periods for removal. The ability of fungi to degrade lignin has been well documented (see Lengeler et al., 1999) and involves a battery of different enzymes, metals, co-substrates, and molecular oxygen. The main enzyme thought to be responsible for the lignolytic activities of fungi is lignin peroxidase, which oxidizes lignin to yield aromatic cation radical intermediates that undergo spontaneous fission reactions.

The breakdown of organic matter by fungi ranges from partial degradation to complete mineralization (conversion of organic to inorganic matter). Partial degradation involves the conversion of organic matter to other pools of organic material (fine particulate organic matter, dissolved organic carbon) – which may be further processed by other microorganisms or transported out of the system. Saprophytic fungi present in streams are particularly important in the conversion of externally-derived (allochthonous) organic matter such as leaf litter to internal (autochthonous) components of the food chain with the lotic system. The saprophytic activities of fungi in the aquatic environment will be considered in relation to colonization and growth, breakdown of leaf litter, and examples of saprophytes in chytrids and deuteromycetes.

8.5.1 Colonization, growth, and fungal succession

The breakdown and decomposition of plant and animal matter is particularly important in benthic regions of lakes and rivers (where the material accumulates) and involves the combined activities of a variety of organisms including bacteria, fungi, and invertebrates. Initial colonization and tissue breakdown is primarily carried out by fungi, with bacteria becoming involved as a second phase of activity. The particular fungi concerned depend to some extent on the substrate. Within the chytrids, for example, *Nowakowskiella ramosa* is found primarily on cellulosic plant material, while another common member of this group, *Chytriomyces hyalinus*, grows saprophytically in freshwater on the exuvae of May flies and fragments of chitin. This fungus can be readily cultured in the laboratory on 0.5 per cent chitin agar, and is an ideal organism for demonstrating the production of extracellular chitinase.

Exploitation of organic substrate by fungi involves three main phases – colonization, growth, and dissemination. Fungal spores can be readily collected from water samples, stained and enumerated. Spore concentrations rise during periods of litter input, and remain high while litter remains in the stream. The ability of spores to colonize fresh substrate can be expressed as the 'inoculum potential' – which relates to spore concentration, attachment efficiency, initial growth rate (dependent on spore size and endogenous reserve), and competitive ability. With some species (e.g., *Clavatospora longibrachiata*), conidial abundance in suspension appears to be the main factor which determines colonization ability, while in others (e.g., *Flagellospora curvula*) quite small numbers of conidia in transport can lead to major colonization of organic detritus. Growth of fungi on individual leaves, which represent a finite food reserve, follows a distinct pattern – with various parameters (fungal growth rate, respiration, spore formation) reaching peaks at approximately the same time (Suberkropp, 1992). As noted previously, the ability to initiate sporulation during growth may be an adaptation to the ephemeral nature of leaf detritus and other substrates in the stream environment. In the development of fungal communities on organic substrates in aquatic environments, fungi which colonize early persist throughout the degradation sequence. Successional replacement of species is not generally observed, and fungi which colonize a leaf soon after it enters a stream have a major and continued impact on the fungal community that develops.

As a result of fungal growth, breakdown of structural polymers and release of nutrients within the substrate renders it more consumable by invertebrate detritivores. Fungi have been shown, for example, to affect both the palatability and food quality of leaves for invertebrate shredders such as amphipods and caddis fly larvae, and are themselves part of the food supply. Caddis fly larvae have the ability to discriminate among fungi of different palatibilities which are colonizing the same leaf, and have been shown to forage selectively on different fungal populations which are distributed over the leaf surface.

8.5.2 Breakdown of leaf litter

Leaf litter provides the most important allochthonous substrate in lotic systems, with microbial breakdown being carried out principally by fungi and bacteria. Degradation of leaf litter is considered initially in comparison with terrestrial systems, then in relation to the time sequence of the process.

Comparison of leaf litter breakdown in terrestrial and aquatic systems

As with terrestrial environments, the accumulation of leaf litter in streams may follow a marked seasonal periodicity, with fungi as the major agent involved in subsequent breakdown. The process of decomposition, however, differs between the two systems in a number of key respects (Suberkropp, 1992).

- In rivers, individual leaves are discrete resource units that are continuously rearranged as they are decomposed. In terrestrial environments, leaves at different stages of decomposition are layered, reflecting the temporal input of leaves into the terrestrial ecosystem.

- In streams, leaf litter tends to occur in discrete patches (leaf packs) separated by leaf-free

regions. Because of this, colonization is primarily by downstream dispersal of conidia rather than by growth of hyphae from one leaf to another.

- Rates of leaf breakdown are typically higher in streams compared with terrestrial environments. This is due to a range of factors, including the constant presence of water, continuous availability of inorganic nutrients from current flowing across the leaf surface, mechanical fragmentation by the force of current, and modulation of diurnal and seasonal temperature changes within the water mass.

Leaf litter in streams will also have a more disturbed existence compared to terrestrial habitats, with individual leaves liable to sink into anaerobic muds, be consumed by invertebrates, or transported out of the system by floods. Under such ephemeral conditions, fungal colonizers of leaves would be expected to have a strategy of rapid growth and reproduction – with spore formation and dispersal representing a major part of resource allocation. Laboratory studies on leaf infection (Suberkropp, 1992) by hyphomycete fungi indicate rapid conversion of leaf to fungal biomass at ambient stream temperatures, with a peak in fungal biomass and metabolic activity approximately 15 days after initial infection (Figure 8.3). The importance of reproduction is emphasized by the fact that sporulation commences shortly after growth and peaks at the same time, indicating that little biomass conversion is required before reproduction is initiated in the stream habitat.

Time sequence of litter decomposition

Leaves are wind-blown or fall directly into streams, often from overhanging foliage, and undergo three main phases in their decomposition sequence (Figure 8.4).

- *Leaching of soluble constituents.* Loss of soluble constituents commences once the leaves have become wetted, and is mostly completed within a few days – releasing dissolved organic matter (DOM) into the stream environment. Up to 25 per

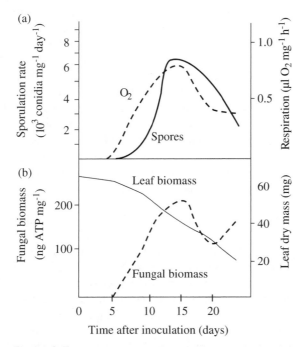

Figure 8.3 Growth and activity of *Tetracladium* (hyphomycete) on leaf discs: (a) fungal respiration closely parallels fungal growth; (b) inoculation of leaf discs leads to a phase (from 5–20 hours) of fungal growth, with a corresponding decline in leaf biomass (based on a figure from Suberkropp, 1992)

cent of the total dry mass of leaves is lost by leaching with the first 24 hours of immersion in the stream environment (Webster and Benfield, 1986). Soluble constituents lost during leaching are mainly polyphenols and carbohydrates (Suberkropp *et al.*, 1976).

- *Microbial degradation.* A period of microbial colonization and growth, involving decomposition by fungi and bacteria, leads to major changes in leaf structure and composition (see below).

- *Mechanical fragmentation by invertebrates.* This usually occurs late in the processing sequence (since it requires prior microbial conditioning) and is complete when no large particles remain. The fine particulate organic material (FPOM) generated by mechanical activity is dispersed

into the water column, being carried downstream and undergoing further microbial decay. Activities of invertebrate grazers in fragmenting detritus and removing microbial populations are discussed in Section 9.13.

Microbial degradation of leaf litter   During the second stage of leaf processing, microbial populations colonize and grow on the leaf substrate. Decrease in dry weight (initially due to loss of soluble constituents) continues, though at a slower rate. Differences in leaf structure and chemistry result in wide differences in microbial decay rates (Webster and Benfield, 1986). Leaves with a high initial nutrient concentration have more rapid decomposition than those with lower nutrient content. Other components, such as lignin and tannins, may have an inhibitory effect. The complexation of proteins to tannins is a principal cause for the slow breakdown of leaf biomass from many broad-leaved woody plants (Allan, 1995).

The activities of fungi and bacteria in the degradation of leaf biomass can be considered in relation to loss of specific chemical constituents, increase in nitrogen concentration, microbial succession, and evidence for the relative roles of fungi and bacteria.

Loss of specific constituents. Studies by Suberkropp *et al.* (1976) on over-winter breakdown of oak and hickory leaves, showed that cellulose and hemicellulose declined at about the same rate as total leaf biomass. Lignin was processed more slowly, but lipids were rapidly degraded.

Increase in nitrogen. During the period of microbial activity, nitrogen typically increases as a percentage of the remaining dry weight, and may increase in absolute terms as well. The increase in nitrogen occurs for two main reasons. First, proteins and polypeptides released during biomass breakdown complex with lignin and become resistant to further breakdown. These nitrogenous compounds thus remain, while other constituents are lost, resulting in a net increase in nitrogen. An increase in organic nitrogen also occurs due to the generation of microbial biomass, with uptake of nitrates from river water.

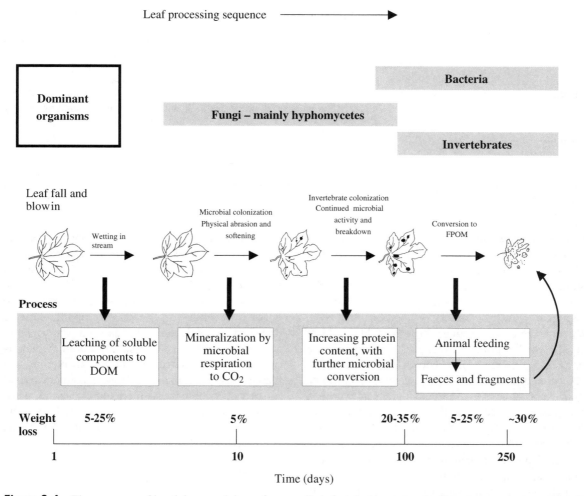

Leaf processing sequence ⟶

Figure 8.4   The sequence of breakdown and decay for a medium-fast deciduous tree leaf in a temperate stream (based on a figure from Allan, 1995)

Microbial succession. Suberkropp and Klug (1976) followed the detailed succession of fungi and bacteria on leaf litter in a north temperate stream from November to June. The first half of the processing period (12–18 weeks) was dominated by fungi, particularly hyphomycetes. Typically, 4–8 fungal species dominate throughout this phase (Barlöcher, 1982), with succession being determined primarily by which fungus arrives first as a waterborne spore. Bacterial numbers gradually increased throughout the decomposition sequence, with domination by these organisms during the terminal processing stage. The entry and activities of bacteria may be

enhanced by the earlier breakdown of leaf biomass by fungi, with the formation of greater substrate surface area and release of labile substances. Propagules of soil fungi were frequently carried into the river environment on leaf surfaces, but appeared to contribute little to leaf decomposition (Suberkropp and Klug, 1976). Studies by Barlöcher (1985) suggest that aquatic fungi are favoured by the colder winter temperatures, with soil fungi possibly being more important during summer.

Relative importance of bacteria and fungi. The importance of microbes to the decomposition process

Table 8.6 Differential influence of fungi and bacteria on the breakdown of elm leaves in stream water (data from Kaushik and Hynes, 1971)

Activity	Fungi and bacteria active	No microbial activity	Bacteria active only	Fungi active only
Treatment	No antibiotics	Antifungal and antibacterial antibiotics	Antifungal antibiotics only	Antibacterial antibiotics only
Loss in mass (%)	21.9	1	9.3	17.5
Final protein content (% final mass)	12.5	5.9	7.3	11.3

15 1-cm leaf discs were placed in 300 ml stream water, enriched with nitrogen and phosphorus, at 10°C. Antibiotics were added as shown and renewed twice weekly. Results show – (a) percentage mass loss 4 weeks after initial submersion of leaves, and (b) final protein content, expressed as a percentage of the final mass. At the beginning of the experiment, mean protein content was 4.3%.

can be assessed by comparing weight loss over 4 weeks, in the presence and absence of antibiotics (Table 8.6). Studies by Kaushik and Hynes (1971) showed that complete suppression of microbial activity blocked leaf breakdown, with no resulting significant loss of biomass or increase in protein content. Inhibition of fungal growth (Table 8.6) had an appreciable effect on biomass loss and increase in protein content, while bacterial inhibition was less effective – suggesting that fungi contribute more to changes in mass and percentage protein than bacteria. Studies on the macerating abilities of five species of aquatic hyphomycetes (Suberkropp and Klug, 1980) suggested that fungi do not require any mechanical fragmentation to carry out their activities and are able to completely degrade leaf tissue (converting 65–75 per cent of biomass) at 10°C, in the absence of macroinvertebrate consumers.

Great diversity occurs in the pattern of leaf breakdown on the stream bottom, where leaves are a mosaic of patches of microbial colonies. This biodiversity in breakdown activities occurs due to variations in the timing of leaf fall (and thus the onset of the colonization process), variation in leaf species-specific rates of degradation by fungi and diversity of fungal species that are present in the river system.

8.5.3 Saprophytic fungi – chytrids and deuteromycetes

Saprophytic nutrition is well represented in the two major groups of aquatic fungi – the chytridiomycetes and deuteromycetes. Within these groups, saprophytic fungi are characterized by the development of an extensive vegetative phase, which is important in substrate penetration and nutrient absorption. In the chytrids, which lack a true mycelium, this vegetative phase is developed as a relatively undifferentiated branched structure, referred to as a pseudomycelium. Deuteromycetes have a true mycelium, which is typically differentiated into two main regions – a filamentous branching structure which colonizes and extends over the organic food supply, and an associated rhizoidal component which extends into the substratum and is involved in anchoring the fungus and in nutrient absorption.

Saprophytic chytrid fungi

The saprophytic habit in chytrid fungi is illustrated by the widely-distributed fungus *Nowakowskiella ramosa*, which occurs throughout Europe, North and South America, and Asia, and lives on decaying plant material in freshwater habitats. In the life cycle of this organism (Figure 8.5), uniflagellate spores act as the main dispersal agent within the aquatic environment and come to rest on the dead organic substrate present in the benthic part of the water body. Zoospores encyst and then germinate by a single germ tube, which begins to branch dichotomously and to develop into a vegetative thallus. This consists of an irregular branched network of filaments lying over the substrate, with fine rhizoids arising from various parts of the network and penetrating the substratum. After an extensive growth phase, filaments

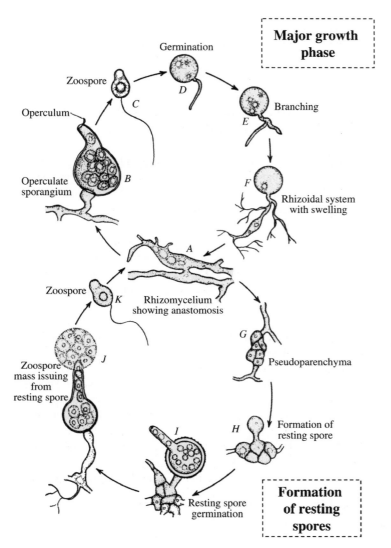

Major growth phase

Germination

Zoospore

Operculum

D

C

Branching

E

Operculate sporangium

B

F

Rhizoidal system with swelling

A

Zoospore

K

Rhizomycelium showing anastomosis

G

Pseudoparenchyma

Zoospore mass issuing from resting spore

J

H

Formation of resting spore

I

Resting spore germination

Formation of resting spores

Figure 8.5 Life cycle of *Nowakowskiella ramosa*: the life cycle consists of two components, a major growth phase (A–F) and a phase of resting spore formation (G–H) with subsequent generation of zoospores (I–K) (figure taken from Alexopoulos *et al.*, 1996, with permission from John Wiley & Sons)

arise from the surface mycelium and develop sporangia which contain between 4 and 40 zoospores. Resting spores develop from pseudoparenchymatous tissue which is formed by cell fusion. Nuclear events involved in the formation of these spores have not been determined. The spores may germinate directly or release zoospores, thus acting as resting zoosporangia.

Saprophytic deuteromycetes – the hyphomycetes

Within the deuteromycetes one particular group, the hyphomycetes, are important saprophytes of flowing waters, occurring worldwide in streams and rivers. They are most abundant on deciduous leaf litter which has entered the system from river bank (riparian) vegetation, but also grow on other types of litter such as wood, grass, and aquatic macrophytes.

Hyphomycetes can be divided into two major groups on the basis of their biology.

• Fungi occurring in well-aerated water, often found growing on submerged leaves and wood. The conidia are produced under water and, after release, are often trapped in surface foam that is generated by water turbulence. Many of these species produce branched, tetra-radiate or multi-radiate conidia. These branches may be important in anchoring the spore to the substratum under turbulent conditions, or entangling the spore within organic debris which subsequently becomes the substratum. Some of these fungi are active at low temperatures, and are primary colonizers of leaf litter in autumn. These fungi are involved in the early stages of leaf litter decomposition, where they release cellulolytic, pectolytic, and proteolytic enzymes. Various studies have also shown that these fungi are important in altering the food quality of the litter for leaf-shredding invertebrates. Caddis flies and amphipods show clear preferences for leaf litter which is inhabited by fungi, and the survival of the animals may depend on the particular fungi concerned (Suberkropp, 1992).

- Fungi present in still and stagnant aquatic environments. These fungi survive in water and mud of low-oxygen content, but apparently need higher oxygen levels for colonization of new substrates. The conidia are often formed at an air–water interface and trap air in cage-like structures. Trapped air makes these conidia buoyant, and the fungi are often referred to as aeroaquatic.

The divergence of hyphomycetes into two separate groups occupying clear water and stagnant aquatic environments parallels the situation seen in aquatic oomycetes, where a similar divergence occurs in the evolution of two ecologically distinct phyla – the Leptomitales and Rhipidiales.

8.6 Parasitic activities of aquatic fungi

Fungi have firmly established themselves as major parasites within the freshwater environment (Table 8.4) attacking other fungi, algae, macrophytes, invertebrates, and fish. All major groups of freshwater fungi contain parasitic species. The importance of fungal parasites extends to all micro-habitats within lakes and rivers, affecting pelagic, benthic, and littoral zones within the water body. In common with viruses, aquatic fungi may be regarded as having a significant impact on the populations of all lake biota and on interactions within aquatic food webs.

The activity and adaptations of fungal parasites range from relatively unspecialized examples within the deuteromycetes to highly specialized parasites of invertebrates and algae within the chytrids.

8.6.1 Parasitic and predatory deuteromycetes: fungi that attack small animals

The majority of deuteromycete fungi are saprophytes, able to break down plant material rich in cellulose and lignin substrates, but relatively low in nitrogen. Acquisition of nitrogen appears to have evolved in some groups by the use of invertebrates and other small animals as a supplementary food source, and show a transition from predation (where the animal is killed prior to nutrient uptake) to parasitism (where at least some nutrient transfer occurs prior to death of the host organism). Parasitic and predatory deuteromycetes attack a wide range of target organisms, including bacteria, copepods, amoebae, other fungi, and nematodes.

Fungi which attack nematodes have been investigated in particular detail. What may have arisen as a supplementary food source for saprophytic fungi has now developed into an obligate feeding habit (Barron, 1992), with corresponding loss of cellulolytic activity and transition from saprophytism to parasitism over evolutionary time. Fungi that attack nematodes have two distinct life styles – endoparasites and predators.

Endoparasites

Some of these fungi attach to the surface of the animal, usually near the mouth parts, germinate, then invade by hyphal extension. In other cases the conidia are ingested and germinate in the animal's gut. In either case, the mycelium grows through the body of the nematode, killing the organism and eventually producing conidia that project from the nematode surface.

Predacious fungi

In addition to endoparasitic species that involve spore germination and attack from within, other fungal species capture their prey then carry out hyphal penetration. These species produce extensive mycelia and capture their prey in various ways – including constricting and non-constricting hyphal rings, adhesive branches, and sticky knobs. These systems have not only developed in the ascomycetes, but have arisen independently in the basidiomycetes and zygomycetes as well.

Dactylella bembicoides and *Arthobotrys dactyloides* are among the species which form constrictive rings. Some species have cells that swell instantly as the nematode passes through, holding the nematode until hyphae penetrate and invade the body.

In other cases, capture is by adhesion – either involving a sticky net of hyphae (*Arthrobotrys oligospora*) or the production of sticky knobs. *Dactylella tylopaga*, *Pagidospora amoebophila*, and species of *Nematoctonus* all produce these knobs and are derived from the basidiomycetes – as indicated by clamp connections or septal pore structures. Trap formation in *Arthrobotrys* can be induced by the presence of nematodes in the locality, or by addition of proteinaceous material to the culture. The fungus mycelium is more attractive to nematodes when traps are present, and the adhesive regions produce carbohydrate-binding proteins (lectins) which attach to specific sites on the nematode surface. Nematode-activating pheromones are also produced by the fungus, attracting nematodes to the mycelial region. Once the animal is trapped, adjacent hyphae invade and digest the body.

8.6.2 Parasitic chytrids: highly specialized parasites of freshwater organisms

Many chytrid fungi have evolved as highly specialized parasites, differing from their saprophytic relatives in the reduction of the pseudomycelium to a highly rudimentary structure. Two examples of parasitic chytrids will be considered in detail – *Coelomomyces* (parasitic on invertebrates) and *Rhizophydium* (infecting phytoplankton). Both parasites have sexual and asexual phases, which in the case of *Coelomomyces* are associated with separate hosts.

Coelomomyces – an important parasite of freshwater invertebrates

This genus of parasitic fungi was initially discovered in 1921, growing in the coelom (body cavity) of a mosquito larva. Studies have since revealed that members of this genus are obligate parasites of chironomid midges and mosquito larvae, with almost 50 species (plus a number of varieties) now recognized. These fungi attracted particular interest as biocontrol agents of mosquitoes due to their high degree of host specificity and lethality, and a considerable amount of information is now known about their biology.

Pioneering studies by Whisler *et al.* (1975) on one particular species, *Coelomomyces psorophorae*, established two key facts

* The fungus has alternation of generations, involving a diploid spore-producing phase (sporothallus) and a haploid gamete-producing phase (gametothallus).

* The fungus is heteroecious, requiring two completely different hosts within the aquatic environment – one a mosquito, the other a copepod – to complete its life cycle. The discovery of heteroecism was of particular interest, since this situation was only known previously in relation to basidiomycete rust fungi.

The complex life cycle of *Coelomomyces psorophorae*, involving copepod and larval mosquito (*Culiseta inornata*) hosts is shown in Figure 8.6. There are two motile infective stages, the first of which involves free swimming haploid (+ and −) spores, produced by meiosis. On coming into contact with the copepod, these spores encyst, penetrate the chitinous cuticle, and give rise to much reduced + and − haploid gametothalli in the host haemocoel. These thalli grow within the host and mature to generate + and − gametes, which are released within the copepod shortly before its death. Fusion of gametes occurs both inside and outside the host, to generate motile diploid zygotes – which constitute the second infective phase. Contact of these zygotes with the mosquito larvae results in

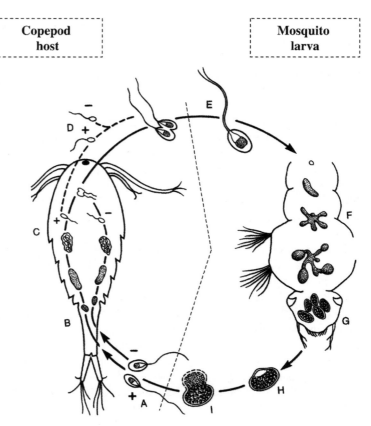

Figure 8.6 Life cycle of *Coelomomyces psorophorae*: the life cycle is divided (dotted line) into a haploid phase (stages A–D) which takes place in the copepod host, and a diploid phase (stages E–I) which occurs in the mosquito larva (figure taken from Alexopoulos *et al.* (1996) with permission from John Wiley & Sons)

encystment, followed by production of an attachment structure (appressorium) and penetration tube. The latter grows through the larval cuticle, enters the underlying epidermal cell and leads to the passage of the fungal protoplast into this cell. The fungal protoplast subsequently develops into a sporothallus which enters the haemocoel, where it multiplies to form discrete units known as hyphal bodies. These grow and develop into a diploid mycelium which eventually produces terminal resting sporangia. These sporangia are produced in such numbers that they almost completely fill the dead or dying larva. Disintegration of larvae results in release of the resting sporangia into the surrounding water. Under appropriate conditions, these sporangia undergo meiosis, leading to the release of numerous haploid motile spores – which are then available for infection of the copepod host.

Rhizophydium – a major parasite of phytoplankton

The parasitic role of aquatic fungi has been investigated particularly in relation to their effect on planktonic algae – where infection typically leads to death of the host cell, and is an example of lethal parasitism (Lund, 1957). These fungi have a major effect on the growth of algae, reducing primary production, and are one of the principal agents that lead to decline of phytoplankton populations.

Chytrid fungi are particularly important as algal parasites, and include the species *Rhizophydium planktonicum* – which infects a wide range of algal species (Van Donk and Bruning, 1992). The ecological success of these *r*-selected parasites (Burdon, 1992) lies in their relatively simple vegetative phase (with rapid exploitation of algal nutrients) and the production of large numbers of

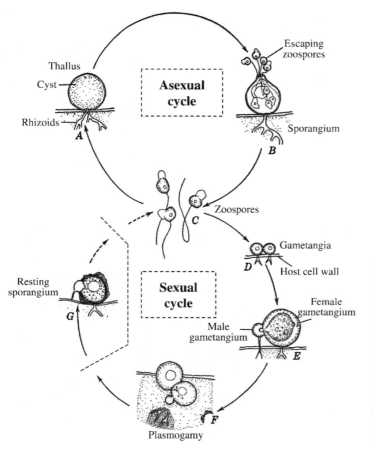

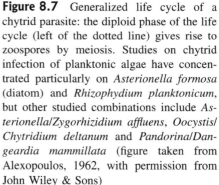

Figure 8.7 Generalized life cycle of a chytrid parasite: the diploid phase of the life cycle (left of the dotted line) gives rise to zoospores by meiosis. Studies on chytrid infection of planktonic algae have concentrated particularly on *Asterionella formosa* (diatom) and *Rhizophydium planktonicum*, but other studied combinations include *Asterionella/Zygorhizidium affluens*, *Oocystis/Chytridium deltanum* and *Pandorina/Dangeardia mammillata* (figure taken from Alexopoulos, 1962, with permission from John Wiley & Sons)

swarming spores (zoospores) that are highly motile and are able to infect new algae at various points in the water column. The strategy of short vegetative phase and enhanced dispersal phase is particularly relevant for infection of planktonic algae, since these organisms often show rapid changes in population which the successful parasite needs to exploit over a short time frame.

The typical life cycle of a chytrid such as *Rhizophydium* (Figure 8.7) consists of two parts, a vegetative phase and a sexual phase. The development and progression of infection throughout the phytoplankton population involves the vegetative phase. Spread of infection takes place by haploid asexual zoospores, which attach to new host cells and develop into a simple vegetative body or thallus. This consists of a globular cyst with rhizoids, which

penetrate the host cell and absorb nutrients. After a period of enlargment the cyst differentiates to form a sporangium, the contents of which divide by mitosis to form numerous zoospores. The sporangium ruptures, liberating the motile zoospores, which migrate through the water medium to infect new host cells. The location of new host cells within mixed phytoplankton populations must involve a precise recognition process, since strains of *Rhizophydium planktonicum* are host-specific. Various features of the vegetative cycle, which are affected by environmental parameters, determine the speed and extent of fungal infection within the phytoplankton population (see Section 8.7).

The sexual phase of the life cycle typically occurs at the end of an epidemic, when the potential for further fungal development has become limited. Zoospore infection leads to the formation of sexual reproductive bodies (gametangia), gamete formation, sexual fusion, and the generation of a resting zygote. The zygote subsequently divides by meiosis to form haploid zoospores, which infect new cells and initiate another epidemic.

8.7 Fungal epidemics in the control of phytoplankton populations

Fungal epidemics may be defined as the exponential and uncontrolled spread of fungal infection through the host population. Because of host specificity, epidemics of planktonic algae typically involve fungal spread within the population of a single species which is dispersed within the mixed phytoplankton population.

8.7.1 Ecological significance

The ecological significance of fungal epidemics in limiting phytoplankton populations was first quantitatively assessed by Canter and Lund (1948, 1951, 1953), who studied chytrid infections of a range of diatoms (*Asterionella formosa, Fragilaria crotonensis, Tabellaria fenestrata*, and *Aulacoseira subarctica*). In all cases, these fungal infestations were observed to cause death of the host cells and to limit the growth of phytoplankton populations. Many of the chytrids were shown to be highly species specific, parasitizing a single species or a small group of related species as host. In addition to diatoms, epidemics involving monoflagellate (chytrid) or

biflagellate (oomycete) fungi have since been reported for planktonic blue-green algae, chrysophycean algae, dinoflagellates, and green algae (Van Donk and Bruning, 1992, 1995).

Fungal parasitism can be an important factor in controlling the seasonal phytoplankton succession (Van Donk, 1989), since infection of one algal species may favour the development of others. Replacement of an infected population of *Asterionella formosa* by *Fragilaria crotonensis* and *Tabellaria fenestrata* was reported, for example, by Canter and Lund (1951) for Esthwaite Water, UK. In other cases, the effect of parasitic infections on interspecific competition may be to alter the balance of populations rather than affecting the seasonal pattern. This is particularly the case for algae such as desmids (Canter and Lund, 1969), which do not dominate the phytoplankton assemblage, and where parasitism reduces their proportional abundance in relation to other algae.

Although the process of infection of algal populations might seem straightforward, little is known about the mechanism of infection from fungal resting spores or how the fungi multiply quickly enough to start an epidemic. Two further complicating factors in the host-parasite interactions are:

- host hypersensitivity to infective zoospores – this algal resistance mechanism involves rapid death of the host phytoplankton cell following zoospore contact, preventing development and spread of infection;

- hyperparasitism, where the fungal parasite of algae is itself parasitized by another fungus – the chytrid *Zygorhizidium affluens*, for example, is frequently infected by another chytrid *Rozella* (Canter, 1969).

Case study 8.1   Chytrid infection of Asterionella during an autumn diatom bloom

Early studies by Lund (1949) demonstrated major limitation of the phytoplankton population by the chytrid fungus *Rhizophydium planktonicum* in Esthwaite Water (UK) (Figure 8.8). An autumn diatom bloom routinely occurs in this eutrophic lake when thermal stratification breaks down at the end of the season and inorganic nutrient concentrations begin to rise.

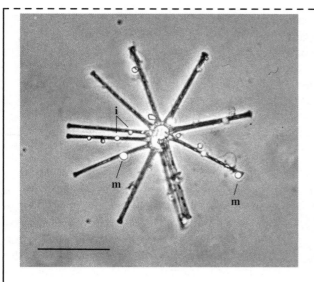

Figure 8.8 Colony of the diatom *Asterionella*, infected with a chytrid fungus (probably *Rhizophydium planktonicum*): immature (i) and mature (m) sporangia can be seen at various points in the colony (scale = 50 μm)

In the sequence shown in Figure 8.9, autumn growth of *Asterionella* commenced in late September. Increase in the count of *Asterionella* cells triggered the onset of fungal infection, with the level of infection rising to a peak (26 per cent of *Asterionella* cells) when the host population reached a maximum level of about 400 cells ml^{-1}. Subsequent decline in the diatom population was accompanied in a decline in the level of live *Rhizophydium* cells ml^{-1} and in the level of diatom infection.

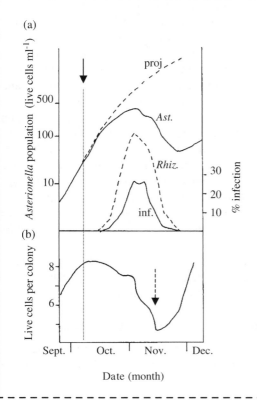

Figure 8.9 Infection of *Asterionella* by *Rhizophidium*.
(a) **Population changes.** Showing changes in the actual population of *Asterionella* (*Ast.*), projected (proj.) population without fungal infection, population of *Rhizophidium* (*Rhiz.*), and percentage infection (inf.) of *Asterionella* by live *Rhizophidium*.
(b) ***Asterionella* colony changes.** Showing the average number of live *Asterionella* cells per colony.

The vertical line (solid arrow) marks the onset of fungal infection in early October. Major flooding (broken arrow) commenced in mid-November (based on a figure from Lund, 1949)

The effect of fungal infection on the diatom population can be seen in two ways – changes in colony size and in the total population. Infection by *Rhizophydium* lead to an early decline in the number of live cells per *Asterionella* colony, suggesting a rapid effect on the growth of host cells. The overall population (live cells ml^{-1}) decreased to 85 ml^{-1} in November, compared with a projected level of 500–1000 ml^{-1} which would have occurred without fungal infection. Seasonal flooding in late November lead to nutrient influx, resulting in a recovery of the *Asterionella* population which was no longer limited by fungal infection.

8.7.2 Net effect of infected and non-infected host cells

The relative impacts of infected and non-infected algal cells on phytoplankton development can be considered in relation to mixed phytoplankton populations and single species populations within these.

Infection within mixed phytoplankton populations

Host–parasite specificity means that the presence and spread of a particular fungal strain is restricted to a single host species within the mixed population. In terrestrial plant communities, the spread of species-specific infection within homogeneous populations (monocultures) is much greater than in mixed stands, due to a reduced proximity of infection source and sink (Sigee, 1993). This principle may apply equally to chytrid infections within mixed phytoplankton populations, in which case the proportion of the total phytoplankton biovolume occupied by the species will be a significant measure of the ability for spread of infection.

Infections of individual species may develop when these occupy quite a low proportion of the total phytoplankton population. Recent field studies by Holfeld (2000) have shown that infections may increase in host populations occupying <1 per cent of the total phytoplankton biovolume, and that in most species examined infected cells were always found when the algal species occupied a higher proportion than this. Two exceptions to this were the diatoms *Stephanodiscis rotula* and *Fragilaria*

crotonensis, where high proportions of biovolume occurred without algal infection. These field studies showed that fungal parasites can exist on their host population even if it comprises only a small proportion of the total phytoplankton biovolume, and that parasites begin to spread when the population balance of a particular species begins to change.

Infection within single species populations

Fungal infection of algal cells impairs physiological processes, inhibits cell growth, and ultimately causes cell death. At any point in time, only a proportion of cells within a single species population are infected, and the net result of fungal infection on the growth of the whole population depends on the balance between net growth of uninfected cells and the loss of infected cells. This situation is shown diagrammatically in Figure 8.10, where respective gain (growth and division) and loss (grazing, sedimentation) factors of infected and uninfected cells separately contribute to overall changes in algal population. Long-term changes in algal population will depend on the future balance between infected and non-infected cells, which relates to the infection rate.

The state of infection at any point in time, and the impact of the fungal parasite on algal growth, can be monitored in two main ways:

- the percentage of infected algal cells,

- the actual biomass of algal population compared with the predicted level of an uninfected population.

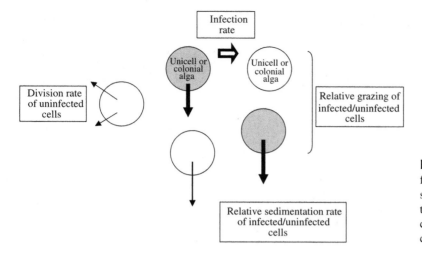

Figure 8.10 Effects of fungal infection on population changes of a single algal species within lake phytoplankton: infected algae (shaded circles) and uninfected algae (open circles)

Assessment in terms of percentage infection is easy to determine, and provides useful information on the spread of infection throughout the algal population. It also correlates with transient changes in both host and parasite populations, as seen in Figure 8.9, for the infection of *Asterionella* by *Rhizophydium*. Although percentage infection can be readily determined, it gives little information, however, on the dynamic state of infection. Comparison of actual and projected algal population provides a more useful overview on the impact of fungal infection, and encompasses all the dynamic activities summarized in Figure 8.10. The projected algal population can be estimated by extrapolation of the algal growth curve prior to fungal infection, and directly compared with the real algal population at any point in time – as seen previously in the case study on infection of *Asterionella* by *Rhizophydium* (Figure 8.8).

8.7.3 Factors affecting the development of fungal infection

Most critical information on the factors which determine the development of fungal infection within algal populations has been obtained by laboratory studies, particularly in relation to the interaction between the diatom *Asterionella formosa* and the chytrid fungus *Rhizophydium planktonicum* (Van

Donk and Bruning, 1992). Field observations have also provided useful information and, in conjunction with the laboratory studies, have identified four interrelated factors which affect the initiation and progression of fungal infection – host population density, relative host/parasite growth rates, fungal reproductive parameters, and environmental effects.

Host population density

Development and growth of parasitic fungal populations within the water column depends on critical thresholds of host algal population (algal cells ml^{-1}). At low levels of algal density, host cells are too far apart for effective zoospore migration and the establishment of new infections that are required to maintain a continued fungal population (Figure 8.11). Under these conditions, fungal multiplication is less than that of the host and the parasite is not able to establish itself. With increase in algal population, a critical threshold is reached (survival threshold) at which the fungal parasite is just able to maintain a stable population. In this situation, at least one out of the total number of zoospores produced per sporangium is able to successfully infect a new host cell. Further increase in algal population leads to a critical point (epidemic threshold) at which the fungal population increases exponentially, resulting in a fungal epidemic (Figure 8.11).

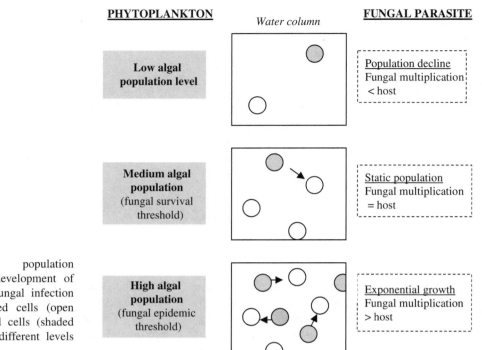

Figure 8.11 Algal population thresholds and the development of fungal populations: fungal infection (arrows) of uninfected cells (open circles) from infected cells (shaded circles) is shown at different levels of algal population

Fungal multiplication substantially succeeds algal growth, and rapid spread occurs throughout the phytoplankton population.

Algal threshold levels are not standard, but vary with different fungi and with the particular growth rates of alga and fungus at any point in time. Although the epidemic threshold values are normally quite high, fungal epidemics can occur at relatively low algal densities. In natural populations of the diatom *Asterionella formosa*, for example, fungal epidemics have been observed at densities down to 10 cells ml^{-1} (Lund, 1957).

Relative host/parasite growth rates

The rate of spread ($^{df}/_{dt}$) of a fungal infection in an algal population depends on the relative growth rates of the host and parasite (Bruning, 1991a), where:

$$^{df}/_{dt} = f(\mu_p - \mu_h) \qquad (8.1)$$

and f = the proportion of algal cells infected, μ_p = specific growth rate of the parasite, μ_h = specific growth rate of the algal host.

The above equation shows that:

- the development of an epidemic requires a higher growth rate of the parasite compared with the host,

- any environmental factors which influence these growth rates will also influence the development of an epidemic.

The importance of host population density in determining fungal growth rates has been noted previously (Figure 8.11). Maximum growth rate of the fungus ($\mu_{p_{max}}$) occurs at 'infinite' host concentration (where separation between host cells has no effect on fungal spread), and is a function of the zoospore production rate of the fungus. When $\mu_h > \mu_{p_{max}}$ the alga will outgrow the fungus, and the development of an epidemic is impossible – even if conditions for fungal infection are highly favourable.

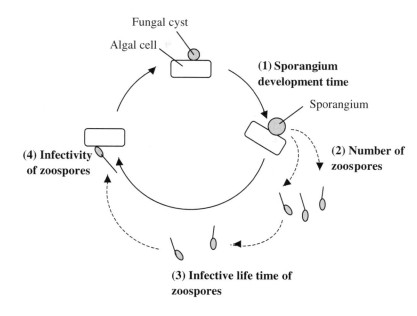

Figure 8.12 Reproductive parameters of a chytrid fungus which affect the development of fungal infection: in the case of *Rhizophydium* infection of *Asterionella*, the mean number of zoospores per sporangium varied from about 30 (2°C, high light conditions) to 2–3 (21°C, low light intensity); the infective lifetime of zoospores was 8 days at 6°C (fungal stages not drawn to scale)

Although the occurrence of an epidemic requires the parasite to 'outgrow' the host, it also requires active growth of the algal population. Recent field studies by Holfield (2000) have shown that an increase in infected phytoplankton cells was usually associated with a growing host population, and that conditions of algal decline often correlated with peak or decreasing levels of infection.

Fungal reproductive parameters

Four major aspects of the chytrid life cycle vary in their timing or the extent to which they promote the spread of fungal infection through the algal population (Figure 8.12). These are: the time taken for development of a mature sporangium from the initial cyst that develops, the average number of zoospores produced per sporangium, the length of time that free zoospores remain infective within the water medium (infective lifetime), and the ability of zoospores to achieve infection once they have contacted the host cell (infectivity).

These reproductive parameters vary with the physiological state of both host and parasite, and are strongly influenced by physical environmental aspects such as light and temperature. Increase

in light intensity, for example, has a major effect on three of these parameters (see next section, Figure 8.13).

Environmental effects on host/parasite interactions

A wide range of environmental factors have an influence on the growth of fungi within phytoplankton populations and the development of fungal epidemics (Van Donk and Bruning, 1992). These include indirect factors which separately affect algal (e.g., nutrient availability, algal competition) and fungal populations (e.g., grazing of fungi), as well as factors which directly influence the host/parasite interaction.

Light intensity, nutrient availability, and temperature all have direct influence on the parasitic process, affecting algal (μ_h) and fungal (μ_p) growth rates, the physiological state of host and parasite and the characteristics of key parameters within the life cycle (see previously).

Light intensity Conditions of low light intensity might be expected to promote the spread of fungal infection within the algal population due to the

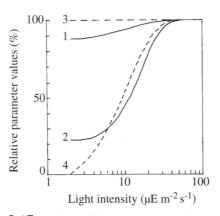

Figure 8.13 Effects of light on the growth parameters of the *Asterionella* parasite *Rhizophydium planktonicum*.

Differences in light intensity have major effects on the number of zoospores produced per sporangium (2) and the infectivity of zoospores (4), but much less influence on sporangium development time (1) and the infective lifetime of zoospores (3). The specific growth rate of the host (*Asterionella*) closely follows the trace of (4) (figure taken from Van Donk and Bruning, 1992, with permission from Biopress Ltd.)

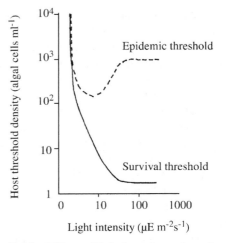

Figure 8.14 Effects of light intensity on host threshold densities for infection by the *Asterionella* parasite *Rhizophydium planktonicum*.

Host threshold densities for fungal survival and epidemic development both vary with light intensity (figure taken from Van Donk and Bruning, 1992, with permission from Biopress Ltd.)

reduction in algal growth rate. Light limitation also reduces the growth rate of the fungus, however, by affecting various stages in the growth cycle (Figure 8.13). Studies on the infection of *Asterionella* by *Rhizophydium* (Bruning, 1991a) have shown that low light causes a marked reduction in the number of zoospores produced per sporangium and in the infectivity of the zoospores, but has little effect on the development time of the sporangium and no apparent influence at all on the infective life time of zoospores (Figure 8.13). The importance of light on zoospore infectivity is particularly interesting. Studies on host-free zoospore suspensions have shown a complete cessation of motility below a critical light intensity, suggesting direct perception of light by these cells. Various authors have also separately reported an absence of zoospore attachment and a failure of germination in complete darkness.

The overall effect of low light intensity on spread of infection will depend on which growth rate has the greatest reduction – μ_h or μ_p. This has been investigated at different light intensities by determining the minimum density of algal (*Asterionella*)

cells necessary for fungal (*Rhizophydium*) survival (survival threshold) and epidemic development (epidemic threshold).

Fungal survival. The population density of algal cells required for fungal survival (survival threshold) falls dramatically as light intensity increases from about 2–50 μE m^{-2} s^{-1} (Figure 8.14).

In terms of fungal survival, reduced light alters the host–parasite balance in favour of the algal host, and appears to relate particularly to light-limitation of the host cells (Bruning, 1991a). The reason why the chytrid fungus requires higher algal levels to survive under light-limiting conditions is not entirely clear, but may relate to lower levels of algal photosynthetic products limiting the key fungal processes of sporangium development and zoospore infectivity noted previously.

The absence of fungal infection below 2 μE m^{-2} s^{-1} is ecologically significant since this light level will still support algal growth. This implies that, under natural conditions, part of the host population can be spatially protected from parasite attack. Deeper layers in the water column may provide an important refuge for algal cells, allowing host and parasite to co-exist in the same water body.

Epidemic development. The effect of light on epidemic development appears to be quite different and more complex than for fungal survival. With the *Asterionella/Rhizophydium* interaction, the epidemic threshold has a minimum value at a light intensity of about 10 μE m^{-2} s^{-1}, increasing with higher light intensity (Figure 8.14). These results suggest that the induction of a fungal epidemic is favoured by conditions of moderate light limitation, even though fungal growth is reduced under such conditions.

Nutrient limitation

Changes in available nutrients can have marked effects on the host–parasite balance. In the case of the *Asterionella/Rhizophydium* interaction (Bruning,1991b), for example, phosphorus deficiency reduces the growth rate of both organisms. Two major effects reduced fungal growth:

- P-limited host cells are less susceptible to infection with zoospores compared with non-limited algal cells;

- sporangia on P-limited algae had the same development time as non-limited host cells, but produced fewer zoospores.

In spite of these limitations to the parasite, zoospore production remained sufficiently high to maintain fungal growth at a higher level than algal growth, and phosphorus limitation of *Asterionella formosa* was found to favour epidemic development of *Rhizophydium planktonicum*. Under such conditions *Asterionella/Rhizophydium* interaction (Bruning,1991b), the host population density at which an epidemic could occur (epidemic threshold) was reduced by a factor of 2.5.

Temperature

Various investigators have reported that temperature may be an important factor in the occurrence and development of fungal parasites. The situation is again complex, since a reduction in temperature below optimum levels reduces the growth rate of both host and parasite.

With infection of *Asterionella* by *Rhizophydium* (Bruning, 1991c), the development time of sporangia increased at low temperatures. This was only partly offset by an increased number of zoospores per sporangium and an increased infective life time of the zoospores. As a result, the maximum growth rate of the fungus was less than that of the alga below 5°C, reducing the potential for epidemic development. Low temperature may also trigger a cessation of parasitic vegetative growth and lead to the formation of resistant asexual or sexual spores.

Inhibition of fungal epidemic development at low temperature has been observed under natural conditions by Van Donk and Ringelberg (1983). Conditions of ice cover (without snow) on Lake Maarsseveen (Holland) coincided with periods of inhibition of the chytrid *Zygorhizidium planktonicum*, a parasite of the diatom *Asterionella formosa*. Under these conditions, *Asterionella* was free of fungal attack but had adequate light, temperature, and nutrient conditions to allow it to grow and outcompete other diatom species.

9

Grazing activities in the freshwater environment: the role of protozoa and invertebrates

At the micro-level, grazing in the freshwater environment involves the removal and consumption of particulate material, including organic debris and living microorganisms. It is carried out by unicellular (protozoa) and multicellular (invertebrates) biota. Grazing is an important aspect of aquatic microbiology since it is a principal factor in the control of microbial populations and is also a major route in the transfer of biomass within the food chain.

This chapter initially considers the biological role of protozoa within the freshwater environment (Part A), including their taxonomic diversity, environmental adaptations, and feeding activities – which primarily involve the consumption of algae and bacteria. The grazing of microorganisms by invertebrates is then discussed, emphasizing the differences that occur between planktonic (Part B) and benthic (Part C) systems.

A. PROTOZOA

9.1 Introduction

Protozoa are important consumers of organic debris and microorganisms in many freshwater bodies, including natural ecosystems of both standing and flowing waters and man-made aquatic systems of economic importance such as wastewater-treatment plants. In most of these systems, protozoa compete with other grazing organisms (multicellular invertebrates) for food supply, and the relative grazing impact of these two major groups can be considered both in terms of biomass transfer and type of food supply.

9.1.1 Relative importance of protozoans, rotifers, and crustaceans in pelagic communities

Although protozoa were often ignored in earlier studies on freshwater ecosystems, their involvement in both pelagic and benthic food webs is becoming increasingly appreciated and the biology of such habitats cannot be properly described unless the protozoan community is taken into account.

In pelagic communities, for example, protozoa have frequently been excluded from grazing studies because rotifers and crustacean zooplankton were

Freshwater Microbiology: Biodiversity and Dynamic Interactions of Microorganisms in the Aquatic Environment   David C. Sigee
© 2005 John Wiley & Sons, Ltd   ISBNs: 0-471-48529-2 (pbk) 0-471-48528-4 (hbk)

simply considered to be more important in terms of population density, total biomass, productivity, rates of grazing, and nutrient regeneration (Pace and Orcutt, 1981). Protozoa also require methods of sampling which differ from those of macro-zooplankton. They are not collected, for example, in standard zooplankton nets (mesh openings of 64 μ or larger) or nets recommended for rotifers (35 μm mesh size). As a result, information about planktonic protozoa has often been incomplete and fragmentary. Many studies on protozoa are limited to brief intervals in the annual cycle or particular depths and are unrelated to the total zooplankton community.

Investigations involving complete analysis of both microzooplankton (protozoa) and macrozooplankton have typically shown just how important protozoan grazing is in the freshwater community. Studies by Pace and Orcutt (1981), for example, showed that protozoa (ciliates and amoebae) dominated the zooplankton community of Lake Oglethorpe (USA) – a warm monomictic lake. During winter mixing, ciliate protozoa were found at population counts of $1–8 \times 10^3 \, 1^{-1}$, and formed up to 32 per cent of the zooplankton community biomass. During summer stratification, populations of crustaceans declined, while abundance of protozoa and rotifers increased. Within the summer water column of this lake, ciliate protozoa reached particularly high bloom populations ($1–2 \times 10^5 \, 1^{-1}$) in the metalimnion – a zone of intense bacterial activity. During the period of these ciliate blooms (July–October), protozoa accounted for 15–62 per cent of the zooplankton biomass and made a significant contribution to grazing of microorganisms within the pelagic ecosystem.

9.1.2   Ecological role of protozoa

As consumers, protozoa are particularly involved in the ingestion of bacteria and algae and are important in a number of interrelated aspects of freshwater systems:

• the microbial loop, linking DOC production and bacterial populations to zooplankton,

• links between organic debris and higher organisms,

• controlling populations (and thus limiting the population increase) of algae and bacteria,

• recycling of inorganic nutrients,

• the carbon cycle.

9.2   Protozoa, algae, and indeterminate groups

Protozoa are unicellular eukaryotes. Together with the unicellular algae and slime moulds they make up the Kingdom Protista (Purves *et al.*, 1997; Table 1.2). They resemble the algae in typically having a complex cell structure, since the functions of the entire organism – nutrition, energy transformation, irritability, motility, and structural support are all contained within a single cell. They also resemble the algae in having wide variation in cell structure, consistent with diverse evolutionary lines within the group. Protozoa differ from algae in a number of general respects.

• *Motility.* Although many protozoa are permanently attached to a substratum (sessile), free-moving (pelagic) protozoa are almost all (but not invariably) actively motile. This is in contrast to pelagic algae, many of which are either non-motile (e.g., diatoms) or are motile via a non-active process (buoyancy: blue-green algae).

• *Cell wall.* Many algae have a hard cell wall, which is important in maintaining cell shape in a hypo-osmotic (freshwater) environment. The majority of protozoa do not have a rigid cell wall, and maintain osmotic integrity via the activity of a contractile vacuole.

• Algae are mainly autotrophic, protozoa are typically heterotrophic.

Protozoa have usually been distinguished from the algae primarily by their heterotrophic nutrition, obtaining energy and nutrients by the uptake of complex organic molecules – either as soluble or particulate matter. On this basis, algae and protozoa can be differentiated by the presence or absence of chlorophyll. This distinction between heterotrophic protozoa and autotrophic algae is not always clear, however, since some groups of organisms, such as dinoflagellates, have members that are mainly heterotrophic and others that are mainly autotrophic. In addition to this, some organisms are able to switch between autotrophic and heterotrophic states, depending on environmental conditions (Section 3.11). As a final complication, some protozoa are able to take up chloroplasts or photosynthetic algae into their cytoplasm and retain them in a functional photosynthetic state as part of a symbiotic relationship. This is particularly characteristic of the ciliate protozoa and is discussed further in Section 9.3.1.

The boundary between protozoa and algae is most problematical in flagellated organisms, where the designation of particular groups containing both green and colourless representatives may be difficult to interpret for the above reasons. In this book, three such groups of flagellates – the dinoflagellates, cryptomonads, and chrysophytes are arbitrarily considered to be algae, largely on the basis that most organisms within these groups are predominantly photosynthetic. Heterotrophic flagellates are thus a diverse group of organisms, including organisms in the above groups which have lost their chlorophyll and become secondarily heterotrophic – plus organisms which are more properly regarded as protozoa, with no recent autotrophic ancestors.

9.3 Taxonomic diversity of protozoa in the freshwater environment

Free-living protozoa can be separated into three main groups – ciliates, flagellates, and amoeboid forms. These are primarily distinguished by their motility (via cilia, flagella, or amoeboid movement) but also differ in terms of internal fine structure, method of feeding, and cell size.

9.3.1 Ciliates

Ciliates are a highly successful group of microconsumers which are present in most freshwater habitats. These protozoa move by cilia, which are internally identical to flagella, but are shorter and tend to occur in larger numbers. Cilia also differ from flagella in their beat pattern, which is a two-phased combination of motor and recovery stroke.

The cilia, which occur in lines (or kineties) at some stage in the life cycle, cause water to move parallel to the cell surface – which is important for both locomotion and feeding. The large size of ciliates compared with other protozoa (the majority are 20–200 µm in length), combined with their widespread occurrence, active motility, and variety of forms, means that these organisms are often the most obvious protozoa that are observed when carrying out microscopical examination of environmental samples.

Ciliate diversity and adaptations to the freshwater environment

Table 9.1 shows some of the major taxonomic groups within the ciliate protozoa. These show different evolutionary strategies within the aquatic environment in terms of microenvironment (planktonic or benthic), life style (motile, free-floating, or sessile) and mode of nutrition (e.g., filter feeders or active predators). These adaptive strategies involve various evolutionary modifications, prominent among which are changes to the ciliary apparatus.

The simplest and possibly archetypal situation in ciliates is where cilia are present over the whole organism and are primarily involved in locomotion. Although many protozoa such as *Paramecium* (Figure 9.1) still retain this feature, they also show various adaptations of their ciliature. These organelles have become modified for more specialized locomotion and for feeding in two main ways as follows.

- Cilia in the vicinity of the mouth (buccal cilia) have developed for feeding. In Oligohymenophora, a line of cilia close to the cytostome (mouth) become developed as an 'undulating membrane,' which is used to intercept particles being carried in currents of water. Cilia on the body surface (somatic cilia) continue to be used for locomotion.

Table 9.1   Major groups of free-living ciliate protozoa

Group	Examples	Motility and microhabitat	Food
Colpopids. Mostly filter-feeders, using tightly packed cilia grouped around the mouth	*Bursaria* *Cyrtolophosis*	Motile – many planktonic	Large food particles such as algae – dinoflagellates, diatoms
Cyrtophores. Have a cluster of strongly developed microtubular rods (nematodesmata) supporting the oral structure	*Nassula* *Pseudo-microthorax*	Motile – planktonic	Often larger algae such as diatoms and filamentous blue-greens
Haptorids. Flattened scavenging or predatory ciliates with killing or holding extrusomes[†] around the mouth region	*Amphileptus* *Loxophyllum*	Motile – many present on benthic benthic substrates	Other protozoa
Oligohymenophora. Specialized buccal ciliature comprising three membranelles and an undulating membrane			
(a) Hymenostomes: short membranelles and undulating membrane	*Colpidium* *Tetrahymena*	Motile – often present in substrate detritus	Typically bacteria
(b) Peritrichs: buccal ciliature forms one or more wreaths around the anterior part of the cell	*Vorticella* *Ophyridium*[*]	Non-motile, sessile – attached via stalk	Typically bacteria
(c) Peniculines: elongate membranelles, weakly developed undulating membrane Star-shaped contractile vacuole complexes	*Paramecium*[*] *Frontonia*[*]	Motile – often present as scavengers in substrate	Bacteria and small algae
Polyhymenophora. Feeding occurs via a band of membranelles (AZM)[**]			
(a) Hypotrichs: usually flattened Walk on substrate or swim using cirri[‡]	*Aspidisca* *Euplotes*	Motile – on benthic substrates	Bacteria and other small particles
(b) Heterotrichs: some of the largest protozoa Move via somatic cilia, clearly arranged in kineties	*Stentor* *Climacostomum*[*]	Motile or non-motile (attached)	Bacteria and other small particles
(c) Oligotrichs: somatic cilia are absent or reduced. AZM[**] used for feeding and locomotion	*Halteria* *Strombidium*	Motile – typically planktonic in open water	Often larger particulate matter such as diatoms
Prostomes. Apical mouth, used for ingestion of detritus	*Urotricha* *Prorodon*[*]	Motile – often in sediments as scavengers	Large detritus – e.g., decomposing animal remains
Suctoria. The main (trophic) stage of life cycle lacks cilia Feeding via arms with extrusomes[†]	*Podophrya* *Tokophrya*	Sessile trophic stage, often with stalk	Motile protozoa such as ciliates

This table gives information on some of the major ciliate groups, but is not meant to be fully comprehensive. Two representative freshwater genera are given for each group, and descriptions of motility and feeding are general to the group (adapted from Patterson, 1996)
[*]Genera with species containing symbiotic green algae.
[†]Extrusome: an organelle with contents which can be extruded to kill or capture prey.
[**]AZM: adoral zone of membranelles, composed of closely grouped cilia.
[‡]Cirrus: a locomotor structure formed from a tight cluster of individual cilia that move as a single entity.

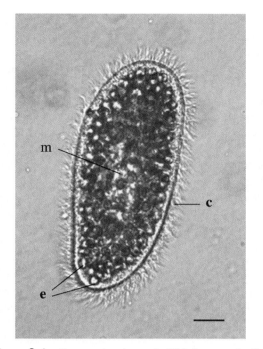

Figure 9.1 *Paramecium bursaria*. This large peniculine ciliate is highly motile, moving over substrata via cilia (c) which cover the entire cell. The organism has dual nutrition, able to ingest food particles (heterotrophy) but also carrying out photosynthesis (autotrophy) via zoo-chlorellae (genus *Chlorella*). These endosymbiotic green algae (e) appear as distinct granules, filling the entire cytoplasm and giving the cell a bright green colour. The algae can be isolated and have been grown in culture for studies on virus infection (Section 7.7). The pale central area indicates the position of a large macronucleus (m) (scale = 25 μm)

- Some groups of cilia have become arranged in tight clumps, which generate more force than groups of separated cilia. These compound organelles have evolved in two main ways – forming 'membranelles' (generating currents towards the mouth) and 'cirri' (walking appendages in the hypotrichs). The role of cilia in filter feeding is discussed later in this section.

In some protozoa the presence and activities of cilia are much reduced. Suctorian protozoa, for example, have entirely lost cilia from the adult phase of the life cycle as an adaptation to a sessile

existence (Section 3). The ciliate identity of these organisms is only evident from the brief presence of cilia in early developmental stages.

Planktonic and benthic ciliates

Some groups have become primarily adapted to the pelagic environment, moving within the main water body in search of food. These include the colpopids, cyrtophores, and oligotrichs (Table 9.1) – all of which are able to consume phytoplankton such as diatoms, dinoflagellates, and blue-green algae. Some of these planktonic ciliates have developed very powerful motility, including the oligotrichs – which have lost their somatic cilia and developed a prominent adoral zone of membranelles (AZM) for both feeding and locomotion.

Benthic ciliates are common in sediments and detritus, where they often show active movement over the substratum as they scavenge for debris, bacteria, and other protozoa. Hypotrichs such as *Euplotes* are particularly active, with the development of cirri as locomotory organelles for use on solid surfaces.

Motile and sessile ciliates

Motile ciliates may be planktonic or benthic, and include both filter feeders and predatory organisms. Some motile ciliates are closely associated with aquatic (e.g., algal) surfaces bearing epiphytic bacteria and debris, and are closely tied to particular microhabitats.

Many sessile ciliates are attached to decomposing detritus, taking advantage of the extensive populations of bacteria which occur in those regions. Such organisms are an important part of the benthic community, and are consumed by various invertebrate predators (Table 9.10) along with the organic substrate to which they are attached. Although sessile protozoa are normally thought of in terms of the benthic environment, they are also well represented within the main water body, where they may be attached to phytoplankton colonies (e.g., the peritrich *Vorticella*) or to freshwater

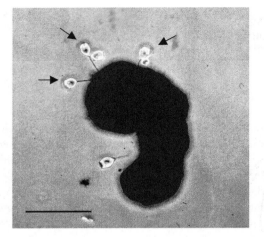

Figure 9.2 Colony of the blue-green alga *Gomphosphaeria* with epiphytic ciliate protozoa (*Vorticella* sp.): attachment to the surface of the alga via an extensile stalk can clearly be seen with three of the protozoa (→) (scale = 100 μm)

fauna such as the gills of the shrimp *Gammarus* (e.g., the suctorian *Dendrocomoetes*). These organisms are able to exploit the advantages of the pelagic environment, escaping benthic predators and gaining access to localized populations of planktonic bacteria, without the requirement for continuous motility and energy expenditure which is necessary for unattached pelagic forms.

Sessile planktonic protozoa such as *Vorticella* are frequently seen associated with colonies of blue-green algae (Figures 9.2 and 6.23), where they are part of the microcosm community which includes diverse populations of epiphytic bacteria, fungi, protozoa, and other algal cells. These protozoa have a complex life cycle which involves alternating sessile and free swimming motile phases. The latter is important for colonization of new algal colonies and dispersal thoughout the planktonic environment.

Filter feeders and predators

Filter feeding has evolved as a particularly successful mode of nutrition in the ciliates, with several groups possessing buccal cilia to generate currents

of water from which suspended particles may be isolated and ingested. In the colpopids, this involves the use of tightly packed cilia around the cytostome (mouth) to remove quite large phytoplankton cells – including diatoms and dinoflagellates. Two other major groups, the oligohymenophora and the polyhymenophora, have evolved more specialized buccal apparatus, and are particularly successful in filtering the extensive bacterial populations which develop in organic-rich detritus and sediments. Oligohymenophora have developed three membranelles near the cystostome, while in polyhymenophora the buccal cilia form a band (the AZM) which extends from the anterior of the cell to the site of food ingestion (the cytostome). The role of the AZM can be seen in Figure 9.3, where cells of the heterotrich ciliate *Climacostomum* orientate their AZM against the organic debris on which they are feeding. Some filter feeders (e.g., colpopids) carry out their activity during active motility, while others (e.g., peritrichs) are sessile, bringing the suspended material to themselves via powerful currents.

Predation, the active pursuit and ingestion of motile prey has evolved in some ciliate groups. The haptorids are a prominent group of scavenging and predatory ciliates, capturing and killing their prey by the use of organelles called extrusomes, which extrude their contents on contact and immobilize the target organism. A type of predation has also evolved in the suctoria, a remarkable group of ciliates which has almost completely suppressed the use of cilia for both feeding and locomotion. The life cycle of these organisms is dominated by a sessile, non-ciliate feeding (trophic) phase, which catches prey via extended arms or tentacles bearing extrusomes.

Symbiotic associations of ciliate protozoa

The uptake of algae by protozoa normally occurs as part of their heterotrophic nutrition, and leads to death and digestion of the photosynthetic organisms. Ciliate protozoa such as *Nassula*, for example, are important consumers of blue-green algae (Case Study 10.4) and are frequently green due to the presence of partly-digested algal remains, which

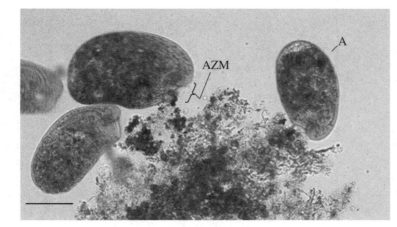

Figure 9.3 Group of ciliate (*Climacostomum*) cells grazing on organic debris. *Climacostomum* belongs to a group of ciliates (spirotrichs) which feed by using an anterior band of fused cilia, the adoral zone of membranelles (AZM). In this micrograph, three *Climacostomum* cells are actively removing particulate matter from a clump of organic debris; one of the organisms (A) is directly associated with the debris via its AZM (scale = 50 μm)

can be clearly seen under the light microscope. With completion of algal digestion, the protozoa return to a colourless state until the next food intake.

In contrast to this temporary greening due to periodic cycles of ingestion and digestion, other protozoa are permanently green due to the continued presence of algae within the cell as part of a symbiotic relationship. One of the best-known examples is *Paramecium bursaria*, which contains photosynthetically-active zoochlorellae (Figure 9.1). These symbiotic algae contribute to the build-up of organic carbon in the protozoon, and benefit themselves from the metabolic coupling and protected environment supplied by their host cell. In these cells the algae are not digested, but persist, enclosed within membranes in the host cytoplasm. Symbiotic associations typically involve green algae (Chlorophyceae) and are particularly common in the ciliates. Some of the genera known to have symbiotic associations are indicated in Table 9.1, and include sessile benthic protozoa (*Ophyridium*), actively motile protozoa living in sediments (*Paramecium*, *Climacostomum*), and planktonic organisms present in the water column (*Strombidium*, *Frontonia*). Despite the permanent presence of photosynthetic algae in their cytoplasm, these protozoa are all active feeders – continuing their heterotrophic mode of nutrition, but benefiting from the presence of autotrophic symbionts.

In some situations, the population of symbiotic protozoa may be ecologically important, making a major contribution to the generation of biomass. Hecky and Kling (1981), for example, reported that in Lake Tanganyika the biomass of *Strombidium viride* equalled the phytoplankton biomass, indicating that this oligotrich ciliate (containing symbiotic algae) was playing a significant role as a primary producer. Symbiotic ciliates may also be ecologically important in benthic environments. The peritrich ciliate *Ophyridium* is commonly found in ponds and shallow lakes, where it is attached to stones and plant surfaces, forming spherical gelatinous colonies which may measure several centimetres across.

9.3.2 Flagellate protozoa

Flagellate protozoa are characterized by the following properties.

- *The possession of one or more flagella.* These motor organelles have a similar internal fine structure to cilia, but are typically fewer in number, attached only at the anterior of the cell and are at least as long as the cell. They occur in multiples of two, one of which may be very short. In addition to motility, flagella may also function in feeding (generation of a water current for particle filtration) and attachment to the substratum.

- *Typically small size.* Many flagellate protozoa have cell diameters <20 µm. This has important implications for feeding activities, limiting the cells to ingestion of fine particulate matter (e.g., bacteria) or to uptake of dissolved organic material over the cell surface.

- *Complex taxonomical relationships.* The taxonomic relationships of flagellate protozoa have proved highly problematical (Patterson, 1996) for various reasons. In addition to the inherent evolutionary complexity of this group, the small size of many of these organisms makes light microscope examination difficult. In some situations, electron microscopy or molecular analysis may be required.

A final source of confusion, as mentioned previously, is the status of flagellate groups such as the euglenids which contain both photosynthetic and colourless forms. Although these are typically designated as algae, the non-photosynthetic members may be a major source of heterotrophic activity in aquatic environments. One such organism, *Peranema* (Figure 9.4), is a colourless member of the euglenid flagellates and is a prominent member of the benthic community. This large and highly active phagotroph is able to manipulate food particles (bacteria, other protozoa, detritus) into the cell via a pair of ingestion rods.

The major groups of free-living colourless flagellates (no photosynthetic members) are shown in Table 9.2. These are almost all benthic protozoa, feeding on bacterial populations in sediments and on bacterial biofilms attached to solid surfaces (Patterson and Larsen, 1991).

Some of the groups (bicosoecids, collar flagellates) are typically sessile, with individuals attached to the substratum enclosed in a protective covering or lorica. Other groups (bodonids, cercomonads, diplomonads) use their flagella to migrate over the substratum, consuming organic debris or bacteria as they travel. Pelobionts are unusual in adopting an amoeboid form of movement, relegating their flagella to a non-motile appendage.

As with the ciliates, food is obtained either by predation or filter feeding. Predation normally

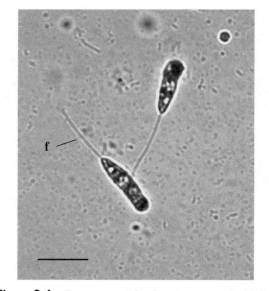

Figure 9.4 *Peranema trichophorum* – a euglenid flagellate. A highly active phagotrophic flagellate, which does not readily swim but glides along the ground using a prominent anterior flagellum (f). *Peranema* also has a short second flagellum which is recurrent and is extremely difficult to see. This species is colourless and feeds on bacteria and small protists (scale = 50 µm)

involves active (motile) location of food sources and uptake of particulate matter (phagocytosis) via cytoplasmic extensions (pseudopodia). Filter feeding is normally carried out by sessile flagellates and involves the creation of filter currents.

Collar flagellates are a particularly common group of filter feeders. The collar is a cytoplasmic extension at the base of the single anterior flagellum and is composed of very fine pseudopodia, acting as a filtration system. Trapped bacteria are subsequently drawn into the cell by pseudopodial activity. Another interesting group are the pedinellids, which alternate between a non-trophic motile state and a trophic sessile state. These organisms usually have a permanent stalk which, in the case of *Actinomonas* (a commonly-occurring freshwater genus), simply trails behind during swimming. This organism enters the sessile state by the attachment of the stalk to the substratum and subsequently grows arms or tentacles around the flagellum. Suspended

Table 9.2 Major groups of free-living colourless flagellates

Group	Examples	Motility and microhabitat[*]	Food
Bicosoecids. Two anterior flagella, one directed backwards for attachment to base of lorica[**]	*Bicoseca*	Sessile, attached by lorica	Suspended bacteria
Bodonids. Small, biflagellated Have kinetoplast[‡]	*Bodo* *Rhynchomonas*	Motile (gliding) or sessile	Typically ingest individual attached bacteria via a discrete mouth
Cercomonads. Two flagella, one anterior (for motility) the other trails behind	*Cercomonas* *Heteromita*	Motile (gliding)	Feed on bacteria by pseudopodial engulfment
Collar flagellates. Single flagellum, surrounded by a basal collar of cytoplasmic fingers Single or colonial Some have a lorica[**]	*Codonosiga* *Monosiga*	Mostly sessile	Filter suspended bacteria or other small particles
Diplomonads. Cells with two nuclei and two clusters of four flagella	*Trepomonas* *Hexamita*	Motile	Consume bacteria by phagocytosis or organic matter by absorption over surface (osmotrophic)
Pedinellids. Single apical flagellum Usually with a stalk	*Actinomonas* *Pteridomonas*	Individual cells alternately motile or sessile	Filter feeding of bacteria when sessile
Pelobionts. Amoeboid cells with a long stiff flagellum (not for motility) *Unassigned genera[†]*	*Mastigamoeba* *Pelomyxa*	Motile, by amoeboid movement	Ingest bacteria and algae by phagocytosis

This table gives information on some of the major colourless flagellate groups, but is not meant to be fully comprehensive. Groups containing photosynthetic flagellates are not included. Two representative freshwater genera are given for each group and descriptions of motility and feeding are general to the group (adapted from Patterson, 1996)

[*]Almost all of the flagellate groups in this table are typical of benthic microenvironments, rich in detritus and bacterial populations.

[†]There are about 70 genera of heterotrophic flagellates which cannot be confidently assigned to any of the recognized groups (Patterson and Larsen, 1991)

[**]Lorica: an organic or inorganic casing or shell, incompletely surrounding an organism and usually loose-fitting.

[‡]Kinetoplast: a large body of mitochondrial DNA.

particles, carried by flagellar currents to the arms, are trapped by extrusome secretions and subsequently ingested. The arms are withdrawn later when the cells enter the swimming phase.

9.3.3 Amoeboid protozoa

Amoeboid protozoa, are common organisms in the freshwater environment, occurring both as planktonic and benthic forms. Although amoebae are a diverse group of organisms (Table 9.3), they have a number of general characteristics:

- they lack cilia or flagella,

- phagocytic feeding and locomotion are typically achieved by cytoplasmic extensions from the cell surface (pseudopodia),

- many amoebae produce a thick external casing, referred to as a test or lorica; this may either be organic (with or without adhering material) or may be secreted as inorganic elements.

As with flagellate protozoa, the demarcation of amoeboid protozoa can be problematic, and

Table 9.3   Major groups of free-living amoeboid protozoa

Group	Examples	Motility and microhabitat[*]	Food
HELIOZOAN AMOEBAE star-like protozoa with radiating arms (stiffened pseudopodia)			
Actinophryids: arms taper from base to tip	*Actinophrys* *Actinosphaerium*	Free-floating, planktonic	Consume motile protozoa, unicellular algae and some metazoa (e.g., rotifers)
Centrohelids: arms thin, do not taper and have extrusomes Microtubular rod supporting the arm terminates on a central granule	*Acanthocystis* *Heterophrys*	Free-floating, planktonic	Algae
Desmothoracids: cells live within a perforated lorica from which arms project	*Clathrulina* *Hedriocystis*	Sessile, benthic	Consume fine particulate material including bacteria
RHIZOPOD AMOEBAE produce temporary pseudopodia			
Euamoebae: have one or more broad pseudopodia, but lack a shell or lorica	*Amoeba* *Mayorella*	Motile, often on sediments and organic matter	Voracious scavengers and predators, consuming detritus and microbes (e.g., bacteria, diatoms)
Heterolobosea: have an amoeboid and flagellated stage in the life cycle A mixed assemblage of organisms	*Naegleria* *Tetramitus*	Motile, planktonic	
Nucleariid filose amoebae: have thin filose pseudopodia arising at any part of the body Pseudopodia not stiffened and without extrusomes	*Nuclearia* *Pompholyxophrys*	Motile, planktonic	Feed on algae, including blue-greens
Vampyrellids: flattened amoebae with delicate filose pseudopodia arising from margins	*Arachnula* *Vampyrella*	Motile, planktonic or on sediments	Attack fungi and algae
Testate amoebae: amoeboid organisms within a shell of organic matter Pseudopodia arise from one or two apertures	*Amphitrema* *Arcella*	Sessile, present in sediments	Range of particulate material

This table gives information on some of the major groups of amoeboid protozoa, but is not meant to be fully comprehensive. Two representative freshwater genera are given for each group and descriptions of motility and feeding are general to the group (adapted from Patterson, 1996)

classification within the group is complex. The absence of cilia and flagella, for example, does not hold for one group normally included in the amoebae – the heterolobosea (Table 9.3). Members of this heterogeneous group, which includes the facultative human parasite *Naegleria*, have both amoeboid and flagellated stages in their life cycle. Some confusion in the identification of actively amoeboid cells may also arise in relation to the acrasid slime moulds, which have amoeboid 'swarmers' as part of their life cycle (Olive, 1975). Similar confusion (in relation to flagellate protozoa) may also occur with myxogastreid (myxomycete) slime moulds, which have flagellated swarmers. With both groups of slime moulds, the swarmers are rarely encountered in the freshwater environment, and are really part of a short-term 'colonial assemblage' rather than a long-term population of individual independent cells.

Evolutionary strategies in the amoeboid protozoa: heliozoan and rhizopod amoebae

Complexity within the amoebae reflects diverse evolutionary lines of adaptation to the aquatic environment. Phylogenetic development of the pseudopodium, the major taxonomic feature of this group of protozoa, has proceeded along two principal lines. This has lead to the evolution of two major groups of organisms, the heliozoan and rhizopod amoebae, each with their own distinct feeding strategy.

Heliozoa: passive diffusion feeders These star-like protozoa have stiffened pseudopodia (axopodia) radiating out from a spherical cell. Light microscopy of large heliozoan protozoa such as *Actinosphaerium*, show the axopodia to be anchored in the central multinucleate region of the cell, which is surrounded by a peripheral vacuolate zone of cytoplasm (Figure 9.5(a)).

Pseudopodia in these organisms are not involved in active motility and have evolved solely for feeding purposes. These protozoa are passive 'diffusion feeders', relying on movements (e.g., drifting or swimming) of the prey to make contact with the pseudopodia. This is then followed by capture and ingestion. The pseudopodia are permanent rod-like structures with an internal microtubular skeleton, and increase the surface area for food contact by extending into the freshwater environment. Small adhesive capture organelles (extrusomes) move along the pseudopodial arms, and are used to hold potential prey.

Evolution of diffusion feeding in amoebae has given rise to three distinct groups – the

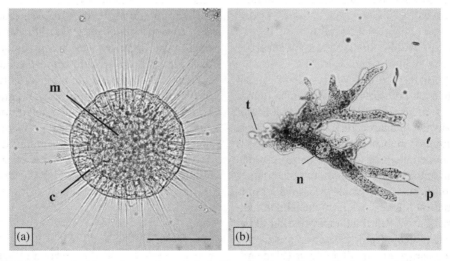

Figure 9.5 Heliozoan and rhizopod amoebae: (**a**) *Actinosphaerium eichornii*. A large heliozoan amoeba. This organism has many arms (axopodia), which are anchored in the central multinucleate zone (m) of the cell. The arms pass through the peripheral vacuolate cytoplasm (c), emerging as elongate structures, tapering from the base to the tip. *Actinosphaerium* consumes motile protozoa, unicellular algae, and small metazoa (e.g., rotifers) which contact, and are immobilized by, the axopodia (scale = 100 μm); (**b**) *Amoeba proteus*. A very large rhizopod amoeba, which is typically active in prey encapsulation and phagocytosis. This specimen is actively moving by extending its pseudopodia (p) to the right, leaving a trailing posterior region (t). The protoplasm is granular and contains a prominent nucleus (n) (scale = 200 μm)

actinophryids, centrohelids, and desmothoracids (Table 9.3). The first two have a typically planktonic life style, drifting within the water column or over bacterial-rich sediments. Desmothoracids are more commonly found as attached organisms in sediments, individually enclosed within a perforated lorica.

Rhizopod amoebae: active predators In rhizopod amoebae, pseudopodia occur as temporary extensions from the cell and are involved in movement and ingestion of food particles. These protozoa have mostly adopted a strategy of active predation, moving within the water column or on sediments in search of food particles. This is seen particularly well in species of *Amoeba* (Figure 9.5(b)) which are highly active in benthic environments in the phagocytosis of bacteria and other fine particulate organic material. Although most amoebae feed on fine particulate matter, some are able to ingest much larger material. The nucleariid and vampyrellid amoebae feed on large planktonic algae, including filamentous blue-green algae, and have potential as biocontrol agents (Chapter 10). One particular species, *Nuclearia delicatula*, attaches to the end of filamentous blue-greens and actively ingests the algal cells, rapidly reducing the algal population within the laboratory culture or within the water column (Figure 10.11)

The ability to form pseudopodia is widespread among eukaryotes, and protozoa which have these structures are not necessarily closely related. Major categories of rhizopod amoebae are distinguished by the shape and number of pseudopodia (Table 9.3) and include some amoebae with one or more broad pseudopodia (euamoebae), while others have larger numbers of thin (filose) pseudopodia (nucleariids and vampyrellids). One group of amoebae, the testate amoebae, have adopted a free-floating existence, living within shells of organic matter on sediments.

9.4 Ecological impact of protozoa: the pelagic environment

Reference to Tables 9.1 to 9.3 emphasizes the contribution of all major groups of protozoa to the planktonic freshwater communities of lakes, reservoirs, and other standing waters. Planktonic protozoa have two important characteristics which are necessary for their continued existence within the main water body.

- They have some mechanism to avoid sedimentation out of the water column. This may involve active motility, buoyancy or attachment to other planktonic biota such as algae and zooplankton. Many freely motile protozoa are able to position themselves at particular depths in the water column.

- They are able to capture and ingest particulate matter in the water column – including organic debris, picoplankton and large algal colonies. The only exception to this are saprotrophic protozoa, which are able to absorb soluble organic compounds through their body surface.

9.4.1 Positioning within the water column

As with motile algae, protozoa are able to position themselves vertically within the water column in relation to optimum environmental conditions. In the case of heterotrophic protozoa, however, positioning is not primarily in relation to light intensity but is a response to other external parameters such as food supply and oxygen concentration.

Many protozoa respond to gradients of oxygen concentration, localized concentrations of which may act either as an attractant or repellant. The ciliate *Loxodes* is a microaerophilic organism, preferring oxygen concentrations of about 5 per cent saturation. This protozoon migrates out of sediments as they become anaerobic at the onset of lake stratification, and positions itself at an appropriate depth below the oxycline (Figure 9.6) as stratification develops. These organisms migrate up or down the water column by negative or positive geotaxis and have well-developed geo-receptors (statoliths) which contain barium particles. The geotactic response varies with oxygen concentration (Figure 9.6), leading to the accumulation of *Loxodes* populations at a particular depth. Movement out of

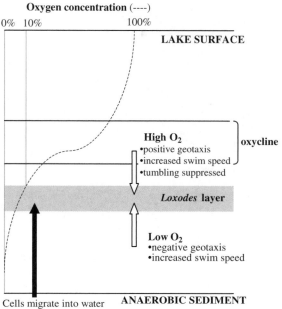

Figure 9.6 Positioning of the micro-aerophilic ciliate protozoon *Loxodes* within the water column: organisms return to the *Loxodes* layer by positive or negative geotaxis when they move into regions of high or low oxygen concentration (figure based on information from Fenchel, 1987)

this region leads to an appropriate geotactic response, with increased motility and reduced random motion. The latter involves periodic tumbling activity, when depolarization of the surface membrane results in an abrupt and random change of direction – and is typical of the movement of many protozoa. The avoidance of even moderate oxygen concentrations by microaerophilic organisms such as *Loxodes* may be explained in terms of the intracellular production of harmful oxygen radicals, which are generated photochemically from molecular oxygen within pigmented species (Finlay, 1981).

Although protozoa are not typically attracted to light, the presence of endosymbiotic algae may promote this. *Paramecium bursaria*, for example, containing symbiotic zoochlorellae (Figure 9.1), is strongly phototactic. This behaviour is lost if the organism no longer contains algae, and may be

mediated in part by oxygen which is photosynthetically produced by the algal symbionts. Some non-symbiotic protozoa, particularly heterotrich ciliates, may also respond to light (Fenchel, 1987). The presence of brightly-coloured flavin and hypericin pigments in these organisms typically correlates with a strong negative reaction to light.

9.4.2   Trophic interactions in the water column

Within the planktonic community, protozoa are often an important component of the food web, where they are particularly involved:

- in the microbial loop (Figure 1.9), ingesting bacteria which have taken up dissolved organic carbon (DOC) derived from phytoplankton and zooplankton;

- as herbivores within the classical food chain (Figure 1.10), able to ingest a wide range of algae, from unicellular to colonial forms; transfer of biomass then occurs with ingestion of protozoa by zooplankton and other invertebrates.

The adaptations and ecological role of planktonic protozoa in standing waters will be illustrated by reference to one particular group of organisms – the heterotrophic nanoflagellates.

9.5   Heterotrophic nanoflagellates: an integral component of planktonic communities

Heterotrophic flagellates are an important ecological assemblage of protozoa and include bacterivores, herbivores (consuming algae), detritivores (consuming non-living organic debris), and saprotrophs (absorbing soluble organic compounds). Two major groups can be recognized:

- microflagellates, with a size range of 20–200 µm, including algal derivatives such as euglenids and dinoflagellates, and

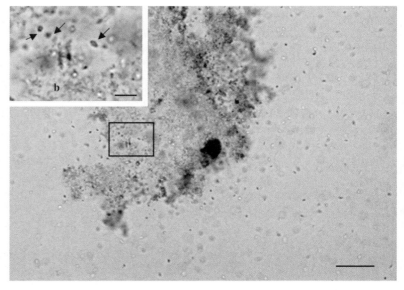

Figure 9.7 Population of *Bodo saltans* – a heterotrophic nanoflagellate: some of the flagellates have penetrated the decomposing organic material, ingesting bacteria within the debris, while others are ingesting bacterial cells which have been released into the general medium. Individual *Bodo* cells have a large posterior flagellum and measure about 5 μm in diameter (scale = 50 μm) *Inset*: detail from inside debris, showing flagellate cells (arrows) adjacent to a bacterial region (b) (Scale = 10 μm)

- nanoflagellates, 2–20 μm, including choanoflagellates and chrysomonads.

Heterotrophic nanoflagellates (HNFs) have received particular attention in recent years for their role in feeding on bacteria (Berninger *et al.*, 1991a, b). This activity is illustrated in Figure 9.7, where a dense population of *Bodo saltans* has accumulated around a piece of organic debris which is a rich source of bacteria. HNFs are able to feed more effectively than other bacterivores on relatively dilute suspensions of bacteria (10^5–10^6 ml^{-1}) and are the major consumers of picoplanktonic microorganisms in a variety of freshwater environments.

The ecological impact of planktonic nanoflagellates will be considered in relation to their population levels, taxonomic diversity, bacterial grazing rates, and distribution within the water column.

9.5.1 Enumeration of nanoflagellate populations in aquatic samples

Recent advances in the methods used to visualize and count nanoflagellates in aquatic samples have indicated much higher levels (number per unit volume) than had been previously realized. Two basic approaches have been used to enumerate the abundance of nanoflagellates – viable counts and total counts.

Viable counts

This technique, also referred to as 'culture counts' and 'most probable number method' (MPN) involves serial dilution of water samples followed by incubation of each dilution with a suspension of food organisms. For heterotrophic nanoflagellates, bacteria have been routinely used as the 'food organism' and have been added directly to the different dilutions to supplement the natural bacterial flora. In theory, any dilution that contains a single flagellate cell will result in flagellate growth and a rise in the flagellate population. The flagellate count in the original water sample can be determined from the proportion of dilutions exhibiting flagellate growth, the dilution interval, and the number of replicate dilutions cultured, using well-established probability tables.

As with the estimation of bacterial populations, viable counts of flagellates suffer from one major drawback – not every organism will grow and multiply in laboratory culture. For this reason, the technique usually provides a low estimate of the flagellate population. The technique also suffers from the disadvantages of being labour-intensive and requires the examination of a large number of laboratory cultures. In spite of these drawbacks, the viable count technique provides the only method to estimate the number of flagellates of particular trophic types (e.g., bacterivores, herbivores). It also provides flagellate mono-cultures for taxonomic and biochemical characterization.

Total counts

Direct microscopical counts using epifluorescence are widely used to enumerate total nanoflagellate populations. In practice, formaldehyde or glutaraldehyde fixatives are added to the water samples immediately after collection to preserve the flagellates, and the samples are kept in cold conditions (0–4°C) to retain natural chlorophyll fluorescence (autofluorescence). Under the fluorescence microscope, heterotrophic flagellates can be visualized (along with other microorganisms) using various nucleic acid-binding fluorochromes (e.g., acridine orange, DAPI) and can then be distinguished from autotrophic organisms by the absence of chlorophyll autofluorescence. Nanoflagellates can be distinguished from microflagellates and other non-chlorophyll containing organisms in terms of size

and shape, and can be enumerated by visual counts or by automated techniques such as computer image analysis, flow cytometry, or microphotometry. Full taxonomic analysis of HNF communities requires examination of these organisms in an unfixed, unstained condition and is best carried out on live samples.

9.5.2 Taxonomic composition of HNF communities

Heterotrophic nanoflagellates are a diverse taxonomic assemblage. A survey of 55 lakes in North Germany (Auer and Arndt, 2001), for example, showed that most nanoflagellates in this geographic region belonged to five major taxonomic categories (Table 9.4), of which chrysophytes and an assemblage of unidentified forms were the most abundant. Other important taxonomic groups which were frequently represented included choanoflagellates, bicosoecids, and bodonids (Figure 9.7).

9.5.3 Abundance and control of flagellate populations

The range of flagellate abundance in freshwater habitats differs between micro- and nanoflagellates. Heterotrophic microflagellates vary in densities up to about 10^2 cells ml^{-1}, while

Table 9.4 Major heterotrophic nanoflagellate groups in temperate lakes (data obtained from analysis of HNF communities in 55 Northern German lakes, taken from Auer and Arndt, 2001)

Taxon	Percentage occurrence in samples	Typical genera	Mean size (μm)
Chrysophytes	93%	*Spumella*	5.2
		***Paraphysomonas**[*]*	**14.2**
Choanoflagellates	57%	*Monosiga*	4.9
		Codonosiga	6.1
Bicosoecids	54%	*Bicosoeca*	5.5
Bodonids	41%	*Bodo*	3.3
Unidentified assemblage	87%	*Kathablepharis*	4.7
		***Quadricilia**[*]*	**9.8**
		***Aulacomonas**[*]*	**16.9**

[*]**Large-bodied HNFs**, showing highest abundance in hypertrophic lakes and in spring.

nanoflagellates are much more prolific, rising to levels of 10^5 cells ml^{-1}.

At any point in time, the population of nanoflagellates within a particular aquatic system depends on the balance between:

- top-down control, grazing pressure from higher trophic levels, particularly ciliate protozoa and zooplankton, and

- bottom-up control, food availability, particularly the supply of bacteria.

Over small time scales and in localized environments, the balance between these two aspects may vary considerably, resulting in substantial oscillations in the nanoflagellate population. The importance of grazing activities has been noted earlier (Case Study 1.1, Figure 1.15) in the food web of Lake Constance (Germany), where the nanoflagellate population has a major role in removing bacteria, but is kept at a relatively low level due to grazing by ciliate protozoa.

Over longer time scales and with comparison of different freshwater environments, clear correlations occur between nanoflagellate populations, bacterial abundance, and concentrations of inorganic nutrients (trophic status). In a range of freshwater systems, Berninger *et al.* (1991b) noted nanoflagellate populations varying from 10^2–10^5 cells ml^{-1}, directly corresponding to differences in trophic status and resulting bacterial abundances of 10^5–10^9 cells ml^{-1}. Nanoflagellate populations increase along a trophic gradient from oligotrophic to eutrophic systems, in line with increased bacterial populations over this sequence (Table 9.5).

Variations in nutrient status can also affect the taxonomic composition of the flagellate population. A survey of North German lakes by Auer and Arndt (2001) showed some variation in taxonomic composition with nutrient status, with increased dominance by chrysophytes in hypertrophic lakes (about 45 per cent of total HNF biomass) compared with mesotrophic ones (30 per cent). The size distribution of HNFs was also strongly influenced by increased lake trophic status, with an increased importance of larger organisms such as *Paraphysomonas*, *Quadricilia* and *Aulacomonas* (Table 9.4). This increase in the proportion of flagellates >10 µm probably reflects an increase in food supply, with many of these larger forms feeding on algae and other flagellates in addition to bacteria.

9.5.4   Nanoflagellate grazing rates and control of bacterial populations

Despite the high growth rates of bacteria, their abundance in individual water bodies tends to remain relatively constant. This constancy implies that bacterial loss processes (bacteriophage attack, grazing by protozoa and zooplankton) are tightly coupled to productivity in most ecosystems, as summarized in Figure 6.13 (Section 6.10.2). In many freshwater sites, grazing of bacteria by heterotrophic flagellates represents one of the major limitations on bacterial population increase. This has been investigated by quantitative comparison of flagellate grazing rates and bacterial productivity.

Table 9.5   Representative abundancies of heterotrophic nanoflagellates in different freshwater environments (adapted from Berninger *et al.*, 1991b)

Environment	Site	Flagellates (cells $\times 10^3$ ml^{-1})
Oligotrophic freshwater	Mountain becks[a]	0.5–1.5
Mesotrophic freshwater	Lake Constance[b]	0.5–8.1
Eutrophic freshwater	Priest pot[a]	50.0–180
Hypereutrophic freshwater	Soda lakes[a]	240–400

References: [a]Berninger *et al.* (1991b); [b]Weisse (1990)

Flagellate grazing rates

Two basic approaches have been used to determine flagellate grazing rates.

- *Grazer interruption experiments.* This involves monitoring changes in bacterial populations after nanoflagellates have been removed or inactivated. Removal of flagellates can be carried out by selective filtration of environmental samples through screens with a mesh sizes that are small enough to collect flagellates but not bacteria. Changes in the bacterial population are then monitored over a period of time (normally about 12 hours), and grazing rates calculated by comparing these changes in filtered and unfiltered samples. This approach has been particularly useful in studying the relative importance of different size ranges of grazing protozoa and for obtaining estimates of grazing activity with mixed protozoon communities.

  A modification of this approach involves the 'serial dilution technique' where aliquots of untreated water samples (containing predator and prey) are mixed with varying amounts of filtered water (prey only). Grazing rates are determined by comparing the growth rates of prey populations in the samples at different levels of grazing (i.e., with different amounts of diluent).

- *Tracer experiments.* These involve measurement of uptake and removal rates of live/dead bacteria

or particles with a similar size range to bacteria, using a suitable label (tracer) to carry out quantitation. Experiments have included the use of radioactively-labelled or fluorescence-labelled bacterial cells, and labelled inert particles such as latex micro-spheres. Using natural particulate material (such as bacteria) has the advantage that – unlike artificial particles – the material is digested by the predator and its disappearance can be monitored.

Table 9.6 shows estimates (Berninger *et al.*, 1991b) of flagellate grazing activities obtained by different workers. These are expressed in much the same way as zooplankton grazing rates (Section 9.8.4), either as rate of food particle uptake (particles cell^{-1} h^{-1}) or as volume of water extracted (nl cell^{-1} h^{-1}). The latter is referred to as the clearance rate, and can be determined as the product of particle uptake and particle concentration. Measured rates of flagellate grazing vary considerably, probably reflecting differences in HNF population, food supply and experimental procedures. The data do suggest, however, that food uptake may be as high as 100 particles flagellate cell^{-1} h^{-1}.

Effects of flagellate grazing on bacterial population increase

In eutrophic waters (HNF populations reaching 10^5 cells ml^{-1}), maximum grazing rates (100 particles cell^{-1} h^{-1}) would result in the removal of

Table 9.6 Flagellate grazing rates in freshwater environments (adapted from Berninger *et al.* (1991b), where references to the varied sources of these data are given)

Method of analysis	Clearance (nl HNF^{-1} h^{-1})	Uptake (food particles HNF^{-1} h^{-1})
Grazer interruption experiments		
Dilution	20.2	37.2
Dilution	0.1–20.0	10.0–100
Two-stage culture	1.0–4.0	2.3–9.2
Selective filtration	0.2–0.4	4.0–7.1
Tracer experiments		
Bacteria	0.2–1.0	10.0–75.0
Fluorescent microspheres	0.1–8.4	0.8–24.0
Fluorescent paint particles	1.0	2.0

Table 9.7   Bacterial production and grazing loss in freshwater environments (adapted from Sanders *et al.*, 1989; in each case the bacterial grazing loss is largely due to heterotrophic flagellates)

Bacterial production ($\times 10^6$ cells ml^{-1} h^{-1})	Grazing loss ($\times 10^6$ cells ml^{-1} h^{-1})	Grazing loss/bacterial production (%)	Sample site
0.12–0.56	0.06–0.51	42–121	Eutrophic Lake Frederiksborg (Denmark)[a]
0.01–0.81	0–0.33	0–200	Eutrophic Lake Oglethorpe (USA)[b]
0.01–0.10	0.01–0.60	11–162	Eutrophic Lake Oglethorpe (USA)[c]
0.11–0.25	0.06–0.29	23–122	Eutrophic Lake Anna (USA)[d]
0.01–0.02	0.01–0.02	76–168	Mesotrophic Lake Biwa (Japan)[e]
0.02–0.04	0.01–0.03	50–147	Meuse River (Belgium)[f]

References: [a]Riemann (1985); [b]Sanders and Porter (1986); [c]Sanders *et al.* (1989); [d]Tremaine and Mills (1987); [e]Nagata (1988); [f]Servais *et al.* (1985).

10^7 bacterial cells ml^{-1} h^{-1}. There is thus potential for massive HNF removal of bacteria from the water column of these aquatic systems.

Studies on bacterial production and HNF grazing levels in different environmental systems (Table 9.7, Sanders *et al.*, 1989) suggest that grazing levels are normally less than this, with values of up to 10^5 bacteria ml^{-1} h^{-1} in eutrophic lakes and up to 10^4 in mesotrophic water bodies. These rates of grazing are of the same magnitude as bacterial productivity in eutrophic (10^5 bacteria ml^{-1} h^{-1}) and mesotrophic (10^4) systems, suggesting that HNF grazing may completely counter-balance any increase in bacterial biomass in these aquatic systems. Calculation of grazing loss as a percentage of bacterial production is highly variable (0–200 per cent) within individual systems. To some extent, this simply reflects sampling and experimental variability within the data, but does suggest that the impact of HNF grazing may also vary according to dynamic conditions within the ecosystem.

9.5.5   Co-distribution of bacteria and protozoa within the water column

The relative importance of bacteria as a food source for heterotrophic nanoflagellates and other protozoa is indicated by their joint distribution within the water column of stratified lakes. During the summer stratification of many lakes, the discrete accumulations of chemolithotrophic and photosynthetic (purple and green) bacteria just below the thermocline are correlated with localized populations of protozoa. In Lake Belovod (Russia), for example, photosynthetic bacteria are predominantly consumed by protozoa, which show a pronounced population peak in this part of the water column (Figure 9.8). The location of protozoa at the top of the hypolimnion also requires a tolerance by these organisms of

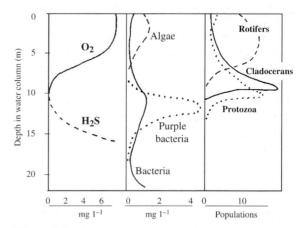

Figure 9.8   Vertical distribution of oxygen, hydrogen sulphide, and microorganisms in Lake Belovod, Russia (based on a figure from Sorokin, 1965); population scales (far right) – rotifers ($\times 10^2$ l^{-1}), cladocerans ($\times 10^4$ l^{-1}), protozoa ($\times 10$ l^{-1})

zero (anoxic) or low (micro-aerobic) conditions. Although cladoceran zooplankton are also major grazers of bacteria, the feeding populations of these organisms are restricted to aerobic parts of the water column and show a distinct peak above the protozoa. Cladocerans have access to chemolithotrophic populations, but not to photosynthetic bacteria – which are strict anaerobes.

9.6 Ecological impact of protozoa: the benthic environment

The bottoms of lakes and rivers are typically covered by a rich array of organic material. This is largely composed of decomposing debris (leaf litter, flocculent detritus), but also includes extensive populations of microorganisms some of which (particularly bacteria) provide important food sources for protozoa. The surface of sediments thus provides a multitude of microhabitats for motile protozoa, a variety of attachment sites for sessile species, and a ready source of complex carbon compounds for these heterotrophic organisms.

Benthic protozoa are an important aspect of most aquatic environments, where high populations and distinctive communities occur in a wide range of micro-niches (Fenchel, 1987). Although most of the protozoa present in sediments are physiologically active, others occur as metabolically inactive stages (Section 1.2.5) – in some cases derived by sedimentation from the water column. Many of these inactive stages occur as resistant cysts (Figure 1.1), able to survive adverse (e.g., winter) conditions present in the water column. All of the major protozoan taxonomic groups are represented in these communities, with many of the most well-known protozoa being typical of benthic environments. These particularly include ciliate (*Paramecium*, *Tetrahymena*, *Stentor*, *Blepharisma*, *Stylonychia*) and amoeboid (*Actinosphaerium*, *Amoeba proteus*) organisms. Flagellates are also well represented in most benthic habitats, with populations of microflagellates in sediments ranging from 10^2–10^5 organisms ml^{-1} and nanoflagellates from 10^5–10^6 organisms ml^{-1} (Alongi, 1991).

In natural habitats, the species composition of the benthic protozoon community varies with local microenvironment and seasonal change. Human activities may also have an impact on benthic organisms as seen, for example, in conditions of organic pollution and in the special case of sewage-treatment plants.

9.6.1 Benthic microenvironments

Benthic microenvironments and their associated microbial communities are defined in relation to localized conditions of hydrology (river flow), nutrient availability, type of substrate, presence of higher plants and macro-algae, oxygen concentration, and many other environmental features. Three particular examples are considered in this section – solid surfaces (typically aerobic), micro-aerophilic conditions, and anaerobic environments.

Protozoon communities associated with solid surfaces

Submerged surfaces of leaves and other parts of higher plants, algal filaments, non-living detritus, and stones often have a rich and diverse community of associated protozoa. The attached organisms are of two main types, with some species being permanently fixed while others move freely over the surface in search of prey.

- *Attached species.* Attached species typically project from the surface via a stalk and carry out active extraction of food particles by filter feeding (e.g., peritrich ciliates), or passive extraction by simply allowing particulate material to come into contact with adhesive arms (e.g., heliozoan amoebae). The peritrich ciliates include both solitary (*Vorticella*) and colonial (*Ophyridium*) genera.

  Many attached ciliates have specialized distribution stages (swarmers or larvae) which are fully ciliated and migrate after release to disperse the species and colonize new surfaces. In one group of ciliates, the Suctoria, the adult stage is permanently fixed to a substratum and has completely lost its cilia. Cilia are only present in the swarmer stage of these organisms, which is thus the only part of the life cycle to reveal the ciliate affinities of this group externally.

Many larger species of attached ciliates have contractile stalks, allowing them to withdraw the main body away from turbulent water, predators such as flatworms and snails, and any other hostile influence.

- *Motile, browsing organisms.* Various protozoa move over submerged surfaces, including hypostome ciliates (with ventral mouths), hypotrich ciliates, and bodonid and euglenoid flagellates. These organisms consume a variety of organic material, including non-living debris and attached organisms such as bacteria, diatoms, blue-greens, and other algae. Such motile protozoa are major consumers within biofilms and epilithon communities.

Microaerophilic environments

Microaerophilic environments have limited oxygen availability. They often occur as localized regions of oxygenation within, or adjacent to, anaerobic conditions and include leaf litter, top layers of sediments in oligotrophic lakes, and regions of the hypolimnion in eutrophic lakes (Figure 9.6). These regions are characterized by obligate and facultative aerobes. They are typically dominated by *Loxodes*, which accumulates at oxygen concentrations of 5 per cent (Section 9.4.1) plus a variety of other protozoa, including the following.

- *Filter feeders.* These include large heterotrich ciliates, with *Spirostomum* and *Blepharisma* being particularly prominent, filter-feeding on large bacteria and small eukaryote cells. Hypotrich ciliates are also a prominent group, including genera such as *Stylonychia* and *Euplotes* feeding on bacteria. Among the colpopids, the giant protozoon *Bursaria* is a filter-feeder which ingests large-sized prey, particularly other ciliates.

- *Active predators*, including carnivorous ciliates such as *Didinium* and *Litonotus* which depend on rotifers or other ciliates for food.

Anoxic environments

Anoxic conditions, where there is a complete absence of oxygen, occur as micro-habitats in both aerobic and anaerobic sediments. In aerobic sediments, oxygen diffuses from the water column into the top of the sediment, where it is consumed by the microbial community. At a certain depth, free oxygen is no longer available for microbial respiration and the localized region becomes anoxic. In anaerobic sediments (typical of eutrophic environments) the lower part of the water column is anoxic, and no oxygen at all reaches the sediment, which is anoxic throughout. Anaerobic protozoa, like bacteria, have to rely on compounds other than oxygen to act as a terminal electron acceptor in such conditions.

A distinct fauna of anaerobic protozoa occurs in anoxic conditions. Some of these, such as pelobiont and diplomonad flagellates are believed to have evolved from times of limiting oxygen availability on Earth, and have survived in anoxic conditions ever since. Most of the ciliates which live in these environments, however, have become secondarily adapted to such conditions by modifying their respiratory activity. Such anaerobic ciliates include a group of distinct genera – *Metopus*, *Caenomorpha*, and *Plagiopyla*, which also characterize marine anaerobic environments.

Protozoa in specialized biological environments

Distinct communities of freshwater protozoa also occur in specialized biological environments, including mats of blue-green algae, fungal colonies, bacterial biofilms, and the surfaces of higher plants. In chalk streams, protozoa are primarily associated with bottom sediments and with the surfaces of the dominant macrophyte, *Ranunculus penicillatus*. Protozoa are located particularly in regions such as older leaf surfaces and nodes, where the current was less strong, and amoebae, flagellates, and cliates can be found. Peritrich ciliates are frequently found attached to the substratum in flowing waters. In streams with a moderate and stable flow regime, protozoa with contractile stalks (e.g., *Vorticella*) predominate. In streams which are liable to more extreme flow conditions, peritrichs with non-contractile stalks (e.g., *Epistylis*) are more common.

9.6.2 Seasonal changes

Marked seasonal changes occur on the benthic fauna of temperate lakes, particularly eutrophic lakes such as Esthwaite Water (UK). At this site, sediments are aerobic from October to May, when the column is fully mixed. During this time, ciliate populations number about 3500 cells ml^{-1}, and are dominated by *Loxodes* and *Spirostomum*. Most lake sediments have a loose, flocculent consistency, and protozoa can penetrate down to a depth of at least 6 cm under aerobic conditions.

With the onset of summer stratification, sediment conditions become anaerobic and aerobic protozoa disappear – either becoming encysted or migrating into the water column. Two species of *Loxodes* migrate into the lower part of the oxycline or upper hypolimnion, where they position themselves within a zone ranging from 0–10 per cent oxygen saturation. In summer, zonation of the sediments becomes much more compressed, with anaerobic protozoa being restricted to the top 1 cm, where conditions are highly reducing and sulphide ions occur at high concentration.

9.6.3 Organic pollution

Freshwater environments which have been contaminated by a transient influx of soluble organic pollutants show a characteristic succession of microorganisms (Figure 9.9) over the following days and weeks (Fenchel, 1987). Pollutant conditions include general leakage of farm effluents, and more specific events such as spillage of plant products, peptone, and milk.

In all cases, the initial response is the development of a substantial bacterial bloom. Where the environmental contamination involves a large amount of organic material, the bacterial bloom is massive, leading to a high biological oxygen demand (BOD), anoxic conditions, and a succession of anaerobic protozoon populations.

Where the addition of organic pollutant is not large, the bacterial bloom that develops is moderate, and the protozoa that develop are aerobic or micro-aerophilic. The succession of organisms which develop involve an initial increase in heterotrophic flagellates which are active in bacterial consumption, have a rapid growth rate, and are also tolerant

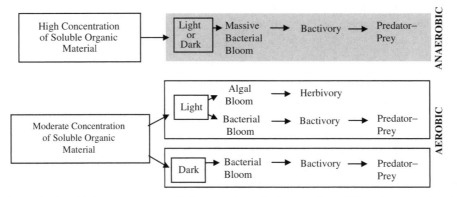

Figure 9.9 General microbial responses to soluble organic pollution – effects of environmental parameters. The concentration of soluble organic pollutant and light/dark conditions affect the general response of the microbial community. With a high concentration of pollutant, the entire aquatic system becomes anaerobic (grey box) and only facultative or obligate anaerobes are able to survive

of temporary conditions of low oxygen and high acidity which result from the bacterial bloom. The flagellates rapidly colonize the high-nutrient environment and reduce the bacterial population considerably, so that oxygen demand is lessened and the water body can become more oxygenated. This reduces some of the physiological constraints on the microbial flora, allowing the community to become more diverse. Rapidly-growing bacterivorous ciliates (e.g., *Colpidium, Glaucoma, Cyclidium*) appear next, together with small amoebae. These are followed a few days later by slow-growing ciliates (e.g., hypotrichs) and amoebae that are often specialized to eat larger and more restricted types of food (filamentous bacteria) or are carnivorous (pleurostomatids and prostomatids). These protozoon populations are eventually followed by rotifers, small crustaceans, and other small metazoa – with a continued trend towards larger sizes. Organic enrichment of freshwater habitats thus triggers a well-defined microbial sequence that involves:

- a bacterial bloom, followed by protozoa then metazoon populations;

- initial colonization and growth by *r*-selected species, followed by domination of the more stable environment by *K*-selected species – the progressive increase in size and reduced growth rate relates to this;

- a transition from low diversity (early heterotrophic flagellate dominance) to high diversity (wide range of ciliate and other protozoa). Low diversity occurs at a time of high temporary stress, when oxygen levels are low and acidity is high. High diversity occurs when the community is more stable, with a complex network of food chains.

In the presence of light, and if the organic material is rich in soluble phosphorus and nitrogen, an algal bloom develops – with increases in populations of blue-green algae, diatoms, and other algae. This bloom supports other protozoon populations in addition to the bacterivorous sequence, giving rise to a very diverse microbial flora.

9.6.4   Sewage-treatment plants: activated sludge

Sewage-treatment plants receive a continuous flow of particulate and dissolved organic matter, and are designed to convert this into a form (flocs or slime) which can then be separated off from a low-nutrient liquor. The solid and liquid waste can then be disposed of without causing major environmental problems. The conversion of raw sewage to disposable waste is carried out by microbial communities – chiefly bacteria and protozoa, in activated sludge and percolating filters.

The protozoon community

Activated sludge and percolating filters represent one of the most extreme types of nutrient-enriched aquatic environment, and develop special communities of protozoa which are sustained by the large-scale production of bacteria. The diverse fauna includes amoebae, heterotrophic flagellates, and ciliates, as well as various small invertebrates.

The protozoon population of activated sludge (Fenchel, 1987) is characterized by:

- high population densities, with values as high as 5×10^4 cells ml^{-1},

- high species diversity, with counts of 50–70 species recorded for both activated sludge and percolating filters,

- high counts of bacterivorous protozoa – ciliates are particularly plentiful and include attached peritrichs, scuticociliates, and tetrahymenine ciliates.

Community dynamics: the effects of operating parameters

Continuous flow of organic material into the system means that a microbial succession (see previously) does not occur. Instead, a steady state is reached, where the succession is terminated at a stage which depends on the flux of material through the system. This in turn depends on two main and interrelated operating parameters – rate of flow and organic loading.

- *Rate of flow.* The passage of fluid through the system continuously removes organisms by simple displacement, so their continued presence depends on the ability for rapid growth to compensate this loss. Sewage-treatment plants with high rates of flow lead to domination of the microbial community by organisms with rapid rates of reproduction, including smaller protozoa (such as flagellates) and small ciliates. In contrast to this, systems with a low rate of flow contain organisms with lower rates or reproduction such as larger ciliates and extending to metazoa.

  High rates of flow can be viewed as imposing an environmental stress on the microbial community, with domination by *r*-selected species and a low biodiversity. Lower rates of flow represent low stress conditions, where the natural succes-

sion can proceed to a more mature stage and the community becomes characterized by *K*-selected species.

- *Rate of organic loading.* Increase in the amount of organic material per unit volume of liquid entering the treatment system (organic loading) generates higher bacterial populations and higher demand for oxygen. Oxygen levels in the activated sludge become depleted, and the whole system may go anoxic – despite procedures to keep it aerated. In this situation, anaerobic protozoa become dominant, including pelobiont and diplomonad flagellates. Heavily loaded systems with low levels of oxygen thus favour these flagellates, plus amoebae and small ciliates which are normally found in aquatic sites with organic pollution.

  As the loading with organic material declines, the activated sludge becomes restored to a more diverse community in which anaerobes are replaced by micro-aerophiles and then by aerobic protozoa.

B. GRAZING OF MICROBIAL POPULATIONS BY ZOOPLANKTON

Target organisms for zooplankton grazing in the planktonic environment are primarily unicellular and colonial algae (phytoplankton), bacteria, and protozoa. Although zooplankton grazers can be separated into three main groups – rotifers, cladocerans crustaceans, and copepod crustaceans, other invertebrates such as insect larvae may also be important.

As with members of the phytoplankton, zooplankton may be divided into holoplankton (present in the water column over a major part of the annual cycle) and meroplankton (restricted to a limited period of the year). Rotifer and crustacean zooplankton are mainly holoplanktonic, while insect (particularly chironomid) larvae occur in the water column for a short summer period and are meroplanktonic. During this time the chironomid larvae may completely dominate the grazing community.

9.7 General features of zooplankton: rotifers, cladocerans and copepods

The grazing activity and biological characteristics of rotifers, cladocerans, and copepods differ in a number of key respects, including morphology and size, feeding methods, reproductive characteristics, and predation by other organisms. These characteristics are shown in Table 9.8, which also makes comparisons with protozoa. Some typical members of the major zooplankton groups are illustrated in Figure 9.10.

9.7.1 Morphology and size

Adult members of the three groups show overlapping size ranges, with rotifers (200–600 µm)

Table 9.8 Feeding parameters and other biological characteristics of grazing organisms

	Protozoa	Rotifera	Cladocera	Copepoda
Example	*Paramecium Amoeba*	*Fiolina Brachionus*	*Daphnia Bosmina*	*Cyclops Diaptomus*
Feeding method	Wide range of feeding strategies	Suspension feeding	Filter feeding	Filter or raptorial feeding
Food size (μm)	1–3	1–20	1–50	5–100
Food type	Bacteria, blue-green algae and small unicells	Bacteria, single-cell algae	Single-cell to colonial algae	Single-cell to colonial algae
Water clearance rate	–	Very low	High	Low
Generation time (days)	1	1–7 parthenogenetic	5–24 parthenogenetic	7–32
Adult size (μm)	2–200	200–600	300–3000	500–5000
Predation by invertebrates	High	High	Moderate	Moderate (adults) High (juveniles)
Predation by vertebrates	Very low	Very low	High	Low

Feeding parameters are indicated by the shaded area.

also showing overlap with some of the larger protozoa.

Rotifers are characterized by the division of their body into three main zones – an anterior region (with one or more circles of cilia, forming the corona), a mid-region, and a contractile foot with an adhesive tip. In the pelagic environment, rotifers which are attached to other organisms (e.g., *Philodina*, Figure 9.10) retain their adhesive capability, while freely-planktonic organisms typically lack a foot. Rotifers are mostly suspension feeders, though some (e.g., *Asplancha*) are active predators.

Crustacean zooplankton are covered in a prominent exoskeleton and have segmentally-arranged limbs which are important in locomotion and feeding. Of the two major groups of crustacean zooplankton, cladocerans may be distinguished (Figure 9.10) by the presence of a distinct head (with compound eye and a smaller ocellus) and a bivalve, cuticular carapace, which covers the body in most genera. Cladocerans are also distinguished by a pair of large, branched second antennae. These are able to propel the cladocerans rapidly though the water by a rather jerky rowing action, giving these organisms their common name – water flea. Mouth parts consist of large, chitinized mandibles which crush the food particles, a pair

of small maxillules that push food between the mandibles and a median labrum which is used to cover the mouth parts. Cladocerans are mostly herbivorous filter feeders (including the well-known genera *Daphnia* and *Bosmina*), but also include carnivorous forms. Organisms such as *Leptodora* and *Polyphemus* are raptorial, feeding on smaller zooplankton and actively capturing their prey. Particle feeders such as *Daphnia* filter their food through long hairs (setae) on their thoracic limbs. These hairs bear finely-spaced setules, separated by gaps of a few μm, which retain small particles from the water current and eventually pass the food to the mouth. Although the correlation between food size and spacing of the setules (Section 9.8) suggests a simple filtration process, the ability of some cladocerans to reject unsuitable particles (toxic or too large to ingest) indicates an additional active component to the feeding process. After filtration, food particles pass along a feeding groove to the mouth, and unsuitable items can be extricated at this stage via a claw on the lower abdomen.

Copepods, the second major group of crustacean zooplankton, have adult members which can grow to a slightly larger size than cladocerans. These organisms lack a carapace and have an anterior

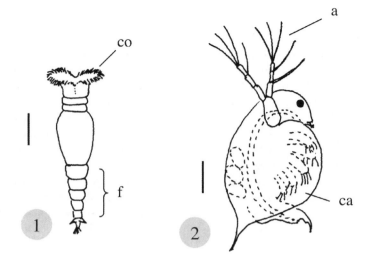

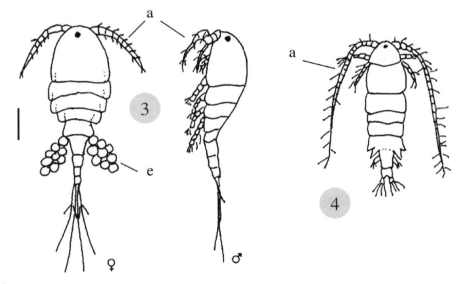

Figure 9.10 Some typical members of the zooplankton, drawn to the same scale.

(1) Rotifer – *Philodina*: this large rotifer is present in the planktonic phase attached to crustaceans. Adhesion is mediated via the foot (f), a mobile contractile organ which terminates in an adhesive tip. The anterior end has a two-wheeled corona (co). Freely planktonic rotifers differ from *Philodina* in typically lacking a foot.

(2) Cladoceran crustacean – *Daphnia*: this common aquatic organism (water flea) has a prominent branched second antenna (a). The filtering limbs are enclosed within a carapace (ca), which also contains an egg pouch with a few eggs.

(3) Copepod crustacean – *Cyclops*: a cyclopoid copepod, showing the female, with paired egg sacs (e), and the male. Antennae are not branched.

(4) Copepod crustacean – *Diaptomus*: a calanoid copepod which differs from cyclopoid copepods in having an elongate antenna, and single, rather than paired, egg sacs (in those females carrying eggs) (scale = 100 μm)

structure (the cephalothorax) which is divided into head region (bearing antennae and mouth parts) and thorax (with six pairs of swimming legs). Behind this, a posterior urosome consists of abdominal segments. Cyclopoid and calanoid sub-groups can be distinguished by a range of morphological features, including the length of their first antennae and the presence of paired or single egg sacs (Figure 9.10). Calanoid and cyclopoid copepods are also distinguished by their mode of feeding. Calanoids ingest small particles and are generally regarded as filter feeders, while cyclopoids are chiefly raptorial – actively capturing prey such as small zooplankters, colonial algae, or aggregated masses of phytoplankton.

9.7.2 Reproduction and generation times

The speed with which zooplankton are able to reproduce is important to the microbiology of freshwater systems since it determines the zooplankton response to increases in food supply, and the dynamics of limitation of bacterial and algal populations. High reproduction rates are also important for rapid replacement of zooplankton populations which are vulnerable to predation.

Rapid population increase in rotifers and cladocerans is achieved by parthenogenesis, with females producing broods of eggs asexually which hatch into females. In rotifers, the eggs are born in external sacs, while in cladocerans they occur in pouches deep in the carapace – emerging as young adults which soon grow large enough to reproduce. The rate of reproduction in rotifers can be very rapid since there is a high rate of egg production (with no requirement for fertilization) and a new reproductive generation produced in only a few days. Each female produces up to 25 young in a life time of 1 to 3 weeks. Cladocerans have a longer generation time (up to 4 weeks) and a longer life time (up to about 12 weeks), producing as many as 700 young per life cycle.

Parthenogenesis is a mechanism for rapid reproduction under good physical conditions and adequate food supply (Smyly, 1979). Under adverse conditions, such as food shortage or approach of winter, males are produced which fertilize the eggs. The sexually-produced zygote becomes thick-walled, divides to produce an embryo and is able to hatch out when conditions improve. New individuals released in spring form the next season's population of parthenogenetic females.

Copepods are not parthenogenetic. Each generation is sexual, and the newly-born progeny have 11 successive moults before mature (reproductive) adulthood is reached. This metamorphosis involves two distinct stages – nauplius larvae (different appearance from the adults, the first six stages), and copepodites (similar appearance to adults, the final five stages). The final copepodite stage develops into a mature adult which has a similar longevity and fecundity to adult Cladocerans.

9.7.3 Predation of zooplankton

As with other lake biota, zooplankton populations are regulated by both top-down (predation) and bottom-up (food supply) control. Predation of zooplankton may be a key determinant of zooplankton seasonal succession (Gliwicz and Pijanowska, 1989). It is also an important aspect of the trophic cascade (Section 1.5.2), since selective removal regulates zooplankton biomass, affecting the grazing pressure exerted by zooplankton filter-feeders and ultimately influencing the populations of planktonic microorganisms such as algae, protozoa, and bacteria. The predation of grazing zooplankton thus has important implications for their target organisms, and is a key factor in lake restoration and the control of algal blooms (see Chapter 10).

Since the 1960s, most studies on predation of zooplankton have focussed on the role of planktivorous fish (Lazzaro, 1987). More recent work has also looked at the importance of other invertebrates, which in some cases have a greater predatory effect than fish (MacKay et al., 1990). In contrast to zooplanktivorous fish, which visually select large prey, invertebrates consume smaller categories of zooplankton such as rotifers, copepod nauplii, and *Bosmina*.

Populations of cladocerans and copepods are vulnerable to predation from within their own communities and from other crustaceans (e.g., mysids), insect larvae (e.g., *Chaoborus*), and fish. Mysids include the genera *Mysis* and *Neomysis*, and are a group of large (up to 2 cm) shrimp-like animals which are predators of other zooplankton.

One of the most important predators of zooplankton is the larva of the Dipteran insect *Chaoborus*. This is present in many lentic systems, from large lakes to small ponds, and is particularly important in tropical waters, where it may have a major influence in water bodies such as Lake Malawi (Irvine, 1997). *Chaoborus* larvae have a particular role in shallow tropical waters (Pagano *et al.*, 2003), where visual predation by zooplanktivorous fish may be limited by low visibility (high turbidity in shallow conditions) and low fish populations. Unlike other invertebrates, *Chaoborus* larvae tend to feed on larger members of the zooplankton. Mesocosm experiments by Pagano *et al.* (2003) showed that *Chaoborus* predation of a mixed zooplankton community removed larger cladocerans and copepods, but had no effect on rotifers or copepod nauplii. Selective predation by this organism shaped the zooplankton community and modified its size structure.

Predation of zooplankton is considerably influenced by their positioning and migration within the water body. Crustacean zooplankton are much more heterogeneously distributed than phytoplankton, occurring in vertical and horizontal patches or shoals. Avoidance of visual location and predation by zooplanktivorous fish is optimized by the location of zooplankton in macrophyte stands (refuges) and by their migration to the upper water column (surface feeding) at night, moving down to avoid predation by day.

It is now widely accepted that zooplankton detection by visual predators is much reduced at night and that this is the main reason for the evolution of diurnal vertical migratory activities in lakes (Gliwicz, 1986), oceans, and coastal water bodies (Holzman and Genin, 2003) – see Section 9.9.3. This diurnal periodicity has further implications in relation to phosphorus cycling (Section 5.7.2, Figure 5.19) and the diurnal migration of phytoplankton (Section 4.10.2).

9.8 Grazing activity and prey selection

In temperate lakes, the grazing activity of zooplankton is highly seasonal (occurring mainly during the spring–autumn period of major phytoplankton growth, Figure 3.14) and is also diurnal (occurring mainly at night, Figure 5.19). During the seasonal cycle, zooplankton grazing activity is normally particularly heavy after the spring diatom bloom, during the 'clear-water' phase, when intense grazing pressure depletes phytoplankton populations and promotes dominance of *r*-selected algae (Section 3.10.2).

The identity of microorganisms which are grazed by zooplankton depends partly on availability and partly on selection by zooplankton from mixed microbial populations. Microbial availability relates particularly to seasonal algal succession (Section 9.8.1), while selectivity depends on the method of feeding and specific selection mechanisms (Sections 9.8.2 and 9.8.3).

9.8.1 Seasonal succession in zooplankton feeding

Seasonal changes in zooplankton food sources depend on the periodic availability of micro-algae, bacteria, and protozoa (Figure 3.13). Phytoplankton form the major food biomass, with temporal changes in different populations of micro-algae being accompanied by (and to some extent determining – DeMott, 1989) changes in different zooplankton populations. This is illustrated by the annual sequence shown in Figure 9.11, which is unusual (compare with Figure 3.14) in that the normal spring diatom bloom is repressed (largely by calanoid copepod feeding) and the 'clear-water phase' has a major growth of cryptomonad algae (grazed mainly by cladocerans and cyclopoid copepods). Within this sequence, particular zooplankton groups are consuming different algae at different times of the year, with a food source at any point in time depending on availability.

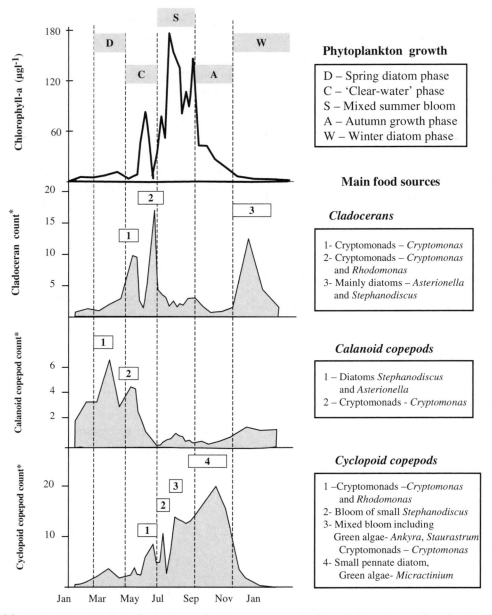

Figure 9.11 Major phytoplankton food sources for crustacean zooplankton during an annual cycle (year 2000) in a temperate eutrophic lake (Rostherne mere, UK). Phytoplankton food sources for the three major groups of crustacean zooplankton are shown for periods of major population increase. This annual cycle was unusual (compare with Figure 3.14) in the lack of a clear spring diatom bloom and the development of a substantiual bloom of cryptomonads in the 'clear-water' phase. Populations of the colonial blue-green algae *Anabaena* (early summer phase), *Aphanizomenon* (mid-summer phase) and *Microcystis* (late summer to autumn phase) were also present, but probably had limited value as a zooplankton food resource. *Population counts for each zooplankton group are expressed as number of individuals l^{-1} (unpublished data, with permission, from Andrew Dean)

9.8.2 Method of feeding

Zooplankton feeding can occur in three main ways as follows.

- *Suspension feeding*, where a current of food particles is directed towards the mouth of the organism, entering the gut. This type of feeding is characteristic of rotifers. This activity is carried out by the anterior circles of cilia, which appear to rotate like a wheel, directing a continuous stream of water with fine suspended particles into the mouth.

- *Filter feeding*, where a current of food particles passes through a filtration apparatus. This is most clearly seen in the cladocerans, where most members of the group are filter-feeders. A water current is created in these organisms by opening and closing of the carapace cavity (Figure 9.12), with subsequent filtration as the water is passed through secondary spines (setules) which constitute the filtering comb of the third and fourth thoracic legs.

- *Raptorial feeding*, where an organism moves in search of prey, makes contact, and ingests it. This

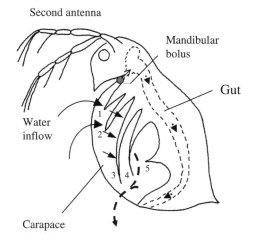

Second antenna

Mandibular
bolus

Gut

Water
inflow

Carapace

Figure 9.12 Water filtration and particle extraction in *Daphnia* (Cladocera). Water is drawn into the carapace cavity and passes through fine hairs on the five pairs of thoracic limbs (1–5). Filtered particles are passed to the mouthparts where they form a mandibular bolus. This passes down the gut and is eventually voided

type of feeding is carried out by some predatory cladocerans (e.g., *Leptodorus*) and cyclopoid copepods such as *Cyclops*.

The process of feeding in copepods is closely connected to the method of movement and the passage of water relative to the feeding parts. Swimming is continuous in the calanoid copepods, due to rotary motion of the long antennae. Although calanoids such as *Diaptomus* and *Eudiaptomus* are sometimes described as filter-feeders, they do not actively strain the particles out of the water. The antennae of these organisms propel a continuous current of water past the body, with selection and removal of food particles by a second set of appendages. This type of feeding is neither filter feeding nor raptorial feeding (since the organisms do not seek out their prey), and can perhaps be regarded as a type of suspension feeding.

Cyclopoid copepods swim with an irregular jerky movement and are adapted to carry out rapid moves towards their prey, which are seized by the predator's mouth parts. The maxillules hold and pierce the prey, forcing the fragments between the mandibles, which oscillate intermittently to macerate some of the food.

Differences in the method of feeding by different groups of zooplankton lead to differences in food selection and ecological impact.

9.8.3 Selection of food by zooplankton

The uptake and consumption of microorganisms from mixed populations within the planktonic environment is highly selective. This has been investigated particularly in relation to phytoplankton grazing, where characteristics of algal size and shape, surface chemistry, presence of mucilage, and secretion of algal toxins are significant. Environmental conditions and zooplankton characteristics (type and size of organisms) are also important.

Particle size and shape

The size of individual phytoplankton units (cells or colonies) is particularly important in determining the pattern of grazing by filter-feeding zooplankton, where food particle uptake occurs over a defined size range (minimum to maximum algal dimensions). Particle size is considered in relation to minimum and maximum dimensions, use of mesocosm experiments, and seasonal changes in the lake environment.

Minimum particle size   In filter-feeders such as *Daphnia*, the minimum size is determined by the spacing between the fine hairs (setules) within the filtration apparatus. The mesh widths of different cladoceran species range from 0.16 to 4.2 μm (Geller and Muller, 1981), with corresponding variation in the minimum size of food particles. Cladocerans with the finest mesh size (e.g., *Diaphanosoma brachyurum*) are able to remove bacteria, while coarse mesh size organisms (e.g., *Holopedium gibberum*, *Sida cristallina*) have a size limit which excludes picoplankton and small microplankton. The majority of *Daphnia* species are in the intermediate range, with a mesh size of about 1.0 μm, which allows them to collect some picoplankton.

Maximum particle size   This is partly determined by the filtration apparatus (opening width of the carapace aperture) and partly by the mouthparts (opening width of the mandibles), both of which tend to vary with species and the overall size of the organism. In the case of *Daphnia* and *Bosmina*, for example, Burns (1968) established a quantitative relationship:

$$S = 22L + 4.87 \qquad (9.1)$$

where S = maximum size of particle ingested (μm) and L = body (carapace) length (mm).

Although examination of the gut contents of *Daphnia* from lake samples suggests that some ingested algae are above the size range predicted by this quantitative relationship, the equation is useful in providing an approximate indication of the fractions of phytoplankton which can be considered as 'edible' and 'inedible'.

The maximum particulate filtration size of cladocerans ranges from about 20 μm in the case of smaller species (corresponding to the upper filtra-

tion limit of rotifers) and to about 50 μm in larger species (comparable to many copepods). Many phytoplankton species appear to have evolved large size as a defence mechanism against grazing, either as large single cells (e.g., dinoflagellates), large colonies (globular blue-greens), or by the formation of spines (e.g., *Staurastrum*) and linear growth form (pennate diatoms). Such adaptations of size and shape do not provide an absolute defence against ingestion, since linear structures can be orientated lengthwise within the filtration mesh, and large colonies such as *Asterionella* can be broken down prior to ingestion.

Mesocosm experiments   The importance of phytoplankton size and shape in terms of food selectivity can be demonstrated experimentally in mesocosm experiments, where removal of particular species by zooplankters such as *Daphnia* can be monitored within a defined environment over a specified time interval. Mesocosm experiments by Sommer (1988) on algal ingestion by *Daphnia magna* led to the designation of selectivity coefficients, which ranged from (maximum) values approaching 1.0 (small diatoms such as *Achnanthes* and *Nitzschia*) to intermediate values (elongate diatoms such as *Synedra*), with low rates of ingestion (coefficient approaching zero) in the case of filamentous and large globular algae. The presence of mucilage is clearly important in promoting a maximum size, and contributes to the success of globular algae such as *Microcystis* (blue-green alga) and *Sphaerocystis* (green alga) in grazing avoidance. Mucilage is also important once ingestion has occurred in protecting the algal cells from digestion (see below).

Seasonal cycle   The significance of algal size in relation to avoidance of zooplankton grazing varies through the seasonal cycle. In temperate lakes, it is particularly important in the late summer bloom, where dominance by *K*-selected organisms (globular blue-greens and dinoflagellates) keeps grazing at a low level. Avoidance of grazing by size does not occur earlier in the year during the clear-water phase, where grazing by zooplankton is very high and *r*-selected algal species dominate.

Surface chemistry

In contrast to filter-feeding zooplankton, which select their food particles purely on the basis of size, other types of zooplankton are able to discriminate on the basis of surface chemistry or 'taste'. This has been demonstrated by feeding zooplankton cultures with equal mixtures of algae and inert polystyrene beads of a similar size and shape (DeMott, 1988). Cladocerans (with the exception of *Bosmina*) do not discriminate, and these filter-feeders consume equal quantities of algae and beads over a range of combined particle concentrations (Figure 9.13). In contrast to the mechanical selection of food particles by cladocerans, herbivorous calanoid copepods such as *Eudiaptomus* also use chemical sensing to select their food. These organisms test the surface chemistry of each particle, rejecting inert beads and consuming only algal cells. The importance of surface chemistry is shown by the fact that if the beads are treated with algal products ('algal-flavoured beads'), they are

then consumed by *Eudiaptomus*. *Eudiaptomus* strongly selected algae over flavoured beads when algae were abundant, but were less selective when algae were scarce – suggesting a further ability to differentiate between dead and living algal cells.

Mucilaginous algae and the avoidance of digestion

The ability of certain mucilaginous algae to avoid digestion within the guts of zooplankton grazers, and to survive as fully viable organisms, was demonstrated in the early studies of Porter (1973) on differential grazing of a natural phytoplankton community. These studies investigated phytoplankton changes within mesocosms under conditions of low grazing pressure (zooplankton removed by filtration), moderate grazing pressure (normal zooplankton community) and high grazing pressure (supplemented zooplankton population). The herbivorous zooplankton included various filter-feeders (*Daphnia*, *Diaptomus*) and the raptorial feeder *Cyclops*. The results showed that phytoplankton species could be divided into three main groups as follows.

- Suppressed species, which were ingested and digested by the zooplankton. Populations of these species became lowest in the high grazing conditions and comprised largely small algae such as flagellates, unicellular diatoms, and various nanoplankton species.

- Unaffected species showed no significant changes in population under conditions of high or low grazing pressure, and were mainly species such as *Cosmarium* (green algae), *Peridinium* (dinoflagellate), and *Anabaena* (blue-green) which were too large (>40 µm) for ingestion.

- Enhanced species, where the algal population was highest under conditions of high grazing pressure compared with low grazing pressure. These members of the phytoplankton were dominated by green algae which were encased in a thick mucilaginous coat, and included species of *Sphaerocystis* and *Elakotothrix*.

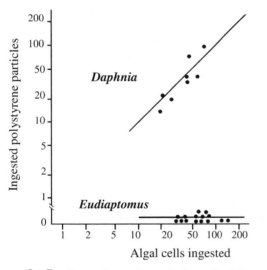

Figure 9.13 Comparison of zooplankton food ingestion in relation to chemical (*Eudiaptomus*) and non-chemical (*Daphnia*) selectivity. Organisms were fed an equal mixture of similar-sized polystyrene beads and unicellular algae; *Daphnia* do not discriminate between the two, but *Eudiaptomus* only consume algal cells (based on a figure from DeMott, 1988)

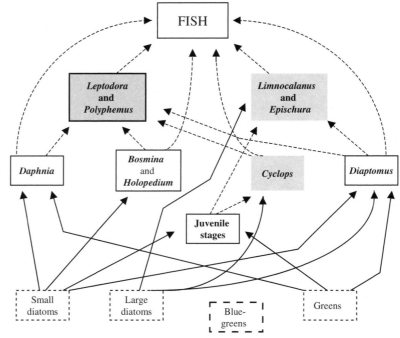

Figure 9.14 Feeding selectivity of zooplankton: incorporation into the Lake Michigan food web model (based on a figure from Canale *et al.*, 1976). In this model, algae are divided into four principle groups (▨), with no ingestion of blue-greens. Zooplankton are of three main types: □ strict herbivores, ▨ combined herbivores and carnivores, ▪ strict carnivores

Examination of zooplankton gut contents showed the digested remains of suppressed species, but intact cells and colonies of enhanced species. Under conditions of high grazing pressure, gelatinous green algae are spared from grazing by viable gut passage and can out-compete both smaller non-mucilaginous species (which are taken up and digested) and larger species. Although larger species are not ingested, they are also typically slow growing and are not able to respond rapidly to the change in environmental situation.

Preferential selection and differential digestion of particular algae by different members of the zooplankton is an important aspect of food webs and is also relevant to the calculation of mass transfer in modelling studies. This is shown in the food web model of Canale *et al.* (1976) for Lake Michigan (USA), where the phytoplankton population is divided into four main groups (Figure 9.14). In this model, blue-green algae are not ingested by zooplankton but small diatoms, large diatoms, and green algae are selectively removed by juvenile stages (nauplius larvae) and adult stages of major

genera. Some of the zooplankters are entirely herbivorous, others combine herbivory and carnivory, while *Leptodora* and *Polyphemus* are entirely carnivorous.

9.9   Grazing rates of zooplankton

Grazing rates of zooplankton are important in relation to the impact of this activity on microbial populations. This section considers the assessment of grazing rates, the effect of intrinsic/seasonal factors, and diurnal variations.

9.9.1   Measurement

The grazing rates of zooplankton can be measured in various ways, and as individual organisms or mixed populations. This has lead to a confusing array of terms.

Individual organisms: ingestion, filtration, and clearance rates

Grazing activities of zooplankton can be monitored either in terms of food uptake (ingestion rate) or in terms of the volume of water (V) which passes through the feeding apparatus. The latter is referred to as the 'filtration rate' even where the organisms concerned are not strictly filter-feeders.

For an individual member of the zooplankton, the ingestion rate (I) can be expressed as the number of particles (N) or the mass of food (M) that is consumed per unit time, and is respectively expressed as N ind^{-1} t^{-1} or M ind^{-1} t^{-1}. Ingestion and filtration rates can also be expressed per unit volume of water or per unit surface area of the lake.

The individual filtration rate (F) can be calculated by dividing the ingestion rate (I) by the particulate food concentration (C), and is expressed as V ind^{-1} t^{-1}. With a defined zooplankton population, such as occurs in laboratory or microcosm conditions, the mean individual filtration rate (F') can be determined from the decrease in particle concentration, where:

$$F' = V \frac{\ln C_2 - \ln C_1}{NT} \qquad (9.2)$$

and C_1 and C_2 = the particle concentrations at the beginning and end of measurement, N = number of animals in the experiment, V = volume of suspension, and T = duration of experiment. The filtration rate is sometimes also referred to as the 'clearance rate', which is defined as the volume of water cleared of particles per unit time.

Mixed zooplankton populations: the community grazing rate

Laboratory studies on feeding rates have usually employed single-species zooplankton populations of a similar size category. In this situation, where there is a uniform population, calculation of a mean filtration rate gives a meaningful value in terms of the individual organisms.

In the aquatic environment, however, zooplankton populations are normally heterogeneous – compris-

ing a mixture of individuals in terms of size, developmental stage, physiological activity, and species. In the more complex situation of mixed zooplankton populations, the community filtration rate is the sum of the filtration rates of all the separate zooplankton categories and is known as the community grazing rate (G) or community grazing index (Thompson *et al.*, 1982), where:

$$G = (F'_1 N_1) + (F'_2 N_2) + \cdots (F'_1 N_1) \qquad (9.3)$$

and F', N are the mean filtration rates and number of individuals in population sub-sets. The units of G are ml filtered l^{-1} d^{-1}. These may also be expressed as the proportion of water filtered daily (e.g., 500 ml l^{-1} = 0.5d^{-1}).

Measurement of the *in situ* grazing rate of natural communities is complicated by zooplankton population heterogeneity. The situation becomes even more complex due to variations within the water column and also with changes during the diurnal cycle. Such measurements require the use of special incubation chambers (Haney, 1973) which allow monitoring under natural conditions of temperature, light, and food supply, with integration of the results over an entire day.

9.9.2   Factors affecting grazing rates

Grazing rates of individual zooplankters and entire populations vary with a wide range of interrelated factors, including internal parameters (physiological state, body size, species-dependent feeding mechanism) and environmental features. Environmental factors include seasonal changes (effects of temperature, light, food availability), location (depth in water column), and diurnal changes in light (Section 9.9.3).

Influence of body size

Filtration rates have been investigated intensively in species of *Daphnia*, which are particularly amenable to laboratory experimentation. Early studies by Burns (1969) demonstrated clear relationships between filtration rate, temperature, and body size in this genus. Other workers have confirmed the importance of body size, in some cases (e.g., Knoechel and Holtby, 1986) demonstrating a linear relationship between filtration rate (F: ml ind^{-1} h^{-1}) and body length (L: mm). These data were obtained from six cladocerans and one calanoid, and the overall relationship between filtration rate and body length was found to depend on particle size, with different equations for bacteria (1 µm diameter) and large (20 µm) algae:

$$\text{Bacteria}: \quad F = 5.105\, L^{2.176}$$
$$\text{Algae}: \quad F = 7.534\, L^{3.002}$$

Algal filtration rates are higher and increase more rapidly with body size than do bacterial filtration rates, suggesting that different processes may be involved in the capture and retention of ultrafine bacterial particles. The general increase in filtration rate with size reflects the larger volume of liquid taken into the carapace cavity in larger animals and results in a greater extraction of microbial biomass with time. Body size is also important in determining the maximum size of particle ingested (Burns, 1968) and may be used to predict which members of the phytoplankton community are available as food for different-sized species of filter-feeding Cladocerans.

Seasonal variations in grazing rate

In temperate lakes, zooplankton community grazing rates would be expected to show marked seasonal variation, depending on the size of the zooplankton population and the availability of suitable microbial biomass. During the early-summer clear-water phase, when zooplankton populations are very high, community grazing rates reach values of 1.0–2.5 d^{-1}. The intensity of filtration at this time is so great that growth of most algae, even at maximum growth rates, is not sufficient to replace algal populations – resulting in a net loss of algal biomass. Such high rates are restricted to this time of year, however, falling to values of 0.2 d^{-1} during late summer (Lampert, 1988).

9.9.3 Diurnal variations in grazing activity

Diel vertical migration (DVM) of zooplankton within the water column of standing waters occurs at particular times of year, and results in diurnal biphasic (high nocturnal, low daytime) grazing activity.

Period of diurnal migration

The seasonal period over which migration takes place varies from lake to lake, being quite long in some cases (e.g., June to November in Lake Constance, Germany) and relatively short in others (May to June in Lake Maarsseveen, Holland). In Lake Maarsseveen (Ringelberg et al., 1991), the onset of *Daphnia* migration is relatively sudden and corresponds to a rapid increase in juvenile perch (*Perca fluviatilis*) – a major predator. It is now established (Nesbitt et al., 1996) that zooplanktivorous fish (and other zooplankton predators) release chemical cues (referred to as 'kairomones') that enhance the light-mediated activation of the circadian rhythm (Figure 9.15). This is also promoted by 'alarm substances' released by injured or partially-eaten prey organisms of the same species. Kairomones have also been shown to cause alterations in zooplankton morphology and life style. Laboratory studies by Krueger and Dodson (1981) showed that a water-soluble factor released by the predatory midge larva *Chaoborus americanus* causes embryos of *Daphnia pulex* to develop into a form (referred to as '*Daphnia minnehaha*') which is less readily eaten by *Chaoborus* compared to the normal embryonic form.

The studies by Ringelberg et al. (1991) showed that upward migration of *Daphnia* commenced well after sunset (coinciding with maximum relative decreases in light intensity), and that the downward movement started about $1\frac{1}{2}$ hours before sunrise (coinciding with the maximum relative increases in light intensity). These movements lead to high rates of nocturnal grazing at the lake surface and low rates of daytime grazing at depth (Figure 5.19). This has implications for the recycling of nitrogen and phosphorus (Section 5.7.2) and for the related growth of phytoplankton and other microorganisms.

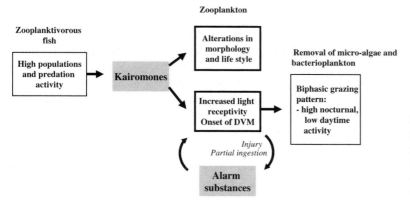

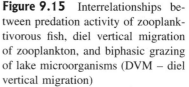

Figure 9.15 Interrelationships between predation activity of zooplanktivorous fish, diel vertical migration of zooplankton, and biphasic grazing of lake microorganisms (DVM – diel vertical migration)

The amplitude of migration is initially low, increasing later with progression of the DVM period. For the rest of the year, zooplankton does not have a diurnal cycle, and in the period immediately before DVM commences organisms are distributed equally within the top 7 m over the diurnal period.

Vertical migration: a strategy for predator avoidance and energy conservation

According to the 'predator avoidance hypothesis' downward migration of zooplankton has evolved as a strategy to avoid predation, by moving to regions of the water column where low light levels limit visual location and consumption. This hypothesis is supported by the discovery of predator kairomones, which promote the onset of the DVM seasonal period. The hypothesis is further supported by the observation (Gliwicz, 1986) that vertical migration of copepods is more pronounced in lakes with long-standing fish populations compared with lakes in which zooplankivorous fish were introduced more recently. The latter would also suggest that DVM is not just a short-term response to high predator populations, but also involves long-term genetic changes in zooplankton populations.

Although the predator-avoidance hypothesis is well established, the adaptive basis for this strategy is not always clear. In Lake Maarsseveen, for example, midday populations of *Daphnia* occur mainly at 8–12 m – well above the 18–21 m depths over which the threshold for visual predation occurs. These lower depths have a light intensity range of 10^{-2} to 10^{-4} μmol m^{-2} s^{-1}, over which the ability of perch to catch *Daphnia* becomes limited (Ringelberg *et al.*, 1991). The predator avoidance response also becomes difficult to interpret in conditions of multiple predation, involving pelagic fish, surface invertebrates (e.g., *Notonecta*), and benthic predators (e.g., *Chaoborus*) – Figure 9.15.

By avoiding daytime fish predation, zooplankton move out of the main phytoplankton zone and into a region of the water column where food availability is limited. This is to some extent compensated by low temperatures and low oxygen concentrations within the hypolimnion, reducing respiration and limiting energy expenditure. This energy conservation occurs at a time when high metabolic activity has no value for food acquisition.

9.10 Effects of algal toxins on zooplankton

The release of toxins by aquatic micro-algae (Section 10.7.3) is primarily regarded as an adaptive measure to reduce predation by zooplankton – although other roles such as antagonistic effects on other algae (competition between species), retrieval of ferric ions from the surrounding medium (siderophore), and immobilization of prey by phagotrophic algae such as *Prymnesium* (Skovgaard and Hansen, 2003) may also be important.

Production of toxins by colonial blue-green algae has particular ecological significance in the freshwater environment. The combination of blue-green algal toxin production and lack of edibility (due to size) are major factors in repressing zooplankton grazing during bloom formation, leading to uncontrolled algal growth and a breakdown in ecosystem homeostasis (Section 10.2.3).

Laboratory studies have shown that algal toxins act by both inhibition of zooplankton feeding and by physiological effects which directly result in mortality. These factors have been investigated by DeMott *et al.* (1991), who studied the effects of purified toxin (microcystin-LR) on the physiological sensitivity and feeding behaviour of selected zooplankton species (Table 9.9). Different species varied in their response to toxin, with the 48 h LC_{50} dose (the concentration of toxin required to kill 50 per cent of the zooplankton population) ranging from under $1\,\mu g\ ml^{-1}$ to over $20\,\mu g\ ml^{-1}$, and feeding being inhibited in one case but not others. These *in vitro* responses to purified toxin were related to survival in the presence of toxic *Microcystis*. Highest rates of survival (*Daphnia pulicaria*) corresponded to a rapid inhibition of ingestion (preventing uptake of toxic cells) and low physiological sensitivity, while the lowest survival (*Diaptomus birgei*) related to a high physiological

sensitivity and lack of inhibition of ingestion. In these experiments, survival of *Daphnia pulicaria* in the presence of *Microcystis* occurs over a relatively short-term (10 day) period. Under field conditions, continued inhibition of feeding would lead to ultimate population decline.

Other studies have also demonstrated physiological and behavioural adaptations of zooplankton to the presence of toxic blue-green algae, including:

- Physiological resistance to toxic blue-green algae. Some species of zooplankton, including the rotifer *Brachionus calyciflorus* (Fulton and Paerl, 1987) and the cladoceran *Bosmina longirostris* (Fulton, 1988) are resistant to toxic algae which cause rapid mortality in other zooplankton species.

- Feeding discrimination by copepods between toxic and non-toxic blue-green algae in mixed populations. This is clearly an adaptive behaviour, limiting the ingestion of toxic cells – but also favouring the survival and evolution of toxic strains of blue-green algae (DeMott and Moxter, 1991). The inhibition of feeding in cladocerans noted earlier may only be temporary. Lampert (1981) noted that *Microcystis*-induced feeding inhibition in *Daphnia pulicaria* quickly recovered when toxic algal cells were removed, demonstrating this adaptive behaviour to be a short-term response.

- Migration within the water column. Field observations by Forsyth *et al.* (1990) suggest that

Table 9.9 Effects of purified algal toxin (microcystin-LR) on the survival of selected zooplankton species (experiment of DeMott *et al.*, 1991)

Species	Physiological sensitivity (48 h LC_{50})[*]	Feeding behaviour	Survival
Diaptomus birgei	0.45–1.00	No inhibition	Poor
Daphnia pulex	9.6	No inhibition	Intermediate
Daphnia hyalina	11.6	ND	ND
Daphnia pulicaria	21.4	Rapid inhibition	Good

[*]Measured as μg toxin ml^{-1}. ND: No data.
Sensitivity and feeding behaviour were tested with purified toxin and related to survival in the presence of live *Microcystis*

populations of zooplankton may avoid depths where toxic blue-green algae (*Anabaena*) are abundant.

9.11 Biomass relationships between phytoplankton and zooplankton populations

Herbivorous zooplankton compete with protozoa in their feeding on bacteria, phytoplankton, and non-living organic debris. In general, protozoa are particularly important in the removal of bacteria and small unicellular algae (see previously), while zooplankton are more involved in grazing larger unicellular and colonial algae. Interactions between phytoplankton and zooplankton populations are of interest in relation to top-down control of phytoplankton (Chapter 10), bottom-up control of zooplankton, and biomass transfer.

9.11.1 Bottom-up control of zooplankton populations

Bottom-up control of zooplankton populations depends on phytoplankton availability, as indicated by a close correlation between phytoplankton and zooplankton biomass.

Various workers have demonstrated a statistical relationship between the major primary producer (phytoplankton) and the main herbivore (zooplankton) populations in lake systems. McCauley and Kalff (1981), for example, found a clear correlation between the two in studies of 20 Canadian Lakes, where:

$$\log_{10}(\text{zooplankton biomass}) = 0.5 \log_{10} \\ (\text{phytoplankton biomass}) + 1.8 \qquad (9.4)$$

The greater mass of herbivore compared to primary producer in these lakes appears to contradict the ecological pyramid noted earlier (Section 1.5.2, Figure 1.10), but can be explained by the high grazing pressure on phytoplankton and the higher turnover rate of algal biomass.

9.11.2 Biomass transfer

The transfer of biomass from phytoplankton to zooplankton is an important aspect of lake microbiology in terms of the general food chain, elemental cycles, natural control of microbial populations, and human implementation of algal control measures under conditions of eutrophication (Chapter 10).

At any point in time, the rate of biomass transfer depends on the standing populations of phytoplankton and zooplankton, and characteristics of grazing activity (Figure 9.16). Many modelling studies have attempted to simulate biomass transfer in aquatic systems (e.g., Jørgensen, 1983; Jørgensen and Bendoricchio, 2001), using quantitative parameters which characterize the above aspects. The conversion of biomass during grazing can be quantified in terms of overall efficiency (assimilation efficiency, Michaelis–Menten constant for grazing), grazing activity of zooplankton (grazing rates, minimum required algal population), and suitability of algal food source (selectivity coefficient) – Jørgensen *et al.* (1978).

Standing populations of phytoplankton and zooplankton are the sum of all the loss (mortality, predation) and gain (growth rate, nutrient input) processes, including biomass transfer. Under conditions of high zooplankton grazing, for example, although phytoplankton productivity may be very high, excessive rates of biomass conversion may result in a low standing phytoplankton population. The high zooplankton population which develops in this situation may ultimately be limited by the 'carrying capacity' of the system. This is an important limit for both zooplankton and phytoplankton populations, and is the maximum stable population reached by a particular species or group of organisms within a particular ecosystem. Although the carrying capacity may be reached after a period of exponential growth, this cannot be sustained due to the limited resources of the environment.

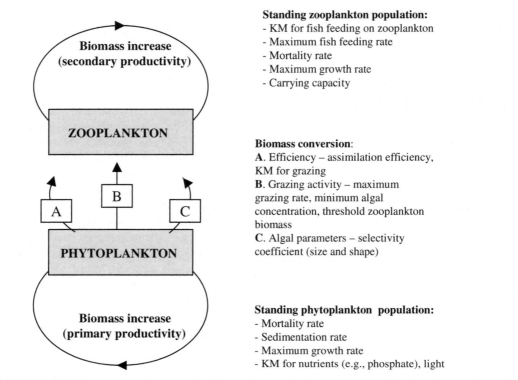

Figure 9.16   Modelling biomass transfer: some of the main quantitative parameters affecting biomass interactions between zooplankton and phytoplankton. Biomass conversion during grazing is affected by interrelated parameters that describe the efficiency of conversion, grazing activity, and algal characteristics. The algal threshold concentration varies with zooplankton species, with values for example of 0.06 μg C ml^{-1} for *Daphnia hyalina* and 0.10 μg C ml^{-1} for *D. galatea* (Geller, 1985). These figures are independent of the population density of *Daphnia*, and are virtually independent of *Daphnia* size and water temperature. The standing populations of zooplankton and phytoplankton reflect the balance between productivity (biomass increase) and removal by predation and parasitism (KM = Michaelis–Menten constant)

C.   GRAZING OF BENTHIC MICROORGANISMS

Benthic microorganisms are important components of both lotic (Steinman, 1996) and lentic (Cuker, 1983) aquatic systems, and include bacterial communities on exposed surfaces (biofilms), fungal and protozoan communities within organic detritus, and algal communities (periphyton) on rocks and stones. Grazing of these microbial populations shows key differences from the pelagic environment, and has important effects on the structure and physiology of the microbial communities.

9.12   Comparison of pelagic and benthic systems

Grazing activities in benthic environments such as river beds (Section 1.8) differ from the pelagic systems seen in the central zone of lakes (Section 1.7) in a number of ways. In benthic systems:

- Microorganisms are typically being removed from exposed surfaces or layers of detritus, so grazing activity occurs primarily in two-dimensional rather than three-dimensional space.

- The grazing organisms include protozoa (as with pelagic systems) but populations of larger grazing organisms are not dominated by crustaceans (zooplankton). While protozoa are particularly important in grazing benthic bacterial populations, removal of algae (periphyton) is largely carried out by mollusca (snails), insects (caddisfly, mayfly and chironomid larvae), amphibia (tadpoles), and fish.

- Primary production involves both algae (periphyton) and vascular macrophytes, so that herbivorous activity is directed against both of these groups of organisms. In the pelagic zone of temperate lakes, herbivory is almost entirely restricted to removal of phytoplankton.

The distinction between pelagic systems (grazing of phytoplankton by zooplankton) and benthic systems (grazing of periphyton by bottom-living invertebrates) becomes less clear in littoral zones and shallow lakes, where benthic organisms are also involved in phytoplankton grazing. Studies on shallow Greenland lakes by Vadeboncoeur *et al.* (2003), for example, demonstrated significant grazing of phytoplankton in oligotrophic lakes by filter feeders (mussels and chironomids), with additional grazing by amphipods, isopods, and snails in eutrophic conditions – where substantial amounts of phytoplankton settle out on the sediments.

Table 9.10 Feeding roles of invertebrate consumers in running waters (adapted from Allan, 1995)

Feeding role	Main food source	Microorganisms	Feeding mechanism	Invertebrate consumers
Shredder	Non-woody CPOM, primarily leaves	Asssociated microbiota, especially fungi, but also protozoa and bacteria	Chewing and mining	Several families of Trichoptera, Plecoptera, and Crustacea Some Diptera, snails
Shredder/gouger	Primarily surface layers of woody CPOM	Asssociated microbiota, especially fungi	As above	Occasional taxa among Diptera, Coleoptera, and Trichoptera
Suspension feeder/ filterer-collector	FPOM released from benthic region, including sloughed periphyton	Protozoa, bacteria, unicellular algae and fragments of filamentous algae	Collect particles using setae, specialized filtering apparatus or nets and secretions	Net-spinning Trichoptera, Simuliidae and other Diptera Some Ephemeroptera
Deposit feeder/ collector gatherer	FPOM and organic microlayer	Bacteria and protozoa	Collect surface deposits, browse on amorphous material, burrow in soft sediments	Many Ephemeroptera, Chironomidae, and Ceratopogonidae
	Surface of algal mats, loosely-attached periphyton	Filamentous algae	Collector feeding structures	
Surface grazer	Low profile periphyton and biofilms	Mainly diatoms and bacteria	Scraping and rasping	Several families of Ephemeroptera and Trichoptera Also snails

CPOM – Coarse particulate organic matter; FPOM – Fine particulate organic matter

9.13   Role of invertebrates in consuming river microorganisms

In running waters, microorganisms may either be directly grazed as part of a dense microbial community (e.g., bacterial biofilm, periphyton) or be consumed within a matrix of extraneous organic matter (e.g., leaf litter). The role of invertebrates in the breakdown of leaf litter is noted in Section 8.5.2. Fragmentation of leaf material by invertebrates occurs as the third major phase of leaf processing (Figure 8.4), and takes place after prior alteration of the leaf tissue by fungal and bacterial activity.

Invertebrates feed in various ways including shredding, gathering, and piercing of bulk material, and scraping and rasping of hard surfaces. Adaptations of feeding mechanism to food resource are shown in Table 9.10, which also emphasizes the diversity of invertebrate grazers (particularly insects) present in the stream environment. Invertebrate feeding on stream periphyton has been inves-

tigated in relation to removal of biomass and changes in algal community structure (Steinman, 1996).

9.13.1   Grazing of periphyton biomass

Interactions between the algal community and the populations of herbivores are complex (Steinman, 1996). Although algal biomass almost always declines in the presence of herbivores, the periphyton community may also be influenced by abiotic factors such as type of substratum, nutrient concentration, light regime, system hydraulics, and disturbance history. These environmental factors all exert direct effects on the growth and development of the algal community, and thus have indirect effects on the algal–grazer interaction.

The identity of the grazing organisms at a particular site and mechanical aspects of the grazing activity depend to a large extent on the structure of the periphyton community. This is shown in

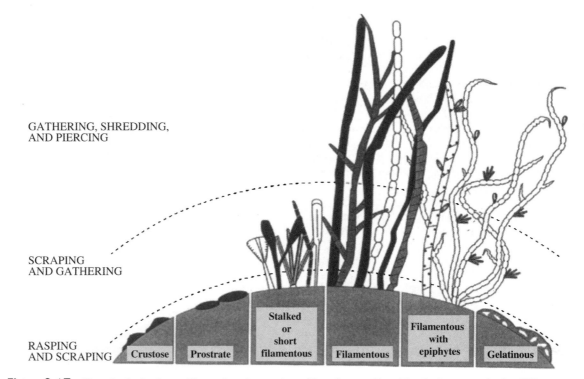

GATHERING, SHREDDING, AND PIERCING

SCRAPING AND GATHERING

RASPING AND SCRAPING

Crustose   Prostrate   Stalked or short filamentous   Filamentous   Filamentous with epiphytes   Gelatinous

Figure 9.17   Hypothetical scheme illustrating the grazing of key forms of benthic algal community by different types of grazers (figure taken from Steinman, 1996, with permission from Elsevier)

Figure 9.17, which divides periphyton into three main types – low profile algae (crustose, prostrate and gelatinous forms), stalked algae (mainly diatoms), and stalked filamentous forms. These categories follow the successional sequence noted previously. These different types of algal community require different types of mechanical grazing activity, with rasping, scraping, gathering, shredding, and piercing all being important in different situations. Some grazers, such as larvae of mayfly species, are particularly adapted for feeding activity at the surface of algal mats or on loosely-attached portions, and have collector feeding structures. In contrast, caddisfly larvae and snails have scraping and rasping mouthparts, respectively, and are better suited to feed on communities of low-profile, tightly attached algae such as diatom biofilms.

9.13.2 Effects of grazing on periphyton community structure

In addition to reducing algal biomass, grazers may also cause alteration to the taxonomic composition and biodiversity of the benthic algal population (Steinman, 1996).

Various studies have noted grazing-induced changes in the taxonomic composition of the periphyton community. In one particular example, larvae of the caddisfly *Leucotrichia pictipes* were reported as removing (but not ingesting) filaments of the blue-green alga *Microcoleus vaginatus* in order to promote the growth of an understory assemblage of diatoms and the blue-green alga *Schizothrix calcicola*. Such 'weeding' behaviour results in the growth of preferred food items for the caddisfly, with major taxonomic alterations to the stream periphyton.

Many grazing studies have noted a decline in periphyton species diversity due to grazing activity. Reduction in species diversity appears to be directly related to grazing pressure, with low to moderate grazing activity having little effect but intense grazing pressure causing a loss of species richness.

In terms of algal physiology, the most notable effect of grazing is to reduce the area-specific primary production (ASPP). This is entirely expected, occurring as a direct consequence of reduction in community biomass. Biomass-specific primary production (BSPP) does not change, however, suggesting that changes in taxonomic composition or community structure do not influence the rate of photosynthesis per unit biomass.

10

Eutrophication: the microbial response to high nutrient levels

Eutrophication may be defined as the inorganic nutrient enrichment of natural waters, leading to an increased production of algae and macrophytes. Many lakes are naturally eutrophic, and in some cases there is a progressive eutrophication as the lake matures (Section 3.7.3). The term 'eutrophication' is more widely known in relation to human activities, where the artificial introduction of plant nutrients (particularly phosphorus and nitrogen) has lead to community changes and a deterioration of water quality in many freshwater systems. This aspect has become increasingly important with increases in human population and more extensive development of agriculture, and eutrophication now ranks with other major anthropogenic effects such as deforestation, global warming, depletion of the ozone layer, and large-scale environmental disturbance in relation to its potentially harmful effect on natural ecosystems.

In terms of aquatic microbiology, eutrophication results in changes in the biomass and taxonomic composition of all groups of microorganisms present in freshwater systems. Some of these effects have already been noted in relation to lake phytoplankton (Section 3.9.1), river diatom biofilms (Section 3.9.2), planktonic bacteria (Section 6.10), and benthic protozoa (Section 9.6). Freshwater microorganisms have a central role in the environmental effects of eutrophication, since it is the microbial response which leads to physico-chemical changes in water quality and can ultimately disrupt ecological balance and system stability.

This chapter considers the origins of eutrophication, the microbial response to increased nutrient levels, and measures that can be taken to control the growth of deleterious microbial populations. These issues are discussed particularly in relation to standing waters (lentic systems) where high nutrient levels can lead to the development of intense algal blooms and the resulting environmental deterioration. Eutrophication of rivers and streams is also an important aspect of freshwater microbiology, but environmental effects are generally less acute than in lentic systems due to the short retention time and lack of stratification (Section 10.1.4).

Freshwater Microbiology: Biodiversity and Dynamic Interactions of Microorganisms in the Aquatic Environment   David C. Sigee
© 2005 John Wiley & Sons, Ltd   ISBNs: 0-471-48529-2 (pbk) 0-471-48528-4 (hbk)

A. ORIGINS OF EUTROPHICATION

10.1 Nutrient status of freshwater environments: from oligotrophic to eutrophic systems

Previous chapters have emphasized the wide-ranging biological effects of physical and chemical parameters in freshwater systems. This chapter focuses on one particular aspect of water chemistry, the impact of variations in inorganic nutrient concentrations. Variations in nutrient level affect the natural ecology of aquatic systems and also cause major problems in human use of water resources.

10.1.1 Eutrophic and oligotrophic lakes: definition of terms

The three inorganic nutrients of major importance in freshwater systems are nitrates, phosphates, and silicates. High concentrations of these nutrients in lake water promote the active growth of phytoplankton, leading to the massive development of algal biomass (high primary productivity) and the resulting growth (high secondary productivity) of all other lake organisms – including bacteria, zooplankton, and fish.

The twin aspects of nutrient concentration and productivity have been used to provide a trophic classification of lakes in temperate climates. Two major categories can be recognized:

- *eutrophic lakes* – high concentrations of nitrates and phosphates, high primary and secondary productivity, and

- *oligotrophic lakes* – low concentrations of either nitrates or phosphates (or both), low primary and secondary productivity.

These two major categories are part of a continuum in terms of water quality, and it is convenient to recognize five main groups – hypertrophic, eutrophic, mesotrophic, oligotrophic, and ultra-oligotrophic in descending order of enrichment and productivity.

Attempts have been made to define these terms in relation to fixed 'boundary' values for both nutrient and productivity water-quality parameters. One particular scheme, developed by the Organization for Economic Cooperation and Development (OECD), provides specific criteria for temperate lakes in terms of the mean annual values of total phosphorus, chlorophyll-a, and Secchi depth (Table 10.1). On this scheme, for example, the mean annual concentration of total phosphorus ranges from 4–10 μg l^{-1} for oligotrophic lakes, and 35–100 μg l^{-1} for eutrophic lakes. Boundary values for the main soluble inorganic nutrients – orthophosphate and dissolved organic nitrogen (nitrate, nitrite, ammonia) – have also been designated (Table 10.1).

Although this scheme provides a useful conceptual framework for defining trophic categories, there are limitations in its practical use for classifying particular lakes. Because of variability within ecosystems, some water bodies can be classified in one or another trophic category, depending on which parameter is used. In an attempt to alleviate this, a more flexible 'open boundary' system has been developed (OECD, 1982), where the status of individual water bodies is determined by statistical fit to more open ranges of the above parameters.

The connection between nutrient status and algal productivity implied in this system is not absolute. Lake Baikal (Russia), for example, is ultra-oligotrophic in terms of physicochemical characteristics but is close to mesotrophic on the basis of primary production and the high productivity of subsequent heterotrophic levels (Mazepova, 1998). In other cases, lakes of high nutrient levels have limited productivity due to conditions of high acidity (acidotrophic lakes), high content of brown humic acids, or high turbidity – all of which may severely limit algal growth.

Climatic differences also bring limitations to the classification system. Tropical lakes show marked differences from temperate lakes, for example, in terms of trophic indicators for the boundary between mesotrophic and eutrophic categories (Ryding and Rast, 1989).

Table 10.1   Trophic classification of temperate freshwater lakes, based on a fixed boundary system

	Trophic category				
	Ultra-oligotrophic	Oligotrophic	Mesotrophic	Eutrophic	Hypertrophic
Nutrient concentration (μg l^{-1})					
Total phosphorus (mean annual value)	<4	4–10	10–35	35–100	>100
Ortho-phosphate[*]		<2	2–5	5–100	>100
DIN[*]		<10	10–30	30–100	>100
Chlorophyll-a concentration (μg l^{-1})					
Mean concentration in surface waters	<1	1–2.5	2.5–8	8–25	>25
Maximum concentration in surface waters	<2.5	2.5–8	8–25	25–75	>75
Secchi depth (m)					
Mean annual value	>12	12–6	6–3	3–1.5	<1.5
Minimum annual value	>6	>3.0	3–1.5	1.5–0.7	<0.7

Lakes are classified according to mean nutrient concentrations and phytoplankton productivity (shaded area). Boundary values are mainly from the OECD classification system (OECD, 1982), with the exception of orthophosphate and dissolved inorganic nitrogen (DIN), which are from Technical Standard Publication (1982).

[*]Orthophosphate and DIN are measured as the mean surface water concentrations during the summer stagnation period.

10.1.2 Determinants of trophic status: location, morphology and hydrology

Location has an important influence on the nutrient status and productivity of lakes in relation to local geological conditions (e.g., phosphate content of their substrata), nature of the catchment area (particularly soil nutrient content), and internal hydrological characteristics (degree of mixing, water retention time). Many oligotrophic lakes occur in mountainous areas, with a distinctive morphology (typically deep lakes) and with an infertile, undisturbed catchment area. In contrast to this, eutrophic lakes are typically shallow lowland water bodies, surrounded by a fertile (often cultivated) catchment. The importance of local conditions is emphasized by variations that occur within a small geographic area such as the English Lake District (UK), where upland oligotrophic lakes (e.g., Wastwater) and lowland eutrophic lakes (e.g., Esthwaite) occur in relatively close proximity (Section 2.3, Figure 2.4).

The high nutrient status of lowland lakes is closely tied to human occupation of lowland areas, with the resulting development and enrichment of agricultural land and the disposal of nutrient, rich human effluent into the freshwater systems. Differences between mountain lakes and those in cultivated lowlands were noted at an early stage in the development of limnology by Naumann (1919), who recognized the connection between geographic location and trophic (nutrient) status, and the importance of the latter in determining phytoplankton productivity.

The potential role of hydrology and residence time in long-term 'natural eutrophication' has been discussed previously in relation to changes in phytoplankton species composition with nutrient status (Section 3.9.1). It has been suggested (Ryding and Rast, 1989) that long-term nutrient enrichment may result in some lakes from the gradual accumulation of inorganic and organic nutrients which enter the lake from external sources. Uptake of nutrients by phytoplankton in the epilimnion leads to deposition in the sediments, with recycling and release back into the water column. Lentic systems are not entirely static, however, and there is some loss of soluble and insoluble nutrients due to water replacement – quantitatively defined as the water residence time. Vollenweider and Kerekes (1980) have shown that if the residence time of the water in a lake is lower than 1 year, the accumulation of organic matter will not be enough to increase its trophic status.

Long-term changes in the trophic status of lakes can be inferred from paleolimnological studies on lake sediments (Battarbee, 1999), where transitions in diatom population reveal gradual alterations in both physical and chemical parameters (Section 3.7.3). In many cases, these changes in diatom population relate to increased nutrient input rather than natural eutrophication processes (Figure 10.3).

With increases in nutrient concentration and productivity the lake enters a steady state of eutrophy, which may last for a long time (thousands of years). Eventually the lake becomes too shallow to support phytoplankton growth or regeneration of nutrients and develops into a closed swamp or fen (Harper, 1992).

10.1.3   Artificial eutrophication: the impact of human activities

Human activities have been a major cause of eutrophication of freshwater systems (Mason, 2002), either by direct discharge of contaminating nutrients into the aquatic system or indirectly, by such processes as deforestation or alteration of drainage patterns. Direct contamination of water sources involves three main types of pollutant – domestic discharges (particularly sewage), industrial effluent, and agricultural waste. The sources of nutrient entry into the freshwater system are of two main types as follows.

- *Point source*: where inflow into the lake or stream is localized. This is typical of sewage and industrial effluent, and is also characteristic of some types of agricultural pollution.

- *Diffuse source*: where entry of organic pollutants occurs over a wide area and includes agricultural seepage, runoff from road systems, and aerial pollution.

In Europe and North America, the effects of human activities on freshwater systems have been particularly acute since the 1930s – when major increases occurred in energy consumption, intensification of land use, introduction of synthetic detergents, and sewage discharge. Various examples of human activities in promoting eutrophication are given later, including river/estuarine systems (Section 10.1.4), an oligotrophic alpine lake (Section 10.3.1), a major wetland area (Case Study 10.4), and fishpond ecosystems (Section 10.4). In the latter case, nutrient enrichment is (unusually) a deliberate policy to increase aquatic productivity.

10.1.4   Eutrophication of rivers and streams

The environmental effects of high nutrient levels are generally less acute in lotic compared with lentic systems. The main reason for this is water flow, which causes continuous displacement of suspended material (low retention time) and prevents the build-up of planktonic biomass in any particular locality. Turbulence and vertical circulation in streams and rivers also prevent stratification, limiting algal growth due to reduced overall light exposure during the circulation cycle and also due to the shading effects of suspended particulate matter. Vertical circulation in lotic systems also prevents the extremes of water column microenvironment (surface O_2 supersaturation, anaerobic hypolimnion) which frequently develop in eutrophic lakes (Figure 5.9). In spite of these growth limitations, algal blooms (with resulting environmental problems) may develop in some rivers.

Algal blooms in lotic systems

Although algal bloom development is generally limited in lotic systems, there are two situations where this does not apply.

- *Large, slow-flowing rivers*. Various studies (e.g., Sellers and Bukaveckas, 2003) have demonstrated substantial riverine phytoplankton development under conditions of low flow and high solar radiation (Section 1.8.2). Serious blue-green algal blooms may develop in response to nutrient enrichment and water extraction (reduced flow), and have been reported in various Australian rivers.

- *River-estuarine systems*. The complex interactions that occur between saline and fresh waters in estuarine systems (Sections 2.12 and 2.13) can result in the development of river blooms of algae which are normally associated with coastal (marine) waters. Nuisance phytoplankton of coastal waters are dominated by dinoflagellates such as *Gonyaulax* and *Gymnodinium* (Section 10.5) – but also include *Pfiesteria piscicida*, which has been implicated in fish kills in estuaries of the middle and southern Atlantic coast of the USA (Silbergeld *et al.*, 2000). *Pfiesteria* outbreaks have been associated with fish kills in the Pocomoke and Chicamacomico Rivers which drain into Chesapeake Bay (USA). These fish kills have been particularly acute in recent years, resulting in closure of the Pocomoke River in 1997. Studies in North Carolina (Burkholder and Glasgow, 1997) have linked *Pfiesteria* abundancy to eutrophication, demonstrating higher populations in waters near sources of organic phosphates, such as sewage outfalls.

Laboratory studies have shown that the presence of fish is critical in promoting a transformation of non-toxic stages of *Pfiesteria* to toxic zoospores (Burkholder and Glasgow, 1997), resulting in the release of toxins which lead to the formation of surface lesions and the ultimate death of fish. High fish populations and eutrophication thus combine to exacerbate the environmental effects of this dinoflagellate. By increasing the concentration (cells ml^{-1}) of *Pfiesteria* in river/estuarine systems, nutrient loads increase the likelihood of a toxic outbreak when adequate numbers of fish are present.

Nutrient concentrations and water quality in lotic systems

The range of nutrient concentrations in rivers extends over a wider scale than lentic systems. In the UK Environment Agency (1998) classification of river systems (Table 10.2), the penultimate (level 5) trophic level ranges from 200–1000 µg l^{-1} for phosphate, and 30–40 mg l^{-1} for nitrate. These values compare to respective eutrophic lake values of 5–100 µg (phosphate) and 0.03–0.1 (nitrate). Although extensive phytoplankton populations do not normally develop in rivers, the growth of aerobic bacterial populations results in some oxygen depletion. Values for dissolved oxygen fall to 20 per cent saturation towards the higher nutrient levels (Table 10.2), with biological oxygen demand (BOD) rising to 15 mg l^{-1} and ammonia concentrations (indicative of anoxic state) rising to 9 mgN l^{-1}. In lakes of an equivalent trophic status, the lower part of the water column would be completely anaerobic.

Biological assessment of water quality (including nutrient status) in streams typically involves a survey of the macro-invertebrate fauna, using a score system which combines measures of taxon richness and

Table 10.2 Eutrophication classification of river systems (table adapted from UK Environment Agency Publication, 1998)

	Trophic category					
	Very low (1)	Low (2)	Moderate (3)	High (4)	Very high (5)	Excessively high (6)
Phosphate (µg l^{-1})	<20	20–60	60–100	100–200	200–1000	>1000
Nitrate (mg l^{-1})	<5	5–10	10–20	20–30	30–40	>40
Dissolved oxygen (% saturation)	80	70	60	50	20	<20
BOD* (mg l^{-1})	2.5	4	6	8	15	N/A
Ammonia	0.25	0.6	1.3	2.5	9	N/A

*BOD – biological oxygen demand.

species sensitivity to nutrient enrichment (Pinder and Farr, 1987). This is in contrast to lentic systems, where microbial indices (particularly phytoplankton) are more commonly used (Section 10.3).

Remedial action

With large slow-flowing rivers, where phytoplankton blooms can become a problem, many of the remedial measures outlined in Sections 10.8–10.10 can be applied to limit the effects of eutrophication.

In addition to long-term enrichment effects, episodic events may also be important. Occasional discharges of inorganic and organic nutrients can occur, for example, resulting in severe deoxygenation due to bacterial growth (Section 9.6.3) and requiring short-term action. One such example occurred in 1987, with the discharge of a large quantity of farm effluent into the Great Stour River (UK). Severe downstream oxygen depletion was avoided by oxygen injection into a stretch of the river, preserving a major part of the fish population (Collingwood, 1987).

B.   ECOLOGICAL EFFECTS OF EUTROPHICATION IN STANDING WATERS

10.2   General biological changes

Natural standing waters range from ultra-oligotrophic to eutrophic, with progressive increase in productivity and related parameters. In addition to such general changes, eutrophication also affects the vertical structure of lakes, with further implications for the biology of freshwater organisms. The transition from eutrophic to hypertrophic status is usually the result of human activities, and ultimately affects the whole ecological balance of the freshwater system.

10.2.1   The progression from oligotrophic to eutrophic waters

Biological changes in standing waters of increasing nutrient content are summarized in Table 10.3. Nutrient concentrations directly affect the population growth of the three main groups of primary producers – phytoplankton, higher plants (macrophytes), and benthic algae (periphyton). In most lakes, phytoplankton forms the largest summer biomass, and eutrophication-linked changes in this group of microorganisms have the greatest impact on lake ecology.

Oligotrophic lakes have clear water and low overall primary productivity, with little development of either phytoplankton or macrophyte populations. The low phytoplankton level means that there is little excretion of carbon (low DOC level) and little microbial productivity (bacteria, viruses, protozoa) in the water column via the microbial loop (Figure 1.9). Low phytoplankton biomass also limits productivity along the main food chain (Figure 10.1), although particulate carbon washed into the system may provide a significant (allochthonous) source of carbon for zooplankton.

Increase to mesotrophic status results in the development of macrophyte stands (with the introduction of diverse microhabitats) and an increase in phytoplankton biomass. This promotes a transition from allochthonous to autochthonous carbon as the main carbon source for secondary productivity, with increased carbon flux through the microbial loop and general food chain. Eutrophic waters support high primary and secondary productivity. At the top end of the trophic range, however, adverse changes begin to appear such as the development of turbidity, decreases in macrophyte development, decreases in biodiversity, and extreme (high to low) levels of oxygen concentration within the water column (Section 10.2.3).

Changes in populations of phytoplankton, macrophytes, and benthic algae are key aspects of the trophic progression.

Table 10.3 Biological changes in temperate standing freshwaters of increasing nutrient content, based on OECD categories

Character	Ultra-oligotrophic	Oligotrophic	Mesotrophic	Eutrophic	Hypertrophic
Water clarity	Clear	Clear	Clear	Often turbid	Turbid
Oxygen saturation	Saturated throughout water column		Variably saturated	Hypersaturated to anaerobic	Very unstable, often no oxygen
Source of C for secondary productivity	Mainly allochthonous		Allochthonous and autochthonous	Mainly autochthonous	
Dissolved organic carbon (DOC) content	Very low	Low	Variable	High	Very high
Total biomass	Very low	Low	Medium	High	Very high
Macrophytes	Low	Low	Variable	High or low	Low
Littoral benthic algae	High	High	Medium	Low	Very low
Phytoplankton Primary productivity	Very low	Low	Medium	High	High, unstable
Typical size	Small, unicellular		Mixture	Blooms of large unicells and colonial	Small, unicellular
Competitive strategy	r-selected		r and K-selected	K-selected	r-selected
Microbial productivity[*]	Very low	Low	Medium	High	High
Zooplankton and fish[†]	Very low	Low	High	High to low	High to low

Differences in populations and productivity of microorganisms are shown for littoral benthic algae (periphyton), phytoplankton (shaded), the microbial loop[*] (including bacteria, viruses, and protozoa), and the main food chain[†]. In shallow lakes, periphyton may occur across the entire area of the water body. A summary of more general changes in oligotrophic and eutrophic lakes is given in Table 2.6.

Phytoplankton

In addition to a general increase in phytoplankton biomass, eutrophication also affects taxonomic composition, competitive strategy, and biodiversity.

Species composition varies considerably in relation to water quality. Desmids and chrysophytes, for example, tend to be characteristic of low-nutrient waters, while colonial blue-greens, chlorococcales (group of green algae), and centric diatoms occur as dominant forms in more nutrient-rich habitats (Section 3.9.1, Table 3.9). The nutrient status of temperate lakes also influences the succession of algal species, both in terms of species content and in the pattern of seasonal phytoplankton succession. Rey-

nolds (1990) identified four generalized cycles relating to the four major nutrient categories, each with their own particular groupings or assemblages of algal species (Table 10.4). In all cases, the season commences with a diatom bloom, but differences occur in the particular species involved and the duration of the bloom. Oligotrophic lakes show a continuation of the diatom bloom over much of the growing season, with the late appearance of dinoflagellates and blue-green algae. The differentiation into discrete diatom bloom, clear-water phase and mixed summer bloom noted previously is more characteristic of mesotrophic and eutrophic lakes. Hypertrophic lakes and ponds (see Section 10.4) tend to have a short diatom (*Stephanodiscus*) bloom,

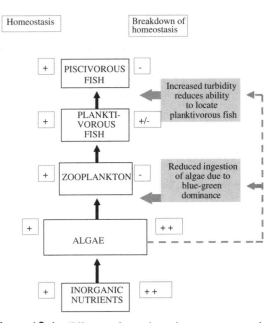

Figure 10.1 Effects of nutrient increase on major populations of pelagic biota in the lake environment. A rise in nutrient (phosphate, nitrate) concentration leads to an increase (+) or decrease (−) in populations of lake organisms. Solid vertical arrows indicate the direction of nutrient and carbon flow. *Left-hand side*: moderate increase in nutrients; homeostasis is maintained, with increased populations of all pelagic biota. *Right-hand side*: high increase in dissolved nutrients causes a breakdown in homeostasis due to limitation at two points (◀━) in the food chain; the balance between pelagic populations is now lost. (This diagram does not include secondary production via the microbial loop, and related pelagic populations of bacteria, viruses and protozoa)

followed by blooms of single-celled rapidly-growing species for the rest of the growth season.

Taxonomic differences reflect differences in algal size and competitive strategy. Oligotrophic sites tend to be dominated over much of the growing season by unicellular phytoplankton such as nanoplanktonic diatoms (Wastwater, UK) and picoplanktonic blue-greens (Lake Baikal, Russia). These *r*-selected algae (Section 3.10.2) are adapted for rapid exploitation of nutrient resources under limiting conditions. In mesotrophic and eutrophic lakes, *r*-selected species are a particular feature of the

clear-water phase, with *K*-selected species (large unicells and colonial blue-greens) forming dominant blooms over the major growth phase. At highest nutrient levels (hypertrophic lakes and ponds), there is a reversion to small, unicellular, rapidly growing species. This is seen in the highly-fertilized fishponds of the Třeboň wetlands (Section 2.8), where *r*-selected species out-compete the more slow-growing dinoflagellates and colonial blue-greens which are prominent in eutrophic waters. These nanoplanktonic organisms form dense blooms of diatoms and green algae as soon as adequate light is available.

Macrophytes

In mesotrophic and eutrophic lakes, higher plants (macrophytes) are present both as emergent plants at the edge (littoral zone) of lakes, and as fully submerged plants in shallow lakes and wetlands (Figure 4.21). Macrophytes are absent or poorly-developed in oligotrophic conditions due to inadequate nutrition. Increase from eutrophic to hypertrophic status typically leads to a loss of submerged macrophytes due to intense competition with phytoplankton (Section 4.8). In the early part of the season, dense algal blooms cause a reduction in light penetration to the young macrophytes on the lake sediments, preventing their growth and development. Long-term survival of macrophytes requires a threshold photon flux density of 45–90 μmol m^{-2} s^{-1} (Sand-Jensen and Borum, 1991). Variations in minimum light requirements between macrophytes depend on their plant-specific carbon value (plant biomass per unit of light-absorbing surface area), indicating that the light requirements of submerged plants are tightly coupled to their ability to harvest light – which relates to their growth form (Middelboe and Markager, 1997).

Changes in macrophyte vegetation have particular significance for the biology of microorganisms in lakes, since they are important sites of epiphytic attachment (filamentous algae, bacterial biofilms), and they act as major refuges for zooplankton and fish – thus influencing the grazing of phytoplankton (Figure 10.10).

Table 10.4 Patterns of phytoplankton succession in water bodies of varying nutrient status, showing some of the main trends in UK lakes and fertilised fish ponds (Czech Republic). Adapted mainly from Reynolds, 1990

Lake type	Spring		Summer	Autumn	Example
Oligotrophic	DIATOMS —————————————→				Wastwater Ennerdale
	Cyclotella —→		DINO *Ceratium* BG *Gomphosphaeria*		
Mesotrophic	DIATOMS —→	CHRYSO —→	DINO —→	DIATOMS	Windermere
	Asterionella	*Mallomonas*	*Ceratium* BG *Gomphosphaeria*	*Asterionella*	Grasmere
Eutrophic	DIATOMS —→	GREEN —→	BG —→ DINO —→	DIATOMS	Rostherne Mere
	Asterionella	*Eudorina*	*Anabaena* *Ceratium* *Aphan.* BG *Microcystis*	*Steph.*	
Hyper-eutrophic	SMALL DIATOMS —→	GREEN —→	GREEN —→	BG	Třeboň fish ponds
	Steph.	*Scenedesmus*	*Pediastrum*	*Aphanocapsa*	

Abbreviations
main groups: DINO – dinoflagellates; BG – blue-green algae; CHRYSO – chrysophytes; Genera: *Steph.* – *Stephanodiscus*; *Aphan.* – *Aphanizomenon*.

Benthic algae

Benthic community responses to lake eutrophication have received much less attention than pelagic systems, and are relatively poorly understood. In most lakes, where sediments are well below the photic zone, benthic algae are restricted to littoral communities and have limited impact on overall productivity. This is not the case for shallow lakes, however, where benthic algae make a major contribution to lake productivity and where the effects of eutrophication on benthic algae are particularly important.

In a recent survey of the effect of eutrophication on phytoplankton and benthic algal communities of a range of Greenland, USA, and Danish lakes, Vadeboncoeur *et al.* (2003) demonstrated an inverse relationship in productivity between pelagic and benthic algal communities. Phytoplankton productivity increased along the phosphorus gradient (total phosphorus [TP]: 2–430 mg m^{-3}), but littoral zone periphyton productivity decreased, substantially limiting the overall increase in primary productivity for the lake as a whole. The depression of benthic algae with increasing nutrient status can be explained in much the same way as the limitation of macrophytes. In both cases, the benthic flora are out-competed by pelagic algal blooms which prevent down-welling of light to the sediments and thus limit photosynthesis.

The effects of nutrient status on the balance between pelagic and benthic communities are particularly acute for shallow lakes. In the above study, benthic productivity of shallow lakes ranged from >80 per cent of whole lake primary productivity in the case of oligotrophic sites down to values of 5 per cent in hypertrophic conditions. In shallow water bodies, eutrophication causes a shift in the site of primary productivity from sediments to water column and uncouples a link between benthic and pelagic communities. Relationships between the pelagic and benthic food webs of lakes are discussed in Case Study 1.2, and competition between pelagic and benthic algae in Section 3.6.1.

10.2.2 Effects of eutrophication on the water column of stratified lakes

Oligotrophic and eutrophic lakes show major differences in the physico-chemistry of their water columns (Section 2.3 and Table 2.6). These differences largely arise due to the high productivity of the eutrophic state, which leads to major changes within the water body. These changes accentuate the distinction between epilimnion and hypolimnion, and include the following.

- Reduced visibility in the epilimnion and light penetration into the hypolimnion due to high phytoplankton levels. The loss of light penetration into the hypolimnion means that photosynthetic activity in this part of the water column no longer occurs, with the loss of deep subsurface chlorophyll maxima due to photosynthetic bacteria (Section 6.7.3) and populations of photo-adaptive algae (Section 4.6.3).

- Increased levels of heterotrophic bacterial activity in the hypolimnion due to increased levels of organic substrate (algal detritus) sedimenting out of the epilimnion. The increased level of bacterial productivity seen with eutrophication may lead to a shift from net autotrophy (carbon uptake mainly by photosynthetic algae) to net heterotrophy (carbon uptake mainly by heterotrophic bacteria) – Section 6.10.5.

- Wider differential in oxygen concentration. At times of high phytoplankton growth (e.g., mid-summer) in temperate climates, the epilimnion typically has elevated oxygen levels while the hypolimnion is often highly reducing. High oxygen levels in the epilimnion are associated particularly with alkaline conditions and the development of surface blooms of blue-green algae. Reducing conditions in the hypolimnion arise due to the oxygen demand of heterotrophic bacteria. These occur at much lower levels in oligotrophic lakes, which are typically aerated throughout the water column.

    The seasonal transition from aerobic to anaerobic sediments seen in many eutrophic lakes is accompanied by vertical migration of various microorganisms up into the water column. This is seen, for example, with the micro-aerophilic protozoon *Loxodes*, which moves into the upper regions of the hypolimnion (Section 9.6.1 and 9.6.2).

- The reducing conditions present in the hypolimnion of eutrophic lakes also have an important influence on the general chemistry of the water column. This is particularly the case for dissolved inorganic nitrogen (Section 5.3 and Figure 5.9), with eutrophic lakes having substantial levels of reduced nitrogen compounds (nitrites and ammonium ions) which are oxidized to nitrate in oligotrophic lakes. Anaerobic conditions also lead to higher levels of phosphate release from sediments (see Section 5.7.2).

- Nutrient availability. The high phytoplankton biomass which develops in the epilimnion of eutrophic lakes leads to a major depletion of nitrates and phosphates in this part of the water column, with much less effect on the hypolimnion. This differential loss of nutrients becomes accentuated towards the end of the growing season, when algal blooms become established (Section 3.7.2 and Figure 3.14)

10.2.3 Major changes in ecological balance: the breakdown of homeostasis

Increasing nutrient concentrations promote growth of both macrophytes and phytoplankton, the relative dominance of which depends on competition (Section 4.8). The response of lake systems to high nutrient levels depends primarily on changes within the water column, and leads to two alternate states – homeostatic and non-homeostatic ecosystems.

Homeostatic ecosystem

Any increase in phytoplankton productivity is matched by higher zooplankton and fish populations as the nutrient flux passes through the main food chain. Under normal conditions, the pelagic ecosystem is therefore able to adjust to increased nutrients and the resulting rise in algal biomass by an increase in the populations of all the other organisms, thus maintaining an ecological balance (Figure 10.1). Macrophytes and associated periphyton communities also occur, and the maintenance of a balanced community within the lake under conditions of increased nutrient loading is an example of homeostasis at the ecological level.

Non-homeostatic ecosystem

An alternative situation may arise at high nutrient levels (external stress), where homeostasis breaks down. This is caused by the development of dense populations of colonial blue-green algae, which have a dual effect on the pelagic ecosystem (Figure 10.1).

- These algae often form very large colonies, which are too big for most of the zooplankton to consume, so there is reduced control by grazing.

- High concentrations of algae in the lake water decrease visibility, so that the fish-eating (piscivorous) fish cannot see and catch the zooplankton-eating (zooplanktivorous) fish. In UK waters, this is particularly the case for perch (*Perca fluviatilis*), which have a strategy of active predation – moving around within the water column in search of their prey. Long-distance visibility is less important for pike (*Esox lucius*), the other main piscivorous fish, which tends to adopt an ambush strategy, locating and attacking its prey in the littoral zone.

Because the plankton-eaters are not being removed the zooplankton are heavily grazed upon – so in turn they exert a reduced control on the algae on which they feed. The growth and activities of zooplankton populations are also inhibited by algal toxins. All these effects combine to reduce the effectiveness of zooplankton grazing. The resulting lack of homeostatic control allows the algae to grow unchecked, causing further ecological problems to occur.

Two stable-state hypothesis   In shallow lakes, the homeostatic state is typically dominated by macrophytes (Section 4.8) and the change to a non-homeostatic state involves a sudden environmental switch to phytoplankton dominance. These two conditions represent alternate stable environmental states, as defined in the 'two stable-state hypothesis' of Scheffer (1998).

The development of these two states typically occurs as a gradual response to long-term eutrophication, and the sudden switch to blue-green algal dominance causes major environmental problems in water bodies such as the Broads Wetland area, UK (Case Study 10.4 and Figure 10.9). This switch is initially reversible, but rapidly becomes irreversible – with uncontrolled growth of blue-green algae and complete suppression of submerged macrophytes.

10.3 Biological assessment of water quality

Although the trophic status of lakes is most directly assessed in terms of nutrient concentrations and phytoplankton productivity (Table 10.1), other biological features may also provide useful supporting information. These follow on from the previous section and include the use of algal indicator groups and determination of indices of species diversity. More broad ecological assessments such as the presence or absence of macrophytes, development of littoral periphyton, physicochemical variations within the water column (oxygen depth profile, dissolved organic carbon (DOC) content, biological oxygen demand (BOD)), and the pattern of seasonal phytoplankton changes (Table 10.4), may also prove useful.

10.3.1 Algal indicator groups

Identification and enumeration of algal taxa from a particular site provides a useful indication of water quality (Table 10.4), and various limnologists have proposed specific quotients for the assessment of lake nutrient status. These quotients use ratios of particular groups of microalgae as a trophic index, and mostly depend on the fact that desmids (Figure 10.2) are typical of low-nutrient waters, while colonial blue-green algae, chlorococcales (green

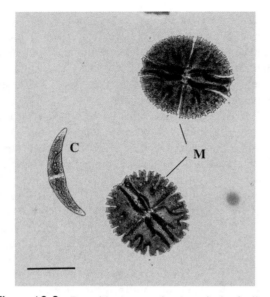

Figure 10.2 Desmids (green algae) typical of oligo-trophic environments, showing single cells of *Micrasterias* (M) and crescent-shaped *Closterium* (C). These placo-derm desmids are frequent members of the phytoplankton in low-nutrient, slightly acidic lakes, and have been used as bio-indicators of oligotrophic conditions (scale = 50 μm)

algae) and centric diatoms are more characteristic of eutrophic conditions. These quotients include the following.

The simple chlorophycean quotient of Rawson (1956), where conditions are eutrophic if:

$$\frac{\text{Number of chlorococcales species}}{\text{Number of desmid species}} > 1$$

The compound microalgal index (CI) of Nygaard (1949) where:

$$CI = \frac{\text{Species of cyanophyceae} + \text{chlorococcales} + \text{centric diatoms}}{\text{Number of desmid species}}$$

and the CI ratio varies from oligotrophy ($<$1), through mesotrophy (1–2.5) to eutrophy ($>$2.5).

The A/C diatom index of Stockner, 1972. The ratio of araphid pennate to centric diatoms (A/C ratio), varies from oligotrophy ($<$1), through mesotrophy (1–2) to eutrophy ($>$2). This ratio has been used, for example, to monitor eutrophication in Lake Tahoe (USA) – see the case study (Case Study 10.1). The   A/C diatom ratio must be used with considerable caution, however, since it may become very misleading in the presence of eutrophic centric diatoms (see below).

In addition to particular indicator groups, individual algal species may also provide diagnostic information on nutrient status of the water body. Within the centric diatoms, for example, *Stephano-discus hantzschii* is frequently found within the water column of enriched rivers and standing waters, and *S. minutula* (Figure 5.27) is common in the plankton of eutrophic lakes.

Case study 10.1   Using the A/C (araphid pennate/centric) diatom ratio to assess eutrophication in Lake Tahoe (USA)

Lake Tahoe (Figure 4.25) is the largest alpine lake in North America, and has undergone a transition from ultraoligotrophic to oligotrophic status in recent years (Carney *et al.*, 1994). This is indicated by changes in lake phytoplankton and particularly by a shift in the species composition of the diatom community. Changes in the A/C diatom ratio, evaluated from mid-lake sediment cores (Byron and Eloranta, 1984), show marked changes in relation to human habitation of lake-shore sites. During the late 1950s, there was a sharp increase in the A/C diatom ratio, from $<$0.1 to 0.7 (Figure 10.3), due particularly to proportional increases in *Fragilaria crotonensis* and *Synedra* species, with decreases in *Cyclotella ocellata* and species of *Melosira*.

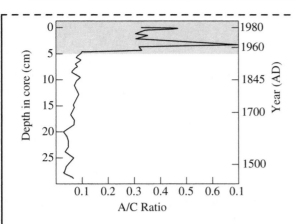

Figure 10.3 Changes in the ratio of araphid pennate to centric diatoms (A/C ratio) as an indicator of gradual eutrophication in Lake Tahoe (USA): the dramatic increase in the A/C ratio during the late 1950s is associated with increasing human population around the lake (based on a figure from Byron and Eloranta, 1984)

These changes in the diatom community parallel other indicators of eutrophication (Carney *et al.*, 1994) such as an increase in primary productivity (annual rise of 5.6 per cent from 1959 to 1990), decrease in Secchi depth, and an increase in available nitrogen (as NO_3). Experimental assays indicate a recent shift from nitrogen limitation to phosphorus limitation of phytoplankton growth, consistent with the onset of early eutrophication caused by human settlement and land use in the watershed area.

10.3.2   Indices of species diversity

Although the species diversity of most groups of freshwater microorganisms probably varies with aquatic nutrient status, there is little general information on this and species diversity has been looked at particularly in relation to the phytoplankton. Standing waters of high nutrient availability do tend to have lower levels of phytoplankton diversity, as demonstrated by Reynolds (1990), where species diversity was assessed using the Margalef index (d_s):

$$d_s = (s - 1)/\log_e N \qquad (10.1)$$

where s = number of species, and N = number of individuals.

In the nutrient-deficient North Basin of Lake Windermere (UK) the value for d_s ranged from 3–6 over the summer-growth period. This compared with values of 2–4 for a nutrient-rich lake (Crose Mere, UK) and 0.2–2 for a hypertrophic water body (fertilized enclosure, Blelham Tarn, UK).

In these and other temperate water bodies, phytoplankton species diversity varies markedly over the summer-growth phase (see also Section 3.13), making quantitative comparison difficult. The develop-

ment of dense algal blooms at high nutrient level does introduce a clear-cut difference, however, since these tend to involve dominance of a few (or even a single) species of algae – resulting in a very low index.

10.4   Problems with intentional eutrophication: destabilization of fishpond ecosystems

Although management of water bodies is normally designed to retain a balance in freshwater biota by avoiding high levels of eutrophication, there are some instances (such as commercial fishponds) where this may not apply.

10.4.1   Promotion of high productivity in fishponds

Application of organic and inorganic nutrients to fishponds may be carried out to enhance fish production by increasing carbon flow through the whole food chain. Increased nutrients promote higher levels of primary productivity (by algae) and heterotrophy (by bacteria and protozoa) – both of which lead to increases in zooplankton and zooplanktivorous fish.

Management of fishponds in the Třeboň wetlands of Central Europe (Pokorny *et al.*, 2002b; Pechar *et al.*, 2002),for example, typically involves:

- intensive stocking with fish such as common carp (*Cyprinus carpio*) – these may reach levels of 600–1000 kg ha^{-1}, and feed on zooplankton and benthos,

- application of lime (supply of carbonate and bicarbonate ions) and nutrients – including organic fertilizers and manure.

10.4.2 Destabilization and restoration of the ecosystem

Although such changes might be expected to lead to the retention of a balanced food chain, with increases in populations of all biota, the combination of fish stocking and increased nutrients can lead to destabilization of the ecosystem (Figure 10.4). High predation pressure of the fish on zooplankton leads to a situation in which phytoplankton growth is uncontrolled by zooplankton, and the rich supply of nutrients promotes growth of the algae.

Intensive growth of phytoplankton early in the season has a number of effects.

- It represses the growth of higher plants (macrophytes), so that more primary productivity is channelled into algal biomass, which promotes higher fish production.

- Increased photosynthesis of the algal population results in a rise of pH to pH10. This may cause gill necrosis and weakening fish stocks.

- Increased biomass and content of organic matter leads to increased populations of heterotrophic organisms (particularly bacteria). This results in increased levels of respiration, reducing oxygen concentrations to levels which can be critical for fish survival.

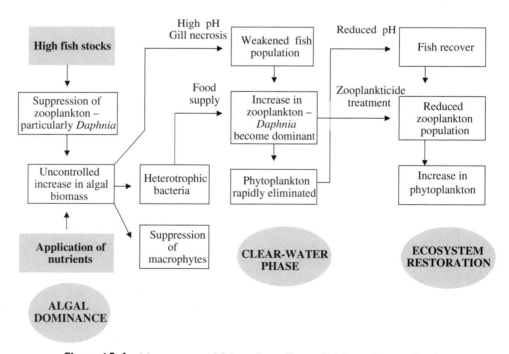

Figure 10.4   Management of fishponds – effects of high-nutrient application

Depleted fish populations lead to an increase in the zooplankton population, which become dominant and largely eliminate the phytoplankton. This clear-water phase can persist for a long time, since the high bacterial and protozoon populations support stable populations of zooplankton, particularly the large *Daphnias*. Continued decomposition of organic matter at the bottom of the pond generates such a high oxygen demand that the negligible photosynthetic activity of algae is not able to compensate, and the low oxygen levels repress growth of the fish population.

The result of intensive nutrient input in these fishponds thus leads to a breakdown of ecosystem homeostasis, as noted previously (Section 10.2.3). The difference here is that nutrient levels are so high that they promote massive early-season growth of green algae, out-competing blue-greens. Unusually for eutrophic systems, extended blooms of blue-greens do not develop, and the resulting environ-mental deterioration follows a very different route which leads to long-term zooplankton dominance.

Restoration

A simple, but rather hazardous way to change the situation (Figure 10.4) has been to add a small quantity of zooplankticide, such as ©SOLDED, which kills the *Daphnias* (Pechar *et al.*, 2002). This restores favourable conditions for the development of phytoplankton, resulting in an increased concentration of oxygen in the water and restoration of a balanced ecosystem. The use of zooplankticides, however, has broad environmental implications beyond short-term ecosystem restoration. The addition of toxins to remove any group of biota should always be regarded with extreme caution, in addition to which any contemplation of their use should first involve a check on legal approval.

C. THE GROWTH AND IMPACT OF ALGAL BLOOMS

10.5 Algal blooms and eutrophication

Algal blooms are simply dense populations of planktonic algae which develop in aquatic systems. They may occur in a wide range of environments, including lakes and rivers, exposed mudflats, and snowpacks – and are part of the normal seasonal development in many ecosystems. In all of these environments, the development of algal blooms can be seen as a balance between the processes of population increase (high growth rate, ability to out-compete other algae) and population loss (effects of grazing, parasitic attack). Increased levels of inorganic nutrients lead to a general increase in primary productivity (see previously) but may also promote algal blooms at different times of the year. In lentic environments, these blooms include the spring diatom bloom, late-spring blooms of green algae, and summer blooms of dinoflagellates and blue-green algae (Figure 3.14). Most of these blooms have no adverse effects on the environment, and the increased algal biomass is transferred to other lake biota via the normal food web (Figure 10.1).

The major problems of eutrophication come with anthropogenic enrichment of the environment and the formation of dense blooms of toxic dinoflagellates (principally marine) and colonial blue-green algae (freshwater).

- *Toxic dinoflagellates.* These organisms are particularly characteristic of marine waters, and are thus largely outside the scope of this volume. They do occur in estuaries, however, and are occasionally seen in major rivers associated with estuaries – so do have some peripheral relevance to freshwater systems. Some of these dinoflagellates are important in the formation of neurotoxins, and are potential hazards in terms of human water contact and food consumption (Silbergeld *et al.*, 2000). Toxic dinoflagellates include *Gambierdiscus toxicus* (CTX toxin), *Gonyaulax Alexandrinum* (STX), *Gymnodinium breve* (brevetoxins), *Diophysis* spp. (Okadaic acid), and *Pfiesteria piscicida* (unknown toxin). *Pfiesteria* has particular significance for freshwater systems since it has been implicated in

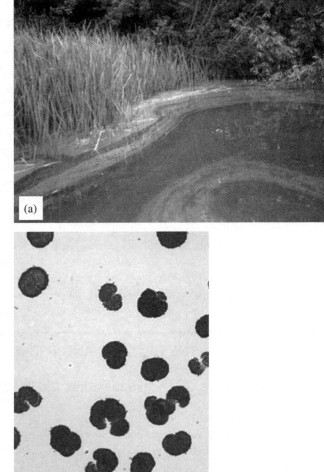

Figure 10.5 Blue-green algal bloom: (a) Edge of Rostherne Mere (UK), a naturally eutrophic lake, showing the development of a blue-green algal bloom in late August. The algal population turns the water bright green and creates a surface scum of decaying organic material. The phyto-plankton of this lake has historically attracted attention due to the large annual blooms of algae, a phenomenon known locally as 'breaking the waters'. (photograph by Elishka Rigney). (b) Low power microscopy of the algal population, the biomass is almost entirely composed of large globular colonies of *Gomphosphaeria* (scale = 100 μm)

eutrophication-related major fish kills in the Pocomoke and Chicamacomico Rivers which are part of the Chesapeake Bay (USA) complex (Silbergeld *et al.*, 2000 – see Section 10.1.4).

• *Colonial blue-green algae*. Colonial blue-greens, not dinoflagellates, form the major nui-sance-algae of freshwater systems and have the potential to cause deterioration in water quality and adverse environmental effects. In many eutrophic environments (e.g., Rostherne Mere, UK; Figure 10.5) and mesotrophic lakes of lesser magnitude, quite dense blue-green blooms occur on an annual basis without any permanent

environmental effects. It is only when these algae form very dense accumulations and totally out-compete other algae that their influence becomes severe.

Four key questions arise in relation to the role of colonial blue-green algae in the environmental effects of freshwater eutrophication.

• What environmental factors lead to bloom for-mation?

• Why is this particular group of algae able to out-compete other algae and form blooms?

- What are the effects of blue-green algae on the freshwater environment?

- How may these algae be controlled?

10.6 Formation of colonial blue-green algal blooms

There has been much written in the past 20 years or so on the ability of blue-green algae to dominate freshwater environments and form blooms. These algae tend to become a prominent feature of northern temperate lakes in midsummer, when their dominance increases and bloom formation occurs.

10.6.1 General requirements for bloom formation

Although the effects of blue-green blooms are often prolonged, their origin may appear to be rapid – with the sudden appearance of dense surface scums in eutrophic lakes and reservoirs (Figure 10.5). The origin of these blooms is partly due to massive growth (population increase) but also the ability to float to the water surface and form dense localized populations. The conditions necessary for the sudden development of surface blooms are summarized in Table 10.5, and include high nutrient concentrations, physicochemical characteristics (high light, temperature, pH) of surface water that promote the dominance of blue-green over other algae, and conditions that lead to large-scale flotation of much of the algal biomass.

The ability of blue-green algae to out-compete other members of the phytoplankton at a time of year when certain environmental aspects (light, temperature) are at an optimum, is a key feature of the success of these organisms in bloom formation.

10.6.2 Competition with other algae

Various hypotheses have been put forward to explain the ability of blue-greens to out-compete other algae (Shapiro, 1990), including their optimum growth at high temperature, low-light tolerance, tolerance of low N/P nutrient ratios, depth regulation by buoyancy, resistance to zooplankton grazing, and tolerance of high pH/low CO_2 concentrations. Most of these features probably contribute to the success of blue-greens, without being individually of sole importance. The dominant success of colonial blue-greens probably results from the sum total of all these characteristics, with an overriding requirement for high nutrient input to achieve high biomass levels.

Optimum growth at high temperature

Blue-green algae have higher growth optima than do green algae and diatoms. The midsummer increase in abundance of these algae in temperate lakes, and their success in tropical lakes, may be the direct result of their ability to grow well in warm water conditions. Chemostat experiments of Tilman *et al.* (1986) have suggested that blue-greens isolated from Lake Michigan and Lake Superior (North America) have maximum growth ability at temperatures exceeding 20°C.

Table 10.5 Requirements for the formation of blue-green algal blooms in lakes and reservoirs

Requirement	Effect
High nutrient concentrations in water (particularly phosphate)	Pre-bloom growth phase and substantial sub-surface population of blue-green algae at the time of bloom formation
High light and temperature High pH	Continued growth of blue-greens at the lake surface: these algae are able to grow and out-compete other algae at high temperature and pH conditions
High proportion of cells with gas vacuoles Stable water column	Short-term floating of cells to the top of the water column leads to surface scum

Although temperature is important, it is secondary to nutrient requirements. Oligotrophic lakes in the same geographic area as eutrophic lakes, and with the same temperature regime, do not develop blue-green blooms. A survey of world lakes lead Robarts and Zohary (1987) to conclude that temperature was of subsidiary importance.

Tolerance of low-light conditions

Various physiological studies (summarized in Zevenboom and Mur, 1980) have suggested that blue-green algae have lower light-energy requirements than green algae and diatoms. Although the ability to grow at low light may seem an unlikely advantage for algae present at the top of the water column in midsummer, the intense self-shading which occurs within turbid bloom populations make this a significant feature. Critical evidence is contradictory, however, with some studies showing that blue-greens are more dominant in turbid conditions, while others show the reverse.

Ability for growth at low N/P ratios

The apparent ability of blue-greens to out-grow other algae at low nitrogen/phosphorus (N/P) ratios, has received much attention. General evidence in support of this is not conclusive, and it may be a feature of secondary importance. Analysis of 17 lakes by Smith (1983) concluded that although the blue-green contribution to the phytoplankton could be high in lakes with low TN/TP ratios (<29/1), they only occurred at low level when ratios were higher. Other studies have been less supportive, and the general importance of low N/P ratios to blue-greens must be queried.

Depth regulation by buoyancy

The ability of vacuolate blue-greens such as *Anabaena* and *Oscillatoria* to regulate their depth by buoyancy is probably important both in the early development of algal populations and in the final phase of dominance.

In terms of population growth, it allows them to adopt an optimum position within the water column in relation to light and CO_2 availability. Regulation of depth is also important in the diurnal migration of these algae to lower (high-nutrient) parts of the water column, allowing them to continue growth at a time of year when epilimnion nutrient levels have reached a low level. During bloom formation, the rapid flotation of these algae to the lake surface leads to changes in the water chemistry and light regime at the lake surface, depressing the growth of other phytoplankton.

Depth regulation by buoyancy is particularly effective in a static water column and confers an advantage on these organisms in such stable conditions. In lakes where the water column remains stratified but without a high degree of stability, changes in the phytoplankton composition (with the emergence of blue-green dominance) still occur – suggesting that buoyancy is not absolutely essential for bloom formation in these algae.

Resistance to zooplankton feeding

Filter-feeding zooplankton often appear to feed ineffectively, if at all, on blue-green algae (Section 9.8.2). Blue-green dominance is further promoted by the elimination of competing green algae and diatoms from mixed phytoplankton populations during grazing activities. Even where blue-green algae are ingested, the presence of a thick outer layer of mucilage often allows them to pass through the zooplankter alimentary canal without being digested.

The relationship between blue-green algal dominance and zooplankton grazing is an important one. The notion that blue-green algae may become dominant in the short term due to a failure of zooplankton grazing (Figure 10.1) is suggested by several studies (Lynch, 1980; Shapiro *et al.*, 1982). These demonstrate that an abundance of filter-feeding grazers does promote dominance by blue-greens such as *Aphanizomenon*. In contrast to this short-term effect, the effectiveness of biomanipulation as a control method depends on the long-term activities of zooplankton such as *Daphnia magna* for controlling blue-greens. Lakes which have acquired these zooplankton populations tend to show a shift from blue-green to green algal dominance (Burns, 1987). The overall relationship between blue-greens and zooplankton grazers at a particular site is thus highly variable, depending on zooplankton biomass, physiological state, species composition, and whether the interaction is being considered short-term or long-term.

Tolerance of high pH and low CO_2 concentrations

In conditions of high light intensity, elevated levels of photosynthesis within the epilimnion lead to pronounced CO_2 uptake resulting in strongly alkaline conditions with low CO_2 availability (Section 4.7.1). It has been suggested that blue-green algae (but not other members of the phytoplankton) have the ability to tolerate these extreme environmental conditions – allowing them to continue active growth at a point when other algae are inhibited, thus out-competing other photosynthetic organisms.

This concept is supported by the fact that blue-green dominance occurs in most lakes only when pH is high. Blue-green dominance is absent from lakes where pH does not rise during the summer – including oligotrophic lakes (low algal biomass, lowphotosynthesis) and lakes with CO_2 sources other than the atmosphere. The importance of CO_2 and pH to blue-green dominance is also supported by enclosure experiments (see below).

Case study 10.2   Use of enclosure experiments to study factors affecting blue-green dominance

Experimental evidence for the importance of CO_2 availability, pH, and concentration of available nitrogen in blue-green dominance have been provided by lake enclosure experiments (Shapiro, 1990). These involved partitioning off discrete volumes of lake water and testing the effect of various additions to the development of mixed populations of blue-green and green algae (Table 10.6).

In control enclosures, with no alteration to the water chemistry, blue-greens became dominant. This was reversed (not strongly) by the single addition of CO_2 but not by the addition of available nitrogen. Respective increase in CO_2 concentration (addition of gaseous CO_2) or maintenance of acid-neutral pH (addition of HCl) in N-enriched enclosures both resulted in a shift to green algal dominance, while maintenance of high pH kept blue-greens dominant. These experiments are consistent with the view that blue-green dominance is determined by multiple rather than single environmental factors and that low CO_2, high pH, and low N/P ratios are all important in this.

10.7   Environmental effects of blue-green blooms

10.7.1   General environmental changes

The effects of dense blooms of blue-green algae are an extension of these features noted earlier (Section 2.3 and Table 2.6) for eutrophic lakes. Very high levels of algal biomass lead to extreme limitation of light penetration, preventing growth of other algae and completely suppressing growth of higher plants. Growth inhibition of other algae results in a dramatic loss of phytoplankton diversity within the water column.

Oxygenation in the water column also reaches new extremes. Active photosynthesis at the water surface leads to high oxygen concentrations (often

Table 10.6 Use of enclosure experiments to investigate factors affecting the dominance of blue-green algae (Shapiro, 1990)

Treatment	Shift in dominance from blue-green to green algae[*]
Control	−
Addition of CO_2 gas	+
Addition of available N (increasing N/P ratio)	−
Addition of CO_2 and available N	+++
pH maintained at 5.5–7.5 with HCl, plus added N	++
pH maintained at 8.5 with HCl, plus added N	−

[*]The shift from blue-green dominance is graded from very marked (+++) to not detectable (−).

supersaturated) and high pH (frequently >pH10). In the lower part of the water column, higher levels of algal detritus and heterotrophic decomposition result in extreme reducing conditions, with oxygen concentrations very low or non-detectable throughout most of the hypolimnion. These reducing conditions permit the accumulation of substantial levels of ammonia and nitrite in the water column, and lead to greater release of phosphate from the lake sediments adding to the already existing eutrophication problems.

10.7.2 Specific effects on water quality

In addition to these rather general environmental effects, blooms of blue-green algae also cause more specific problems in relation to water quality and human use of the water body:

- Production of high concentrations of small molecular weight toxins which are poisonous to a wide range of animals – these are discussed more fully in Section 10.7.3.

- Fish kills: towards the end of an algal bloom, acute anoxia may develop throughout most of the water column due to large-scale death of the algal cells, resulting in a decrease in oxygen evolution by photosynthesis and an increase in oxygen uptake by heterotrophic bacteria metabolizing the algal breakdown products. Under these sudden and extreme conditions of oxygen depletion, massive loss of fish populations may occur, with long-term effects on the ecology of the freshwater environment.

- Problems with water extraction and treatment for domestic use: the filtration systems of water-treatment works can become blocked with algae, affecting the efficiency and maintenance costs of the extraction process. Mucopolysaccharides produced by algal breakdown can chelate the iron or aluminium coagulants added to the water during the treatment process, leading to an increase in the metal complexes entering the water supply. The presence of algae can also lead to unpleasant changes in the odour and taste of the water. Final problems come with the collapse of the bloom, which may lead to the accumulation of ammonia – affecting the oxidation and disinfection capacity of chlorine and converting iron and manganese to soluble forms that can lead to discoloration of water.

- Impacts on agriculture and fisheries: blue-green blooms can have adverse effects on agriculture through damage to livestock (toxin consumption), with an increased risk of flooding to farmland. In the case of fish farming, blue-green blooms can cause changes in fish species composition, and result in fish kills (see above).

- Loss of recreation amenity: development of intense algal blooms frequently leads to a ban on human access for swimming, sailing, and other recreational activities (Case study 10.6).

10.7.3 Production of toxins

In the freshwater environment, toxins are produced and secreted in quantity by one major group of algae – the blue-greens (Carmichael, 1994). These toxins are part of a wide range of bioactive secondary metabolites produced by these organisms, including polyketones, alkaloids, and peptides. Toxins reach particularly high concentrations under conditions of bloom formation, when they contaminate the water supplies of wild and domestic animals and can also harm humans. Toxin production by other algae may occur in some situations. Toxic populations of the eukaryote alga *Prymnesium parvum* (Prymnesiophyceae), for example, may develop in the brackish waters of tidal or coastal freshwater systems causing dramatic changes in fish communities (Skovgaard and Hansen, 2003). In the marine environment, toxin production by dinoflagellates assumes greatest significance – contaminating sea water, but also causing shellfish to become poisonous due to ingestion and bioconcentration by these organisms.

General aspects of blue-green toxins

Although blue-green toxins have harmful effects on cattle and other large vertebrates via their drinking supply, they appear to have evolved primarily as a defence against invertebrate predators – reducing population loss by grazing (Section 9.10). On the basis of their physiological effects on vertebrates, algal toxins have been divided into two main groups – neurotoxins (inducing neuromuscular dysfunction) and hepatotoxins (causing liver damage). The chemical structure of some of the more important toxins is shown in Figure 10.6.

Neurotoxins These include anatoxin-a (a secondary amine alkaloid produced by some strains of *Anabaena* and other blue-green algae), anatoxin-a(s) (an organophosphate produced by *Anabaena* and *Oscillatoria*), and saxitoxin/neosaxitoxin (produced by both freshwater blue-greens and marine dinoflagellates).

These toxins act at neural synapses or neuromuscular junctions, causing loss of function by acting in various ways. Anatoxin-a is a neuromuscular blocking agent, mimicking the effect of the neurotransmitter acetylcholine and promoting prolonged muscular spasm. Anatoxin–a(s) achieves a similar effect by inhibiting acetylcholine esterase, thus preventing breakdown of the neurotransmitter. Saxitoxin and neosaxitoxin are both sodium channel-blocking neurotoxins, preventing or delaying the normal post-synaptic processes.

Hepatoxins These toxins are of two main types – microcystins (cyclic heptapeptides) and nodularins (cyclic pentapeptides). They are produced by a wide range of freshwater blue-green algae, causing liver damage by promoting tumour development and specifically inhibiting protein phosphatases.

Microcystin has been the most intensively-studied of all the blue-green toxins, with information obtained on its molecular synthesis, structural diversity, and mode of action (Dittman *et al.*, 2001). The toxin is secreted by species belonging to the genera *Anabaena*, *Oscillatoria*, *Nostoc*, and *Microcystis* – of which *Microcystis aeruginosa* is the most widely distributed. In addition to microcystin, *Microcystis aeruginosa* also synthesizes other peptides, including depsipeptides (cyanopeptolin, micropeptin), tricyclic microviridins, and the linear microginins and aeruginosins. Most strains of *Microcystis* secrete at least one of these compounds, which act as protease inhibitors.

The production and secretion of microcystin may be partially controlled by microcystin-related proteins (MrpA and MrpB), synthesis of which requires the presence of microcystin and is promoted by blue light (Figure 3.4). Close proximity of other *Microcystis* cells, detected by quorum sensing, may also be important and may lead to enhanced levels of toxin under bloom conditions.

Field populations of individual species of blue-green algae are composed of a mixture of genotypes (Section 3.3.1), not all of which encode the protein machinery for toxin production. In natural populations of *Microcystis* sp., for example, Kurmayer and Kutzenberger (2003) demonstrated that only 1–38 per cent of genotypes contained microcystin (*mcy*) genes. In these particular samples (from Lake Wannsee, Germany), the mean proportion of microcystin genotypes was stable from winter to summerwith no variation in relation to species population change.

In addition to neurotoxins and hepatoxins, lipopolysaccharide (LPS) endotoxins are also produced by various blue-green algae. These resemble the LPS toxins produced by a range of Gram-negative bacteria such as *Salmonella*, causing fevers and inflammation.

NEUROTOXINS

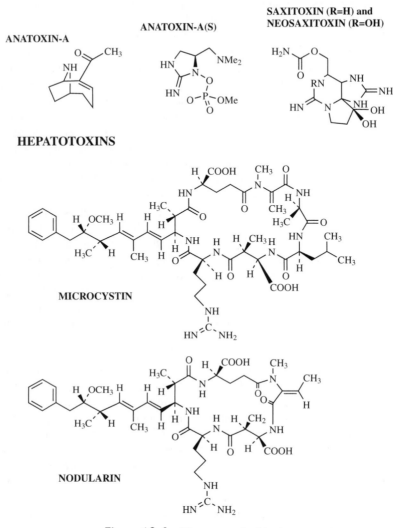

Figure 10.6 Blue-green algal toxins

D. CONTROL OF BLUE-GREEN ALGAE

10.8 Strategies for the control of blue-green algae

Control of algal blooms has become a major issue in the management of many freshwater bodies, where human activities have resulted in a shift in the composition and biomass of phytoplankton populations – creating problems in water usage. The most effective procedures to limit or control the algae vary with the particular water body concerned (see below) and require an understanding of microbial interactions in the freshwater environment. One question of particular relevance, raised in Chapter 1,

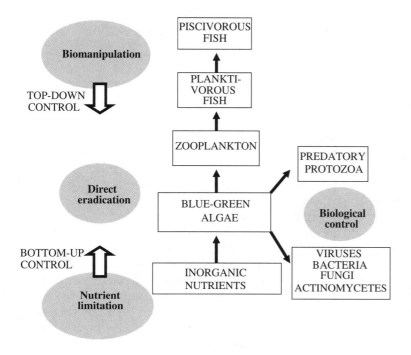

Figure 10.7 Major approaches to the control of blue-green algae in freshwater environments: the diagram shows four major types of control procedure in relation to the major pelagic food chain. Solid arrows indicate the direction of nutrient and carbon flow. Details of the microbial loop, with ingestion of protozoa by zooplankton, and bacteria by protozoa and zooplankton, are not included

is what environmental factors actually control planktonic algal populations? Is grazing by zooplankton (top-down control) of prime importance, or does nutrient limitation (bottom-up control) take precedence?

There are three main approaches currently used to control blue-green algae in the freshwater environment – nutrient limitation, direct eradication, and biomanipulation. These are considered in this section, while biological control, a further strategy with considerable potential, is discussed in Section 10.9. Two of these procedures, nutrient limitation and biomanipulation, involve bottom-up or top-down alterations respectively to the major trophic sequence, while direct eradication and biological control do not fit clearly into either category and can be viewed as acting at the same level as the algae within the main food chain (Figure 10.7). In practice, the effects of these measures (single or combined) varies from simple reduction in blue-green dominance to complete eradication of blue-greens from the system concerned.

10.8.1   Nutrient limitation (bottom-up control)

With many freshwater systems, reduction in the inflow of dissolved inorganic (phosphates, nitrates) and organic nutrients (the external nutrient loading) removes the root cause of the problem and signals a return towards the original ecosystem and the restoration of water quality. The first step in bottom-up control is to identify the major point or diffuse sources of nutrient inflow, then reduce the external loading by diverting the inflow to other sites, sewage treatment, or by chemical stripping of phosphates by precipitation using aluminium or ferric salts. More extensive measures may be required where a site has been contaminated with nutrients long-term. Phosphate pollution gives particular problems with the formation of phosphorus-rich sediments, which need to be removed if long-term reduction in nutrients is to be achieved.

Limitation in nutrient input is always the first measure in attempting to redress the effects of eutrophication, and is an essential prerequisite if other procedures are to prove successful. The importance of nutrient limitation in the control of

algal blooms is illustrated by the case studies on Lake Washington (USA) and the Broads Wetlands Area (UK).

Case study 10.3   Restoration of water quality in Lake Washington, North West USA

Lake Washington (USA) provides a classic example of bottom-up control of blue-green algae, where dramatic reduction in external nutrient loading restored a polluted water body to its previous water quality level (Edmondson, 1991). The lake lies close to the major city of Seattle, acting as an important source of water for the city and as a key wildlife and recreation area. Lake Washington receives inflow from Cedar river and has two main outflows to Puget Sound and Lake Hammamish respectively.

Between 1900 and 1936, raw sewage was pumped directly into the lake. The problem of eutrophication was recognized towards the end of this period, and in 1936 some diversion of sewage occurred into Puget Sound. Discharge of treated sewage continued into the lake, and feeder streams also continued to contribute to the eutrophication process with contamination from septic tanks. By 1963, the level of eutrophication had increased to unacceptable levels, with high winter levels of phosphorus and nitrogen, maximum chlorophyll-a concentrations $>40 \, \mu g l^{-1}$, minimum Secchi depth values < 1 m, and extensive growth of toxic blue-green algae.

In 1963, drastic remedial action was instigated, with 99 per cent of treated sewage being diverted from the lake to Puget Sound by 1967, and an improvement being made to the quality of all effluent discharges into the lake. The response of the lake to these changes was rapid, with a marked decline in phosphate levels after 1965, a slower decline in nitrogen levels (which are derived mainly from agricultural sources), and a dramatic fall in algal biomass and the occurrence of blue-green algae.

The lake is now in excellent condition for a water body in an urban area. Water transparency (Secchi depth) is almost always >5 m, and although small numbers of blue-green algae do occur, they are never a nuisance. Lakeside bathing beaches are no longer closed for reasons of pollution.

10.8.2   Direct eradication

This includes all those procedures which can be used for the direct removal, destruction, or growth inhibition of blue-green algae.

- *Removal.* In some cases, blue-green algae can be directly removed from the aquatic environment. This can become commercially viable if the alga can be used for agricultural purposes. The blue-green alga *Spirulina* of tropical and sub-tropical lakes, for example, can be harvested and used as cattle feed.

    One effective way of removing blue-green algae from some lakes and reservoirs is to flush them out of the system. In practical terms, this involves draining the lake or reservoir or reducing the water retention time by altering the hydrology to increase the inflow and outflow rates.

- *Destruction of blue-greens: the use of algicides.* The use of chemicals such as Cu salts to destroy blue-green algae has been widely used for the rapid and effective elimination of these organisms from small water bodies such as ponds and small lakes in urban parks. The drawback of using chemical treatments is that they are non-selective, with a wide range of other freshwater biota being destroyed in addition to the blue-green algae. Although the effect of algicides is short-term, their toxic influence may persist over a long period of time.

- *Inhibition of algal growth rate.* The growth rate of blue-green algae and the development of blooms can be very effectively controlled in small water bodies by hypolimnetic aeration. This involves injecting a stream of air into the hypolimnion, breaking stratification and circulating water within the entire water column. In this situation, blue-green algae are no longer able to accumulate at the water surface, but are cycled throughout the water column. The reduced exposure to light results in lower rates of photosynthesis and much reduced growth rates. The blue-green algae are also not able to develop the conditions of low CO_2 concentration and high pH which are so important for their ability to out-compete other algae.

10.8.3 Top-down control of blue-green algae: the use of biomanipulation

Biomanipulation is a long-term ecological strategy which involves the removal of zooplanktivorous fish to increase populations of zooplankton, particularly Cladocera. This long-term management strategy limits the growth of blue-greens by promoting the grazing activity of the larger Cladoceran species such as *Daphnia magna*, which are able to ingest the colonial algae. Piscivorous fish such as perch and pike are not removed, and where successful, biomaniopulation leads to the restoration of a stable food chain dominated by these carnivores.

The maintenance of large populations of zooplankton is further promoted by the re-establishment of beds of large aquatic plants (macrophytes), which act as important refuges for the zooplankton (Figure 10.10). These refuges are typical of littoral and other shallow regions of lakes and wetlands, and provide regions where zooplankton can obtain some protection from zooplanktivorous fish.

Biomanipulation has been widely used throughout the northern hemisphere in the rehabilitation of natural ecosystems and the restoration of water quality. The Broads Wetland Area (UK) provides a good example of this.

Case study 10.4   Top-down and bottom-up control of algal populations in the Broads Wetland Area (UK)

In combination with nutrient limitation (bottom-up control), biomanipulation (top-down control) has become an important tool for the rehabilitation of the Broads Wetland Area in Eastern England, UK (Madgwick, 1999). This lowland region consists of approximately 40 shallow lakes (broads) which originated as peat diggings, interconnected by a complex system of waterways (Figure 10.8).

Figure 10.8 Broads Wetland (UK): view from a shallow lake area (broad) to a connecting channel – extensive areas of reed-swamp, dominated by *Phragmites communis*, occur on either side

The eutrophication problem

In 1965, a report by the UK Nature Conservancy drew attention to a pattern of major changes in the wetland area – including a catastrophic loss of water plants, decline of the fishery, and incidence of intense blue-green algal blooms (particularly *Oscillatoria*). These changes result from nutrient increases caused by increases in indigenous and tourist population and agricultural practices. Erosion of the reed swamp margins of the lakes and channels was also evident, caused by increased boating activities.

The Broads may be grouped into three main ecological phases (Figure 10.9), which relate to a gradient of nutrient concentrations within the Broads system (Madgwick, 1999). With increasing nutrient status, a

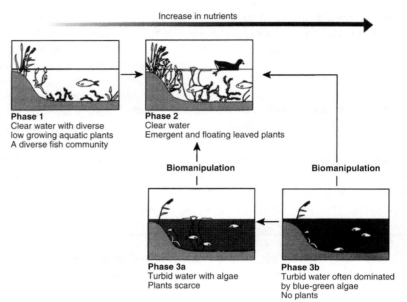

Figure 10.9 Three major phases in the ecological state of shallow lakes in the Broads Wetland Area (UK) (figure taken from Madgwick, 1999, with permission from Kluwer Academic Publishers). The phases follow an increase in nutrient status (range 50–100 mg l^{-1} total phosphorus), with a transition from clear (Phases 1 and 2) to turbid (Phases 3a and 3b) water. Restoration of water quality by biomanipulation can be viewed as a reverse switch from Phase 3 to Phase 2

critical point is reached at which the water body is dominated either by macrophytes (Phase 2) or by phytoplankton (Phase 3a). These two-states represent alternative stable conditions (Section 10.2.3). The transition from macrophyte to a blue-green dominated community corresponds to the switch from homeostatic control to lack of control noted earlier (Figure 10.1) and involves a change from clear to turbid water conditions. Further eutrophication leads to complete blue-green dominance, with no submerged macrophytes.

Lake restoration

The Broads Authority has established four essential steps in the restoration of water quality from the Phase 3 situation:

- reduction in external and internal nutrient loading conditions,

- removal of zooplanktivorous fish (biomanipulation),

- re-establishment of macrophyte beds,

- stabilization of the lake ecosystem, including the re-establishment of appropriate fish populations.

Reduction in external and internal nutrient loading

Phosphorus is the nutrient of most concern in promoting algal growth in the broadland area, entering the rivers and water column of the shallow lakes from both external and internal sources.

External loading   Most of the broads are distant from the river system and receive nutrients from their agricultural catchment area. Diffuse-source input of phosphates from arable land and numerous small treated-sewage inflows is becoming increasingly important. Potential control measures include a reduction in the application of fertilizers or poultry litter to farmland, and the introduction of buffer zones or created wetland areas to intercept and contain particulate phosphorus runoff.

The presence of 15 key sewage works makes a major contribution to eutrophication of the wetlands area. These point sources cause nutrient-enrichment of the river system and also some of the broads via river inflow. This has lead to a phased programme of phosphorus reduction, with the objective of limiting the total phosphorus concentration to 1mg l^{-1} in the sewage effluent.

Internal loading   Long-term eutrophication of the Broads area has lead to phosphorus-rich sediments, and the continuous release of phosphates from these represents the major source of internal loading. This has been controlled with limited success (Pitt *et al.,* 1996) by sediment removal (suction dredging) or phosphate precipitation (dosing with ferric salts). Removal of sediment by suction dredging is currently the only effective means of reducing the available sediment nutrient supply. The main practical constraints of this procedure are cost and the local availability of suitable spoil disposal sites.

With both external and internal loading, the major objective of nutrient limitation is to bring soluble concentrations down to such a level that biomanipulation can be used as the main procedure for restoration of water quality.

Biomanipulation

Since the Broads are interconnected by water channels, long-term reduction in the population of zooplanktivorous fish also requires the creation of fish barriers to isolate individual lakes where biomanipulation is being carried out. The retention of piscivorous fish such as tench and pike during biomanipulation is not only important for the restoration process but allows angling to continue.

With several Broads, reduction in nutrient loading coupled with biomanipulation has proved highly successful – leading to the development of increased zooplankton populations, suppression of blue-green algae, and restoration of clear-water conditions. Biomanipulation may be viewed as a 'reverse switch' (Figure 10.9) in the transition of dominance noted previously from Phase 2 (macrophyte dominance) to Phase 3 (algal dominance).

Re-establishment of macrophyte beds

Although the growth of macrophytes in biomanipulated Broads is no longer repressed by algal shading effects, the development of seedlings and re-growth of these plants is typically slow and erratic. The reasons for this are not entirely clear, but stabilization of lake sediments may be a key factor in promoting the re-establishment of macrophyte populations.

Two major ecological groups of large aquatic plants (macrophytes) are important in the Broads wetlands:

- submerged macrophytes, present in open water and attached to the lake sediment, including *Ceratophyllum demersum, Potamogeton* spp. *Elodea* spp., and Chara spp..

- Littoral vegetation at the edge of the lake, composed mainly of floating reedswamp.

Beds of submerged macrophytes play a key role in maintaining the natural ecological balance, once this has become re-etablished by biomanipulation. This is shown in Figure 10.10, where the importance of the submerged vegetation is emphasized in relation to three key aspects within the pelagic food chain – balance within fish populations, grazing of zooplankton, and competition between algae and macrophytes.

Figure 10.10 Importance of submerged macrophytes in maintaining fish dominance and suppression of algal blooms in a shallow lake or wetland area (e.g., Norfolk Broads, UK). Submerged macrophytes have an important stabilizing effect on all major phases (fish predation, zooplankton herbivory, and algal primary productivity) of the pelagic food chain. This figure also emphasizes the link between benthic and pelagic organisms, and the importance of higher plants (macrophytes) on the development of microorganisms (algae) in the freshwater system. Block arrows indicate carbon flow within the main food chain, solid arrows -macrophyte interactions

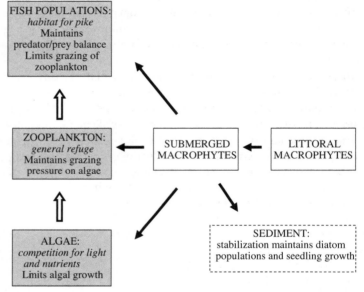

Predator/prey balance within the fish population In conjunction with littoral vegetation, submerged macrophytes present an important habitat for pike (Esox lucius), one of the top carnivores in the wetlands system. In the presence of well-developed beds of submerged vegetation, there is a tendancy for continued pike dominance, ensuring maintenance of a normal predator/prey balance within the fish community and the promotion of a diverse fish population. Once the equilibrium between piscivorous/zooplanktivorous fish is stabilized, zooplanktivorous fish can be allowed to occur without removal, and restocking with piscivores should not be necessary.

Zooplankton refuge Macrophyte beds are important as 'buffer zones', limiting the ability of planktivorous fish to graze zooplankton during that period of their diurnal cycle when they occur within the submerged vegetation.

Phytoplankton growth Submerged macrophytes limit algal growth by competing for light, CO_2, and inorganic nutrients. In shallow aquatic systems, macrophytes are major primary producers and important competitors of phytoplankton (see Section 4.8).

In addition to these effects on the pelagic food chain, submerged macrophytes are important (with benthic diatoms) in stabilizing the sediment against wind-induced re-suspension and current erosion. This stability is important in the development of diatom populations and in the establishment of macrophyte plants from seedlings. In wetland areas such as the Norfolk Broads, the littoral vegetation (composed mainly of floating reedswamp) also has an important role in the development of submerged macrophytes, providing sheltered bays for the establishment of this open water vegetation.

Field observations from broads of different nutrient status suggest that the development and stability of the macrophyte vegetation may vary with nutrient conditions (Madgwick, 1999) – in much the same way that the pelagic community was affected. In moderately eutrophic broads (50–100 µg l^{-1} total phosphorus), the macrophyte flora is dominated by tall, leafy plants such as *Ceratophyllum demersum*, *Potamogeton* spp.n, and *Elodea* spp.. These communities are relatively unstable, showing cyclic changes in relation to weather alterations and other destabilizing events. In contrast to this, charophyte-dominated lakes, which are invariably associated with low-nutrient (<50 µg l^{-1} total phosphorus), tend to form much more stable macrophyte communities with little variation from one year to the next.

Long-term stabilization of the lake ecosystem

Once successful biomanipulation has been carried out, lake stability is maintained by continued limitation in nutrient input and by the 'top-down' control exerted by piscivores such as pike.

In line with earlier comments, field observations show that the continued effectiveness of this top-down control is dependent on the presence of substantial areas of macrophyte beds of varying density, interspersed by patches of water, and with well-developed littoral margins (Perrow and Jowitt, 1997).

10.9 Biological control of blue-green algae

Biological control (or biocontrol) is a short-term strategy to eliminate or reduce the growth of blue-green algae by the addition of naturally-occurring microorganisms (biological control agents) which act as antagonists of blue-greens in the aquatic environment (Sigee *et al.*, 1999b). In the simplest situation, control agent inoculum would be added to the surface waters of the lake or reservoir where the algal bloom was developing or had already developed.

The introduction of organisms which are a natural part of the freshwater community means that biological control can be viewed simply as an alteration in the balance of existing biota within the freshwater environment. Once these organisms have been introduced into the freshwater environment they will limit the development of blue-green algae, but will also themselves become subject to the whole range of environmental pressures present in the ecosystem – and their population will tend to decline to the level normally present. One of the major potential advantages of biological control (in contrast to chemical control) is that the agent can be used to target specific 'nuisance organisms', such as blue-green algae, without having any general adverse effect on other lake organisms.

In recent years, the addition of plant litter (e.g., barley straw) to lakes and ponds has also been used to control the development of blue-green algal populations in freshwater systems. This is an extension of the biological control concept, where a substrate (rather than a microbial agent) is being added to promote the microbial release of anti-algal substances, and is discussed in Case Study 10.5.

10.9.1 Biological control agents

A wide range of freshwater microorganisms can be used as biological control agents. These act on blue-green algae in three main ways:

- Parasitism – fungi and viruses,

- Predation – protozoa,

- Competitive antagonism – secreting lytic enzymes or antibiotics, and including a range of fungi, bacteria, and actinomycetes.

Two examples of organisms which have been used as biological control agents (Sigee *et al.*, 1999b) are shown in Figure 10.11 (the protozoon *Nuclearia*) and Figure 10.12 (the actinomycete *Streptomyces*). *Nuclearia* is highly active in the ingestion of filamentous blue-greens such as *Oscillatoria*, while *Streptomyces* releases a lysozyme compound which rapidly destroys both filamentous and globular colonial blue-greens.

The potential effectiveness of different biological agents for the control of blue-green algae varies considerably and relates to their natural ecological

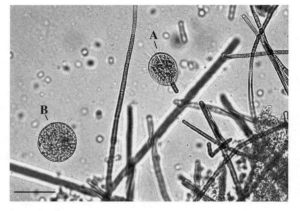

Figure 10.11 *Nuclearia delicatula* – a potential protozoon biocontrol agent. Filaments of the blue-green alga *Oscillatoria* are being grazed by the amoeba *Nuclearia*, which makes attachment via fine pseudopodia (filopodia). Ingestion occurs from the end of the algal filament, as seen with organism (A), which has a short length of remaining filament poking out of the cell. Organism (B) has the typical spherical shape of a cell not involved in ingestion, and appears to be laterally attached to algal cells (see Figure 10.14 for biocontrol activity, scale = 30 μm)

Figure 10.12 *Streptomyces exfoliatus* – a potential actinomycete biocontrol agent. (a) Colony of *Streptomyces* (arrow) growing on a monolayer of the blue-green alga *Microcystis aeruginosa* that had been inoculated with soil water. The colony has a clear area (plaque) of algal inhibition due to the production of lytic enzymes which attack and destroy the algal cells. (b) colony of *Streptomyces* in a liquid culture of the actinomycete obtained from the colony shown in (a) (scale = 100 μm) (photographs in collaboration with RF Glenn and HA Epton)

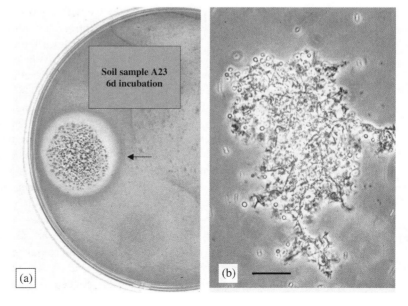

Table 10.7 Major characteristics of different freshwater microorganisms as potential biological control agents

Group	General ecological features	Effectiveness as biological control agents
Fungi		
(a) Chytrids *Rhizophydium*	Obligate parasites within phytoplankton Major influence in terminating algal blooms	Potentially very useful with dense algal blooms, but limited by: • high specificity • difficulties in large-scale culture
(b) Non-chytrid fungi *Acremonium* *Emericellopsis*	Saprophytes Not normally present in phytoplankton No major role in reducing planktonic algal populations	Limited use, though extracellular products (including β-lactam antibiotics) can destroy blue-greens
Viruses		
LPP-1 Cyanophage	Obligate parasites Major effect in limiting algal growth and bloom control in freshwater environment	Potentially very useful since they have rapid generation time and high burst size. Limited by: • high degree of host specificity • development of resistant host mutants • difficulties in producing large amounts of inoculum
Protozoa *Nassula* (ciliate) *Ochromonas* (flagellate) *Nuclearia* (amoeba) *Acanthamoeba*	Remove algae by predation Major ecological role in reduction of algal populations by grazing	Depends on: • protozoon growth and feeding rates • specificity of predation • predation of protozoa by copepods and cladocerans
Bacteria Include bacteria acting by: general lysis – *Bacillus, Flexibacter* contact lysis – *Myxobacter* entrapment lysis – *Myxococcus*	Act by lytic enzyme secretion Readily isolated during bloom formation, where: • numbers correlate with high algal biomass • associated with sudden decline in algal population	Bacteria such as *Myxococcus* are potentially very useful: • ability to search out prey • rapid digestion of algae • rapid growth and multiplication • survive low prey densities • wide host range • overcome host resistance
Actinomycetes *Streptomyces*	Act by lytic enzyme secretion Not normal components of plankton Very little ecological impact on phytoplankton populations	Give good results in laboratory experiments, but limited in environment: • rapidly sink out of epilimnion • lytic products rapidly diluted/degraded

role (Table 10.7). Organisms such as chytrid fungi, viruses, protozoa, and lytic bacteria, which are already part of the plankton population and have a natural role in bloom termination, have the greatest potential for biological control.

In this context, the ideal biological control agent should have the following characteristics.

• A broad specificity to a range of bloom-forming blue-green algae. This limits the usefulness of chytrid fungi and viruses, which are typically specific to single algal species or small species groups. Protozoa such as *Nassula* and *Nuclearia* are also limited since they tend to ingest filamentous algae (e.g., *Anabaena* and *Oscillatoria*) but not globose algae such as *Microcystis*.

• Needs to be effective at high algal density and to increase rapidly within the bloom population. All of the organisms identified as being involved in natural bloom control (see above) have this capacity.

- Needs to persist and be effective for as long as possible within the epilimnion. Biological control agents are themselves liable to rapid loss from the environment by degradation (viruses), protozoon grazing (fungi, bacteria), zooplankton grazing (fungi, bacteria, and protozoa), and loss by sedimentation (all agents). Lytic enzymes become diluted or destroyed with time and the effectiveness of parasitic organisms rapidly decreases when the blue-green population drops below a critical level.

- Is easy to culture in the laboratory and to produce large amounts of inoculum. Parasitic organisms (chytrid fungi and viruses) are difficult in this respect since they need blue-green hosts for their culture (which may be difficult to separate off from the agent) and might be difficult to store in an infective stage.

10.9.2 Protocol for the development of biological control agents

As with biological control of other target pest organisms (e.g., plant pathogenic bacteria, see Sigee

(1993)), a protocol for the biological control of blue-green algae involves four main stages – isolation of the agent from environmental samples, laboratory testing, field trials, and use in lake management (Figure 10.13).

Isolation of biological control agent

Biological control agents can be potentially isolated from any freshwater site. One simple way of isolating the agent is to pipette small (2 ml) aliquots of the water sample onto an algal monolayer overlying an agar base. Antagonistic organisms present in the environmental water sample can be detected and isolated from colourless plaques in the algal monolayer, which develop where the algae are being destroyed or removed. These antagonists can be subcultured and reapplied to algal lawns as part of the procedure for purifying the isolates.

Laboratory analysis

Laboratory analysis is an essential part of the selection process, and is important for:

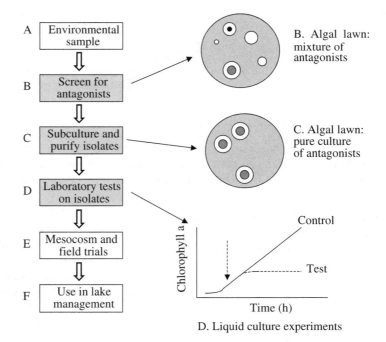

Figure 10.13 General scheme for the testing and development of biological control agents to be used against planktonic blue-green algae. The central part of the scheme (laboratory work) is illustrated to the right-hand side of the flow diagram. In D, the time of addition of antagonists to the test culture is indicated by a vertical arrow (figure derived from Sigee *et al.*, 1999b, with permission from Kluwer Academic Publishers)

- identification and detailed characterization of the potential biological control agent;

- testing the effectiveness of the agent against a range of blue-green algae, concentrating particularly on bloom-formers and largely involving liquid culture experiments;

- determination of the mode of action and the parameters of algal/antagonist interaction. These tests would include the effects of different physical conditions, use of environmental algal isolates and analysis of altering the control agent/ algal biomass ratios.

Field trials

Biological control agents which look promising from laboratory tests can be carried on to small-scale field trials. These may involve the use of lake enclosures (mesocosms) containing a small (e.g., 5–10 litres) volume of lake water. Addition of a control agent to a test system within the mesocosms ensures that the agent can be tested under quasi-natural conditions, without being released into the general aquatic environment.

Testing potential biological control agents, such as protozoa and actinomycetes in small-scale field trials (Sigee *et al.*, 1999b) demonstrates their continued activity under environmental conditions, but with reduced effectiveness. The lake environment, involving varied conditions of light/dark regime, environmental populations of blue-green algae, and the presence of other lake biota is clearly much more complex than the laboratory situation, and the activity of the biocontrol agent is much less predictable.

Use in lake management

Although experiments with biological control agents have met with some success (e.g., Sigee *et al.*, 1999b), none appear to have yet gone to the stage of commercial development and use in lake management. Commercial development would require extensive toxicity testing in relation to human water supplies and the development of techniques for mass production, packaging, and storage. Use in lake management would require the product to be cost-effective and would need the development of a strategy for field application.

Case study 10.5   Potential protozoon control agents

Laboratory studies on protozoon control agents have shown them to be effective against a range of filamentous bloom-forming blue-green algae (e.g., *Anabaena*, *Oscillatoria*) but not globular forms such as *Microcystis*. Figure 10.14 shows the results of laboratory studies using one of these potential biological control agents, *Nuclearia delicatula* (Figure 10.11), where the grazing efficiency on *Anabaena* is being tested by varying the absolute and relative populations of algae and protozoa. Experiments such as this give some idea of the protozoon population that would be required to control different blue-green algal populations in the lake environment. At the lowest algal population tested, which was equivalent to pre-bloom conditions, all levels of protozoon population (100, 500, and 1000 cells ml^{-1}) were able to control the growth of the algal population. At the highest algal population tested (in excess of normal bloom conditions), only the highest protozoon population level succeeded.

10.9.3   Application of plant litter to control blue-green algae

Addition of plant litter, such as barley straw and deciduous leaf matter, to standing waters leads to aerobic decomposition, with the release of metabolites which can suppress nuisance algae (Ridge *et al.*, 1999). Barley straw was the first type of plant litter shown to inhibit algal growth and is still the main type in practical use. It is typically used in

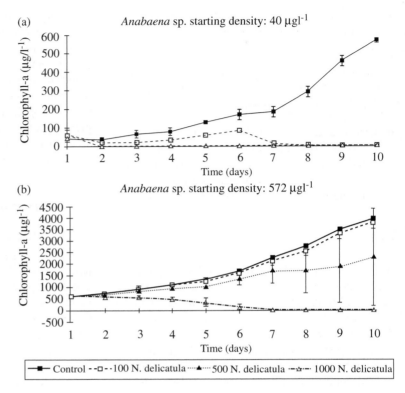

Figure 10.14 Predator-prey experiments on blue-green algae using laboratory cultures of an amoeboid protozoon. The ability of different inoculum levels of *Nuclearia delicatula* (see Figure 10.11) to control the growth of *Anabaena* sp. in laboratory culture is shown for two starter levels of algal population, with respective chlorophyll-a concentrations of (a) 40 µg ml^{-1} and (b) 572 µg ml^{-1}. In each case, protozoa were added to give overall initial population levels of 100, 500 and 1000 cells ml^{-1}. Algal population levels are given as the mean chlorophyll-a concentration (+) SD of three replicate flasks (taken from Sigee *et al.*, 1999b, with permission from Kluwer Academic Publishers.)

small, enclosed volumes of water such as ponds and shallow lakes (including trout farms and reservoirs), where the released inhibitory compounds are not dissipated or lost within a large volume of water. Application of barley straw to eutrophic waters is particularly effective at promoting the switch from algal to macrophyte domination.

Release of inhibitory compounds

The generation and release of inhibitory compounds is the result of microbial decomposition of the plant material. Oxidizing conditions are essential, and the application of plant litter for algal control is only effective if the material is maintained submerged and in an aerobic state. Decomposition of plant litter

leads to the formation of a wide range of organic compounds, including simple phenolics and long chain fatty acids.

Although the exact nature of the inhibitory compounds is not known, they are thought to be derived from oxidized polyphenolics which originatemainly from lignin. The release of these compounds from plant litter (such as oak leaves) appears to be biphasic (Ridge *et al.*, 1999), involving two main phases.

- An early phase (4–90 days), with the formation of soluble, relatively stable inhibitors which are probably derived from oxidized tannins.

- A late phase (120–900 days), where the inhibitors are relatively unstable in solution. These are mainly associated with fine particulate organic material (FPOM), and are probably derived by oxidative breakdown of lignin.

Laboratory and field studies have shown that the inhibitory compounds are effective against a wide range of algae, including both filamentous attached (e.g., *Cladophora*) and planktonic forms. The latter include green algae (*Chlorella*), diatoms (*Asterionella*, *Tabellaria*), and blue-greens involved in toxic bloom formation (*Microcystis*, *Anabaena*). Reports from various applications (see Ridge *et al.*, 1999) suggest that decomposing litter has no adverse effects on freshwater invertebrates, amphibians (frog, newt), or fish (use in trout farms). The inhibitory effect of decomposing plant litter only lasts for 6–12 months, after which fresh material needs to be added for continued control.

Field trials

Many field trials with barley straw have been apparently successful in limiting the growth of nuisance algae. In one of these, Barrett *et al.* (1996) applied barley straw to a reservoir in the Grampian Region of Scotland (area $25\,000\,m^2$, normal capacity $250\,000\,m^3$) which had a perennial problem with intensive growths of diatoms and blue-green algae. Addition of barley straw at a level of $38\,g\,m^{-3}$ in spring resulted in reduced populations of algae throughout the summer, with no typical early-autumn *Anabaena* bloom. Further application of barley straw in December lead to a much-reduced diatom bloom the following spring, with no subsequent blue-green algal bloom later in the growth season.

10.10 Strategies for the control of blue-green algae in different water bodies

In the previous section, different practical approaches were discussed in relation to the control of blue-green algae (Figure 10.7) with examples of

SHORT-TERM:
Prevention: hypolimnetic aeration
Rapid limitation or removal: biological control, flushing the system

MEDIUM- TO LONG-TERM:
Prevention: use of barley straw
Reduction of external nutrient loading: use of buffer strips, diversion to other sites, P-stripping
Reduction of internal nutrient loading: suction-dredging of sediment, ferric chloride dosing

LONG-TERM:
Reversal of a deteriorated environment: biomanipulation, with measures to ensure the long-term stability of the ecosystem

Figure 10.15 Short-term to long-term practical measures for the control of blue-green algae and the limitation of algal blooms

their application in various case studies. These approaches range from short-term to long-term strategies (Figure 10.15), depending partly on the urgency of the required control and whether they are implemented to achieve prevention, limitation, or reversal of algal blooms.

In practice, the management of freshwater sites extends beyond these interventionist procedures to include other practical, theoretical, and administrative aspects forming an integrated management policy. Within this framework, different approaches apply to different aquatic systems.

10.10.1 Integrated management policy

Integrated management of aquatic bodies and regional aquatic systems (Constanza and Voinov, 2000) also includes continuous monitoring of the environment, computer modelling to predict adverse ecological events, and the use of legal constraints to minimize or avoid nutrient pollution.

Environmental monitoring

Environmental monitoring involves the regular collection and analysis of freshwater samples to assess the trophic status of the water body, using the criteria previously described. This activity can provide various levels of information about the freshwater system, including:

- long-term changes in eutrophication,

- medium-term measures required for general remediation and restoration of water quality,

- short-term information on the imminent development of blue-green blooms and the necessary response.

Analysis of a time sequence of environmental samples can, like paleolimnological studies on lake sediments, provide useful information on long-term changes in lakes. Because of climatic oscillations, however, data sets should extend over a relatively long (>10 years) time interval if the long-term effects of climate or environmental change on eutrophication are to be elucidated.

In the medium-term, environmental monitoring is important for the assessment of trophic status in relation to potential remediation. This is emphasized in a review by Van der Molen and Boers (1999) on eutrophication in The Netherlands, where monitoring and restoration of shallow eutrophic lakes is envisaged as taking place over the next decade and beyond.

Short-term environmental monitoring for the early detection of an impending bloom of blue-green algae is a routine management activity for many water bodies. This activity is important to give an early blue-green warning to the public and also to implement short-term control measures at a time when the algal population is just beginning to develop.

Mathematical modelling

Mathematical models attempt to relate the various physical, chemical, and biological processes that occur within a particular ecosystem in quantitative terms. The extent of the modelling process varies from quite simple models (e.g., relating phytoplankton biomass to lake nutrients) to more complex models, where a wider range of parameters is taken into consideration. In practice, no model can encompass all the factors which determine a particular event, so all models present a simplified synthesis of ecological events.

In general, ecological mathematical models can have two general properties:

- simulation, with an accurate description of a particular data set, and

- prediction, involving the ability to make reliable predictions of future hydrological, chemical, and biological events in lake aquatic systems.

The major value of such models is that they are useful in understanding the complex, multivariate processes which determine ecological events such as build-up of phytoplankton populations. This involves the need to identify and quantitate the principal regulatory processes in operation. Many (site-specific) models are developed in relation to a particular ecosystem (Krivtsov *et al.*, 1999), and their applicability to other systems requires subsequent testing. The predictive nature of simulation models also means that they are potentially useful tools in freshwater management, where future adverse events (e.g., development of blue-green algal blooms) can be anticipated and corrective measures taken (Reynolds, 1999).

Nutrient pollution – legal constraints

Legal constraints are important world-wide as a first line of defence against the promotion of eutrophication by human activities. In Europe, there is no single piece of legislation which comprehensively deals with the problem of eutrophication in freshwater systems. The laws which apply fall into three categories relating to conservation, water quality, and agricultural intensification. Readers are referred to a review by Wilson (1999) for a detailed analysis of European legislation on eutrophication.

Case study 10.6 Environmental monitoring at Hollingworth Lake, Greater Manchester (UK)

Hollingworth Lake is a shallow upland lake within an urban conurbation, and is an important recreational centre (Figure 10.16). Although this lake is dominated by diatoms for most of the annual cycle (Section 2.3.1), the lake has experienced a number of outbreaks of blue-green algae (mainly *Oscillatoria*) in recent years – on each occasion disrupting public access and recreational activities. At the present time, the only method of algal control in this lake is the use of barley straw. The lake is administered by a local management group, with overall responsibility for sampling and water quality in the hands of a combined regional authority (referred to here as the 'water authority').

Figure 10.16 Hollingworth Lake, Greater Manchester (UK): a shallow upland lake and important recreation centre, this lake was the site of major blue-green algal blooms in 1996 and 1999, limiting public access and activities at these times

Regular monitoring is carried out at Hollingworth, using a protocol closely similar to the one summarized in Figure 10.17. Early warning signs of a possible blue-green development are provided by lake surveillance, regular phytoplankton and water chemical analysis, and a prolonged fall in the lake water level (static water column, no flushing of lake contents). Positive feed-back for any these parameters triggers a more intensive phytoplankton analysis and the implementation of a blue-green control measure – the application of barley straw. Confirmation that the blue-green algae has reached a critical level (>100 colonies ml^{-1}) promotes a third wave of activity with intensive testing for blue-green toxins, frequent phytoplankton sampling to record the development of blue-green populations, erection of warning notices, and the inception of a blue-green action algal group. The above activities are downgraded once the blue-green count has consistently (over a 1 month period) dropped to under 100 colonies ml^{-1}.

The use of barley straw in this context is probably of limited control value, since this procedure is most effective as a medium-to long-term preventative measure rather than one that can be used in responsive mode.

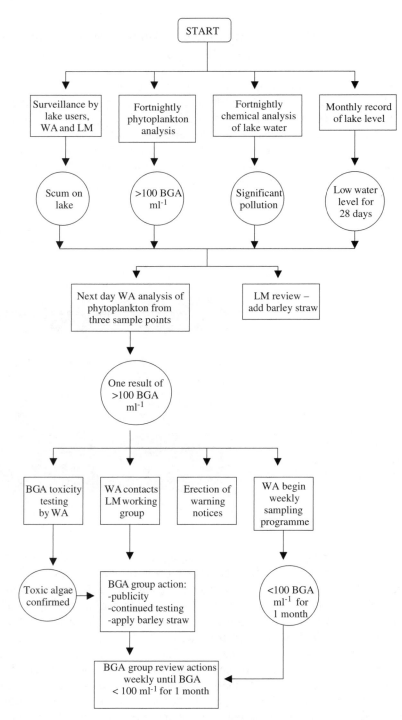

Figure 10.17 Reactive monitoring programme for blue-green algae. WA – water authority, LM – local lake management, BGA – blue-green algae (counted as colonies ml^{-1})

10.10.2   Specific remedial measures in different freshwater systems

Although the basic causes of anthropogenic eutrophication (from agricultural, domestic, or industrial sources) are universally important, the individual circumstances of particular water bodies vary considerably (Carvalho and Moss, 1995). Some of these differences are shown in Table 10.8, where individual details of nutrient enrichment, bloom development, and necessary remedial measures for different aquatic systems differ markedly. Various environmental and hydrological factors can have a major impact on this situation, including the following.

Infrequent mixing of the water column

In stratified lakes of temperate climates, the summer development of a low-nutrient epilimnion and high-nutrient hypolimnnion is normally terminated by complete mixing of the water column during the autumn overturn. This annual restoration of surface nutrients does not occur in deep subalpine lakes of the southern Italian alps, where complete mixing of the water column only occurs in harsh winters (Salmaso et al., 1999). These historically oligotrophic lakes, with a total volume of $124 \times 10^9\,\mathrm{m}^3$, constitute one of the largest freshwater supplies in Europe and have experienced progressive

Table 10.8   Blue-green algal blooms in different freshwater systems: specific eutrophication problems and remedies

Water body (and major problems)	Direct cause of eutrophication	Remedial measures
Deep oligotrophic lakes (intermittent mixing) (*e.g., Lake Garda, Italian Alps*) Salmaso et al., 1999	Increased external loading $\longrightarrow$ high P levels in hypolimnion Intermittent mixing $\longrightarrow$ periodic entry of P into upper layers	Water treatment removal of P in lake inflows Reduce sewage discharge into deep layers of lake
Lakes with variable climate (*e.g., Lake Sanabria, Spain*) Hoyos and Comin, 1999	Variable rainfall $\downarrow$ Periods of high retention time)	Develop computer model to predict climate effects Stringent control of P input at critical times
Wetland area (effects of agriculture and tourism) (*e.g., Norfolk Broads, UK*) Maggs, 1999	External P loading from DE and AE sources[*] Internal P loading from high P sediments	Range of measures: – reduce external P loading – sediment removal – biomanipulation – restore macrophytes
Shallow reservoir (nutrient inflow) (*e.g., Foxcote Reservoir, UK*) (Daldorph, 1999)	Increased P loading (DE and AE) via defined inflow	Reduce nutrient entry by ferric sulphate dosing of inflow
Quays and dockland areas (hydrology) (*e.g., Salford Quays, UK*)	High retention time High nutrients Static water column	Direct control of algal blooms by hypolimnetic aeration
Shallow urban lakes (hydrology and public access) (Birch and McCaskie, 1999)	High retention time Waterfowl excreta P-rich sediment	Range of measures: – flush system – dredge sediment – reduce waterfowl – restore macrophytes

[*]DE – domestic effluent, AE – agricultural effluent
Publications quoted in this table are taken from the volume edited by Harper et al. (1999).

eutrophication in recent decades – showing periodic outbreaks of blue-green blooms. In the absence of regular mixing, these lakes (referred to as oligomictic) undergo nutrient accumulation in the hypolimnion, but not the epilimnion – which remains depleted. When mixing does occur, the surface layers become nutrient-rich and algal blooms develop later in the season. The only long-term solution for lakes with this topology is a reduction in the nutrient loads. Construction of water-treatment plants has led to an improvement of water quality in some of the lakes, and there is a need to keep the discharge of excess sewage into the deep waters to a minimum.

Climatic variation

Lake Sanabria, the largest lake in Spain, provides a good example of the effects of climatic variation on water retention time, nutrient accumulation and the development of phytoplankton populations (Hoyos and Comin, 1999). This lake is located in a geographic area which receives the North Atlantic Climatic Oscillation (NACO) and experiences high inter-annual variability – particularly in relation to rainfall. The phytoplankton of this oligotrophic lake is dominated by Cryptophyta and Chlorophyta, with a range of blue-green algae also routinely present. In years of high rainfall, when water residence time is low (<0.5 years), populations of blue-green algae are limited to summer growth and are restricted within the water column to the thermocline. Years of low rainfall increase water residence time to a period of 1 year or more, leading to nutrient accumulation and a quite different phytoplankton development. Blue-greens increase in population, occupying the whole water column and dominate into the autumn period and early spring.

In this lake, the most effective management strategy to avoid excessive nutrient accumulation and blue-green growth is intensive monitoring of rainfall and lake development over the summer (with the development of a predictive computer model for rainfall effects). This allows a stringent control of phosphorus input to be put in place at the onset of low rainfall periods.

Lake hydrology

With some water bodies, hydrological characteristics lead to a high retention time and result in nutrient build-up and the accumulation of algal biomass. Freshwater systems in this category typically have limited (or no) inflow/outflow, and include urban lakes and quay areas.

Urban lakes   Urban lakes tend to be relatively small, shallow water bodies in areas of high human population. They are typically artificial features, with no provision for water through-flow, and the fundamental problem of high retention time is compounded by various routes of nutrient loading – including fertilizer runoff and faecal contamination from wildfowl populations. These lakes are liable to extreme problems of eutrophication, as indicated by a 1992 survey of parkland lakes in Wandsworth, London (UK) – where problems encountered included the development of extensive algal blooms (including toxic blue-greens), loss of all higher aquatic plants, and fish kills due to anaerobic conditions during hot weather (Birch and McCaskie, 1999). With these lakes a range of measures can be implemented to restore water quality, including flushing the system (to remove algae and contaminated water), dredging the sediment, reducing the wildfowl populations (to cut down on nutrient entry), and re-establishment of macrophytes. Fish kills due to anaerobiosis were eliminated by the installation of aerators. Fish management was also important, reducing the overall level of fish, and particularly removing carp and bream. These fish are known to be benthic feeders, stirring up the sediment and promoting nutrient release and re-suspension of particulate matter. Aquatic plants were also uprooted.

GLOSSARY

Actinomycetes A group of aerobic, Gram-positive prokaryotes that form branching filaments and asexual spores. See 'fungal-like' microorganisms.

Acidophile An organism that lives in acid conditions (growth range – pH0–5.5).

Aerobe An organism that grows in the presence of atmospheric oxygen.

Aquifer Underground aquatic system.

Algae A phylogenetically-diverse group of simple (mainly unicellular to colonial) organisms that have chlorophyll-a as their main photosynthetic pigment and lack a sterile covering of cells around reproductive cells. Algae carry out oxygenic photosynthesis, and include eukaryote and prokaryote organisms.

Algal bloom A dense population of planktonic algae, resulting from natural or anthropogenic causes.

Algal indicators Particular species or groups of algae with distinctive ecological preferences. Their environmental presence indicates aspects such as high pH, low inorganic nutrient level and high turbulence.

Algal lawn Laboratory preparation, an algal monolayer overlying agar within a culture dish. Used to detect and assess the presence of algal antagonists (e.g. viruses, lytic bacteria) within water samples.

Algal trophic index The ratio (quotient) of particular algal species or groups, providing a measure of the trophic status of the aquatic environment.

Algicide An agent that kills algae.

Alkalophile An organism that lives in alkaline conditions (growth range – pH8.5–12.0).

Allochthonous material Soluble or particulate matter in an aquatic system that is derived from an external source.

Amoeboid protozoa A diverse group of protozoa, characterised by the formation of pseudopodia.

Amoeboid movement Moving by cytoplasmic flow and the formation of pseudopodia.

Anaerobe An organism that grows in the absence of free oxygen.

Anabolic Biochemical process or processes leading to an increase in potential energy, involving synthesis of macromolecules or energy-rich bonds.

Anoxic Without oxygen

Anoxygenic photosynthesis Photosynthetic process in which electron donors (e.g.organic matter, sulphide) are used which do not result in oxygen evolution. Typical of photosynthetic bacteria.

Antagonist An organism that limits or prevents the growth of another organism e.g Parasitic fungi are antagonists of algal hosts.

Antibiotic A microbial product or its derivative that kills susceptible organisms or limits their growth

Archaea A major taxonomic group (domain) of prokaryote organisms that have isoprenoid glycerol diether or diglycerol tetraether lipids in their membranes

Freshwater Microbiology: Biodiversity and Dynamic Interactions of Microorganisms in the Aquatic Environment David C. Sigee
© 2005 John Wiley & Sons, Ltd ISBNs: 0-471-48529-2 (pbk) 0-471-48528-4 (hbk)

and archaeal rRNA. Many live in extreme environments. See also 'Bacteria,' and 'Eukarya.'

Atomic Force Microscopy (AFM) A new type of high-resolution microscopy that can be used to investigate surface structure.

Autochthonous material Soluble or particulate matter in an aquatic system that is derived internally from within the water body.

Ascomycetes (Phylum Ascomycota) A relatively minor group of freshwater (true) fungi that have a septate mycelium and produce ascospores.

Autotroph An organism that uses inorganic carbon (CO_2, HCO_3^-) as its sole source of carbon.

Auxotroph A mutant organism that has lost the ability to synthesis an essential nutrient and must obtain it as a precursor from its surroundings.

Bacillus A rod-shaped bacterium.

Bacteria A major taxonomic group (domain) of prokaryote organisms that contain primarily diacyl glycerol diesters in their membranes and have bacterial rRNA. See also 'Archaea,' 'Eukarya.'

Bactericide An agent that kills bacteria.

Bacteriochlorophyll A modified chlorophyll present in photosynthetic bacteria.

Bacteriophage (phage) A virus that is specific to a bacterial host. See also 'cyanophage.'

Basidiomycetes (Phylum Basidiomycota) A relatively minor group of freshwater (true) fungi that have a septate mycelium and produce basidiospores.

Batch culture Microbial laboratory culture, grown in a fixed volume of growth medium.

Benthos Organisms living at the bottom of an aquatic system, associated with the sediments.

Biodiversity The range of taxonomic, phenotypic or genetic characteristics within a population or community.

Biofilm A community of microorganisms occurring at a physical (e.g. water/solid) interface, typically present within a layer of extracellular polysaccharide that is secreted by the community.

Biological control (biocontrol) A short-term strategy to limit the growth of nuisance organisms (e.g. blue-green algae) in the environment by the addition of naturally-occurring (or derived) antagonists.

Biological oxygen demand (BOD) Experimentally determined oxygen requirement for microorganisms within a sealed water sample, usually measured over a five day period. BOD is determined by the heterotrophic bacterial population and organic nutrient content, and is a measure of trophic status.

Biomanipulation A long-term management strategy to maintain or restore the ecological balance in waters where nutrient input may promote deleterious algal blooms. The strategy operates by promoting high zooplankton populations and maintaining grazing pressure on phytoplankton (See 'top-down' control).

Biomass Total mass of all living organisms or a set of organisms, usually expressed as dry weight.

Biome The community of organisms living within an ecosystem or distributed over a wide climatic area.

Biosilification The conversion of soluble to insoluble silica that occurs in diatom cell wall formation.

Biosphere The part of the planet containing living organisms. The living world.

Biota Living organisms

Biovolume The volume occupied by a single organism (unit biovolume) or an entire species population (population biovolume). The latter is normally expressed as a % of the total community biovolume.

Blue-green algae (Class Cyanophyta) A major group of prokaryotic organisms that contain chlorophyll-a and carry out oxygenic photosynthesis. Also referred to as 'cyanobacteria.'

Bottom-up control Restricted population growth of a group of organisms due to limitations in nutrient availability.

Boundary layer A region of restricted water movement associated with the surface of solid objects (surface drag) that are exposed to laminar flow.

Brown algae (Class Phaeophyta) A group of attached brown-pigmented eukaryote algae, poorly represented in freshwater environments.

Calvin cycle   The major pathway for reduction and incorporation (fixation) of CO_2 into organic material by photoautotrophs during photosynthesis. Also carried out by chemolithoautotrophs.

Capsid   A protein coat that surrounds a virion's nucleic acid.

Carbon flux   Carbon flow, within food webs.

Catabolic   Biochemical process or processes involving the conversion of potential to kinetic energy, by the breakdown of macromolecules or energy-rich bonds.

Chemolithotrophic autotroph   An organism that oxidises reduced inorganic compounds to derive both energy and electrons, and uses CO_2 as its carbon source.

Chemoorganic heterotroph   An organism that uses organic compounds as a source of energy, hydrogen, electrons and carbon for biosynthesis.

Chemotaxis   Response of motile microorganisms to a chemical stimulus, by movement towards or away from the source of the chemical.

Chemotrophs   Organisms that obtain energy from the oxidation of chemical compounds.

Chitin   A highly resistant nitrogen-containing polysaccharide forming the wall of certain fungi and other microorganisms.

Chrysophytes (Class Chrysophyta)   A group of golden-brown eukaryote algae.

Chytrids (Phylum Chytridiomycota)   An important group of freshwater parasitic fungi, living on invertebrate and phytoplankton hosts.

Ciliates   A major group of motile (*via cilia*) free-living protozoa.

Circadian rhythm   Metabolic or behavioural periodicity with a cycle of about 24h. See also 'diel periodicity,' and 'diurnal activity'.

Cladocerans (Phylum Cladocera)   A major component of the zooplankton, important filter feeders of algae.

Clear-water phase   A time period within the seasonal cycle of temperate lakes, when phytoplankton biomass is much reduced due to high levels of zooplankton grazing.

Climax community   The group of organisms that finally develops, and is maintained, under constant (steady-state) environmental conditions.

Co-aggregation   The spatial association of particular bacteria on a surface due to cell-to-cell recognition.

Colonial   Term used for multicellular algae and protozoa which lack the cell differentiation and division of labour seen in higher organisms.

Commensalism   Association between two organisms where there is no direct metabolic cooperation.

Community   A naturally-occurring assemblage of organisms living within a particular habitat.

Community filtration rate   The sum of the filtration rates of all the separate organisms within a heterogeneous zooplankton community.

Compensation point   The depth in the water column at which the evolution of oxygen due to photosynthesis (determined by light intensity) is balanced by the uptake of oxygen due to respiration.

Competition   Interaction between individuals of the same species or different species for habitat resources such as food, space or physical parameters (e.g. light).

Conjugation   A process of gene transfer and recombination in bacteria that requires direct cell-to-cell contact.

Constitutive   Gene expression which occurs all the time and is not initiated by a specific external stimulus (non-induced).

Copepods (Phylum Copepoda)   A major component of zooplankton, important filter and raptorial feeders of algae.

Cryophilic   Organisms typical of snow and ice environments.

Cryptomonads (Class Cryptophyta)   A group of unicellular motile eukaryote algae.

Cyanobacteria   See 'blue-green algae.'

Cyanophage   A virus that is specific to a blue-green algal host. See also 'bacteriophage.'

Cyclic AMP   Derivative of adenosine monophosphate that is biochemically important in signal transduction and control of gene expression.

Cytokinesis   Cytoplasmic division, during mitosis or meiosis. Follows on from nuclear division (karyokinesis).

Denitrification   Environmental conversion of nitrate to nitrogen.

Detritus   Small pieces of dead and decomposing organic material.

Deuteromycetes (Phylum Deuteromycota)   A taxonomically-mixed group of 'true' fungi that appear to lack a sexual stage in their life cycle.

Diatoms (Class Bacillariophyta)   A group of non-motile eukaryote algae, with silica cell walls. See 'frustule.'

Diel periodicity   Occurring at 24h intervals. See also 'circadian rhythm', 'diurnal activity.'

Diffusion feeding   Feeding strategy in which the predator relies on the movements of the prey organism to make contact, as in heliozoan and suctorian protozoa.

Dinoflagellates (Class Dinophyta)   A group of unicellular, highly motile eukaryote algae.

Diurnal activity   Occurring every day, as part of a sequence of activities with a 24h time frame. See also 'circadian rhythm,' 'diel periodicity.'

Dissolved organic carbon (DOC)   The soluble fraction (able to pass through a 0.2μm filter membrane) of organic matter present in the water medium, measured as carbon. Derived mainly from internal (autochthonous, e.g. algae) or external (allochthonous) sources.

Dissolved organic matter (DOM)   The soluble fraction (see above) of organic matter present in the water medium, corresponding to DOC but measured as dry weight.

Domain   1. Major taxonomic assemblage. The highest level of biological classification. 2. A functional DNA sequence.

Early mRNA   Messenger RNA produced early in virus infection and codes for proteins required to take over host cell activities and manufacture viral nucleic acids.

Eclipse period   The earliest phase of viral infection in which the host cells do not contain any complete virions.

Ecological dominance   The tendency for one group of organisms or species to out-compete other biota and form a dense homogeneous population (low biodiversity).

Ecological niche   The functional position of an organism in its local environment. The term is normally used to describe a particular role within a community (e.g. a secondary coloniser of biofilms) or the occurrence of a particular microenvironment to which organisms are adapted (e.g. part of water column containing microaerophile protozoa).

Ecological pyramid   The steady state balance between primary producers, herbivores and carnivores in relation to population counts, biomass or energy values.

Ecosystem   A self-regulating biological community living in a defined habitat.

Electron transport chain   A series of electron carriers that transfer electrons along a succession of electron donors and acceptors.

Endemic species   Organisms restricted to a particular geographic or local region.

Entrainment   The process by which a free-running endogenous rhythm is synchronised to a periodic environmental stimulus.

Epidemic   Uncontrolled spread of parasitic infection within a host population – e.g. infection of phytoplankton by chytrid fungi. See also 'fungal epidemic threshold.'

Epilimnion   The surface layer of a stratified water body.

Epilithic   Occurring on gravel, pebbles and large rocks.

Epipelic   Occurring on fine sediments such as mud.

Epiphytic   Associated with, or attached to, plant surfaces – including macrophytes, phytoplankton and benthic algae. See also 'phycosphere.'

Epipsammic   Occurring on sand.

Eukarya The major taxonomic group (domain) that contains eukaryote organisms, with cells that have primarily glycerol fatty acyl diesters in their membranes and eukarytic rRNA. See also 'Bacteria,' 'Archaea.'

Euglenoids (Class Euglenophyta) A group of unicellular motile eukaryote algae.

Eukaryote Unicellular or multicellular organism, with cells that contain a membrane-bound nucleus and characteristic cytoplasmic membrane-bound organelles. See also 'Eukarya,' 'prokaryote.'

Eutrophic An aquatic system rich in inorganic nutrients, with high primary productivity.

Eutrophication Environmental increase in soluble inorganic plant nutrients such as phosphates and nitrates.

External loading Entry of inorganic nutrients into the water column of an aquatic system from an outside source (e.g. river inflow).

Extreme environment An environment in which physicochemical factors such as irradiation, pH, salinity and temperature are outside the normal range for growth of most microorganisms.

Extremophile A microorganism that is adapted to grow in an extreme environment.

Femtoplankton The smallest size category ($<0.2\mu m$ diameter) of planktonic organisms. Mainly viruses and small bacteria. Part of the DOC.

Fermentation An energy-yielding process in which an energy substrate is oxidised without an exogenous electron acceptor. Usually organic molecules serve as both electron donors and acceptors.

Filter feeding Removal of suspended food particles from the aquatic medium by directing a current of water through a filter device. Typical of some protozoa and zooplankton.

Filtration rate (clearance rate) The volume of water cleared of particles per unit time by a filter feeding organism. See also 'community filtration rate.'

Flagellates A phylogenetically diverse assemblage of motile (via flagella) unicellular microorganisms, with links to both algae and protozoa. See also 'heterotrophic nanoflagellates.'

Food chain A linear progression of feeding from one major group of organisms to another, describing the main sequence of biomass transfer.

Food web An interconnected system of feeding relationships.

Free energy (ΔG_o^1) The maximum amount of kinetic energy released during a chemical reaction, expressed as kJ mol^{-1} substrate.

Frustule The silicified cell wall of diatoms.

Fungal epidemic threshold The critical phytoplankton population density above which a fungal epidemic occurs.

Fungal survival threshold The critical phytoplankton population density that is required to maintain a static infective fungal population.

Fungi A phylogenetically diverse group of heterotrophic microorganisms with a typically branched filamentous (mycelial) body organisation. See also 'true' fungi.

Gas vacuole A gas-filled vacuole, found in some aquatic blue-green algae and bacteria, that increases buoyancy. It is composed of gas vesicles, which are made of protein.

Generation time The time required for a microbial population to double in number.

Grazing Strictly, the consumption of green plant material by herbivores. In this book, the term is more broadly used to describe the consumption of algae and other microorganisms by zooplankton, benthic invertebrates and protozoa.

Gram stain A differential staining procedure that separates bacteria into Gram-positive and Gram-negative organisms on the basis of their ability to retain crystal violet dye when decolorised by an organic solvent such as ethanol.

Green algae (Class Chlorophyta) A major group of eukaryote algae that have predominantly green coloration.

Gross productivity The underlying rate of biomass synthesis in a population of organisms, without any deduction of biomass loss. See also 'productivity' and 'net productivity.'

Habitat   The living place of an organism or community, characterised by its physicochemical or biotic properties.

Halophile   An organism that is adapted to grow in a saline environment, ranging from 2–20% of a saturated salt (NaCl) solution.

Helophytes   Macrophytes present in the upper part of the littoral zone, with their foliage mainly out of the water.

Heterocyst   A specialised cell produced by some blue-green algae that is involved in nitrogen fixation.

Heterotroph   An organism that uses organic molecules as its principal source of carbon.

Heterotrophic nanoflagellates (HNFs)   A diverse group of non-photosynthetic flagellates (protozoa) that have a small size range (2–20µm) and are important consumers of bacteria.

Histone   A small basic protein containing large amounts of lysine and arginine that is associated with eukaryotic nuclear DNA.

Holoplanktonic   Aquatic organisms which are present in the water column over most of the annual cycle.

Homeostasis   Maintenance of the internal environment. A broad biological concept ranging from molecular systems to single cells, whole organisms and ecosystems.

Host   An organism or species that is liable to infection by a particular parasite in a specific parasitic relationship.

Hydraulic water retention time   The average time taken to refill a lake basin with water if it were emptied.

Hydrology   All aspects of water flow connected with an aquatic system, including inflow and outflow of water.

Hydrophytes   Macrophytes that are totally submerged.

Hypha   A tubular filament, forming the mycelium that is typical of most fungi.

Hyphomycetes   An important group of aquatic fungi belonging to the phylum Deuteromycota, typical of flowing waters.

Hypolimnion   The lower region of water column in a stratified water body, between the metalimnion and sediments.

Infection   The invasion of a host by a microbial agent, with subsequent establishment and multiplication of the microorganism.

Internal loading   Entry of inorganic nutrients into the water column of an aquatic system from the sediments.

Irradiance   The amount of incident light energy per unit area and per unit time.

Kairomones   Chemical signals released by zooplankton predators that enhance the light-mediated activation of the zooplankton circadian rhythm, promoting the onset of vertical diurnal migration.

K-strategist (K-selected organism)   An organism adapted to survive, grow and reproduce in a crowded environment, where it is subjected to high levels of competition from other biota.

Lag phase   The period following introduction of microorganisms into fresh culture medium, when there is no increase in biomass or population count.

Latent period   The initial infection phase in the one-step viral growth experiment (e.g. infection of blue-green algae by cyanophages), during which no viruses are released.

Latent virus   A non-virulent virus, that integrates into the host chromosome (prophage) and remains dormant.

Late mRNA   Messenger RNA produced during the final phase of viral infection (e.g. of eukaryotic algae by phycoviruses). The mRNA codes for proteins involved in capsid construction and virus release.

Lentic system   A standing freshwater body.

Limnology   The study of aquatic systems contained within continental boundaries.

Lithotroph   An organism that uses inorganic compounds as its source of electrons.

Littoral zone   The peripheral shoreline at the edge of lakes and rivers.

Log phase (exponential phase) The phase of growth in laboratory batch culture when the microbial population is growing at a constant and maximum rate.

Lotic system A flowing freshwater body.

Luxury consumption The ability of aquatic microorganisms to take up and store inorganic nutrients under conditions of excess availability.

Lysogeny A state of viral infection in which the phage genome (prophage) remains within the host cell in a non-virulent condition.

Lysis The physical disintegration or rupture of a cell.

Lytic cycle A virus infection cycle that results in death of the host cell (lytic host).

Macrophyte A large aquatic plant (e.g. water-lily - *Nuphar*), as opposed to smaller plants such as duckweed (*Lemna*), macroscopic algae (e.g. *Cladophora*) and phytoplankton.

Macroplankton Multicellular or colonial planktonic organisms with sizes $> 200\mu m$.

Meiofauna Invertebrate animals inhabiting the bottom of a lake or river – including copepods, nematodes and rotifers.

Meroplanktonic Aquatic organisms that have only a limited planktonic existence in the water column. Most of the annual cycle is spent on sediments (e.g. algae) or outside the aquatic system (e.g. chironomid insects).

Mesocosm An environmental enclosure, used for experimental purposes. Lake mesocosms can be used, for example, to test the effect of nutrient addition on phytoplankton composition within a restricted volume of environment, and range from quite small containers (~ 101) up to much larger enclosures such as Lund tubes ($>10^4 l$).

Mesotrophic An aquatic system with moderate levels of inorganic nutrients and moderate primary productivity. An intermediate state between oligotrophic and eutrophic,

Mesophile An organism with a growth range between 15–45°C.

Metalimnion The middle layer of the water column in a stratified water body, between the epilimnion and hypolimnion. Characterised by a steep temperature gradient (thermocline).

Metaphyton Planktonic filamentous algae that have become detached from inorganic substrate or higher plant surfaces.

Methanogen A strictly anaerobic bacterium that derives energy by converting CO_2, H_2, formate, acetate and other organic compounds to either methane or to methane and CO_2.

Michaelis-Menten constant (KM or K_m) A kinetic constant for an enzyme reaction, defined as the substrate concentration required for the enzyme to operate at half maximal capacity. Michaelis-Menten kinetics can also be applied to other dynamic activities involving substrates, such as algal nutrient uptake and zooplankton grazing.

Microaerophile A microorganism that requires low levels of oxygen (2–10% saturation) for growth, but is damaged by normal oxygen levels.

Microalgae A term generally applied to all microscopic algae, planktonic or benthic, with sizes up to about $200\mu m$.

Microbial loop A major route of biomass retrieval in the freshwater environment, involving the uptake of carbon (from DOC) by heterotrophic bacteria with subsequent transfer to the main food chain. The DOC is either autochthonous (e.g. eutrophic standing waters) or allochthonous (most rivers and streams).

Microcell A very small ($0.2\mu m$ or smaller) bacterial cell that has adopted the miniaturised form as a response to low nutrient conditions. Such ultramicrobacteria are able to increase in size when high nutrient conditions return.

Microcosm A limited or localised community of organisms, either occurring naturally (e.g. phycosphere community) or as part of an experiment (e.g. small volume containers in a lake).

Microenvironment A small-scale habitat occurring within a general ecosystem.

Micronutrient Trace element

Microorganism An organism not clearly visible to the naked eye, requiring a microscope for detailed observation.

Microplankton   Unicellular and multicellular planktonic organisms in the size range 20–200µm.

Mineralization   1. The release or conversion of inorganic nutrients from organic matter due to microbial activity. 2. The conversion of soluble to insoluble silica during diatom cell wall formation. See also 'biosilification.'

Mixotrophy   The ability of organisms to combine autotrophic (using inorganic electron sources) and heterotrophic (organic carbon sources) nutrition. The term is sometimes used more specifically in phycology to describe algae that can carry out photosynthesis and phagotrophy.

Mortality rate   The proportion of a population dying over a period of time, calculated as the number of deaths divided by the initial number of live individuals.

Mycelium   A branched filamentous (hyphal) growth form typical of actinomycetes, oomycetes and 'true' fungi.

Nanoplankton   Unicellular planktonic organisms in the size range 2–20µm. Typically eukaryotic.

Nekton   Strongly swimming pelagic biota – e.g. fish.

Net productivity   The observed rate of biomass synthesis in a population of organisms, equal to gross productivity minus intrinsic mass loss. Intrinsic mass loss occurs due to internal metabolic processes (e.g. respiration, excretion) and is separate from mass loss due to external (extrinsic) factors such as grazing and parasitism. See also 'productivity' and 'gross productivity.'

Neuston   The distinct community of organisms living at the air/surface water interface of water bodies.

Neutrophile   Organism living under neutral pH conditions (growth range pH5.5 – 8.5).

Niche   See 'ecological niche.'

Nitrification   Environmental oxidation of ammonia to nitrate.

Nitrogen fixation   The metabolic reduction of atmospheric nitrogen to ammonia, catalysed by the nitrogenase enzyme.

Nutrient cycling   The sequence of nutrient transitions and transfers that occur between different ecological compartments in an aquatic system.

Nutrient re-cycling   The regeneration of biologically-available nutrients as part of the general cycling process.

Nutrient spiralling   The linear progression and completion of nutrient cycling over a distance of water flow in a lotic system.

Oceanography   The study of aquatic systems between continents.

Oligotrophic   An aquatic system poor in inorganic nutrients, with low primary productivity.

One-step growth experiment   A laboratory experiment to investigate the kinetics of virus infection, in which one round of virus reproduction begins and ends with the lysis of the microbial host cells.

Oomycetes (Phylum Oomycota)   Water moulds. A group of eukaryotic microorganisms with a mycelial growth form that have a separate phylogenetic origin from other 'fungal-like' organisms.

Organotroph   1. An organism that uses reduced organic compounds as its source of electrons. 2. An organism that carries out organotrophy (osmotrophy).

Osmotrophy   Direct uptake of soluble organic compounds over the cell surface by an aquatic organism. Also referred to as organotrophy, saprotrophy.

Oxidation-reduction (redox) reactions   Chemical reactions involving electron transfers, in which a reductant molecule donates electrons to an oxidant.

Oxycline   The zone of steep oxygen gradient in the water column of a stratified lake, corresponding approximately to the thermocline.

Oxygenic photosynthesis   Photosynthesis that oxidises water to generate oxygen, typical of eukaryote algae and blue-green algae.

Parasite   An organism that derives its food from a living organism of another species (host) by direct association. Transfer or conversion of biomass occurs over a period of time. Ecologically-important parasites in the freshwater environment include chytrid fungi (phytoplankton host) and bacteriophages (bacterial host).

Pathogen An infective agent that causes disease.

Pelagic Organisms normally present in the water column of aquatic systems, divided into plankton and nekton.

Pelagic zone The main central part of a lake.

Peptidoglycan (mucopeptide) A large cell wall polymer typical of bacteria and blue-green algae, composed of long chains of alternating N-acetylglucosamine and N-acetylmuramic acid residues.

Periphyton A community of mainly plant-like organisms (including algae, bacteria and fungi) present on underwater substrata.

Phagotrophy Ingestion and intracellular digestion of particulate organic matter.

Photoinhibition The damaging effects of high light intensity on metabolic activities.

Photon flux density (PFD) A measure of light intensity. The number of incident photons per unit area and time.

Photoadaptation The physiological responses of aquatic algae to temporary or long-term limitations in light energy.

Photo-organotrophy The ability to obtain carbon from either CO_2 (photosynthesis) or soluble organic compounds (organotrophy). Energy is obtained from light.

Photoprotection Internal shielding (within a multicellular plant body or individual cell) from harmful radiation by carotenoid pigments (photoprotectants).

Photoreactivation Light-induced repair of damaged DNA.

Photosynthetically-available radiation (PAR) The range of wavelengths used by most phototrophic organisms – including algae and higher plants.

Photosystem II complex The functional association of pigments and protein molecules present on thylakoid membranes that catalyses the transfer of electrons from water to plastoquinone.

Phototroph An organism that uses solar energy to manufacture organic compounds by photosynthesis.

Phycovirus A virus that is specific to a eukaryote algal host. See also 'bacteriophage' and 'cyanophage.'

Phycosphere The microenvironment that occurs within and immediately around the surface of algae. See also 'microcosm.'

Phytoplankton Free-floating photosynthetic microorganisms, including algae and bacteria.

P-I curves Photosynthesis-irradiance plots for laboratory-cultured algae.

Picoplankton Unicellular planktonic organisms in the size range 0.2–2µm. Mainly prokaryotes.

Piscivorous Fish-eating. Examples of piscivorous fish in temperate lakes include Perch (*Perca*) and Pike (*Esox*).

Placoderm desmids Unicellular algae in the family Desmidiaceae. Consist of two separate halves (semicells) joined by connection zone (isthmus). Typical of unpolluted, low nutrient waters with a slightly acidic pH.

Plankton Free-floating pelagic organisms including bacterioplankton, phytoplankton, zooplankton and virioplankton.

Plantae The kingdom of multicellular photosynthetic organisms (plants) within the domain Eukarya.

Plaque A clear area in a monolayer of cultured cells, indicating destruction of cells by a single infective microorganism. Used to make counts of virulent bacteria or viruses as plaque forming units (PFU) in a plaque assay.

Pleuston The community of floating (unattached) higher plants on the water surface of standing waters.

Polymerase chain reaction (PCR) An *in vitro* technique used to synthesise large quantities of specific nucleotide sequences from small amounts of DNA.

Predation Capture and killing of other organisms for food. Includes herbivory (e.g. consumption of algal cells) and carnivory (e.g. consumption of zooplankton).

Primary production Synthesis of biomass by photosynthetic organisms – higher plants, eukaryotic algae, blue-green algae and anoxygenic photosynthetic bacteria. The first stage of biomass formation in freshwater ecosystems.

Productivity The intrinsic rate of increase in biomass (growth rate) in a population of organisms. Can be expressed as net or gross productivity. See also 'primary' and 'secondary' production.

Prokaryote Unicellular or colonial organism, with cells that lack a membrane-bound nucleus and do not have the membrane-bound organelles seen in eukaryotes.

Prophage The latent form of a temperate phage (virus) that is non-virulent and is typically integrated into the host chromosome.

Protista (Protoctista) The kingdom of unicellular and colonial microorganisms (including algae and protozoa) within the domain Eukarya.

Protozoa (protozoans) A subkingdom of heterotrophic unicellular and colonial eukaryotes within the kingdom Protista, Domain Eukarya.

Pseudomonad A member of the genus *Pseudomonas*, a widely occurring and ecologically-important group of freshwater bacteria.

Psychrophile An organism that grows well at $0°C$, with an optimum temperature of $15°C$ or lower and a temperature maximum around $20°C$.

Quorum sensing The process by which microorganisms such as bacteria monitor their own population density by sensing the levels of signals released within the community. At a particular population density, when the population has reached a critical level or quorum, quorum-dependent genes are expressed.

Raptorial feeding A type of feeding in which the consumer moves round in active search and capture of food organisms.

Red algae (Class Rhodophyta) A group of red-pigmented eukaryote algae that are typically attached (non-planktonic).

Remineralisation Environmental conversion of organically-bound mineral material back to a soluble inorganic component – e.g. conversion of phosphorus-containing biomass to phosphate. See also 'recycling,' 'mineralisation.'

Restoration (rehabilitation) Human intervention to reverse the deterioration of a polluted water body, with remedial activity to return the aquatic system to an earlier (pre-pollution) state.

Rotifers (phylum Rotifera) A major component of zooplankton, important as suspension feeders of bacteria.

r-strategists (r-selected organisms) Organisms adapted to an uncrowded environment, with low competition from other biota. Typically pioneer organisms.

Saline Aquatic environments with a high ionic content, typically dominated by sodium and chloride ions

Saprotroph (saprophyte) An organism that obtains its nutrients from dead organic matter.

Secchi depth The depth within a water column at which a suspended sectored plate (Secchi disk) can no longer just be seen. Provides a useful measure of water turbidity, and under appropriate conditions, of phytoplankton biomass.

Secondary metabolism Biochemical processes that occur after growth has been completed. Typical of the stationary phase of laboratory batch cultures.

Secondary production Synthesis of biomass by heterotrophic organisms such as bacteria, protozoa, fungi and zooplankton. Follows on from primary production, and involves biomass conversion along the food chain.

Sedimentation Descent of non-motile plankton in the water column due to gravitational forces.

Senescence A series of structural and metabolic changes leading to death.

Siderophore A small organic molecule that is secreted by microorganisms and complexes with ferric ions present in the surrounding medium. Important in the uptake of iron from the environment and in competition between microorganisms.

Sigma factor (transcription factor) A key protein required for the transcription of bacterial mRNA.

Species diversity index A numerical evaluation of population composition in relation to species content, providing a measure of species richness and dominance. Provides useful information on the trophic and pollution status of lakes (e.g. using algal species index) and rivers (invertebrate species index).

Stationary phase The end phase of growth in laboratory batch culture when population growth ceases and the growth curve levels off.

Stratification The vertical structuring of static or very slow moving water bodies into three distinct layers – epilimnion, metalimnion and hypolimnion. Determined by temperature and circulatory differences within the water column.

Stress factors Environmental changes that impair biological function at the level of ecosystems, individual organisms and molecular systems.

Succession The temporal sequence of organisms that occurs in a developing community such as a biofilm or lake pelagic community. See also 'climax community.'

Suspension feeding Uptake and consumption of suspended food particles includes diffusion and filter feeding by protozoa and zooplankton.

Symbiosis Mutual association between living organisms, where both parties gain metabolic benefit by the exchange of nutrients – e.g. association of bacteria with blue-green algal cells.

Temperate virus A non-virulent virus that can infect its host and establish a lysogenic relationship rather than cause immediate lysis. See also 'lysogeny' and 'lytic cycle.'

Thermophile An organism that is able to grow at temperatures $>55°C$, with a growth optimum typically between $55–65°C$.

Thermocline See 'Metalimnion.'

Top-down control Limitation in the population growth of a particular group of organisms by the activities of predators, parasites or anatagonists. Occurs both as a natural ecological phenomenon and as a result of human intervention in aquatic systems. See also 'biomanipulation' and 'bottom-up control.'

Toxin A microbial product or component that can cause injury to other organisms.

Trace elements Biologically-essential elements that are required by aquatic organisms in very small amounts.

Transduction Bacteriophage-mediated gene transfer between bacterial cells.

Transformation Process of gene transfer between bacteria in which a fragment of free bacterial DNA is taken up by a recipient cell and integrated into the bacterial chromosome.

Trophic Connected with nutrition and feeding. The term is used to describe the inorganic nutrient status of different water bodies (oligotrophic to eutrophic), and the feeding relationships (trophic interactions) of freshwater biota.

Trophic cascade The effect of a top predator (e.g. fish) or intermediate predator (e.g. macroinvertebrate) on the zooplankton community and their grazing on phytoplankton.

True fungi A distinctive group of eukaryote microorganisms with a typical mycelial growth form that is part of the diverse assemblage of 'fungal-like' organisms. See also 'actinomycetes' and 'oomycetes.'

Ultramicrobacteria Very small ($0.2\mu m$ or smaller) bacteria that occur under low nutrient conditions. Represent dormant stages that can regain metabolic activity and larger size when high nutrient conditions return. See 'microcell'.

Ultraviolet (UV) light Short ($10–400nm$) wavelength radiation that is important in photoinhibition.

Viral unit An arbitrary measure of viral abundance, where one unit is the total population of a virus measured as total count, viable count or rate of incorporation of ^{32}P-labelled orthophosphate into viral DNA. Changes in viral population over a period of time can be expressed in terms of 'viral units.'

Virion The extracellular phase of the virus life cycle, present as a free virus particle within the water medium and typically consisting of a protein capsid surrounding a single nucleic acid molecule. Virions are directly released by lysis of infected cells and are subsequently able to infect new host cells.

Virioplankton   The aquatic population of free particulate viruses (virions). An important component of the femtoplankton and also the DOC.

Water activity (a$_w$)   A measure of water availability to aquatic organisms in saline environments, relating the vapour pressure of the saline solution to the vapour pressure of pure water.

Zooplankton   A diverse assemblage of invertebrate planktonic organisms.

Zooplankton refuge   Localised regions of the aquatic environment (e.g. macrophyte beds) where zooplankton populations are less exposed to predation. A buffer zone.

Zooplanktivorous   Zooplankton-eating. The Alewife (*Alosa aestavalis*) is an example of an ecologically-important fish consumer of zooplankton in North America.

Zygomycetes (Phylum Zygomycota)   A group of 'true' fungi that typically have a well-developed mycelium and form sexual resting spores.

References

Abraham-Peskir, J. *et al.* (1997). Seasonal changes in whole-cell metal levels in protozoa of activated sludge. *Ecotoxicology and Environmental Safety* **38**: 272–280.

Alberts, B. *et al.* (1962) *Molecular Biology of the Cell.* Garland Publishing Inc., New York, USA.

Alexopoulos, C. (1962). *Introductory Mycology.* John Wiley and Sons, New York, USA.

Alexopoulos, C. *et al.* (1996). *Introductory Mycology, Second Edition.* John Wiley and Sons, New York, USA.

Allan, J. (1995). *Stream Ecology.* Chapman and Hall, London, UK.

Allison, D. *et al.*, Eds. (2000). *Community Structure and Co-operation in Biofilms.* Society for General Microbiology Symposia. Cambridge University Press, Cambridge, UK.

Alongi, D. (1991). Flagellates of benthic communities: characteristics and methods of study. In *The Biology of Free-living Heterotrophic Flagellates*, (Patterson D. and Larsen, J., Eds.), pp. 57–75. Clarendon Press, Oxford, UK.

Amin, M. and Day, M. (1988). *Donor and Recipient Effects on Transduction Frequency in situ.* REGEM 1 Program and Abstracts, REGEM, Cardiff, UK.

Amy, P. and Morita, R. (1983). Starvation-survival patterns of 16 freshly-isolated open-ocean bacteria. *Applied and Environmental Microbiology* **45**: 1109–1115.

Anderson, W. *et al.* (1996). Evidence of recent warming and El Nino-related variations in ice breakup of Wisconsin lakes. *Limnology and Oceanography* **41**: 815–821.

Andrews, J. (1992). Fungal life histories. In *The Fungal Community*, (Carroll G. and Wicklow D., Eds), pp. 119–145. Marcel Dekker, New York, USA.

Angyal, S. (1972). Complex formation between sugars and metal ions. In *Carbohydrate Chemistry – VI*, (Doane W., Ed) Butterworths, London, UK.

Ashton, P. (1979). Nitrogen fixation in a nitrogen-limited impoundment. *Journal of Water Pollution Control Federation* **51**: 570–579.

Auclair, J. (1995). Implications of increased UV-B induced photoreduction: iron (II) enrichment stimulates picocyanobacterial growth in the microbial food web in clear water acidic Canadian Shield lakes. *Canadian Journal of Fisheries and Aquatic Science* **52**: 782–787.

Auer, B. and Arndt, H. (2001). Taxonomic composition and biomass of heterotrophic flagellates in relation to lake trophy and season. *Freshwater Biology* **46**: 959–972.

Baker, P. (1991). Identification of common noxious Cyanobacteria. Part I- Nostocales. *Urban Water Research Association of Australia Research Reports* **29**.

Baker, P. (1992). Identification of common noxious Cyanobacteria. Part II - Chroococcales and Oscillatoriales. *Urban Water Research Association of Australia Research Reports* **46**.

Bale, M. *et al.* (1988). Transfer and occurrence of large mercury resistance plasmids in river epilithon. *Applied and Environmental Microbiology* **54**: 972–978.

Barlöcher, F. (1982). On the ecology of Ingoldian fungi. *Bioscience* **32**: 581–586.

Barlöcher, F. (1985). The role of fungi in the nutrition of stream invertebrates. *Botanical Journal of the Linnean Society* **91**: 83–94.

Barrett, P. *et al.* (1996). The control of diatom and cyanobacterial blooms in reservoirs using barley straw. *Hydrobiologia* **340**: 307–311.

Barron, G. (1992). Lignolytic and cellulolytic fungi as predators and parasites. In *The Fungal Community: its Organisation and Role in the Ecosystem*, (Carroll, G. and Wicklow, D., Eds), pp. 311–326. Marcel Dekker, New York, USA.

Battarbee, R. (1999). The importance of paleolimnology to lake restoration. *Hydrobiologia* **395/396**: 149–159.

Begon, M. *et al.* (1996). *Ecology.* Blackwell, Oxford, UK.

Bell, W. and Sakshaug, E. (1980). Bacterial utilization of algal extracellular products. 2. A kinetic study of natural populations. *Limnology and Oceanography* **25**: 1021–1033.

Bell, W. *et al.* (1974). Selective stimulation of marine bacteria by algal extracellular products. *Limnology and Oceanography* **19**: 833–839.

Beltrami, E. and Carroll, T. (1994). Modeling the role of viral disease in recurrent phytoplankton blooms. *Journal of Mathematical Biology* **32**: 857–863.

Berendse, F. (1993). Ecosystem stability, competition and nutrient cycling. In *Biodiversity and Ecosystem Function*, (E.-D. Schulze and H. A. Mooney, Eds). pp. 409-. Springer Verlag, Berlin, Germany

Bergh, O. *et al.* (1989). High abundance of viruses found in aquatic environments. *Nature* **340**: 467–468.

Berner, E. and Berner, R. (1987). *The Global Water Cycle.* Prentice-Hall, Englewood Cliffs, New Jersey, USA.

Berninger, U. *et al.* (1991a). Protozoon control of bacterial abundances in freshwater. *Limnology and Oceanography* **36**: 139–147.

Berninger, U. *et al.* (1991b). Heterotrophic flagellates of planktonic communities, their characteristics and methods of study. In *The Biology of Free-living Heterotrophic Flagellates*, (Patterson, D. and Larsen, J., Eds), pp. 38–56. Clarendon Press, Oxford, UK.

Bidigare, R. *et al.* (1993). Evidence for a photoprotective function for secondary carotenoids of snow algae. *Journal of Phycology* **29**: 427–434.

Biggs, B. J. F. (1996). Patterns in benthic algae of streams. In *Algal Ecology*, (Stevenson, R. J., Bothwell, M. and Lowe, R. L., Eds), pp. 31–56. Academic Press, New York, USA.

Biggs, B. J. F. *et al.* (1998). Subsidy and stress responses of stream periphyton to gradients in water velocity as a function of community growth form. *Journal of Phycology* **34**: 598–607.

Billen, G. *et al.* (1990). Dynamics of bacterioplankton in oligotrophic and eutrophic aquatic environments: bottom-up or top-down control? *Hydrobiologia* **207**: 37–42.

Birch, S. and McCaskie, J. (1999). Shallow urban lakes: a challenge for lake management. In *The Ecological Bases for Lake and Reservoir Management*, (Harper, D. *et al.*, Eds), pp. 365–377. Kluwer, Dordrecht, The Netherlands.

Bird, D. and Kalff, J. (1984). Empirical relationships between bacterial abundance and chlorophyll concentration in fresh and marine waters. *Canadian Journal of Fisheries and Aquatic Science* **41**: 1015–1023.

Bohringer, H. *et al.* (1995). UDP-glucose is a potential intracellular signal molecule in the control of expression of sigma S and sigma S-dependent genes in *Escherichia coli*. *Journal of Bacteriology* **177**: 413–422.

Bolch, C. J. S. *et al.* (1996). Genetic characterization of strains of cyanobacteria using PCR-RFLP of the cpcBA intergenic spacer and flanking regions. *Journal of Phycology* **32**: 445–451.

Booth, K. *et al.* (1987). Studies on the occurrence and elemental composition of bacteria in freshwater phytoplankton. *Scanning Microscopy* **1**: 2033–2042.

Borchardt, M. (1996). Nutrients. In *Algal Ecology*, (Stevenson, R. J. Bothwell M. and Lowe, R. L., Eds), pp. 183–228. Academic Press, New York, USA.

Boston, H. and Hill, W. (1991). Photosynthesis-light reactions of stream periphyton communities. *Limnology and Oceanography* **36**: 644–656.

Bothwell, M. *et al.* (1989). Evidence of dark avoidance by phototrophic periphytic diatoms in lotic systems. *Journal of Phycology* **25**: 85–94.

Boynton, W. and Kemp, W. (2000). Influence of river flow and nutrient loads on selected ecosystem processes. In *Estuarine Science* (Hobbie, J., Ed). pp. 269–298. Island Press, Washington DC, USA.

Bratbak, G. (1987). Carbon flow in an experimental microbial ecosystem. *Marine Ecology Progress Series* **36**: 267–276.

Bratbak, G. *et al.* (1990). Viruses as partners in spring bloom microbial trophodynamics. *Applied and Environmental Microbiology* **56**: 1400–1405.

Bratbak, G. *et al.* (1993). Viral mortality of the marine alga *Emiliania huxleyi* (Haptophyceae) and termination of algal blooms. *Marine Ecology Progress Series* **93**: 39–48.

Bratbak, G. *et al.* (1994). Viruses and the microbial loop. *Microbial Ecology* **49**: 1488–1493.

Brinkmeyer, R. *et al.* (2003). Diversity and structure of bacterial communities in Arctic versus Antarctic pack ice. *Applied and Environmental Microbiology* **69**: 6610–6619.

Brisou, J. (1995). *Biofilms*. CRC Press, Boca Raton, USA.

Brook, A. *et al.* (1988). A study of barium accumulation in desmids using the Oxford scanning proton microprobe (SPM). *Nuclear Instruments and Methods in Physics Research* **B30**: 372–377.

Brown, E. and Button, D. (1979). Phosphate-limited growth kinetics of *Selenastrum capricornutum* (Chlorophyceae). *Journal of Phycology* **15**: 305–311.

Brown, R. M. *et al.* (1964). Airborne algae: their abundance and heterogeneity. *Science* **143**: 583–585.

Brunberg, A.-K. (1999). Contribution of bacteria in the mucilage of *Microcystis* spp. (Cyanobacteria) to benthic and pelagic bacterial production in a hypereutrophic lake. *FEMS Microbiology Ecology* **29**: 13–22.

Brunberg, A.-K. and Blomqvist, P. (2003). Recruitment of *Microcystis* (Cyanophyceae) from lake sediments: the importance of littoral inocula. *Journal of Phycology* **39**: 58–65.

Bruning, K. (1991a). Infection of *Asterionella* by a chytrid. II. Effects of light on survival and epidemic development of the parasite. *Journal of Plankton Research* **13**: 119–129.

Bruning, K. (1991b). Effects of phosphorus limitation on the epidemiology of a chytrid phytoplankton parasite. *Freshwater Biology* **25**: 409–417.

Bruning, K. (1991c). Effect of temperature and light on population dynamics of the *Asterionella-Rhizophydium* association. *Journal of Plankton Research* **13**: 707–719.

Brussaard, C. and Riegman, R. (1998). Influence of bacteria on phytoplankton cell mortality with phosphorus or nitrogen as the algal-growth-limiting nutrient. *Aquatic Microbial Ecology* **14**: 271–280.

Burdige, D. and Zheng, S. (1998). The biogeochemical cycling of dissolved organic nitrogen in estuarine sediments. *Limnology and Oceanography* **43**: 1796–1813.

Burdon, J. (1992). The growth and regulation of pathogenic fungal populations. In *The Fungal Community*, (Carroll, G. and Wicklow, D., Eds), pp. 173–181. Marcel Dekker, New York, USA.

Burkert, U. *et al.* (2003). Members of a readily enriched beta-proteobacterial clade are common in surface waters of a humic lake. *Applied and Environmental Microbiology* **69**: 6550–6559.

Burkholder, J. and Glasgow, H. (1997). Trophic controls on stage transformations of a toxic ambush-predator dinoflagellate. *Journal of Eukaryotic Microbiology* **44**: 200–205.

Burney, C. (1994). Seasonal and diel changes in particulate and dissolved organic matter. In *The Biology of Particles in Aquatic Systems*, (R. Wotton, Ed), pp. 97–136. Lewis Publishers, London, UK.

Burnham, J. C. *et al.* (1981). Entrapment and lysis of the cyanobacterium *Phormidium luridum* by aqueous cultures of *Myxococcus xanthus* PCO2. *Archives of Microbiology* **129**: 285–294.

Burns, C. (1968). The relationship between body size of filter-feeding cladocera and the maximum size of particle ingested. *Limnology and Oceanography* **13**: 675–678.

Burns, C. (1969). Relation between filtering rate, temperature and body size in four species of *Daphnia*. *Limnology and Oceanography* **14**: 693–700.

Burns, C. (1987). Insights into zooplankton – cyanobacteria interactions derived from enclosure studies. *New Zealand Journal of Freshwater and Marine Research* **21**: 477–482.

Byron, E. and Eloranta, P. (1984). Recent historical changes in the diatom community of Lake Tahoe, California-Nevada, U.S.A. *Verhandlungen der internationale Vereinigung für theoretische und angewandte Limnologie* **22**: 1372–1376.

Cain, J. and Trainor, F. (1976). Regulation of gametogenesis in *Scenedesmus obliquus* (Chlorophyceae). *Journal of Phycology* **12**: 383–390.

Caiola, M. and Pellegrini, S. (1984). Lysis of *Microcystis aeruginosa* by *Bdellovibrio*-like bacteria. *Journal of Phycology* **20**: 471–475.

Cairns, J. (2002). Biotic community response to stress. In *Biological Response Signatures*, (Simon, T., Ed), pp. 13–21. CRC Press Boca Raton USA.

Caljon, A. and Cocquyt, C. (1992). Diatoms from the surface sediments of the northern part of Lake Tanganyika. *Hydrobiologia* **230**: 135–156.

Campbell, D. *et al.* (1993). Electron transport regulates cellular differentiation in the filamentous cyanobacterium *Calothrix*. *Plant Cell* **5**: 451–463.

Canale, R. *et al.* (1976). A plankton-based food web model for Lake Michigan. In *Modeling Biochemical Processes in Aquatic Ecosystems*, (Canale, R., Ed), pp. 33–74. Ann Arbor Science, Ann Arbor, USA.

Canter, H. (1969). Studies on British Chytrids XXIX. A taxonomic revision of certain fungi found on the diatom *Asterionella*. *Botanical Journal of the Linnean Society* **62**: 267–278.

Canter, H. and Lund, E. (1951). Studies on plankton parasites III. Examples of interaction between parasitism and other factors determining the growth of diatoms. *Annals of Botany* **15**: 359–371.

Canter, H. and Lund, E. (1953). Studies on plankton parasites II. The parasitism of diatoms with special reference to the English Lake District. *Transactions of the British Mycological Society* **36**: 13–37.

Canter, H. and Lund, E. (1969). The parasitism of planktonic desmids by fungi. *Österreischer Botanischen Zeitung* **116**: 351–377.

Canter, H. and Lund, J. (1948). Studies on plankton parasites. 1. Fluctuations in the numbers of *Asterionella formosa* Hass. in relation to fungal epidemics. *New Phytologist* **47**: 238–261.

Carmichael, W. (1994). The toxins of cyanobacteria. *Scientific American* **January**: 64–70.

Carney, H. *et al.* (1994). *Seasonal, Interannual, and Long-term Dynamics of Plankton Diatoms in Oligotrophic Lake Tahoe*. 11th International Diatom Symposium, San Francisco, California, USA, pp. 621–629. California Academy of Sciences California, USA.

Caron, D. A. *et al.* (1999). Molecular phylogenetic analysis of the heterotrophic chrysophyte genus *Paraphysomonas* (Chrysophyceae), and the design of rRNA-targetted oligonucleotide probes for two species. *Journal of Phycology* **35**: 824–837.

Carpenter, S. R. and Kitchell, J. F. (1993). *The Trophic Cascade in Lakes*. Cambridge University Press, Cambridge, UK.

Carrell, E. (1969). Studies on chloroplast development and replication in *Euglena*. 1. Vitamin B12 and chloroplast replication. *Journal of Cell Biology* **47**: 525–530.

Carvalho, L. and Moss, B. (1995). The current status of a sample of English Sites of Special Scientific Interest subject to eutrophication. *Aquatic Conservation: Marine and Freshwater Ecosystems* **5**: 191–204.

Castlebury, L. (1994). Small-subunit ribosomal gene phylogeny of *Plasmodiophora brassicae*. Fifth International Mycological Congress, Vancouver BC, Canada, pp. 32.

Catteneo, A. *et al.* (1998). Organisms' response in a chronically polluted lake supports hypothesized link between stress and size. *Limnology and Oceanography* **43**: 1938–1943.

Chen, F. and Suttle, C. (1995). Amplification of DNA polymerase gene fragments from viruses infecting microalgae. *Applied and Environmental Microbiology* **61**: 1274–1278.

Chesnick, J. *et al.* (1996). Utility of the mitochondrial *nas4L* gene for algal and protistan phylogenetic analysis. *Journal of Phycology* **32**: 452–456.

Chiovitti, A. *et al.* (2003). The complex polysaccharides of the raphid diatom *Pinnularia viridis* (Bacillariophyceae). *Journal of Phycology* **39**: 543–554.

Clay, S. *et al.* (1991). X-ray microanalytical studies of freshwater biota: changes in the elemental composition of *Anabaena spiroides* during blooms of 1988 and 1989. *Scanning Microscopy* **5**: 207–217.

Clegg, M. *et al.* (2003). The effect of photon irradiance on the behavioural ecology and potential niche separation of freshwater phytoplanktonic flagellates. *Journal of Phycology* **39**: 650–662.

Cole, D. *et al.* (2003). Evaluation of F + RNA and DNA coliphages as source-specific indicators of fecal contamination in surface waters. *Applied and Environmental Microbiology* **69**: 6507–6514.

Cole, J. *et al.* (1988). Bacterial production in fresh and saltwater ecosystems: a cross-system overview. *Marine Ecology Progress Series* **43**: 1–10.

Coles, J. F. and Jones, R. C. (2000). Effect of temperature on photosynthesis response and growth of four phytoplankton species isolated from a tidal freshwater river. *Journal of Phycology* **36**: 7–16.

Collingwood, M. (1987). Injecting new life into the Great Stour. *Water and Waste Treatment* **30**: 12.

Costanza, R. and Voinov, A. (2000). Integrated ecological economic regional modeling. In *Estuarine Science*, (Hobbie, J. Ed). pp. 461–506. Island Press, Washington, USA.

Cota, G. (1985). Photoadaptation of high Arctic ice algae. *Nature* **315**: 219–221.

Cottrell, M. and Suttle, C. (1991). Wide-spread occurrence and clonal variation in viruses which cause lysis of a cosmopolitan, eukaryotic marine phytoplankter, *Micromonas pusilla*. *Marine Ecology Progress Series* **78**: 1–9.

Cottrell, M. and Suttle, C. (1995). Dynamics of a lytic virus infecting the photosynthetic marine picoflagellate *Micromonas pusilla*. *Limnology and Oceanography* **40**: 730–739.

Coulombe, A. and Robinson, G. (1981). Collapsing *Aphanizomenon flos-aquae* blooms: Possible contributions of photo-oxidation, O2 toxicity and cyanophages. *Canadian Journal of Botany* **59**: 1277–1284.

Cross, T. (1982). Aquatic actinomycetes: a critical survey of the occurrence, growth and role of actinomycetes in aquatic habitats. *Journal of Applied Bacteriology* **50**: 397–423.

Cuker, B. (1983). Grazing and nutrient interactions in controlling the activity and composition of the epilithic algal community of an arctic lake. *Limnology and Oceanography* **28**: 133–141.

Czeczuga, B. (1968a). Primary production of the green hydrosulphuric bacteria, *Chlorobium limicola* (Chlorobacteriaceae). *Photosynthetica* **2**: 11–15.

Czeczuga, B. (1968b). Primary production of the purple sulphuric bacteria, *Thiopedia rosea* (Thiorhodaceae). *Photosynthetica* **2**: 161–166.

Daft, M. J. and Stewart, W. (1973). Light and electron microscope observations on algal lysis by bacterium CP-1. *New Phytologist* **72**: 799–808.

Daft, M. J. *et al.* (1975). Ecological studies on algal-lysing bacteria in fresh waters. *Freshwater Biology* **5**: 577–596.

Damerval, T. *et al.* (1991). Gas vesicle synthesis in the cyanobacterium *Pseudanabaena* sp.: occurrence of a single photoregulated gene. *Molecular Microbiology* **5**: 657–664.

Davies, D. (2000). Physiological events in biofilm formation. In *Community Structure and Co-operation in Biofilms*, (Allison, D. *et al.*, Eds), pp. 37–52. Cambridge, Cambridge University Press, Cambridge, UK.

Deacon, J. (1997). *Modern Mycology*, Blackwell Science, Oxford, UK.

Dean, J. (1999). *Lange's Handbook of Chemistry*. McGraw-Hill, New York, USA.

Debacon, M. and McIntyre, A. (1991). Taxonomic structure of phytoplankton assemblages in Crater Lake, Oregon, USA. *Freshwater Biology* **25**: 95–104.

DeHaan, H. (1993). Solar UV-light penetration and photodegradation of humic substances in peaty lake water. *Limnology and Oceanography* **38**: 1072–1076.

DeMott, W. R. (1988). Discrimination between algae and artificial particles by freshwater and marine copepods. *Limnology and Oceanography* **33**: 397–408.

DeMott, W. R. (1989). The role of competition in zooplankton succession. In *Plankton Ecology*, (Sommer, U., Ed), pp. 195–252. Springer-Verlag, Berlin, Germany.

DeMott, W. R. and F. Moxter (1991). Foraging on cyanobacteria by copepods: Responses to chemical defenses and resource abundance. *Ecology* **72**: 1820–1834.

DeMott, W. R. *et al.* (1991). Effects of toxic cyanobacteria and purified toxins on the survival and feeding of a copepod and three species of *Daphnia*. *Limnology and Oceanography* **36**: 1346–1357.

Dent, R. M. *et al.* (2001). Functional genomics of plant photosynthesis in the fast lane using *Chlamydomonas reinhardtii*. *Trends in Plant Science* **6**: 364–371.

Descy, J.-P. *et al.* (2000). Pigment ratios and phytoplankton assessment in northern Wisconsin lakes. *Journal of Phycology* **36**: 274–286.

Dick, M. (1990). *Key to* Pythium. University of Reading Press, Reading, UK.

Dillon, J. G. and Castenholz, R. W. (1999). Scytonemin, a cyanobacterial sheath pigment, protects against UVC radiation: implications for early photosynthetic life. *Journal of Phycology* **35**: 673–681.

Dittmann, E. *et al.* (2001). Altered expression of two light-dependent genes in a microcystin-lacking mutant of *Microcystis aeruginosa* PCC 7806. *Microbiology* **147**: 3113–3119.

Dobson, M. and Frid, C. (1998). *Ecology of Aquatic Systems*. Addison Wesley Longman Ltd., Harlow, UK.

Dodds, W. K. *et al.* (1999). Photosynthesis-irradiance patterns in benthic microalgae: variations as a function of assemblage thickness and community structure. *Journal of Phycology* **35**: 42–53.

Doers, M. and Parker, D. (1988). Properties of *Microcystis aeruginosa* and *M. flos-aquae* (Cyanophyta) in culture: taxonomic implications. *Journal of Phycology* **24**: 502–508.

Donk, E. V. (1989). The role of fungal parasites in phytoplankton succession. In *Plankton Ecology*, (Sommer, U., Ed), pp. 171–195. Springer-Verlag Berlin, Germany.

Duarte, C. M. *et al.* (1990). Size plasticity of freshwater phytoplankton: implications for community structure. *Limnology and Oceanography* **35**: 1846–1851.

Dubinsky, Z. and Berman, T. (1981). Light utilisation by phytoplankton in Lake Kinneret (Israel). *Limnology and Oceanography* **24**: 652–663.

Ducklow, H. and Carlson, C. (1992). Oceanic bacterial production. *Advances in Microbial Ecology* **12**: 113–181.

Dugdale, R. (1967). Nutrient limitation in the sea: dynamics identification and significance. *Limnology and Oceanography* **12**: 685–695.

Dunbar, M. (1973). The ecosystem as a unit of natural selection. *Transactions of the Connecticut Academy of Arts and Sciences* **44**: 111–130.

Edgar, L. and Zavortink, M. (1983). The mechanism of diatom locomotion II. Identification of actin. *Proceedings of the Royal Society of London. Series B.* **218**: 345–348.

Edmondson, W. (1991). *The Uses of Ecology: Lake Washington and Beyond*. University of Washington Press, Seattle, USA.

Egger, K. (1992). Analysis of fungal population structure using molecular techniques. In *The Fungal Community*, (Carroll, G. and Wicklow, D., Eds), pp. 193–208. Marcel Dekker, New York, USA.

Ehlers, L. (2000). Gene transfer in biofilms. In *Community Structure and Co-operation in Biofilms*, (Allison, D. *et al.*, Eds), pp. 215–257. Cambridge University Press, Cambridge, UK.

Eichler, B. and Pfennig, N. (1990). Seasonal development of anoxygenic phototrophic bacteria in a holomictic drumlin lake (Schleinsee, F.R.G.). *Archives of Hydrobiology* **119**: 369–392.

Ellis, H. (1984). *Book of Data*. Longman, Harlow, UK.

Elton, C. (1958). *The Ecology of Invasion by Plants and Animals*. Methuen, London, UK.

Elwood, J. *et al.* (1981). The role of microbes associated with organic and inorganic substrates in phosphorus spiralling in a woodland stream. *Verhandlungen der internationale Vereinigung für theoretische und angewandte Limnologie* **21**: 850–856.

Ensign, J. (1992). Introduction to the Actinomycetes. In *The Prokaryotes*, (Balows, A. *et al.*, Eds), pp. 811–814. Springer-Verlag, New York, USA.

Environment Agency, (1998). The State of the Environment of England and Wales The Stationary Office, London, UK.

Escoubas, J. *et al.* (1995). Light intensity regulation of calogene transcription is signaled by the redox state of the plastoquinone pool. *Proceedings of the National Academy of Sciences* **92**: 10237–10241.

Evans, J. and Prepas, E. (1997). Relative importance of iron and molybdenum in restricting phytoplankton biomass in high phosphorus saline lakes. *Limnology and Oceanography* **42**: 461–472.

Ewert, D. and Paynter, M. (1980). Enumeration of bacteriophages and host bacteria in sewage and the activated sludge treatment process. *Applied and Environmental Microbiology* **39**: 576–583.

Fairchild, G. *et al.* (1985). Algal periphyton growth on nutrient-diffusing substrates: an *in situ* bioassay. *Ecology* **66**: 465–472.

Falkowski, P. G. (2000). Rationalizing elemental ratios in unicellular algae. *Journal of Phycology* **36**: 3–6.

Falkowski, P. G. and La Roche, J. (1991). Acclimation to spectral irradiance in algae. *Journal of Phycology* **27**: 8–14.

Fallowfield, H. J. and Daft, M. J. (1988). The extracellular release of dissolved organic carbon by freshwater cyanobacteria and algae and the interaction with *Lysobacter* CP-1. *British Phycological Journal* **23**: 317–326.

Federici, B. (1981). Mosquito control by the fungi *Culinomyces*, *Lagenidium* and *Coelomyces*. In *Microbial Control of Pests and Plant Diseases 1970–1980*, (Burges, H., Ed), pp. 555–572. Academic Press, New York, USA.

Fenchel, T. (1987). *Ecology of Protozoa*. Springer-Verlag, Berlin, Germany.

Fenchel, T. (1991). Flagellate design and function. In *The Biology of Free-living Heterotrophic Flagellates*, (Patterson, D. and Larsen, J., Eds), pp. 7–19. Clarendon Press, Oxford, UK.

Fennel, K. and Boss, E. (2003). Subsurface maxima of phytoplankton and chlorophyll: steady-state solutions from a simple model. *Limnology and Oceanography* **48**: 1521–1534.

Fields, S. and Rhodes, R. (1991). Ingestion and retention of *Chroomonas* spp. (Cryptopyhyceae) by *Gymnodinium acidotum* (Dinophyceae). *Journal of Phycology* **27**: 525–529.

Findlay, S. *et al.* (2003). Metabolic and structural response of hypopheic microbial communities to variations in supply of dissolved organic matter. *Limnology and Oceanography* **48**: 1608–1617.

Finlay, B. (1981). Oxygen availability and seasonal migrations of ciliated protozoa in a freshwater lake. *Journal of General Microbiology* **123**: 173–178.

Fischer, H. *et al.* (1999). Targeting and covalent modification of cell wall and membrane proteins heterologously expressed in the diatom *Cylindrotheca fusiformis* (Bacillariophyceae). *Journal of Phycology* **35**: 113–120.

Flower, R. *et al.* (1987). The recent paleolimnology of acid lakes in Galloway, south-west Scotland: diatom analysis, pH trends and the role of afforestation. *Journal of Ecology* **75**: 797–824.

Forsyth, D. *et al.* (1990). Alteration of seasonal and diel patterns in vertical migration of zooplankton by *Anabaena* and planktivorous fish. *Archives of Hydrobiology* **117**: 385–404.

Francko, D. and Wetzel, R. (1982). The isolation of cyclic adenosine $3':5'$-monophosphate (cAMP) from lakes of differing trophic status: correlation with planktonic metabolic variables. *Limnology and Oceanography* **27**: 27–38.

Frederickson, J. *et al.* (1989). Lithotrophic and heterotrophic bacteria in deep subsurface sedimenta and their relation to sediment properties. *Geomicrobiological Journal* **7**: 53–66.

Freifelder, G. (1987). *Microbial Genetics*. Jones and Bartlett Publishers Inc., Boston, USA.

Fukumoto, R.-h. *et al.* (2003). Cloning and characterisation of a cDNA encoding a sexual cell division-inducing pheromone from a unicellular green alga *Closterium ehrenbergii* (Chlorophyta). *Journal of Phycology* **39**: 931–936.

Fulton, R. (1988). Resistance to blue-green toxins by *Bosmina longirostris*. *Journal of Plankton Research* **10**: 771–778.

Fulton, R. and H. W. Paerl (1987). Toxic and inhibitory effects of the blue-green alga *Microcystis aeruginosa* in herbivorous zooplankton. *Journal of Plankton Research* **9**: 837–855.

Gardner, W. *et al.* (1998). Nitrogen cycling rates and light effects in tropical Lake Maracaibo, Venezuela. *Limnology and Oceanography* **43**: 1814–1825.

Gasol, J. and Vaque, D. (1993). Lack of coupling between heterotrophic nanoflagellates and bacteria: a general phenomenon across aquatic systems? *Limnology and Oceanography* **38**: 657–665.

Geider, R. J. and MacIntyre, H. L. (1996). A dynamic model of photoadaptation in phytoplankton. *Limnology and Oceanography* **41**: 1–15.

Geider, R. J. and Platt, T. (1986). A mechanistic model of photoadaptation in microalgae. *Marine Ecology Progress Series* **30**: 85–92.

Gell, P. and Gasse, F. (1990). *Relationships between salinity and diatom flora from some Australian saline lakes.*

11th International Diatom Symposium California Academy of Sciences, pp. 631–641. San Francisco, USA.

Geller, W. and Muller, H. (1981). The filtration apparatus of cladocera: filter mesh sizes and their implications on food selectivity. *Oecologia* **49**: 316–321.

Geyer, W. *et al.* (2000). Interaction between physical processes and ecosystem structure. In *Estuarine Science*, (Hobbie, J., Ed), pp. 177–209. Island Press, Washington DC, USA.

Ginzburg, M. and Ginzburg, B. (1981). Interrelationships of light, temperature, sodium chloride and carbon source in growth of halotolerant and halophilic strains of *Dunaliella*. *British Phycological Journal* **16**: 313–324.

Gliwicz, Z. M. and Pijanowska, J. (1989). The role of predation in zooplankton succession. In *Plankton Ecology*, (Sommer, U., Ed), pp. 253–296. Springer-Verlag, Berlin, Germany.

Golden, S. (1995). Light-responsive gene expression in cyanobacteria. *Journal of Bacteriology* **177**: 1651–1654.

Goldman, C. R. (1960). Primary productivity and limiting factors in three lakes of the Alaskan Peninsula. *Ecological Monographs* **30**: 207–270.

Goldman, C. R. *et al.* (1972). Comparative study of the limnology of two small lakes on Ross Island, Antarctica. In *Antarctic Terrestrial Biology*, (Llano, G., Ed). pp. 1–50. American Geophysical Union, Washington, USA.

Goldsborough, L. and Robinson, G. (1996). Pattern in wetlands. In *Algal Ecology*, (Stevenson, R. J., Bothwell, M. and Lowe, R., Eds), pp. 77–117. Academic Press, New York, USA.

Gorham, P. *et al.* (1964). Isolation and culture of toxic strains of *Anabaena flos-aquae* (Lyngb.) de Breb. *Verhandlungen der internationale Vereinigung fur theoretische und angewandte Limnologie* **15**: 796–804.

Graham, D. and Phillips, M. (1979). Proteins at liquid surfaces.1. Kinetics of adsorption and surface denaturation. *Journal of Colloid Interface Science* **70**: 403–414.

Graham, J. *et al.* (1985). Light, temperature and photo period as factors controlling reproduction in *Ulothrix zonata* (Ulvophyceae). *Journal of Phycology* **21**: 235–239.

Grobbelaar, N. *et al.* (1986). Dinitrogen-fixing endogenous rhythm in *Synechococcus* RF-1. *FEMS Microbiology Letters* **37**: 173–177.

Groisman, P. and Davies, T. (2001). Snow cover and the climate system. In *Snow Ecology*, (Jones, H. *et al.*, Eds), pp. 1–44. Cambridge University Press, Cambridge, UK.

Groth, P. (1971). Untersuchungen uber einige Spurenelementente in Seen. *Archives of Hydrobiology* **68**: 305–375.

Güde, H. (1989). The role of grazing on bacteria in plankton succession. In *Plankton Ecology*, (Sommer, U., Ed), pp. 337–364. Springer-Verlag, Berlin, Germany.

Guillard, R. and Cassie, V. (1963). Minimum cyanocabalamin requirements of some marine centric diatoms. *Limnology and Oceanography* **8**: 161–165.

Häder, D.-P. (1995). Influence of ultraviolet radiation on phytoplankton ecosystems. In *Algae, Environment and Human Affairs*, (Weissner, W., Schnepf, E. and Starr, R., Eds), pp. 41–55. Biopress Ltd., Bristol, UK.

Häder, D.-P. and M. Häder (1989). Effects of solar radiation on photomovement and motility in photosynthetic and colorless flagellates. *Environmental Experimental Botany* **29**: 273–282.

Hairston, N. *et al.* (1960). Community structure, population control and competition. *The American Naturalist* **94**: 421–425.

Hall, D. and Walmsley, R. (1991). Effect of temperature on germination of *Rhizoclonium riparium* (Siphonocladales, Chlorophyta) akinetes and zoospores. *Journal of Phycology* **27**: 537–539.

Handfield, M. *et al.* (1992). Seasonal fluctuation pattern of the microflora on Agassiz ice sheet, Ellesmere Island, Canadian Arctic. *Musk-ox* **39**: 119–123.

Haney, J. (1973). An *in situ* examination of the grazing activities of natural zooplankton communities. *Archives of Hydrobiology* **72**: 87–132.

Happey-Wood, C. (1988). Ecology of freshwater planktonic algae. In *Growth and Reproductive Strategies of Freshwater Phytoplankton*, (Sandgren, C., Eds). pp. 175–226. Cambridge University Press, Cambridge, UK.

Hardy, J. and Apts, C. (1989). Photosynthetic carbon reduction: high rates in the sea-surface microlayer. *Marine Biology* **101**: 411–417.

Harper, D. (1992). *Eutrophication of Freshwaters: principles, Problems and Restoration*. Chapman and Hall, London, UK.

Harper, D. *et al.*, Eds. (1999). *The Ecological Bases of Lake and Reservoir Management*. Kluwer Academic Publishers, Dordrecht, The Netherlands.

Harris (1980). The measurement of photosynthesis in natural populations of phytoplankton. In *The Physiological Ecology of Phytoplankton*, (Morris, I., Ed), pp. 129–187. Blackwell Science Publications, Oxford, UK.

Hart, D. and Finelli, C. (1999). Physical-biological coupling in streams: the pervasive effects of flow on benthic organisms. *Annual Review of Ecological Systems* **30**: 363–395.

Havens, K. and Carlson, R. (1998). Functional complimentarity in plankton communities along a gradient of acid stress. *Environmental Pollution* **101**: 427–436.

Hayakawa, K. *et al.* (2002). Fatty acid composition as an indicator of physiological condition of the cyanobacterium *Microcystis aeruginosa*. *Limnology* **3**: 29–35.

Hecky, R. and Kling, H. (1981). The phytoplankton and protozooplankton of the euphotic zone of Lake Tanganyika: species composition, biomass, chlorophyll content, and spatio-temporal distribution. *Limnology and Oceanography* **26**: 548–564.

Heinonen, P. *et al.* (2000). *Hydrological and Limnological Aspects of Lake Monitoring*. John Wiley and Sons Ltd, Chichester, UK.

Held, A. *et al.* (1969). *Blastocladia* and *Aqualinderella*: fermentative water molds with high carbon dioxide optima. *Science* **165**: 706–709.

Heldt, H.-W. (1997). *Plant Biochemsitry and Molecular Biology*. Oxford University Press, Oxford, UK.

Hellebust, J. (1971). Glucose uptake by *Cyclotella cryptica*: dark induction period and light inactivation of transport system. *Journal of Phycology* **7**: 345–349.

Hermansson, M. *et al.* (1982). Hydrophobic and electrostatic characterisation of surface structures of bacteria and its relationship to adhesion at an air–water interface. *Archives of Microbiology* **131**: 308–312.

Hessen, D. *et al.* (1995). Growth responses, P-uptake and loss of flagellae in *Chlamydomonas reinhardtii* exposed to UV-B. *Journal of Plankton Research* **17**: 17–27.

Hewson, I. *et al.* (2001). Virus-like particle distribution and abundance in sediments and overlying waters along eutrophication gradients in two subtropical waters. *Limnology and Oceanography* **46**: 1734–1746.

Higgins, M. *et al.* (2003). Probing the surface of living diatoms with atomic force microscopy: The nanostructure and nanomechanical properties of the mucilage layer. *Journal of Phycology* **39**: 722–734.

Hildebrand, M. *et al.* (1997). A gene family of silicon transporters. *Nature* **385**: 688–689.

Hildebrand, M. *et al.* (1998). Characterisation of a silicon transporter gene family in *Cylindrotheca fusiformis*. *Molecular Genetics and Genomics* **260**: 480–486.

Hill, W. (1996). Effects of light. In *Algal Ecology*, (Stevenson, R., Bothwell, M. and Lowe, R. L., Eds), pp. 121–149. Academic Press, New York, USA.

Hobbie, J. (2000). *Estuarine Science*. Island Press, Washington DC, USA.

Hofer, J. and Sommaruga, R. (2001). Seasonal dynamics of viruses in an alpine lake: importance of filamentous forms. *Aquatic Microbial Ecology* **26**: 1–11.

Hoham, R. (1975). Optimum temperatures and optimum temperature ranges for growth of snow algae. *Arctic Alpine Research* **7**: 13–24.

Hoham, R. and Duval, B. (2001). *Microbial Ecology of Snow and Freshwater Ice with Emphasis on Snow Algae*. Cambridge University Press, Cambridge, UK.

Hoham, R. *et al.* (1993). Snow algae and other microbes in Several alpine areas in New England. *50th Annual Eastern Snow Conference*, pp. 165–173.

Holfeld, H. (2000). Relative abundance, rate of increase, and fungal infections of freshwater phytoplankton. *Journal of Plankton Research* **22**: 987–995.

Holland, A. *et al.* (1974). Quantitative evidence concerning the stabilization of sediments by marine benthic diatoms. *Marine Biology* **27**: 191–196.

Holt, J. Eds. (1994). *Bergey's Manual of Systematic Bacteriology*. Williams and Wilkins Company, Baltimore, USA.

Holzman, R. and Genin, A. (2003). Zooplanktivory by a nocturnal coral-fish: effects of light, flow and prey density. *Limnology and Oceanography* **48**: 1367–1375.

Horn, B. and Lichtwardt, R. (1986). Studies of the nutritional relationships of larval *Aedes aegypti* (Diptera: Culicidae) with *Smittium culisetae* (Trichomycetes). *Mycologia* **73**: 724–740.

Horne, A. (1975). *The Ecology of Clear Lake Phytoplankton* Clear Lake Algal Research Unit, Lakeport, USA.

Horne, A. and Goldman, C. R. (1972). Nitrogen fixation in Clear Lake, California.1. Seasonal variation and the role of heterocysts. *Limnology and Oceanography* **17**: 678–692.

Horne, A. and Goldman, C. R. (1994). *Limnology*. McGraw-Hill, New York, USA.

Hoyos, C. and Comin, F. (1999). The importance of interannual variability for management. In *The Ecological Bases for Lake and Reservoir Management*, (Harper, D. *et al.*, Eds), pp. 281–289. Kluwer Academic, Dordrecht, The Netherlands.

Huckauf, J. *et al.* (2000). Stress responses of *Synechocystis* sp. strain PCC 6803 mutants impaired in genes encoding putative alternative sigma factors. *Microbiology* **146**: 2877–2889.

Huntsman, S. and W. Sunda (1980). The role of trace metals in regulating phytoplankton growth. In *The Physiological Ecology of Phytoplankton*, (Morris, I., Ed.), pp. 285–328. Blackwell, Oxford, UK.

Hutchinson, G. (1961). The paradox of the plankton. *American Naturalist* **95**: 137–146.

Iqbal, S. and Webster, J. (1973). Trapping of aquatic hyphomycete spores by air bubbles. *Transactions of the British Mycological Society* **61**: 331–346.

Irvine, K. (1997). Food selectivity and diel vertical distribution of *Chaoborus edulis* (Diptera, Chaoboridae) in Lake Malawi. *Freshwater Biology* **37**: 605–620.

Ishiura, M. *et al.* (1998). Expression of a gene cluster *kai*ABC as a circadian feedback process in cyanobacteria. *Science* **281**: 1519–1523.

Janse, I. *et al.* (2003). High-resolution differentiation of cyanobacteria by using rRNA-internal transcribed spacer denaturing gradient gel electrophoresis. *Applied and Environmental Microbiology* **69**: 6634–6643.

Jansen, M. A. K. *et al.* (1998). Higher plants and UV-B radiation: balancing damage, repair and acclimation. *Trends in Plant Science* **3**: 131–135.

Jenik, J. *et al.* (2002). Plant life in an endangered mire: cervene blato bog. In *Freshwater Wetlands and their Sustainable Future*, (Kvet, J., Jenik, J. and Soukupova, L., Eds), pp. 399–408. The Parthenon Publishing Group, New York, USA.

Jewson, D. (1977). Light penetration in relation to phytoplankton content of the euphotic zone of Lough Neagh, N. Ireland. *Oikos* **28**: 74–83.

Jobling, M. *et al.* (1988). Plasmid-borne mercury resistance in aquatic bacteria. *FEMS Microbiology Letters* **49**: 31–37.

John, D. *et al.* (2002). *The Freshwater Algal Flora of the British Isles*. Cambridge University Press, Cambridge, UK.

Johnson, C. and Golden, S. (1999). Circadian programs in cyanobacteria: adaptiveness and mechanism. *Annual Revue of Microbiology* **53**: 389–409.

Jones, A. K. (1986). Eukaryotic algae – antimicrobial systems. In *Natural Antimicrobial Systems*, (Gould, G. W. *et al.*, Eds), pp. 232–256. Bath University Press, Bath, UK.

Jones, H. *et al.*, Eds. (2001). *Snow Ecology*. Cambridge University Press, Cambridge, UK.

Jones, J. (1971). Studies on freshwater bacteria: factors which influence the population and its activity. *Journal of Ecology* **59**: 593–613.

Jones, J. *et al.* (1983). Bacterial uptake of algal extracellular products: an experimental approach. *Journal of Applied Bacteriology* **54**: 355–365.

Jørgensen, S. (1983). Ecological modeling of lakes. In *Mathematical Modeling of Water Quality – Streams, Lakes and Reservoirs*, (Orlob, G., Ed), pp. 337–393. John Wiley and Sons Ltd., Chichester, UK.

Jørgensen, S. and Bendoricchio, G. (2001). *Fundamentals of Ecological Modelling*. Elsevier, Amsterdam, The Netherlands.

Jørgensen, S. *et al.* (1978). Examination of a lake model. *Ecological Modelling* **4**: 253–278.

Kalff, J. (2002). *Limnology*. Prentice Hall, Upper Saddle River, New Jersey, USA.

Kalff, J. *et al.* (1972). Pigment cycles in two high-arctic Canadian lakes. *Verhandlungen der internationale Vereinigung für theoretische und angewandte Limnologie* **18**: 250–256.

Kaneko, T. *et al.* (1996). Sequence analysis of the genome of the unicellular cyanobacterium *Synechocystis* sp. strain PCC6803. II. Sequence determination of the entire genome and assignment of potential coding regions. *DNA Research* **3**: 109–136.

Kapuscinski, R. and Michell, R. (1980). Processes controlling virus inactivation in coastal waters. *Water Research* **14**: 363–371.

Karentz, D. (1999). Evolution and ultraviolet light tolerance in algae. *Journal of Phycology* **35**: 629–630.

Karentz, D. *et al.* (1991). Cell survival characteristics and molecular responses of Antarctic phytoplankton to ultraviolet-B radiation. *Journal of Phycology* **27**: 326–341.

Karp-Boss, L. and Jumars, P. (1998). Motion of diatom chains in steady shear flow. *Limnology and Oceanography* **43**: 1767–1773.

Kato, T. *et al.* (1991). Allozyme divergence in *Microcystis* (Cyanophyceae) and its taxonomic inference. *Algological Studies* **64**: 129–140.

Kaushik, N. and Hynes, H. (1971). The fate of autumn-shed leaves that fall into streams. *Archives of Hydrobiology* **68**: 59–63.

Kazumi, J. and Capone, D. (1994). Heterotrophic microbial activity in shallow aquifer sediments of Long Island, New York. *Microbial Ecology* **28**: 19–37.

Kelley, D. (1997). Convection in ice-covered lakes: effects on algal suspension. *Journal of Plankton Research* **19**: 1859–1880.

Ketchum, B. and Redfield, A. (1949). Some physical and chemical characteristics of algae growth in mass cultures. *Journal of Cellular and Comparative Physiology* **13**: 373–381.

Kirchman, D. *et al.* (2003). Diversity and abundance of uncultured *Cytophaga*-like bacteria in the Delaware Estuary. *Applied and Environmental Microbiology* **69**: 6587–6596.

Kirk, J. (1994). *Light and Photosynthesis in Aquatic Ecosystems*. Cambridge University Press, Cambridge, UK.

Kitayama, T. *et al.* (1999). Subcellular localisation of iron and manganese superoxide dismutase in *Chlamydomonas reinhardtii* (Chlorophyceae). *Journal of Phycology* **35**: 136–142.

Klaveness, D. (1988). Ecology of the Cryptomonadida: a first review. In *Growth and Reproductive Strategies of Freshwater Phytoplankton*, (Sandgren, C., Ed), pp. 105–133. Cambridge University Press, Cambridge, UK.

Klotz, R. and Matson, E. (1978). Dissolved organic carbon fluxes in the Shetucket River of Eastern Connecticut, USA. *Freshwater Biology* **8**: 347–355.

Knoechel, R. and Holtby, L. (1986). Cladoceran filtering rate: body length relationships for bacterial and large algal particles. *Limnology and Oceanography* **31**: 195–200.

Kol, E. (1968). Kryobiologie, biologie und limnologie des Schnees und eases. I. Kryovegetation. In *Die Binnen gewasser*. (Elster, H. and Ohle, W., Eds). Schweizerbart'sche Verlagsbuch-handlung, Stuttgart, Germany.

Kolber, Z. and Falkowski, P. G. (1993). Use of active fluorescence to estimate phytoplankton photosynthesis *in situ*. *Limnology and Oceanography* **38**: 1646–1665.

Komarek, J. (1991). A review of water-bloom forming *Microcystis* species, with regard to populations from Japan. *Archives of Hydrobiology* **64**: 115–127.

Kornberg, A. (1995). Inorganic polyphosphate: toward making a forgotten polymer unforgettable. *Journal of Bacteriology* **177**: 491–495.

Kozhov, M. (1963). *Lake Baikal and its Life*. W. Junk, The Hague, The Netherlands.

Kozhova, O. M. and L. Izmest'eva, Eds. (1998). *Lake Baikal, Evolution and Biodiversity*. Backhuys Publications, Leiden, The Netherlands.

Kozhova, O. M. and Matveev, A. A. (1998). Lake Baikal. In *A Water Quality Assessment of the Former Soviet Union*, (Kimstach, V., Meybeck, M. and Baroudy, E., Eds), pp. 481–529. E and F.N. Spon, London, UK.

Kreuger, D. and Dodson, S. (1981). Embryological induction and predation ecology in *Daphnia pulex*. *Limnology and Oceanography* **26**: 219–223.

Krivtsov, V. *et al.* (1999). Examination of the phytoplankton of Rostherne Mere using a simulation mathematical model. *Hydrobiologia* **414**: 71–76.

Krivtsov, V. *et al.* (2002). Elemental concentrations and correlations in winter micropopulations of *Stephanodiscus rotula*: an autecological study over a period of cell size reduction and restoration. *European Journal of Phycology* **37**: 27–36.

Kroger, N. *et al.* (1999). Polycationic peptides from diatom biosilica that direct silica nanosphere formation. *Science* **286**: 1129–1132.

Kurath, G. and Morita, R. (1983). Starvation-survival physiological studies of a marine *Pseudomonas* sp. *Applied and Environmental Microbiology* **45**: 1206–1211.

Kurmeyer, R. and Kutzenberger, T. (2003). Application of real-time PCR for quantification of microcystin genotypes in a population of the toxic cyanobacterium *Microcystis* sp. *Applied and Environmental Microbiology* **69**: 6723–6730.

Kuznetsov, S. (1970). *Microflora of Lakes and their Geochemical Activities [In Russian]*. Izdatel'stvo Nauka, Leningrad, Russia.

Kvet, J. *et al.* Eds. (2002). *Freshwater Wetlands and their Sustainable Use*. The Parthenon Publishing Group, New York, USA.

Laflamme, M. and Lee, R. (2003). Mitochondrial genome conformation among CW-group chlorophycean algae. *Journal of Phycology* **39**: 213–220.

Lamothe, G. *et al.* (2003). Reverse transcription-PCR analysis of bottled and natural waters for the presence of noroviruses. *Applied and Environmental Microbiology* **69**: 6541–6549.

Lampert, W. (1981). Toxicity of blue-green *Microcystis aeruginosa*: effective defense mechanism against grazing pressure by *Daphnia*. *Internationale Vereinigung für theoretische und angewandte Limnologie Verhandlung* **21**: 1436–1440.

Lampert, W. (1988). The relationship between zooplankton biomass and grazing: a review. *Limnologica* **19**: 11–20.

Larsson, T. *et al.* (1985). Distribution of reducing power between photosynthetic carbon and nitrogen assimilation in *Scenedesmus*. *Planta* **163**: 69–76.

Laurion, I. and Vincent, W. (1998). Cell size versus taxonomic composition as determinants of UV-sensitivity in natural phytoplankton communities. *Limnology and Oceanography* **43**: 1774–1779.

Lazzaro, X. (1987). A review of planktivorous fishes: their evolution, feeding behaviours, selectivities, and impacts. *Hydrobiologia* **146**: 283–293.

Leavitt, P. *et al.* (1997). Past ultraviolet environments in lakes derived from fossil pigments. *Nature* **388**: 457–459.

Lee, D.-Y. and Rhee, G.-Y. (1999). Circadian rhythm in growth and death of *Anabaena flos-Aquae* (Cyanobacteria). *Journal of Phycology* **35**: 694–699.

Lee, R. (1997). *Phycology*. Cambridge University Press, Cambridge, UK.

Lengeler, J. *et al.* (1999). *Biology of the Prokaryotes*. Blackwell Science, New York, USA.

Lesack, L. *et al.* (1984). Transport of carbon, nitrogen, phosphorus and major solutes in the Gambia River, West Africa. *Limnology and Oceanography* **29**: 816–830.

Levado, E. (2001). Studies on phytoplankton diversity within the water column of two freshwater lakes –

Rostherne Mere (UK) and Lake Glubokoe (Russia). Ph.D Thesis *School of Biological Sciences*. University of Manchester, Manchester, UK.

Lewis, W. M. (1976). Surface/volume ratio: implications for phytoplankton morphology. *Science* **192**: 885–887.

Li, A. *et al.* (2000). Mixotrophy in *Gyrodinium galatheanum* (Dinophyceae): grazing responses to light intensity and inorganic nutrients. *Journal of Phycology* **36**: 33–45.

Lin, C. and Blum, J. (1977). Recent invasion of a red alga (*Bangia atropurpurea*) in Lake Michigan. *Journal of the Fisheries Research Board of Canada*: 2413–16.

Litaker, R. *et al.* (2003). Identification of *Pfiesteria piscicida* (Dinophyceae) and *Pfiesteria*-like organisms using internal transcribed spacer-specific PCR assays. *Journal of Phycology* **39**: 754–761.

Liu, Y. *et al.* (1995). Bacterial luciferase as a reporter of circadian gene expression in cyanobacteria. *Journal of Bacteriology* **177**: 2080–2086.

Lock, M. A. *et al.* (1984). River epilithon: toward a structural-functional model. *Oikos* **42**: 10–22.

Loewen, P. and Hengge-Aronis, R. (1994). The role of the sigma factor sigmas (KatF) in bacterial global regulation. *Annual Revue of Microbiology* **56**: 412–429.

Lowe, R. L. (1996). Periphyton patterns in lakes. In *Algal Ecology*, (Stevenson, R. J., Bothwell, M. and Lowe, R. L., Eds), pp. 57–76. Academic Press, New York, USA.

Lund, J. (1949). Studies on *Asterionella*. I. The origin and nature of cells producing the spring maximum. *Journal of Ecology* **37**: 389–419.

Lund, J. (1954). The seasonal cycle of the plankton diatom *Melosira italica* (Ehr.) Kutz. subsp. *subarctica* O.Mull. *Journal of Ecology* **42**: 151–179.

Lund, J. (1957). Fungal diseases of planktonic algae. In *Biological Aspects of the Transmission of Diseases*, (Horton-Smith, C., Ed). pp. 19–24. Oliver and Boyd, Edinburgh, UK.

Lupton, F. S. and Marshall, K. C. (1981). Specific adhesion of bacteria to heterocysts of *Anabaena* spp. and its ecological significance. *Applied and Environmental Microbiology* **42**: 1085–1092.

Lynch, M. (1980). Alternate control and cultivation by *Daphnia pulex*. In *Evolution and Ecology of Zooplankton Populations*, (Kerfoot, W., Ed). pp. 299–304. University Press, New England, USA.

Lynn, S. G. *et al.* (2000). Effect of nutrient availability on the biochemical and elemental stoichiometry in the freshwater diatom *Stephanodiscus minutulus* (Bacillariophyceae). *Journal of Phycology* **36**: 510–522.

MacArthur, R. (1955). Fluctuations in animal populations and a measure of community stability. *Ecology* **36**: 533–536.

MacArthur, R. and Wilson, E. (1967). *The Theory of Island Biogeography*. Princeton University Press, Princeton, USA.

MacKay, N. *et al.* (1990). The impact of two *Chaoborus* species on a zooplankton community. *Canadian Journal of Zoology* **68**: 981–985.

Madgwick, F. (1999). Strategies for conservation management of lakes. In *The Ecological Bases for Lake and Reservoir Management*, (Harper, D. *et al.*, Eds), pp. 309–323. Kluwer Academic, Dordrecht, The Netherlands.

Madronich, S. *et al.* (1998). Changes in biologically active ultraviolet radiation reaching the Earth's surface. *Photochemistry and Photobiology* **46**: 5–19.

Maki, J. and Hermansson, M. (1994). The dynamics of surface microlayers in aquatic environments. In *The Biology of Particles in Aquatic Systems*, (Wotton, R., Ed), pp. 161–182. Lewis Publishers, London, UK.

Maranger, R. and Bird, D. (1995). Viral abundance in aquatic systems: a comparison between marine and fresh waters. *Marine Ecology Progress Series* **121**: 217–226.

Maranger, R. and Bird, D. (1996). High concentrations of viruses in the sediments of Lac Gilbert, Quebec. *Microbial Ecology* **31**: 141–151.

Margalef, R. (1958). Temporal succession and spatial heterogeneity in phytoplankton. In *Perspectives in Marine Biology*, (Buzzati-Traverso, A., Ed). pp. 323–349. University of California Press, Berkeley, USA.

Margolin, P. (1987). Generalised transduction. In *Escherichia coli and Salmonella typhimurium. Cellular and Molecular Biology*, (Neidhardt, F. *et al.*, Eds), pp. 1154–1168. American Society for Microbiology, Washington DC, USA.

Markaki, Z. *et al.* (2003). Atmospheric deposition of inorganic phosphorus in the Levantine Basin, eastern Mediterranean. Spatial and temporal variability and its role in seawater productivity. *Limnology and Oceanography* **48**: 1557–1568.

Martens, K. (1997). Speciation in ancient lakes. *TREE* **12**: 177–182.

Martin, E. and Benson, R. (1988). Phages of cyanobacteria. In *The Bacteriophages*, (Calendar, R., Eds), pp. 607–645. Plenum Press, New York, USA.

Martin, E. and Kokjohn, T. (1999). Cyanophages. In *Encyclopaedia of Virology*, (Granoff, A. and Webster, R., Eds), pp. 324–332. Academic Press, San Diego, USA.

Martin-Jézéquel, V. *et al.* (2000). Silicon metabolism in diatoms: implications for growth. *Journal of Phycology* **36**: 821–840.

Mason, C. (2002). *Biology of Freshwater Pollution*. Pearson Education Ltd., Harlow, UK.

Mazepova, G. (1998). The role of copepods in the Baikal ecosystem. *Journal of Marine Systems* **15**: 113–120.

McCauley, E. and Kalff, J. (1981). Empirical relationships between phytoplankton and zooplankton biomass in lakes. *Canadian Journal of Fisheries and Aquatic Science* **38**: 458–463.

McCormick, P. V. and Stevenson, R. J. (1991). Mechanisms of benthic algal succession in lotic environments. *Ecology* **72**: 1835–1848.

McLean, *et al.* (1997). Evidence of autoinducer activity in naturally occurring biofilms. *FEMS Microbiology Letters* **154**: 259–263.

McManus, J. *et al.* (2003). Hypolimnetic oxidation rates in Lake Superior: role of dissolved organic material on the lake's carbon budget. *Limnology and Oceanography* **48**: 1624–1632.

McNaughton, S. J. (1993). Biodiversity and function of grazing systems. In *Biodiversity and Ecosystem Function*, (Schulze, E.-D. and Mooney, H. A., Eds), pp. 361–383. Springer-Verlag, Berlin, Germany.

McNight, D. M. *et al.* (2000). Phytoplankton dynamics in a stably stratified antarctic lake during winter darkness. *Journal of Phycology* **36**: 852–861.

Megard, R. *et al.* (1979). Attenuation of light and daily integral rates of photosynthesis attained by planktonic algae. *Limnology and Oceanography* **24**: 1038–1050.

Meints, R. *et al.* (1986). Assembly site of the virus PBCV-1 in a *Chlorella*-like green alga: ultrastructure studies. *Virology* **154**: 240–245.

Menge, B. and Sutherland, J. (1987). Community regulation: variation in disturbance, competition, and predation in relation to environmental stress and recruitment. *American Naturalist* **130**: 730–757.

Menzel, D. and Ryther, J. (1970). Distribution and cycling of organic matter in the oceans. In *Organic Matter in Natural Waters*, (Hood, D., Ed), pp. 31–54. Institute of Marine Science Publication, Alaska, USA.

Meyer, J. (1994). The microbial loop in flowing waters. *Microbial Ecology* **28**: 195–199.

Middelboe, A. and Markager, S. (1997). Depth limits and minimum light requirements of freshwater macrophytes. *Freshwater Biology* **37**: 553–568.

Middelboe, M. *et al.* (2003). Distribution of viruses and bacteria in relation to diagenetic activity in an estuarine sediment. *Limnology and Oceanography* **48**: 1447–1456.

Mill, A. (1980). Colloidal and macromolecular forms of iron in natural waters. A review. *Environmental Technology Letters* **1**: 987–108.

Miller, R. and Sayler, G. (1992). Bacteriophage–host interactions in aquatic systems. In *Genetic Interactions among Microorganisms in the Natural Environment*, (Wellington, E. and Van Elsas, J., Eds), pp. 176–193. Pergamon Press, Oxford, UK.

Miller, R. *et al.* (1990). Environmental and molecular characterisation of systems which affect genome alteration in *Pseudomonas aeruginosa*. In *Pseudomonas '89: Biotransformation, Pathogenesis, and Evolving Biotechnology*, (Silver, S. *et al.*, Eds), pp. 252–268. American Society for Microbiology, Washington DC, USA.

Minshall, G. (1978). Autotrophy in stream ecosystems. *BioScience* **28**: 767–771.

Mitchell, S. and Galland, A. (1981). Phytoplankton photosynthesis, eutrophication and vertical migration of dinoflagellates in a New Zealand reservoir. *Verhandlungen der internationale Vereinigung fur theoretische und angewandte Limnologie* **21**: 1017–1020.

Moebus, K. (1987). Ecology of marine bacteriophages. In *Phage Ecology*, (Goyal, S., Gerba, C. and Bitton, G., Eds). pp. 137–156. John Wiley and Sons, New York, USA.

Molin, S. *et al.* (2000). Molecular ecology of biofilms. In *Biofilms*, (Bryers, J., Ed). pp. 89–120. John Wiley and Sons, New York, USA.

Moorhead, D. and Reynolds, J. (1992). Modeling the contributions of decomposer fungi in nutrient cycling. In *The Fungal Community*, (Carroll, G. and Wicklow, D., Eds), pp. 691–714. Marcel Dekker, New York, USA.

Morel, A. and Bricaud, A. (1981). Theoretical results concerning light absorption in a discrete medium, and application to specific absorption of phytoplankton. *Deep-Sea Research* **28**: 1375–1393.

Morita, R. (1982). Starvation-survival of heterotrophs in the marine environment. *Advanced Microbial Ecology* **6**: 171–198.

Morita, R. (1997). Oligotrophic environments. In *Bacteria in Oligotrophic Environments*, (Morita, R., Ed), pp. 1–36. Chapman and Hall, New York, USA.

Morrison, W. *et al.* (1979). Frequency of F116-mediated transduction of *Pseudomonas aeruginosa*. *Applied and Environmental Microbiology* **36**: 724–730.

Mortimer, C. (1941). The exchange of dissolved substances between mud and water in lakes (Parts I and II). *Journal of Ecology* **29**: 280–329.

Mosser, J. *et al.* (1977). Photosynthesis in the snow: the alga *Chlamydomonas nivalis* (Chlorophyceae). *Journal of Phycology* **13**: 22–27.

Mulholland, P. (1996). Role of nutrient cycling in streams. In *Algal Ecology*, (Stevenson, R. J., Bothwell, M. and Lowe, R. L., Eds), pp. 609–640. Academic Press, New York, USA.

Münster, U. (1984). Distribution, dynamic and structure of free dissolved carbohydrates in the Pluss-see, a north German eutrophic lake. *Verhandlungen der internationale Vereinigung für theoretische und angewandte Limnologie* **22**: 929–935.

Murray, A. (1995). Phytoplankton exudation: exploitation of the microbial loop as a defence against viruses. *Journal of Plankton Research* **17**: 1079–1094.

Murray, A. and Jackson, G. (1992). Viral dynamics: a model of effects of size, shape, motion and abundance of single-celled planktonic organisms and other particles. *Marine Ecology Series* **89**: 103–116.

Musgrave, A. (1993). Mating in *Chlamydomonas*. *Progress in Phycological Research* **9**: 193–237.

Nagata, T. (1988). The microflagellate–picoplankton food linkage in the water column of Lake Biwa. *Limnology and Oceanography* **33**: 504–517.

Nagata, T. *et al.* (1994). Autotrophic picoplankton in Southern lake Baikal:abundance, growth and grazing mortality during summer. *Journal of Plankton Research* **16**: 945–959.

Nalewajko, C. *et al.* (1980). Significance of algal extracellular products to bacteria in lakes and cultures. *Microbial Ecology* **6**: 199–207.

Natvig, D. (1987). *Aqualinderella fermentans*. In *Zoosporic Fungi in Teaching and Research*, (Fuller, M. and Jaworski, A., Eds), pp. 78–79. Southeastern, Athens, GA, USA.

Naumann, E. (1919). Nagra synpunkter Angaende Limnoplanktons Okologi med Sarskild Hansyn till Fytoplankton. *Svensk Botanische Tidschrift (English translation by the Freshwater Biological Association #49)* **13**: 129–163.

Nedwell, D. B. *et al.* (1999). Nutrients in estuaries. In *Advances in Ecological Research*, (Nedwell, D. B. and Raffaelli, D. G., Eds). pp. 43–92. Academic Press, New York, USA.

Nesbitt, L. *et al.* (1996). Opposing predation pressures and induced vertical migration responses in *Daphnia*. *Limnology and Oceanography* **41**: 1306–1311.

Noble, R. and Fuhrman, J. (1998). Use of SYBR-Green I for rapid epifluorescence counts of marine viruses and bacteria. *Aquatic Microbial Ecology* **14**: 113–118.

Nüsslein, B. *et al.* (2003). Stable isotope biogeochemistry of methane formation in profundal sediments of Lake Kinneret (Israel). *Limnology and Oceanography* **48**: 1439–1446.

Nygaard, G. (1949). Hydrobiological studies in some ponds and lakes. Part II. The quotient hypothesis and some new or little known phytoplankton organisms. *Kengelige Danske Videnskabernes Selskab. Biologiske Skrifter* **7**: 1–293.

Ochoa de Alda, J. and Houmard, J. (2000). Genomic survey of cAMP and cGMP signalling components in the cyanobacterium *Synechocystis* PCC 6803. *Microbiology* **146**: 3183–3194.

Odum, E. (1962). Relationships between structure and function in the ecosystem. *Japanese Journal of Ecology* **12**: 108–118.

Odum, E. (1971). *Fundamentals of Ecology*. WB Saunders, Philadelphia, USA.

Odum, E. (1985). Trends expected in stressed ecosystems. *BioScience* **35**: 419–422.

Odum, E. (1997). *Ecology. A Bridge Between Science and Society*. Sinauer Associates Inc., Sunderland, USA.

OECD (1982). Eutrophication of waters, assessment and control. Final report. OECD Cooperative Programme on monitoring inland waters (eutrophication control). OECD (Organisation for Economic Cooperation and Development), Paris, France.

Ogunseitan, O. *et al.* (1990). Dynamic interactions of *Pseudomonas aeruginosa* and bacteriophages in lake water. *Microbial Ecology* **19**: 171–156.

Okamoto, O. and Hastings, J. (2003). Novel dinoflagellate clock-related genes identified through microarray analysis. *Journal of Phycology* **39**: 519–530.

Olive, L. (1975). *The Mycetozoans*. Academic Press, New York, USA.

O'Morchoe, S. *et al.* (1988). Conjugal transfer of R68.45 and FP5 between *Pseudomonas aeruginosa* strains in a freshwater environment. *Applied and Environmental Microbiology* **54**: 1923–1929.

Ouyang, Y. *et al.* (1998). Resonating circadian clocks enhance fitness in cyanobacteria. *Proceedings of the National Academy of Sciences* **95**: 8660–8664.

Overbeck, J. (1979). Dark CO_2 uptake – biochemical background and its relevance to *in situ* bacterial production. *Archiv für Hydrobiologie Beih. Ergebnisse der Limnologie* **12**: 38–47.

Ovreas, L. *et al.* (2003). Characterisation of microbial diversity in hypersaline environments by melting profiles and reassociation kinetics in combination with terminal restriction fragment length polymorphism (T-RFLP). *Microbial Ecology* **46**: 291–301.

Pace, M. and Cole, J. (1994). Primary and bacterial production in lakes: are they coupled over depth? *Journal of Plankton Research* **16**: 661–672.

Pace, M. and Orcutt, J. (1981). The relative importance of protozoans, rotifers, and crustaceans in a freshwater zooplankton community. *Limnology and Oceanography* **26**: 822–830.

Padan, E. and Shilo, M. (1967). Isolation of 'cyanophages' from freshwater ponds and their interaction with *Plectonema boryanum*. *Virology* **32**: 234–246.

Padan, E. and Shilo, M. (1969). Distribution of cyanophages in natural habitats. *Verhandlungen der internationale Vereinigung fur theoretische und angewandte Limnologie* **17**: 747–751.

Padan, E. and M. Shilo (1973). Cyanophages – viruses attacking blue-green algae. *Bacteriological Reviews* **3**: 343–370.

Padan, E. *et al.* (1970). The reproductive cycle of cyanophage LPP1-G in *Plectonema boryanum* and its dependence on photosynthetic and respiratory systems. *Virology* **40**: 514–521.

Paerl, H. W. (1988). Growth and reproductive strategies of freshwater blue-green algae (cyanobacteria). In *Growth and Reproductive Strategies of Freshwater Phytoplankton*, (Sandgren, C., Ed), pp. 261–315. Cambridge University Press, Cambridge, UK.

Paerl, H. W. (1992). Epi- and endobiotic interactions of cyanobacteria. In *Algae and Symbioses: Plants, Animals, Fungi, Viruses, Interactions Explored*, (Reisser, W., Ed), pp. 537–565. Biopress, Bristol, UK.

Paerl, H. W. *et al.* (1983). Carotenoid enhancement and its role in maintaining blue-green algal (*Microcytsis aeruginosa*) surface blooms. *Limnology and Oceanography* **28**: 847–857.

Pagano, M. *et al.* (2003). An experimental study of the effects of nutrient supply and *Chaoborus* predation on zooplankton communities of a shallow tropical reservoir (Lake Brobo, Cote d'Ivoire). *Freshwater Biology* **48**: 1379–1395.

Palinska, K. *et al.* (2000). *Prochlorococcus marinus* strain PCC 9511, a picoplanktonic cyanobacterium, synthesises the smallest urease. *Microbiology* **146**: 3099–3107.

Pant, A. and Fogg, G. (1976). Uptake of glycollate by *Skeletonema costatum* (Grev) Cleve in bacterial culture. *Journal of Experimental Marine Biological Ecology* **22**: 227–234.

Park, C. (1980). *Ecology and Environmental Management*, Dawson Westview Press.

Paterson, D. M. and Black, K. S. (1999). Water flow, sediment dynamics and benthic biology. *Advances in Ecological Research*, **29**: 155–193. Academic Press, New York, USA.

Patrick, S. and Holding, A. (1985). The effect of bacteria on the solubilisation of silica in diatom frustules. *Journal of Applied Bacteriology* **59**: 7–16.

Patterson, D. (1996). *Free-living Freshwater Protozoa*. Manson Publishing Ltd., London, UK.

Patterson, D. and Larsen, J. (1991). *The Biology of Free-living Heterotrophic Flagellates*. Oxford University Press, Oxford, UK.

Paul, J. (1979). Osmoregulation in the marine diatom *Cylindrotheca fusiformis*. *Journal of Phycology* **15**: 280–284.

Paul, J. (1989). Production of extracellular nucleic acids by genetically altered bacteria in aquatic-environment microcosms. *Applied and Environmental Microbiology* **55**: 1865–1869.

Paul, J. *et al.* (1988). Seasonal and diel variability in dissolved DNA and in microbial biomass and activity in a sub-tropical estuary. *Applied and Environmental Microbiology* **54**: 718–727.

Paul, J. *et al.* (1991). Concentration of viruses and dissolved DNA from aquatic environments by vortex flow filtration. *Applied and Environmental Microbiology* **57**: 2197–2204.

Pechar, L. *et al.* (2002). Hydrobiological evaluation of Třeboň fishponds since the end of the nineteenth century. In *Freshwater Wetlands and their Sustainable Future*, (Kvet, J., Jenik, J. and Soukupova, J., Eds), pp. 31–62. The Parthenon Publishing Group, New York, USA.

Pechlaner, R. *et al.* (1972). The production processes in two high mountain lakes (Vorder and Hinterer Finstertaler See), Kuhtai, Austria. In *Productivity Problems of Freshwaters*, (Kajak, Z. and Hillbricht-Ilkowska, A., Eds), pp. 239–269. PWN (Polish Scientific Publishers), Warsaw-Krakow, Poland.

Pedros-Alio, C. (1989). Toward an autoecology of bacterioplankton. In *Plankton Ecology*, (Sommer, U., Ed), pp. 297–336. Springer-Verlag, Berlin, Germany.

Perrow, M. and Jowitt, A. (1997). Factors affecting water plant recovery – the influence of macrophytes on the structure and function of fish communities. In *Restoration of the Norfolk Broads – Final Report to the EU Life Programme*, (Madgwick, F. and Phillips, G., Eds). Broads Authority, Norwich, UK.

Petersen, J. *et al.* (1999). Cloning and characterisation of a class II DNA photolyase from *Chlamydomonas*. *Plant Molecular Biology* **40**: 1063–1071.

Petersen, J. and Small, G. (2001). A gene required for the novel activation of a class II DNA photolyase in *Chlamydomonas*. *Nucleic Acids Research* **29**: 4472–4481.

Pfennig, N. (1989). Ecology of phototrophic purple and green sulfur bacteria. In *Autotrophic Bacteria*, (Schlegel, H. and Bowien, B., Eds), pp. 97–116. Springer-Verlag, New York, USA.

Pfennig, N. (1967). Photosynthetic bacteria. *Annual Review of Microbiology* **21**: 285–324.

Pick, F. R. (1991). The abundance and composition of freshwater picocyanobacteria in relation to light penetration. *Limnology and Oceanography* **36**: 1457–1462.

Pickup, R. (1992). Detection of gene transfer in aquatic environments. In *Genetic Interactions among Microorganisms in the Natural Environment*, (Wellington, E. and Van Elsas, J., Eds), pp. 145–164. Pergamon Press, Oxford, UK.

Pinder, L. and Farr, I. (1987). Biological surveillance of water quality 3: the influence of organic enrichment on the macro-invertebrate fauna of small chalk streams. *Archives of Hydrobiology* **109**: 619–637.

Pitt, J. *et al.* (1996). Control of nutrient release from sediment. In *Restoration of the Norfolk Broads*, (Madgwick, F. and Phillips, G., Eds). Broads Authority and Environment Agency, Norwich, UK.

Platt, T. *et al.* (1980). Photoinhibition of photosynthesis in natural assemblages of marine phytoplankton. *Journal of Marine Research* **38**: 687–701.

Plude, J. *et al.* (1991). Chemical characterisation of polysaccharide from the slime layer of the cyanobacterium *Microcystis flos-aquae* C3–40. *Applied and Environmental Microbiology* **57**: 1696–1700.

Pokorny, J. *et al.* (2002a). The role of wetlands in energy and material flows in the landscape. In *Freshwater Wetlands and their Sustainable Future*, (Kvet, J., Jenik, J. and Soukupova, L., Eds), pp. 445–462. The Parthenon Publishing Group, New York, USA.

Pokorny, J. *et al.* (2002b). Role of macrophytes and filamentous algae in fishponds. In *Freshwater Wetlands and their Sustainable Future*, (Kvet, J., Jenik, J. and Soukupova, L., Eds), pp. 97–124. The Parthenon Publishing Group, New York, USA.

Pollingher, U. (1988). Freshwater armored dinoflagellates: growth, reproduction strategies, and population dynamics. In *Growth and Reproductive Strategies of Freshwater Phytoplankton*, (Sandgren, C., Ed), pp. 134–174. Cambridge University Press, Cambridge, UK.

Pomati, F. *et al.* (2000). The freshwater cyanobacterium *Planktothrix* sp. FP1: molecular identification and detection of paralytic shellfish poisoning toxins. *Journal of Phycology* **36**: 553–562.

Pomeroy, J. and Brun, E. (2001). Physical properties of snow. In *Snow Ecology*, (Jones, A. K. *et al.*, Eds), pp. 45–126. Cambridge University Press, Cambridge, UK.

Porta, D. *et al.* (2003). Physiological characterisation of a *Synechococcus* sp. (cyanophyceae) strain PCC 7942 iron-dependent bioreporter for freshwater environments. *Journal of Phycology* **39**: 64–79.

Porter, K. (1973). Selective grazing and differential digestion by zooplankton. *Nature* **244**: 179–180.

Pradeep Ram *et al.* (2003). Seasonal shift in net ecosystem production in a tropical estuary. *Limnology and Oceanography* **48**: 1601–1607.

Prescott, L. *et al.* (2002). *Microbiology*. McGraw Hill, Boston, USA.

Proctor, L. *et al.* (1993). Calibrating estimates of phage-induced mortality in marine bacteria: ultrastructural studies of marine bacteriophage development from one-step experiments. *Microbial Ecology* **25**: 161–182.

Purves, W. *et al.* (1997). *Life: the Science of Biology*. Sinauer Associates, Sunderland, USA.

Rabalais, N. *et al.* (2000). Gulf of Mexico biological system responses to nutrient changes in the Mississippi river. In *Estuarine Science*, (Hobbie, J., Ed). pp. 241–268. Island Press, Washington DC, USA.

Raven, J. A. (1984). *Energetics and Transport in Aquatic Plants*. Alan Liss Inc., New York, USA.

Raven, J. A. (1997). Phagotrophy in phototrophs. *Limnology and Oceanography* **42**: 198–205.

Rawson, D. (1956). Algal indicators of trophic lake types. *Limnology and Oceanography* **1**: 18–25.

Reed, R. (1985). Osmoacclimation in *Bangia atropurpurea* (Rhodophyta, Bangiales): the osmotic role of floridoside. *British Phycological Journal* **20**: 211–218.

Reed, R. *et al.* (1985). The osmotic role of mannitol in the Phaeophyta: an appraisal. *Phycologia* **24**: 35–47.

Reeve, C. *et al.* (1984). Role of protein synthesis in the survival of carbon-starved *Escherichia coli*. *Journal of Bacteriology* **160**: 1041–1046.

Regier, H. and Cowell, E. (1972). Applications of ecosystem theory, succession, diversity, stability, stress and conservation. *Biological Conservation* **4**: 83–88.

Reisser, W. (1992). Interactions of eukaryotic algae and viruses. In *Algae and Symbioses: Plants, Animals, Fungi, Viruses, Interactions Explored*, (Reisser, W., Ed), pp. 531–536. Biopress Ltd., Bristol, UK.

Reisser, W. (1995). Phycovirology: aspects and prospect of a new phycological discipline. In *Algae, Environment and Human Affairs*. (Weissner, W., Schnepf, E. and Starr, R., Eds), pp. 143–158. Biopress Ltd., Bristol, UK.

Reith, M. (1999). Chloroplast genome diversity: from large to small circles. *Journal of Phycology* **35**: 893–895.

Reynolds, C. (1990). *The Ecology of Freshwater Phytoplankton.* Cambridge University Press, Cambridge, UK.

Reynolds, C. (1997). *Vegetation Processes in the Pelagic: a Model for Ecosystem Theory.* Ecology Institute, Nordbunte, Germany.

Reynolds, C. (1999). Modelling phytoplankton dynamics and its application to lake management. In *The Ecological Bases for Lake and Reservoir Managament,* (Harper, D. *et al.*, Eds), pp. 123–131. Kluwer Academic Publishers, Dordrecht, The Netherlands.

Ricciardi-Rigault, M. *et al.* (2000). Changes in sediment viral and bacterial abundances with hypolimnetic oxygen depletion in a shallow eutrophic Lac Brome (Quebec, Canada). *Canadian Journal of Fisheries and Aquatic Science* **56**: 1284–1290.

Richardson, L. *et al.* (1988). Manganese oxidation in pH and O_2 microenvironments produced by phytoplankton. *Limnology and Oceanography* **33**: 352–363.

Rickard, A. *et al.* (2000). Coaggregation between aquatic bacteria is mediated by specific-growth-phase-dependent lectin-saccharide interactions. *Applied and Environmental Microbiology* **66**: 431–434.

Rickard, A. *et al.* (2002). Phylogenetic relationships and coaggregation ability of freshwater biofilm bacteria. *Applied and Environmental Microbiology* **68**: 3644–3650.

Ridge, I. *et al.* (1999). Algal growth control by terrestrial leaf litter: a realistic tool? In *The Ecological Bases for Lake and Reservoir Management,* (Harper, D., *et al.*, Eds), pp. 173–191. Kluwer Academic Publishers, Dordrecht, The Netherlands.

Rieman, B. (1985). Potential importance of fish predation and zooplankton grazing on natural populations of freshwater bacteria. *Applied and Environmental Microbiology* **50**: 187–193.

Ringelberg, J. *et al.* (1991). Diel vertical migration of *Daphnia hyalina* (*sensu latiori*) in Lake Maarsseveen, Part 1. Aspects of seasonal and daily timing. *Archives of Hydrobiology* **121**: 129–145.

Robarts, R. D. and Zohary, T. (1984). *Microcystis aeruginosa* and underwater light attenuation in a hypertrophic lake (Hartbeesport Dam, South Africa). *Journal of Ecology* **72**: 1001–1017.

Robarts, R. D. and Zohary, T. (1987). Temperature effects on photosynthetic capacity, respiration and growth rates of bloom-forming cyanobacteria. *New Zealand Journal of Marine and Freshwater Research* **21**: 391–399.

Roberts, D. and Boylen, C. (1988). Patterns of epipelic algal distribution in an Adirondack lake. *Journal of Phycology* **24**: 146–152.

Rodo, X. *et al.* (1997). Variations in seasonal rainfall in Southern Europe during the present century: relationships with the North Atlantic Oscillation and the El-Nino Southern Oscillation. *Climate Dynamics* **13**: 275–284.

Round, F. E. (1993). *A Review and Methods for the Use of Epilithic Diatoms for Detecting and Monitoring Changes in River Quality.* HMSO, London, UK.

Ryding, S.-O. and Rast, W. (1989). *The Control of Eutrophication of Lakes and Reservoirs.* The Parthenon Publishing Group, Paris, France.

Sabater, S. and Munoz, I. (2000). *Nostoc verrucosum* (Cyanobacteria) colonized by a chironomid larva in a mediterranean stream. *Journal of Phycology* **36**: 59–61.

Sabater, S. *et al.* (2003). Structure and function of benthic algal communities in an extremely acid river. *Journal of Phycology* **39**: 481–492.

Safferman, R. and Morris, M. (1964). Control of algae with viruses. *Journal of the American Water Works Association* **56**: 1217–1224.

Safferman, R. and Morris, M. (1967). Observations on the occurrence, distribution and seasonal incidence of blue-green algal viruses. *Applied Microbiology* **15**: 1219–1222.

Sag, Y. *et al.* (1998). The simultaneous biosorption of Cu(II) and Zn on *Rhizopus arrhizus*: application of adsorption models. *Hydrometall* **50**: 297–314.

Salmaso, N. *et al.* (1999). Understanding deep oligotrophic lakes for efficient management. In *The Ecological Bases for Lake and Reservoir Management,* (Harper, D. *et al.*, Eds), pp. 253–263. Kluwer Academic, Dordrecht, The Netherlands.

Sanders, H. (1968). Marine benthic diversity: a comparative study. *American Naturalist* **102**: 243–282.

Sanders, R. (1991). Trophic strategies among heterotrophic flagellates. In *The Biology of Free-living Heterotrophic Flagellates,* (Patterson, D. and Larsen, J., Eds), pp. 21–38. Clarendon Press, Oxford, UK.

Sanders, R. and Porter, K. (1986). Use of metabolic inhibitors to estimate protozooplankton grazing and bacterial production in a monomictic eutrophic lake with an anaerobic hypolimnion. *Applied and Environmental Microbiology* **52**: 101–107.

Sanders, R. *et al.* (1992). Relationship between bacteria and heterotrophic nanoplankton in marine and freshwaters: an inter-ecosystem comparison. *Marine Ecology Progress Series* **86**: 1–14.

Sanders, R. *et al.* (1989). Seasonal patterns of bacterivory by flagellates, ciliates, rotifers, and cladocerans in a freshwater planktonic community. *Limnology and Oceanography* **34**: 673–687.

Sandgren, C. (1988a). The ecology of chrysophyte flagellates: their growth and perennation strategies as freshwater phytoplankton. In *Growth and Reproductive Strategies of Freshwater Phytoplankton*, (Sandgren, C., Ed), pp. 9–104. Cambridge University Press, Cambridge, UK.

Sandgren, C., Ed. (1988b). *Growth and Reproductive Strategies of Freshwater Phytoplankton*. Cambridge University Press, Cambridge, UK.

Sand-Jensen, K. and Borum, J. (1991). Interactions among phytoplankton, periphyton and macrophytes in temperate freshwaters and estuaries. *Aquatic Botany* **41**: 137–175.

Saye, D. *et al.* (1987). Potential for transduction of plasmids in a natural freshwater environment: effect of donor concentration and the natural community on transduction in *Pseudomonas aeruginosa. Applied and Environmental Microbiology* **53**: 987–995.

Saye, D. *et al.* (1990). Transduction of linked chromosomal genes between *Pseudomonas aeruginosa* strains during incubation *in situ* in a freshwater habitat. *Applied and Environmental Microbiology* **56**: 140–145.

Scheffer, M. (1998). *Ecology of Shallow Lakes*. Chapman and Hall, London, UK.

Schindler, D. (1987). Detecting ecosystem responses to anthropogenic stress. *Canadian Journal of Fisheries and Aquatic Science* **44**: 6–25.

Schneegurt, M. A. *et al.* (2000). Metabolic rhythms of a diazotrophic cyanobacterium, *Cyanothece* sp. strain ATCC 51142, heterotrophically grown in continuous phase. *Journal of Phycology* **36**: 107–117.

Scholin, C. *et al.* (1997). Detection and quantitation of *Pseudo-nitzschia australis* in cultured and natural populations using LSU rRNA-targeted probes. *Limnology and Oceanography* **42**: 1265–1272.

Seckbach, J., Ed. (2000). *Journey to Diverse Microbial Worlds*. Kluwer Academic, Dordrecht, The Netherlands.

Seckbach, J. and Oren. A. (2000). A vista into the diverse microbial world. In *Journey to Diverse Microbial Worlds*, (Seckbach, J., Ed). Kluwer Academic, Dordrecht, The Netherlands.

Sellers, T. and Bukaveckas, P. (2003). Phytoplankton production in a large, regulated river: a modeling and mass balance assessment. *Limnology and Oceanography* **48**: 1476–1487.

Semenova, E. and Kuznedelov, K. (1998). A study of the biodiversity of Baikal picoplankton by comparative analysis of the 16s rRNA gene 5'-terminal regions. *Molecular Biology* **32**: 754–760.

Semovski, S. *et al.* (2000). Lake Baikal: analysis of AVHRR imagery and simulation of under-ice phytoplankton bloom. *Journal of Marine Systems* **27**: 117–130.

Sengbusch, P. and Muller, U. (1983). Distribution of glycoconjugates at algal surfaces as monitored by FITC-conjugated lectins. *Protoplasma* **114**: 103–113.

Senger, H. and Fleischhacker, P. (1978). Adaptation of the photosynthetic apparatus of *Scenedesmus obliquus* to strong and weak light conditions. *Physiologia Plantarum* **43**: 35–42.

Servais, P. *et al.* (1985). Rate of bacterial mortality in aquatic environments. *Applied and Environmental Microbiology* **49**: 1448–1454.

Shapiro, J. (1990). Current beliefs regarding dominance by blue-greens: the case for the importance of CO_2 and pH. *Verh. Int. Verein. Limnol* **24**: 38–54.

Shapiro, J. *et al.* (1982). Experiments and experiences in biomanipulation: studies of biological ways to reduce algal abundance and eliminate blue-greens. *U.S. EPA 600/3-82-096.*

Shearer, C. (1993). The Freshwater Ascomycetes. *Nova Hedwig* **56**: 1–33.

Sherman, D. M. *et al.* (2000). Heterocyst development and localisation of cyanophycin in N_2-fixing cultures of *Anabaena* sp. PCC 7120 (Cyanobacteria). *Journal of Phycology* **36**: 932–941.

Sickman, J. O. *et al.* (2003). Evidence for nutrient enrichment of high-elevation lakes in the Sierra Nevada, California. *Limnology and Oceanography* **48**: 1885–1892.

Sigee, D. (1984). Some observations on the structure, cation content and possible evolutionary status of dinoflagellate chromosomes. *Botanical Journal of the Linnean Society* **88**: 127–147.

Sigee, D. (1993). *Bacterial Plant Pathology: Cell and Molecular Aspects*. Cambridge University Press, Cambridge, UK.

Sigee, D. and Levado, E. (2000). Cell surface elemental composition of *Microcystis aeruginosa*: high-Si and low-Si sub-populations within the water column of a eutrophic lake. *Journal of Plankton Research* **22**: 2137–2153.

Sigee, D. *et al.* (1999b). Biological control of cyanobacteria: principles and possibilities. In *The Ecological Bases of Lake and Reservoir Management*, (Harper, D. *et al.*, Eds), pp. 161–172. Kluwer Academic, Dordrecht, The Netherlands.

Sigee, D. *et al.* (1998). Elemental concentrations, correlations and ratios in micropopulations of *Ceratium hirundinella* (Pyrrophyta):L an X-ray microanalytical study. *European Journal of Phycology* **33**: 155–164.

Sigee, D. *et al.* (1999a). Elemental composition of depth samples of *Ceratium hirundinella* (Pyrrophyta) within a stratified lake: an X-ray microanalytical study. *Aquatic Microbial Ecology* **19**: 177–187.

Sigee, D. *et al.* Eds. (1993). *X-ray Microanalysis in Biology: Experimental Techniques and Applications.* Cambridge University Press, Cambridge, UK.

Sigee, D. *et al.* (1999c). Elemental composition of the cyanobacterium *Anabaena flos-aquae* collected from different depths within a stratified lake. *European Journal of Phycology* **34**: 477–486.

Sigee, D. *et al.* (2002). Fourier-transform infrared spectroscopy of Pediastrum duplex: characterisation of a micropopulation isolated from a eutrophic lake. *European Journal of Phycology* **37**: 19–26.

Silbergeld, E. *et al.* (2000). *Pfeisteria*: harmful algal blooms as indicators of human-ecosystem interactions. *Environmental Research* **82**: 97–105.

Simenstad, C. *et al.* (2000). Habitat–biotic interactions. In *Estuarine Science*, (Hobbie, J., Ed). pp. 427–455. Island Press, Washington DC, USA.

Simon, T. (2002). Biological response signatures: toward the detection of cause-and-effect and diagnosis in environmental disturbance. In *Biological Response Signatures*, (Simon, T., Ed), pp. 3–11. CRC Press, Boca Raton, USA.

Sinclair, N. A. and Stokes, J. L. (1965). Obligately phsychrophilic yeasts from the polar region. *Canadian Journal of Microbiology* **11**: 259–270.

Skovgaard, A. (1996). Mixotrophy in *Fragilidium subglobosum* (Dinophyceae): growth and grazing responses as functions of light intensity. *Marine Ecology Progress Series* **143**: 247–253.

Skovgaard, A. and Hansen, P. (2003). Food uptake in the harmful alga *Prymnesium parvum* mediated by excreted toxins. *Limnology and Oceanography* **48**: 1161–1166.

Sleigh, M. (1973). *The Biology of Protozoa*. William Clowes and Son, London, UK.

Smith, D. J. and Underwood, G. J. C. (2000). The production of extracellular carbohydrates by estuarine benthic diatoms: the effects of growth phase and light and dark treatment. *Journal of Phycology* **36**: 321–333.

Smith, R. C. *et al.* (1973). Optical properties and color of Lake Tahoe and Crater Lake. *Limnology and Oceanography* **18**: 189–199.

Smith, V. (1983). Low nitrogen to phosphorus ratios favor dominance by blue-green algae in lake phytoplankton. *Science* **221**: 669–671.

Smyly, W. (1979). Population dynamics of *Daphnia hyalina* Leydig (Crustacea: *Cladophora*) in a productive and unproductive lake in the English Lake District. *Hydrobiologia* **64**: 269–278.

Sødergren, A. (1984). Small-scale temporal changes in the biological and chemical composition of surface microlayers in an eutrophic lake. *Verhandlungen der internationale Vereinigung für theoretische und angewandte Limnologie* **22**: 765–771.

Søndergaard, M. (1984). Dissolved organic carbon in Danish Lakes: concentration, composition and lability. *Verhandlungen der internationale Vereinigung für theoretische und angewandte Limnologie* **22**: 780–784.

Sommer, U. (1988). Growth and reproductive strategies of planktonic diatoms. In *Growth and Survival Strategies of Freshwater Phytoplankton*, (Sandgren, C., Ed), pp. 227–260. Cambridge University Press, Cambridge, UK.

Sorokin, Y. (1965). On the trophic role of chemosynthesis and bacterial biosynthesis in water bodies. *Memorie. Istituto Italiano di Idrobiologia Dottore Marce de March.* **18 Suppl.**: 187–205.

Steemann Nielsen, E. (1952). The use of radioactive carbon (^{14}C) for measuring organic production in the sea. *Journal du Conseel International pour L'Exporation de la Mer* **18**: 117–140.

Steinberg, C. E. W. and Geller, W. (1993). Biodiversity and Interactions within pelagic nutrient cycling and productivity. In *Biodiversity and Ecosystem Function*, (Schulze, E.-D. and Mooney, H. A., Eds), pp. 43–64. Springer Verlag, Berlin, Germany.

Steinman, A. (1996). Effects of grazers on freshwater benthic algae. In *Algal Ecology*, (Stevenson, R. J., Bothwell, M. and Lowe, R. L., Eds), pp. 341–374. Academic Press, New York, USA.

Stent, G. (1963). *Molecular Biology of Bacterial Viruses*. Freeman and Company, San Francisco, USA.

Stevenson, R. J. (1990). The stimulation and drag of current. In *Algal Ecology: Freshwater Benthic Ecosystems*, (Stevenson, R. J., Bothwell, M. and Lowe, R. L., Eds), pp. 321–341. Academic Press, New York, USA.

Stevenson, R. J. (1996). An introduction to algal ecology in freshwater benthic habitats. In *Algal Ecology*, (Stevenson, R. J., Bothwell, M. and Lowe, R., Eds), pp. 3–30. Academic Press, New York, USA.

Stevenson, R. J. *et al.*, Eds. (1996). *Algal Ecology*. Academic Press, New York, USA.

Stewart, G. (1992). Transformation in natural environments. *Genetic Interactions among Microorganisms in the Natural Environment*. (Wellington, E. and Van Elsas, J., Eds), pp. 216–234. Oxford University Press, Oxford, UK.

Stockner, J. (1972). Paleolimnology as a means of assessing eutrophication. *Verhandlungen der internationale Vereinigung für theoretische und angewandte Limnologie.* **18**: 1018–1030.

Stoltze, H. *et al.* (1969). The influence of the mode of nutrition on the digestion system of *Ochromonas malhamensis. Journal of Cell Biology* **43**: 396–409.

Straskrabova, V. and Komarkova, J. (1979). Seasonal changes of bacterioplankton in a reservoir related to algae. 1. Numbers and biomass. *Int. Rev. ges. Hydrobiol.* **64**: 285–302.

Stübing, D. *et al.* (2003). On the use of lipid biomarkers in marine food web analyses: an experimental case study on the Antarctic krill, *Euphausia superba. Limnology and Oceanography* **48**: 1685–1700.

Stulp, B. K. and Stam, W. T. (1984). Genotypic relationships between strains of *Anabaena* (Cyanophyceae) and their correlation with morphological affinities. *British Phycological Journal* **19**: 287–301.

Stumm, W. and Morgan, J. (1970). *Aquatic Chemistry.* John Wiley and Sons, New York, USA.

Suberkropp, K. (1992). Aquatic hyphomycete communities. In *The Fungal Community: its Organisation and Role in the Ecosystem*, (Carroll, G. and Wicklow, D., Eds). pp. 729–747. MarcelDekker, New York, USA.

Suberkropp, K. and Klug, M. (1976). Fungi and bacteria associated with leaves during processing in a woodland stream. *Ecology* **57**: 707–719.

Suberkropp, K. and Klug, M. (1980). The maceration of deciduous leaf litter by aquatic hyphomycetes. *Canadian Journal of Botany* **58**: 1025–1031.

Suberkropp, K. *et al.* (1976). Changes in the chemical composition of leaves during processing in a woodland stream. *Ecology* **57**: 720–727.

Suh, S. *et al.* (1999). Effect of *rpo*S mutation on the stress response and expression of virulence factors in *Pseudomonas aeruginosa. Journal of Bacteriology* **181**: 3890–3897.

Sullivan, C. (1976). Diatom mineralisation of silicic acid. 1. Si(OH)$_4$ transport characteristics in *Navicula pelliculosa. Journal of Phycology* **12**: 390–396.

Sullivan, C. (1977). Diatom mineralisation of silicic acid. 1. Si(OH)$_4$ transport characteristics in *Navicula pelliculosa. Journal of Phycology* **13**: 86–91.

Suttle, C. (1992). Inhibition of photosynthesis in phytoplankton by the submicron size fraction concentrated from seawater. *Marine Ecology Progress Series* **87**: 105–112.

Suttle, C. (1994). The significance of viruses to mortality in aquatic microbial communities. *Microbial Ecology* **28**: 237–243.

Suttle, C. (2000). Cyanophages and their role in the ecology of cyanobacteria. In *The Ecology of Cyanobacteria*, (Whitton, B. and Potts, M., Eds), pp. 563–589. Kluwer, Amsterdam, The Netherlands.

Suttle, C. and Chen, F. (1992). Mechanisms and rates of decay of marine viruses in sea water. *Applied and Environmental Microbiology* **58**: 3721–3729.

Suttle, C. *et al.* (1990). Infection of phytoplankton by viruses and reduction of primary productivity. *Nature* **347**: 467–469.

Swift, D. (1980). Vitamins and phytoplankton growth. In *The Physiological Ecology of Phytoplankton*, (Morris, I., Ed), pp. 329–368. Blackwell, Oxford, UK.

Takao, M. *et al.* (1989). Expression of *Anacystis nidulans* photolyase gene in *Escherichia coli*, functional complementation and modified action spectrum of photoreactivation. *Photochemistrry and Photobiology* **50**: 633–637.

Talling, J. (1960). Self-shading effects in natural populations of a planktonic diatom. *Wetterinid Leben* **12**: 235–242.

Tansley, A. (1935). The use and abuse of vegetational concepts and terms. *Ecology* **16**: 284–307.

Technical Standard Publication (1982). *Utilisation and Protection of Waterbodies. Standing Inland Waters. Classification.* Technical standard 27885/01. Berlin, Germany.

Terauchi, K. and Ohmori, M. (1999). An adenylate cyclase, Cya1, regulates cell motility in the cyanobacterium *Synechocystis* sp. PCC 6803. *Plant Cell Physiology* **40**: 248–251.

Thompson, J. *et al.* (1982). Natural filtration rates of zooplankton in a closed system: the derivation of a community grazing index. *Journal of Plankton Research* **4**: 545–556.

Thompson, P. *et al.* (1994). Storage of phosphorus in nitrogen-fixing *Anabaena flos-aquae* (Cyanophyceae). *Journal of Phycology* **30**: 267–273.

Tidona, C. and Darai, G. Eds. (2002). *The Springer Index of Viruses.* Springer Verlag, Berlin, Germany.

Tien, C.-J. *et al.* (2002). Occurrence of cell-associated mucilage and soluble extracellular polysaccharides in Rostherne Mere and their possible significance. *Hydrobiologia* **485**: 245–252.

Tilman, D. (1977). Resource competition between planktonic algae: an experimental and theoretical approach. *Ecology* **58**: 338–348.

Tilman, D. and Kilham, S. S. (1976). Phosphate and silicate growth and uptake kinetics of the diatoms *Asterionella formosa* and *Cyclotella meneghiniana* in batch and semicontinuous culture. *Journal of Phycology* **12**: 375–383.

Tilman, D. *et al.* (1986). Green, blue-green and diatom algae: taxonomic differences in competitive ability for phosphorus, silica and nitrogen. *Archives of Hydrobiology* **106**: 473–486.

Tilzer, M. M. (1983). The importance of fractional light absorption by photosynthetic pigments for phytoplankton productivity in Lake Constance. *Limnology and Oceanography* **28**: 833–846.

Tilzer, M. M. and Goldman, C. R. (1978). Importance of mixing, thermal stratification and light adaptation for phytoplankton productivity in Lake Tahoe (California-Nevada). *Ecology* **59**: 810–821.

Torsvik, V. *et al.* (2000). Molecular biology and genetic diversity of microorganisms. In *Journey to Diverse Microbial Worlds*, (Seckbach, J., Ed), pp. 45–57. Kluwer Academic, Dordrecht, The Netherlands.

Tranter, M. and Jones, H. (2001). The chemistry of snow: processes and nutrient cycling. In *Snow Ecology*, (Jones, H. *et al.*, Eds), pp. 127–167. Cambridge University Press, Cambridge, UK.

Tremaine, S. and Mills, A. (1987). Inadequacy of the eucaryote inhibitor cycloheximide in studies of protozoon grazing on bacteria at the freshwater – sediment interface. *Applied and Environmental Microbiology* **53**: 1969–1972.

Tuchman, N. (1996). The role of heterotrophy in algae. In *Algal Ecology*, (Stevenson, R. J., Eds), pp. 299–318. Academic Press, New York, USA.

Tuji, A. (2000). The effect of irradiance on the growth of different forms of freshwater diatoms: implications for succession in attached diatom communities. *Journal of Phycology* **36**: 659–661.

Underwood, G. J. C. and Kromkamp, J. (1999). Primary production by phytoplankton and microphytobenthos in estuaries. In *Advances in Ecological Research*, (Nedwell, D. B. and Raffaelli, D. G., Eds), pp. 93–153. Academic Press, New York, USA.

Urban, N. *et al.* (1990). Geochemical processes controlling concentrations of Al, Fe, and Mn in Nova Scotia lakes. *Limnology and Oceanography* **35**: 1516–1534.

Vadeboncoeur, Y. *et al.* (2003). From Greenland to green lakes: cultural eutrophication and the loss of benthic pathways in lakes. *Limnology and Oceanography* **48**: 1408–1418.

Van den Hoek *et al.* (1995). *Algae: an Introduction to Phycology*. Cambridge University Press, Cambridge, UK.

Van der Molen, D. and Boers, P. (1999). Eutrophication control in the Netherlands. In *The Ecological Bases for Lake and Reservoir Management*, (Harper, D. *et al.*, Eds), pp. 403–409. Kluwer Academic, Dordrecht, The Netherlands.

Van Donk, E. and Bruning, K. (1992). Ecology of aquatic fungi in and on algae. In *Algae and Symbioses: Plants, Animals, Fungi, Viruses, Interactions Explored*, (Reisser, W., Ed), pp. 567–592. Biopress Ltd., Bristol, UK.

Van Donk, E. and Bruning, K. (1995). Effects of fungal parasites on planktonic algae and the role of environmental factors in the fungus–alga relationship. In *Algae, Environment and Human Affairs*, (Weissner, W., Schnepf, E. and Starr, R., Eds), pp. 223–224. Biopress, Bristol, UK.

Van Donk, E. and Ringelberg, J. (1983). The effect of fungal parasitism on the succession of diatoms in Lake Maarsseveen (The Netherlands). *Freshwater Biology* **13**: 241–251.

Van Elsas, J. (1992). Antibiotic resistance gene transfer in the environment: an overview. In *Genetic Interactions among Microorganisms in the Natural Environment*, (Wellington, E. and Van Elsas, J., Eds). pp. 17–39. Oxford University Press, Oxford, UK.

Van Etten, J. (2002a). Phycodnaviridae: Chlorovirus. In *The Springer Index of Viruses*, (Tidona, C. and Darai, G., Eds), pp. 724–733. Springer-Verlag, Berlin, Germany.

Van Etten, J. (2002b). Chlorella virus history. *WEBSITE* (http://www.ianr.unl.edu/plantpath/facilities/Virology/)

Van Etten, J. and Meints, R. (1999). Giant viruses infecting algae. *Annual Reviews of Microbiology* **53**: 447–494.

Van Etten, J. *et al.* (1991). Viruses and virus-like particles of eukarytic algae. *Microbiological Reviews* **55**: 586–620.

Vannote, R. *et al.* (1980). The river continuum concept. *Canadian Journal of Fisheries and Aquatic Science* **37**: 130–137.

Villareal, T. and Lipschultz, F. (1995). Internal nitrate concentrations in single cells of large phytoplankton from the Sargasso Sea. *Journal of Phycology* **31**: 689–696.

Vincent, W. (1988). *Microbial Ecosystems in Antarctica*. Cambridge University Press, Cambridge, UK.

Vinebrooke, R. and Leavitt, P. (1998). Direct and interactive effects of allochthonous dissolved organic matter, inorganic nutrients, and ultraviolet radiation on an alpine littoral food web. *Limnology and Oceanography* **43**: 1065–1081.

Vollenweider, R. and Kerekes. J. (1980). The loading concept as a basis for controlling eutrophication philosophy and preliminary results of the OECD programme on eutrophication. *Progress in Water Technology* **12**: 5–38.

Vorosmarty, C. and Peterson, B. (2000). Macro-scale models of water and nutrient flux to the coastal zone.

In *Estuarine Science*, (Hobbie, J., Ed). pp. 43–79. Island Press, Wasjington DC, USA.

Vrieling, E. G. *et al.* (1999). Silicon deposition in diatoms: Control by the pH inside the silicon deposition vesicle. *Journal of Phycology* **35**: 548–559.

Vrieling, E. G. *et al.* (2000). Nanoscale uniformity of pore architecture in diatomaceous silica: a combined small and wide angle X-ray scattering study. *Journal of Phycology* **36**: 146–159.

Waaland, J. and Branton, D. (1969). Gas vacuole development in a blue-green alga. *Science* **163**: 1339–1341.

Waaland, J. *et al.* (1971). Gas vacuoles. Light shielding in a blue-green alga. *Journal of Cell Biology* **48**: 212–215.

Walsby, A. (1974). The extracellular products of *Anabaena cylindrica* (Lemm). 1. Isolation of a macromolecular pigment-peptide complex and other components. *British Phycological Journal* **9**: 371–381.

Walsby, A. (1994). Gas vesicles. *Microbiological Reviews* **58**: 94–144.

Wangersky, P. (1976). The surface film as a physical environment. *Annual Review of Ecological Systems* **7**: 161–176.

Warr, S. *et al.* (1987). Low-molecular-weight carbohydrate biosynthesis and the distribution of cyanobacteria (blue-green algae) in marine environments. *British Journal of Phycology* **22**: 175–180.

Webster, J. and Benfield, E. (1986). Vascular plant breakdown in freshwater ecosystems. *Annual Review of Ecological Systems* **17**: 567–594.

Wehr, J. and Sheath, R. (2003). *Freshwater Algae of North America*. Academic Press, Amsterdam, The Netherlands.

Wehr, J. and Thorp, J. (1997). Effects of navigation dams, tributaries, and littoral zones on phytoplankton communities in the Ohio River. *Canadian Journal of Fisheries and Aquatic Science* **54**: 378–395.

Weinbauer, M. and Höfle, M. (1998). Significance of viral lysis and flagellate grazing as factors controlling bacterioplankton production in a eutrophic lake. *Applied and Environmental Microbiology* **64**: 431–438.

Weisse, T. (1990). Trophic interactions among heterotrophic microplankton, nanoplankton, and bacteria in lake Constance. *Hydrobiologia* **191**: 111–122.

Weisse, T. *et al.* (1990). Response of the microbial loop to the phytoplankton spring bloom in a large prealpine lake. *Limnology and Oceanography* **35**: 781–794.

Wellburn, A. (1998). *Air Pollution and Climate Change*. Pearson Education Ltd., Harlow, UK.

Wellington, E. and Van Elsas, J. Eds. (1992). *Genetic Interactions among Microorganisms in the Natural Environment*. Pergamon Press, Oxford, UK.

Werner, D., Ed. (1977). *Biology of Diatoms*. University of California Press, Berkeley, USA.

Wetherbee, R. *et al.* (1998). The first kiss, establishment and control of initial adhesion by raphid diatoms. *Journal of Phycology* **34**: 9–15.

Wetzel, R. (1973). Productivity investigations of interconnected marl lakes. 1.The eight lakes of the Oliver and Walters chains, northestern Indiana. *Hydrobiological Studies* **3**: 91–143.

Wetzel, R. (1983). *Limnology*. Harcourt Brace College Publishers, Fort Worth, USA.

Wetzel, R. (2001). *Limnology: Lake and River Systems*. Academic Press, San Diego, USA.

Wharton, R. *et al.* (1983). Distribution, species composition and morphology of algal mats in Antarctic dry valley lakes. *Phycologia* **22**: 355–365.

Whisler, H. *et al.* (1975). Life History of *Coelomomyces psorophorae*. *Proceedings of the National Academy of Sciences* **72**: 693–696.

Wiebe, W. and Smith, D. (1977). Direct measurement of dissolved organic carbon release by phytoplankton and incorporation by microheterotrophs. *Marine Biology* **42**: 213–223.

Wilcox, L. and Wedemayer, G. (1991). Phagotrophy in the freshwater, photosynthetic dinoflagellate *Amphidinium cryophilum*. *Journal of Phycology* **27**: 600–609.

Wilhelm, S. *et al.* (1996). Growth, iron requirements, and siderophore production in iron-limited *Synechococcus* PCC 7002. *Limnology and Oceanography* **41**: 89–97.

Williams, P. (1993). On the definition of plankton production terms. *Measurements of Primary Production from the Molecular to the Global Scale*, (Li, W. and Maestrini, S., Eds). International Council for the Exploration of the Sea MSS 197, pp. 9–19. Copenhagen, Denmark.

Williams, T. *et al.* Eds. (2002). *Phytoplankton Productivity. Carbon Assimilation in Marine and Freshwater Ecosystems* Blackwell, Oxford, UK.

Wilson, H. (1999). Legislative challenges for lake eutrophication control in Europe. In *The Ecological Bases for Lake and River Management*, (Harper, D. *et al.*, Ed), pp. 389–401. Kluwer Academic, The Netherlands.

Wilson, J. (1990). Mechanisms of species coexistence: twelve explanations for Hutchinson's 'Paradox of the Plankton': Evidence from New Zealand plant communities. *New Zealand Journal of Ecology* **13**: 17–42.

Winkelmann (1991). Specificity of iron transport in bacteria and fungi. In *Handbook of Microbial Chelates*, pp. 65–105. CRC Press.

Winstanley, C. *et al.* (1992). The development of detection systems for pseudomonads released into lake water. In *Genetic Interactions Among Microorganisms in the Natural Environment*, (Wellington, E. and van Elsas, J., Eds), pp. 165–175. Oxford University Press, Oxford, UK.

Wipat, A. *et al.* (1992). Detection systems for streptomycetes. *Genetic Interactions Among Microorganisms in the Natural Environment*, (Wellington, E. and Van Elsas, J., Eds). pp. 83–90. Oxford University Press, Oxford, UK.

Wohl, D. and McArthur, J. (1998). Actinomycete-flora associated with submerged freshwater macrophytes. *FEMS Microbiology Ecology* **26**: 135–140.

Wommak, K. and Colwell, R. (2000). Virioplankton: viruses in aquatic ecosystems. *Microbiology and Molecular Biology Reviews* **64**: 69–114.

Wright, S. J. L. (1986). Microbial antagonism of cyanobacteria. In *Natural Antimicrobial Systems*, (Gould, G. W. *et al.*, Eds), pp. 257–276. Bath University Press, Bath, UK.

Wright, S. J. L. and Thompson, R. J. (1985). *Bacillus* volatiles antagonize bacteria. *FEMS Microbiology Letters* **30**: 263–267.

Wright, S. J. L. *et al.* (1991). Isoamyl alcohol (3-methyl-1-butanol), a volatile anti-cyanobacterial and phytotoxic product of some *Bacillus* spp. *Letters in Applied Microbiology* **13**: 130–132.

Wurtsbaugh, W. and Horne, A. (1983). Iron in eutrophic Clear Lake, California: its importance for algal nitrogen fixation and growth. *Canadian Journal of Fisheries and Aquatic Science* **40**: 1419–1429.

Yamomoto, Y. *et al.* (1998). Distribution and identification of actinomycetes lysing bacteria in a eutrophic lake. *Journal of Applied Phycology* **10**: 391–397.

Yoder, C. and Rankin, E. (1995). Biological response signatures and the area of degradation value: new tools for interpreting multimetric data. In *Biological Assessment and Criteria: Tools for Water Resource Planning and Decision Making*, (David, W. and Simon, T., Eds), pp. 263–286. Lewis Publishers, Boca Raton, USA.

Yoshii, K. *et al.* (1999). Stable isotope analysis of the pelagic food web in Lake Baikal. *Limnology and Oceanography* **44**: 502–511.

Young, J. (1992). The role of gene transfer in bacterial evolution. In *Genetic Interactions among Microorganisms in the Natural Environment*, (Wellington, E. and Van Elsas, J., Eds), pp. 3–13. Pergamon Press, Oxford, UK.

Zambrano, M. and Kolter, R. (1996). GASPing for life in stationary phase. *Cell* **86**: 181–184.

Zemskaya, T. (2001). Ecophysiological characteristics of the mat-forming bacterium Thioploca in bottom sediments of the Frolikha Bay, Northern Baikal. *Microbiology* **70**: 335–341.

Zenova, G. and Zvyagintsev, D. (2002). Actinomycetes of the genus *Micromonospora* in a meadow ecosystem. *Microbiology* **71**: 570–574.

Zevenboom, W. and Mur, L. (1980). N_2-fixing cyanobacteria: why they do not become dominant in Dutch hypertrophic lakes. In *Hypertrophic Ecosystems*, (Barica, J. and Mur, L., Eds). Junk, The Hague, The Netherlands.

Zirbel, M. J. *et al.* (2000). The reversible effect of flow on the morphology of *Ceratocorys horrida* (Peridiniales, Dinophyta). *Journal of Phycology* **36**: 46–58.

Index

Freshwater Microbiology: Biodiversity and Dynamic Interactions of Microorganisms in the Aquatic Environment   David C. Sigee
© 2005 John Wiley & Sons, Ltd   ISBNs: 0-471-48529-2 (pbk) 0-471-48528-4 (hbk)